ELECTRONIC PRINCIPLES

Other McGraw-Hill Books by Albert P. Malvino
Experiments for Electronic Principles
Digital Computer Electronics
Semiconductor Circuit Approximations
Experiments for Semiconductor Circuit Approximations
Electronic Instrumentation Fundamentals
Basic Electronics: A Text-Lab Manual (with P. Zbar and M. Miller)
Digital Principles and Applications (with D. Leach)

ELECTRONIC PRINCIPLES

Fourth Edition

Albert Paul Malvino, Ph.D., E.E.

Glencoe/McGraw-Hill Educational Division
A Macmillan/McGraw-Hill Company
Westerville, OH   Mission Hills, CA   Peoria, IL

Sponsoring Editor: Brian Mackin
Editing Supervisor: Kelly A. Warsak
Design and Art Supervisor: Caryl Valerie Spinka
Production Supervisor: Catherine Bokman

Text Designer: Delgado Design Inc.
Cover Designer: Delgado Design Inc.
Cover Photographer: Ken Karp
Technical Art Studio: Fine Line Inc. and
                 Monotype Composition Company, Inc.

Library of Congress Cataloging-in-Publication Data

Malvino, Albert Paul.
    Electronic principles / Albert Paul Malvino. — 4th ed.
       p.   cm.
    Includes index.
    ISBN 0-07-039957-3
    1. Electronics.   I. Title.
TK7816.M25 1989
621.381—dc 19   88-13383
                                             CIP

Electronic Principles, Fourth Edition

3 4 5 6 7 8 9 10 11 12 13 14 15 — 00 99 98 97 96 95 94 93 92 91 90

ISBN 0-07-039957-3

CONTENTS

PREFACE

Five years ago, I thought I knew everything about electronics. After all, I had received my Ph.D. in electrical engineering from Stanford University, worked at Hewlett-Packard for many years, been a technician for nine years and an engineer for twenty-four years. Then, I discovered what electronics really is. It is not a discipline with complex formulas, big words, and hard analysis. It is not a rigid science with only one right answer to every problem or only one method or formula that can be used to solve the problem.

Electronics is an art as well as a science. The meaning of this important idea cannot be given in a few words or sentences. But the meaning will become very clear when you read this book. In preparation for a new approach to learning electronics, read Chapter 1 because it sets the tone for the entire book and tells you how to get started in your quest for the real electronics. After reviewing some important ideas about basic electricity, Chapter 1 goes on to discuss recent brain research that has major importance for teaching and learning electronics. This chapter also explains what I mean when I say electronics is an art as well as a science.

This new edition updates the preceding one in many ways. To begin with, the material is now divided into basic and optional topics. Coverage of basic topics appears at the beginning of each chapter. These are the topics that I classify as essential for a modern technician. They are the fundamental principles behind electronics. I don't think I could survive for five minutes in the everyday world of electronics without these topics.

Following the basic topics, you will find an "Optional Topics" section. These continue the discussion at a more specialized or advanced level. It is not necessary to cover any of these topics because they are not used in other chapters. But these topics will be valuable to many teachers who want additional discussions in certain areas. Also, you may find these topics very valuable when you get to industry.

To see what is covered in this book, look at the Contents. There you will find classical topics such as diodes, transistors, amplifiers, FETs, and op amps. These topics are important, but equally important is the way they are discussed. The best textbook I ever used was a microwave text by Dr. H. H. Skilling of Stanford University. His book proved to me that textbooks don't have to be boring, hard to read, and nearly impossible to understand. I have tried my best to emulate Dr. Skilling. I have broken some rules, invented new viewpoints, simplified my language, and done everything I could think of to make this book understandable and valuable to a beginning student.

When I was a student, I always liked the study aids at the end of each chapter because they gave me a chance to find out what I had learned. This new edition includes many study aids such as summaries, vocabularies, important equations, important processes, multiple-choice questions, and homework problems. Some of the chapters include two new features never before in print: the Software Engine™ and the T-Shooter™. Both of these are textbook versions of software I wrote for computer-assisted instruction. The Software Engine allows you to practice up-down thinking (described elsewhere), while the T-Shooter allows you to trouble-shoot basic circuits. Besides teaching you things you can learn in no other way, these new textbook features are a lot of fun to work with.

As before, this book is for a student taking a first course in linear electronics. The prerequisites are a dc-ac course, algebra, and some trigonometry. In many schools it will be possible to take the ac and trigonometry courses concurrently.

In addition to this textbook, a correlated laboratory manual, *Experiments for Electronic Principles*, is available. It contains over 50 experiments including optional exercises in troubleshooting and design. An extensive instructor's guide is also available.

A final point. Einstein once said, "Make things as simple as possible but no simpler." It is a very good point. Many books go out of their way to make things as difficult as possible. Others oversimplify the material and leave only sawdust. Very few books can sail the narrow channel between being too hard and too easy. I hope this is one of those rare books with just the right touch. I think it may be because I did more than write this text; I bled over it. My soul is in this book.

Albert Paul Malvino

ACKNOWLEDGMENTS

As a preparation for writing this latest edition of *Electronic Principles*, I had the help of a reviewer-consultant team. During a period of several months, I would send them questions and they would respond. Their answers helped me to see clearly what had to be done. I could have written this book without their help, but it would not have turned out as well as it did. They provided the positive and negative feedback that gave this book much of its quality. As Ruskin said, "Quality is never an accident; it is always the result of intelligent effort." Because of their intelligent effort, I feel this is my best book yet. I want to thank all of the following colleagues sincerely for their help and guidance.

Robert Abrams, Alabama Technical College
Esin Ayen, NYC Technical College
James Bentley, Mayo Area Vocational Technical Institute
Richard Berube, Community College of Rhode Island
William Blanton Jr., Delta-Ouachita Vocational Technical
    Institute
Frank Gergelyi, Metropolitan Technical Institute
John Gonzales, Cape Fear Technical Institute
Don Grob, Salina Area Vocational Technical Institute
Herbert Hall Jr., Lakeland College
Reginald Hamer, DeAnza Community College
Fred Harding, Tunkhannock Area High School
Michael Herndon, Pima Community College
Hassie Holmes, Asnuntuck Community College
Richard Honeycutt, Davidson Community College
Thomas E. Hopkins, AT & T Training
Lottie Johnson, State Area Vocational Technical—Harriman
O. M. Kuritza, College of DuPage
Dan Landiss, Forest Park Community College
Paul Lecoq, Spokane Falls College
Steven Linley, Herbert Henry Dow High School
Charles Miller, Ann Arundel Community College
Carl Morgan, Miami University
Harry Partin, Hinds Junior College—Raymond Campus
Stephen Purpura, West Virginia Northern Community
    College
Dr. Lee Rosenthal, Fairleigh Dickinson University
Bernard Rudin, Community College of Philadelphia
E. R. Shackley, Central Carolina Technical College
Serge Silbey, North Seattle Community College
James Splitstone, Belleville Area College
Richard Sturtevant, Springfield Technical Community
    College
Joseph Warren, Sheridan Vocational Technical Center
William Yule, Alpena Community College
Ulrich Zeisler, Utah Technical College at Salt Lake

To Joanna
My brilliant and beautiful wife
without whom I would be nothing.
She always comforts and consoles,
never complains or interferes,
asks nothing and endures all,
and writes my dedications.

1 INTRODUCTION

One of the prerequisites for reading this book is a course in dc circuit theory in which Ohm's law, Kirchhoff's laws, and other circuit theorems have been discussed. This first chapter reviews some basic concepts needed to understand electronics. We will also discuss the structure of the human brain and other ideas that affect the way you learn.

1-1 CONVENTIONAL AND ELECTRON FLOW

Which way do electric charges flow? Murphy's law states that the number of deeply held beliefs is equal to the number of possibilities, no matter how ridiculous. Fortunately, there are only two possible directions for current: plus to minus, or minus to plus. Some people insist that current flows from positive to negative, while others will take the opposite position. Which position is correct? They both are, because it depends on what you mean by current.

The Fluid Theory

Franklin (1750) made an outstanding contribution with his fluid theory of electricity. He visualized electricity as an invisible fluid. If a body had more than its normal share of this fluid, he said it had a positive charge; if the body had less than a normal share, its charge was considered negative. On the basis of this theory, Franklin concluded that electric fluid flowed from positive (excess) to negative (deficiency).

The fluid theory was easy to visualize and agreed with all experiments conducted in the eighteenth and nineteenth centuries. As a result, everybody accepted the notion that charges were flowing from positive to negative (now called *conventional flow*). Between 1750 and 1897, a large number of concepts and formulas based on conventional flow came into existence. During this period, the scientific community became committed to conventional flow as a way of life.

Even today, the bulk of engineering literature continues to use conventional flow. Somebody (usually an engineer or scientist) who invents a new device tends to insert arrows on the device that point in the direction of conventional current.

The Electron

In 1897, Thomson discovered the electron and proved that it had a negative charge. Nowadays, the planetary concept of matter is well known. Matter is made up of atoms. Each atom is a positively charged nucleus surrounded by orbiting electrons. The outward push of centrifugal force on each electron is exactly balanced by the inward pull of the nucleus. Therefore, electrons travel in stable orbits, in a manner similar to the motion of the planets around the sun.

A copper atom has 29 protons and 29 electrons. Of the 29 electrons, 28 travel in tight orbits around the nucleus; because of their small orbits, these electrons are locked into the atom by the strong pull of the nucleus. However, the 29th electron travels in a very large orbit, since it is relatively far from the nucleus and feels almost no nuclear attraction. As a result, it is called a *free electron* because it can easily wander from one copper atom to the next.

Electron Flow

In a piece of copper wire, the only physical charges that flow are the free electrons. Under the influence of an electric field, these free electrons flow out of the negative terminal of a battery through the wire to the positive terminal. This is the exact opposite of conventional flow, which creates a problem. Everybody now agrees that charges actually flow from negative to positive in a piece of copper wire, but not everyone is willing to discard the use of conventional flow.

Why the resistance to change? Because once you get above the atomic level, it makes no difference whether you visualize charges flowing from negative to positive or vice versa. Mathematically, you get the same answers either way. Therefore, even though *electron flow* is the truth, the whole truth, and nothing but the truth, conventional flow preserves the mathematical foundations of almost 200 years of circuit theory.

What it comes down to is this: It is convenient for engineers to use both conventional and electron flow, rather than choosing one or the other. At the atomic level, they use electron flow to explain what is actually happening. Above the atomic level, they pretend that a hypothetical positive charge flows, rather than an electron. Maybe someday the engineering community will change to electron flow when analyzing circuits mathematically, but at this time, the consensus is that such a change is not worth the hassle.

Either Flow Valid

Some people insist that you must use electron flow because it agrees with physical reality. Other people insist that you must use conventional flow because the engineering community uses it. Both of these claims are false. You don't have to choose between one or the other. You can use either flow. In fact, many people use electron flow at one level of understanding and conventional flow at another level. When you have mastered the art of electronics, you will be comfortable with either approach. The direction of flow doesn't make any difference when you really know what you are doing.

If you are a beginning student, you should settle on one type of flow and use it consistently. Eventually, as you gain experience, you will begin to realize that electronics is an art as well as a science. Because of this, conventional flow is just as valid as electron flow. In other words, we are not dealing with a rigid subject where there is only one right answer. We are dealing with a flexible subject where many right answers are possible.

1-2 VOLTAGE SOURCES

For any electronic circuit to work, there has to be a source of energy. An energy source is either a voltage source or a current source. This section discusses the voltage source, and the next section is about the current source.

Ideal Voltage Source

An *ideal* or perfect voltage source produces an output voltage that does not depend on the value of load resistance. The simplest example of an ideal voltage source is a perfect battery, one whose internal resistance is zero. For instance, the battery of Fig. 1-1*a* produces an output voltage of 12 V across a load resistance of 10 kΩ; Ohm's law tells us that the load current is 1.2 mA. If we reduce the load resistance to 30 Ω, as shown in Fig. 1-1*b*, the load voltage is still 12 V; the load current, however, increases to 0.4 A. Figure 1-1*c* shows an adjustable load resistance (rheostat). The ideal voltage source will always produce 12 V across the load resistance, regardless of what value it is adjusted to. Therefore, the load voltage is constant; only the load current changes.

Real Voltage Source

An ideal voltage source cannot exist in nature. It can exist only in our minds as a theoretical device. It is not hard to see why. Suppose the load resistance of Fig. 1-1*c* approaches zero; then the load current approaches infinity. No *real voltage source* can produce infinite current because every real voltage source has some internal resistance. For instance, a flashlight battery has an internal resistance of less than 1 Ω, a car battery has an internal resistance of less than 0.1 Ω, and an electronic voltage source may have an internal resistance of less than 0.01 Ω. Finally, an ideal voltage source has an internal resistance of zero.

The load current has to flow through the internal resistance of the voltage source. Because of this, some voltage is dropped across the internal resistance of the source. This means the load voltage is always less than the ideal voltage. When the load resistance is large compared to the source resistance, the voltage across the internal resistance of the source is so small that we don't notice it. In other words, the load voltage approximately equals the ideal voltage when the load resistance is large compared to the internal resistance of the source.

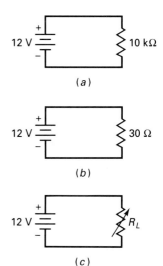

FIGURE 1-1
Voltage source.

Stiff Voltage Source

How small does the internal resistance of the voltage source have to be before we can treat the source as ideal? In this book, we will ignore the internal resistance when it is at least 100 times smaller than the load resistance:

$$R_S < 0.01R_L \tag{1-1}$$

Any source that satisfies this condition is called a *stiff voltage source*. Unless otherwise stated, all voltage sources in this book are assumed to be stiff. In other words, a stiff voltage source is equivalent to an ideal voltage source when an error of less than 1 percent is acceptable.

For instance, the load resistance is adjustable in Fig. 1-2. Over what range of load resistance does the voltage source appear stiff? Multiply by 100 to get

$$R_L = 100(0.06 \ \Omega) = 6 \ \Omega$$

As long as the load resistance is greater than 6 Ω, we can safely ignore the internal resistance of 0.06 Ω in our calculations for load voltage and current. This is equivalent to saying we can treat the stiff voltage source as equivalent to an ideal voltage source.

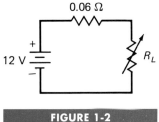

FIGURE 1-2

Load current.

EXAMPLE 1-1

Suppose a voltage source has an ideal voltage of 15 V and an internal resistance of 0.2 Ω. For what values of load resistance will the voltage source appear stiff?

SOLUTION

Multiply by 100 to get

$$R_L = 100(0.2 \ \Omega) = 20 \ \Omega$$

As long as the load resistance is greater than 20 Ω, the voltage source is stiff. This means we can ignore the 0.2 Ω and visualize all of the 15 V across the load resistance.

1-3 CURRENT SOURCES

A voltage source has a very small internal resistance. A *current source* is different; it has a very large internal resistance. Furthermore, a current source produces an output current that does not depend on the value of load resistance.

The simplest example of a current source is the combination of a battery and a large source resistance, as shown in Fig. 1-3a. In this circuit, the load current is

$$I_L = \frac{V_S}{R_S + R_L} \tag{1-2}$$

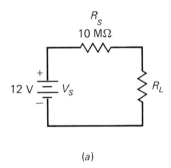

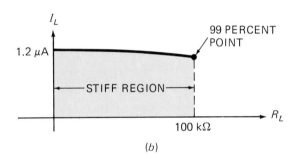

(a)

(b)

FIGURE 1-3   Current source.

Because R_S is 10 MΩ, small load resistances have almost no effect on the load current. For instance, when R_L is 10 kΩ, the load current is

$$I_L = \frac{12\text{ V}}{10\text{ M}\Omega + 10\text{ k}\Omega} = \frac{12\text{ V}}{10.01\text{ M}\Omega} = 1.2\ \mu\text{A}$$

Figure 1-3*b* shows a graph of load current versus load resistance. As you can see, the load current is approximately constant. When the load resistance equals 100 kΩ, the load current is 99 percent of the ideal value. This is equivalent to saying that R_S is at least 100 times greater than R_L. For future discussions, a *stiff current source* is one whose internal resistance is at least 100 times greater than the load resistance:

$$R_S > 100R_L \tag{1-3}$$

Notice that this is the exact opposite of the condition for a stiff voltage source. A current source works best when it has a very high internal resistance, while a voltage source works best when it has a very low internal resistance.

You can now forget about using a battery and a large resistor to build a current source. This method is never used in practice because the currents are too small. How are current sources built? Primarily with transistors. As you will see in a later chapter, a transistor is a device that acts like a current source. Norton's theorem (discussed in basic circuits books) uses the symbol of Fig. 1-4*a* for an ideal current source, one whose internal resistance is infinite. A device like this produces a constant current of I_S. The internal resistance of a real current source is in parallel with an ideal current source as shown in Fig. 1-4*b*.

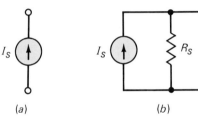

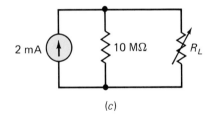

(a)

(b)

(c)

FIGURE 1-4   Symbol for current source.

EXAMPLE 1-2

Figure 1-4c shows a current source of 2 mA with an adjustable load resistance. For the current source to appear stiff, what is the largest acceptable value of load resistance?

SOLUTION

The current source is stiff when the load resistance is 100 times smaller than the internal resistance. If you divide the internal resistance by 100, you get the maximum allowable load resistance:

$$R_{L(\text{max})} = \frac{10 \text{ M}\Omega}{100} = 100 \text{ k}\Omega$$

This means the current source will appear ideal to any load resistance between 0 and 100 kΩ. Stated another way, the load current will be 2 mA for any load resistance between 0 and 100 kΩ. As long as the load resistance is less than 100 kΩ, we can ignore the internal resistance of 10 MΩ and visualize the current source as ideal.

1-4 THEVENIN'S THEOREM

Every once in a while, somebody makes a big breakthrough in engineering and carries all of us to a new high. M. L. Thevenin made one of these quantum jumps when he discovered the circuit theorem named after him; *Thevenin's theorem.* You probably covered this theorem in a basic dc circuits course and no doubt were told it was important. But it's impossible to tell a beginner how important the theorem is. It's the kind of tool that gives you new eyes. With Thevenin's theorem, you can see things that are impossible to see in any other way. It truly separates the professional from the amateur. Now is a good time to rededicate yourself to learning this incredible theorem. It's a gold mine to those who know how to use it.

What a Theorem Is

A theorem is a statement that can be proved mathematically. Thevenin's theorem is one of the most powerful theorems in electronics. Experienced engineers and technicians use it constantly in circuit analysis and design. This method of analysis far outperforms the solutions of Kirchhoff loop equations. The Thevenin theorem allows us to do things that are either difficult or impossible with other circuit methods. As a wise man once said, "Thevenin's theorem did not revise circuit analysis; it created it." This may be an overstatement, but you won't find many practicing engineers and technicians who disagree with it.

Basic Idea

In Fig. 1-5a, what is the load current for each of these values of R_L: 1.5, 3, and 4.5 kΩ? Before Thevenin came along, the classical engineering

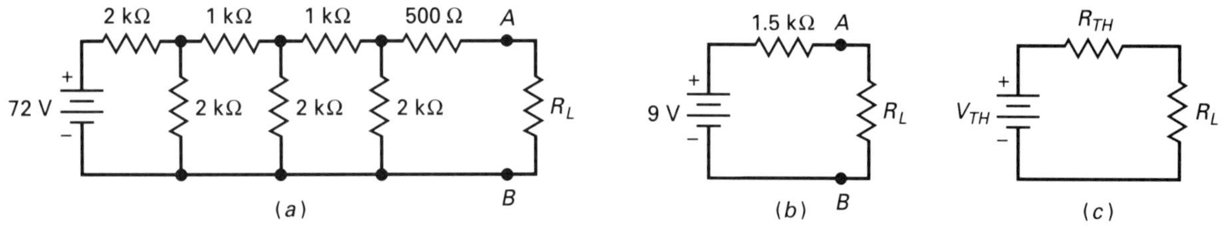

FIGURE 1-5 Applying Thevenin's theorem.

solution was to write and solve four Kirchhoff loop equations. Assuming that you know how to solve four simultaneous loop equations, you can grind your way to the answer for a load resistance of 1.5 kΩ. Then you have to repeat the process for 3 and for 4.5 kΩ. Half an hour later (more or less), you may have worked out all three load currents.

There is another way. When R_L = 1.5 kΩ, you can add 500 Ω to get the total resistance of the right branch, which is 2 kΩ. This 2 kΩ is in parallel with 2 kΩ to give 1 kΩ. Continuing like this, you can work toward the source until you have the total resistance seen by the source. Then you calculate the total current and work back toward the load until you find the load current. After you have the load current for 1.5 kΩ, you can repeat the whole boring process for 3 and 4.5 kΩ.

When Thevenin tried to analyze a circuit like Fig. 1-5a, he was dissatisfied with things as they were. He did not like the "simultaneous equations" solution, and he did not like the "work toward the source and back again" solution. Here are the thoughts that ran through his mind when he was on the verge of his breakthrough:

> **Why should I bother finding all the currents and voltages in the circuit? All I want is the load current. Something tells me there may be a shortcut here, a way to solve this problem quickly and easily.**

Breakthroughs usually begin with dissatisfaction—when something is too hard to do in the old ways we have become accustomed to. All the known methods required that Thevenin calculate every current in the circuit, even though all he wanted was the current through the load resistance. His dissatisfaction with the status quo led him to think along these lines:

> **I wonder if there is any way to replace all of the circuit to the left of the load resistance by a simpler circuit? The simplest circuit I can think of is a battery and an internal resistance. Is it possible that all of the circuit left of the AB terminals might be represented by a single battery and a series resistance?**

This intuitive jump at a possible right answer turned out to be correct. Thevenin was able to prove mathematically that all of the circuit on the left of the AB terminals could be replaced by a single battery and series resistor as shown in Fig. 1-5b.

In Fig. 1-5b, the load resistance may be 1.5, 3, or 4.5 kΩ. When the load resistance is 1.5 kΩ, the load current is

$$I_L = \frac{9\,\text{V}}{3\,\text{k}\Omega} = 3\,\text{mA}$$

In a similar way, you can calculate load currents of 2 mA for 3 kΩ and 1.5 mA for 4.5 kΩ. Why is this new circuit so much easier to analyze than the original? Because the new circuit has only one loop, compared with four loops in the original circuit. Anybody can solve a one-loop problem because all it takes is Ohm's law. A four-loop circuit requires a simultaneous solution or other involved methods.

You can have a nightmare of a circuit, but it still can be reduced to a single-loop circuit (see Prob. 1-28 if you like nightmares). Thevenin's theorem is the great simplifier. It takes big, complicated circuits and turns them into simple one-loop circuits like Fig. 1-5c. This is why practicing engineers and technicians love Thevenin's theorem. It allows them to analyze multiloop circuits quickly and easily. It allows them to fly instead of crawl.

The general idea is this: Whenever you're after the load current in a circuit with more than one loop, think Thevenin, or at least consider it as a possible way to go. More often than not, Thevenin's theorem will offer the most efficient way to solve the problem, especially if the load resistance takes on several values.

In this book, to *thevenize* a circuit means to apply Thevenin's theorem to a circuit, that is, to reduce a multiloop circuit with load resistance to an equivalent single-loop circuit with the same load resistance. In the Thevenin equivalent circuit, the load resistor sees a single source resistance in series with a voltage source. How much easier can life be than this?

Thevenin Voltage and Resistance

Recall the following ideas about Thevenin's theorem from earlier courses. The Thevenin voltage is the voltage that appears across the load terminals when you disconnect the load resistor. Because of this, the Thevenin voltage is sometimes called the *open-circuit* voltage. The Thevenin resistance is the resistance looking back into the load terminals when all sources have been reduced to zero. This means replacing voltage sources by short circuits and current sources by open circuits.

When a circuit is already built, you can measure its Thevenin voltage as follows. Disconnect the load resistor. Then use a voltmeter to measure the voltage across the load terminals. The reading you get is the Thevenin voltage. This assumes no voltmeter loading error, which is equivalent to saying the voltmeter's input resistance is much larger than the Thevenin resistance.

Next, measure the Thevenin resistance as follows. Reduce all sources to zero. This means physically replacing voltage sources with short circuits and physically opening or removing current sources. After the sources have been reduced to zero, use an ohmmeter to measure the resistance between the load terminals. This is the Thevenin resistance.

For instance, suppose you have breadboarded the unbalanced Wheatstone bridge shown in Fig. 1-6a. To thevenize the circuit, you disconnect the load resistance and measure the voltage between *A* and *B* (the load terminals). Assuming no measurement error, you will read 2 V. Next, you replace the battery by a short circuit and measure the resistance between *A* and *B*; you should read 4.5 kΩ. Now you can draw the Thevenin equivalent of Fig. 1-6*b*. With this equivalent circuit, you can easily and quickly calculate the load current for any value of load resistance.

Take a moment and let the significance of this example sink in. The Wheatstone bridge is not balanced. An unbalanced Wheatstone bridge is a nightmare if you don't use Thevenin's theorem. Because the bridge is unbalanced, a current exists in the load resistor. Rather than try to find the current in old ways, we can replace the Wheatstone bridge by the battery and resistance of Fig. 1-6*b*. This new circuit will produce exactly the same load current as the original circuit. If you had to calculate the load current for 10 different load resistances, which circuit would you prefer to work with: Fig. 1-6*a* or 1-6*b*?

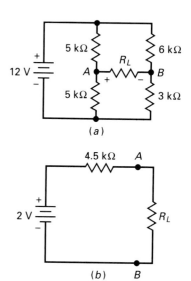

FIGURE 1-6

Wheatstone bridge.

EXAMPLE 1-3

Calculate the Thevenin voltage and resistance of Fig. 1-6*a*.

SOLUTION

If the circuit is not built, you have to use your head instead of a VOM (volt-ohm-milliammeter) to find the Thevenin voltage and resistance. Given the unbalanced Wheatstone bridge of Fig. 1-6*a*, you *mentally* disconnect the load resistor. If you are visualizing this correctly, you then see a voltage divider on the left side and a voltage divider on the right side. The one on the left produces 6 V, and the one on the right produces 4 V, as shown in Fig. 1-7*a*. The Thevenin voltage is the voltage between nodes *A* and *B*, which is

$$V_{TH} = 6\text{ V} - 4\text{ V} = 2\text{ V}$$

Next, mentally replace the battery by a short circuit to get Fig. 1-7*b*. By redrawing the circuit, you can get the two parallel circuits shown in Fig. 1-7*c*. Now, the first parallel circuit has 2.5 kΩ and the second has 2 kΩ. Since these parallel circuits are in series, the total resistance between *A* and *B* is

$$R_{TH} = 2.5\text{ kΩ} + 2\text{ kΩ} = 4.5\text{ kΩ}$$

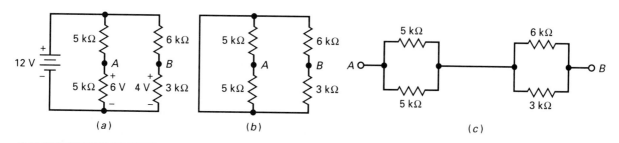

FIGURE 1-7   Calculating Thevenin voltage and resistance.

Thevenin's theorem is useful because it works in the laboratory or on paper. That is, you can measure the Thevenin quantities when the circuit is built, or you can calculate these quantities if the circuit is still a dream on your scratch pad. This gives it a double advantage that many other theorems don't have. It's both practical and theoretical. In terms of usefulness, it has to be ranked close to Ohm's law.

1-5 NORTON'S THEOREM

The Norton theorem is closely related to the Thevenin theorem. Given a Thevenin circuit like that in Fig. 1-8a, the Norton theorem says that you can replace it by the equivalent circuit of Fig. 1-8b. The Norton circuit has an ideal current source in parallel with a source resistance. Notice that the current source produces a fixed current of

$$I_N = \frac{V_{TH}}{R_{TH}} \qquad (1\text{-}4)$$

Also notice that the Norton resistance has the same value as the Thevenin resistance:

$$R_N = R_{TH} \qquad (1\text{-}5)$$

Incidentally, the arrow in the Norton current source points in the direction of conventional current; this is because the inventor was an engineer. This is the first of many devices in which the arrow will point in the direction of conventional current. If you prefer electron flow, your training in reverse thinking starts now. When you see the arrow in a current source, it means that electrons go the other way.

The Norton current is sometimes called the *shorted-load current* because it equals the current that would flow if the load resistance were zero. The Norton resistance should be easy to remember because it equals the Thevenin resistance. For instance, if the Thevenin resistance is 2 kΩ, the Norton resistance is 2 kΩ. The only difference is that the Norton resistance always appears in parallel with a current source, while the Thevenin resistance always appears in series with a voltage source.

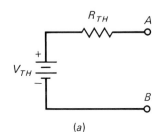

(a)

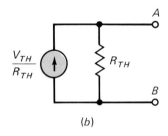

(b)

FIGURE 1-8

(a) Thevenin circuit;
(b) Norton circuit.

EXAMPLE 1-4

Figure 1-9a shows a Thevenin circuit. Convert this to a Norton circuit.

SOLUTION

First, short the load terminals as shown in Fig. 1-9b and calculate the load current, which is

$$I_N = \frac{10 \text{ V}}{2 \text{ k}\Omega} = 5 \text{ mA}$$

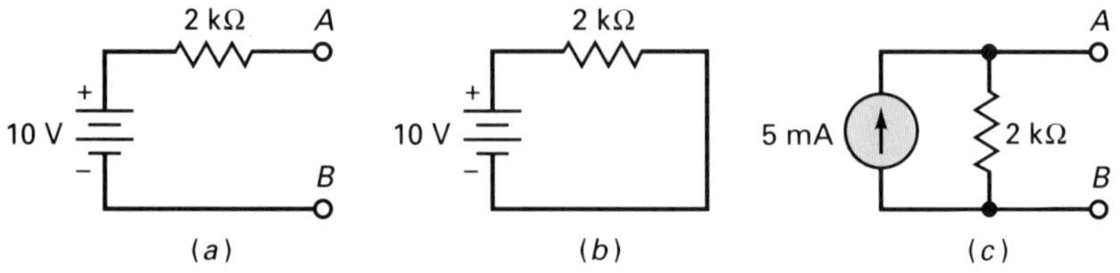

FIGURE 1-9   Deriving Norton's circuit from Thevenin's circuit.

This shorted-load current equals the Norton current. The Norton resistance equals the Thevenin resistance:

$$R_N = 2 \text{ k}\Omega$$

Second, draw the Norton circuit of Fig. 1-9c. The Norton current equals the shorted-load current (5 mA) and the Norton resistance equals the Thevenin resistance (2 kΩ). That's all there is to it.

1-6   TROUBLESHOOTING

Troubleshooting means finding out why a circuit is not doing what it is supposed to do. Every engineer and technician does a lot of troubleshooting when designing or testing electronic circuits. The most common troubles are opens and shorts. Devices like transistors can become open or shorted in a number of ways. One way to destroy any transistor is by exceeding its maximum-power rating. On data sheets for various devices, manufacturers carefully list the absolute maximum ratings for their devices. Do not take these ratings lightly. If you exceed the maximum-power rating, for example, you will almost always have to buy a new device. Sometimes, exceeding the maximum-power rating burns out the interior of the device, leaving an empty space called an *open circuit*. At other times, excessive power dissipation melts the interior of the device, producing an expensive *short circuit*.

Resistors become open circuits when their power dissipation is excessive. But you can get a shorted resistor indirectly as follows. During the stuffing and soldering of printed-circuit boards, an undesirable splash of solder may connect two nearby conducting lines. Known as a *solder bridge*, this effectively shorts any device between the two conducting lines. On the other hand, a poor solder connection usually means no connection at all. This is known as a *cold-solder joint* and implies that the device appears to be open. Perhaps the best starting strategy in troubleshooting is to look for open circuits first, then for short circuits.

Beyond opens and shorts, anything is possible. For instance, temporarily applying too much heat to a resistor may permanently change the resistance by several percent. If the value of resistance is critical, the circuit may not work properly after the heat shock. In this case, it may be a bit more difficult to isolate the faulty resistor in a built-up circuit. In this book, we will keep our troubleshooting practice limited primarily to open and shorted devices. This will go a long way toward training you to think like a troubleshooter.

An Open Device

To be effective as a troubleshooter, you have to know what effect an open or a short will have on the circuit operation. Here is something that will help. Always remember these two facts about an open device:

 The current through an open device is zero.
The voltage is unknown.

The first statement is true because an open device has infinite resistance. No current can exist in an infinite resistance. The second statement is true because of Ohm's law:

$$V = IR = (0)(\infty)$$

In this equation, zero times infinity is mathematically indeterminate, meaning the answer can be anything. You have to figure out what the voltage is by looking at the rest of the circuit. How this is done will be discussed soon.

A Shorted Device

A shorted device is exactly the opposite. Always remember these two statements about a shorted device:

 The voltage across a shorted device is zero.
The current is unknown.

The first statement is true because a shorted device has zero resistance. No voltage can exist across zero resistance. The second statement is true because of Ohm's law:

$$I = \frac{V}{R} = \frac{0}{0}$$

Zero divided by zero is mathematically meaningless. That is, the answer is indeterminate. You have to figure out what the current is by looking at the rest of the circuit.

Table of Troubles

Troubleshooting is a lot of fun. You get to play detective because the circuit gives you clues and you then figure out what is probably causing the trouble. Normally, you measure voltages with respect to ground. From these measurements and your knowledge of basic electricity, you can usually deduce the most likely trouble. After you have isolated a component as the top suspect, you can unsolder or disconnect the component and use an ohmmeter or other instrument for a confirmation.

The best way to learn how to troubleshoot is to do it. Let's get started. In Fig. 1-10, a voltage divider consisting of R_1 and R_2 drives resistors R_3 and R_4 in series. Before you can troubleshoot this circuit, you have to know what the normal voltages are. The first thing to do, therefore, is to work out the values of V_A and V_B. The first is the voltage between node A and ground. The second is the voltage between node B and ground. Because R_1 and R_2 are much smaller than R_3 and R_4 (10 Ω versus 100 kΩ), the voltage at node A is approximately $+6$ V. Furthermore, since R_3 and R_4 are equal, the voltage at node B is approximately $+3$ V. When this circuit is trouble free, you will measure 6 V between node A and ground, and 3 V between node B and ground. These two voltages are the first entry of Table 1-1.

When R_1 is open, what do you think happens to the node voltages? Since no current can flow through the open R_1, no current can flow through R_2. Ohm's law then tells us the voltage across R_2 is zero. Therefore, it follows that $V_A = 0$ and $V_B = 0$ as shown in Table 1-1 for R_1 open.

When R_2 is open, what happens to the node voltages? Since no current can flow through the open R_2, the voltage at node A is pulled up toward the supply voltage. Since R_1 is much smaller than R_3 and R_4, the voltage at node A is approximately 12 V. Since R_3 and R_4 are equal, the voltage at node B becomes 6 V. This is why $V_A = 12$ V and $V_B = 6$ V as shown in Table 1-1 for an R_2 open.

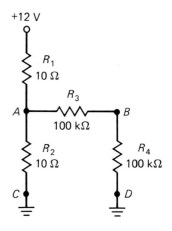

FIGURE 1-10
Troubleshooting example.

Table 1-1	Troubles and Clues	
Trouble	V_A	V_B
Circuit OK	6 V	3 V
R_1 open	0	0
R_2 open	12 V	6 V
R_3 open	6 V	0
R_4 open	6 V	6 V
C open	12 V	6 V
D open	6 V	6 V
R_1 shorted	12 V	6 V
R_2 shorted	0	0
R_3 shorted	6 V	6 V
R_4 shorted	6 V	0

If ground C is open, no current can pass through R_2. This is equivalent to an open R_2. This is why the trouble C open has $V_A = 12$ V and $V_B = 6$ V in Table 1-1.

These examples give you an idea of how to think when you are troubleshooting. You have to visualize the circuit modified by an open or a shorted resistor. Then you use Ohm's law to work out approximate voltages at the nodes. You should work out all of remaining entries in Table 1-1, making sure that you understand why each voltage exists for the given trouble. This basic exercise will go a long way toward teaching you fundamental troubleshooting. For additional practice in troubleshooting, work with the T-shooter in the homework problems. The T-shooter problems for Chap. 1 are found on page 25.

EXAMPLE 1-5

Look at the T-shooter shown in Fig. 1-18 (at the end of the homework problems). It has a circuit at the top, similar to the one summarized in Table 1-1. The circuit values are different, so the voltages will be different. How do you use the T-shooter to measure the node voltages at A, B, and E when the circuit has no troubles?

SOLUTION

In Fig. 1-18, the first box is labeled "Circuit OK." This is the one to use when you want to measure normal voltages. To measure the voltage at node A, read the entry adjacent to V_A, which is $B5$. This entry $B5$ is called a *token*. Go to the large box labeled "Voltages" and read the value of the token, $B5$. Find row B and column 5. You should read 4, meaning 4 V. This is the voltage at node A.

Similarly, V_B has a token of $E2$. The large box then gives a voltage of 2 V. Finally, the supply voltage V_E has a token of $C4$. The corresponding voltage is 12 V.

That's all there is to measuring voltages with the T-shooter. The homework problems at the end of this chapter will use the T-shooter to continue your troubleshooting practice. Also, many later chapters will include T-shooters to improve your troubleshooting skill with more advanced circuits.

1-7 SPERRY'S MODEL OF THE HUMAN BRAIN

Recent discoveries indicate that the human brain processes information on different levels of consciousness. Because of this, learning is far more complicated than most people realize. This section looks at some recent brain research that has enormous implications for teaching and learning electronics.

When we look at the human brain, it appears to be a single organ. But a closer examination reveals that it is two separate hemispheres joined by a bundle of nerve fibers called the *corpus callosum* (see Fig. 1-11). Looking at the two brain halves, a philosopher might ask "Is there one mind here or are there two?" The traditional answer is "one mind." But the one-mind model turns out to be as short-sighted as the idea that the world is flat. As you will see, a new Christopher Columbus claims that we have two minds, not one.

Dr. Roger Sperry (California Institute of Technology) won a Nobel prize for the research he did on the human brain. Here is what happened. During the 1960s, the standard treatment for severe epilepsy was to cut the corpus callosum. With this drastic surgery, the epilepsy disappears but the brain is split into two isolated halves because the left half is no

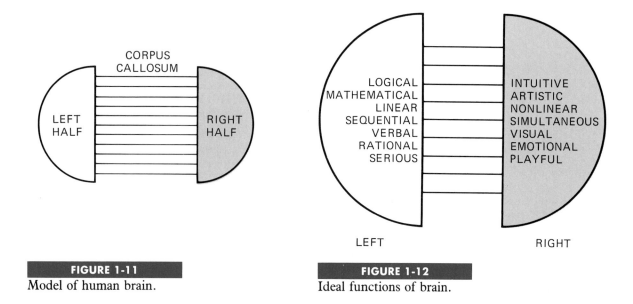

FIGURE 1-11

Model of human brain.

FIGURE 1-12

Ideal functions of brain.

longer connected to the right half. Over several years, Sperry examined dozens of split-brain people and arrived at these four conclusions:

1. The left half of the brain thinks in words and numbers.

2. The right half of the brain thinks in pictures and other nonverbal images.

3. The two brain halves process information so differently that it is more accurate to speak of a left brain and a right brain, rather than a single brain.

4. It is impossible to describe in words how the right brain works.

Figure 1-12 shows some adjectives that describe the two brains. Ideally, the left brain is logical, mathematical, linear, etc. The right brain is intuitive, artistic, nonlinear, etc. Some researchers identify the left brain with Freud's conscious mind and the right brain with the unconscious mind.

Some occupations are primarily left-brain occupations such as accounting, law, and mathematics. Other occupations are primarily right-brain occupations like art, music, and sports. Electronics is both an art (right brain) and a science (left brain). To be really great at electronics, you will have to learn to use both of your brains. In the simplest terms, you have two brains to teach, not one. You teach the left brain with words, numbers, and logic. You teach the right brain with visual aids, what-if questions, and metaphors. The goal is to use both brains to produce a superior type of learning now identified as *whole-brain learning*.

To become a whole-brain learner, here are a few things you can do to get started. First, accept the duality of the human mind—the notion that you have two minds or modes of thinking. Once you realize in your bones that there are two complementary ways of thinking, you will make much better progress in learning electronics.

Second, one of the left-brain traps of formal education is the idea that every problem has only one right answer. Anyone who has worked in industry for at least five minutes knows that most problems have many right answers. Often, the best solution to a real-life problem is the second or third right answer that you can find. So, accept the idea that most problems have many right answers.

Third, another left-brain trap that is even more dangerous is the idea that you must solve problems in a particular way, with exactly the right formula. Absolutely false! There are many ways to solve problems because electronics is an art as well as a science. This gives you the freedom to do things your way, provided you don't violate any of the basic laws of electricity. This is why the overuse of formulas is a mistake. It gets you in the rut of trying to solve problems in just the right way. As you will see, you don't have to rely on formulas all of the time. There are alternative ways to solve many problems using only Ohm's law. Therefore, accept the idea that problems can be solved in many ways.

Fourth, some people scoff at the idea of *intuition*. But Sperry's research proves the right brain does process information in nonverbal ways at another level of consciousness. We cannot verbally explain intuition, because it is a thinking process that occurs in the right brain. In other words, your right brain can solve problems without your left brain being aware of it. When someone has a hunch, a dream, or other inspiration that solves a difficult problem, it is an answer being transferred from the right brain to the left brain. So, accept the idea that you have intuition, a talent for knowing things without having logically deduced them. In time, your intuition will become one of your most valuable tools for solving problems. If you have any reservations about intuition, listen to what Einstein had to say about it: "The really valuable thing is intuition. Without it, I could not see how to begin."

Finally, try to develop *what-if thinking* (right brain) as well as *sequential thinking* (left brain). Sequential thinking is what we usually do. It is the kind of thinking where the result of each step is used in the next step. It is logical and mathematical. It is neat and clean and unforgiving. It is what a computer does. What-if thinking is different. In this kind of thinking, anything goes and all things are possible. What-if thinking searches for more than one right answer. It is sometimes illogical, it breaks rules, it makes mistakes, it is playful, it is sometimes foolish, and it is creative.

1-8 WHERE FORMULAS COME FROM

A formula is a compact mathematical summary of how quantities are related. You will see many formulas in this book. Unless you know where each one comes from, you may become confused and discouraged as they accumulate. Fortunately, there are only three ways in which formulas come into existence. Knowing what they are makes all the difference in studying electronics. In fact, knowing the origin of a formula tells you why it is true or valid.

Defining Formula

In Gaelic (the ancient language of Ireland), *fear* means man, *bean* stands for woman, and *paiste* is child. To understand Gaelic, you have to memorize these definitions:

fear = man
bean = woman
paiste = child

and so on. The trick to learning a foreign language is to memorize the definitions of the words. Then you will know what someone means when they use a foreign word.

To understand electricity and electronics you have to realize you are dealing with a foreign language. This is why you have to memorize the meanings of new words such as current, voltage, and resistance. But a verbal definition of these words is not enough. Why? Because your idea of current must be mathematically identical to everyone else's. The only way to get this identical mathematical meaning is with a *defining formula* (sometimes called a *definition*). If you recall, current was defined as the number of coulombs, Q, passing through the cross section during time, t. The defining formula for current is

$$I = \frac{Q}{t} \qquad (1\text{-}6)$$

A defining formula like this tells you exactly what the quantity, I, equals. If Q is 6 C and $t = 2$ s, then I is 3 C/s or 3 A. Everyone using this defining formula will get the same answer for current.

A defining formula is an equation that explains a new quantity in terms of known quantities. Before you saw Eq. (1-6) for the first time, you did not know the mathematical meaning of current. After this defining formula was given and you memorized it, you knew the mathematical definition of current. Because defining formulas are starting points, you don't have to ask where they come from. You just memorize them, the same as you would memorize the words of a foreign language.

Experimental Formulas

An *experimental formula* (often called a *law*) is different. It summarizes a relation between quantities that are already defined. Here is an example of an experimental formula:

$$f = \frac{Q_1 Q_2}{K d^2} \qquad (1\text{-}7)$$

This is called Coulomb's law. It gives the force between two charges. Where does it come from? By ingenious experiments, Coulomb proved that the force was directly proportional to each charge and inversely proportional to the square of the distance between the charges. It is an example of a relation that already exists in nature. Before a discovery, someone may have a vague feeling that such a relation exists. After a number of experiments, he or she is able to write a mathematical equation that summarizes the discovery. After enough people verify the truth of the discovery through separate experiments, the experimental formula is called a law.

Derived Formula

Given an equation like

$$y = 3x$$

we can add 5 to both sides to get

$$y + 5 = 3x + 5$$

The new equation is valid because both sides are still mathematically equal. There are many other mathematical operations such as subtraction, multiplication, division, factoring, and substitution, that preserve the equality of both sides of the equation. For this reason, we can get all kinds of new formulas by mathematical juggling.

A *derived formula* is one that you create with mathematics. This means you start with one or more formulas, and by different mathematical operations you arrive at a new formula not in your original set of formulas. When this happens, you have come up with a derived formula. For instance, the original form of Ohm's law was

$$R = \frac{V}{I} \tag{1-8}$$

Transposing R and I gives

$$I = \frac{V}{R} \tag{1-9}$$

This is a derived formula. It is the original form of Ohm's law rearranged into another equivalent form.

The Answer

When you don't know what kind of formulas you are dealing with, your understanding of electronics is built on sand. The question, "Where did the equation come from?" has three possible answers and you should know which one applies. To build your understanding of electronics on solid ground, always remember these distinctions:

1. Defining formulas are invented or made up. They cannot be proved experimentally or mathematically. The defined quantity appears for the first time in a formula. Before this time, no other formulas have used the quantity being defined. When you encounter a defining formula, all you have to do is memorize it, much the same as you would memorize the words of a foreign language.

2. Experimental formulas represent mathematical summaries of existing relations in nature. They cannot be proved mathematically.

They can be proved only by experiment or observation. All the quantities in an experimental formula have already been defined.

3. Derived formulas are the only ones that can be proved mathematically, that is, by rearranging other formulas using valid mathematical operations that preserve the equality of both sides of an equation. Derived formulas allow us to see things in a new light or from another angle. Most formulas in this book are derived formulas.

1-9 APPROXIMATIONS

Mathematics is a marvelous tool. Without it, we could do little to improve the world. But like all good things, it can be overdone. The greatest folly of early mathematical training is the idea that you must get an *exact answer* to every problem. This is a left-brain trap because there are many situations where exact answers do more harm than good. For instance, exact answers are wrong when troubleshooting electronics equipment. In fact, exact answers would probably get most troubleshooters fired because it takes too long to get exact answers. Because electronics is an art and a science, you will have to learn to use approximations instead of exact answers.

Did you know that 1 ft of AWG 22 wire that is 1 in from a chassis has a resistance of 0.016 Ω, an inductance of 0.24 μH, and a capacitance of 3.3 pF? If we had to include the effects of resistance, inductance, and capacitance in every calculation for current, we would spend a disproportionate amount of time on calculations. This is why everybody ignores the resistance, inductance, and capacitance of connecting wires in most situations.

The *ideal approximation* (sometimes called the *first approximation*) of a device is the simplest equivalent circuit of the device. In the case of a connecting wire, the ideal approximation is a conductor of zero resistance. The ideal approximation includes only one or two basic ideas of how a device works. It is a rock-bottom approximation beyond which you cannot go without losing the meaning of the device. For instance, a resistor has resistance, inductance, and capacitance. The ideal approximation of a resistor is a pure resistance.

The *second approximation* includes some extra features to improve the analysis. Usually, this is as far as many engineers and technicians go in daily work. For instance, the ideal approximation of a flashlight battery is a voltage source of 1.5 V. The second approximation is a voltage source of 1.5 V in series with a Thevenin resistance of approximately 1 Ω.

The *third approximation* includes other effects of lesser importance. Only the most demanding applications require this level of approximation. When we study diodes and transistors, you will see some examples of the third approximation.

The approximation to use depends on what you are trying to do. If you are troubleshooting, you will find the ideal approximation is usually adequate. If you are sending a man to the moon, you will need the third approximation because the greatest precision is required. For most situations, the second approximation is the best compromise.

STUDY AIDS

The following study aids will help to reinforce the ideas discussed in this chapter. For best results, use these study aids within 6 hours of reading the earlier material. Then review these study aids a week later and a month later to ensure that the concepts remain in your long-term memory.

SUMMARY

Sec. 1-1 Conventional and Electron Flow

Conventional flow assumes charges flow from positive to negative. This hypothetical view of current still dominates at the engineering level, even though it is a mathematical abstraction rather than a physical reality. Electrons are the physical carriers in a piece of wire. They flow from negative to positive. Either conventional flow or electron flow is a valid viewpoint of current when you know what you are doing. A beginner should settle on using one type of flow until he or she has more experience.

Sec. 1-2 Voltage Sources

An ideal voltage source produces a constant voltage, no matter what the value of load resistance. This is equivalent to saying an ideal voltage source has an internal resistance of zero. A real voltage source acts like an ideal voltage source and a series resistance. A stiff voltage source has an internal resistance that is at least 100 times smaller than the load resistance. When an error of less than 1 percent is acceptable, we can treat all stiff voltage sources as ideal voltage sources.

Sec. 1-3 Current Sources

An ideal current source produces a constant current, no matter what the value of load resistance. A stiff current source is one whose internal resistance is at least 100 times larger than the load resistance. When an error of less than 1 percent is acceptable, we can treat all stiff current sources as ideal current sources.

Sec. 1-4 Thevenin's Theorem

Thevenin's theorem is a bread-and-butter tool used by many technicians and engineers. Any circuit facing a load resistance can be replaced by an ideal voltage source and a series resistance. The Thevenin voltage equals the load voltage when the load resistor is disconnected. The Thevenin resistance is the equivalent resistance facing the load resistor.

Sec. 1-5 Norton's Theorem

Norton's theorem is the dual or alternative to Thevenin's theorem. Any circuit facing a load resistance can be replaced by an ideal current source and a parallel resistance. The Norton current equals the load current when the load resistor is shorted. The Norton resistance equals the Thevenin resistance.

Sec. 1-6 Troubleshooting

The most common troubles are shorts and opens. Whenever you exceed maximum power ratings, you may short or open a device. Also, solder splashes may short components, and cold-solder joints may create opens.

Sec. 1-7 Sperry's Model of the Human Brain

The left half of the brain thinks in words and numbers, while the right half thinks in pictures and other nonverbal images. The two brain halves process information so differently that we refer to them as the left brain and the right brain. It is impossible to describe in words how the right brain works. Electronics is an art and a science. Because of this, it requires the use of the left brain (logical, mathematical, linear, etc.) and the right brain (intuitive, artistic, nonlinear, etc.).

Sec. 1-8 Where Formulas Come From

Formulas originate in three ways. They may be defining formulas, experimental formulas, or derived formulas. Defining formulas give the meaning of a quantity in terms of other defined quantities. The quantity being defined for the first time has not appeared in earlier formulas. Experimental formulas are mathematical summaries of relations that already exist in nature. Derived formulas are mathematical rearrangements of existing formulas.

Sec. 1-9 Approximations

The ultimate left-brain trap is our desire for exact answers. It originates with our early mathematical training. In many applications, exact answers are a waste of time. Practicing engineers and technicians use approximations. The ideal, or first, approximation is the simplest equivalent circuit of a device; it is used a lot in troubleshooting. The second approximation includes some extra features to improve accuracy; it is used a lot in everyday work. The third approximation is highly accurate but rarely used.

VOCABULARY

Without words, your left brain cannot work. This is why you must know the meaning of important words introduced in this book. Explain to yourself what each of the following terms means. Keep your answers short and to the point. Verify your answer by rereading the appropriate discussion or by looking at the end-of-book Glossary.

current source	right brain
defining formula	second approximation
derived formula	stiff current source
experimental formula	stiff voltage source
ideal approximation	third approximation
intuition	voltage source
left brain	

IMPORTANT RELATIONS

Here are some important relations. Say each of the following in symbols, then say each in words. Try to explain what the relation means and how it is used. Then read the description that follows.

Eq. 1-1 Stiff Voltage Source

$$R_S < 0.01R_L$$

This is how you can recognize a stiff voltage source. Its internal resistance is at least 100 times smaller than the load resistance. When this condition is satisfied, more than 99 percent of the ideal voltage appears across the load resistor. When errors of less than 1 percent are acceptable, we can treat all stiff voltage sources as ideal voltage sources.

Eq. 1-3 Stiff Current Source

$$R_S > 100R_L$$

This is how you can recognize a stiff current source. Its internal resistance is at least 100 times larger than the load resistance. When this condition is satisfied, more than 99 percent of the ideal current flows through the load resistor. When errors of less than 1 percent are acceptable, we can treat all stiff current sources as ideal current sources.

Eqs. 1-4 and 1-5 Norton and Thevenin Theorems

$$I_N = \frac{V_{TH}}{R_{TH}}$$

and

$$R_N = R_{TH}$$

It is often necessary to look at things from an alternative viewpoint. These equations allow you to convert a Thevenin circuit to a Norton circuit. Notice that the Norton and Thevenin resistances are equal in value, but different in their physical locations. The Thevenin resistance is always in series with a voltage source, while the Norton resistance is always in parallel with a current source.

STUDENT ASSIGNMENTS

QUESTIONS

The following questions may have more than one right answer. Select the best answer. This is the one that is always true, or covers more situations, or fits the context, etc.

1. The right way to visualize current in a wire is with
    a. Electron flow
    b. Conventional flow
    c. Either flow
    d. Neither flow

2. An ideal voltage source has
   a. Zero internal resistance
   b. Infinite internal resistance
   c. A load-dependent voltage
   d. A load-dependent current

3. A real voltage source has
   a. Zero internal resistance
   b. Infinite internal resistance
   c. A small internal resistance
   d. A large internal resistance

4. If a load resistance is 1 kΩ, a stiff voltage source has a resistance of
   a. At least 10 Ω
   b. Less than 10 Ω
   c. More than 100 kΩ
   d. Less than 100 kΩ

5. An ideal current source has
   a. Zero internal resistance
   b. Infinite internal resistance
   c. A load-dependent voltage
   d. A load-dependent current

6. A real current source has
   a. Zero internal resistance
   b. Infinite internal resistance
   c. A small internal resistance
   d. A large internal resistance

7. If a load resistance is 1 kΩ, a stiff current source has a resistance of
   a. At least 10 Ω
   b. Less than 10 Ω
   c. More than 100 kΩ
   d. Less than 100 kΩ

8. The Thevenin voltage is the same as the
   a. Shorted-load voltage
   b. Open-load voltage
   c. Ideal source voltage
   d. Norton voltage

9. The Thevenin resistance is equal in value to the
   a. Load resistance
   b. Half the load resistance
   c. Internal resistance of a Norton circuit
   d. Open-load resistance

10. To get the Thevenin voltage, you have to
    a. Short the load resistor
    b. Open the load resistor
    c. Short the voltage source
    d. Open the voltage source

11. To get the Norton current, you have to
    a. Short the load resistor
    b. Open the load resistor
    c. Short the voltage source
    d. Open the voltage source

12. The Norton current is sometimes called the
    a. Shorted-load current
    b. Open-load current
    c. Thevenin current
    d. Thevenin voltage

13. A solder bridge may cause
    a. A short
    b. An open
    c. A fire
    d. A Norton current

14. A cold-solder joint may cause
    a. A short
    b. An open
    c. A fire
    d. A Thevenin voltage

15. An open resistor has
    a. Infinite current through it
    b. Zero voltage across it
    c. Infinite voltage across it
    d. Zero current through it

16. A shorted resistor has
    a. Infinite current through it
    b. Zero voltage across it
    c. Infinite voltage across it
    d. Zero current through it

17. Problems should be solved
    a. In only one way
    b. Only with formulas
    c. To get one right answer
    d. Any way that works

18. What-if thinking
    a. Searches for more than one right answer
    b. Breaks rules
    c. Makes mistakes
    d. All of the above and more

19. If a quantity is being described for the first time, the kind of formula used is a
    a. Defining formula
    b. Experimental formula
    c. Derived formula
    d. Quadratic formula

20. An experimental formula
    a. Defines a quantity
    b. Represents an existing relation in nature
    c. Can be derived mathematically
    d. Is like a foreign word

21. An ideal voltage source and an internal resistance is an example of the
    a. Ideal approximation
    b. Second approximation
    c. Third approximation
    d. Exact model

22. Treating a connecting wire as a conductor with zero resistance is an example of the
    a. Ideal approximation
    b. Second approximation
    c. Third approximation
    d. Exact model

23. Conventional flow assumes charges flow from
    a. Positive to negative
    b. Negative to positive
    c. Deficiency to excess
    d. Positive to positive

24. The voltage out of an ideal voltage source
    a. Is zero
    b. Is constant
    c. Depends on the value of load resistance
    d. Depends on the internal resistance

25. The current out of an ideal current source
    a. Is zero
    b. Is constant
    c. Depends on the value of load resistance
    d. Depends on the internal resistance

26. Thevenin's theorem replaces a complicated circuit facing a load by an
    a. Ideal voltage source and parallel resistor
    b. Ideal current source and parallel resistor
    c. Ideal voltage source and series resistor
    d. Ideal current source and series resistor

27. Norton's theorem replaces a complicated circuit facing a load by an
    a. Ideal voltage source and parallel resistor
    b. Ideal current source and parallel resistor
    c. Ideal voltage source and series resistor
    d. Ideal current source and series resistor

28. One way to short a device is
    a. With a cold-solder joint
    b. With a solder bridge
    c. By disconnecting it
    d. By opening it

29. Practicing engineers and technicians use
    a. Exact methods of calculation only
    b. Approximations
    c. Computers for everything
    d. Precise methods only

30. Electronics is
    a. A cold and hard science
    b. A rigid and precise discipline
    c. Always logical and exact
    d. An art and a science

BASIC PROBLEMS

Sec. 1-2 Voltage Sources

1-1. Suppose a voltage source has an ideal voltage of 12 V and an internal resistance of 0.5 Ω. For what values of load resistance will the voltage source appear stiff?

1-2. A load resistance may vary from 270 Ω to 100 kΩ. If a stiff voltage source drives this load resistance, what is internal resistance of the source?

1-3. A flashlight battery has an internal resistance of 1 Ω. For what values of load resistance does the flashlight battery appear stiff?

1-4. A car battery has an internal resistance of 0.06 Ω. For what values of load resistance does the car battery appear stiff?

1-5. The internal resistance of a voltage source equals 0.05 Ω. How much voltage is dropped across this internal resistance when the current through it equals 2 A?

1-6. In Fig. 1-13, the ideal voltage is 9 V and the internal resistance is 0.4 Ω. If the load resistance is zero, what is the load current?

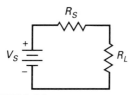

FIGURE 1-13

Sec. 1-3 Current Sources

1-7. Suppose a current source has an ideal current of 10 mA and an internal resistance of 20 MΩ. For what values of load resistance will the current source appear stiff?

1-8. A load resistance may vary from 270 Ω to 100 kΩ. If a stiff current source drives this load resistance, what is the internal resistance of the source?

1-9. A current source has an internal resistance of 100 kΩ. What is the largest load resistance if the current source must appear stiff?

1-10. In Fig. 1-14, the ideal current is 10 mA and the internal resistance is 100 kΩ. If the load resistance equals zero, what does the load current equal?

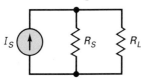

FIGURE 1-14

1-11. The ideal current is 5 mA and the internal resistance is 250 kΩ in Fig. 1-14. If the load resistance is 10 kΩ, what is the load current? Is this a stiff current source?

Sec. 1-4 Thevenin's Theorem

1-12. What is the Thevenin voltage in Fig. 1-15? The Thevenin resistance?

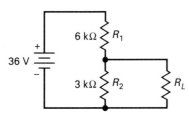

FIGURE 1-15

1-13. Calculate the load current in Fig. 1-15 for each of these load resistances: 0, 1 kΩ, 2 kΩ, 3 kΩ, 4 kΩ, 5 kΩ, and 6 kΩ.

1-14. The voltage source of Fig. 1-15 is decreased to 12 V. What happens to the Thevenin voltage? To the Thevenin resistance?

1-15. All resistances are doubled in Fig. 1-15. What happens to the Thevenin voltage? To the Thevenin resistance?

Sec. 1-5 Norton's Theorem

1-16. A circuit has a Thevenin voltage of 15 V and a Thevenin resistance of 3 kΩ. What is the Norton circuit?

1-17. A circuit has a Norton current of 10 mA and a Norton resistance of 10 kΩ. What is the Thevenin circuit?

1-18. What is the Norton circuit for Fig. 1-15?

Sec. 1-6 Troubleshooting

1-19. Suppose the load voltage of Fig. 1-15 is 36 V. What is wrong with R_1?

1-20. The load voltage of Fig. 1-15 is zero. The battery and the load resistance are all right. Suggest two possible troubles.

1-21. If the load voltage is zero in Fig. 1-15 and all resistors are normal, where does the trouble lie?

1-22. The load voltage is 12 V in Fig. 1-15. What is the probable trouble?

ADVANCED PROBLEMS

1-23. A voltage source is temporarily shorted. If the ideal voltage is 6 V and the shorted-load current is 150 A, what is the internal resistance of the source?

1-24. In Fig. 1-13, the ideal voltage is 10 V and the load resistance is 75 Ω. If the load voltage equals 9 V, what does the internal resistance equal? Is the voltage source stiff?

1-25. Somebody hands you a black box with a 2-kΩ resistor connected across the exposed load terminals. How can you measure the Thevenin voltage?

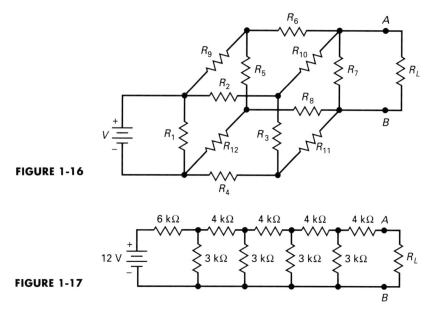

FIGURE 1-16

FIGURE 1-17

1-26. The black box in Prob. 1-25 has a knob on it that allows you to reduce all internal voltage and current sources to zero. How can you measure the Thevenin resistance?

1-27. Solve Prob. 1-13. Then solve the same problem without using Thevenin's theorem. After you are finished, comment on what you have learned about the Thevenin theorem.

1-28. You are in the laboratory looking at a circuit like Fig. 1-16. Somebody challenges you to find the Thevenin circuit driving the load resistor. Describe an experimental procedure for measuring the Thevenin voltage and the Thevenin resistance.

1-29. Design a hypothetical current source using the battery and resistor. The current source must meet the following specifications: It must supply a stiff 1 mA of current to any load resistance between 0 and 10 kΩ.

1-30. Design a voltage divider (similar to Fig. 1-15) that meets these specifications: Ideal source voltage is 30 V, open-load voltage is 15 V, and Thevenin resistance is equal to or less than 2 kΩ.

1-31. Design a voltage divider like Fig. 1-15 so that it produces a stiff 10 V to all load resistances greater than 1 MΩ. Use an ideal voltage of 30 V.

1-32. Somebody hands you a D cell flashlight battery and a volt-ohm-milliammeter (VOM). You have nothing else to work with. Describe an experimental method for finding the Thevenin equivalent circuit of the flashlight battery.

1-33. You have a D cell flashlight battery, a VOM, and a box of different resistors. Describe a method that uses one of the resistors to find the Thevenin resistance of the battery.

1-34. Calculate the load current in Fig. 1-17 for each of these load resistances: 0, 1 kΩ, 2 kΩ, 3 kΩ, 4 kΩ, 5 kΩ, and 6 kΩ.

T-SHOOTER PROBLEMS

Use Fig. 1-18, on page 26, for the remaining problems. If you haven't already done so, read Example 1-5 before attempting these problems. The T-shooter is a simplified textbook version of software written for computer-assisted instruction. You will find the T-shooter useful for building your troubleshooting skills. You can measure voltages in any order; for instance, V_E first, V_A second, and V_B third, or whatever. These voltages are the clues to the trouble. After measuring one or more voltages, try to figure out what the trouble is. The possible troubles are an open resistor, a shorted resistor, an open ground, and no supply voltage.

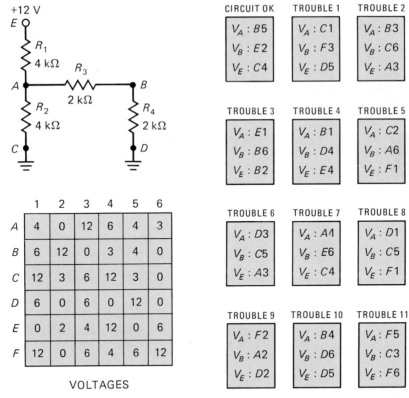

FIGURE 1-18 T-Shooter™. (*Patent pending: Courtesy of Malvino Inc.*)

1-35. What is causing Trouble 1?

1-36. What is causing Trouble 2?

1-37. What is causing Troubles 3 through 6?

1-38. What is causing Troubles 7 through 11?

SEMICONDUCTORS

To understand how diodes, transistors, and integrated circuits work, you first have to study semiconductors: materials that are neither conductors nor insulators. Semiconductors contain some free electrons but what makes them unusual is the presence of holes. In this chapter you will learn about semiconductors, holes, and other related topics.

2-1 CONDUCTORS

Copper is a good conductor. The reason for this is clear when we look at its atomic structure as shown in Fig. 2-1. The nucleus or center of the atom contains 29 protons (positive charges). When a copper atom has a neutral charge, 29 electrons (negative charges) circle the nucleus like planets around the sun. Notice that 2 electrons are in the first orbit, 8 in the second, 18 in the third, and 1 in the fourth.

Stable Orbits

The positive nucleus of Fig. 2-1 attracts the planetary electrons. The reason these electrons are not pulled into the nucleus is because of the centrifugal or outward force created by their orbital motion. When an electron is in a stable orbit, the centrifugal force exactly equals the inward attraction of the nucleus. The centrifugal force decreases for slower electrons. This is why the electron in the largest orbit travels more slowly than an electron in a smaller orbit. Less centrifugal force is needed to offset the attraction of the nucleus.

The Core and the Free Electron

In Fig. 2-1, the nucleus and the inner electrons are of little interest in the study of electronics. Our focus throughout most of this book will be the outer orbit, also called the *valence orbit*. This outer orbit controls the electrical properties of the atom. To emphasize the importance of the outer orbit, we can define the core of an atom as the nucleus and all the

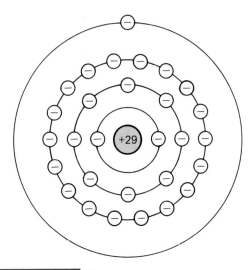

FIGURE 2-1   Copper atom

inner orbits. For a copper atom, the core is the nucleus ($+29$) and the first three orbits (-28).

The core of a copper atom has a net charge of $+1$ because it contains 29 protons and 28 inner electrons. Since the valence electron is in a large orbit around a core with a net charge of only $+1$, the inward pull felt by the outer electron is very small. In fact, the electron in the outer orbit barely feels any force of attraction. Because the attraction is so weak, the outer electron is often called a *free electron*.

Main Idea

The most important idea for you to remember about a copper atom is this: because the valence electron is only weakly attracted by the core, an outside force can easily dislodge this free electron from the copper atom. This is why copper is a good conductor. The slightest voltage can cause the free electrons in a copper wire to flow from one atom to the next. The best conductors (silver, copper, and gold) have a single valence electron.

EXAMPLE 2-1

Suppose the free electron moves away from the copper atom of Fig. 2-1. What is the net charge of the copper atom?

SOLUTION

Since the nucleus has 29 protons and the inner orbits have 28 electrons, the copper atom has a net charge of $+1$. Whenever a neutral atom loses one or more of its electrons, it becomes a positively charged atom and is referred to as a positive ion. If a neutral atom gains electrons, it becomes negatively charged and is called a negative ion.

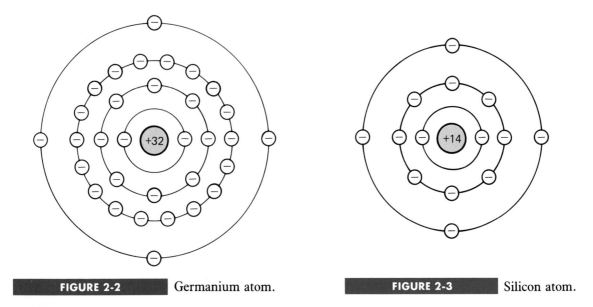

FIGURE 2-2 Germanium atom.   **FIGURE 2-3** Silicon atom.

2-2 SEMICONDUCTORS

A semiconductor is an element with a valence of four. This means an isolated atom of the material has four electrons in its outer or valence orbit. The number of electrons in the valence orbit is the key to electrical conductivity. Conductors have one valence electron, semiconductors have four valence electrons, and insulators have eight valence electrons.

Germanium

Germanium is an example of a semiconductor, a material whose resistivity is between that of a conductor and that of an insulator. Figure 2-2 shows a germanium atom. In the center is a nucleus with 32 protons. This time, the orbiting electrons distribute themselves as follows: 2 electrons in the first orbit, 8 in the second orbit, and 18 in the third. The last four electrons are in the outer or valence orbit. The main thing to remember about the germanium atom is that it has four valence electrons. This property applies to all semiconductors. This is how you can tell if a substance is a semiconductor. If it is a semiconductor, it will have four electrons in the outer or valence orbit.

Silicon

The most widely used semiconductor material is silicon. An isolated silicon atom has 14 protons and 14 electrons. As shown in Fig. 2-3, the first orbit contains two electrons and the second orbit contains eight electrons. The four remaining electrons are in the outer orbit.

In Fig. 2-3, the nucleus and the first two orbits are the core of the silicon atom. This core has a net charge of $+4$ because of the 14 protons in the nucleus and 10 electrons in the first two orbits. Notice the four electrons in the outer or valence orbit. This tells us that silicon is a semiconductor.

EXAMPLE 2-2

What is the net charge of the silicon atom of Fig. 2-3 if it loses one of its valence electrons? What is its net charge if it loses all four valence electrons?

SOLUTION

Initially, the silicon atom is neutral because it has 14 protons and 14 electrons. If it loses one valence electron, it becomes a positive ion with a charge of $+1$. If the silicon atom loses all four of its valence electrons, it becomes a positive ion with a charge of $+4$.

2-3 SILICON CRYSTALS

When silicon atoms combine to form a solid, they automatically arrange themselves into an orderly pattern called a *crystal*. Each silicon atom shares its electrons with the other silicon atoms in such a way as to have eight electrons in the valence orbit as shown in Fig. 2-4. Whenever an atom has eight electrons in its valence orbit as shown here, it becomes chemically stable. The shaded circles represent the silicon cores. Although the central atom originally has four electrons in its valence orbit, it now has eight electrons in this orbit.

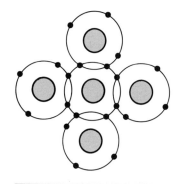

FIGURE 2-4

Covalent bonds.

Covalent Bonds

Each neighboring atom shares an electron with the central atom. In this way, the central atom appears to have four additional electrons, giving it a total of eight electrons in its valence orbit. Actually, the electrons no longer belong to any single atom; they are now shared by adjacent atoms.

In Fig. 2-4, each core has a charge of $+4$. Look at the central core and the one to its right. These two cores attract the pair of electrons between them with equal and opposite forces. This pulling in opposite directions is what holds the silicon atoms together. The idea is similar to tug-of-war teams pulling on a rope. As long as both teams pull with equal and opposite forces, they remain bonded together.

Since each shared electron in Fig. 2-4 is being pulled in opposite directions, the electron is a bond between the opposite cores. This type of chemical bond is known as a *covalent bond*. In a silicon crystal, there are billions of silicon atoms, each with eight valence electrons. These valence electrons are the covalent bonds that hold the crystal together, that give it solidity.

No More Than Eight Valence Electrons

Each atom in a silicon crystal has eight electrons in its valence orbit. These eight electrons produce a chemical stability that results in a solid piece of silicon material. There are advanced mathematical equations that partially explain why eight electrons produce chemical stability in different materials, but no one really knows the ultimate reason why the number eight is so special. It is one of those experimental laws like the law of gravity. Physicists can observe it and prove it by experiments, but they cannot give you the ultimate reason for its existence.

The valence orbit can hold no more than eight electrons. Because of this, it is described as filled or saturated when it contains eight electrons. Furthermore, the eight valence electrons are called bound electrons because they are tightly held by the atoms. Because of these bound electrons, a silicon crystal is almost a perfect insulator at room temperature (approximately 25°C).

Heat Energy Can Create a Hole

The ambient temperature is the temperature of the surrounding air. When the ambient temperature is above absolute zero (-273°C), the heat energy of the surrounding air causes the atoms in a silicon crystal to vibrate back and forth inside the silicon crystal. The higher the ambient temperature, the stronger the mechanical vibrations of these atoms. If you pick up a warm object, the warmth you feel is caused by its vibrating atoms.

The vibrations of the silicon atoms can occasionally dislodge an electron from the valence orbit. When this happens, the released electron gains enough energy to go into a larger orbit as shown in Fig. 2-5. In this larger orbit, the electron is a free electron. Furthermore, the departure of the electron leaves a vacancy in the valence orbit that is called a hole. This hole behaves like a positive charge in the sense that it will attract and capture any electron in the immediate vicinity.

The number of free electrons and holes created by heat energy is very small in a silicon crystal. At room temperature, there are so few free electrons and holes that a silicon crystal acts almost like a perfect insulator.

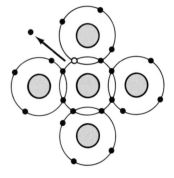

FIGURE 2-5

Heat energy produces free electron.

Recombination and Lifetime

In a pure silicon crystal, equal numbers of free electrons and holes are created by thermal (heat) energy. The free electrons move randomly throughout the crystal. Occasionally, a free electron will approach a hole, feel its attraction, and fall into it. This merging of a free electron and a hole is called *recombination*.

The amount of time between the creation and disappearance of a free electron is called the lifetime. It varies from a few nanoseconds to several microseconds, depending on how perfect the crystal is and other factors.

Main Ideas

At any instant, the following is taking place inside a silicon crystal:

1. Some free electrons and holes are being created by thermal energy.

2. Other free electrons and holes are recombining.

3. Some free electrons and holes exist in an in-between state; they were previously created and have not yet recombined.

EXAMPLE 2-3

If a pure silicon crystal has 1 million free electrons inside it, how many holes does it have? What happens to the number of free electrons and holes if the ambient temperature increases?

SOLUTION

Look at Fig. 2-5. When heat energy creates a free electron, it automatically creates a hole at the same time. Therefore, a pure silicon crystal always has the same number of holes and free electrons. If there are 1 million free electrons, there are 1 million holes.

A higher temperature increases the vibrations at the atomic level, which means that more free electrons and holes are created. But no matter what the temperature is, a pure silicon crystal has the same number of free electrons and holes.

2-4   INTRINSIC SEMICONDUCTORS

An *intrinsic* semiconductor is a pure semiconductor. A silicon crystal is an intrinsic semiconductor if every atom in the crystal is a silicon atom. At room temperature, a silicon crystal acts approximately like an insulator because it has only a few free electrons and holes produced by thermal energy. In this section we will look at how an intrinsic silicon crystal conducts.

Flow of Free Electrons

Figure 2-6 shows part of a silicon crystal between charged metallic plates. Assume that thermal energy has produced a free electron and a hole. The free electron is in a large orbit at the right end of the crystal. Because of the negatively charged plate, the free electron is repelled to the left. This free electron can move from one large orbit to the next until it reaches the positive plate.

Flow of Holes

Notice the hole at the left of Fig. 2-6. This hole attracts the valence electron at point A. This causes the valence electron to move into the hole. This action is not the same as recombination where a free electron falls into a hole. Instead of a free electron, we have a valence electron moving into a hole.

When the valence electron at point A moves to the left, it creates a new hole at point A. The effect is the same as moving the original hole to the right. The new hole at point A can then attract and capture another valence electron. In this way, valence electrons can travel along the path shown by the arrows. This means the hole can move the opposite way, along path A-B-C-D-E-F.

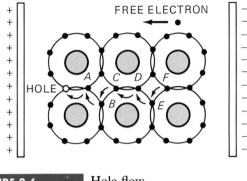

| FIGURE 2-6 | Hole flow.

A semiconductor has two distinct types of flow. It has the flow of free electrons in large orbits, and it has the flow of holes in small orbits. This second type of flow occurs only in semiconductors and is what makes them useful in radio, television, and computers. You will learn why in later chapters.

2-5 TWO TYPES OF FLOW

Figure 2-7 shows an intrinsic semiconductor. Notice that it has the same number of free electrons and holes. This is because thermal energy produces free electrons and holes in pairs. The applied voltage will force the free electrons to flow left and the holes to flow right. When the free electrons arrive at the left end of the crystal, they enter the external wire and flow to the positive battery terminal. On the other hand, the free electrons at the negative battery terminal will flow to the right end of the crystal. At this point, they enter the crystal and recombine with holes that arrive at the right end of the crystal. In this way, a steady flow of free electrons and holes occurs inside the semiconductor.

In Fig. 2-7, the free electrons and holes move in opposite directions. From now on, we will visualize the current in a semiconductor as the combined effect of the two types of flow: the flow of free electrons in one direction and the flow of holes in the other direction. Free electrons and holes are often called *carriers* because they carry a charge from one place to another.

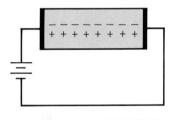

| FIGURE 2-7 |

Intrinsic semiconductor.

2-6 DOPING A SEMICONDUCTOR

One way to increase conductivity of a semiconductor is by doping. This means deliberately adding impurity atoms to an intrinsic crystal to alter its electrical conductivity. A doped semiconductor is called an *extrinsic* semiconductor.

Increasing the Free Electrons

How does a manufacturer dope a silicon crystal? The first step is to melt a pure silicon crystal. This breaks the covalent bonds and changes the

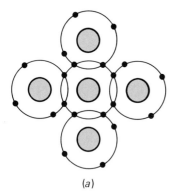

● FREE ELECTRON

(a)

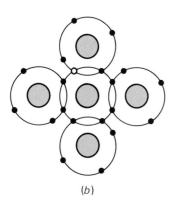

(b)

FIGURE 2-8

(a) Increasing free electrons; (b) increasing holes.

silicon from a solid to a liquid. To increase the number of free electrons, pentavalent atoms are added to the molten silicon. Pentavalent atoms have five electrons in the valence orbit. Examples of pentavalent atoms include arsenic, antimony, and phosphorus. Because these materials will donate an extra electron to the silicon crystal, they are often referred to as *donor* impurities.

Figure 2-8*a* shows how the doped silicon crystal appears after it cools down and reforms its solid crystal structure. A pentavalent atom is in the center surrounded by four silicon atoms. As before, the neighboring atoms share an electron with the central atom. But this time, there is an extra electron left over. Remember that each pentavalent atom has five valence electrons. Since only eight electrons can fit into the valence orbit, the extra electron remains in a larger orbit. In other words, it is a free electron.

Each pentavalent or donor atom in a silicon crystal produces one free electron. This is how a manufacturer controls the conductivity of a doped semiconductor. The more impurity that is added, the greater the conductivity. In this way, a semiconductor may be lightly or heavily doped. A lightly doped semiconductor has a high resistance, while a heavily doped semiconductor has a low resistance.

Increasing the Number of Holes

How can we dope a pure silicon crystal to get an excess of holes? By using a trivalent impurity, one whose atoms have only three valence electrons. Examples include aluminum, boron, and gallium.

Figure 2-8*b* shows a trivalent atom in the center. It is surrounded by four silicon atoms, each sharing one of its valence electrons. Since the trivalent atom originally had only three valence electrons and each neighbor shares one electron, only seven electrons are in the valence orbit. This means a hole exists in the valence orbit of each trivalent atom. A trivalent atom is also called an *acceptor* atom because each hole it contributes can accept a free electron during recombination.

Points to Remember

Before manufacturers can dope a semiconductor, they must produce it as an absolutely pure crystal. Then by controlling the amount of impurity, they can precisely control the properties of the semiconductor. Historically, pure germanium crystals were easier to produce than pure silicon crystals. This is why the earliest semiconductor devices were made of germanium. Eventually, manufacturing techniques improved and pure silicon crystals became available. Because of its advantages, silicon has emerged as the most popular and useful semiconductor material. (One big advantage is that heat energy produces fewer free electrons and holes in silicon than germanium. A later discussion tells you why this is important.)

EXAMPLE 2-4

A doped semiconductor has 10 billion silicon atoms and 15 million pentavalent atoms. If the ambient temperature is 25°C, how many free electrons and holes are there inside the semiconductor?

SOLUTION

Each pentavalent atom contributes one free electron. Therefore, the semiconductor has 15 million free electrons produced by doping. There will be almost no holes by comparison because the only holes in the semiconductor are those produced by heat energy.

2-7 TWO TYPES OF EXTRINSIC SEMICONDUCTORS

A semiconductor can be doped to have an excess of free electrons or an excess of holes. Because of this, there are two types of doped semiconductors.

n-type Semiconductor

Silicon that has been doped with a pentavalent impurity is called an n-type semiconductor, where the n stands for negative. Figure 2-9 shows an n-type semiconductor. Since the free electrons outnumber the holes in an n-type semiconductor, the free electrons are called the *majority carriers* and the holes are called the *minority carriers*.

Because of the applied voltage, the free electrons inside the semiconductor move to the left and the holes move to the right. When a hole arrives at the right end of the crystal, one of the free electrons from the external circuit enters the semiconductor and recombines with the hole. Stated another way, the free electron from the external circuit enters the semiconductor and becomes a valence electron.

The free electrons shown in Fig. 2-9 flow to the left end of the crystal where they enter the wire and flow on to the positive terminal of the battery. In addition to the free electrons, a valence electron occasionally leaves the left end of the crystal. The departure of this valence electron creates a hole at the left end of the crystal.

In the n-type semiconductor of Fig. 2-9, there are many free electrons at room temperature and only a few holes. Because of this, the flow of the free electrons (majority carriers) has a much greater effect in an n-type semiconductor. The only reason for mentioning the holes (minority carriers) is because they become important in later discussions.

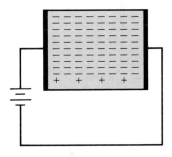

FIGURE 2-9

n-type semiconductor.

p-type Semiconductor

Silicon that has been doped with a trivalent impurity is called a p-type semiconductor, where the p stands for positive. Figure 2-10 shows a p-type semiconductor. Since holes outnumber free electrons, the holes are referred to as the majority carriers and the free electrons are known as the minority carriers.

Because of the applied voltage, the free electrons move to the left and the holes move to the right. In Fig. 2-10, the holes arriving at the right end of the crystal will recombine with free electrons from the external circuit. Stated another way, free electrons leave the negative end of the

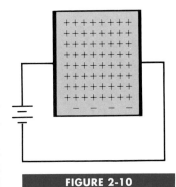

FIGURE 2-10

p-type semiconductor.

battery and arrive at the right end of the semiconductor. At this point, they enter the right end of the semiconductor and become valence electrons. These valence electrons flow to the left end of the crystal, equivalent to holes flowing to the right. When the valence electrons leave the left end of the crystal, they become free electrons in the external circuit.

There is also a flow of minority carriers in Fig. 2-10. The free electrons inside the semiconductor flow from right to left. Because there are so few of the minority carriers, they have almost no effect in this circuit.

2-8 THE UNBIASED DIODE

By itself, a piece of *n*-type semiconductor is about as useful as a carbon resistor; the same can be said for a *p*-type semiconductor. But when a manufacturer dopes a crystal so that one half of it is *p*-type and the other half is *n*-type, something new comes into existence.

The border between *p*-type and *n*-type is called the *pn* junction. The *pn* junction has such useful properties that it has led to all kinds of inventions including diodes, transistors, and integrated circuits. Understanding the *pn* junction enables you to understand all kinds of semiconductor devices. This is why we now will discuss the *pn* junction in detail.

The Unbiased Diode

As discussed in the preceding chapter, each pentavalent atom in a silicon crystal produces one free electron. For this reason, we can visualize a piece of *n*-type semiconductor as shown on the right side of Fig. 2-11. Each circled plus sign represents a pentavalent atom, and each minus sign is the free electron that it contributes to the semiconductor.

Similarly, we can visualize the trivalent atoms and holes of a *p*-type semiconductor as shown on the left side of Fig. 2-11. Each circled minus sign is the trivalent atom and each plus sign is the hole in its valence orbit. Notice that each piece of semiconductor material is electrically neutral because the number of pluses and minuses is equal.

A manufacturer can produce a single crystal with *p*-type material on one side and *n*-type on the other side as shown in Fig. 2-12. The junction is the border where the *p*-type and the *n*-type regions meet, and *junction*

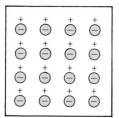

FIGURE 2-11 Separate pieces of semiconductor.

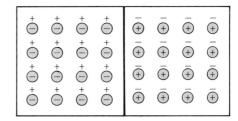

FIGURE 2-12 Crystal.

diode is another name for a *pn* crystal. The word diode is a contraction of two electrodes, where "di" stands for two.

The Depletion Layer

Because of their repulsion for each other, the free electrons on the *n* side of Fig. 2-12 tend to diffuse (spread) in all directions. Some of the free electrons diffuse across the junction. When a free electron enters the *p* region, it becomes a minority carrier. With so many holes around it, this minority carrier has a short lifetime. Soon after entering the *p* region, the free electron falls into a hole. When this happens, the hole disappears and the free electron becomes a valence electron.

Each time an electron diffuses across a junction, it creates a pair of ions. When an electron leaves the *n* side, it leaves behind a pentavalent atom that is short one negative charge; this pentavalent atom becomes a positive ion. After the migrating electron falls into a hole on the *p* side, it makes a negative ion out of the trivalent atom that captures it.

Figure 2-13 shows these ions on each side of the junction. The circled plus signs are the positive ions, and the circled minus signs are the negative ions. The ions are fixed in the crystal structure because of covalent bonding and cannot move around like free electrons and holes.

Each pair of positive and negative ions at the junction is called a dipole. The creation of a dipole means one free electron and one hole have been taken out of circulation. As the number of dipoles builds up, the region near the junction is emptied of carriers. We call this charge-empty region the *depletion layer*.

Barrier Potential

Each dipole has an electric field between the positive and negative ions. Therefore, if additional free electrons enter the depletion layer, the electric field tries to push these electrons back into the *n* region. The strength of the electric field increases with each crossing electron until equilibrium is reached. To a first approximation, this means the electric field eventually stops the diffusion of electrons across the junction.

In Fig. 2-13, the electric field between the ions is equivalent to a difference of potential called the *barrier potential*. At 25°C, the barrier potential approximately equals 0.3 V for germanium diodes and 0.7 V for silicon diodes.

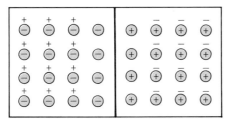

FIGURE 2-13 Depletion layer.

2-9 FORWARD BIAS

Figure 2-14 shows a dc source across a diode. The negative source terminal is connected to the *n*-type material, and the positive terminal is connected to the *p*-type material. This connection is called *forward bias*.

Flow of Free Electrons

Current flows easily in a circuit like Fig. 2-14. Why? Because the battery will force free electrons and holes to flow toward the junction. When the free electrons move toward the junction, positive ions are created at the right end of the crystal. These positive ions pull electrons into the crystal from the external circuit. In this way, free electrons can leave the negative source terminal and flow into the right end of the crystal.

When we look at Fig. 2-14, here is what we see. Electrons are pouring into the right end of the crystal, while the bulk of electrons in the *n* region moves toward the junction. The left edge of this moving group disappears as it hits the junction (the electrons fall into holes). In this way, there is a continuous stream of electrons from the negative source terminal toward the junction.

Flow of Valence Electrons

What happens to the free electrons that disappear at the junction? They become valence electrons. As valence electrons, they move through holes in the *p* region. In other words, the valence electrons on the *p* side move toward the left end of the crystal, equivalent to holes moving toward the junction. When the valence electrons reach the left end of the crystal, they enter the external circuit and flow to the positive terminal of the source.

Recap

Here is what happens to an electron in Fig. 2-14:

1. After leaving the negative source terminal, it enters the right end of the crystal.

2. It travels through the *n* region as a free electron.

3. At the junction it recombines with a hole and becomes a valence electron.

4. It travels through the *p* region as a valence electron.

5. After leaving the left end of the crystal, it flows into the positive source terminal.

What to Remember

Current flows easily in a forward-biased silicon diode. As long as the applied voltage is greater than the barrier potential, there will be a large continuous current in the circuit. This is worth repeating: *Current flows easily in a forward-biased diode.* In other words, if the source voltage is greater than 0.7 V, a silicon diode produces a continuous current in the forward direction.

EXAMPLE 2-5

Suppose the battery current is 1 mA in Fig. 2-14. What does this tell you?

SOLUTION

Ohm's and Kirchhoff's laws are the universal laws of electricity. Even though we are in new territory, we can continue to apply these basic laws. In Fig. 2-14, we have a series circuit. Kirchhoff's current law tells us the current is the same in all parts of a series circuit. Therefore, a battery current of 1 mA means the current is 1 mA through the p side, 1 mA through the junction, 1 mA through the n side, etc.

2-10 REVERSE BIAS

Turn the dc source around and you reverse-bias the diode as shown in Fig. 2-15. This time, the negative battery terminal is connected to the p side, and the positive battery terminal to the n side. This particular connection is called *reverse bias*.

Depletion Layer Widens

The negative battery terminal attracts the holes, and the positive battery terminal attracts the free electrons. Because of this, holes and free electrons flow away from the junction. Therefore, the depletion layer gets wider.

How wide does the depletion layer get? When the holes and electrons move away from the junction, the newly created ions increase the difference of potential across the depletion layer. The wider the depletion layer, the

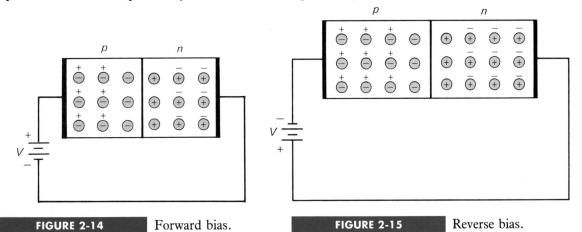

| **FIGURE 2-14** | Forward bias. | **FIGURE 2-15** | Reverse bias. |

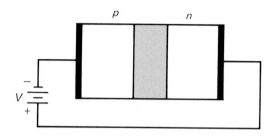

FIGURE 2-16   Depletion layer.

greater the difference of potential. The depletion layer stops growing when its difference of potential equals the applied reverse voltage. When this happens, electrons and holes stop moving away from the junction.

Sometimes the depletion layer is shown as a shaded region like that of Fig. 2-16. The width of this shaded region is proportional to the reverse voltage. Therefore, as the reverse voltage increases, the depletion layer gets wider. This phenomenon is used to produce the varactor diode. (Varactor diodes are used to set the resonant frequency of the tuning circuits that select a particular radio station or TV channel.)

Minority-Carrier Current

Is there any current at all after the depletion layer settles down? Yes. A small current exists with reverse bias. Recall that thermal energy continuously creates pairs of free electrons and holes. This means a few minority carriers exist on both sides of the junction. Most of these recombine with the majority carriers. But those inside the depletion layer may live long enough to get across the junction. When this happens, a small current flows in the external circuit.

Figure 2-17 illustrates the idea. Assume that thermal energy has just created a free electron and hole near the junction. The depletion layer pushes the free electron to the right, forcing one electron to leave the right end of the crystal. The hole in the depletion layer is pushed to the left. This extra hole on the p side lets one electron enter the left end of

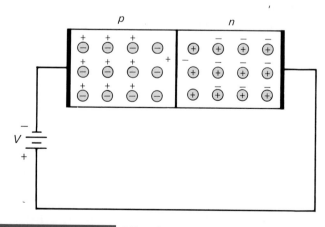

FIGURE 2-17   Minority carrier current.

the crystal and fall into a hole. Since thermal energy is continuously producing electron-hole pairs inside the depletion layer, we will get a small continuous current in the external circuit.

The reverse current caused by the thermally-produced minority carriers is called the *saturation current*. In equations, the saturation current is symbolized by I_S. The name *saturation* reminds us that we cannot get more minority-carrier current than is produced by the thermal energy. In other words, increasing the reverse voltage will not increase the number of thermally created minority carriers. This is a function of temperature alone.

The higher the junction temperature, the greater the saturation current. As mentioned earlier, silicon has come to dominate the semiconductor industry. One of the main reasons is because there are fewer minority carriers in silicon diodes than in germanium diodes. In other words, a silicon diode has a much smaller I_S than a germanium diode of the same size and shape.

Surface-Leakage Current

Besides the thermally produced minority-carrier current, does any other current flow in a reverse-biased diode? Yes. A small current flows on the surface of the crystal. Known as the *surface-leakage current*, it is caused by surface impurities and imperfections in the crystal structure. (If you want to know more about this, see "Optional Topics," on page 43.)

What to Remember

The total reverse current in a diode consists of a minority-carrier current (very small and dependent on temperature) and a surface-leakage current (very small and directly proportional to voltage). In most applications, the reverse current in a silicon diode is so small that you don't even notice it. The main idea to remember is this: *Current is approximately zero in a reverse-biased silicon diode*. In fact, a silicon diode acts approximately the same as an open switch when it is reverse-biased.

2-11 BREAKDOWN

You already know that a resistor has a maximum power rating. Diodes also have maximum power ratings, but they also have maximum voltage ratings. In other words, there is a limit to how much reverse voltage a diode can withstand before it is destroyed. This section tells you why such a limit exists.

Avalanche Effect

Keep increasing the reverse voltage and you eventually reach the breakdown voltage. For rectifier diodes (those manufactured to conduct better one way than the other), the breakdown voltage is usually greater than 50 V. Once the breakdown voltage is reached, a large number of the minority carriers suddenly appears in the depletion layer and the diode conducts heavily.

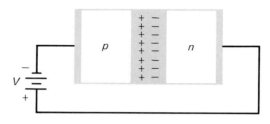

FIGURE 2-18   Avalanche.

Where do the carriers come from? They are produced by the *avalanche effect* (see Fig. 2-18), which occurs at higher reverse voltages. Here is how it works. As usual, there is a small reverse minority-carrier current. When the reverse voltage increases, it forces the minority carriers to move more quickly. These minority carriers will collide with the atoms of the crystal. When these minority carriers have enough energy, they can knock valence electrons loose, that is, produce free electrons. These new minority carriers can then join the existing minority carriers to collide with other atoms. The process is geometric because one free electron liberates one valence electron to get two free electrons. These two free electrons then free two more electrons to get four free electrons. The process continues until the reverse current is huge.

Figure 2-19 shows a magnified view of the depletion layer. The reverse bias forces the free electron to move to the right. As it moves, the electron gains speed. The larger the reverse bias, the faster the electron moves. If the high-speed electron has enough energy, it can bump the valence electron of the first atom into a larger orbit. This results in two free electrons. Both of these can then accelerate and go on to dislodge two

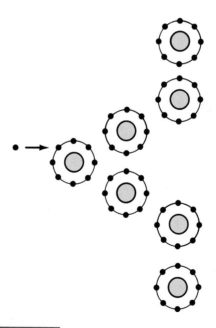

FIGURE 2-19   How avalanche occurs.

more electrons. In this way, the number of minority carriers may become quite large and the diode can conduct heavily.

The breakdown voltage of a diode depends on how heavily doped the diode is. With rectifier diodes (the most common type), the breakdown voltage is usually greater than 50 V. When using a rectifier diode, you should not exceed the breakdown voltage. The diode manufacturer specifies the largest reverse voltages the diodes can handle. We will discuss these maximum ratings in the next chapter.

Zener Effect

When a diode is heavily doped, the depletion layer is very narrow. Because of this, the electric field across the depletion layer (voltage divided by distance) is very intense. When the field strength reaches approximately 300,000 V/cm, the field is intense enough to pull electrons out of their valence orbits. The creation of free electrons in this way is called the *zener effect* (also known as *high-field emission*). This is distinctly different from the avalanche effect, which depends on high-speed minority carriers knocking valence electrons loose. The zener effect depends only on the intensity of the electric field.

The zener effect appears for breakdown voltages less than 4 V, whereas the avalanche effect requires a reverse voltage of at least 6 V before it appears. Between 4 and 6 V, both effects may be present. In a later chapter, we will discuss special devices optimized for use in the breakdown region. Known as *zener diodes*, these devices are widely used to regulate voltage. More will be said about this later.

OPTIONAL TOPICS

The following material continues the earlier discussions at a more advanced and specialized level. All the topics are optional because they are not used in any of the basic discussions in later chapters. This section will be a useful reference when you are in industry because then you will probably want more advanced viewpoints.

2-12 ENERGY LEVELS

To a good approximation, we can identify the total energy of an electron with the size of its orbit. That is, we can think of each radius of Fig. 2-20a as equivalent to an energy level in Fig. 2-20b. Electrons in the smallest orbit are on the first energy level; electrons in the second orbit are on the second energy level, and so on.

Higher Energy in Larger Orbit

Since the electron is attracted by the nucleus, extra energy is needed to lift the electron into a larger orbit. When an electron is moved from the first to the second orbit, it gains potential energy with respect to the nucleus. Some of the external forces that can lift an electron to higher energy levels are heat, light, and voltage.

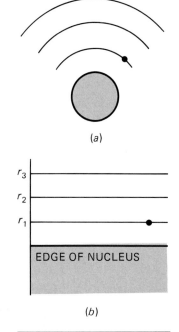

(a)

(b)

FIGURE 2-20

Energy levels.

For instance, assume an outside force lifts the electron from the first to the second orbit. This electron has more potential energy because it is further from the nucleus. The idea is similar to an object above the earth. The higher the object, the greater its potential energy with respect to the earth. If the object is released, it can fall further and do more work when it hits the earth.

Falling Electrons Radiate Light

After an electron has been moved into a larger orbit, the electron may fall back to its original energy level. If it does, the electron gives up its extra energy in the form of heat, light, and other radiation. The energy lost by a falling electron equals the amount of energy that is radiated out of the atom. Since different elements have different energy levels, the color of the radiated light (red, green, orange, etc.) depends on the material being used.

The principle behind the light-emitting diode (LED) is based on energy levels. With this type of device, the applied voltage lifts the electrons to higher energy levels. When these electrons fall back to their original energy levels, they give off light. Depending on the material used, the light is red, green, orange, blue, etc.

Energy bands

At absolute zero temperature, the semiconductor of Fig. 2-21 has no free electrons. One way to understand this is the concept of energy levels. No two electrons in a crystal can be at exactly the same energy level. Because of this, all electrons in the first orbit have slightly different energy levels. This is why the first energy level of Fig. 2-21 is shown as a band of energy levels rather than a single horizontal line. In a similar way, the second-orbit electrons lie within the second band, and the valence electrons lie in the valence band.

At −273°C (absolute zero temperature) all of the valence electrons are tightly held in the valence band of energy. But at room temperature,

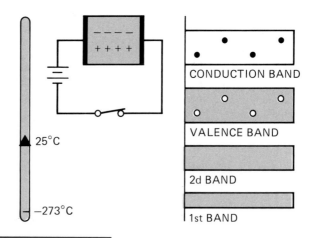

FIGURE 2-21

Energy bands for intrinsic semiconductor.

thermal energy can occasionally lift a valence electron into the conduction band. The additional energy allows a few electrons to move from the valence band to the conduction band as shown in Fig. 2-21. The free electrons are in conduction-band orbits, while the holes are in valence-band orbits. The current consists of free electrons flowing through the conduction band and holes flowing through the valence band. Because thermal energy produces only a few carriers in an intrinsic semiconductor, the current is too small for most applications. What we need for semiconductors to be practical is a way of increasing either the free electrons or the holes.

n-type Energy Bands

What do the energy bands look like for an *n*-type semiconductor? At absolute zero temperature, the conduction band has many free electrons, but the valence band has no holes. At room temperature, however, thermal energy produces a few minority carriers. Figure 2-22 summarizes the situation. The conduction band has many free electrons, while the valence band has only a few holes.

You can see why free electrons are the majority carriers in an *n*-type semiconductor. They are the only available carriers, except for an occasional hole in the valence band. The free electrons travel in large conduction-band orbits. Occasionally, a free electron recombines with a hole.

p-type Energy Bands

What do the energy bands look like for a *p*-type semiconductor? At absolute zero temperature, the conduction band has no free electrons, while the valence band has many holes produced by doping. At room temperature, however, thermal energy produces a few minority carriers. Figure 2-23 summarizes the situation. The conduction band has only a few free electrons, while the valence band has many holes.

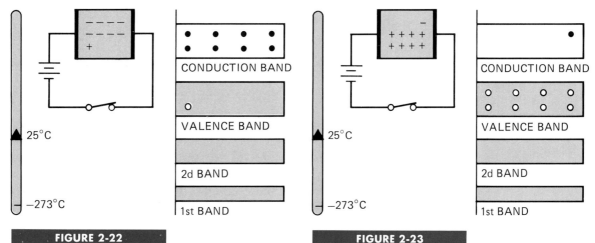

FIGURE 2-22
Energy bands for *n*-type semiconductor.

FIGURE 2-23
Energy bands for *p*-type semiconductor.

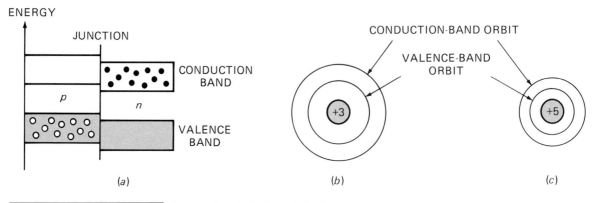

FIGURE 2-24 Energy bands before diffusion.

2-13 THE ENERGY HILL

To understand more advanced types of semiconductor devices, you will need to know how energy levels control the action of a *pn* junction.

Before Diffusion

Assuming an abrupt junction (one that suddenly changes from *p* to *n* material), what does the energy diagram look like? Figure 2-24*a* shows the energy bands before electrons have diffused across the junction. The *p* side has many holes in the valence band, and the *n* side has many electrons in the conduction band. But why are the *p* bands slightly higher than the *n* bands?

The *p* side has trivalent atoms with a core charge of +3, shown in Fig. 2-24*b*. On the other hand, the *n* side has pentavalent atoms with a core charge of +5 (Fig. 2-24*c*). A +3 core attracts an electron less than a +5 core. Therefore, the orbits of a trivalent atom (*p* side) are slightly larger than those of a pentavalent atom (*n* side). This is why the *p* bands of Fig. 2-24*a* are slightly higher.

An abrupt junction like that of Fig. 2-24*a* is an idealization because the *p* side cannot suddenly end where the *n* side begins. A manufactured diode has a gradual change from one material to the other. For this reason, Fig. 2-25*a* is a more realistic energy diagram of a junction diode.

At Equilibrium

When the diode is first formed, there is no depletion layer (Fig. 2-25*a*). In this case, free electrons will diffuse across the junction. In terms of energy levels, this means the electrons near the top of the *n* conduction band move across the junction as previously described. Soon after crossing the junction, a free electron will recombine with a hole. In other words, the electron will fall from the conduction band to the valence band. As it does, it emits heat, light, and other radiation. This recombination not only creates the depletion layer, it also changes the energy levels at the junction.

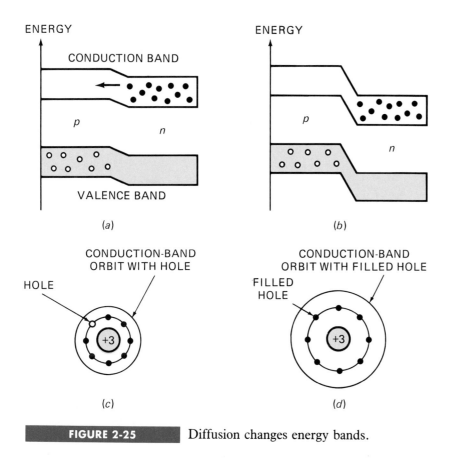

ENERGY

CONDUCTION BAND

p

n

VALENCE BAND

(a)

ENERGY

p

n

(b)

CONDUCTION-BAND
ORBIT WITH HOLE

HOLE

(c)

CONDUCTION-BAND
ORBIT WITH FILLED HOLE

FILLED
HOLE

(d)

FIGURE 2-25 Diffusion changes energy bands.

Figure 2-25*b* shows the energy diagram after the depletion layer is created. The *p* bands have moved up with respect to the *n* bands. As you can see, the bottom of each *p* band is level with the top of the corresponding *n* band. This means electrons on the *n* side no longer have enough energy to get across the junction. What follows is a simplified explanation of why the *p* band moves up.

Figure 2-25*c* shows a conduction-band orbit around one of the trivalent atoms before diffusion has occurred. When an electron diffuses across the junction, it falls into the hole of a trivalent atom (Fig. 2-25*d*). This extra electron in the valence orbit will push the conduction-band orbit farther away from the trivalent atom as shown in Fig. 2-25*d*. Therefore, any new electrons coming into this area will need more energy than before to travel in a conduction-band orbit. Stated another way, the larger conduction-band orbit means the energy level has increased. This is equivalent to saying the *p* bands move up with respect to the *n* bands after the depletion layer has built up.

At equilibrium, conduction-band electrons on the *n* side travel in orbits not quite large enough to match the *p* side orbits (Fig. 2-25*b*). In other words, electrons on the *n* side do not have enough energy to get across the junction. To an electron trying to diffuse across the junction, the path it must travel looks like a hill, an energy hill (see Fig. 2-25*b*). The electron cannot climb this hill unless it receives energy from an outside voltage source.

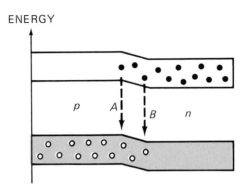

| FIGURE 2-26 | Energy bands with forward bias. |

Forward Bias

Forward bias lowers the energy hill (see Fig. 2-26). In other words, the battery increases the energy level of the free electrons, equivalent to forcing the *n* band upward. Because of this, free electrons have enough energy to enter the *p* region. Soon after entering the *p* region, each electron falls into a hole (path *A*). As a valence electron, it continues its journey toward the left end of the crystal.

A conduction-band electron may fall into a hole even before it crosses the junction. In Fig. 2-26 a valence electron may cross the junction from right to left; this leaves a hole just to the right of the junction. This hole does not live long. A conduction-band electron soon falls into it (path *B*).

Regardless of where the recombination takes place, the result is the same. A steady stream of free electrons moves toward the junction and falls into holes near the junction. The captured electrons (now valence electrons) move left in a steady stream through the holes in the *p* region. In this way, we get a continuous flow of electrons through the diode.

Incidentally, when free electrons fall from the conduction band to the valence band, they radiate their excess energy in the form of heat and light. With an ordinary diode, the radiation is heat energy, which serves no useful purpose. But with a light-emitting diode (LED), the radiation is a colored light such as red, green, blue, or orange. LEDs are widely used as visual indicators on electronic instruments, computer keyboards, consumer equipment, and so forth.

❑ 2-14 BARRIER POTENTIAL AND TEMPERATURE

The *ambient* temperature is the temperature of the surrounding air. The *junction* temperature is the temperature inside the diode, right at the junction of the *p*- and *n*-type materials. The ambient and junction temperatures may be different.

The value of barrier potential depends on the junction temperature. A higher temperature creates more free electrons and holes. These extra electrons and holes reduce the width of the depletion layer, equivalent to decreasing the barrier potential. Many people use this guideline for estimating the change in barrier potential: *For either germanium or silicon diodes, the barrier potential decreases 2 mV for each Celsius degree rise.*

EXAMPLE 2-6

What is the barrier potential of a silicon diode when the junction temperature is 100°C?

SOLUTION

If the junction temperature increases to 100°C, the barrier potential decreases by

$$(100 - 25)2 \text{ mV} = 150 \text{ mV} = 0.15 \text{ V}$$

and the barrier potential becomes

$$V_B = 0.7 \text{ V} - 0.15 \text{ V} = 0.55 \text{ V}$$

Temperature is one of the most important factors in the design of semiconductor circuits. Changes in temperature usually affect the performance of electronic circuits. Even if you don't design circuits, remember the basic idea: The barrier potential decreases when the temperature increases.

2-15 REVERSE-BIASED DIODE

Let's discuss a few advanced ideas about a reverse-biased diode. To begin with, you know that the depletion layer changes in width when the reverse voltage changes. Let us see what this implies.

Transient Current

When the reverse voltage increases, holes and electrons move away from the junction. As the free electrons and holes move away from the junction, they leave positive and negative ions behind. Therefore, the depletion layer gets wider. The greater the reverse bias, the wider the depletion layer becomes. While the depletion layer is adjusting to its new width, a current flows in the external circuit. This transient current drops to zero after the depletion layer stops growing.

The amount of time the transient current flows depends on the *RC* time constant of the external circuit. It typically happens in a matter of nanoseconds. Because of this, you can ignore the effects of the transient current below approximately 10 MHz. In other words, for most circuits you can disregard the current produced by the change in the width of the depletion layer.

Reverse Saturation Current

Figure 2-27, on the next page, illustrates saturation current in terms of energy bands. Assume an electron-hole pair appears in the junction area *A* and *B*. The free electron at *A* falls down the energy hill, pushing an electron out the right end of the conduction band. Similarly, a valence

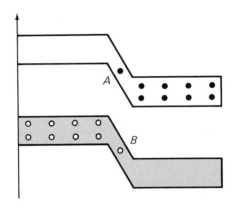

FIGURE 2-27   Reverse saturation current.

electron falls down the hill into the hole at B. The falling valence electron leaves a hole behind. This extra hole on the p side lets an electron enter the left end of the crystal.

The greater the reverse voltage, the steeper the energy hill. A conduction-band electron falling down this hill can gain a high velocity. The importance of this is brought out when we discuss breakdown voltage.

The higher the junction temperature, the greater the saturation current. A useful approximation to remember is this: I_S doubles for each 10°C rise. This is the same as a 7 percent increase in saturation current for each Celsius degree increase.

Surface-Leakage Current

Here is a simplified explanation of why surface-leakage current exists. Suppose the atoms at the top and bottom of Fig. 2-28a are atoms on the surface of the crystal. Since these atoms have no neighbors, they have broken covalent bonds (holes). Visualize these holes along the surface of the crystal shown in Fig. 2-28b. In effect, the skin of a crystal is like a p-type semiconductor. Because of this, electrons can enter the left end of the crystal, travel through the surface holes, and leave the right end of the crystal. In this way, we get a small reverse current along the surface. Unlike the reverse minority-carrier current which is independent of reverse voltage, the surface-leakage current is directly proportional to the reverse voltage.

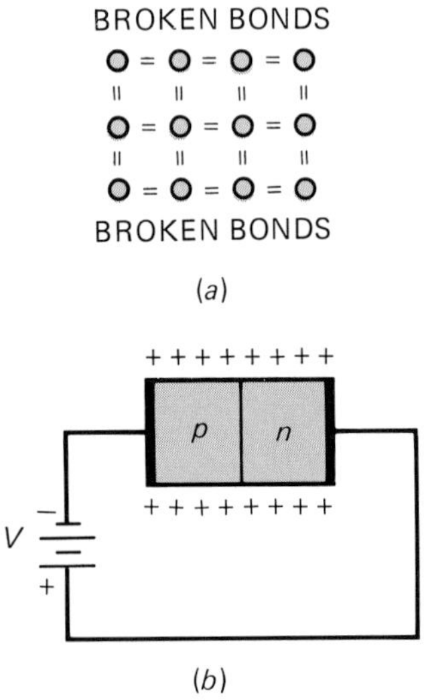

BROKEN BONDS

O = O = O = O

O = O = O = O

O = O = O = O

BROKEN BONDS

(a)

+ + + + + + +

p n

+ + + + + + +

V

(b)

FIGURE 2-28

Surface-leakage current.

EXAMPLE 2-7

A silicon diode has a saturation current of 5 nA at 25°C. Estimate the saturation current at 100°C.

SOLUTION

The saturation current doubles for each increase of 10°C. Therefore, it equals 10 nA at 35°C, 20 nA at 45°C, 40 nA at 55°C, 80 nA at 65°C, 160 nA at 75°C, 320 nA at 85°C, 640 nA at 95°C, and 1.28 µA at 105°C. So, it is roughly 1 µA at 100°C.

Remember that the rule is only an approximation to begin with, so you don't have to work out a really accurate answer. But if you do want a more accurate answer, you can use the equivalent rule: 7 percent rise per degree increase. In this case,

$$I_S = (1.07)(1.07)(1.07)(1.07)(1.07)640 \text{ nA} = 898 \text{ nA}$$

In fact, a mathematician would solve the problem like this: there are seven steps of 10° and five steps of 1° between 25°C and 100°C. Therefore,

$$I_S = (2^7)(1.07^5)(5 \text{ nA}) = 898 \text{ nA}$$

EXAMPLE 2-8

A silicon diode has a saturation current of 5 nA and a surface-leakage current of 10 nA when the reverse voltage is 15 V. What is the total reverse current when the reverse voltage is doubled to 30 V?

SOLUTION

The saturation current stays at 5 nA because temperature is the only factor that can change it. But the surface-leakage current will double because it obeys Ohm's law. Therefore, the total reverse current is

$$I_R = 5 \text{ nA} + 20 \text{ nA} = 25 \text{ nA}$$

STUDY AIDS

The following study aids will help to reinforce the ideas discussed in this chapter. For best results, use these study aids within 6 hours of reading the earlier material. Then review these study aids a week later and a month later to ensure that the concepts remain in your long-term memory.

SUMMARY

Sec. 2-1 Conductors

A neutral copper atom has only one electron in its outer orbit. Since this single electron can be easily dislodged from its atom, it is called a free electron. Copper is a good conductor because the slightest voltage causes free electrons to flow from one atom to the next.

Sec. 2-2 Semiconductors

Silicon is the most widely used semiconductor material. An isolated silicon atom has four electrons in its outer or valence orbit. The number of electrons in the valence orbit is the key to conductivity. Conductors have one valence electron, semiconductors have four valence electrons, and insulators have eight valence electrons.

Sec. 2-3 Silicon Crystals

Each silicon atom in a crystal has its four valence electrons plus four more electrons that are shared by the neighboring atoms. At room temperature, a pure silicon crystal has only a few thermally-produced free electrons and holes. The amount of time between the creation and recombination of a free electron and a hole is called the lifetime.

Sec. 2-4 Intrinsic Semiconductors

An intrinsic semiconductor is a pure semiconductor. When an external voltage is applied to the intrinsic semiconductor, the free electrons flow toward the positive battery terminal and the holes flow toward the negative battery terminal.

Sec. 2-5 Two Types of Flow

Two types of carrier flow exist in an intrinsic semiconductor. First, there is the flow of free electrons through larger orbits (conduction band). Second, there is the flow of holes through smaller orbits (valence band).

Sec. 2-6 Doping a Semiconductor

Doping increases the conductivity of a semiconductor. A doped semiconductor is called an *extrinsic semiconductor*. When an intrinsic semiconductor is doped with pentavalent (donor) atoms, it has more free electrons than holes. When an intrinsic semiconductor is doped with trivalent (acceptor) atoms, it has more holes than free electrons.

Sec. 2-7 Two Types of Semiconductors

In an *n*-type semiconductor the free electrons are the majority carriers, while the holes are the minority carriers. In a *p*-type semiconductor the holes are the majority carriers, while the free electrons are the minority carriers.

Sec. 2-8 The Unbiased Diode

An unbiased diode has a depletion layer at the *pn* junction. The ions in this depletion layer produce a barrier potential. At room temperature, this barrier potential is approximately 0.7 V for a silicon diode.

Sec. 2-9 Forward Bias

When an external voltage opposes the barrier potential, the diode is forward-biased. If the applied voltage is greater than the barrier potential, the current is large. In other words, current flows easily in a forward-biased diode.

Sec. 2-10 Reverse Bias

When an external voltage aids the barrier potential, the diode is reverse-biased. The width of the depletion layer increases when the reverse voltage increases. The current is approximately zero. In other words, a reverse-biased diode acts approximately like an open switch.

Sec. 2-11 Breakdown

Too much reverse voltage will produce either avalanche or zener effect. Then, the large breakdown current destroys the diode. In general, diodes are never operated in the breakdown region. The only exception is the zener diode, a special-purpose diode discussed in a later chapter.

VOCABULARY

In your own words, explain what each of the following terms means. Keep your answers short and to the point. If necessary, verify your answer by rereading the appropriate discussion or by looking at the end-of-book Glossary.

avalanche effect	majority carrier
barrier potential	minority carrier
breakdown voltage	*n*-type
diode	*p*-type
doping	recombination
forward bias	reverse bias
free electron	surface leakage
hole	zener effect
lifetime	

STUDENT ASSIGNMENTS

QUESTIONS

The following may have more than one right answer. Select the best answer. This is the one that is always true, or covers more situations, or fits the context, etc.

1. The nucleus of a copper atom contains how many protons?
   a. 1
   b. 4
   c. 18
   d. 29

2. The net charge of a neutral copper atom is
   a. 0
   b. +1
   c. −1
   d. +4

3. Assume the valence electron is removed from a copper atom. The net charge of the atom becomes
   a. 0
   b. +1
   c. −1
   d. +4

4. The valence electron of a copper atom experiences what kind of attraction toward the nucleus?
   a. None
   b. Weak
   c. Strong
   d. Impossible to say

5. How many valence electrons does a silicon atom have?
   a. 0
   b. 1
   c. 2
   d. 4

6. Which is the most widely used semiconductor?
   a. Copper
   b. Germanium
   c. Silicon
   d. None of the above

7. How many protons does the nucleus of a silicon atom contain?
   a. 4
   b. 14
   c. 29
   d. 32

8. Silicon atoms combine into an orderly pattern called a
   a. Covalent bond
   b. Crystal
   c. Semiconductor
   d. Valence orbit

9. An intrinsic semiconductor has some holes in it at room temperature. What causes these holes?
   a. Doping
   b. Free electrons
   c. Thermal energy
   d. Valence electrons

10. Each valence electron in an intrinsic semiconductor establishes a
    a. Covalent bond
    b. Free electron
    c. Hole
    d. Recombination

11. The merging of a free electron and a hole is called
    a. Covalent bonding
    b. Lifetime
    c. Recombination
    d. Thermal energy

12. At room temperature an intrinsic silicon crystal acts approximately like
    a. A battery
    b. A conductor
    c. An insulator
    d. A piece of copper wire

13. The amount of time between the creation of a hole and its disappearance is called
    a. Doping
    b. Lifetime
    c. Recombination
    d. Valence

14. The valence electron of a conductor is also called a
    a. Bound electron
    b. Free electron
    c. Nucleus
    d. Proton

15. A conductor has how many types of flow?
    a. 1
    b. 2
    c. 3
    d. 4

16. A semiconductor has how many types of flow?
    a. 1
    b. 2
    c. 3
    d. 4

17. When a voltage is applied to a semiconductor, holes will flow
    a. Away from the negative potential
    b. Toward the positive potential
    c. In the external circuit
    d. None of the above

18. At room temperature a conductor has how many holes?
    a. Many
    b. None
    c. Only those produced by thermal energy
    d. Same number as free electrons

19. In an intrinsic semiconductor, the number of free electrons
    a. Equals the number of holes
    b. Is greater than the number of holes
    c. Is less than the number of holes
    d. None of the above

20. Absolute zero temperature equals
    a. −273°C
    b. 0°C
    c. 25°C
    d. 50°C

21. At absolute zero temperature an intrinsic semiconductor has
    a. A few free electrons
    b. Many holes
    c. Many free electrons
    d. No holes or free electrons

22. At room temperature an intrinsic semiconductor has
    a. A few free electrons and holes
    b. Many holes
    c. Many free electrons
    d. No holes

23. The number of free electrons and holes in an intrinsic semiconductor increases when the temperature
    a. Decreases
    b. Increases
    c. Stays the same
    d. None of the above

24. The flow of valence electrons to the left means that holes are flowing to the
    a. Left
    b. Right
    c. Either way
    d. None of the above

25. Holes act like
    a. Atoms
    b. Crystals
    c. Negative charges
    d. Positive charges

26. How many types of flow are there in a conductor?
    a. 0
    b. 1
    c. 2
    d. 3

27. How many types of flow are there in a semiconductor?
    a. 0
    b. 1
    c. 2
    d. 3

28. Trivalent atoms have how many valence electrons?
    a. 1
    b. 3
    c. 4
    d. 5

29. A donor atom has how many valence electrons?
    a. 1
    b. 3
    c. 4
    d. 5

30. If you wanted to produce a p-type semiconductor, which of these would you use?
    a. Acceptor atoms
    b. Donor atoms
    c. Pentavalent impurity
    d. Silicon

31. Holes are the minority carriers in which type of semiconductor?
    a. Extrinsic
    b. Intrinsic
    c. n-type
    d. p-type

32. How many free electrons does a p-type semiconductor contain?
    a. Many
    b. None
    c. Only those produced by thermal energy
    d. Same number as holes

33. Silver is the best conductor. How many valence electrons do you think it has?
    a. 1
    b. 4
    c. 18
    d. 29

34. Suppose an intrinsic semiconductor has 1 billion free electrons at room temperature. If the temperature changes to 75°C, how many holes are there?
    a. Less than 1 billion
    b. 1 billion
    c. More than 1 billion
    d. Impossible to say

35. An external voltage source is applied to a p-type semiconductor. If the left end of the crystal is positive, which way do the majority carriers flow?
    a. Left
    b. Right
    c. Neither
    d. Impossible to say

36. Which of the following belongs least?
    a. Conductor
    b. Semiconductor
    c. Four valence electrons
    d. Crystal structure

37. Which of the following is approximately equal to room temperature?
    a. 0°C
    b. 25°C
    c. 50°C
    d. 75°C

38. How many electrons are there in the valence orbit of a silicon atom within a crystal?
    a. 1
    b. 4
    c. 8
    d. 14

39. Positive ions are atoms that have
    a. Gained a proton
    b. Lost a proton
    c. Gained an electron
    d. Lost an electron

40. Which of the following describes an n-type semiconductor?
    a. Neutral
    b. Positively charged
    c. Negatively charged
    d. Has many holes

41. A p-type semiconductor contains holes and
    a. Positive ions
    b. Negative ions
    c. Pentavalent atoms
    d. Donor atoms

42. Which of the following describes a p-type semiconductor?
    a. Neutral
    b. Positively charged
    c. Negatively charged
    d. Has many free electrons

43. Which of the following cannot move?
    a. Holes
    b. Free electrons
    c. Ions
    d. Majority carriers

44. What causes the depletion layer?
    a. Doping
    b. Recombination
    c. Barrier potential
    d. Ions

45. What is the barrier potential of a silicon diode at room temperature?
    a. 0.3 V
    b. 0.7 V
    c. 1 V
    d. 2 mV per degree Celsius

46. To produce a large forward current in a silicon diode, the applied voltage must be greater than
    a. 0
    b. 0.3 V
    c. 0.7 V
    d. 1 V

47. In a silicon diode the reverse current is usually
    a. Very small
    b. Very large
    c. Zero
    d. In the breakdown region

48. Surface-leakage current is part of the
    a. Forward current
    b. Forward breakdown
    c. Reverse current
    d. Reverse breakdown

49. The voltage where avalanche occurs is called the
 a. Barrier potential
 b. Depletion layer
 c. Knee voltage
 d. Breakdown voltage

50. Diffusion of free electrons across the junction of an unbiased diode produces
 a. Forward bias
 b. Reverse bias
 c. Breakdown
 d. The depletion layer

51. When the reverse voltage increases from 5 to 10 V, the depletion layer
 a. Becomes smaller
 b. Becomes larger
 c. Is unaffected
 d. Breaks down

52. When a diode is forward-biased, the recombination of free electrons and holes may produce
 a. Heat
 b. Light
 c. Radiation
 d. All of the above

53. A reverse voltage of 20 V is across a diode. What is the voltage across the depletion layer?
 a. 0 V
 b. 0.7 V
 c. 20 V
 d. None of the above

BASIC PROBLEMS

2-1. What is the net charge of a copper atom if it gains three electrons?

2-2. What is the net charge of a silicon atom if it loses all of its electrons?

2-3. Classify each of the following as a conductor or semiconductor:
 a. Germanium
 b. Silver
 c. Silicon
 d. Gold

2-4. A diode is forward-biased. If the current is 5 mA on the n side, what is the current in each of the following:
 a. p side
 b. External connecting wires
 c. Junction

2-5. Classify each of the following as n-type or p-type semiconductors:
 a. Doped by acceptor atoms
 b. Crystal with pentavalent impurities
 c. Majority carriers are holes
 d. Donor atoms were added to crystal
 e. Minority carriers are free electrons

ADVANCED PROBLEMS

2-6. A designer will be using a silicon diode over a temperature range of 0 to 75°C. What are the minimum and maximum values of barrier potential?

2-7. A silicon diode has a saturation current of 10 nA at 25°C. If it operates over a temperature range of 0 to 75°C, what are the minimum and maximum values of saturation current?

2-8. A diode has a surface-leakage current of 10 nA when the reverse voltage is 10 V. What is the surface-leakage current if the reverse voltage is increased to 50 V?

DIODE THEORY

This chapter is about diode approximations. We need these diode approximations for troubleshooting, analysis, and design. The approximation we will use in any situation depends on what we are trying to do. If we are troubleshooting, the ideal approximation is usually adequate. If we are designing, the third approximation may be needed. Most of the time, the second approximation is the best compromise between simplicity and accuracy.

3-1 THE SCHEMATIC SYMBOL

Some electronic devices are *linear*, meaning that their current is directly proportional to their voltage. The reason they are called linear is because a graph of current versus voltage turns out to be a straight line. The simplest example of a linear device is an ordinary resistor. If you graph its current versus voltage, you will get a straight line.

A diode is different. Because of the built-in barrier potential, a diode does not act like a resistor. Therefore, you can hardly expect it to have the same kind of graph as a resistor. As you will see, a plot of the current versus voltage for a diode produces a nonlinear graph.

Figure 3-1 shows the schematic symbol of a rectifier diode. The *p* side is called the anode, and the *n* side the cathode. The diode symbol looks like an arrow that points from the *p* side to the *n* side, from the anode to the cathode. Because of this, the diode arrow is a reminder that conventional current flows easily from the *p* side to the *n* side. If you prefer using electron flow, you will have to reverse your thinking. In this case, the easy direction for electron flow is against the diode arrow. Stated another way, you can visualize the diode as pointing to where the free electrons are coming from.

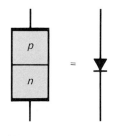

FIGURE 3-1

Schematic symbol of a rectifier diode.

3-2 THE DIODE CURVE

When a manufacturer optimizes a diode to convert alternating current to direct current, the diode is called a *rectifier* diode. This is the oldest and most widely used type of diode. One of its main applications is in *power*

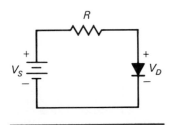

FIGURE 3-2

Forward bias.

supplies—circuits that convert the alternating voltage from the power line into direct voltage for electronics equipment.

Figure 3-2 shows one of the simplest diode circuits. How do you go about analyzing a circuit like this? A forward-biased diode conducts well, and a reverse-biased one conducts poorly. Therefore, when you are analyzing diode circuits, one of the things to decide is whether the diode is forward- or reverse-biased. This is not always easy to do. But here is something that helps. Ask yourself this question: Is the external circuit trying to push conventional current in the direction of the diode arrow or in the opposite direction? If conventional current is in the same direction as the diode arrow, the diode is forward-biased. On the other hand, if conventional current tries to flow in the opposite direction of the arrow, the diode is reverse-biased.

If you prefer electron flow, then ask yourself the same question restated for the flow of free electrons. Is the external circuit trying to push free electrons in the opposite direction of the arrow? If so, the diode conducts easily.

3-3 THE FORWARD REGION

We have the main idea behind a rectifier diode: it conducts better one way than the other. This is a start. We can now sharpen our understanding by graphing diode current versus diode voltage.

Figure 3-2 is a circuit that you can set up in the laboratory. After you connect this circuit, you can measure the voltage across the diode and the current through it. This gives you corresponding pairs of I and V to use in your graph.

What do you think the graph will look like? Since the dc source tries to set up conventional current in the same direction as the diode arrow, the diode is forward-biased. The greater the applied voltage, the larger the diode current. By varying the applied voltage, you can measure the diode current (use a series ammeter) and the diode voltage (a voltmeter in parallel with the diode). By plotting the corresponding currents and voltages, you can get a graph of the diode current versus diode voltage.

Knee Voltage

Figure 3-3 shows how the graph looks for a forward-biased silicon diode. It is customary to plot voltage along the horizontal axis because voltage is the independent variable. Each value of diode voltage produces a particular current. The current is the dependent variable and is plotted along the vertical axis.

What does the graph tell us? To begin with, the diode doesn't conduct well until the applied voltage overcomes the barrier potential. This is why the current is small for the first few tenths of a volt. As we approach 0.7 V, free electrons start crossing the junction in larger numbers. This is the reason the current starts increasing rapidly. Above 0.7 V, the slightest increase in diode voltage produces a large increase in current.

The voltage where the current starts to increase rapidly is called the *knee voltage* of the diode. For a silicon diode, the knee voltage equals the

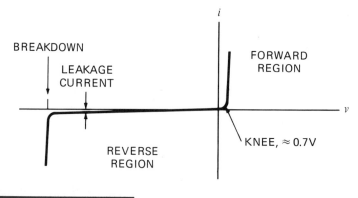

FIGURE 3-3 Diode curve.

barrier potential, approximately 0.7 V. A germanium diode, on the other hand, has a knee voltage of about 0.3 V.

Nonlinear Device

A diode is a *nonlinear device* because its current is not directly proportional to its voltage. Below 0.7 V, the diode has very little current. Just above 0.7 V, the current increases rapidly. This action is quite different from an ordinary resistor where the current increases in direct proportion to the voltage. The reason the diode is different is because it has a built-in barrier potential produced by its depletion layer. It is as though a small battery of 0.7 V were built into the diode. Until the applied voltage can offset this built-in battery, the diode current remains very small.

A nonlinear device like a diode always has a current-voltage graph that is curved. The size and shape of the nonlinear graph depend on the particular device. But you can count on the fact that the graph will not be a straight line when a device is nonlinear.

Bulk Resistance

Above the knee voltage, the diode current increases rapidly. This means small increases in the diode voltage cause large increases in diode current. The reason is this: after overcoming the barrier potential, all that impedes the current is the resistance of the p and n regions. Since any conductor has some resistance, so too does the p region have a certain amount of resistance. The n region also has some resistance. The sum of these resistances is called the *bulk resistance* of the diode. In symbols,

$$r_B = r_P + r_N$$

The value of bulk resistance depends on the doping level, and the size of the p and n regions. Typically, the bulk resistance of rectifier diodes is less than 1 Ω.

Maximum DC Forward Current

If the current in a diode is too large, excessive heat will destroy the diode. Even approaching the burnout value without reaching it can shorten diode life and degrade other properties. For this reason, a manufacturer's data sheet specifies the maximum current a diode can safely handle without shortening its life or degrading its characteristics.

The *maximum forward current* is one of the maximum ratings given on a data sheet. This current may be listed as $I_{F(\max)}$, I_O, etc. depending on the manufacturer. For instance, a 1N456 has a maximum rating of 135 mA. This means it can safely handle a continuous forward current of 135 mA.

Current-Limiting Resistor

This brings us to why a resistor is almost always used in series with a diode. In Fig. 3-2, the resistor is referred to as a *current-limiting* resistor. The larger this resistance is, the smaller the diode current for a particular applied voltage. The exact size of R will depend on what the designer is trying to do with the circuit. But at a minimum, the current-limiting resistance has to keep the diode current less than the maximum rating.

The diode current is given by

$$I = \frac{V_S - V_D}{R} \tag{3-1}$$

where V_S is the source voltage and V_D is the voltage across the diode. This equation is nothing more than Ohm's law applied to the current-limiting resistor. In other words, the voltage across the resistor equals $V_S - V_D$. Dividing this voltage by the resistance gives the current through the resistor. Since this is a series circuit, the current is the same in all parts, which means the diode current has the same value as the current through the resistor.

Maximum Power Dissipation

Closely related to the maximum dc forward current is the *maximum power dissipation*. Like a resistor, a diode has a power rating. This rating tells how much power the diode can safely dissipate without shortening its life or degrading its properties. You can calculate the power dissipation as follows. When the diode current is a direct current, the product of diode voltage and current equals the power dissipated by the diode.

With rectifier diodes, the maximum power rating is not normally used because all of the burnout information is already contained in the maximum current rating. For instance, the data sheet of a 1N4001 lists a maximum forward current I_O of 1 A. The power rating is not listed because you don't need it. As long as you keep the maximum forward current under 1 A, you will not blow out the diode.

EXAMPLE 3-1

Suppose a diode has $r_P = 0.13\ \Omega$ and $r_N = 0.1\ \Omega$. What is its bulk resistance?

SOLUTION

Add the individual resistances to get the bulk resistance:

$$r_B = 0.13 \ \Omega + 0.1 \ \Omega = 0.23 \ \Omega$$

The bulk resistance is sometimes called the ohmic resistance of a diode because it is nothing more than the resistance of the material to the flow of charges.

EXAMPLE 3-2

In Fig. 3-2, on page 58, $V_S = 10$ V, $V_D = 0.7$ V, and $R = 1$ kΩ. What does the diode current equal?

SOLUTION

The voltage across the resistor is equal to 9.3 V. Now, use Ohm's law to find the current in the circuit:

$$I = \frac{9.3 \ \text{V}}{1 \ \text{k}\Omega} = 9.3 \ \text{mA}$$

Don't memorize Eq. (3-1). All you have to remember is that Ohm's law applies to the current-limiting resistor. This should jog your memory enough to make you realize that you have to subtract the diode voltage from the source voltage to get the voltage across the current-limiting resistor.

EXAMPLE 3-3

The voltage across a 1N4001 is 0.93 V when the current is 1 A. What is the power dissipation of the diode for these values?

SOLUTION

$$P = (0.93 \ \text{V})(1 \ \text{A}) = 0.93 \ \text{W}$$

This is an application of the basic power formula: $P = VI$. The power dissipated by the diode equals the voltage across the diode times the current through the diode. The power creates heat, which raises the junction temperature of the diode.

3-4 THE REVERSE REGION

When you reverse-bias a diode, you get only a small leakage current. By measuring diode current and voltage, you can plot the reverse curve; it will look something like the reverse region of Fig. 3-3. There are no surprises here; diode current is very small for all reverse voltages less than the breakdown voltage. At breakdown, the current increases rapidly for small increases in voltage. As already mentioned, with a rectifier diode you always try to keep the applied voltage less than the breakdown voltage specified on a manufacturer's data sheet.

At normal frequencies (those for which the rectifier diode is intended), Fig. 3-3 on page 59 applies. This curve tells us the value of diode current for a particular diode voltage. Even though we measure the current and voltage in the circuit of Fig. 3-2, on page 58, the curve of Fig. 3-3 can be used for any diode circuit. Why? Because the relation between diode current and voltage is the same, no matter how the diode is connected. In other words no matter what circuit we use a diode in, a given diode voltage produces exactly the same current indicated by Fig. 3-3.

3-5 THE IDEAL DIODE

Typical circuits use resistors with tolerances of at least ±5 percent. Other devices such as capacitors, diodes, and transistors, often have tolerances of ±10 percent or more. Because of this, each circuit design usually results in voltages and currents that have tolerances of more than ±5 percent. For much of electronics work, therefore, it makes sense to use approximations of complicated devices like diodes. After all, what is the point of spending a lot of time working out exact answers if you don't need exact answers? For instance, when troubleshooting a circuit, it may be enough to know that some voltage should be roughly 9 V. Then if you measure only 2 V, you will know something is wrong.

For the remainder of this chapter, we will discuss three approximations that are widely used for silicon diodes. Each is useful under certain conditions that you will learn about. We will begin with the simplest approximation, which is called the *ideal diode*. This approximation is used to get ballpark answers. It is ideal for troubleshooting because it usually produces answers that are accurate enough to guide your troubleshooting efforts.

What does a rectifier diode do? It conducts well in the forward direction and poorly in the reverse direction. Boil this down to its essence, and this is what you get. Ideally, a rectifier diode acts like a perfect conductor (zero resistance) when forward-biased and like a perfect insulator (infinite resistance) when reverse-biased.

Figure 3-4 shows the current-voltage graph of an ideal diode. It echoes what we just said: zero resistance when forward-biased and infinite resistance when reverse-biased. True, it is impossible to build such a device, but this is what manufacturers would produce if they could. Furthermore, when the supply voltage is high enough and the diode current is small enough, a real silicon diode almost behaves like an ideal diode.

Is there any real device that acts like an ideal diode? Almost. An ordinary switch has zero resistance when closed and infinite resistance when open. Therefore, an ideal diode acts like an intelligent switch that has the sense to close when forward-biased and to open when reverse-biased. Figure 3-5 summarizes the intelligent switch idea. This is rock bottom; we cannot simplify beyond this point without losing the main idea of a diode.

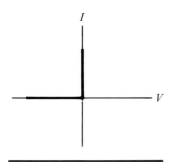

FIGURE 3-4

Ideal diode curve.

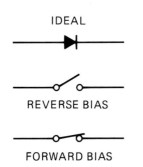

FIGURE 3-5

Ideal diode acts like a switch.

Extreme as the ideal-diode approximation seems at first, it gives usable answers for many diode circuits. And of course, a lot depends on what you are trying to do. If you are troubleshooting, it may be fine. If you are trying to send an astronaut to the moon, you may need a more accurate approximation. There will be times when the ideal-diode approximation produces terrible answers. For this reason, we need a second and third approximation. But for all preliminary analysis of rectifier-diode circuits, the ideal diode is an excellent starting point.

EXAMPLE 3-4

Use the ideal-diode approximation to calculate the load current, load voltage, load power, diode power, and total power in Fig. 3-6.

SOLUTION

The source voltage forward-biases the diode. Visualize the diode replaced by a closed switch. Then, we have a simple series circuit with a source voltage of 10 V and a load resistance of 1 kΩ. With Ohm's law,

$$I = \frac{10\,\text{V}}{1\,\text{k}\Omega} = 10\,\text{mA}$$

Because of the closed switch, all of the source voltage appears across the load resistor and

$$V_L = 10\,\text{V}$$

Next, use the VI product to get the powers as follows:

$$P_L = (10\,\text{V})(10\,\text{mA}) = 100\,\text{mW}$$

$$P_D = (0\,\text{V})(10\,\text{mA}) = 0$$

The total power is the sum of the individual powers:

$$P_T = P_D + P_L = 0 + 100\,\text{mW} = 100\,\text{mW}$$

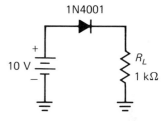

FIGURE 3-6
Example.

Everything here is straightforward. Notice how the ideal diode has reduced the problem down to an analysis of a simple series circuit. This is because we replaced the diode (a new device) with a switch (an old device). We replaced the unknown by the known.

The one thing worth noting here is that the diode power is zero. In a real diode, there will be some power dissipation. But the ideal diode gives us a hint about the power in a real diode. The power probably will be low, approaching zero.

3-6 THE SECOND APPROXIMATION

Because of the barrier potential, the applied voltage has to be more than 0.7 V before a silicon diode really conducts well. When the source voltage is much greater than 0.7 V, the ideal diode approximation produces only

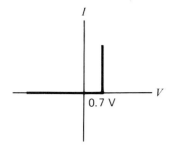

FIGURE 3-7

Diode curve for second approximation.

2D APPROXIMATION

REVERSE BIAS

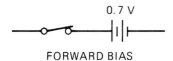

FORWARD BIAS

FIGURE 3-8

Second approximation acts like switch and battery.

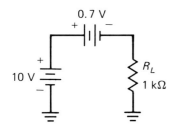

FIGURE 3-9

Example.

a small error. But when the source voltage is not large, the barrier potential becomes important. As a guide, if the source voltage is 7 V (this is 10 times the barrier potential), the error in using the ideal diode is roughly 10 percent. If this kind of error is too much for the work you are doing, then you will have to use a better approximation.

Figure 3-7 shows the graph of current versus voltage for the *second approximation*. The graph says no current exists until 0.7 V appears across the diode. At this point the diode turns on. Thereafter, only 0.7 V can appear across the diode, no matter what the current. (Use 0.3 V for germanium diodes.)

Figure 3-8 shows the equivalent circuit for the second approximation. We think of the diode as a switch in series with a barrier potential of 0.7 V. If the source voltage is at least 0.7 V, the switch will close. In this case, the voltage across the device is 0.7 V. Because the barrier potential is fixed at 0.7 V, the total diode drop remains at 0.7 V for any value of forward current.

On the other hand, if the source voltage is less than 0.7 V or if the source voltage is negative (reversed polarity), the switch is open. Then the barrier potential has no effect at all, and you can think of the diode as an open circuit.

EXAMPLE 3-5

Use the second approximation to calculate the load current, load voltage, load power, diode power, and total power in Fig. 3-6.

SOLUTION

Visualize the diode replaced by a closed switch and a barrier potential of 0.7 V. Then, we have a series circuit with two series-opposing batteries as shown in Fig. 3-9. The opposing voltages subtract and Ohm's law gives

$$I = \frac{10\,\text{V} - 0.7\,\text{V}}{1\,\text{k}\Omega} = \frac{9.3\,\text{V}}{1\,\text{k}\Omega} = 9.3\,\text{mA}$$

The load voltage equals

$$V_L = I_L R_L = (9.3\,\text{mA})(1\,\text{k}\Omega) = 9.3\,\text{V}$$

An alternative way to calculate load voltage is by subtracting the diode drop from the source voltage:

$$V_L = V_S - V_D = 10\,\text{V} - 0.7\,\text{V} = 9.3\,\text{V}$$

Next, use the *VI* product to get the powers as follows:

$$P_L = (9.3\,\text{V})(9.3\,\text{mA}) = 86.5\,\text{mW}$$

$$P_D = (0.7\,\text{V})(9.3\,\text{mA}) = 6.51\,\text{mW}$$

The total power is the sum of the individual powers:

$$P_T = P_D + P_L = 6.51\,\text{mW} + 86.5\,\text{mW} = 93\,\text{mW}$$

Notice how the diode power is no longer zero. But diode power is still quite low when compared to the load power. This makes sense because the same current flows through the diode and the load, but the diode voltage is much smaller than the load voltage. This is the way a well-designed circuit should work. The barrier potential of the diode should be much smaller than the load voltage.

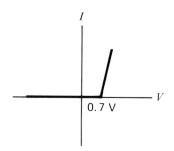

FIGURE 3-10
Diode curve for third approximation.

3-7 THE THIRD APPROXIMATION

In the *third approximation* of a diode, we include the bulk resistance r_B. Figure 3-10 shows the effect that r_B has on the diode curve. After the silicon diode turns on, the voltage increases linearly or proportionally with an increase in current. The greater the current, the larger the voltage because we have to add the *IR* drop across r_B to the total diode voltage.

The equivalent circuit for the third approximation is a switch in series with a barrier potential of 0.7 V and a resistance of r_B (see Fig. 3-11). When the applied voltage is larger than 0.7 V, the diode conducts. The total voltage across the diode equals

$$V_D = 0.7 + I_D r_B \qquad (3-2)$$

(For a germanium diode, use 0.3 V instead of 0.7 V.)

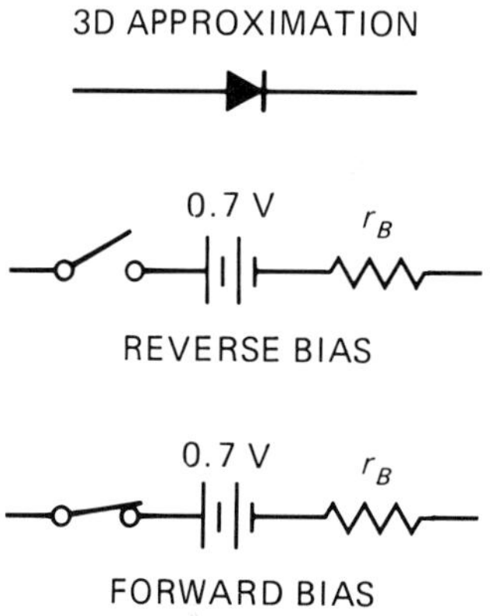

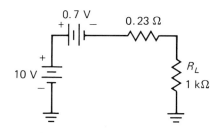

FIGURE 3-11
Equivalent circuit for third approximation.

EXAMPLE 3-6

Use the third approximation to calculate the load current, load voltage, load power, diode power, and total power in Fig. 3-6. A 1N4001 has a bulk resistance of 0.23 Ω.

SOLUTION

This solution is similar to Example 3-5, except that we will include the barrier potential and the bulk resistance of a 1N4001. Visualize the diode replaced by a closed switch, a barrier potential of 0.7 V, and a bulk resistance of 0.23 Ω. Then, we have a series circuit with two series-opposing batteries and two resistances as shown in Fig. 3-12. The opposing voltages subtract, and the two resistances add (series connection). Therefore, Ohm's law gives

$$I = \frac{10\,\text{V} - 0.7\,\text{V}}{1\,\text{k}\Omega + 0.23\,\Omega} = \frac{9.3\,\text{V}}{1\,\text{k}\Omega} = 9.3\,\text{mA}$$

Notice what happened in the denominator. The bulk resistance (0.23 Ω) is so small compared to the load resistance (1 kΩ) that the sum is still 1 kΩ after rounding off to three significant digits.

FIGURE 3-12
Example of series circuit.

Because of this, all the other calculations produce almost the same results as before:

$$V_L = I_L R_L = (9.3 \text{ mA})(1 \text{ k}\Omega) = 9.3 \text{ V}$$

$$P_L = (9.3 \text{ V})(9.3 \text{ mA}) = 86.5 \text{ mW}$$

$$P_D = (0.702 \text{ V})(9.3 \text{ mA}) = 6.53 \text{ mW}$$

$$P_T = P_D + P_L = 6.53 \text{ mW} + 86.5 \text{ mW} = 93 \text{ mW}$$

The only quantity that is slightly different is the diode power. This happens because the diode voltage is a bit larger because of the additional drop across the bulk resistance:

$$V_D = 0.7 \text{ V} + I_L r_B = 0.7 \text{ V} + (9.3 \text{ mA})(0.23 \text{ }\Omega) = 0.702 \text{ V}$$

The conclusion? The third approximation isn't worth the effort when the load resistance is much larger than the bulk resistance. What does "much larger" mean? Here are some guidelines. When the load resistance is 100 times larger than the bulk resistance, the calculation error is less than 1 percent when you ignore the bulk resistance. When the load resistance is 10 times larger than the bulk resistance, the calculation error is less than 10 percent.

Table 3-1 Error when Ignoring 0.7 V

V_S	Ideal Diode
3.5 V	20%
7 V	10%
14 V	5%
28 V	2.5%
70 V	1%

Table 3-2 Error when Ignoring Bulk Resistance

R_L/r_B	Ideal or Second Approximation
X 5	20%
X 10	10%
X 20	5%
X 40	2.5%
X 100	1%

3-8 SELECTING AN APPROXIMATION

Which approximation should you use? If you are troubleshooting or making a preliminary analysis, large errors are often acceptable. There is no need to waste time if all you want is a basic idea of how a circuit works. On the other hand, if your circuit uses precision resistors with a tolerance of ±1 percent, you may want to use the third approximation. But most of the time, the second approximation turns out to be the best compromise and is the one that is used most often in circuit analysis.

The guiding equation that tells you which approximation to use is this:

$$I_F = \frac{V_S - 0.7}{R_L + r_B} \tag{3-3}$$

This equation for forward current pinpoints the effect of the barrier potential and the bulk resistance. There are two sources of error. First, you have to consider how large the source voltage, V_S, is compared to 0.7 V. If V_S is equal to 7 V, ignoring the barrier potential produces a calculation error of 10 percent as shown in Table 3-1. If V_S is equal to 14 V, the calculation error is 5 percent, and so on.

Similarly, when the load resistance is 10 times the bulk resistance, ignoring the bulk resistance produces a calculation error of 10 percent. When the load resistance is 20 times greater, the error drops to 5 percent, as shown in Table 3-2.

Calculation errors are additive. For instance, if the source voltage is 7 V and the load resistance is 10 times the bulk resistance, the total calculation error is the sum of the two errors, or about 20 percent when you use an ideal diode. Under the same conditions, the second approximation produces a calculation error of approximately 10 percent.

Most rectifier diodes have bulk resistances under 1 Ω, which means that the second approximation produces less than 5 percent error with load resistances greater than 20 Ω. This covers just about all the practical circuits you may encounter. This is why the second approximation is an excellent compromise to use whenever you are in doubt about which approximation to use.

3-9 TROUBLESHOOTING

You can quickly check the condition of a diode with an ohmmeter. Measure the dc resistance of the diode in either direction, and then reverse the leads and measure the dc resistance again. The forward current will depend on which ohmmeter range is used, which means you get different readings on different ranges. The main thing to look for, however, is a high ratio of reverse to forward resistance. How high is high? For typical silicon diodes used in electronics work, the ratio should be higher than 1000:1.

Using an ohmmeter to check diodes is an example of go/no-go testing. You're really not interested in the exact dc resistance of the diode; all you want to know is whether the diode is acting approximately like a one-way conductor or not, that is, if it has a low resistance in the forward direction and a high resistance in the reverse direction. Diode troubles are indicated for any of the following: extremely low resistance in both directions (diode shorted); high resistance in both directions (diode open); somewhat low resistance in the reverse direction (called a *leaky diode*).

The test is usually done with the diode out of a circuit. But even when the diode is in a circuit, an ohmmeter check (turn the circuit power off first) should indicate resistance lower in one way than the other.

A final point: Some ohmmeters can produce enough current on low ranges to destroy a small-signal diode. For this reason, you should test small-signal diodes on ranges greater than R X 10. On these higher scales, the internal resistance of the ohmmeter prevents excessive diode current.

EXAMPLE 3-7

Figure 3-13, on the next page, shows the diode circuit analyzed earlier. Suppose something causes the diode to burn out. What kind of symptoms will you get?

SOLUTION

When a diode burns out, it becomes an open circuit. In this case, the current drops to zero. Therefore, if you measure the load voltage, the voltmeter will indicate zero.

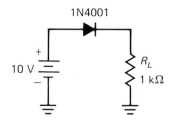

FIGURE 3-13   Example of diode circuit.

EXAMPLE 3-8

Suppose the circuit of Fig. 3-13 is not working. You measure the load voltage and get zero. What is the trouble?

SOLUTION

Many troubles are possible. First, the diode could be open. Second, the supply voltage could be zero. Third, one of the connecting wires could be open.

How do you find the trouble? Measure the voltages to isolate the defective component. Then disconnect any suspected component and test its resistance. For instance, you could measure the load voltage first and the source voltage second. If there is source voltage but no load voltage, the diode is highly suspicious. An ohmmeter test will tell. If the diode passes the ohmmeter test, then check the connections because there's nothing else to account for having source voltage but no load voltage.

If there is no source voltage, the power supply is defective or a connection between the supply and the diode is open. Power-supply troubles are very common. Often, when electronics equipment is not working, the trouble is in the power supply. This is why most troubleshooters start by measuring the voltage out of the power supply.

3-10   UP-DOWN THINKING

Numbers are nice, but there is nothing quite like *up-down* thinking to help you understand circuits. The idea is this. Any circuit has independent variables (like source voltages and branch resistances) and dependent variables (like voltages across resistors, currents, powers, etc.). When an independent variable increases, each of the dependent variables will usually respond by increasing or decreasing. If you really understand how the circuit works, you will be able to predict whether a dependent variable will increase or decrease. To do up-down thinking, you don't have to calculate how much a quantity changes; only whether it goes up (U), goes down (D), or shows no change (N).

Up-down thinking is very useful in troubleshooting, analysis, and design. It really helps you to get your bearings and to use formulas intelligently rather than blindly. When you know beforehand how a dependent variable should respond to changes in an independent variable, you are less likely to make dumb mistakes with formulas. In short, up-down thinking gives you perspective, insight, and control over the situation.

Here is how it works for Fig. 3-13. A voltage V_S of 10 V is applied to a diode in series with a load resistance R_L of 1 kΩ. In the second approximation of a diode, there are three independent variables for this circuit: V_S, R_L, and V_K. We are including the knee voltage, V_K, as an independent variable because it may be slightly different from the ideal value of 0.7 V. There are five dependent variables as follows: V_L, I_L, P_D, P_L, and P_T. These are the load voltage, load current, diode power, load power, and total power.

Suppose the source voltage V_S increases slightly, say 10 percent. How will each of the dependent variables respond? Will each go up (U), go down (D), or show no change (N). To solve this problem, the first thing to do is throw away your calculator. Up-down thinking works best when you use only your head. Here are some of the thoughts that might pass through your mind as you solve this problem:

In the second approximation, the diode has a voltage drop of 0.7 V. If the source voltage increases slightly, the diode drop is still 0.7 V, which means the load voltage has to increase. If the load voltage increases, Ohm's law tells me the load current increases. An increase in load current means a larger *VI* product for the diode and for the load. Therefore, the diode power and load power both increase. Finally, the total power is the sum of diode power and load power, so total power must also increase.

The first entry of Table 3-3 summarizes the effect of a small increase in source voltage. As you can see, each dependent variable increases. What do you think happens when the load resistance of Fig. 3-13 increases slightly? Since the diode voltage is constant in the second approximation, the load voltage shows no change, but the load current will go down. In turn, this implies less diode power, load power, and total power. The second entry of Table 3-3 summarizes this case. Finally, consider the effect of knee voltage. If the knee voltage increases slightly, the dependent variables decrease, except for the diode power, as shown in the third entry of Table 3-3.

Table 3-3 Up-down Thinking

	V_L	I_L	P_D	P_L	P_T
V_S increase	U	U	U	U	U
R_L increase	N	D	D	D	D
V_K increase	D	D	U	D	D

Practice up-down thinking with new circuits and you will become a better troubleshooter, analyzer, or designer. It cuts across all of these areas because up-down thinking requires true understanding of what is going on in a circuit. For additional practice in up-down thinking, work with the Software Engine Problems at the end of the chapter.

EXAMPLE 3-9

Look at the Software Engine of Fig. 3-21 (in the homework problems at the end of the chapter). How do you use this to find dependent changes?

SOLUTION

The idea is similar to the T-shooter introduced in Chap. 1. To avoid confusion, we will always use a slight *increase* in the independent variable and find the response of each dependent variable. The first box shows the response to a small increase in source voltage V_S. The response of V_A is given by the token $C3$. In the large box labeled "Responses," the token $C3$ gives U, meaning up or increase.

The way you can practice up-down thinking for the circuit is by selecting one independent variable (V_S, R_1, R_2, R_3, or V_K). Next, you select any dependent variable in the box (V_A, V_B, V_C, I_1, etc.). Then try to figure out if the dependent variable goes up, goes down, or shows no change. To check your answer, read the token and then the response.

For instance, how does an increase in knee voltage affect the current in R_3? In Fig. 3-21, a stiff voltage divider drives the diode in series with the 100 kΩ. Therefore, a slight increase in knee voltage will decrease the voltage across the 100 kΩ. Then, Ohm's law tells us I_3 should decrease. To check this answer, look at the box labeled V_K. I_3 has a token of $A3$. Then, $A3$ translates into a D, which means down. Our answer is correct.

The homework problems at the end of this chapter will use the Software Engine to continue your training in up-down thinking. Later chapters include additional exercises in this important type of thinking.

3-11 READING A DATA SHEET

The Appendix includes data sheets for some widely used semiconductor devices. You can refer to these data sheets to learn more about the devices in this book. Much of the information on a manufacturer's data sheet is obscure and of use only to circuit designers. For this reason, we will discuss only those entries on the data sheet that describe quantities in this book.

Reverse Breakdown Voltage

Let us start with the data sheet for a 1N4001, a very popular rectifier diode used in power supplies (circuits that convert ac voltage to dc voltage). In the Appendix you will find a data sheet for the 1N4001 to 1N4007 series of diodes: seven diodes that have the same forward characteristics but differ in their reverse characteristics. We are interested in learning how to read the data sheet for the 1N4001 member of this family. The first entry under "Maximum Ratings" is this:

	Symbol	1N4001
Peak Repetitive Reverse Voltage	V_{RRM}	50 V
Working Peak Reverse Voltage	V_{RWM}	50 V
DC Blocking Voltage	V_R	50 V

These three different breakdown symbols specify breakdown under certain conditions of operation. All you have to know is that the breakdown voltage for this diode is 50 V, no matter how the diode is being used. This breakdown occurs because the diode goes into avalanche where huge number of carriers suddenly appear in the depletion layer. With a rectifier diode like the 1N4001, breakdown is always destructive.

With a 1N4001, a reverse voltage of 50 V represents a destructive level that a designer must be careful to avoid under all operating conditions. This is why a designer includes a *safety factor*. There is no absolute rule on how large to make the safety factor because it depends on too many design factors. A conservative design would use a safety factor of 2, which means never allowing a reverse voltage of more than 25 V across the 1N4001. A less conservative design might allow as much as 40 V across the 1N4001.

Maximum Forward Current

Another entry of interest is average rectified forward current, which appears like this on the data sheet:

	Symbol	Value
Average Rectified Forward Current (single-phase, resistive-load, 60 Hz, $T_A = 75°C$)	I_O	1 A

This entry tells us that the 1N4001 can handle up to 1 A in the forward direction when used as a rectifier. You will learn more about average rectified forward current in the next chapter. For now, all you need to know is that 1 A is the level of forward current where the diode burns out because of excessive power dissipation. (Sometimes a diode may be able to withstand the maximum rating without burning out. But the device's characteristics are usually degraded so that it doesn't work as specified by the manufacturer.)

Again, a designer looks upon 1 A as the absolute maximum rating of the 1N4001, a level of forward current that should not even be approached. This is why a safety factor would be included; possibly a factor of 2. In other words, a reliable design would ensure that the forward current is less than 0.5 A under all operating conditions. Failure studies of devices show that the lifetime of a device decreases the closer you get to the maximum rating. This is why some designers have been known to use a safety factor of as much as 10:1. A really conservative design would keep the maximum forward current of a 1N4001 at 0.1 A or less.

Forward Voltage Drop

Under "Electrical Characteristics" in the Appendix, the first entry shown is this:

Characteristic and Condition	Symbol	Typical Values	Maximum Values
Maximum Instantaneous Forward Voltage Drop ($i_F = 1.0$ A, $T_J = 25°C$)	v_F	0.93 V	1.1 V

These measurements are made with an ac signal, which is why the word instantaneous appears in the specification. The typical 1N4001 has a forward voltage drop of 0.93 V when the current is 1 A and the junction temperature is 25°C. If you test thousands of 1N4001s, you will find that few have as much as 1.1 V across them when the current is 1 A. Typical values are useful in the early stages of a design. But in the final stages, the designer has to use the maximum values to make sure the circuit works in the worst case. This is where bulk resistance comes in. The foregoing data allows a designer to calculate the typical bulk resistance and the maximum (worst-case) bulk resistance. If you are interested in seeing how this is done, refer to "Optional Topics," on the next page.

Maximum Reverse Current

Another entry on the data sheet that is worth discussing is this one:

	Symbol	Typical Value	Maximum Value
Maximum Reverse Current	I_R		
$T_J = 25°C$		0.05 μA	10 μA
$T_J = 100°C$		1.0 μA	50 μA

This is the reverse current at the dc rated voltage (50 V for a 1N4001). At 25°C, the typical 1N4001 has a reverse current of 0.05 μA. But notice how it increases to 1 μA at 100°C. In the worst case, the reverse current is 10 μA at 25°C and 50 μA at 100°C. Remember this reverse current includes thermally-produced current and surface-leakage current. You can see from these numbers that temperature might be important. A design that is based on a reverse current of 0.05 μA will work fine at 25°C with a typical 1N4001, but may fail in mass production if the circuit has to work in environments where the junction temperature reaches 100°C.

OPTIONAL TOPICS

The following material continues the earlier discussions at a more advanced and specialized level. All the topics are optional because they are not used in any of the basic discussions in later chapters. This section will be a useful reference when you are in industry because then you will probably want more advanced viewpoints.

□ **3-12** **LINEAR DEVICES**

Ohm's law tells us the current through an ordinary resistor is proportional to the voltage across the resistor. This means a graph of resistor current versus resistor voltage is linear. For instance, given a resistor of 500 Ω, its *I-V* graph looks like Fig. 3-14. Notice the sample points. The current is 1 mA for a voltage of 0.5 V and 2 mA for 1 V. In either case, the ratio of voltage to current equals 500 Ω. Reversing the voltage has no effect on the linearity of the graph. A reverse current of -1 mA exists for a reverse voltage of -0.5 V; the current increases to -2 mA for -1 V.

An ordinary resistor is often called a *linear device* because the graph of its current versus voltage is a straight line similar to Fig. 3-14. An ordinary resistor is also referred to as a *passive device* because all it does is dissipate power; it cannot generate power. A battery, on the other hand, is an *active device* because it can generate power.

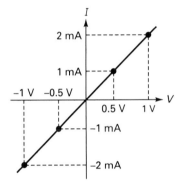

FIGURE 3-14

Linear resistance.

□ **3-13** **HOW TO CALCULATE BULK RESISTANCE**

When you are trying to analyze a diode circuit accurately, you will need to know the bulk resistance of the diode. Manufacturer's data sheets do not usually list the bulk resistance separately, but they do give enough information to allow you to calculate it. Here is the formula for bulk resistance:

$$r_B = \frac{V_2 - V_1}{I_2 - I_1} \tag{3-4}$$

where V_1 and I_1 are the voltage and current at some point at or above the knee voltage; V_2 and I_2 are the voltage and current at some higher point on the diode curve.

For instance, the data sheet of a 1N4001 (see the Appendix) gives a forward voltage of 0.93 V for a current of 1 A. Since this is a silicon diode, it has a knee voltage of approximately 0.7 V and a current of approximately zero. Therefore, the values to use are $V_2 = 0.93$ V, $I_2 = 1$ A, $V_1 = 0.7$ V, and $I_1 = 0$. Substituting these values into equation, we get a bulk resistance of

$$r_B = \frac{V_2 - V_1}{I_2 - I_1} = \frac{0.93\,\text{V} - 0.7\,\text{V}}{1\,\text{A} - 0\,\text{A}} = \frac{0.23\,\text{V}}{1\,\text{A}} = 0.23\,\Omega$$

Incidentally, the reason that you don't use the total voltage divided by the total current when calculating the bulk resistance is because of the built-in barrier potential that every diode has. The bulk resistance is the ohmic resistance of only the p and n regions without the depletion layer. Therefore, to calculate the bulk resistance, we have to eliminate the effect of the barrier potential. The way to do this is by using two points on the diode curve that are above the knee voltage.

3-14 DC RESISTANCE OF A DIODE

If you take the ratio of total diode voltage to total diode current, you get the *dc resistance* of the diode. In the forward direction, this dc resistance is symbolized by R_F; in the reverse direction, it is designated R_R.

Forward Resistance

Because the diode is a nonlinear resistance, its dc resistance varies with the current through it. For example, here are some pairs of forward current and voltage for a 1N914: 10 mA at 0.65 V, 30 mA at 0.75 V, and 50 mA at 0.85 V. At the first point, the dc resistance is

$$R_F = \frac{0.65 \text{ V}}{10 \text{ mA}} = 65 \text{ } \Omega$$

At the second point,

$$R_F = \frac{0.75 \text{ V}}{30 \text{ mA}} = 25 \text{ } \Omega$$

And at the third point,

$$R_F = \frac{0.85 \text{ mA}}{50 \text{ mA}} = 17 \text{ } \Omega$$

Notice how the dc resistance decreases as the current increases. In any case, the forward resistance is low.

Reverse Resistance

Similarly, here are two sets of reverse current and voltage for a 1N914: 25 nA at 20 V; 5 μA at 75 V. At the first point, the dc resistance is

$$R_R = \frac{20 \text{ V}}{25 \text{ nA}} = 800 \text{ M}\Omega$$

At the second point,

$$R_R = \frac{75 \text{ V}}{5 \text{ μA}} = 15 \text{ M}\Omega$$

Notice how the dc resistance decreases as we approach the breakdown voltage (75 V). Nevertheless, the reverse resistance of the diode is still high, well into the megohms.

3-15 LOAD LINES

This section is about the *load line*, a tool used to find the exact value of diode current and voltage. Load lines are especially useful with transistors, so a detailed explanation will be given later in the transistor discussions.

Equation for the Load Line

How can we find the exact diode current and voltage in Fig. 3-15? In this series circuit, a voltage source V_S forward-biases the diode through a current-limiting resistor R_S. The voltage from the left end of the resistor to ground is V_S, the source voltage. The voltage from the right end of the resistor to ground is V, the diode voltage. Therefore, the difference of potential across the resistor is $V_S - V$, and the current is

$$I = \frac{V_S - V}{R_S} \tag{3-5}$$

Because of the series circuit, this current is the same throughout the circuit.

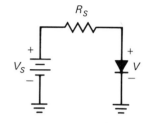

FIGURE 3-15
Diode circuit.

An Example

If the source voltage and current-limiting resistance are given, only the diode current and voltage are unknown. For instance, if the source voltage is 2 V and the current-limiting resistance is 100 Ω, then Eq. (3-5) becomes

$$I = \frac{2 - V}{100} \tag{3-6}$$

Equation (3-6) is a linear relation between current and voltage. If we plot this equation, we will get a straight line. For instance, let V equal zero. Then

$$I = \frac{2\,\text{V} - 0\,\text{V}}{100\,\Omega} = 20\,\text{mA}$$

Plotting this point ($I = 20$ mA, $V = 0$) gives the point on the vertical axis of Fig. 3-16. This point is called *saturation* because it represents maximum current.

Here's how to get another point. Let V equal 2 V. Then Eq. (3-6) gives

$$I = \frac{2\,\text{V} - 2\,\text{V}}{100\,\Omega} = 0$$

When we plot this point ($I = 0$, $V = 2$ V), we get the point shown on the horizontal axis (Fig. 3-16). This point is called *cutoff* because it represents minimum current.

By selecting other voltages, we can calculate and plot additional points. Because Eq. (3-6) is linear, all points will lie on the straight line shown

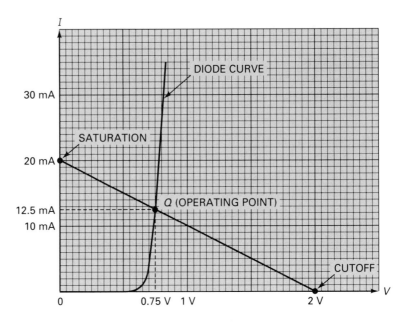

FIGURE 3-16

Load line.

in Fig. 3-16. (Try plotting other points if you don't believe it.) The straight line is called the *load line*.

The *Q* Point

Figure 3-16 shows the graph of a load line and a diode curve. The point of intersection represents a simultaneous solution. In other words, the coordinates of point Q are the values of diode current and voltage for a source voltage of 2 V and a current-limiting resistance of 100 Ω. By reading the coordinates of the Q point, we get a diode current of approximately 12.5 mA and a diode voltage of 0.75 V. The Q point is referred to as the *operating point* because it represents the current through the resistor and the diode.

STUDY AIDS

The following study aids will help to reinforce the ideas discussed in this chapter. For best results, use these study aids within 6 hours of reading the earlier material. Then review these study aids a week later and a month later to ensure that the concepts remain in your long-term memory.

SUMMARY

Sec. 3-1 The Schematic Symbol

The *p* side is called the anode, and the *n* side is called the cathode. The diode symbol looks like an arrow that points in the easy direction of conventional flow. The opposite way is the easy direction for electron flow.

Sec. 3-2 The Diode Curve

When a manufacturer optimizes a diode to convert alternating current to direct current, the diode is called a rectifier diode. One of its main uses is in power supplies; circuits that convert alternating voltage from the power line into direct voltage for electronics equipment.

Sec. 3-3 The Forward Region

The knee voltage of a diode is where the forward curve turns upward. This voltage is approximately equal to the barrier potential of the diode. The diode is called a nonlinear device because the graph of its current versus voltage is not a straight line. A current-limiting resistor is always used with a diode to prevent the current from exceeding the maximum rating.

Sec. 3-4 The Reverse Region

Only a small current exists in a reverse-biased diode. To a first approximation, this current is zero because a reverse-biased diode is like an open switch.

Sec. 3-5 The Ideal Diode

The ideal diode is the first approximation of a diode. The idea is to visualize the diode as an intelligent switch that automatically closes when forward-biased and opens when reverse-biased.

Sec. 3-6 The Second Approximation

In this approximation we visualize a silicon diode as a switch in series with a battery of 0.7 V. The switch closes when the source voltage is equal to or greater than 0.7 V. The switch opens when the source voltage is less than 0.7 V.

Sec. 3-7 The Third Approximation

In this approximation, the bulk resistance of the diode is in series with the switch and the battery. Because of this, the total forward voltage of a conducting silicon diode is the sum of 0.7 V and the voltage across the bulk resistance.

Sec. 3-8 Selecting an Approximation

The ideal diode is perfect for troubleshooting and preliminary circuit analysis. The third approximation is used mostly by designers in the later stages of a design. The second approximation is an excellent compromise for troubleshooting and design. Knowing which one to use in a given situation comes with experience.

Sec. 3-9 Troubleshooting

When you suspect that a diode is the trouble, use an ohmmeter to check its resistance in each direction. You should get a low resistance in one direction and a high resistance in the other. This test may work even when the diode is in the circuit and the power is off. But you can be fooled. The only sure way is to disconnect the diode from the circuit before testing it with an ohmmeter.

Sec. 3-10 Up-down Thinking

Up-down thinking is very useful in troubleshooting, analysis, and design. It really helps you to get your bearings and to use formulas intelligently rather than blindly. When you know beforehand how a dependent variable should respond to changes in an independent variable, you are less likely to make mistakes with formulas. In short, up-down thinking gives you perspective, insight, and control over the situation.

Sec. 3-11 Reading a Data Sheet

Data sheets specify the characteristics of semiconductor devices. The data sheet of the 1N4001 contains this useful information: breakdown voltage, maximum forward current, forward voltage drop, and maximum reverse current.

VOCABULARY

In your own words, explain what each of the following terms means. Keep your answers short and to the point. If necessary, verify your answer by rereading the appropriate discussion or by looking at the end-of-book Glossary.

approximation	ideal diode
bulk resistance	knee voltage
current-limiting resistor	leakage current
linear	power dissipation
maximum forward current	rectifier diode
nonlinear device	safety factor

IMPORTANT EQUATIONS

Formulas are dangerous things. Improperly used, they make bridges fall down and circuits go up in smoke. The following formulas are useless if you don't know what they mean in words. Suggestion: Look at each formula, then read the words to find out what it means. Your chances of learning and remembering are much better if you concentrate on words rather than formulas.

Eq. 3-1 Current-limiting Resistor

$$I = \frac{V_S - V_D}{R}$$

This is Ohm's law for the current through the current-limiting resistor. It says the current equals the voltage across the resistor divided by the resistance.

Eq. 3-2 The Third Approximation

$$V_D = 0.7 + I_D r_B$$

This is a combination of Ohm's law and Kirchhoff's voltage law. This is the equation for the total voltage across the diode when you are using the third approximation. The diode voltage equals the barrier potential (0.7 V) plus the voltage across the

bulk resistance (the diode current times the bulk resistance).

Eq. 3-3 Selecting an Approximation

$$I_F = \frac{V_S - 0.7}{R_L + r_B}$$

This is as complicated as it gets, but is still simple if you understand what is going on. Again, it's Ohm's law in disguise. The numerator is the net voltage in the circuit: the difference between the source voltage and the barrier potential. This net voltage appears across the total series resistance. The denominator is the total series resistance: the sum of load resistance and bulk resistance. The equation says the forward current equals the net voltage divided by the total resistance.

STUDENT ASSIGNMENTS

QUESTIONS

The following may have more than one right answer. Select the best answer. This is the one that is always true, or covers more situations, or fits the context, etc.

1. When the graph of current versus voltage is a straight line, the device is referred to as
    a. Active
    b. Linear
    c. Nonlinear
    d. Passive

2. What kind of device is a resistor?
    a. Unilateral
    b. Linear
    c. Nonlinear
    d. Bipolar

3. What kind of a device is a diode?
    a. Bilateral
    b. Linear
    c. Nonlinear
    d. Unipolar

4. How is a nonconducting diode biased?
    a. Forward
    b. Inverse
    c. Poorly
    d. Reverse

5. When the diode current is large, the bias is
    a. Forward
    b. Inverse
    c. Poor
    d. Reverse

6. The knee voltage of a diode is approximately equal to the
    a. Applied voltage
    b. Barrier potential
    c. Breakdown voltage
    d. Forward voltage

7. The leakage current consists of minority-carrier current and
    a. Avalanche current
    b. Forward current
    c. Surface-leakage current
    d. Zener current

8. How much voltage is there across the second approximation of a silicon diode when it is forward-biased?
    a. 0
    b. 0.3 V
    c. 0.7 V
    d. 1 V

9. How much current is there through the second approximation of a silicon diode when it is reverse-biased?
   a. 0
   b. 1 mA
   c. 300 mA
   d. None of the above

10. How much diode voltage is there with the ideal-diode approximation?
   a. 0
   b. 0.7 V
   c. more than 0.7 V
   d. 1 V

11. The bulk resistance of a 1N4001 is
   a. 0
   b. 0.23 Ω
   c. 10 Ω
   d. 1 kΩ

12. If the bulk resistance is zero, the graph above the knee becomes
   a. Horizontal
   b. Vertical
   c. Tilted at 45°
   d. None of the above

13. The ideal diode is usually adequate when
   a. Troubleshooting
   b. Doing precise calculations
   c. The source voltage is low
   d. The load resistance is low

14. The second approximation works well when
   a. Troubleshooting
   b. Load resistance is high
   c. Source voltage is high
   d. All of the above

15. The only time you have to use the third approximation is when
   a. Load resistance is low
   b. Source voltage is high
   c. Troubleshooting
   d. None of the above

16. How much load current is there in Fig. 3-17 with the ideal diode?
   a. 0
   b. 14.3 mA
   c. 15 mA
   d. 50 mA

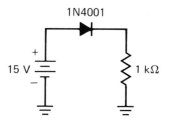

FIGURE 3-17 Diode circuit.

17. How much load current is there in Fig. 3-17 with the second approximation?
   a. 0
   b. 14.3 mA
   c. 15 mA
   d. 50 mA

18. How much load current is there in Fig. 3-17 with the third approximation?
   a. 0
   b. 14.3 mA
   c. 15 mA
   d. 50 mA

19. If the diode is open in Fig. 3-17, the load voltage is
   a. 0
   b. 14.3 V
   c. 20 V
   d. − 15 V

20. If the resistor is ungrounded in Fig. 3-17, the voltage measured between the top of the resistor and ground is
   a. 0
   b. 14.3 V
   c. 20 V
   d. − 15 V

21. The load voltage measures zero in Fig. 3-17. The trouble may be
   a. A shorted diode
   b. An open diode
   c. An open load resistor
   d. Too much supply voltage

BASIC PROBLEMS

Sec. 3-3 The Forward Region

3-1. A diode is in series with 220 Ω. If the voltage across the resistor is 4 V, what is the current through the diode?

3-2. A diode has a voltage of 0.7 V and a current of 50 mA. What is the diode power?

3-3 Two diodes are in series. The first diode has a voltage of 0.75 V and the second has a voltage of 0.8 V. If the current through the first diode is 500 mA, what is the current through the second diode?

Sec. 3-5 The Ideal Diode

3-4. In Fig. 3-18a, calculate the load current, load voltage, load power, diode power, and total power.

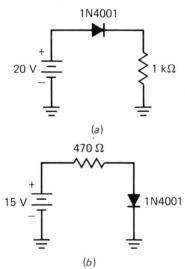

(a)

(b)

FIGURE 3-18

3-5. If the resistor is doubled in Fig. 3-18a, what is the load current?

3-6. In Fig. 3-18b, calculate the load current, load voltage, load power, diode power, and total power.

3-7. If the resistor is doubled in Fig. 3-18b, what is the load current?

3-8. If the diode polarity is reversed in Fig. 3-18b, what is the diode current? The diode voltage?

Sec. 3-6 The Second Approximation

3-9. In Fig. 3-18a, calculate the load current, load voltage, load power, diode power, and total power.

3-10. If the resistor is doubled in Fig. 3-18a, what is the load current?

3-11. In Fig. 3-18b, calculate the load current, load voltage, load power, diode power, and total power.

3-12. If the resistor is doubled in Fig. 3-18b, what is the load current?

3-13. If the diode polarity is reversed in Fig. 3-18b, what is the diode current? The diode voltage?

Sec. 3-7 The Third Approximation

3-14. In Fig. 3-18a, calculate the load current, load voltage, load power, diode power, and total power.

3-15. If the resistor is doubled in Fig. 3-18a, what is the load current?

3-16. In Fig. 3-18b, calculate the load current, load voltage, load power, diode power, and total power.

3-17. If the resistor is doubled in Fig. 3-18b, what is the load current?

3-18. If the diode polarity is reversed in Fig. 3-18b, what is the diode current? The diode voltage?

Sec. 3-9 Troubleshooting

3-19. Suppose the voltage across the diode of Fig. 3-19a is 5 V. Is the diode open or shorted?

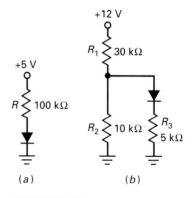

(a)

(b)

FIGURE 3-19

3-20. Something causes R to short in Fig. 3-19a. What will the diode voltage be? What will happen to the diode?

3-21. You measure 0 V across the diode of Fig. 3-19a. Next you check the source voltage and it reads +5 V with respect to ground. What is wrong with the circuit?

3-22. In Fig. 3-19*b*, you measure a potential of +3 V at the junction of R_1 and R_2. (Remember, potentials are always with respect to ground.) Next you measure 0 V at the junction of the diode and the 5-kΩ resistor. Name some possible troubles.

3-23. You measure 0 V at the junction of R_1 and R_2 in Fig. 3-19*b*. What are some of the things that can be wrong with this circuit?

Sec. 3-11 Reading Data Sheets

3-24. Which diode would you select in the 1N4001 series if it has to withstand a peak repetitive reverse voltage of 700 V?

3-25. The data sheet shows a band on one end of the diode. What is the name of this band? Does the diode arrow of the schematic symbol point toward or away from this band?

3-26. Boiling water has a temperature of 100°C. If you drop a 1N4001 into a pot of boiling water, will it be destroyed or not? Explain your answer.

ADVANCED PROBLEMS

3-27. Here are some diodes and their worst-case specifications:

Diode	I_F	I_R
1N914	10 mA at 1 V	25 nA at 20 V
1N4001	1 A at 1.1 V	10 μA at 50 V
1N1185	10 A at 0.95 V	4.6 mA at 100 V

Calculate the forward and the reverse resistance for each of these diodes.

3-28. In Fig. 3-19*a*, what value should R be to get a diode current of approximately 10 mA?

3-29. What value should R_2 be in Fig. 3-19*b* to set up a diode current of 0.25 mA?

3-30. A silicon diode has a forward current of 50 mA at 1 V. Use the third approximation to calculate its bulk resistance.

3-31. Given a silicon diode with a reverse current of 5 μa at 25°C and 100μA at 100°C, calculate the surface leakage current.

3-32. The power is turned off and the upper end of R_1 is grounded in Fig. 3-19*b*. Now you use an ohmmeter to read the forward and reverse resistance of the diode. Both readings are identical. What does the ohmmeter read?

3-33. Some systems, like burglar alarms, and computers, use battery backup just in case the main source of power should fail. Describe how the circuit of Fig. 3-20 works.

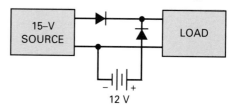

FIGURE 3-20

SOFTWARE ENGINE PROBLEMS

Use Fig. 3-21 for the remaining problems. If you haven't already done so, read Example 3-9 on page 70 before attempting these problems. The Software Engine is a simplified textbook version of software written for computer-assisted instruction. You will find the Software Engine useful for improving your troubleshooting and design skills. You can check circuit responses in any order—for instance, V_C first, P_2 second, and I_1 third. After selecting the independent and dependent variables, try to figure out what the response will be to a small *increase* in the independent variable. The labels at the top of each box are the independent variables. The box contents are the dependent variables and their tokens. The token values are given in the large box labeled "Responses."

A final point. Assume increases of approximately 10 percent in the independent variable and use the second approximation of a diode. Also, use U for up, D for down, and N for no change. A response should be an N if the change is so small in your

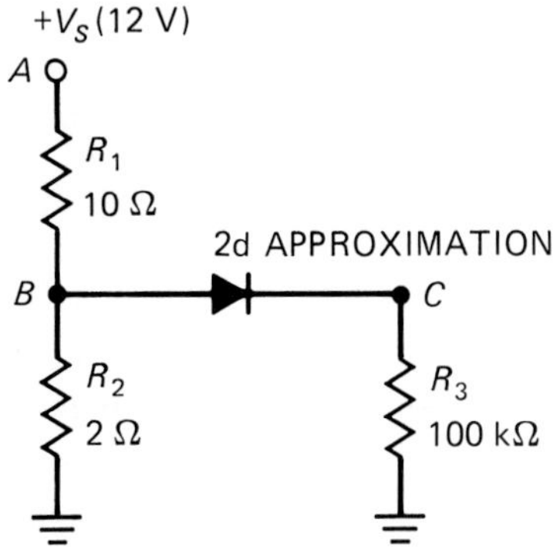

	1	2	3	4	5	6
A	U	U	D	D	U	D
B	U	D	U	U	N	D
C	D	N	U	D	N	N
D	D	U	N	U	D	U
E	U	U	N	D	D	U
F	N	N	D	U	N	U

RESPONSES

V_S	R_1	R_2	R_3	V_K
V_A : *C*3	V_A : *C*2	V_A : *C*6	V_A : *F*1	V_A : *E*3
V_B : *A*5	V_B : *B*6	V_B : *A*1	V_B : *B*5	V_B : *F*1
V_C : *E*6	V_C : *F*3	V_C : *D*6	V_C : *F*2	V_C : *C*4
I_1 : *B*1	I_1 : *A*3	I_1 : *B*2	I_1 : *F*5	I_1 : *D*3
I_2 : *E*2	I_2 : *C*1	I_2 : *C*4	I_2 : *D*3	I_2 : *F*5
I_3 : *D*4	I_3 : *A*6	I_3 : *F*4	I_3 : *E*4	I_3 : *A*3
P_1 : *A*2	P_1 : *E*5	P_1 : *D*5	P_1 : *C*2	P_1 : *C*2
P_2 : *B*4	P_2 : *D*1	P_2 : *B*3	P_2 : *F*1	P_2 : *E*3
P_3 : *F*6	P_3 : *A*4	P_3 : *E*1	P_3 : *D*5	P_3 : *A*4

FIGURE 3-21 Software Engine ™. (*Patent pending: Courtesy of Malvino Inc.*)

opinion that you would have difficulty measuring it. For instance, you can easily see a change of 10 percent in a variable, but you would find it difficult to see a change of less than 1 percent.

3-34. Try to predict the response of each dependent variable in the box labeled V_S. Check your answers. Then, answer the following question as simply and directly as possible. What effect does an increase in source voltage have on the dependent variables of the circuit?

3-35. Predict the response of each dependent variable in the box labeled R_1. Check your answers. Then summarize your findings in one or two sentences.

3-36. Predict the response of each dependent variable in the box labeled R_2. Check your answers. List the dependent variables that decrease. Explain why these variables decrease, using Ohm's law or similar basic ideas.

3-37. Predict the response of each dependent variable in the box labeled R_3. List the dependent variables that show no change. Explain why these variables show no change.

3-38. Predict the response of each dependent variable in the box labeled V_K. List the dependent variables that decrease. Explain why these variables decrease.

4

DIODE CIRCUITS

A rectifier diode is ideally a closed switch when forward-biased and an open switch when reverse-biased. Because of this, it is useful for converting alternating current to direct current. This chapter discusses three basic rectifier circuits called the *half-wave rectifier*, the *full-wave rectifier*, and the *bridge rectifier*.

4-1 THE INPUT TRANSFORMER

Power companies in the United States supply a nominal line voltage of 115 V rms at a frequency of 60 Hz. The actual voltage coming out of a power outlet may vary from 105 V to 125 V rms, depending on the time of day, locality, and other factors. Recall that the relation between the rms value and the peak value of a sine wave is given by

$$V_{rms} = 0.707V_p \tag{4-1}$$

This equation says that the rms voltage equals 70.7 percent of the peak voltage. Recall what rms value means. This is the equivalent dc voltage that would produce the same amount of power over one complete cycle.

Basic Equation

Line voltage is too high for most of the devices used in electronics equipment. This is why a transformer is commonly used in almost all electronics equipment. This transformer steps the ac voltage down to lower levels that are more suitable for use with devices like diodes and transistors.

Figure 4-1 shows an example of a transformer. The left coil is called the *primary winding* and the right coil is called the *secondary winding*. The number of turns on the primary winding is N_1, and the number of turns on the secondary winding is N_2. The vertical lines between the primary and secondary windings indicate that the turns are wrapped on an iron core.

With this type of transformer, the coefficient of coupling k approaches one, which means tight coupling exists. In other words, all the flux

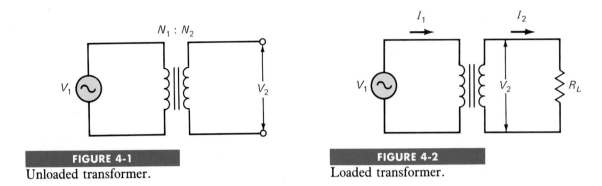

FIGURE 4-1

Unloaded transformer.

FIGURE 4-2

Loaded transformer.

produced by the primary winding cuts through the secondary winding. The voltage induced in the secondary winding is given by

$$V_2 = \frac{N_2}{N_1} V_1 \qquad (4\text{-}2)$$

The voltages in this equation may be either rms or peak voltages. Just be consistent and use rms for both, or peak for both.

Step-Up Transformer

When the secondary winding has more turns than the primary winding, more voltage is induced in the secondary than in the primary. In other words, when N_2/N_1 is greater than one, the transformer is referred to as a *step-up* transformer. If $N_1 = 100$ turns and $N_2 = 300$ turns, the same flux cuts through three times as many turns in the secondary as in the primary winding. This is why the secondary voltage is three times as large as the primary voltage.

Step-Down Transformer

When the secondary winding has fewer turns than the primary winding, less voltage is induced in the secondary than in the primary. In this case, the turns ratio, $N_2{:}N_1$, is less than one, and the transformer is called a *step-down* transformer. If $N_1 = 100$ turns and $N_2 = 50$ turns, the same flux cuts through half as many turns in the secondary as in the primary winding, and the secondary voltage is half the primary voltage.

Effect on Current

Figure 4-2 shows a load resistor connected across the secondary winding. Because of the induced voltage across the secondary winding, a load current exists. If the transformer is ideal ($k = 1$ and no power is lost in the windings or the core), the output power equals the input power:

$$P_2 = P_1$$

or

$$V_2 I_2 = V_1 I_1$$

We can rearrange the foregoing equation as follows:

$$\frac{I_1}{I_2} = \frac{V_2}{V_1}$$

But Eq. (4-2) implies that $V_2/V_1 = N_2/N_1$. Therefore,

$$\frac{I_1}{I_2} = \frac{N_2}{N_1}$$

or

$$I_1 = \frac{N_2}{N_1} I_2 \qquad (4\text{-}3)$$

An alternative way to write the foregoing equation is

$$I_2 = \frac{N_1}{N_2} I_1 \qquad (4\text{-}4)$$

Notice the following. For a step-up transformer, the voltage is stepped up but the current is stepped down. On the other hand, for a step-down transformer, the voltage is stepped down but the current is stepped up.

EXAMPLE 4-1

Suppose the voltage from a power outlet is 120 V rms. What is the peak voltage?

SOLUTION

Using algebra, we can rewrite Eq. (4-1) in this equivalent form:

$$V_p = \frac{V_{\text{rms}}}{0.707}$$

Now, substitute the rms voltage and calculate the peak voltage:

$$V_p = \frac{120 \text{ V}}{0.707} = 170 \text{ V}$$

This tells us that the sinusoidal voltage out of the power outlet has a peak value of 170 V.

EXAMPLE 4-2

A step-down transformer has a turns ratio of 5:1. If the primary voltage is 120 V rms, what is the secondary voltage?

SOLUTION

Divide the primary voltage by 5 to get the secondary voltage:

$$V_2 = \frac{120 \text{ V}}{5} = 24 \text{ V}$$

EXAMPLE 4-3

Suppose a step-down transformer has a turns ratio of 5:1. If the secondary current is 1 A rms, what is the primary current?

SOLUTION

With Eq. (4-3),

$$I_1 = \frac{1\,A}{5} = 0.2\,A$$

As a check on this answer, use your common sense as follows. This is a step-down transformer, which means the current is stepped up going from primary to secondary, equivalent to saying the current is stepped down as we go from the secondary to the primary. This means the primary current is five times smaller than the secondary current. Whenever possible, you should check that your answers are logical because it is easy to make a mistake with equations.

4-2   THE HALF-WAVE RECTIFIER

The simplest circuit that can convert alternating current to direct current is the *half-wave* rectifier, shown in Fig. 4-3. Line voltage from an ac power outlet is applied to the primary winding of the transformer. Usually, the power plug has a third prong to ground the equipment. Because of the turns ratio, the peak voltage across the secondary winding is

$$V_{p2} = \frac{N_2}{N_1} V_{p1}$$

Recall the dot convention used with transformers. The dotted ends of a transformer have the same polarity of voltage at any instant in time. When the upper end of the primary winding is positive, the upper end of the secondary winding is also positive. When the upper end of the primary winding is negative, the upper end of the secondary winding is also negative.

Here is how the circuit works. On the positive half cycle of primary voltage, the secondary winding has a positive half sine wave across it. This means the diode is forward-biased. However, on the negative half cycle of primary voltage, the secondary winding has a negative half sine wave. Therefore, the diode is reverse-biased. If you use the ideal-diode approximation for an initial analysis, you will realize that the positive half cycle appears across the load resistor, but not the negative half cycle.

For instance, Fig. 4-4 shows a transformer with a turns ratio of 5:1. The peak primary voltage is

$$V_{p1} = \frac{120\,V}{0.707} = 170\,V$$

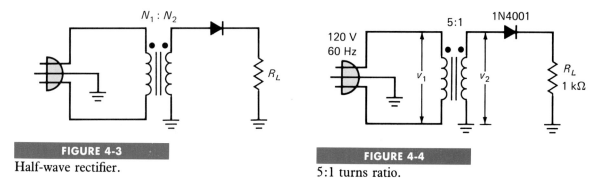

FIGURE 4-3

Half-wave rectifier.

FIGURE 4-4

5:1 turns ratio.

The peak secondary voltage is

$$V_{p2} = \frac{170\text{ V}}{5} = 34\text{ V}$$

With the ideal-diode approximation, the load voltage has a peak value of 34 V.

Figure 4-5 shows the load voltage. This type of waveform is called a *half-wave signal* because the negative half cycles have been clipped off or removed. Since the load voltage has only a positive half cycle, the load current is unidirectional, meaning that it flows only in one direction. Therefore, the load current is a pulsating direct current. It starts at zero at the beginning of the cycle, then increases to a maximum value at the positive peak, then decreases to zero where it sits for the entire negative half cycle.

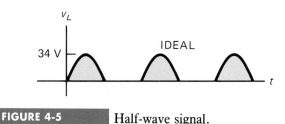

FIGURE 4-5   Half-wave signal.

Period

The frequency of the half-wave signal is still equal to the line frequency, which is 60 Hz. (In Europe, line frequency is 50 Hz.) Recall that the period, T, equals the reciprocal of the frequency. Therefore, the half-wave signal has a period of

$$T = \frac{1}{f} = \frac{1}{60\text{ Hz}} = 0.0167\text{ s} = 16.7\text{ ms}$$

This is the amount of time between the beginning of a positive half cycle and the start of the next positive half cycle. This is what you would measure if you looked at a half-wave signal with an oscilloscope.

DC or Average Value

If you connect a dc voltmeter across the load resistor of Fig. 4-4, it will indicate a dc voltage of V_p/π, which may be written as

$$V_{dc} = 0.318V_p \qquad (4-5)$$

where V_p is the peak value of the half-wave signal across the load resistor. For instance, if the peak voltage is 34 V, the dc voltmeter will read

$$V_{dc} = 0.318(34 \text{ V}) = 10.8 \text{ V}$$

This dc voltage is sometimes called the *average* value of the half-wave signal because the voltmeter reads the average voltage over one complete cycle. The needle of the voltmeter cannot follow the rapid variations of the half-wave signal, so the needle settles down on the average value, which is 31.8 percent of the peak value. (The 31.8 percent can be proved with calculus.)

Approximations

Because the secondary voltage is much greater than the knee voltage, using the second approximation will improve the analysis only slightly. If we use the second approximation, the half-wave signal has a peak of 33.3 V. Furthermore, since the bulk resistance of a 1N4001 is only 0.23 Ω compared to a load resistance of 1 kΩ, there is no increase in accuracy when using the third approximation. In conclusion, either the ideal diode or the second approximation is adequate in analyzing this circuit.

EXAMPLE 4-4

In Europe, a half-wave rectifier has an input voltage of 240 V rms with a frequency of 50 Hz. If the step-down transformer has a turns ratio of 8:1, what is the load voltage?

SOLUTION

You can divide 240 V by 0.707 to get the answer. Here is an alternative way to get the peak voltage. Since the rms voltage is twice as large as previous examples, the peak voltage is twice as large as before:

$$V_{p1} = 2(170 \text{ V}) = 340 \text{ V}$$

Because of the 8:1 step down, the secondary voltage has a peak value of

$$V_{p2} = \frac{340 \text{ V}}{8} = 42.5 \text{ V}$$

Ignoring the diode drop means that the load voltage is a half-wave signal with a peak value of 42.5 V.

The period of the rectified output voltage is slightly longer:

$$T = \frac{1}{50\text{ Hz}} = 0.02\text{ s} = 20\text{ ms}$$

This is what you would measure with an oscilloscope.

4-3 THE FULL-WAVE RECTIFIER

Figure 4-6 shows a *full-wave rectifier*. Notice the grounded center tap on the secondary winding. Because of this center tap, the circuit is equivalent to two half-wave rectifiers. The upper rectifier handles the positive half cycle of secondary voltage, while the lower rectifier handles the negative half cycle of secondary voltage. In other words, D_1 conducts on the positive half cycle and D_2 conducts on the negative half cycle. Because of this, the rectified load current flows during both half cycles. Furthermore, this load current flows in one direction only.

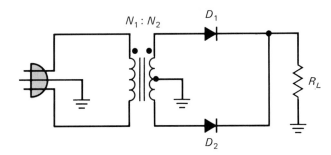

FIGURE 4-6 Full-wave rectifier.

For instance, Fig. 4-7 shows a transformer with a turns ratio of 5:1. The peak primary voltage is still equal to

$$V_{p1} = \frac{120\text{ V}}{0.707} = 170\text{ V}$$

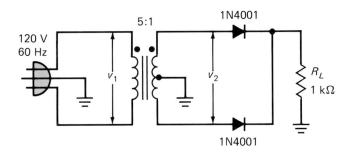

FIGURE 4-7 Example of full-wave rectifier.

The peak secondary voltage is

$$V_{p2} = \frac{170 \text{ V}}{5} = 34 \text{ V}$$

Because of the grounded center tap, each half of the secondary winding has a sinusoidal voltage with a peak of only 17 V. Therefore, the load voltage has an ideal peak value of only 17 V instead of 34 V. This factor-of-two reduction is a characteristic of all full-wave rectifiers. It is a direct result of using a grounded center tap on the secondary winding.

Figure 4-8 shows the load voltage. This type of waveform is called a *full-wave signal*. It is equivalent to inverting or flipping the negative half cycles of a sine wave to get positive half cycles. Because of Ohm's law, the load current is a full-wave signal with a peak value of

$$I_p = \frac{17 \text{ V}}{1 \text{ k}\Omega} = 17 \text{ mA}$$

DC or Average Value

If you connect a dc voltmeter across the load resistor of Fig. 4-7, it will indicate a dc voltage of $2V_p/\pi$, which is equivalent to

$$V_{dc} = 0.636V_p \tag{4-6}$$

where V_p is the peak value of the half-wave signal across the load resistor. For instance, if the peak voltage is 17 V, the dc voltmeter will read

$$V_{dc} = 0.636(17 \text{ V}) = 10.8 \text{ V}$$

This dc voltage is the average value of the full-wave signal because the voltmeter reads the average voltage over one complete cycle.

Output Frequency

The frequency of the full-wave signal is double the input frequency. Why? Recall how a complete cycle is defined. A waveform has a complete cycle when it repeats. In Fig. 4-8, the rectified waveform begins repeating after one half cycle of the primary voltage. Since line voltage has a period, T_1, of

$$T_1 = \frac{1}{f} = \frac{1}{60 \text{ Hz}} = 0.0167 \text{ s} = 16.7 \text{ ms}$$

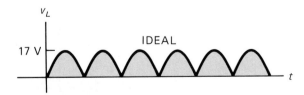

FIGURE 4-8 Full-wave signal.

The rectified load voltage has a period, T_2, of

$$T_2 = \frac{16.7 \text{ ms}}{2} = 8.33 \text{ ms}$$

The frequency of the load voltage therefore equals

$$f_2 = \frac{1}{T_2} = \frac{1}{8.33 \text{ ms}} = 120 \text{ Hz}$$

This says the output frequency equals two times the input frequency. In symbols,

$$f_{\text{out}} = 2f_{\text{in}} \qquad\qquad (4\text{-}7)$$

This doubling of the frequency is a characteristic of all full-wave rectifiers. It is a direct result of using two diodes, one to rectify the positive half cycle of input voltage and the other to rectify the negative half cycle of input voltage. Visually, the effect is to invert the negative half of the input voltage to get a full-wave signal.

Again, notice the following about the use of diode approximations. Because the secondary voltage is much greater than the knee voltage, the second approximation results in a full-wave output voltage with a peak value of 16.3 V instead of 17 V. Once more, the small bulk resistance of a 1N4001 has almost no effect. In conclusion, either the ideal diode or the second approximation is adequate in analyzing most full-wave circuits. The only time you would consider using the third approximation is when the load resistance is small. (Refer to Table 3-2 if you need guidance on what "small" means.)

EXAMPLE 4-5

Suppose the full-wave rectifier of Fig. 4-7 has an input voltage of 240 V rms with a frequency of 50 Hz. If the step-down transformer has a turns ratio of 8:1, what is the load voltage?

SOLUTION

The peak primary voltage is the same as the previous example:

$$V_{p1} = 340 \text{ V}$$

The peak secondary voltage has the same peak value as before:

$$V_{p2} = 42.5 \text{ V}$$

The center tap reduces this voltage by a factor of 2. In other words, the entire secondary winding has a sine wave across it with a peak value of 42.5 V. Therefore, each half of the secondary winding has a sine wave with only half this peak value, or approximately 21.2 V. Ignoring the diode drop means that the load voltage is a full-wave signal with a peak value of 21.2 V.

Also, the rectified output signal has a frequency of twice the input frequency. In this case, the output frequency is

$$f = 2(50 \text{ Hz}) = 100 \text{ Hz}$$

4-4 THE BRIDGE RECTIFIER

Figure 4-9 shows a *bridge rectifier*. By using four diodes instead of two, this clever design eliminates the need for a grounded center tap. The advantage of not using a center tap is that the rectified load voltage is twice what it would be with the full-wave rectifier.

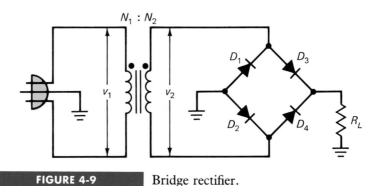

FIGURE 4-9   Bridge rectifier.

Here is how it works. During the positive half cycle of line voltage, diodes D_2 and D_3 conduct; this produces a positive half cycle across the load resistor. During the negative half cycle of line voltage, diodes D_1 and D_4 conduct; this produces another positive half cycle across the load resistor. The result is a full-wave signal across the load resistor.

For instance, Fig. 4-10 shows a transformer with a turns ratio of 5:1. The peak primary voltage is still equal to

$$V_{p1} = \frac{120 \text{ V}}{0.707} = 170 \text{ V}$$

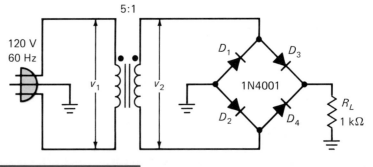

FIGURE 4-10   Example of bridge rectifier.

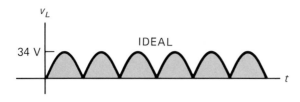

FIGURE 4-11   Full-wave signal.

The peak secondary voltage is still

$$V_{p2} = \frac{170 \text{ V}}{5} = 34 \text{ V}$$

Because the full secondary voltage is applied to the conducting diodes in series with the load resistor, the load voltage has an ideal peak value of 34 V, twice that of the full-wave rectifier discussed earlier.

Figure 4-11 shows the ideal load voltage. As you see, the shape is identical to that of a full-wave rectifier. Therefore, the frequency of the rectified signal equals 120 Hz, twice the line frequency. Because of Ohm's law, the load current is a full-wave signal with a peak value of

$$I_p = \frac{34 \text{ V}}{1 \text{ k}\Omega} = 34 \text{ mA}$$

There is a new factor to consider when using the second approximation with a bridge rectifier: there are two conducting diodes in series with the load resistor during each half cycle. Therefore, we must subtract two diode drops instead of only one. This means the peak voltage with the second approximation is

$$V_p = 34 \text{ V} - 2(0.7 \text{ V}) = 32.6 \text{ V}$$

The additional voltage drop across the second diode is one of the few disadvantages of the bridge rectifier. Also, there are two bulk resistances in series with the load resistance. But the effect is again negligible with the circuit values shown in Fig. 4-10. Unless you are designing a bridge rectifier, you will not normally use the third approximation because the bulk resistance is usually much smaller than the load resistance.

Most designers feel that having two diode drops and two bulk resistances is only a minor disadvantage. The advantages of the bridge rectifier include a full-wave output, an ideal peak voltage equal to the peak secondary voltage, and no center tap on the secondary winding. These advantages have made the bridge rectifier the most popular rectifier design. Most equipment uses a bridge rectifier to convert the ac line voltage to a dc voltage suitable for use with semiconductor devices.

EXAMPLE 4-6

Suppose the bridge rectifier of Fig. 4-9 has an input voltage of 240 V rms with a frequency of 50 Hz. If the step-down transformer has a turns ratio of 8:1, what is the load voltage?

SOLUTION

The peak primary voltage is the same as the previous example:

$$V_{p1} = 340 \text{ V}$$

The peak secondary voltage has the same peak value as before:

$$V_{p2} = 42.5 \text{ V}$$

This time, the entire secondary voltage is across two conducting diodes in series with the load resistor. Ignoring the diode drop means that the load voltage is a full-wave signal with a peak value of 42.5 V. Also, the frequency of the rectified output voltage is 100 Hz.

4-5 THE CAPACITOR-INPUT FILTER

The load voltage out of a rectifier is pulsating rather than steady. For instance, look at Fig. 4-11. Over one complete output cycle, the load voltage increases from zero to a peak, then decreases back to zero. This is not the kind of dc voltage needed for most electronic circuits. What is needed is a steady or constant voltage similar to what a battery produces. To get this type of rectified load voltage, we need to use a *filter*.

Half-wave Filtering

The most common type of filter is the *capacitor-input* filter shown in Fig. 4-12. To simplify the initial discussion of filters, we have represented an ideal diode by a switch. As you can see, a capacitor has been inserted in parallel with the load resistor. Before the power is turned on, the capacitor is uncharged; therefore, the load voltage is zero. During the first quarter cycle of the secondary voltage, the diode is forward-biased. Ideally, it looks like a closed switch. Since the diode connects the secondary winding directly across the capacitor, the capacitor charges to the peak voltage, V_p.

Just past the positive peak, the diode stops conducting, which means the switch opens. Why? Because the capacitor has V_p volts across it. Since

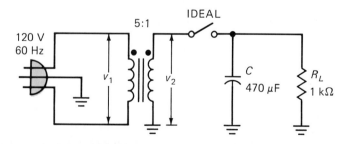

FIGURE 4-12 Capacitor-input filter.

the secondary voltage is slightly less than V_p, the diode goes into reverse bias. With the diode now open, the capacitor discharges through the load resistance. But here is the key idea behind the capacitor-input filter: by deliberate design, the discharging time constant (the product of R_L and C) is much greater than the period, T, of the input signal. Because of this, the capacitor will lose only a small part of its charge during the off time of the diode as shown in Fig. 4-13a.

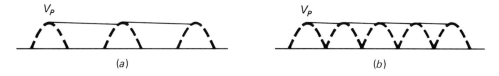

(a)   (b)

| FIGURE 4-13 | (a) Half-wave filtering; (b) full-wave filtering. |

When the source voltage again reaches its peak, the diode conducts briefly and recharges the capacitor to the peak voltage. In other words, after the capacitor is initially charged during the first quarter cycle, its voltage is approximately equal to the peak secondary voltage. This is why the circuit is sometimes called a *peak detector*.

The load voltage is now almost a steady or constant dc voltage. The only deviation from a pure dc voltage is the small ripple caused by charging and discharging the capacitor. The smaller the ripple is, the better. One way to reduce this ripple is by increasing the discharging time constant, which equals $R_L C$.

Full-Wave Filtering

Another way to reduce the ripple is to use a full-wave rectifier or bridge rectifier; then the ripple frequency is 120 Hz instead of 60 Hz. In this case, the capacitor is charged twice as often and has only half the discharge time (see Fig. 4-13b). As a result, the ripple is smaller and the dc output voltage more closely approaches the peak voltage. From now on, our discussion will emphasize the bridge rectifier driving a capacitor-input filter because this is the most commonly used circuit.

Brief Conduction of Diode

In the unfiltered rectifiers discussed earlier, each diode conducts for half a cycle. In the filtered rectifiers we are now discussing, each diode conducts for much less than half a cycle. When the power switch is first turned on, the capacitor is uncharged. Ideally, it takes only a quarter of a cycle to charge the capacitor to the peak secondary voltage. After this initial charging, the diodes turn on only briefly near the peak and are off during the rest of the cycle. In terms of degrees, the diodes turn on for only a couple of degrees during each cycle (half a cycle is 180°).

An Important Formula

Whether you are troubleshooting, analyzing, or designing, you have got to know how to estimate the size of the ripple. Normally, the ripple is small compared to the peak secondary voltage. For most applications, the ripple is considered small when it is less than 10 percent of the load voltage. For instance, if the load voltage is 15 V, the ripple in most filtered rectifiers will be less than 1.5 V peak-to-peak.

Here is the formula for ripple expressed in terms of easily measured circuit values:

$$V_R = \frac{I}{fC} \tag{4-8}$$

where V_R = peak-to-peak ripple voltage
I = dc load current
f = ripple frequency
C = capacitance

The proof of Eq. (4-8) is too lengthy and complicated to show in this book. But the derivation assumes that the peak-to-peak ripple is less than 20 percent of the load voltage. Beyond this point, you cannot use Eq. (4-8) without encountering a lot of error. But as was already discussed earlier, the whole point of the capacitor-input filter is to produce a steady or constant dc voltage. For this reason, most designers deliberately select circuit values to keep the ripple less than 10 percent of the load voltage. In the circuits you encounter, you will find that the ripple is usually less than 10 percent of the load voltage.

DC Voltage

To be successful in electronics, you have to learn the following basic idea: approximations are the rule, not the exception. Why? Because electronics is not an exact science like pure mathematics. The idea that you must always get exact answers is a false idea, a left-brain trap. For most of the work in electronics, approximate answers are adequate and even desirable.

The situation is like an artist painting a picture. The best artist starts with the largest brush when beginning a painting. The artist then switches to a medium-sized brush to improve the picture, and, finally, may use the smallest brush to get the finest detail. No good artist ever uses a small brush all of the time.

The three diode approximations are like an artist's brushes. You should start with the ideal diode to get the big picture. In many cases (troubleshooting, for instance), this will be all you need. Often, you will want to improve your analysis by using the second approximation (a lot of everyday work is done with this one). Finally, the third approximation may be best in some situations (if the circuit uses 1 percent resistors, for example).

With the foregoing in mind, here is how the diode approximations affect the value of the load voltage. For an ideal diode and no ripple, the dc load voltage out of a filtered bridge rectifier equals the peak secondary voltage:

$$V_{dc} = V_{p2}$$

This is what you want to remember when you are troubleshooting or making a preliminary analysis of a filtered bridge rectifier.

With the second approximation of a diode, we have to allow for the 0.7 V across each diode. Since there are two conducting diodes in series with the load resistor, the dc load voltage with no ripple out of a filtered bridge rectifier is

$$V_{dc} = V_{p2} - 1.4 \text{ V}$$

In the third approximation, two bulk resistances are in the charging path of the capacitor. This complicates the analysis because the diode conducts briefly only near the peak. Fortunately, bulk resistances of rectifier diodes are typically less than 1 Ω. Because of this, they usually have little or no effect on the load voltage. Unless you are designing a filtered bridge rectifier, you will not need to consider the effect of bulk resistance. (If you are designing the circuit, you will need to use advanced mathematics because you have to deal with an exponential function. The alternative is to build the circuit and arrive at circuit values by experiment. The main rule here is to keep the load resistance as large as possible compared to the bulk resistance.)

There is one more improvement that we can use. We can include the effect of the ripple as follows:

$$V_{dc(\text{with ripple})} = V_{dc(\text{without ripple})} - \frac{V_R}{2}$$

The idea here is to subtract half the peak-to-peak ripple to refine the answer slightly. Since peak-to-peak is usually less than 10 percent, the improvement in the answer is less than 5 percent.

A Basic Guideline

The resistors used in typical electronic circuits have tolerances of ± 5 percent. Sometimes, you will see precision resistors of ± 1 percent used in critical applications. And sometimes, you will see resistors of ± 10 percent used. But if we take 5 percent as the usual tolerance, then one guideline for selecting an approximation is this: Ignore a quantity if it produces an error of less than 5 percent. This means we can use the ideal diode if it produces less than 5 percent error. If the ideal diode results in 5 percent or more error, switch to the second approximation. Also, ignore the effect of ripple when it is less than 10 percent of the load voltage. (Remember: the peak-to-peak ripple is divided by two before subtracting from the load voltage. Therefore, a 10 percent ripple produces only a 5 percent error in load voltage.)

The foregoing guideline will be of some help in deciding which approximation to use, but don't lean on this guideline too heavily. You may have a situation where a 5 percent guideline is not suitable. Remember the artist's brushes. The job may require a smaller or larger brush. It is impossible to give you a rule for every situation because real life is too messy and has too many exceptions. But don't be discouraged. That's what makes electronics more interesting than accounting. Use the basic guideline given here, but be ready to abandon it if you feel it doesn't apply to your situation.

EXAMPLE 4-7

Suppose a bridge rectifier has a dc load current of 10 mA and a filter capacitance of 470 μF. What is the peak-to-peak ripple out of a capacitor-input filter?

SOLUTION

Use Eq. (4-8) to get

$$V_R = \frac{10\text{ mA}}{(120\text{ Hz})(470\ \mu\text{F})} = 0.177\text{ V}$$

This assumes the input frequency is 60 Hz, which is the nominal line frequency in the United States.

EXAMPLE 4-8

Assume we have a filtered bridge rectifier with a line voltage of 120 V rms, a turns ratio of 9.45, a filter capacitance of 470 μF, and a load resistance of 1 kΩ. What is the dc load voltage?

SOLUTION

Start by calculating the rms secondary voltage:

$$V_2 = \frac{120\text{ V}}{9.45} = 12.7\text{ V}$$

This is what you would measure with an ac voltmeter connected across the secondary winding.

Next, calculate the peak secondary voltage:

$$V_{p2} = \frac{12.7\text{ V}}{0.707} = 18\text{ V}$$

With an ideal diode and ignoring the ripple, the dc load voltage equals the peak secondary voltage:

$$V_{dc} = 18\text{ V}$$

This answer would be adequate if you were troubleshooting a circuit like this. The dc load voltage is the approximate value you would read with a dc voltmeter across the load resistor. If there were trouble in such a circuit, the dc voltage probably would be much lower than 18 V.

The second approximation improves the answer by including the effect of the two-diode voltage drops:

$$V_{dc} = 18\text{ V} - 1.4\text{ V} = 16.6\text{ V}$$

This is more accurate, so let us use it in the remaining calculations.

To calculate the ripple, we need the value of dc load current:

$$I = \frac{16.6\,\text{V}}{1\,\text{k}\Omega} = 16.6\,\text{mA}$$

Now, we can use Eq. (4-8):

$$V_R = \frac{16.6\,\text{mA}}{(120\,\text{Hz})(470\,\mu\text{F})} = 0.294\,\text{V}$$

This is the peak-to-peak ripple and is what you would see if you looked at the load voltage with the ac input of an oscilloscope.

This ripple has little effect on the dc load voltage:

$$V_{\text{dc(with ripple)}} = 16.6\,\text{V} - \frac{0.294\,\text{V}}{2} = 16.5\,\text{V}$$

This gives you the basic idea of how to calculate the dc load voltage and ripple.

4-6 CALCULATING OTHER QUANTITIES

Besides the load voltage and the peak-to-peak ripple, there are other quantities that you may want to calculate for a filtered bridge rectifier (Fig. 4-14). Let us begin with the diode current. During the positive half cycle, D_2 and D_3 conduct. During the negative half cycle, these diodes are off. Therefore, the dc or average current through these diodes equals half the dc load current:

$$I_D = 0.5 I_L \qquad (4\text{-}9)$$

By a similar argument, D_1 and D_4 conduct during the negative half cycles. Therefore, each of them has a diode current given by Eq. (4-9). This diode current must be less than the maximum dc current rating specified on the data sheet for the diode. For instance, a 1N4001 has a current rating of 1 A. Therefore, the load current must be less than 2 A to avoid damaging the diode.

Another thing to watch out for is the peak inverse voltage (PIV) across the nonconducting diodes. In Fig. 4-14, diode D_1 is off and diode D_2 is

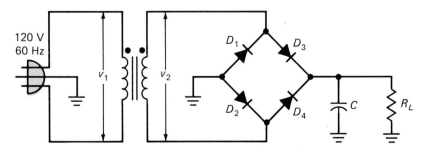

FIGURE 4-14 Bridge rectifier with filter.

on during the positive half cycle. Visualize diode D_2 as a closed switch. If you are doing this correctly, you will see that the entire secondary voltage is across diode D_1. At the positive peak of secondary voltage, therefore, diode D_1 has to withstand a reverse voltage of V_{p2}. This is the peak inverse voltage across the diode:

$$PIV = V_{p2} \tag{4-10}$$

By a similar argument each diode has to withstand the same maximum voltage given by Eq. (4-10). For instance, a 1N4001 has a PIV rating (same as breakdown voltage) of 50 V. This means it will work with a bridge rectifier that produces up to 50 V of load voltage.

A final quantity you may need is the primary current. Here is how you can get it:

$$I_1 = \frac{N_2}{N_1} I_L \tag{4-11}$$

This is almost identical to Eq. (4-3), which was discussed earlier. An alternative equation which amounts to the same thing is this:

$$I_1 = \frac{P_L}{V_1} \tag{4-12}$$

This current has to be less than the fuse rating; otherwise, the fuse blows out.

EXAMPLE 4-9

If the load voltage is 16.5 V and the load current is 16.5 mA, what is the diode current, peak inverse voltage, and primary current in Fig. 4-14? Use a peak secondary voltage of 18 V.

SOLUTION

$$I_D = 0.5 I_L = 0.5(16.5 \text{ mA}) = 8.25 \text{ mA}$$

$$PIV = V_{p2} = 18 \text{ V}$$

$$I_1 = \frac{P_L}{V_1} = \frac{(16.5 \text{ V})(16.5 \text{ mA})}{120 \text{ V}} = 2.27 \text{ mA}$$

These are straightforward calculations based on the preceding discussion. Here, you see that diode current is half of the load current, peak inverse voltage equals peak secondary voltage, and primary current equals load power divided by input voltage.

4-7 SURGE CURRENT

Before the power is turned on, the filter capacitor is uncharged in Fig. 4-14. At the instant the power is applied, the uncharged capacitor looks like a short. Therefore, the initial charging current may be quite large.

In the worst case, the circuit may be energized at the instant when line voltage is at its maximum. This means $V_{2(PEAK)}$ is across the secondary winding with the filter capacitor uncharged. The only thing that impedes current is the resistance of the windings and the bulk resistance of the diodes. For this reason, the initial current is very large. As the capacitor charges, the current decreases to lower levels.

The sudden gush of current when the power is first turned on it called the *surge current*. A designer has to make sure the diode he uses can withstand the surge current, brief as it is. The key to surge current is the size of the filter capacitor. If the filter capacitor is less than 1000 μF, the surge current is usually too brief in duration to damage the diodes. But when the filter capacitor is larger than 1000 μF, it takes many cycles to charge the capacitor. In this case, diode damage may occur. If you want to learn more about this design problem, see the "Optional Topics" section on page 105.

4-8 TROUBLESHOOTING

Let's wrap up our discussion of the capacitor-input filter by talking about troubleshooting. Almost every piece of electronics equipment has a power supply, typically a rectifier driving a capacitor-input filter followed by a voltage regulator (discussed later). This power supply provides dc voltages needed by transistors and other devices. If the equipment is not working properly, the first thing to test is the dc voltage out of the power supply.

You can check out the rectifier and capacitor-input filter as follows. Use a floating VOM to measure the secondary voltage (ac range). This reading is the rms voltage across the secondary winding. This rms voltage multiplied by 1.414 gives the peak value. Use a calculator or better yet, mentally estimate the peak voltage. For instance, here is one way a troubleshooter can estimate the peak voltage in Fig. 4-14 after measuring an rms secondary voltage of 12.7 V.

> **All I need is a rough estimate of the peak value. 12.7 V rounds off to 13 V. The peak value is about 40 percent higher than this. When I multiply 13 V by 0.4, I get 5.2 V, which rounds off to 5 V. After I add this to 13 V, I have 18 V. This is the approximate peak voltage across the secondary winding.**

After you have a value for the peak secondary voltage, you can measure the dc load voltage. This dc voltage should be in the vicinity of the estimated peak secondary voltage. If the peak voltage is distinctly different or if you are suspicious of its value, then look at the dc load voltage with an oscilloscope, checking the ripple. A peak-to-peak ripple around 10 percent of the ideal load voltage is reasonable. The ripple may be somewhat more or less than this, depending on the design. Furthermore, the ripple frequency should be 120 Hz for a full-wave rectifier or a bridge rectifier.

Here are some common troubles that arise and the symptoms they produce in bridge rectifiers with capacitor-input filters. If a diode is defective, the dc load voltage will be somewhat lower than it should be and the ripple frequency will equal 60 Hz instead of 120 Hz. If the filter capacitor is open, the dc load voltage will be low, equal to the average value instead of the peak value because the output will be an unfiltered full-wave signal. On the other hand, if the filter capacitor is shorted, one or more diodes may be ruined and the transformer may be damaged. Sometimes the filter capacitor becomes leaky with age and this reduces the dc load voltage. Occasionally, shorted windings in the transformer reduce the dc output voltage. Besides these troubles, you can also have the solder bridges, cold-solder joints, etc.

Successful troubleshooting starts with knowing how a circuit should work. When you know the normal voltages for a circuit, you can measure these voltages with a voltmeter or oscilloscope. Voltages that are much lower or much higher than normal are the clues that you can use to figure out what the trouble is.

With the foregoing ideas in mind, there isn't much left to do except to start troubleshooting. The only way you will master this important skill is through experience with circuits that have troubles. The T-shooter at the end of the homework problems will go a long way toward helping you learn how to troubleshoot. Before you attempt to use the T-shooter, read the three examples that follow.

EXAMPLE 4-10

When the circuit of Fig. 4-15 is working normally, it has an rms secondary voltage of 12.7 V, a load voltage of 18 V, and a peak-to-peak ripple of 318 mV. If the filter capacitor is open, what are some of the voltages that this trouble produces?

SOLUTION

With an open filter capacitor, the circuit reverts to an ordinary bridge rectifier with no filter capacitor. Suppose you use a floating ac voltmeter to measure the rms secondary voltage. This would indicate 12.7 V rms because the trouble is beyond the second winding.

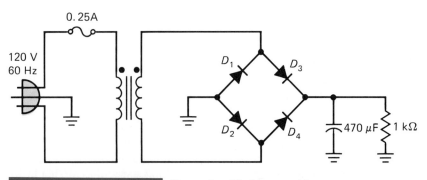

FIGURE 4-15 Example of bridge rectifier.

Because there is no filtering, the load voltage is a full-wave signal with a peak of 18 V. If a dc voltmeter were connected across the load, it would indicate 11.4 V (this is 63.6 percent of 18 V).

An oscilloscope across the load could be used to check the ripple. In this case, it would display an unfiltered full-wave rectified sine wave with a peak value of approximately 18 V. The average or dc value for this full-wave signal is 63.6 percent of 18 V, which is 11.4 V. This is why the dc voltmeter indicates 11.4 V across the load resistor.

EXAMPLE 4-11

Suppose the load resistor of Fig. 4-15 is shorted. Describe the symptoms. (Use the normal voltages of the previous example.)

SOLUTION

A short across the load will increase the current to an extremely high value. This will blow out the fuse. Furthermore, it is possible that one or more diodes will be destroyed before the fuse blows. If you are troubleshooting a circuit with a shorted load, you will measure zero for all voltages because of the blown fuse. After you test the fuse with an ohmmeter, don't simply replace it and turn on the power. You should check the diodes with an ohmmeter to see if any of them have been destroyed. After you replace any defective diode, you should measure the load resistance with an ohmmeter. If it measures zero, you have more troubles to find.

The trouble could be a solder bridge across the load resistor, a miswire, any number of possibilities. In general, remember this: whenever you find a blown fuse, there is probably something else causing the trouble. So, don't turn the power on until you find the trouble. Fuses do occasionally blow out without a short elsewhere in the circuit, but usually there is another trouble that causes them to blow.

EXAMPLE 4-12

Find Trouble 8 in the T-shooter of Fig. 4-32 (on page 123 at the end of the homework problems).

SOLUTION

The use of this T-shooter is similar to the T-shooter of Chap. 1. If you haven't read the introductory material on the T-shooter, go back to Chap. 1 and read Sec. 1-6 and Example 1-5 before proceeding.

The circuit is shown at the top of Fig. 4-32. It is a bridge rectifier and a capacitor-input filter, the most common type of power supply. The box labeled "OK" gives the measurements for the circuit when it is working properly.

We are going to troubleshoot the circuit for Trouble 8. Go to the box labeled "T8." This contains all the measurements you need for Trouble 8. You can start with any measurement you want. For instance, to measure the load voltage, read the token for V_L and you get $F4$. Go to the big box labeled "Measurements" and read the value of this token, which is 0. This means that the load voltage is zero.

Let's try V_2 next. Its token is $C1$, which translates into 0. So, there is no secondary voltage. All right, let us measure the resistance of fuse, F_1. Its token is $B3$, which has a value of ∞. This means the fuse is open. Either this is an isolated trouble or a short elsewhere in the circuit caused the fuse to blow.

How about measuring R_L next? Its token is $E4$, which translates to 1 kΩ. So, the load resistor is okay. Let's try the filter capacitor C_1. Its token is $B5$, equivalent to zero. Bingo! The filter capacitor is shorted. That's why the fuse blew and why there is no secondary voltage.

Now, you have the idea of how to use the T-shooter for troubleshooting the power-supply circuit. You can gain a great deal of simulated troubleshooting experience in this way. The T-shooter is easy, fast, and exposes you to a wide variety of troubles. The homework problems will continue your experiences in troubleshooting.

4-9 READING A DATA SHEET

Refer to the data sheet of a 1N4001 in the Appendix. It should make more sense to you now that you know how a rectifier circuit works. To begin with, the peak repetitive reverse voltage, designated V_{RRM} on the data sheet, is the same as the peak inverse voltage discussed earlier. The data sheet says the 1N4001 can withstand a voltage of 50 V in the reverse direction.

Also, the average rectified forward current I_O now has more meaning. This is the dc or average current through the diode. As you know, it equals half the dc load current for either a full-wave rectifier or a bridge rectifier. For a half-wave rectifier, the diode current equals the dc load current. The data sheet says a 1N4001 can have a dc current of 1 A, which means the dc load current can be as much as 2 A in a bridge rectifier.

Notice also the surge-current rating I_{FSM}. As indicated on the data sheet, a 1N4001 can withstand as much as 30 A when the power is first turned on. But it can withstand this large current only for one cycle, which means the filter capacitor has to charge within one cycle. So, the size of the filter capacitor becomes very important in determining whether or not a diode can handle the surge current. This is a design problem, so we won't say anything more about it here. (See "Optional Topics," on the next page, if you want more.)

□ **OPTIONAL TOPICS**

The following material continues the earlier discussions at a more advanced and specialized level. All the topics are optional because they are not used in any of the basic discussions in later chapters. This section will be a useful reference when you are in industry because then you will probably want more advanced viewpoints.

□ **4-10 FUSES**

In an ideal transformer, the currents are given by

$$\frac{I_1}{I_2} = \frac{N_2}{N_1}$$

You can use this equation to work out the size of a fuse. For instance, if the load current is 1.5 A rms and the turns ratio is 9:1, then

$$\frac{I_1}{1.5\,\text{A}} = \frac{1}{9}$$

or

$$I_1 = \frac{1.5\,\text{A}}{9} = 0.167\,\text{A rms}$$

This means that the fuse must have a value greater than 0.167 A, plus 10 percent in case the line voltage is high, plus approximately another 10 percent for transformer losses (these produce extra primary current). The next higher standard fuse size of 0.25 A (slow-blow in case of line surges) would probably be satisfactory. The purpose of the fuse is to prevent excessive damage in case the load resistance is accidentally shorted.

□ **4-11 REAL TRANSFORMERS**

The transformers you buy from a parts supplier are not ideal because the windings have resistance which produces power losses. Furthermore, the laminated core has eddy currents which produce additional power losses. Because of these unwanted power losses, a real transformer is a difficult device to specify fully. The data sheets for transformers rarely list the turns ratio, the resistances of windings, and other transformer quantities. Usually, all you get is the secondary voltage at a rated current. For instance, the F25X is an industrial transformer whose data sheet gives only the following specifications: for a primary voltage of 115 V ac, the secondary voltage is 12.6 V ac when the second current is 1.5 A. If the secondary current is less than 1.5 A, the secondary voltage will rise slightly because of lower IR drops across the winding resistance.

When it is necessary to know the primary current, you can estimate the turns ratio of a real transformer by using Eq. (4-2) and you can find the primary current with Eq. (4-3).

4-12 DESIGN GUIDELINES

If you are designing a capacitor-input filter, you need to choose a capacitor that is large enough to keep the ripple small. How small is small? This depends on how bulky a capacitor you're willing to use. As the ripple decreases, the capacitor becomes larger and it also becomes more expensive.

As a compromise between small ripple and large capacitance, many designers use the 10 percent rule, which states that you should select a capacitor that will hold the peak-to-peak ripple at approximately 10 percent of the peak voltage. For instance, if the peak voltage is 15 V, then select a capacitor that makes the peak-to-peak ripple around 1.5 V. This may sound like too much ripple, but it is not. As you will see later, additional filtering is usually done with electronic circuits called *voltage regulators*.

Incidentally, the derivation of Eq. (4-8) assumed small ripple; but we will use it to estimate large ripple as well. Even though the answers we get will contain some error, they are useful in practice. Why are we using answers that will contain errors? Because the filter capacitor is typically an electrolytic capacitor with a tolerance of ± 20 percent or more, so exact answers are unnecessary.

4-13 SURGE CURRENT

Before the power is turned on, the filter capacitor is uncharged. At the instant the circuit is energized, the capacitor looks like a short; therefore, the initial charging current may be quite large. This sudden gush of current is called the surge current.

In the worst case, the circuit may be energized at the instant that line voltage is at its maximum. This means $V_{2(\text{peak})}$ is across the secondary winding, and the capacitor is uncharged. The only thing that impedes current is the resistance of the windings and the bulk resistance of the diodes. We can symbolize this resistance as R_{TH}, the Thevenin resistance looking from the capacitor back toward the rectifier. Therefore, in the worst case,

$$I_{\text{surge}} = \frac{V_{2(\text{peak})}}{R_{TH}} \tag{4-13}$$

For instance, suppose the secondary voltage is 12.6 V ac and the Thevenin resistance facing the capacitor is 1.5 Ω. As found earlier, $V_{2(\text{peak})} = 17.8$ V, which means a maximum surge current of

$$I_{\text{surge}} = \frac{17.8\,\text{V}}{1.5\,\Omega} = 11.9\,\text{A}$$

This current starts to decrease as soon as the capacitor charges. If the capacitor is extremely large, however, the surge current can remain at a high level for a while and may cause diode damage.

Large Capacitor Means Longer Surge

Here is something to give you more insight into the problem. The secondary voltage has a period of

$$T = \frac{1}{f} = \frac{1}{60\ \text{Hz}} = 16.7\ \text{ms}$$

For a Thevenin resistance of 1 Ω, a capacitor of 1000 μF produces a time constant of 1 ms. This means the capacitor can charge within a few milliseconds, a fraction of a cycle. This usually is not long enough to damage the diode.

When the capacitance is much larger than 1000 μF, the time constant becomes very long and it may take many cycles to get the capacitor fully charged. If the surge current is too high, diode damage can easily occur, as can capacitor damage from heating and gas formation in the electrolyte.

Data Sheets

Data sheets list the surge-current rating as I_{surge}, $I_{FM(\text{surge})}$, I_{FSM}, etc. You have to read the fine print here because this rating depends on the number of cycles needed to charge the filter capacitor. For instance, the surge-current rating of a 1N4001 is 30 A for one cycle, 24 A for two cycles, 18 A for four cycles, and so on. Most of the designs in this book will charge the filter capacitance within a fraction of a cycle. In fact, whenever the filter capacitor is less than 1000 μF, the capacitor usually charges in less than one cycle.

Design Hints

Suppose you are designing a rectifier circuit with a capacitor-input filter. What do you do about surge current? As before, you select a capacitance to produce a ripple of about 10 percent of the dc load voltage. If the capacitance is less than 1000 μF, usually you can ignore the surge current because it's unlikely that it will damage the rectifier diodes of a typical circuit.

On the other hand, if the filter capacitance is greater than 1000 μF, you may need to use winding resistance and bulk resistance to calculate the surge current using Eq. (4-13). You can measure the winding resistances with an ohmmeter. And here is how to estimate the bulk resistance:

$$r_B = \frac{V_F - 0.7}{I_F} \tag{4-14}$$

The quantities V_F and I_F are listed on data sheets. After you calculate the surge current, you select a diode whose surge-current rating is larger than your estimated surge current.

4-14 *RC* AND *LC* FILTERS

With the 10 percent rule, we get a dc load voltage with a peak-to-peak ripple of around 10 percent. Before the 1970s, passive filters were connected between the filter capacitor and the load to reduce the ripple to less than 1 percent. The whole idea was to get an almost perfect dc voltage, similar to what you get from a battery. You rarely see passive filters used in new circuit designs, but occasionally there are special applications where they may still be viable. What follows is a brief discussion to give you the basic idea.

RC Filter

Figure 4-16*a* shows two *RC* filters between the input capacitor and the load resistor. By deliberate design, R is much greater than X_C at the ripple frequency. Therefore, the ripple is dropped across the series resistors instead of across the load resistor. Typically, R is at least 10 times greater than X_C; this means that each section attenuates (reduces) the ripple by a factor of at least 10. The main disadvantage of the *RC* filter is the loss of dc voltage across each R. This means that the *RC* filter is suitable only for light loads (small load current or large load resistance).

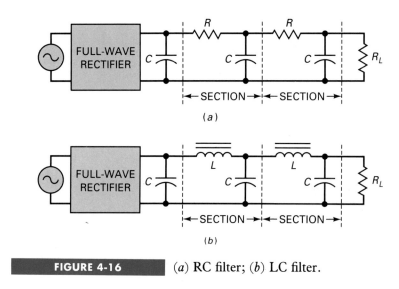

(*a*)

FIGURE 4-16 (*a*) RC filter; (*b*) LC filter.

LC Filter

When the load current is large, the *LC* filters of Fig. 4-16*b* are an improvement over *RC* filters. Again, the idea is to drop the ripple across the series components; in this case, the inductors. This is accomplished by making X_L much greater than X_C at the ripple frequency. In this way, the ripple can be reduced to extremely low levels. Furthermore, the dc voltage drop across the inductors is much smaller because only the winding resistance is involved.

The *LC* filter was quite popular at one time. Now, it's becoming obsolete in typical power supplies because of the size and cost of inductors. For low-voltage power supplies, the *LC* filter has been replaced by *IC* voltage regulators, active filters that reduce ripple and hold the final dc voltage constant.

4-15 VOLTAGE MULTIPLIERS

A *voltage multiplier* is two or more peak detectors or peak rectifiers that produce a dc voltage equal to a multiple of the peak input voltage ($2V_P$, $3V_P$, $4V_P$, and so on). These power supplies are used for high voltage/low current devices like cathode-ray tubes (the picture tubes in TV receivers, oscilloscopes, and computer displays).

Half-Wave Voltage Doubler

Figure 4-17a is a *voltage doubler*. At the peak of the negative half cycle, D_1 is forward-biased and D_2 is reverse-biased. Ideally, this charges C_1 to the peak voltage, V_P, with the polarity shown in Fig. 4-17b. At the peak of the positive half cycle, D_1 is reverse-biased and D_2 is forward-biased. Because the source and C_1 are in series, C_2 will try to charge toward $2V_P$. After several cycles, the voltage across C_2 will equal $2V_P$, as shown in Fig. 4-17c.

By redrawing the circuit and connecting a load resistance, we get Fig. 4-17d. Now it's clear that the final capacitor discharges through the load resistor. As long as R_L is large, the output voltage equals $2V_P$ (ideally). That is, provided the load is light (long time constant), the output voltage is double the peak input voltage. This input voltage normally comes from the secondary winding of a transformer.

For a given transformer, you can get twice as much output voltage as you get from a standard peak rectifier. This is useful when you are trying to produce high voltages (several hundred volts or more). Why? Because higher secondary voltages result in bulkier transformers. At some point, a designer may prefer to use voltage doublers instead of bigger transformers.

The circuit is called a half-wave doubler because the output capacitor, C_2, is charged only once during each cycle. As a result, the ripple

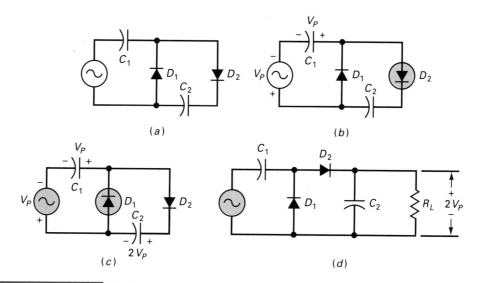

(a)　　　　　(b)

(c)　　　　　(d)

FIGURE 4-17　　Half-wave voltage doubler.

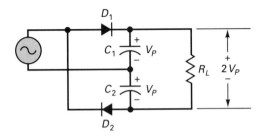

FIGURE 4-18 Full-wave voltage doubler.

frequency is 60 Hz. Sometimes you will see a surge resistor in series with C_1.

Full-Wave Voltage Doubler

Figure 4-18 shows a full-wave voltage doubler. On the positive half cycle of the source, the upper capacitor charges to the peak voltage with the polarity shown. On the next half cycle, the lower capacitor charges to the peak voltage with the indicated polarity. For a light load, the final output voltage is approximately $2V_P$.

The circuit is called a full-wave voltage doubler because one of the output capacitors is being charged during each half cycle. Stated another way, the output ripple is 120 Hz. This ripple frequency is an advantage because it is easier to filter. Another advantage of the full-wave doubler is that the PIV rating of the diodes need only be greater than V_P.

The disadvantage of a full-wave doubler is the lack of a common ground between input and output. In other words, if we ground the lower end of the load resistor in Fig. 4-18, the source is floating. In the half-wave doubler of Fig. 4-17d, grounding the load resistor also grounds the source, an advantage in some applications.

Voltage Tripler

By connecting another section, we get the *voltage tripler* of Fig. 4-19a. The first two peak rectifiers act like a doubler. At the peak of the negative

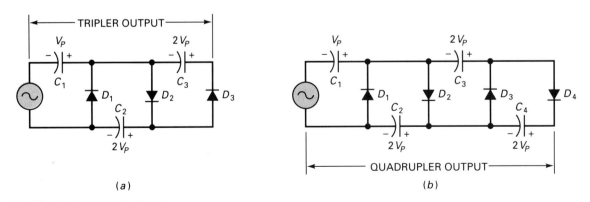

(a)   (b)

FIGURE 4-19 (*a*) Voltage tripler; (*b*) voltage quadrupler.

half cycle, D_3 is forward-biased. This charges C_3 to $2V_P$ with the polarity shown in Fig. 4-19a. The tripler output appears across C_1 and C_3.

The load resistance is connected across the tripler output. As long as the time constant is long, the output approximately equals $3V_P$.

Voltage Quadrupler

Figure 4-19b is a *voltage quadrupler* with four peak rectifiers in cascade (one after another). The first three are a tripler, and the fourth makes the overall circuit a quadrupler. The first capacitor charges to V_P; all others charge to $2V_P$. The quadrupler output is across the series connection of C_2 and C_4. As usual, a large load resistance (long time constant) is needed to have an output of approximately $4V_P$.

Theoretically, we can add sections indefinitely; however, the ripple gets worse as additional sections are added. This is why voltage multipliers are not used in low-voltage supplies, which are the supplies you encounter the most. As stated earlier, voltage multipliers are almost always used to produce high voltages, well into the hundreds or thousands of volts.

4-16 THE LIMITER

The diodes used in power supplies are *rectifier diodes;* those with a power rating greater than 0.5 W and optimized for use at 60 Hz. In the remainder of the chapter, we will be using *small-signal diodes;* these have power ratings of less than 0.5 W (with current in milliamperes rather than amperes) and are typically used at frequencies much greater than 60 Hz.

The first small-signal circuit to be discussed is the limiter; it removes signal voltages above or below a specified level. This is useful not only for signal shaping but also for protecting circuits that receive the signal.

Positive Limiter

Figure 4-20 shows a *positive limiter* (sometimes called a *clipper*), a circuit that removes positive parts of the signal. As shown, the output voltage has all positive half cycles clipped off. The circuit works as follows: During the positive half of input voltage, the diode turns on. Ideally, the output voltage is zero; to a second approximation, it is approximately $+0.7$ V.

During the negative half cycle, the diode is reverse-biased and looks open. In many limiters, the load resistor, R_L, is at least 100 times greater than series resistor, R. For this reason, the source is stiff and the negative half cycle appears at the output.

Figure 4-20 shows the output waveform. The positive half cycle has been clipped off. The clipping is not perfect. To a second approximation, a conducting silicon diode drops approximately 0.7 V. Because the first 0.7 V is used to overcome the barrier potential, the output signal is clipped near $+0.7$ V rather than 0 V.

If you reverse the polarity of the diode in Fig. 4-20, you get a negative limiter that removes the negative half cycle. In this case, the clipping level is near -0.7 V.

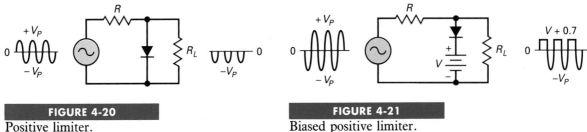

FIGURE 4-20
Positive limiter.

FIGURE 4-21
Biased positive limiter.

Biased Limiter

With the *biased limiter* of Fig. 4-21, you can move the clipping level to $V + 0.7$. When the input voltage is greater than $V + 0.7$, the diode conducts and the output is held at $V + 0.7$. When the input voltage is less than $V + 0.7$, the diode opens and the circuit becomes a voltage divider. As before, load resistance should be much greater than the series resistance; then the source is stiff and all the input voltage reaches the output.

You can combine biased positive and negative limiters as shown in Fig. 4-22. Diode D_1 turns on when the input voltage exceeds $V_1 + 0.7$; this is the positive clipping level. Similarly, diode D_2 conducts when the input is more negative than $-V_2 - 0.7$; this is the negative clipping level. When the input signal is large, that is, when V_P is much greater than the clipping levels, the output signal resembles a square wave like that of Fig. 4-22.

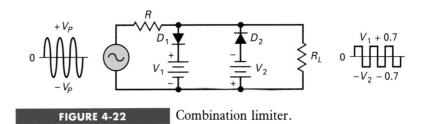

FIGURE 4-22   Combination limiter.

Variations

Using batteries to set the clipping level is impractical. One approach is to add more silicon diodes because each produces an *offset* of 0.7 V. For instance, Fig. 4-23*a* shows two diodes in a positive limiter. Since each diode has an offset of around 0.7 V, the pair of diodes produce a clipping level of approximately $+1.4$ V. Figure 4-23*b* extends this idea to four diodes; this results in a clipping level of approximately $+2.8$ V. There is no limit on the number of diodes used, and this is practical because diodes are inexpensive.

Limiters are sometimes used to protect the load against excessive voltage. For instance, Fig. 4-23*c* shows a 1N914 protecting the load (not shown) against excessively large input voltages. The 1N914 conducts when the input exceeds $+5.7$ V. In this way, a destructively large input voltage like $+100$ V never reaches the load because the diode clips at $+5.7$ V, the maximum voltage to reach the load.

Incidentally, a circuit like Fig. 4-23c is often called a *diode clamp* because it holds the signal at a fixed level. It literally clamps the output voltage at $+5.7$ V when the input voltage exceeds this level. A typical use for a diode clamp is to protect the load.

Sometimes a variation like Fig. 4-23d is used to remove the offset of limiting diode D_1. Here is the idea: Diode D_2 is biased slightly into forward conduction, so that it has approximately 0.7 V across it. This 0.7 V is applied to 1 kΩ in series with D_1 and 100 kΩ. This means diode D_1 is on the verge of conduction. Therefore, when a signal comes in, diode D_1 conducts near 0 V.

4-17 THE DC CLAMPER

The diode clamp is a variation of the limiter discussed in the preceding section. The *dc clamper* is different, so don't confuse the similar-sounding words. A dc clamper adds a dc voltage to the signal. For instance, if the incoming signal swings (varies) from -10 V to $+10$ V, a positive dc clamper will produce an output that ideally swings from 0 to $+20$ V. (A negative dc clamper would produce an output between 0 and -20 V.)

Positive Clamper

Figure 4-24a shows a positive dc clamper. Ideally, here is how it works. On the first negative half cycle of input voltage, the diode turns on, as shown in Fig. 4-24b. At the negative peak, the capacitor must charge to V_P with the polarity shown.

Slightly beyond the negative peak, the diode shuts off, as shown in Fig. 4-24c. The $R_L C$ time constant is deliberately made much greater than the period, T, of the incoming signal. For this reason, the capacitor remains almost fully charged during the off time of the diode. To a first approximation, the capacitor acts like a battery of V_P volts. This is why the output voltage in Fig. 4-24a is a positively clamped signal.

Figure 4-24d shows the circuit as it is usually drawn. Since the diode drops 0.7 V when conducting, the capacitor voltage does not quite reach V_P. For this reason, the dc clamping is not perfect, and the negative peaks are at -0.7 V.

Negative Clamper

What happens if we turn the diode in Fig. 4-24d around? The polarity of capacitor voltage reverses, and the circuit becomes a negative clamper. Both positive and negative clampers are widely used. Television receivers, for instance, use a dc clamper to add a dc voltage to the video signal. In television work, the dc clamper is usually called a dc restorer.

To remember which way the dc level of a signal moves, look at Fig. 4-24d. Notice that the diode arrow points upward, the same direction as the dc shift. In other words, when the diode points upward, you have a positive dc clamper. When the diode points downward, the circuit is a negative dc clamper.

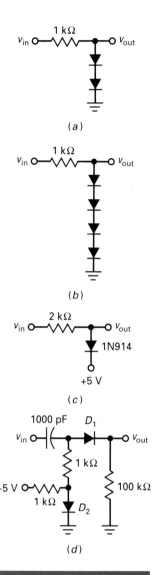

(a)

(b)

(c)

(d)

FIGURE 4-23

Limiters. (a) Two-diode offset; (b) four-diode offset; (c) diode-clamp; (d) biased near zero.

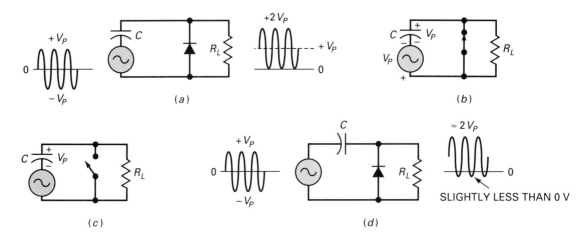

(a)

(b)

(c)

(d)

SLIGHTLY LESS THAN 0 V

FIGURE 4-24 Positive dc clamper.

4-18 THE PEAK-TO-PEAK DETECTOR

If you cascade a dc clamper and a peak detector (same as a peak rectifier), you get a *peak-to-peak detector* (see Fig. 4-25). The input sine wave is positively clamped; therefore, the input to the peak detector has a peak value of $2V_P$. This is why the output of the peak detector is a dc voltage equal to $2V_P$.

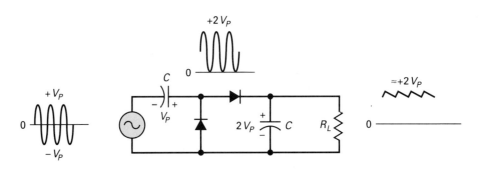

FIGURE 4-25 Peak-to-peak detector.

As usual, the discharge time constant $R_L C$ must be much greater than the period of the incoming signal. By satisfying this condition, you get good clamping action and good peak detection. The output ripple will therefore be small.

Where are peak-to-peak detectors used? Sometimes, the output of a peak-to-peak detector is applied to a dc voltmeter. The combination acts like a peak-to-peak ac voltmeter. For instance, suppose that a signal swings from -20 to $+50$ V. If you try to measure this with an ordinary ac voltmeter, you will get an incorrect reading. If you use a peak-to-peak detector in front of a dc voltmeter, you will read 70 V for the peak-to-peak value of the signal.

❑ 4-19 THE DC RETURN

One of the most baffling things that may happen in the laboratory is this: You connect a signal source to a circuit and for some reason the circuit will not work, yet nothing is defective in the circuit or the signal source. As a concrete example, Fig. 4-26a shows a sine-wave source driving a half-wave rectifier. When you look at the output with an oscilloscope, you see no signal at all; the rectifier refuses to work. To add to the confusion, you may try another sine-wave source and find a normal half-wave signal across the load (Fig. 4-26b).

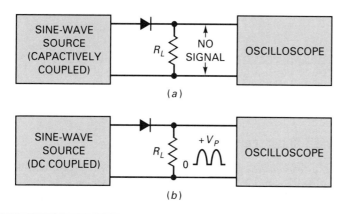

(a)

(b)

FIGURE 4-26   The dc-return problem.

The phenomenon just described is a classic in electronics; it occurs again and again in practice. It may happen with diode circuits, transistor circuits, integrated circuits, etc. Unless you understand why one kind of source works and another doesn't, you will be confused and possibly discouraged every time the problem arises.

Types of Coupling

The signal source of Fig. 4-27a is *capacitively coupled;* this means that it has a capacitor in the signal path. Many commercial signal generators use a capacitor to dc-isolate the source from the load. The idea of a capacitively coupled source is to let only an ac signal pass from source to load.

The *dc-coupled* source of Fig. 4-27b is different. It has no capacitor; therefore, it provides a path for both alternating and direct currents. When you connect this kind of source to the load, it is possible for the load to force a direct current through the source. As long as this direct current is not too large, no damage to the source occurs. Many commercial signal generators are dc-coupled like this.

Sometimes, a signal source is *transformer-coupled* like Fig. 4-27c. The advantage is that it passes the ac signal and at the same time provides a dc path through the secondary winding.

All circuits discussed earlier in this chapter work with dc-coupled and transformer-coupled sources. It is only with the capacitively-coupled sources that the trouble may arise.

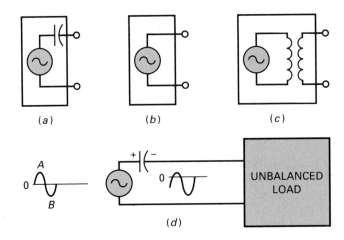

(a)   (b)   (c)

(d)

FIGURE 4-27   (a) Capacitively coupled source; (b) direct-coupled source; (c) transformer-coupled source; (d) unbalanced load causes unequal charging currents.

Unbalanced Diode Circuits

An *unbalanced load* is one that has more resistance on one half cycle than on the other. Figure 4-27d shows an unbalanced load. If the current is greater on the positive half cycle, the capacitor charges with the polarity shown. As we saw with dc clampers, a charged capacitor causes the dc level of the signal to shift.

Now we know why a half-wave rectifier won't work when connected to a capacitively coupled source. In Fig. 4-28a the capacitor charges to V_P during the first few cycles. Because of this, the signal coming from the source is negatively clamped, and the diode cannot turn on after the first few cycles. This is why we see no signal on the oscilloscope.

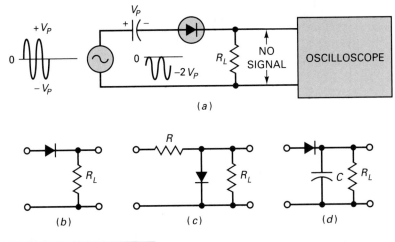

(a)

(b)   (c)   (d)

FIGURE 4-28   Capacitively coupled source produces unwanted clamping.

Among the diode circuits discussed earlier, the following are unbalanced loads: the half-wave rectifier, the limiter, the peak detector, the dc clamper, and the peak-to-peak detector. The last two are supposed to dc clamp the signal; therefore, they work fine with a capacitively coupled source. But the half-wave rectifier, the limiter, and the peak detector of Fig. 4-28b, c, and d will not work with a capacitively coupled source because of unwanted dc clamping.

DC Return

Is there a remedy for unwanted dc clamping? Yes. You can add a *dc-return* resistor across the input to the unbalanced circuit (see Fig. 4-29a). This resistor, R_D, allows the capacitor to discharge during the off time of the diode. In other words, any charge deposited on the capacitor plates is removed during the alternate half cycle.

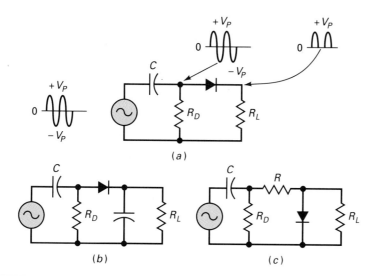

FIGURE 4-29   Dc return eliminates unwanted clamping.

The size of R_D is not critical. The main idea in preventing unwanted dc clamping is to keep the discharging resistance, R_D, less than or equal to the charging resistance in series with the diode. In Fig. 4-29a, this means

$$R_D < R_L$$

When this condition is satisfied, only a slight shift occurs in the dc level of the signal. The same rule applies to Fig. 4-29b. (For best results, make R_D less than one-tenth of R_L. This gives a highly balanced load with negligible dc shift.)

The limiter of Fig. 4-29c is slightly different. When the diode is conducting, the charging resistance in series with the diode equals R instead of R_L. Therefore, in Fig. 4-29c, the rule is

$$R_D < R$$

When possible, R_D should be less than one-tenth of R.

STUDY AIDS

The following study aids will help to reinforce the ideas discussed in this chapter. For best results, use these study aids within 6 hours of reading the earlier material. Then review these study aids a week later and a month later to ensure that the concepts remain in your long-term memory.

SUMMARY

Sec. 4-1 The Input Transformer
The input transformer is usually a step-down transformer. In this type of transformer, the voltage is stepped down and the current is stepped up. One way to remember this is by remembering that the output power equals the input power in a lossless transformer.

Sec. 4-2 The Half-wave Rectifier
The half-wave rectifier has a diode in series with a load resistor. The load voltage is a half-wave rectified sine wave with a peak value approximately equal to the peak secondary voltage. The dc or average load voltage equals 31.8 percent of the peak load voltage.

Sec. 4-3 The Full-wave Rectifier
The full-wave rectifier has a center-tapped transformer with two diodes and a load resistor. The load voltage is a full-wave rectified sine wave with a peak value approximately equal to half of the peak secondary voltage. The dc or average load voltage equals 63.6 percent of the peak load voltage. The ripple frequency equals two times the input frequency.

Sec. 4-4 The Bridge Rectifier
The bridge rectifier has four diodes. The load voltage is a full-wave rectified sine wave with a peak value approximately equal to peak secondary voltage. The dc or average load voltage equals 63.6 percent of the peak load voltage. The ripple frequency equals two times the line frequency.

Sec. 4-5 The Capacitor-input Filter
This is a capacitor across the load resistor. The idea is to charge the capacitor to the peak voltage and let it supply current to the load when the diodes are nonconducting. With a large capacitor, the ripple is small and the load voltage is almost a pure dc voltage.

Sec. 4-6 Calculating Other Quantities
In a full-wave or bridge rectifier, the diode current is half the load current and the peak inverse voltage equals the peak secondary voltage. In any kind of rectifier, the primary current approximately equals the load power divided by the primary voltage.

Sec. 4-7 Surge Current
Because the filter capacitor is uncharged before the power is turned on, the initial charging current is quite high. If the filter capacitor is less than 1000 μF, the surge current is usually too brief to damage the diodes.

Sec. 4-8 Troubleshooting
The basic measurements you can make on a rectifier circuit include a floating ac voltmeter across the secondary winding to measure the rms secondary voltage, a dc voltmeter across the load resistor to measure the dc load voltage, and an oscilloscope across the load resistor to measure the peak-to-peak ripple.

Sec. 4-9 Reading a Data Sheet
The three most important specifications on the data sheet of a diode are the peak inverse voltage, the maximum diode current, and the maximum surge current.

VOCABULARY

In your own words, explain what each of the following terms mean. Keep your answers short and to the point. If necessary, verify your answer by rereading the appropriate discussion or by looking at the end-of-book Glossary.

bridge rectifier	peak value
capacitor-input filter	rectifier diode
dc value	ripple
full-wave rectifier	rms value
half-wave rectifier	step-down transformer
line voltage	surge current
peak inverse voltage	

IMPORTANT EQUATIONS

The following formulas are useless if you don't know what they mean in words. Suggestion: Look at each formula, then read the words to find out what the formula means. Your chances of learning and remembering are much better if you concentrate on words rather than formulas.

Eq. 4-1 RMS Voltage

$$V_{rms} = 0.707V_p$$

This equation relates the heating effect of a dc voltage to an ac voltage. In effect, it converts a sine wave with a peak value of V_p to a dc voltage with a value of V_{rms}. It says a sine wave with a peak value of V_p produces the same amount of heat or power as a dc voltage with a value of V_{rms}. The magic number 0.707 comes from a calculus derivation. There's not much else you can do here except memorize the relation.

Eq. 4-5 DC Voltage from a Half-wave Rectifier

$$V_{dc} = 0.318V_p$$

One of the things you can do with calculus is work out the average value of a time-varying signal. If you really want to know where the number 0.318 comes from, you will have to learn calculus. Otherwise, just memorize the equation. It says the dc or average value of a half-wave rectified sine wave equals 31.8 percent of the peak voltage.

Eq. 4-6 DC Voltage from a Full-wave Rectifier

$$V_{dc} = 0.636V_p$$

Because the full-wave signal has twice as many cycles as a half-wave signal, the average voltage is twice as much. The equation says that the dc voltage equals 63.6 percent of the peak voltage of the full-wave rectified sine wave.

Eq. 4-7 Frequency of Full-wave Voltage

$$f_{out} = 2f_{in}$$

This applies to full-wave and bridge rectifiers. It says the ripple frequency equals two times the line frequency. If line frequency is 60 Hz, the ripple frequency is 120 Hz. Very important for troubleshooting. Remember it.

Eq. 4-8 Ripple out of Capacitor-input Filter

$$V_R = \frac{I}{fC}$$

This equation is the key to the value of ripple, something a troubleshooter or designer needs to know. It says that the peak-to-peak ripple equals the dc load current divided by the ripple frequency times the filter capacitance.

Eq. 4-9 Diode Current

$$I_D = 0.5I_L$$

This applies to the full-wave and bridge rectifiers. The equation says that the dc current in any diode equals half the dc load current.

Eq. 4-10 Peak Inverse Voltage

$$PIV = V_{p2}$$

This applies to the full-wave and bridge rectifiers. It says that the peak inverse voltage across a nonconducting diode equals the peak secondary voltage.

STUDENT ASSIGNMENTS

QUESTIONS

The following may have more than one right answer. Select the best answer. This is the one that is always true, or covers more situations, or fits the context, etc.

1. If $N_1/N_2 = 2$, and the primary voltage is 120 V, what is the secondary voltage?
   a. 0 V
   b. 36 V
   c. 40 V
   d. 60 V

2. In a step-down transformer, which is larger?
   a. Primary voltage
   b. Secondary voltage
   c. Neither
   d. No answer possible

3. A transformer has a turns ratio of 4:1. What is the peak secondary voltage if 115 V rms is applied to the primary winding?
   a. 40.7 V c. 163 V
   b. 64.6 V d. 170 V

4. With a half-wave rectified voltage across the load resistor, load current flows for what part of a cycle?
   a. 0°   c. 180°
   b. 90°   d. 360°

5. Suppose line voltage may be as low as 105 V rms or as high as 125 rms in a half-wave rectifier. With a 5:1 step-down transformer, the maximum peak load voltage is closest to
   a. 21 V   c. 29.6 V
   b. 25 V   d. 35.4 V

6. The voltage out of a bridge rectifier is
   a. Half-wave signal
   b. Full-wave signal
   c. Bridge-rectified signal
   d. Sine wave

7. If the line voltage is 115 V rms, a turns ratio of 5:1 means the rms secondary voltage is closest to
   a. 15 V   c. 30 V
   b. 23 V   d. 35 V

8. What is the peak load voltage in a full-wave rectifier if the secondary voltage is 20 V rms?
   a. 0 V   c. 14.1 V
   b. 0.7 V d. 28.3 V

9. We want a peak load voltage of 40 V out of a bridge rectifier. What is the approximate rms value of secondary voltage?
   a. 0 V   c. 28.3 V
   b. 14.4 V d. 56.6 V

10. With a full-wave rectified voltage across the load resistor, load current flows for what part of a cycle?
    a. 0°   c. 180°
    b. 90°   d. 360°

11. What is the peak load voltage out of a bridge rectifier for a secondary voltage of 15 V rms? (Use second approximation.)
    a. 9.2 V   c. 19.8 V
    b. 15 V   d. 24.3 V

12. If line frequency is 60 Hz, the output frequency of a half-wave rectifier is
    a. 30 Hz   c. 120 Hz
    b. 60 Hz   d. 240 Hz

13. If line frequency is 60 Hz, the output frequency of a bridge rectifier is
    a. 30 Hz   c. 120 Hz
    b. 60 Hz   d. 240 Hz

14. With the same secondary voltage and filter, which has the most ripple?
    a. Half-wave rectifier
    b. Full-wave rectifier
    c. Bridge rectifier
    d. Impossible to say

15. With the same secondary voltage and filter, which produces the least load voltage?
    a. Half-wave rectifier
    b. Full-wave rectifier
    c. Bridge rectifier
    d. Impossible to say

16. If the filtered load current is 10 mA, which of the following has a diode current of 10 mA?
    a. Half-wave rectifier
    b. Full-wave rectifier
    c. Bridge rectifier
    d. Impossible to say

17. If the load current is 5 mA and the filter capacitance is 1000 μF, what is the peak-to-peak ripple out of a bridge rectifier?
    a. 21.3 pV   c. 21.3 mV
    b. 56.3 nV   d. 41.7 mV

18. The diodes in a bridge rectifier each have a maximum dc current rating of 2 A. This means the dc load current can have a maximum value of
    a. 1 A   c. 4 A
    b. 2 A   d. 8 A

19. What is the PIV across each diode of a bridge rectifier with a secondary voltage of 20 V rms?
    a. 14.1 V   c. 28.3 V
    b. 20 V   d. 34 V

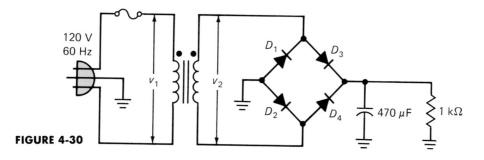

FIGURE 4-30

20. If the secondary voltage increases in a bridge rectifier with a capacitor-input filter, the load voltage will
   a. Decrease   c. Increase
   b. Stay the same d. None of these

21. If the filter capacitance is increased, the ripple will
   a. Decrease   c. Increase
   b. Stay the same d. None of these

22. In Fig. 4-30, the filter capacitor is open. What will the load voltage look like on an oscilloscope?
   a. Horizontal line at 0 V
   b. Horizontal line at normal output
   c. Half-wave signal
   d. Full-wave signal

23. Something is shorting out the load resistor of Fig. 4-30. After you remove the short, you should check the condition of the
   a. Fuse
   b. Odd-numbered diodes
   c. Even-numbered diodes
   d. All of the foregoing

24. In Fig. 4-30, the secondary voltage has an rms value of 12.7 V. If a dc voltmeter indicates a load voltage of 11.4 V, the trouble is probably
   a. An open filter capacitor
   b. Blown fuse
   c. Open secondary winding
   d. No center tap

25. The dc load voltage of Fig. 4-30 seems normal, but the ripple is 60 Hz. Which of these is a possible trouble:
   a. An open filter capacitor
   b. Blown fuse
   c. Open secondary winding
   d. Open diode

BASIC PROBLEMS

Sec. 4-1 The Input Transformer

4-1. Suppose the peak value of a sinusoidal voltage is 50 V. What is the rms value?

4-2. Line voltage may vary from 105 to 125 V rms. Calculate the peak value for low-line voltage and high-line voltage.

4-3. A step-up transformer has a turns ratio of 1:4. If the line voltage is 115 V rms, what is the peak secondary voltage?

4-4. A step-down transformer has a primary voltage of 110 V rms and a secondary voltage of 12.7 V rms. What is the turns ratio?

4-5. A transformer has a primary voltage of 120 V rms and a secondary voltage of 25 V rms. If the secondary current is 1 A rms, what is the primary current?

Sec. 4-2 The Half-wave Rectifier

4-6. During the day the line frequency varies slightly from its nominal value of 60 Hz. Suppose the line frequency is 61 Hz. What is the period of the rectified output voltage from a half-wave rectifier?

4-7. A step-down transformer with a turns ratio of 3:1 is connected to a half-wave rectifier. If the line voltage is 115 V rms, what is the peak load voltage? Give the two answers: one for an ideal diode, and another for the second approximation.

Sec. 4-3 The Full-wave Rectifier

4-8. During the day, the line frequency drops down to 59 Hz. What is the frequency out of a full-wave rectifier for this input frequency? What is the period of the output?

4-9. Refer to Fig. 4-7. Suppose the line voltage varies from 105 V rms to 125 V rms. What is the peak load voltage for the two extremes? (Use ideal diodes.)

4-10. If the turns ratio of Fig. 4-7 is changed to 6:1, what is the dc load current?

Sec. 4-4 The Bridge Rectifier

4-11. Refer to Fig. 4-10. If the load resistance is changed to 3.3 kΩ, what is the dc load current? Give answers for two cases: ideal diode and second approximation.

4-12. If in Fig. 4-10, the turns ratio is changed to 6:1 and the load resistance to 820 Ω, what is the dc load current? (Give ideal- and second-approximation answers.)

Sec. 4-5 The Capacitor-input Filter

4-13. A bridge rectifier has a dc load current of 20 mA and a filter capacitance of 680 μF. What is the peak-to-peak ripple out of a capacitor-input filter?

4-14. In the previous problem, the rms secondary voltage is 15 V. What is the dc load voltage? Give three answers: one based on ideal diodes, another based on the second approximation, and a third based on the effect of ripple.

Sec. 4-6 Calculating Other Quantities

4-15. The rms secondary voltage of Fig. 4-30 is 12.7 V. Use the ideal diode and ignore the effect of ripple on dc load voltage. Work out the values of each of these quantities: dc load voltage, dc load current, dc diode current, rms primary current, peak inverse voltage, and turns ratio.

4-16. Repeat Prob. 4-15, but use the second approximation and include the effect of ripple on the dc load voltage.

4-17. Draw the schematic diagram of a bridge rectifier with a capacitor-input filter and these circuit values: $V_2 = 20$ V, $C = 1000$ μF, $R_L = 1$ kΩ. What is the load voltage and peak-to-peak ripple?

Sec. 4-8 Troubleshooting

4-18. You measure 24 V rms across the secondary of Fig. 4-30. Next you measure 21.6 V dc across the load resistor. What is the most likely trouble?

4-19. The dc load voltage of Fig. 4-30 is too low. Looking at the ripple with a scope, you discover it has a frequency of 60 Hz. Give some possible causes.

4-20. There is no voltage out of the circuit of Fig. 4-30. Give some possible troubles.

4-21. Checking with an ohmmeter, you find all diodes in Fig. 4-30 open. You replace the diodes. What else should you check before you power up?

ADVANCED PROBLEMS

4-22. You are designing a bridge rectifier with a capacitor-input filter. The specifications are a dc load voltage of 15 V and a ripple of 1 V for a load resistance of 680 Ω. How much rms voltage should the secondary winding produce for a line voltage of 115 V rms? What size should the filter capacitor be? What are the minimum I_O and PIV ratings for diodes?

4-23. Design a full-wave rectifier using a 48 V rms center-tapped transformer that produces a 10 percent ripple across a capacitor-input filter with a load resistance of 330 Ω. What are the minimum I_O and PIV ratings of the diodes?

4-24. Design a power supply to meet the following specifications: The secondary voltage is 12.6 V rms and the dc output is approximately 17.8 V at 120 mA. What are the minimum I_O and PIV ratings of the diodes?

4-25. A full-wave signal has a dc value of 0.636 times the peak value. With your calculator or a table of sine values, you can derive the average value of 0.636. Describe how you would do it.

4-26. The secondary voltage in Fig. 4-31 is 25 V rms. With the switch in the upper position, what is the output voltage?

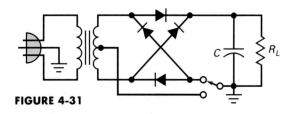

FIGURE 4-31

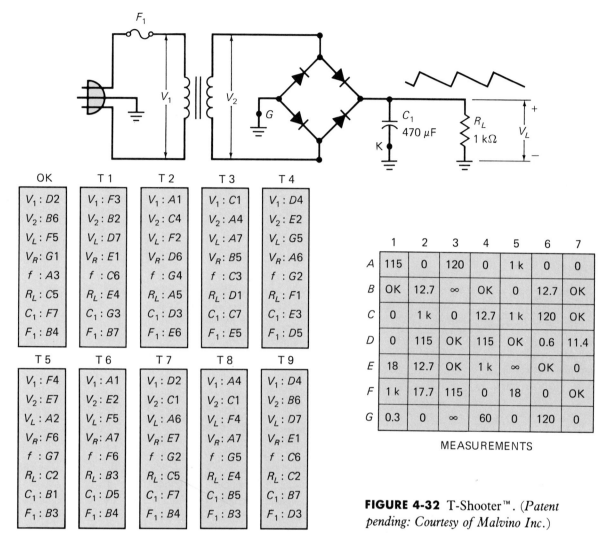

	1	2	3	4	5	6	7
A	115	0	120	0	1 k	0	0
B	OK	12.7	∞	OK	0	12.7	OK
C	0	1 k	0	12.7	1 k	120	OK
D	0	115	OK	115	OK	0.6	11.4
E	18	12.7	OK	1 k	∞	OK	0
F	1 k	17.7	115	0	18	0	OK
G	0.3	0	∞	60	0	120	0

MEASUREMENTS

OK

$V_1 : D2$	
$V_2 : B6$	
$V_L : F5$	
$V_R : G1$	
$f : A3$	
$R_L : C5$	
$C_1 : F7$	
$F_1 : B4$	

T 1

$V_1 : F3$
$V_2 : B2$
$V_L : D7$
$V_R : E1$
$f : C6$
$R_L : E4$
$C_1 : G3$
$F_1 : B7$

T 2

$V_1 : A1$
$V_2 : C4$
$V_L : F2$
$V_R : D6$
$f : G4$
$R_L : A5$
$C_1 : D3$
$F_1 : E6$

T 3

$V_1 : C1$
$V_2 : A4$
$V_L : A7$
$V_R : B5$
$f : C3$
$R_L : D1$
$C_1 : C7$
$F_1 : E5$

T 4

$V_1 : D4$
$V_2 : E2$
$V_L : G5$
$V_R : A6$
$f : G2$
$R_L : F1$
$C_1 : E3$
$F_1 : D5$

T 5

$V_1 : F4$
$V_2 : E7$
$V_L : A2$
$V_R : F6$
$f : G7$
$R_L : C2$
$C_1 : B1$
$F_1 : B3$

T 6

$V_1 : A1$
$V_2 : E2$
$V_L : F5$
$V_R : A7$
$f : F6$
$R_L : B3$
$C_1 : D5$
$F_1 : B4$

T 7

$V_1 : D2$
$V_2 : C1$
$V_L : A6$
$V_R : E7$
$f : G2$
$R_L : C5$
$C_1 : F7$
$F_1 : B4$

T 8

$V_1 : A4$
$V_2 : C1$
$V_L : F4$
$V_R : A7$
$f : G5$
$R_L : E4$
$C_1 : B5$
$F_1 : B3$

T 9

$V_1 : D4$
$V_2 : B6$
$V_L : D7$
$V_R : E1$
$f : C6$
$R_L : C2$
$C_1 : B7$
$F_1 : D3$

FIGURE 4-32 T-Shooter™. (*Patent pending: Courtesy of Malvino Inc.*)

4-27. A rectifier diode has a forward voltage of 1.2 V at 2 A. The winding resistance is 0.3 Ω. If the secondary voltage is 25 V rms, what is the surge current in a bridge rectifier?

T-SHOOTER PROBLEMS

Use Fig. 4-32 for the remaining problems. If you haven't already done so, read Example 4-12 before attempting these problems. You can measure voltages in any order; for instance, V_2 first, V_L second, and V_R third, or whatever. These voltages are the clues to the trouble. After measuring a voltage, try to figure out what to measure next. Troubleshooting has so many possibilities that it is impractical to try to give rules for every situation. The best approach is to measure something, then think about

what this tells you. Usually, the measurement gives you an idea of what you should measure next. Keep making measurements until you have enough clues to logically figure out what the trouble is.

The possible troubles are open or shorted components (diodes, resistors, capacitors, etc.). Besides voltage measurements, there are other measurements as follows: f for ripple frequency, R_L for load resistance, C_1 for capacitor resistance, and F_1 for fuse resistance.

4-28. Find Trouble 1.

4-29. Find Troubles 2 and 3.

4-30. Find Troubles 4 and 5.

4-31. Find Troubles 6 and 7.

4-32. Find Troubles 8 and 9.

5

SPECIAL-PURPOSE DIODES

Rectifier diodes are the most common type of diode. They are used in power supplies to convert ac voltage to dc voltage. But rectification is not all that a diode can do. Now we will discuss diodes used in other applications. The chapter begins with the zener diode, which is optimized for its breakdown properties. Zener diodes are very important because they are the key to voltage regulation. The chapter also covers optoelectronic diodes. Schottky diodes, varactors, and other diodes.

5-1   THE ZENER DIODE

Small-signal and rectifier diodes are never intentionally operated in the breakdown region because this may damage them. A *zener diode* is different; it is a silicon diode that the manufacturer has optimized for operation in the breakdown region. In other words, unlike ordinary diodes that never work in the breakdown region, zener diodes work best in the breakdown region. Sometimes called a *breakdown diode,* the zener diode is the backbone of voltage regulators, circuits that hold the load voltage almost constant despite large changes in line voltage and load resistance.

I-V Graph

Figure 5-1a shows the schematic symbol of a zener diode; Fig. 5-1b is an alternative symbol. In either symbol, the lines resemble a "z," which stands for zener. By varying the doping level of silicon diodes, a manufacturer can produce zener diodes with breakdown voltages from about 2 to 200 V. These diodes can operate in any of three regions: forward, leakage, and breakdown.

Figure 5-1c shows the *I-V* graph of a zener diode. In the forward region, it starts conducting around 0.7 V, just like an ordinary silicon diode. In the leakage region (between zero and breakdown) it has only a small reverse current. In a zener diode, the breakdown has a very sharp knee, followed by an almost vertical increase in current. Note that the

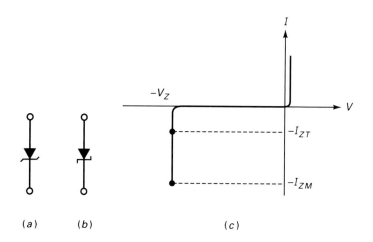

FIGURE 5-1 Zener diode. (*a*) Symbol; (*b*) alternative symbol; (*c*) diode curve.

voltage is almost constant, approximately equal to V_Z over most of the breakdown region. Data sheets usually specify the value of V_Z at a particular test current I_{ZT}.

Do not let the use of the minus signs confuse you. Minus signs need to be included with graphs because you are simultaneously showing forward and reverse values. But you don't have to use minus signs in other discussions if the meaning is clear without them. For instance, it is preferable to say that a zener diode has a breakdown voltage of 10 V, rather than to say it has a breakdown voltage of -10 V. Anyone who knows how a zener diode works already knows it has to be reverse-biased. A pure mathematician might prefer to say a zener diode has a breakdown voltage of -10 V, but a practicing engineer or technician will prefer to say it has a breakdown voltage of 10 V.

Zener Resistance

Because all diodes have some bulk resistance in the *p* and *n* regions, the current through a zener diode produces a small voltage drop in addition to the breakdown voltage. To state it another way, when a zener diode is operating in the breakdown region of Fig. 5-1*c*, an increase in current produces a slight increase in voltage. The increase is very small, typically a few tenths of a volt. This may be important in design work, but not for troubleshooting and preliminary analysis. Unless otherwise indicated, our discussions will ignore the zener resistance.

Zener Regulator

A zener diode is sometimes called a *voltage-regulator* diode because it maintains a constant output voltage even though the current through it changes. For normal operation, you have to reverse-bias the zener diode as shown in Fig. 5-2*a*. Furthermore, to get breakdown operation, the source voltage V_S must be greater than the zener breakdown voltage V_Z. A series resistor R_S is always used to limit the zener current to less than

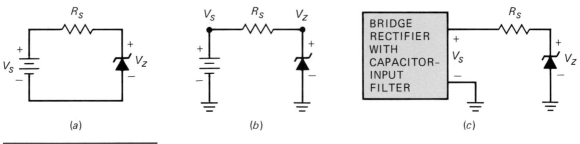

FIGURE 5-2 Zener regulator.

its maximum current rating. Otherwise, the zener diode will burn out like any device with too much power dissipation.

Figure 5-2*b* shows an alternative way to draw the circuit with grounds. Whenever a circuit has grounds, it is usually best to measure node voltages with respect to ground. In fact, if you are using a voltmeter with a power plug, its common terminal may be grounded. In this case, it is necessary to measure node voltages to ground.

For instance, suppose you want to know the voltage across the series resistor of Fig. 5-2*b*. Here is the usual way to find it when you have a built-up circuit. First, measure the voltage from the left end of R_S to ground. Second, measure the voltage from the right end of R_S to ground. Third, subtract the two voltages to get the voltage across R_S. This indirect method is necessary because the common lead of many plug-in voltmeters is grounded. (*Note:* If you have a floating VOM, you can connect directly across the series resistor.)

Figure 5-2*c* shows the output of a power supply connected to a series resistor and a zener diode. This circuit is used when you want a dc output voltage that is less than the output of the power supply. A circuit like this is called a *zener voltage regulator*, or simply a *zener regulator*.

Ohm's Law Again

In Fig. 5-2, the voltage across the series resistor equals the difference between the source voltage and the zener voltage. Therefore, the current through the resistor is

$$I_S = \frac{V_S - V_Z}{R_S} \tag{5-1}$$

Don't memorize this equation. It is nothing more than Ohm's law applied to the series resistor. The series current equals the voltage across the series resistor divided by the resistance. The only thing you have to remember is that the voltage across the series resistor is the difference between the source voltage and the zener voltage. In fact, you don't even have to remember that because the circuit itself contains this information. When you look at Fig. 5-2, you can see at a glance that the voltage across the series resistor equals V_S minus V_Z.

Once you have the value of series current, you also have the value of zener current. Why? Because Fig. 5-2 is a series circuit and you know that current is the same in all parts of a series circuit.

Ideal Zener Diode

For troubleshooting and preliminary analysis, we can approximate the breakdown region as vertical. Therefore, the voltage is constant even though the current changes, which is equivalent to ignoring the zener resistance. Figure 5-3a shows the ideal approximation of a zener diode. This means that a zener diode operating in the breakdown region ideally acts like a battery. In a circuit, it means that you can mentally replace a zener diode by a voltage source of V_Z, provided the zener diode is operating in the breakdown region.

Second Approximation

Figure 5-3b shows the second approximation of a zener diode. A zener resistance (relatively small) is in series with an ideal battery. This resistance produces a voltage drop equal to the product of the current and the resistance.

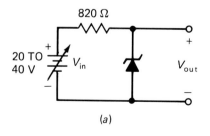

FIGURE 5-3

Zener approximation. (a) Ideal; (b) second approximation.

EXAMPLE 5-1

Suppose the zener diode of Fig. 5-4a has a breakdown voltage of 10 V. What are the minimum and maximum zener currents?

SOLUTION

The applied voltage may vary from 20 to 40 V. Ideally, a zener diode acts like the battery shown in Fig. 5-4b. Therefore, the output voltage is 10 V for any source voltage between 20 and 40 V.

The minimum current occurs when the source voltage is minimum. Visualize 20 V on the left end of the resistor and 10 V on the right end. Then you can see that the voltage across the resistor is 20 V − 10 V, or 10 V. The rest is Ohm's law:

$$I_S = \frac{10 \text{ V}}{820 \text{ }\Omega} = 12.2 \text{ mA}$$

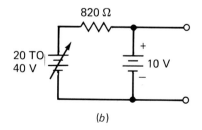

FIGURE 5-4   Example.

The maximum current occurs when the source voltage is 40 V. In this case, the voltage across the resistor is 30 V, which gives a current of

$$I_S = \frac{30 \text{ V}}{820 \text{ }\Omega} = 36.6 \text{ mA}$$

In a voltage regulator like Fig. 5-4a, the output voltage is held constant at 10 V, despite the change in source voltage from 20 to 40 V. The larger source voltage produces more zener current, but the output voltage holds rock-solid at 10 V. (If the zener resistance is included, the output voltage increases slightly when the source voltage increases.)

5-2 THE LOADED ZENER REGULATOR

Figure 5-5a shows a loaded zener regulator, and Fig. 5-5b shows the same circuit in a practical form. This circuit is more complicated than the unloaded zener regulator analyzed in the previous section, but the basic idea is the same: The zener diode operates in the breakdown region and holds the load voltage constant. Even if the source voltage changes or the load resistance varies, the load voltage will remain fixed and equal to the zener voltage.

Breakdown Operation

Always remember this: The zener diode has to operate in the breakdown region to hold the load voltage constant. To put it another way, the zener diode cannot regulate if the load voltage is less than the zener voltage.

How can you tell if the zener diode of Fig. 5-5 is operating in the breakdown region? The designer of the circuit usually takes care of this. Here is the formula that applies:

$$V_{TH} = \frac{R_L}{R_S + R_L} V_S \tag{5-2}$$

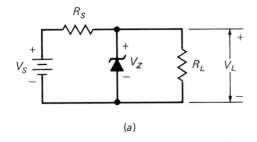

(a)

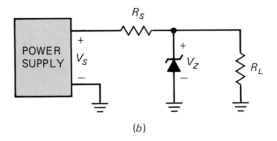

(b)

FIGURE 5-5 Zener regulator.

This is the voltage that exists when the zener diode is disconnected from the circuit. This voltage has to be greater than the zener voltage; otherwise, breakdown cannot occur.

Here is where the equation comes from. When the zener diode is disconnected from the circuit, all that's left is a voltage divider consisting of R_S in series with R_L. The current through this voltage divider is

$$I = \frac{V_S}{R_S + R_L}$$

The load voltage without the zener diode equals the previous current times the load resistance. When you multiply the current by the load resistance, you get the right side of Eq. (5-2), where V_{TH} stands for the Thevenin voltage. This is the voltage with the zener diode out of the circuit.

Series Current

Unless otherwise indicated, in all subsequent discussions we assume the zener diode is operating in the breakdown region. In Fig. 5-5, the current through the series resistor is given by

$$I_S = \frac{V_S - V_Z}{R_S} \qquad (5\text{-}3)$$

This is Ohm's law applied to the current-limiting resistor. It is the same whether or not there is a load resistor. In other words, if you disconnect the load resistor, the current through the series resistor still equals the voltage across the resistor divided by the resistance.

Load Current

Ideally, the load voltage equals the zener voltage because the load resistor is in parallel with the zener diode. As an equation,

$$V_L = V_Z \qquad (5\text{-}4)$$

This allows us to use Ohm's law to calculate the load current:

$$I_L = \frac{V_L}{R_L} \qquad (5\text{-}5)$$

Zener Current

With Kirchhoff's current law,

$$I_S = I_Z + I_L$$

This should be clear from your study of series-parallel circuits. The zener diode and the load resistor are in parallel. The sum of their currents has to equal the total current, which is the same as the current through the series resistor.

We can rearrange the foregoing equation to get this important formula:

$$I_Z = I_S - I_L \qquad (5\text{-}6)$$

This tells you that the zener current no longer equals the series current, as it does in an unloaded zener regulator. Because of the load resistor, the zener current now equals the series current minus the load current.

Process

Troubleshooters, designers, and other professionals don't blindly plug numbers into formulas, hoping to get the right answer. Professionals know the meaning of each step they take when they solve a problem. Knowing what you are doing is a lot better than relying on formulas.

If professionals don't use formulas, what do they use? Something called a *process*. A process is a step-by-step routine used to solve problems. When professionals solve a problem, they work out the values of different quantities, using Ohm's law in a logical sequence. Occasionally, a complicated formula may be necessary, but that is the exception rather than the rule. Often, problems in electronics are simply Ohm's law and other basic ideas applied over and over to the different components and devices in the circuit.

Here is a three-step process for finding the zener current:

1. Calculate the current through the series resistor.
2. Calculate the load current.
3. Calculate the zener current.

These steps can be abbreviated to

1. Series current
2. Load current
3. Zener current

or symbolically,

1. I_S
2. I_L
3. I_Z

This is what professionals remember. You get the series current first, the load current second, and the zener current third. And you use Ohm's and other basic ideas in the process. The details of the calculations are automatically remembered, at least most of the time.

If you can remember the three quantities in the process, your mind usually takes care of the rest of the details. If you do get stuck, look at the formulas to jog your memory. But don't use formulas blindly. Reread the discussion or examples if you can't remember the details of some step in the process. In general, don't memorize any formula unless you expect to use it a few thousand times. Ohm's law is an example of a formula to memorize. The equations of this chapter are examples of formulas you do not memorize because most of them are rewrites of Ohm's law.

Ripple across the Load Resistor

In Fig. 5-5b, the output of a power supply drives a zener regulator. As you know, the power supply produces a dc voltage with a ripple. Ideally, the zener regulator reduces the ripple to zero because the load voltage is constant and equal to the zener voltage. As an example, suppose the power supply produces a dc voltage of 20 V with a peak-to-peak ripple of 2 V. Then the supply voltage is swinging from 19 V minimum to 21 V maximum. Variations in supply voltage will change the zener current, but they have almost no effect on the load voltage.

If you take into account the small zener resistance, you will find that there is a small ripple across the load resistor. But this ripple is much smaller than the original ripple coming out of the power supply. In fact, you can estimate the new ripple with this equation:

$$V_{R(\text{out})} = \frac{R_Z}{R_S + R_Z} V_{R(\text{in})} \qquad (5\text{-}7)$$

This is an accurate approximation of peak-to-peak output ripple. If it reminds you of a voltage divider, you are right on target. It comes from visualizing the zener diode replaced by its second approximation. With respect to the ripple, the circuit acts like a voltage divider formed by R_S in series with R_Z.

Temperature Coefficient

One final point: Raising the *ambient* (surrounding) temperature changes the zener voltage slightly. On data sheets, the effect of temperature is listed under the *temperature coefficient*, which is the percentage change per degree change. A designer needs to calculate the change in zener voltage at the highest ambient temperature. But even a troubleshooter should know that temperature can change the zener voltage.

For zener diodes with breakdown voltages less than 5 V, the temperature coefficient is negative. For zener diodes with breakdown voltages of more than 6 V, the temperature coefficient is positive. Between 5 and 6 V, the temperature coefficient changes from negative to positive; this means that you can find an operating point for a zener diode at which the temperature coefficient is zero. This is important in some applications where a solid zener voltage is needed over a large temperature range.

EXAMPLE 5-2

Figure 5-6 has these circuit values: $V_S = 18$ V, $V_Z = 10$ V, $R_S = 270\ \Omega$, and $R_L = 1\ \text{k}\Omega$. Is the zener diode operating in the breakdown region?

SOLUTION

Use Eq. (5-2), or better still, use your head. Mentally disconnect the zener diode. Then all that is left is a voltage divider with 270 Ω in series with 1 kΩ. Therefore, the current through the voltage divider is

$$I = \frac{18\ \text{V}}{1.27\ \text{k}\Omega} = 14.2\ \text{mA}$$

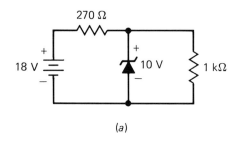

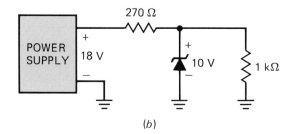

(a)   (b)

FIGURE 5-6   Example.

Multiply this current by the load resistance to get the Thevenin voltage:

$$V_{TH} = (14.2 \text{ mA})(1 \text{ k}\Omega) = 14.2 \text{ V}$$

Since this voltage is greater than the zener voltage (10 V), the zener diode will operate in the breakdown region when it is reconnected to the circuit.

Naturally, you can plug the values directly into Eq. (5-2) as follows:

$$V_{TH} = \frac{1 \text{ k}\Omega}{1.27 \text{ k}\Omega} 18 \text{ V} = 14.2 \text{ V}$$

The result is the same, so either method is acceptable.

The advantage of the first method is that you are more likely to remember it because it is Ohm's law applied twice. Also, the first method requires you to think logically about what is happening in the circuit. But either method is valid, so use whichever you prefer.

EXAMPLE 5-3

What does the zener current equal in Fig. 5-6b?

SOLUTION

You are given the voltage on both ends of the series resistor. Subtract the voltages, and you can see that 8 V is across the series resistor. Then Ohm's law gives

$$I_S = \frac{8 \text{ V}}{270 \text{ }\Omega} = 29.6 \text{ mA}$$

Since the load voltage is 10 V, the load current is

$$I_L = \frac{10 \text{ V}}{1 \text{ k}\Omega} = 10 \text{ mA}$$

The zener current is the difference of the two currents:

$$I_Z = 29.6 \text{ mA} - 10 \text{ mA} = 19.6 \text{ mA}$$

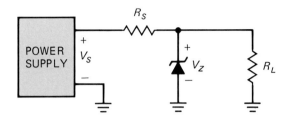

FIGURE 5-7 Zener regulator with the load resistor.

EXAMPLE 5-4

The data sheet of a 1N961 gives a zener resistance of 8.5 Ω. Suppose this zener diode is used in Fig. 5-7 with a series resistance of 270 Ω. What is the load ripple if the supply ripple is 2 V?

With Eq. (5-7),

$$V_{R(\text{out})} = \frac{8.5\ \Omega}{278.5\ \Omega}(2\ \text{V}) = 0.061\ \text{V} = 61\ \text{mV}$$

The final output is a dc voltage of 10 V with a peak-to-peak ripple of only 61 mV.

EXAMPLE 5-5

What does the circuit of Fig. 5-8 do?

SOLUTION

This is an example of a preregulator (the first zener diode) driving a zener regulator (the second zener diode). First, notice that the preregulator has an output voltage of 20 V. This is the input to the second zener regulator, whose output is 10 V. The basic idea is to provide the second regulator with a well-regulated input, so that the final output is extremely well regulated.

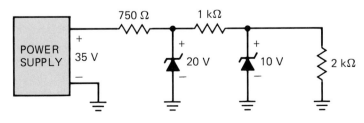

FIGURE 5-8 Example.

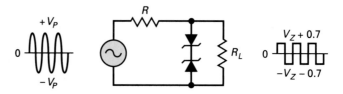

FIGURE 5-9 Zener diode used in combination limiter.

EXAMPLE 5-6

What does the circuit of Fig. 5-9 do?

SOLUTION

In most applications, zener diodes are used in voltage regulators where they remain in the breakdown region. But there are exceptions. Sometimes zener diodes are used in waveshaping circuits like Fig. 5-9.

Notice the back-to-back connection of two zener diodes. On the positive half-cycle, the upper diode conducts and the lower diode breaks down. Therefore, the output is clipped as shown. The clipping level equals the zener voltage (broken-down diode) plus 0.7 V (forward-biased diode).

On the negative half-cycle, the action is reversed. The lower diode conducts, and the upper diode breaks down. In this way, the output is almost a square wave. The larger the input sine wave, the better looking the output square wave.

5-3 OPTOELECTRONIC DEVICES

Optoelectronics is the technology that combines optics and electronics. This exciting field includes many devices based on the action of a *pn* junction. Examples of optoelectronic devices are light-emitting diodes (LEDs), photodiodes, optocouplers, etc. Our discussion begins with the LED.

Light-Emitting Diode

Figure 5-10a shows a source connected to a resistor and a LED. The outward arrows symbolize the radiated light. In a forward-biased LED, free electrons cross the junction and fall into holes. As these electrons fall from a higher to a lower energy level, they radiate energy. In ordinary diodes, this energy goes off in the form of heat. But in a LED, the energy is radiated as light. LEDs have replaced incandescent lamps in many applications because of their low voltage, long life, and fast on-off switching.

Ordinary diodes are made of silicon, an opaque material that blocks

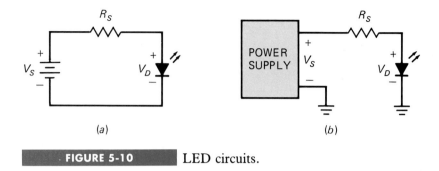

FIGURE 5-10 LED circuits.

the passage of light. LEDs are different. By using elements like gallium, arsenic, and phosphorus, a manufacturer can produce LEDs that radiate red, green, yellow, blue, orange, or infrared (invisible). LEDs that produce visible radiation are useful with instruments, calculators, etc. The infrared LED finds applications in burglar alarm systems and other areas requiring invisible radiation.

LED Voltage and Current

The resistor of Fig. 5-10 is the usual current-limiting resistor that prevents the current from exceeding the maximum current rating of the diode. Since the resistor has a node voltage of V_S on the left and a node voltage of V_D on the right, the voltage across the resistor is the difference between the two voltages. With Ohm's law, the series current is

$$I_S = \frac{V_S - V_D}{R_S} \qquad (5\text{-}8)$$

For most of the commercially available LEDs, the typical voltage drop is from 1.5 to 2.5 V for currents between 10 and 50 mA. The exact voltage drop depends on the LED current, color, tolerance, etc. Unless otherwise specified, we will use a nominal drop of 2 V when troubleshooting or analyzing the LED circuits in this book. If you get into design work, consult the data sheets for the LEDs you are using.

Seven-Segment Display

Figure 5-11a shows a *seven-segment display*. It contains seven rectangular LEDs (A through G). Each LED is called a segment because it forms part of the character being displayed. Figure 5-11b is a schematic diagram of the seven-segment display. External series resistors are included to limit the currents to safe levels. By grounding one or more resistors, we can form any digit from 0 through 9. For instance, by grounding A, B, and C, we get a 7. Grounding A, B, C, D, and G produces a 3.

A seven-segment display can also display capital letters A, C, E, and F, plus lowercase letters b and d. Microprocessor trainers often use seven-segment displays that show all digits from 0 through 9, plus A, b, C, d, E, and F.

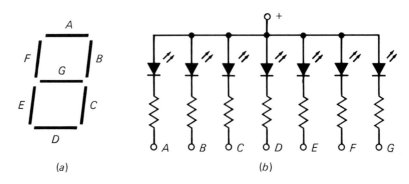

(a)   (b)

FIGURE 5-11 (*a*) Seven-segment indicator; (*b*) schematic diagram.

The seven-segment indicator of Fig. 5-11*b* is referred to as the *common-anode* type because all anodes are connected together. Also available is the *common-cathode* type where all cathodes are connected together.

Photodiode

As previously discussed, one component of reverse current in a diode is the flow of minority carriers. These carriers exist because thermal energy keeps dislodging valence electrons from their orbits, producing free electrons and holes in the process. The lifetime of the minority carriers is short, but while they exist they can contribute to the reverse current.

When light energy bombards a *pn* junction, it can dislodge valence electrons. The more light striking the junction, the larger the reverse current in a diode. A *photodiode* is one that has been optimized for its sensitivity to light. In this diode, a window lets light pass through the package to the junction. The incoming light produces free electrons and holes. The stronger the light, the greater the number of minority carriers and the larger the reverse current.

Figure 5-12 shows the schematic symbol of a photodiode. The arrows represent the incoming light. Especially important, the source and the series resistor reverse-bias the photodiode. As the light becomes brighter, the reverse current increases. With typical photodiodes, the reverse current is in the tens of microamperes.

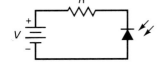

FIGURE 5-12

Photodiode.

Optocoupler

An optocoupler (also called an *optoisolator* or an *optically coupled isolator*) combined a LED and a photodiode in a single package. Figure 5-13 shows an optocoupler. It has a LED on the input side and a photodiode on the output side. The left source voltage and the series resistor set up a current through the LED. Then the light from the LED hits the photodiode, and this sets up a reverse current in the output circuit. This reverse current produces a voltage across the output resistor. The output voltage then equals the output supply voltage minus the voltage across the resistor.

When the input voltage is varying, the amount of light is fluctuating. This means that the output voltage is varying in step with the input voltage. This is why the combination of a LED and a photodiode is called

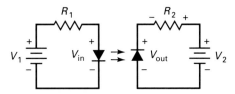

FIGURE 5-13   Optocoupler.

an optocoupler. The device can couple an input signal to the output circuit.

The key advantage of an optocoupler is the electrical isolation between the input and output circuits. With an optocoupler, the only contact between the input and the output is a beam of light. Because of this, it is possible to have an insulation resistance between the two circuits in the thousands of megohms. Isolation like this comes in handy in high-voltage applications where the potentials of the two circuits may differ by several thousand volts.

EXAMPLE 5-7

In Fig. 5-10 the source voltage is 10 V, and the series resistance is 680 Ω. What is the LED current?

SOLUTION

Use a nominal LED drop of 2 V. Then the series resistor has 10 V on the left end and 2 V on the right end. This means the voltage across the resistor is 8 V. Finish off the problem with Ohm's law:

$$I = \frac{8\,V}{680\,\Omega} = 11.8\,mA$$

5-4 THE SCHOTTKY DIODE

At lower frequencies, an ordinary diode can easily turn off when the bias changes from forward to reverse. But as the frequency increases, the diode reaches a point where it cannot turn off fast enough to prevent noticeable current during part of the reverse half-cycle. This effect is known as *charge storage*. It places a limit on the useful frequency of ordinary rectifier diodes.

What happens is this. When a diode is forward-biased, some of the carriers in the depletion layers have not yet recombined. If the diode is suddenly reverse-biased, these carriers can flow in the reverse direction for a little while. The greater the lifetime, the longer these charges can contribute to reverse current.

The time it takes to turn off a forward-biased diode is called the *reverse recovery time*. The reverse recovery time is so short in small-signal diodes that you don't even notice its effect at frequencies below 10 MHz or so. It's only when you get well above 10 MHz that it becomes important.

The solution is a special-purpose device called a *Schottky diode*. This type of diode has no depletion layer, which eliminates the stored charges at the junction. The lack of charge storage means the Schottky diode can switch off faster than an ordinary diode. In fact, a Schottky diode can easily rectify frequencies above 300 MHz.

The most important application of Schottky diodes is in digital computers. The speed of computers depends on how fast their diodes and transistors can turn on and off. This is where the Schottky diode comes in. Because it has no charge storage, the Schottky diode has become the backbone of low-power Schottky TTL, a group of widely used digital devices.

A final point: In the forward direction, a Schottky diode has a barrier potential of only 0.25 V. Therefore, you may see Schottky diodes used in a low-voltage bridge rectifiers because you have to subtract only 0.25 instead of the usual 0.7 V for each diode.

5-5 THE VARACTOR

The varactor (also called the *voltage-variable capacitance, varicap, epicap,* and *tuning diode*) is widely used in television receivers, FM receivers, and other communications equipment. Here is the basic idea. In Fig. 5-14*a*, the depletion layer is between the *p* region and the *n* region. The *p* and *n* regions are like the plates of a capacitor, and the depletion layer is like the dielectric. When a diode is reverse-biased, the width of the depletion layer increases with the reverse voltage. Since the depletion layer gets wider with more reverse voltage, the capacitance becomes smaller. It's as though you moved apart the plates of a capacitor. The key idea is that capacitance is controlled by voltage.

Figure 5-14*b* shows the equivalent circuit for a reverse-biased diode. At higher frequencies, the varactor acts the same as a variable capacitance. Figure 5-14*d* shows how the capacitance varies with reverse voltage. This graph shows that the capacitance gets smaller when the reverse voltage gets larger. The really important idea here is that reverse voltage controls capacitance. This opens the door to remote control.

Figure 5-14*c* shows the schematic symbol for a varactor. How is this device used? You can connect a varactor in parallel with an inductor to get a resonant circuit. Then you can change the reverse voltage to change the resonant frequency. This is the principle behind tuning in a radio station, a TV channel, etc.

5-6 VARISTORS

Lightning, power-line faults, etc., can pollute the line voltage by superimposing dips, spikes, and other transients on the normal 115 V rms. *Dips* are severe voltage drops lasting microseconds or less. *Spikes* are short overvoltages of 500 to more than 2000 V. In some equipment, filters are

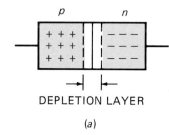

p *n*

DEPLETION LAYER

(a)

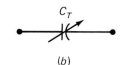

C_T

(b)

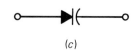

(c)

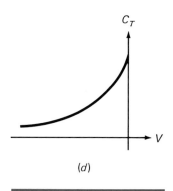

C_T

V

(d)

FIGURE 5-14

Varactor. (*a*) Structure; (*b*) equivalent circuit; (*c*) schematic symbol; (*d*) graph.

used between the power line and the primary of the transformer to eliminate the problems caused by line transients.

One of the devices used for line filtering is the *varistor* (also called a *transient suppressor*). This semiconductor device is like two back-to-back zener diodes with a high breakdown voltage in both directions. For instance, a V130LA2 is a varistor with a breakdown voltage of 184 V (equivalent to 130 V rms) and a peak current rating of 400 A. Connect one of these across the primary winding, and you don't have to worry about spikes. The varistor will clip all spikes at the 184-V level and protect your equipment.

5-7 READING A DATA SHEET

The Appendix shows the data sheet for the 1N746 series of zener diodes. This data sheet also covers the 1N957 series and the 1N4370 series. Refer to these data sheets during the following discussion. Again, most of the information on a data sheet is for designers, but there are a few items that even troubleshooters and testers will want to know about.

Maximum Power

The power dissipation of a zener diode equals the product of its voltage and current:

$$P_Z = V_Z I_Z \qquad (5\text{-}9)$$

For instance, if $V_Z = 12$ V and $I_Z = 10$ mA, then

$$P_Z = (12 \text{ V})(10 \text{ mA}) = 120 \text{ mW}$$

As long as P_Z is less than the power rating, the zener diode can operate in the breakdown region without being destroyed. Commercially available zener diodes have power ratings from ¼ to more than 50 W.

For example, the data sheet for the 1N746 series lists a maximum power rating of 400 mW. A safe design includes a safety factor to keep the power dissipation well below this 400-mW maximum. As mentioned elsewhere, safety factors of 2 or more are used for conservative designs.

Maximum Current

Data sheets usually include the *maximum current* a zener diode can handle without exceeding its power rating. This maximum current is related to the power rating as follows:

$$I_{ZM} = \frac{P_{ZM}}{V_Z} \qquad (5\text{-}10)$$

where I_{ZM} = maximum rated zener current
P_{ZM} = power rating
V_Z = zener voltage

For example, the 1N759 has a zener voltage of 12 V. Therefore, it has a maximum current rating of

$$I_{ZM} = \frac{400\,\text{mW}}{12\,\text{V}} = 33.3\,\text{mA}$$

The data sheet gives two maximum current ratings: 30 and 35 mA. Notice these values bracket our theoretical answer of 33.3 mA. The data sheet gives you two values because of the tolerance in the zener voltage.

If you satisfy the current rating, you automatically satisfy the power rating. For instance, if you keep the maximum zener current less than 33.3 mA, you are also keeping the maximum power dissipation less than 400 mW. If you throw in the safety factor of 2, you don't have to worry about a marginal design blowing the diode.

Tolerance

Note 1 on the data sheet shows these tolerances:

1N4370 series: ±10 percent, suffix A for ±5 percent units
1N746 series: ±10 percent, suffix A for ±5 percent units
1N957 series: ±20 percent, suffix A for ±10 percent units, suffix B for ±5 percent units

For instance, a 1N758 has a zener voltage of 10 V with a tolerance of ±10 percent, while the 1N758A has the same zener voltage with a tolerance of ±5 percent. The 1N967 has a zener voltage of 18 V with a tolerance of ±20 percent. The 1N967A has the same zener voltages with a tolerance of ±10 percent, and the 1N967B has the same voltage with a tolerance of ±5 percent.

Zener Resistance

The zener resistance (also called *zener impedance*) may be designated R_{ZT} or Z_{ZT}. For instance, the 1N961 has a zener resistance of 8.5 Ω measured at a test curent of 12.5 mA. As long as the zener current is above the knee of the curve, you can use 8.5 Ω as the approximate value of the zener resistance. But note how the zener resistance increases at the knee of the curve (700 Ω). The point is this: Operation should be at or near the test current, if at all possible. Then you know the zener resistance is relatively small.

The data sheet contains a lot of additional information, but it is primarily aimed at designers. If you do get involved in design work, then you have to read the data sheet carefully, including the notes that specify how quantities were measured. Data sheets vary from one manufacturer to the next, so you have read between the lines if you want to get to the truth.

Derating

The *derating factor* shown on a data sheet tells you how much you have to reduce the power rating of a device. For instance, the 1N746 series has a power rating of 400 mW for a lead temperature of 50°C. The derating factor is given as 3.2 mW/°C. This means that you have to subtract 3.2 mW for each degree above 50°C. Even though you may not be involved in design, you have to be aware of the effect of temperature. If it is known that the lead temperature will be above 50°C, the designer has to derate or reduce the power rating of the zener diode.

5-8   TROUBLESHOOTING

Figure 5-15 shows a zener regulator. When the circuit is working properly, the voltage between node A and ground is $+18$ V, the voltage between node B and ground is $+10$ V, and the voltage between node C and ground is $+10$ V.

Now, let's discuss what can go wrong with the circuit. When a circuit is not working as it should, a troubleshooter usually starts by measuring node voltages. These voltage measurements give clues that help isolate the trouble. For instance, suppose he or she measures these node voltages:

$$V_A = +18 \text{ V} \qquad V_B = +10 \text{ V} \qquad V_C = 0$$

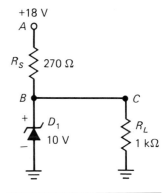

+18 V

FIGURE 5-15

Zener regulator.

When you are trying to figure out what causes incorrect voltages, trial and error is useful. That is, you play the what-if game. Here is what may go through a troubleshooter's mind after measuring the foregoing node voltages:

> What if the load resistor were open? No, the load voltage would still be $+10$ V. What if the load resistor were shorted? No, that would pull nodes B and C down to ground, producing 0 V. All right, what if the connecting wire between nodes B and C were open? Yes, that would do it. That's got to be it.

This trouble produces unique symptoms. The only way you can get this set of voltages is with an open connection between nodes B and C.

Not all troubles produce unique symptoms. Sometimes, two or more troubles produce the same set of voltages. Here is an example. Suppose the troubleshooter measures these node voltages:

$$V_A = +18 \text{ V} \qquad V_B = 0 \qquad V_C = 0$$

What do you think the trouble is? Think about this for a few minutes. When you have an answer, read what follows.

Here is a way that a troubleshooter might find the trouble. The thinking goes like this:

> I've got voltage at A, but not at B and C. What if the series resistor were open? Then no voltage could reach node B or node C, but I would still measure $+18$ V between node A and ground. Yes, the series resistor is probably open.

At this point, the troubleshooter would disconnect the series resistor and measure its resistance with an ohmmeter. Chances are that it would be open. But suppose it measures okay. Then the troubleshooter's thinking continues like this:

> That's strange. Well, is there any other way I can get +18 V at node A and 0 V at nodes B and C? What if the zener diode were shorted? What if the load resistor were shorted? What if a solder splash were between node B or node C and ground? Any of these will produce the symptoms I'm getting.

Now, the troubleshooter has more possible troubles to check out. Eventually, she or he will find the trouble.

When components burn out, they usually become open, but not always. Some semiconductor devices can develop internal shorts, in which case, they are like zero resistances. Other ways to get shorts include a solder splash between traces on a printed-circuit board, a solder ball touching two traces, etc. Because of this, you must include what-if questions in terms of shorted components, as well as open components.

EXAMPLE 5-8

Assume an ideal zener diode and work out the node voltages for all possible shorts and opens in Fig. 5-15.

SOLUTION

In working out the voltages, remember this. A shorted component is equivalent to a resistance of zero, while an open component is equivalent to a resistance of infinity. If you have trouble calculating with 0 and ∞, then use 0.001 Ω and 1000 MΩ. In other words, use a very small resistance for a short and a very large resistance for an open.

To begin, the series resistor R_S may be shorted or open. Let us designate these R_{SS} and R_{SO}, respectively. Similarly, the zener diode may be shorted or open, symbolized by D_{1S} and D_{1O}. Also, the load resistor may be shorted or open, R_{LS} and R_{LO}. Finally, the connecting wire between B and C may be open, designated BC_O.

If the series resistor were shorted, +18 V would appear at nodes B and C. This would destroy the zener diode and possibly the load resistor, but the voltage would remain at +18 V. Then a troubleshooter would measure $V_A =$ +18 V, $V_B =$ +18 V, and $V_C =$ +18 V. This trouble and its voltages are shown in Table 5-1.

If the series resistor were open, then the voltage could not reach node B. In this case, nodes B and C would have zero voltage. Continuing like this, we can get the remaining entries shown in Table 5-1.

In Table 5-1, the comments indicate troubles that might occur as a direct result of the original short circuits. For instance, a shorted R_S will destroy the zener diode and may also burn out the load resistor. It depends on the power rating of the load resistor. A shorted R_S means there's 18 V across 1 kΩ. This produces a power of 0.324 W. If the load resistor is rated at only 0.25 W, it will burn out.

Study the table. You can learn a lot from it. Also, use the T-shooter at the end of this chapter to practice troubleshooting a zener regulator.

Table 5-1 Zener Regulator Troubles and Symptoms

Trouble	V_A, V	V_B, V	V_C, V	Comments
None	18	10	10	No trouble.
R_{SS}	18	18	18	D_1 and R_L may be blown.
R_{SO}	18	0	0	
D_{1S}	18	0	0	R_S may be blown.
D_{1O}	18	14.2	14.2	
R_{LS}	18	0	0	R_S may be blown.
R_{LO}	18	10	10	
BC_O	18	10	0	
No supply	0	0	0	Check power supply.

❑ OPTIONAL TOPICS

The following material continues the earlier discussions at a more advanced and specialized level. All the topics are optional because they are not used in any of the basic discussions in later chapters. This section will be a useful reference when you are in industry because then you will probably want more advanced viewpoints.

❑ 5-9 LOAD LINES

The current through the zener diode of Fig. 5-16a is given by

$$I_Z = \frac{V_S - V_Z}{R_S} \tag{5-11}$$

This says the zener current equals the voltage across the series resistor divided by the resistance. Equation (5-11) can be used to construct the load line as previously discussed. For instance, suppose $V_S = 20$ V and $R_S = 1$ kΩ. Then the foregoing equation reduces to

$$I_Z = \frac{20 - V_Z}{1000}$$

As before, we get the saturation point (vertical intercept) by setting V_Z equal to zero and solving for I_Z to get 20 mA. Similarly, to get the cutoff point (horizontal intercept), we set I_Z equal to zero and solve for V_Z to get 20 V.

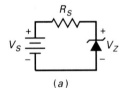

(a)

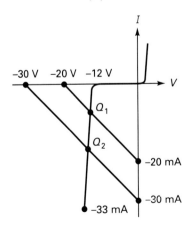

(b)

FIGURE 5-16

Zener diode circuit.

Alternatively, you can get the ends of the load line as follows. Visualize Fig. 5-16a with $V_S = 20$ V and $R_S = 1$ kΩ. With the zener diode shorted, the maximum diode current is 20 mA. With the diode open, the maximum diode voltage is 20 V.

Suppose the zener diode has a breakdown voltage of 12 V. Then its graph appears as shown in Fig. 5-16b. When we plot the load line for $V_S = 20$ V and $R_S = 1$ kΩ, we get the upper load line with an intersection point of Q_1. The voltage across the zener diode will be slightly more than the knee voltage at breakdown because the curve slopes slightly.

To understand how voltage regulation works, assume the source voltage changes to 30 V. Then the zener current changes to

$$I_Z = \frac{30 - V_Z}{1000}$$

This implies that the ends of the load line are 30 mA and 30 V, as shown in Fig. 5-16b. The new intersection is at Q_2. Compare Q_2 with Q_1, and you can see there is more current through the zener diode but approximately the same zener voltage. Therefore, even though the source voltage has changed from 20 to 30 V, the zener voltage is still approximately equal to 12 V. This is the basic idea of voltage regulation; the output voltage has remained almost constant even though the input voltage has changed by a large amount.

5-10 SECOND APPROXIMATION

Figure 5-17 shows the second approximation of a zener diode. A zener resistance (relatively small) is in series with an ideal battery. This resistance produces a voltage drop equal to the product of the current and the resistance. For instance, the voltage at Q_1 (Fig. 5-16b) is

$$V_1 = I_1R_Z + V_Z$$

and the voltage at Q_2 is

$$V_2 = I_2R_Z + V_Z$$

The change in voltage is

$$V_2 - V_1 = (I_2 - I_1)R_Z$$

This is usually written as

$$\Delta V_Z = \Delta I_Z R_Z \tag{5-12}$$

where ΔV_Z = change in zener voltage
ΔI_Z = change in zener current
R_Z = zener resistance

This tells us the change in zener voltage equals the change in zener current times the zener resistance. Usually, R_Z is small, so the voltage change is slight.

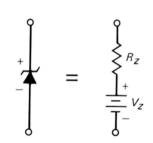

FIGURE 5-17

Second approximation.

EXAMPLE 5-9

Suppose the zener diode has a breakdown voltage of 10 V and a zener resistance of 8.5 Ω. What is the additional voltage when the current is 20 mA?

SOLUTION

Use Eq. (5-12) to get

$$\Delta V_Z = (20 \text{ mA})(8.5 \text{ Ω}) = 0.17 \text{ V}$$

This means the zener voltage is 10.17 instead of 10 V.

☐ 5-11 RIPPLE

As discussed earlier, a zener regulator like Fig. 5-18*a* reduces the ripple. How much reduction is there in ripple? Visualize the zener diode replaced by its second approximation as shown in Fig. 5-18*b*. At the beginning of the capacitor discharge, the current through the series resistor is

$$I_{S(\text{max})} = \frac{V_{S(\text{max})} - V_Z}{R_S}$$

At the end of the discharge,

$$I_{S(\text{min})} = \frac{V_{S(\text{min})} - V_Z}{R_S}$$

Subtracting these equations gives

$$I_{S(\text{max})} - I_{S(\text{min})} = \frac{V_{S(\text{max})} - V_{S(\text{min})}}{R_S}$$

which is usually written as

$$\Delta I_S = \frac{\Delta V_S}{R_S}$$

Rearranging gives

$$\Delta V_S = \Delta I_S R_S$$

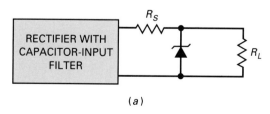

(a)

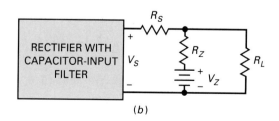

(b)

FIGURE 5-18   Effect on ripple.

This says the peak-to-peak input ripple equals the change in series current times the series resistance.

Earlier, we derived the change in zener voltage:

$$\Delta V_Z = \Delta I_Z R_Z$$

If these are the maximum changes in Fig. 5-18b, the peak-to-peak ripple across the zener diode equals the change in zener current times the zener resistance. Taking the ratio of output ripple to input ripple gives

$$\frac{\Delta V_Z}{\Delta V_S} = \frac{\Delta I_Z R_Z}{\Delta I_S R_S}$$

For a constant load resistance, the change in zener current equals the change in source current, so the foregoing ratio reduces to

$$\frac{\Delta V_Z}{\Delta V_S} = \frac{R_Z}{R_S} \tag{5-13}$$

where ΔV_Z = output ripple
$\quad \Delta V_S$ = input ripple
$\quad R_Z$ = zener resistance
$\quad R_S$ = series resistance

This is a useful equation because it tells you at a glance how the output ripple and input ripple are related. The equation says the ratio of output to input ripple equals the ratio of zener to series resistance. For instance, if the zener resistance is 7 Ω and the series resistance is 700 Ω, then the output ripple is ¹⁄₁₀₀ of the input ripple.

❏ 5-12 ZENER DROP-OUT POINT

For a zener regulator to hold the output voltage constant, the zener diode must remain in the breakdown region under all operating conditions; this is equivalent to saying there must be zener current for all source voltages and load currents. The worst case occurs for minimum source voltage and maximum load current because the zener current drops to a minimum. For this case,

$$I_{S(\text{min})} = \frac{V_{S(\text{min})} - V_Z}{R_{S(\text{max})}}$$

which can be rearranged as

$$R_{S(\text{max})} = \frac{V_{S(\text{min})} - V_Z}{I_{S(\text{min})}} \tag{5-14}$$

As shown earlier,

$$I_Z = I_S - I_L$$

In the worst case, this is written as

$$I_{Z(\text{min})} = I_{S(\text{min})} - I_{L(\text{max})}$$

The critical point occurs when the maximum load current equals the minimum series current:

$$I_{L(\text{max})} = I_{S(\text{min})}$$

At this point, the zener current drops to zero, and regulation is lost.

By substituting $I_{L(\text{max})}$ for $I_{S(\text{min})}$ in Eq. (5-14), we get this useful design relationship:

$$R_{S(\text{max})} = \frac{V_{S(\text{min})} - V_Z}{I_{L(\text{max})}} \qquad (5\text{-}15)$$

where $R_{S(\text{max})}$ = critical value of series resistance
$V_{S(\text{min})}$ = minimum source voltage
V_Z = zener voltage
$I_{L(\text{max})}$ = maximum load current

The *critical resistance* $R_{S(\text{max})}$ is the maximum allowable series resistance. The series resistance R_S must always be less than the critical value; otherwise, breakdown operation is lost, and the regulator stops working. In this case, we lose the constant load voltage, and the output ripple becomes almost as large as the input ripple.

EXAMPLE 5-10

A zener regulator has an input voltage from 15 to 20 V and a load current from 5 to 20 mA. If the zener voltage is 6.8 V, what value should the series resistor have?

SOLUTION

The worst case occurs for minimum source voltage and maximum load current. With Eq. (5-15), the critical series resistance is

$$R_{S(\text{max})} = \frac{15 \text{ V} - 6.8 \text{ V}}{20 \text{ mA}} = 410 \ \Omega$$

Therefore, the series resistor must be less than 410 Ω for the zener diode to operate in the breakdown region under the worst-case condition.

5-13 LED DESIGN GUIDELINE

The brightness of a LED depends on the current. Ideally, the best way to control brightness is by driving the LED with a current source. The next best thing to a current source is a large supply voltage and a large series resistance. In this case, the LED current is given by

$$I = \frac{V_S - V_{\text{LED}}}{R_S}$$

The larger the source voltage, the less effect V_{LED} has. In other words, a large V_S swamps out the variation in LED voltage.

For example, a TIL222 is a green LED with a minimum drop of 1.8 V and a maximum drop of 3 V for a current of approximately 25 mA. If you drive a TIL222 with a 20-V source and a 750-Ω resistor, the current varies from 22.7 to 24.3 mA. This implies a brightness that is essentially the same for all TIL222s. But suppose your design uses a 5-V source and a 120-Ω resistor. Then the current varies from about 16.7 to 26.7 mA; this results in a noticeable change in brightness. Therefore, to get approximately constant brightness with LEDs, use as large a supply voltage as possible.

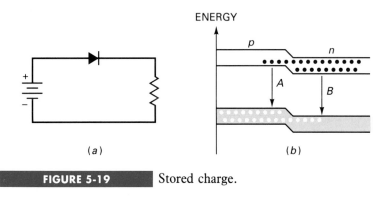

(a)   (b)

FIGURE 5-19   Stored charge.

5-14   REVERSE RECOVERY TIME

Figure 5-19*a* shows a forward-biased diode, and Fig. 5-19*b* illustrates the energy bands. As you can see, conduction-band electrons have diffused across the junction and traveled into the *p* region before recombining (path *A*). Similarly, holes have crossed the junction and traveled into the *n* region before recombination occurs (path *B*). If the lifetime equals 1 μs, free electrons and holes exist for an average of 1 μs before recombination takes place. Because of the lifetime of minority carriers, the charges in a forward-biased diode are temporarily stored in different energy bands near the junction. The greater the forward current, the larger the number of stored charges. This effect is referred to as *charge storage*.

Charge storage is important when you try to switch a diode from on to off. Why? Because if you suddenly reverse-bias a diode, the stored charges can flow in the reverse direction for a while. The greater the lifetime, the longer these charges can contribute to reverse current. For example, suppose a forward-biased diode is suddenly reverse-biased as shown in Fig. 5-20*a*. Then a large reverse current can exist for a while because of the stored charges shown in Fig. 5-20*b*. Until the stored charges either cross the junction or recombine, the reverse current can continue.

The time it takes to turn off a forward-biased diode is called the *reverse recovery time* t_{rr}. The conditions for measuring t_{rr} vary from one manufacturer to the next. As a guide, t_{rr} is the time it takes for the reverse current to drop to 10 percent of the forward current. For instance, the 1N4148

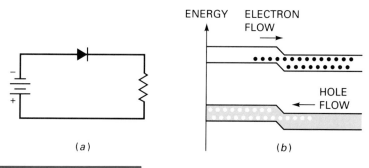

FIGURE 5-20 Stored charges can flow in reverse direction.

has a t_{rr} of 4 ns. If this diode has a forward current of 10 mA and it is suddenly reverse-biased, it will take approximately 4 ns for the reverse current to decrease to 1 mA. Reverse recovery time is so short in small-signal diodes that you don't even notice its effect at frequencies below 10 MHz or so. It's only when you get well above 10 MHz that you have to take t_{rr} into account.

What effect does reverse recovery time have on rectification? Take a look at the half-wave rectifier shown in Fig. 5-21a. At low frequencies the output is well behaved because it is the classic half-wave rectified signal shown in Fig. 5-21b. As the frequency increases well into megahertz, however, the output signal begins to deviate from its normal shape as shown in Fig. 5-21c. As you see, some conduction is noticeable near the beginning of the reverse half-cycle. The reverse recovery time is now becoming a significant part of the period. For instance, if t_{rr} = 4 ns and the period is 50 ns, then the early part of the reverse half-cycle will have a wiggle in it similar to Fig. 5-21c.

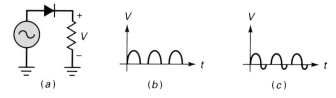

FIGURE 5-21 (a) Half-wave rectifier; (b) normal output; (c) distorted output with stored charge.

5-15 THE SCHOTTKY DIODE

A Schottky diode uses a metal such as gold, silver, or platinum on one side of the junction and doped silicon (typically *n*-type) on the other side. When a Schottky diode is unbiased, free electrons on the *n* side are in smaller orbits than free electrons on the metal side. This difference in orbit size is called the *Schottky barrier*. When the diode is forward-biased, free electrons on the *n* side can gain enough energy to travel in larger orbits. Because of this, free electrons can cross the junction and enter the metal, producing a large forward current. Since the metal has no holes, there is no charge storage and no reverse recovery time. The absence of charge storage means the reverse recovery time approaches zero. Because of this, a Schottky diode can switch off faster than an ordinary diode. When it is used in a circuit like Fig. 5-21a, the Schottky diode produces a perfect half-wave signal like Fig. 5-21b even at frequencies above 300 MHz.

5-16 VARACTOR CHARACTERISTICS

Varactors are silicon diodes optimized for their variable capacitance (Fig. 5-22a). Because the capacitance is voltage-controlled, varactors have replaced mechanically tuned capacitors in many applications such as television receivers and automobile radios. Data sheets for varactors list a reference value of capacitance measured at a specific reverse voltage, typically -4 V. For instance, the data sheet of a 1N5142 lists a reference capacitance of 15 pF at -4 V.

Besides the reference value of capacitance, data sheets give a tuning range and a voltage range. For example, along with the reference value of 15 pF, the data sheet of a 1N5142 shows a tuning range of 3:1 for a voltage range of -4 to -60 V. This means that the capacitance decreases from 15 to 5 pF when the voltage varies from -4 to -60 V.

The tuning range of a varactor depends on the doping level. For instance, Fig. 5-22b shows the doping profile for an abrupt-junction diode (the ordinary type of diode). Notice that the doping is uniform on both sides of the junction; this means that the number of holes and free electrons is equally distributed. The tuning range of an abrupt-junction diode is between 3:1 and 4:1.

To get larger tuning ranges, some varactors have a hyperabrupt junction, one whose doping profile looks like Fig. 5-22c. This profile tells us that the density of charges increases as we approach the junction. The heavier concentration leads to a narrower depletion layer and a larger capacitance. Furthermore, changes in reverse voltage have more pronounced effects on capacitance. A hyperabrupt varactor has a tuning range of about 10:1, enough to tune an AM radio through its frequency range (535 to 1605 kHz).

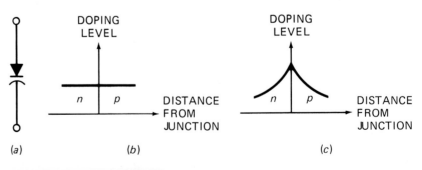

(a)   (b)   (c)

FIGURE 5-22   (a) Symbol; (b) abrupt doping profile; (c) hyperabrupt doping profile.

5-17   OTHER DIODES

Besides the special-purpose diodes discussed so far, there are a few others you should know about. Because these are so specialized, only a brief description follows. The intention is to make you aware of their existence, just in case you want to research them in greater detail.

Constant-Current Diodes

These are diodes that work in a way exactly opposite to zener diodes. Instead of holding the voltage constant, these diodes hold the current constant. Known as *constant-current diodes* (and also as *current-regulator diodes*), these devices keep the current through them fixed when the voltage changes. For example, the 1N5305 is a constant-current diode with a typical current of 2 mA over a voltage range of 2 to 100 V.

Step-Recovery Diodes

The step-recovery diode has an unusual doping profile because the density of carriers decreases near the junction. This unusual distribution of carriers causes a phenomenon called *reverse snap-off*. During the positive half-cycle, the diode conducts like any silicon diode. But during the negative half-cycle, reverse current exists for a while because of the stored charges, then suddenly drops to zero. The snap-off current of a step-recovery diode is rich in harmonics and can be filtered to produce a sine wave of a higher frequency. Because of this, step-recovery diodes are useful in frequency multipliers, circuits whose output frequency is a multiple of the input frequency.

Back Diodes

Zener diodes normally have breakdown voltages greater than 2 V. By increasing the doping level, we can get the zener effect to occur near zero. Forward conduction still occurs around $+0.7$ V, but now reverse conduction (breakdown) starts at approximately -0.1 V. A diode like this is called a *back diode* because it conducts better in the reverse than in the forward direction. Back diodes are occasionally used to rectify weak signals whose peak amplitudes are between 0.1 and 0.7 V.

Tunnel Diodes

By increasing the doping level of a back diode, we can get breakdown to occur at 0 V. Diodes like this are called *tunnel diodes*. This type of diode exhibits a phenomenon known as *negative resistance*. This means that an increase in forward voltage produces a decrease in forward current, at least over part of the forward curve. The negative resistance of tunnel diodes is useful in high-frequency circuits called *oscillators*. These circuits are able to convert dc power to ac power because they create a sinusoidal signal.

STUDY AIDS

The following study aids will help to reinforce the ideas discussed in this chapter. For best results, use these study aids within 6 hours of reading the earlier material. Then review these study aids a week later and a month later to ensure that the concepts remain in your long-term memory.

SUMMARY

Sec. 5-1 The Zener Diode
This is a special diode optimized for operation in the breakdown region. Its main use is in voltage regulators, circuits that hold the load voltage constant. Ideally, a zener diode is like a perfect battery. To a second approximation, it has a bulk resistance that produces a small additional voltage.

Sec. 5-2 The Loaded Zener Regulator
When a zener diode is in parallel with a load resistor, the current through the current-limiting resistor equals the sum of the zener current and the load current. The process for analyzing a zener regulator consists of finding the series current, load current, and zener current (in that order).

Sec. 5-3 Optoelectronic Devices
The LED is widely used as an indicator on instruments, calculators, and other electronic equipment. By combining seven LEDs in a package, we get a seven-segment indicator. Another important optoelectronic device is the optocoupler, which allows us to couple a signal between two isolated circuits.

Sec. 5-4 The Schottky Diode
The reverse recovery time is the time it takes a diode to shut off after it is suddenly switched from forward to reverse bias. This time may only be a few nanoseconds, but it places a limit on how high the frequency can be in a rectifier circuit. The Schottky diode is a special diode with almost zero reverse recovery time. Because of this, the Schottky diode is useful at high frequencies where short switching times are needed.

Sec. 5-5 The Varactor
The width of the depletion layer increases with the reverse voltage. This is why the capacitance of a varactor can be controlled by the reverse voltage. This leads to remote tuning of radio and television sets.

Sec. 5-6 Varistors
These protective devices are used across the primary winding of a transformer to prevent voltage spikes from damaging or otherwise polluting the input voltage to the equipment.

Sec. 5-7 Reading a Data Sheet
The most important quantities on the data sheet of zener diodes are the zener voltage, the maximum power rating, the maximum current rating, and the tolerance. Designers also need the zener resistance, the derating factor, and a few other items.

Sec. 5-8 Troubleshooting
Troubleshooting is an art and a science. Because of this, you can only learn so much from a book. The rest has to be learned from direct experience with circuits in trouble. Because troubleshooting is an art, you have to ask What if? often and feel your way to a solution.

VOCABULARY

In your own words, explain what each of the following terms mean. Keep your answers short and to the point. If necessary, verify your answer by rereading the appropriate discussion or by looking at the end-of-book Glossary.

light-emitting
  diode (LED)

open

optocoupler

photodiode

process

Schottky diode

short

temperature coefficient

varactor

varistor

voltage regulator

zener resistance

zener voltage

IMPORTANT EQUATIONS

The following formulas are useless if you don't know what they mean in words. *Suggestion:* Look at each formula, then read the words to find out what it means. Your chances of learning and remembering are much better if you concentrate on words rather than formulas.

Eq. 5-1 Current through Series Resistor

$$I_S = \frac{V_S - V_Z}{R_S}$$

This is an equation that you do not have to memorize. It says the current through the series resistor equals the voltage across the series resistor divided by the resistance. It is another example of Ohm's law, where the voltage is the difference of the node voltages on the ends of a resistor.

Eq. 5-2 Thevenin Voltage

$$V_{TH} = \frac{R_L}{R_S + R_L} V_S$$

This is the voltage across the load resistor when the zener diode is disconnected. One way to remember it is this: V_S divided by $R_S + R_L$ is the load current. Multiply this load current by R_L and you get V_{TH}. The value of V_{TH} has to be larger than the zener voltage to get voltage regulation.

Eq. 5-6 Zener Current

$$I_Z = I_S - I_L$$

This is a disguised form of Kirchhoff's current law. It says the zener current equals the difference between the series current and the load current. To use it, you must already have carried out the two preceding steps in the process: 1. Find I_S 2. Find I_L.

Eq. 5-7 Zener Power

$$P_Z = V_Z I_Z$$

The zener power equals the zener voltage times the zener current. This power has to be less than the maximum power rating listed on the data sheet. Otherwise, you may burn out or seriously degrade the characteristics of the zener diode.

Eq. 5-8 LED Current

$$I_S = \frac{V_S - V_D}{R_S}$$

This gives you the current through a resistor in series with a LED. It says the current equals the voltage across the series resistor divided by the resistance. Use 2 V for the value of V_D, unless you have a more accurate value for the voltage across the LED.

STUDENT ASSIGNMENTS

QUESTIONS

The following may have more than one right answer. Select the best answer. This is the one that is always true, or covers more situations, etc.

1. What is true about the breakdown voltage in a zener diode?
   a. It decreases when current increases.
   b. It destroys the diode.
   c. It equals the current times the resistance.
   d. It is approximately constant.

2. Which of these is the best description of a zener diode?
   a. It is a diode.
   b. It is a constant-voltage device.
   c. It is a constant-current device.
   d. It works in the forward region.

3. A zener diode
   a. Is a battery
   b. Acts like a battery in the breakdown region
   c. Has a barrier potential of 1 V
   d. Is forward-biased

4. The voltage across the zener resistance is usually
   a. Small
   b. Large
   c. Measured in volts
   d. Subtracted from the breakdown voltage

5. If the series resistance decreases in an unloaded zener regulator, the zener current
   a. Decreases
   b. Stays the same
   c. Increases
   d. Equals the voltage divided by the resistance

6. In the second approximation, the total voltage across the zener diode is the sum of the breakdown voltage and the voltage across the
   a. Source
   b. Series resistor
   c. Zener resistance
   d. Zener diode

7. The load voltage is approximately constant when a zener diode is
   a. Forward-biased
   b. Reverse-biased
   c. Operating in the breakdown region
   d. Unbiased

8. In a loaded zener regulator, which is the largest current?
   a. Series current
   b. Zener current
   c. Load current
   d. None of these

9. If the load resistance decreases in a zener regulator, the zener current
   a. Decreases
   b. Stays the same
   c. Increases
   d. Equals the source voltage divided by the series resistance

10. If the load resistance decreases in a zener regulator, the series current
    a. Decreases
    b. Stays the same
    c. Increases
    d. Equals the source voltage divided by the series resistance

11. When the source voltage increases in a zener regulator, which of these currents remains approximately constant?
    a. Series current
    b. Zener current
    c. Load current
    d. Total current

12. If the zener diode in a zener regulator is connected with the wrong polarity, the load voltage will be closest to
    a. 0.7 V
    b. 10 V
    c. 14 V
    d. 18 V

13. At high frequencies, ordinary diodes don't work properly because of
    a. Forward bias
    b. Reverse bias
    c. Breakdown
    d. Charge storage

14. The capacitance of a varactor diode increases when the reverse voltage across it
    a. Decreases
    b. Increases
    c. Breaks down
    d. Stores charges

15. Breakdown does not destroy a zener diode, provided the zener current is less than the
    a. Breakdown voltage
    b. Zener test current
    c. Maximum zener current rating
    d. Barrier potential

16. To display the digit 8 in a seven-segment indicator,
    a. C must be lighted
    b. G must be off
    c. F must be on
    d. All segments must be lighted

17. A photodiode is normally
    a. Forward-biased
    b. Reverse-biased
    c. Neither forward- nor reverse-biased
    d. Emitting light

18. When the light increases, the reverse minority-carrier current in a photodiode
    a. Decreases
    b. Increases
    c. Is unaffected
    d. Reverses direction

19. The device associated with voltage-controlled capacitance is a
    a. LED
    b. Photodiode
    c. Varactor diode
    d. Zener diode

20. If the depletion layer gets wider, the capacitance
    a. Decreases
    b. Stays the same
    c. Increases
    d. Is variable

21. When the reverse voltage increases, the capacitance
    a. Decreases
    b. Stays the same
    c. Increases
    d. Has more bandwidth

22. The varactor is usually
    a. Forward-biased
    b. Reverse-biased
    c. Unbiased
    d. In the breakdown region

BASIC PROBLEMS

Sec. 5-1 The Zener Diode
5-1. An unloaded zener regulator has a source voltage of 20 V, a series resistance of 330 Ω, and a zener voltage of 12 V. What is the zener current?

5-2. If the source voltage in Prob. 5-1 varies from 20 to 40 V, what is the maximum zener current?

5-3. If the series resistor of Prob. 5-1 has a tolerance of ± 10 percent, what is the maximum zener current?

Sec. 5-2 The Loaded Zener Regulator
5-4. If the zener diode is disconnected in Fig. 5-23, what is the load voltage?

5-5. Assume the supply voltage of Fig. 5-23 decreases from 20 to 0 V. At some point along the way, the zener diode will stop regulating. Find the supply voltage where regulation is lost.

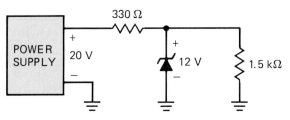

FIGURE 5-23

5-6. Calculate all three currents in Fig. 5-23.

5-7. Assuming a tolerance of ± 10 percent in both resistors of Fig. 5-23, what is the maximum zener current?

5-8. Suppose the supply voltage of Fig. 5-23 can vary from 20 to 40 V. What is the maximum zener current?

5-9. What is the power dissipation in the resistors and zener diode of Fig. 5-23?

5-10. The zener diode of Fig. 5-23 is replaced with a 1N961. What are the load voltage and the zener current?

5-11. The zener diode of Fig. 5-23 has a zener resistance of 11.5 Ω. If the power supply has a ripple of 1 V, what is the ripple across the load resistor?

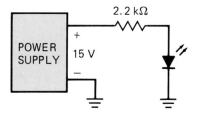

FIGURE 5-24

5-12. Draw the schematic diagram of a zener regulator with a supply voltage of 25 V, a series resistance of 470 Ω, a zener voltage of 15 V, and a load resistance of 1 kΩ. What are the load voltage and the zener current?

Sec. 5-3 Optoelectronic Devices
5-13. What is the current through the LED of Fig. 5-24?

5-14. If the supply voltage of Fig. 5-24 increases to 40 V, what is the LED current?

5-15. If the resistor is decreased to 1 kΩ, what is the LED current in Fig. 5-24?

5-16. The resistor of Fig. 5-24 is decreased until the LED current equals 13 mA. What is the value of the resistance?

Sec. 5-7 Reading a Data Sheet

5-17. A zener diode has a voltage of 10 V and a current of 20 mA. What is the power dissipation?

5-18. A 1N968 has 5 mA through it. What is the power?

5-19. The zener diode of Fig. 5-23 is a 1N963B. What is the minimum zener voltage? The maximum?

5-20. What is the maximum current rating of a 1N758? Find two answers. First, divide the maximum power rating of 400 mW by the zener voltage. Second, take the average of the two maximum currents listed on the data sheet.

Sec. 5-8 Troubleshooting

5-21. In Fig. 5-23, what is the load voltage for each of these conditions?
  a. Zener diode shorted
  b. Zener diode open
  c. Series resistor open
  d. Load resistor shorted

5-22. If you measure approximately 16.4 V for the load voltage of Fig. 5-23, what do you think the trouble is?

5-23. You measure 20 V across the load of Fig. 5-23. An ohmmeter indicates the zener diode is open. Before replacing the zener diode, what should you check for?

5-24. In Fig. 5-25, the LED does not light. Which of the following are possible troubles?
  a. V130LA2 is open.

  b. Ground between two left bridge diodes is open.
  c. Filter capacitor is open.
  d. Filter capacitor is shorted.
  e. Load resistor is open.
  f. Load resistor is shorted.

ADVANCED PROBLEMS

5-25. The zener diode of Fig. 5-23 has a zener resistance of 11.5 Ω. What is the load voltage if you include R_Z in your calculations?

5-26. The zener diode of Fig. 5-23 is a 1N963. If the load resistance varies from 1 to 10 kΩ, what is the minimum load voltage? The maximum load voltage? (Use the second approximation.)

5-27. Design a zener regulator to meet these specifications: load voltage is 6.8 V, source voltage is 20 V, and load current is 30 mA.

5-28. A TIL312 is a seven-segment indicator. Each segment has a voltage drop between 1.5 and 2 V at 20 mA. The supply voltage is +5 V. Design a seven-segment display circuit controlled by on-off switches that has a maximum current drain of 140 mA.

5-29. The secondary voltage of Fig. 5-25 is 12.6 V rms when the line voltage is 115 V rms. During the day the power line varies ±10 percent. The resistors have tolerances of ±5 percent. The 1N753 has a tolerance of ±10 percent and a zener resistance of 7 Ω. If R_2 equals 560 Ω, what is the maximum possible value of the zener current at any instant during day?

5-30. In Fig. 5-25, the secondary voltage is 12.6 V rms, and diode drops are 0.7 V

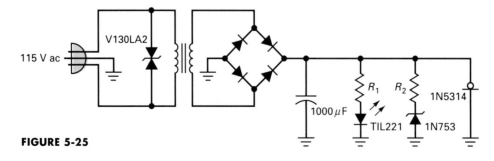

FIGURE 5-25

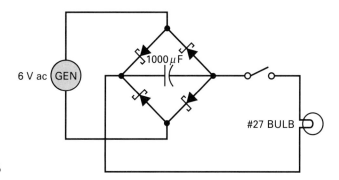

FIGURE 5-26

each. The 1N5314 is a constant-current diode with a current of 4.7 mA. The LED current is 15.6 mA, and the zener current is 21.7 mA. The filter capacitor has a tolerance of ±20 percent. What is the maximum peak-to-peak ripple?

5-31. Figure 5-26 shows part of a bicycle lighting system. The diodes are Schottky diodes. Use the second approximation to calculate the voltage across the filter capacitor.

T-SHOOTER PROBLEMS

Use Fig. 5-27 for the remaining problems. If you haven't already done so, read Example 4-11 before

attempting these problems. The possible troubles are open or shorted components, open leads, open grounds, etc. The box labeled OK gives measurements for the zener regulator when it is trouble-free. The remaining boxes $T1$ through $T8$ are the measurements for different troubles. The big box "Measurements" converts the tokens to measured values.

5-32. Find Trouble 1.

5-33. Find Trouble 2.

5-34. Find Troubles 3 and 4.

5-35. Find Troubles 5 and 6.

5-36. Find Troubles 7 and 8.

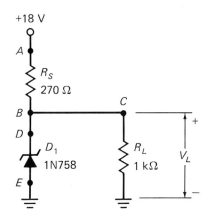

	1	2	3	4	5	6	7
A	14.2	OK	10.5	0	18	14.2	OK
B	18	0	OK	14.2	0	10.5	18
C	0	18	14.2	∞	10.3	0	14.2
D	18	14.2	OK	18	OK	OK	18
E	OK	18	14.2	0	14.2	10.3	0
F	10.3	0	0	0	18	0	10
G	18	0	10.5	14.2	0	18	OK

MEASUREMENTS

OK	T1	T2	T3	T4	T5	T6	T7	T8
V_A : C2	V_A : D7	V_A : A5	V_A : D1	V_A : E2	V_A : B2	V_A : B1	V_A : B7	V_A : G6
V_B : F1	V_B : F4	V_B : B4	V_B : A6	V_B : D4	V_B : C6	V_B : A3	V_B : C3	V_B : A4
V_C : C5	V_C : C1	V_C : E3	V_C : G4	V_C : F5	V_C : B5	V_C : B6	V_C : C7	V_C : F3
V_D : E6	V_D : F2	V_D : F2	V_D : D2	V_D : G1	V_D : F6	V_D : G3	V_D : E5	V_D : G2
D_1 : A7	D_1 : B3	D_1 : A2	D_1 : C4	D_1 : C4	D_1 : G7	D_1 : E1	D_1 : D6	D_1 : E7

FIGURE 5-27 T-Shooter™. (*Patent pending: Courtesy of Malvino Inc.*)

6

BIPOLAR TRANSISTORS

The radio or TV signal received by an antenna is so weak that it cannot adequately drive a loudspeaker or TV tube. This is why we have to amplify the weak signal until it has enough power to be useful. Before 1951, the vacuum tube was the main device used for amplifying weak signals. Although it amplifies quite well, the vacuum tube has several disadvantages. First, it has an internal filament or heater which requires 1 W or more of power. Second, it only lasts a few thousand hours before its filament burns out. Third, it takes up a lot of space. Fourth, it gives off heat that raises the internal temperature of the electronics equipment.

In 1951, Shockley invented the first junction transistor, a semiconductor device that can amplify radio and TV signals. The advantages of a transistor overcome the disadvantages of the vacuum tube. First, a transistor has no filament or heater; therefore, it requires much less power. Second, since it is a semiconductor device, the transistor can last indefinitely. Third, since it is so small, it takes up much less space. Fourth, since the transistor produces much less heat, electronics equipment can run at lower internal temperatures.

The transistor has led to many other inventions including the integrated circuit (IC), a small device that contains thousands of transistors. Because of the IC, modern computers and other electronic miracles are possible. This chapter discusses bipolar transistors, the kind that use both free electrons and holes. (The word *bipolar* is an abbreviation for two polarities.)

6-1 THE UNBIASED TRANSISTOR

A transistor has three doped regions as shown in Fig. 6-1. The bottom region is called the *emitter*, the middle region is the *base*, and the top region is the *collector*. This particular transistor is an *npn* device. Transistors are also manufactured as *pnp* devices. In this chapter, we emphasize *npn* transistors. Later chapters include discussions of *pnp* transistors.

Emitter and Collector Diodes

The transistor of Fig. 6-1 has two junctions: one between the emitter and the base and another between the base and the collector. Because of this, a transistor is similar to two diodes. The emitter and the base form one of the diodes, while the collector and the base form the other diode.

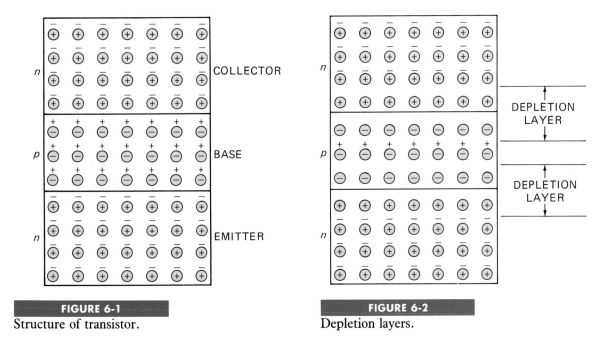

FIGURE 6-1
Structure of transistor.

FIGURE 6-2
Depletion layers.

From now on, we refer to these diodes as the *emitter diode* (the lower one) and the *collector diode* (the upper one).

Before and After Diffusion

Figure 6-1 shows the transistor regions before diffusion has occurred. As discussed in Chap. 2, free electrons in the *n* region diffuse across the junction and recombine with the holes in the *p* region. Visualize the free electrons in each *n* region crossing the junction and recombining with holes. The result is two depletion layers as shown in Fig. 6-2. For each of these depletion layers, the barrier potential is approximately 0.7 V at 25°C. As before, we emphasize silicon devices because they are more widely used than germanium devices.

6-2 THE BIASED TRANSISTOR

An unbiased transistor is like two back-to-back diodes. Each diode has a barrier potential of approximately 0.7 V. When you connect external voltage sources to the transistor, you will get some new and unexpected results. These results are what make the transistor the great invention that it is. Let's take a look at how a transistor works.

Emitter Electrons

Figure 6-3 shows a biased transistor. The minus signs represent free electrons. Not shown but still present are the ions in each region. (The drawing gets too cluttered if we try to include them.) In Fig. 6-3, the emitter is heavily doped; its job is to emit or inject free electrons into the

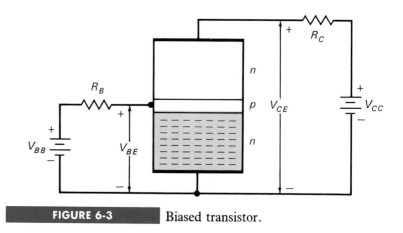

FIGURE 6-3   Biased transistor.

base. The base is lightly doped and very thin; it passes most of the emitter-injected electrons on to the collector. The doping level of the collector is between the heavy doping of the emitter and the light doping of the base. The collector is so named because it collects or gathers electrons from the base.

The left source of Fig. 6-3 forward-biases the emitter diode, while the right source reverse-biases the collector diode. At the instant that forward bias is applied to the emitter diode, electrons in the emitter have not yet entered the base region.

Base Electrons

If V_{BB} is greater than the barrier potential, emitter electrons will enter the base region, as shown in Fig. 6-4. These free electrons can flow in either of two directions. First, they can flow to the left and out the base, passing through R_B on the way to the positive source terminal. Second, the free electrons can flow into the collector.

Which way do most of the free electrons go? Most will continue on to the collector. Why? There are two reasons. First, the base is lightly doped. Because of this, free electrons have a long lifetime in the base

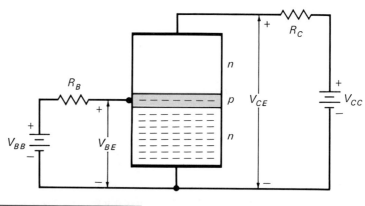

FIGURE 6-4   Electrons enter base.

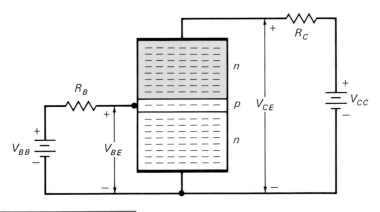

FIGURE 6-5   Electrons enter collector.

region. This gives them the time needed to reach the collector. Second, the base is very thin. This also gives the electrons a better chance of reaching the collector.

In other words, to flow out of the base into the external resistor, the free electrons must first recombine with holes in the base. Then, as valence electrons, they can flow to the left until they leave the base and enter the external connecting wire. Since the base is lightly doped and very thin, very few electrons manage to recombine and escape into the external base lead.

Collector Electrons

Almost all the free electrons go into the collector as shown in Fig. 6-5. Once they are in the collector, they feel the attraction of the V_{CC} source voltage. Because of this, the free electrons flow through the collector and through R_C until they reach the positive terminal of the collector supply voltage.

Here's a summary of what's going on. In Fig. 6-5, V_{BB} forward-biases the emitter diode, forcing the free electrons in the emitter to enter the base. The thin and lightly doped base gives almost all these electrons enough time to diffuse into the collector. These electrons flow through the collector, through R_C, and into the positive terminal of the V_{CC} voltage source. In most transistors, more than 95 percent of the emitter electrons flow to the collector; less than 5 percent flow out the external base lead.

6-3 TRANSISTOR CURRENTS

Figure 6-6 shows the schematic symbol for a transistor. (If you prefer conventional flow, use Fig. 6-6a. If you prefer electron flow, use Fig. 6-6b.) In Fig. 6-6, there are three different currents in a transistor: emitter current I_E, base current I_B, and collector current I_C. Because the emitter is the source of the electrons, it has the largest of the three currents. Almost all the emitter electrons flow into the collector; therefore, the collector current approximately equals the emitter current. The base

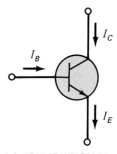

(a) CONVENTIONAL

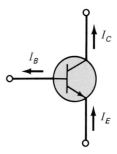

(b) ELECTRON

FIGURE 6-6

Schematic symbol of transistor.

current is very small compared to the two other currents. In low-power transistors, the base current is usually less than 1 percent of the collector current.

Recall Kirchhoff's current law. It says the sum of all currents into a point or junction equals the sum of all currents out of the point or junction. When applied to a transistor, Kirchhoff's current law gives us this important relation among the three transistor currents:

$$I_E = I_C + I_B \tag{6-1}$$

This says the emitter current is the sum of the collector current and the base current. Since the base current is much smaller than the collector current, the equation is screaming this idea at us: To a close approximation, the collector current is equal to the emitter current.

One thing that makes the transistor useful is that the collector current is much larger than the base current. In fact, the current gain β_{dc} of a transistor is defined as the collector current divided by the base current. In symbols, it looks like this:

$$\beta_{dc} = \frac{I_C}{I_B} \tag{6-2}$$

This is the kind of equation that you will use hundreds of times, so you should memorize it. For low-power transistors, the current gain is typically 100 to 300. Even high-power transistors have current gains of 20 to 100. In general, this means that 95 percent or more of the emitter electrons pass on to the collector, while less than 5 percent escape through the base lead.

Equation (6-2) may be rearranged into two equivalent forms. First, when you know the value of β_{dc} and I_B, you can calculate the collector current with this equation:

$$I_C = \beta_{dc}I_B \tag{6-3}$$

To get this, you multiply both sides of Eq. (6-2) by I_B. Second, when you have the value of the collector current and β_{dc}, you can calculate the base current with

$$I_B = \frac{I_C}{\beta_{dc}} \tag{6-4}$$

You get this by multiplying both sides of Eq. (6-2) by I_B and then dividing both sides by β_{dc}. Remember what to remember. Remember Eq. (6-2), and be able to derive Eqs. (6-3) and (6-4) with algebra. All three equations are important in the analysis and design of transistor circuits.

EXAMPLE 6-1

In Fig. 6-6, the transistor has a collector current of 10 mA and a base current of 40 μA. What is the current gain of the transistor?

SOLUTION

Divide the collector current by the base current to get

$$\beta_{dc} = \frac{10 \text{ mA}}{40 \text{ μA}} = 250$$

EXAMPLE 6-2

The transistor has a current gain of 175 in Fig. 6-6. If the base current is 0.1 mA, what is the collector current?

SOLUTION

Multiply the current gain by the base current to get

$$I_C = 175(0.1 \text{ mA}) = 17.5 \text{ mA}$$

EXAMPLE 6-3

The transistor of Fig. 6-6 has a collector current of 2 mA. If the current gain is 135, what is the base current?

SOLUTION

Divide the collector current by the current gain to get

$$\beta_{dc} = \frac{2 \text{ mA}}{135} = 14.8 \text{ μA}$$

6-4 THE CE CONNECTION

In Fig. 6-7, the common or ground side of each voltage source is connected to the emitter. Because of this, the circuit is referred to as a *common-emitter* (CE) connection. Notice that the circuit has two loops. The left loop is called the *base circuit*, and the right loop is called the *collector*

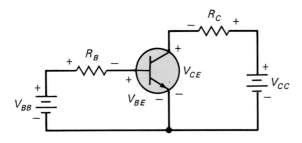

FIGURE 6-7   *CE* connection.

circuit. As you will see, the base circuit controls the collector circuit. This is what makes a transistor useful.

In the base circuit, there is a supply voltage of V_{BB} and a current-limiting resistance of R_B. The voltage between the base and the emitter is symbolized as V_{BE}. As a guideline, V_{BB} must be greater than V_{BE} to turn on the emitter diode. Typically, V_{BB} is in the range of 5 to 15 V for most low-power applications. By using different values of V_{BB} and/or R_B, we can control the base current. As you will see later, the base current controls the collector current. Therefore, any change in the base current produces a change in the collector current. This leads to applications in remote control, amplification, etc.

In the collector circuit, there is a supply voltage of V_{CC} and a current-limiting resistance of R_C. The voltage between the collector and the emitter is symbolized V_{CE}. The supply voltage V_{CC} must reverse-bias the collector diode, or else the transistor won't work properly. This is usually satisfied when V_{CE} is greater than 1 V. A typical range of V_{CE} is 1 to 15 V in low-power circuits.

One way to summarize the operation of a transistor is with graphs that show its currents and voltages. These graphs are more complicated than those of a diode because a transistor has three currents (I_E, I_C, and I_B) and two voltages (V_{BE} and V_{CE}). As it turns out, we need two graphs to summarize the CE circuit: a graph of base current versus base voltage and a graph of collector current versus collector voltage.

6-5 THE BASE CURVE

Let us start with the base current in Fig. 6-7. What do you think the graph of I_B versus V_{BE} looks like? It looks like the graph of an ordinary rectifier diode as shown in Fig. 6-8. And why not? We are talking about the current and voltage of the emitter diode, so we would expect to see a diode curve of current versus voltage. What this means is that we can use any of the three diode approximations discussed earlier.

For instance, if you are troubleshooting a transistor circuit, you can treat the base-emitter part of the transistor as an ideal diode. This will allow you to quickly estimate the currents and voltages. But if you are involved in a precise design, you may want to include the bulk resistance of the emitter diode in your calculations.

Most of the time, whether you are troubleshooting or designing, you will find that the second approximation is the best compromise between speed of the ideal diode and accuracy of the third approximation. All you need to remember for the second approximation is that V_{BE} is 0.7 V as shown in Fig. 6-8.

Here is how to calculate the base current in Fig. 6-7. Focus your attention on the base circuit. It consists of a voltage source V_{BB}, a resistor of R_B, and V_{BE} across the base-emitter part of the transistor. The voltage across the base resistor equals the difference between the source voltage V_{BB} and the base-emitter voltage V_{BE}. Apply Ohm's law to the base resistor to find the base current:

$$I_B = \frac{V_{BB} - V_{BE}}{R_B} \tag{6-5}$$

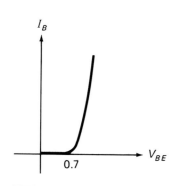

FIGURE 6-8

Diode curve.

Don't memorize this equation. It's not necessary. If you can remember how to get the voltage across the base resistor, then Ohm's law takes care of the rest. As always, the voltage across a resistor is the difference between the node voltages on each end of the resistor.

Just one word of caution. Because of the barrier potential of the emitter diode, the base supply voltage has to be greater than 0.7 V for silicon transistors. If this condition is not satisfied, the emitter diode is nonconducting and the base current is zero. In other words, if V_{BB} is less than V_{BE}, the numerator is negative, which means you cannot use the equation. In most transistor circuits, however, V_{BB} will be greater than V_{BE}, which gives a positive numerator. In this case, Ohm's law applies and you get a base current that is greater than zero.

Incidentally, voltages with single subscripts (V_C, V_E, V_B) refer to the voltage of a transistor terminal with respect to ground. Double subscripts (V_{BE}, V_{CE}, V_{CB}) refer to the voltage between two transistor terminals. You can calculate a double-subscript voltage by subtracting the corresponding single-subscript voltages. For instance, to get V_{CE}, subtract V_E from V_C.

$$V_{CE} = V_C - V_E$$

To get V_{CB}, subtract V_B from V_C:

$$V_{CB} = V_C - V_B$$

To get V_{BE}, subtract V_E from V_B:

$$V_{BE} = V_B - V_E$$

EXAMPLE 6-4

In Fig. 6-7, $V_{BB} = 10$ V and $R_B = 100$ kΩ. What is the base current?

SOLUTION

Unless otherwise stated, you can always assume a silicon transistor and the second approximation. This means you can use a V_{BE} of 0.7 V. The base resistor has a node voltage of 10 V on the left end and 0.7 V on the right end. Therefore, it has 9.3 V across it. This gives a base current of

$$I_B = \frac{9.3 \text{ V}}{100 \text{ k}\Omega} = 93 \text{ } \mu\text{A}$$

6-6 COLLECTOR CURVES

You can vary V_{BB} and V_{CC} in Fig. 6-7 to set up different voltages and currents. By measuring I_C and V_{CE}, you can get data for a graph of I_C versus V_{CE}. For instance, suppose you set I_B to 10 μA. Then you can

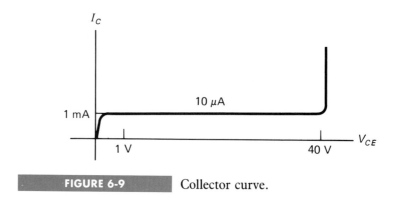

FIGURE 6-9   Collector curve.

vary V_{CC} and measure the resulting I_C and V_{CE}. Plotting the data gives the graph shown in Fig. 6-9.

There is nothing mysterious about this graph. It is nothing more than a compact summary of many ideas discussed earlier. When V_{CE} is zero, the collector diode is not reverse-biased; therefore, the collector current is zero. For V_{CE} between 0 and approximately 1 V, the collector current rises sharply and then becomes almost constant. This is tied in with the idea of reverse-biasing the collector diode. It takes approximately 0.7 V to reverse-bias the collector diode. Once you reach this level, the collector is gathering all electrons that reach its depletion layer.

Above 0.7 V, the exact value of V_{CE} is not too important because even a little bit of reverse bias is enough to collect all the available electrons from the base. This is why the graph becomes flat or horizontal when V_{CE} is greater than 1 V or so. The horizontal graph means that the collector current is constant and equal to approximately 1 mA for any collector voltage between 1 and 40 V. (Incidentally, this graph is for a 2N3904, a widely used low-power transistor.)

If V_{CE} is greater than 40 V, the collector diode breaks down and normal transistor action is lost. The transistor is not intended to operate in the breakdown region. For this reason, one of the maximum ratings to look for on a transistor data sheet is the collector-emitter breakdown voltage.

Collector Voltage and Power

Kirchhoff's voltage law says the sum of all voltages around a loop or closed path is equal to zero. When applied to the collector circuit of Fig. 6-7, on page 163, Kirchhoff's voltage law gives us this important equation:

$$V_{CE} = V_{CC} - I_C R_C \qquad (6\text{-}6)$$

This says the collector-emitter voltage equals the collector supply voltage minus the voltage across the collector resistor. Memorize either the equation or the basic process used to find the collector-emitter voltage. This equation is essential in troubleshooting and design.

In Fig. 6-7, the transistor has a power dissipation of approximately

$$P_D = V_{CE} I_C \qquad (6\text{-}7)$$

This says the transistor power equals the collector-emitter voltage times the collector current. This power is what causes the junction temperature of the collector diode to increase. The higher the power, the higher the junction temperature. Transistors will burn out when the junction temperature is between 150 and 200°C. One of the most important pieces of information on a data sheet is the maximum power rating $P_{D(\text{max})}$. The power dissipation given by Eq. (6-7) must be less than $P_{D(\text{max})}$ to avoid destroying the transistor.

Three Regions of Operation

The curve of Fig. 6-9 has three regions or areas where the operation of the transistor is different. First, there is the region in the middle where V_{CE} is between approximately 1 and 40 V. This is the most important region because it represents normal operation of the transistor. In this region, the emitter diode is forward-biased, and the collector diode is reverse-biased. Furthermore, the collector is gathering almost all the electrons that the emitter has sent into the base. This is why changes in collector voltage have no effect on the collector current. This region is called the *active region*. Graphically, the active region is the horizontal part of the curve.

Another distinct region of operation is the *breakdown region*. The transistor should never operate in this region because it very likely will be destroyed or degraded. Unlike the zener diode which is optimized for breakdown operation, a transistor is not intended for operation in the breakdown region.

Finally, there is the rising part of the curve, where V_{CE} is between 0 and approximately 1 V. This sloping part of the curve is called the *saturation region*. In this region, the collector diode is not reverse-biased. (*Note:* With low-power transistors, the curve may become horizontal well below 1 V. For instance, the curve of a 2N3904 reaches 1 mA at only 0.3 V.)

In summary, the curve of Fig. 6-9 has a saturation region, an active region, and a breakdown region. A transistor can operate safely in either the saturation region or the active region, but not the breakdown region. In applications where the transistor amplifies weak radio and TV signals, it will always be operating in the active region.

More Curves

If we measure I_C and V_{CE} for $I_B = 20 \ \mu\text{A}$, we can plot the second curve of Fig. 6-10. The curve is similar to the first curve, except that the collector current is 2 mA in the active region. Again, the collector current is essentially constant in the active region.

When we plot several curves on a piece of rectangular coordinate paper, we get a set of collector curves like Fig. 6-10. Another way to get this set of curves is with a curve tracer (a test instrument with a video display). In the active region of Fig. 6-10, each collector current is 100 times greater than the corresponding base current. For instance, the top curve has a collector current of 7 mA and a base current of 70 μA. This gives a current gain of

$$\beta_{dc} = \frac{7 \text{ mA}}{70 \ \mu\text{A}} = 100$$

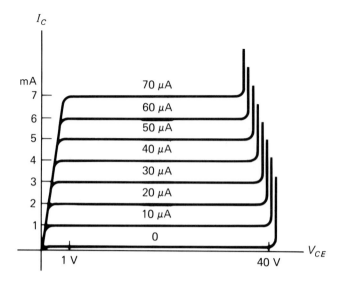

FIGURE 6-10   Set of collector curves.

If you check any other curve, you get the same result: a current gain of 100.

With other transistors, the current gain may be different from 100, but the shape of the curves will be the same. All transistors have an active region, a saturation region, and a breakdown region. The active region is the most important because amplification is possible in the active region.

Cutoff Region

Figure 6-10 has an unexpected curve, the one on the bottom. Notice that the base current is zero, but there still is some collector current. On a curve tracer, this current is usually so small that you cannot see it. We have exaggerated the bottom curve by drawing it larger than usual. This bottom curve is called the *cutoff region* of the transistor, and the small collector current is called the *collector cutoff current.*

Why does the collector cutoff current exist when there is no base current? Because the collector diode is a diode, and like any diode it has reverse minority-carrier current and surface-leakage current. You can ignore the collector cutoff current when the collector current is much larger. For instance, a 2N3904 has a collector cutoff current of 50 nA. Applying the 20-to-1 rule, you can ignore this 50 nA when the ordinary collector current is at least 20 times 50 nA, or 1 mA. This ensures a calculation error of less than 5 percent.

Recap

The transistor has four distinct operating regions: active, cutoff, saturation, and breakdown. Transistors operate in the active region when they are used as amplifiers, circuits that amplify weak signals. Sometimes, these circuits are called linear circuits because changes in the input signal produce proportional changes in the output signal. The saturation and cutoff regions are useful in digital circuits and other computer circuits.

EXAMPLE 6-5

What is the collector-emitter voltage in Fig. 6-7, on page 163, if the collector current is 1 mA, the collector resistance is 3.6 kΩ, and the collector supply voltage is 10 V?

SOLUTION

Subtract the IR drop across the collector resistor from the collector supply voltage as follows:

$$V_{CE} = 10 \text{ V} - (1 \text{ mA})(3.6 \text{ k}\Omega) = 6.4 \text{ V}$$

This type of calculation is useful in troubleshooting because you often have to measure the collector-emitter voltage of a suspicious circuit.

EXAMPLE 6-6

What is the transistor power in Example 6-5?

SOLUTION

$$P_D = (6.4 \text{ V})(1 \text{ mA}) = 6.4 \text{ mW}$$

This is the collector power, the product of the collector-emitter voltage and the collector current. This answer is quite accurate. The only error is not including the base power, the product of V_{BE} and I_B. Most designers ignore the base power because it is much smaller than the collector power.

6-7 TRANSISTOR APPROXIMATIONS

Transistors are complicated devices. Exact analysis is out of the question, unless you have access to a computer. So let us inject some reality into our discussion. There are many approximations for a transistor. To be effective as a troubleshooter, designer, etc., you have to develop an inner sense of which approximations to use. Without simple approximations, a troubleshooter could spend hours trying to locate a trouble in transistor circuits. Without advanced approximations, a designer would create inferior transistor circuits.

Exact Answers

The ultimate left-brain trap is our desire for exact answers. It originates with our early mathematical training, and it gets worse after we buy our first pocket calculator. But we have to remember that electronics is not an exact science. Because of its complicated and imperfect devices, electronics forces us to compromise as we fight our way to a solution. It forces us to use all our brain instead of just the left half.

The search for exact answers is an old one. Even the early Greeks learned to compromise in their search for truth. Aristotle had this to say about exact answers:

> It is the mark of an instructed mind to rest satisfied with that degree of precision which the nature of the subject admits, and not to seek exactness where only an approximation of the truth is possible.

So try to understand what this game of electronics is all about. It is not about getting exact answers. It is about getting the best answers for the job you are doing. The best answer is the one that takes the least time and is still accurate enough for the job at hand. This section will discuss three approximations: the ideal transistor, the second approximation, and the third approximation.

The Ideal Transistor

If a manufacturer could produce an ideal or a perfect transistor, here is what would happen to the curves of Fig. 6-10. First, there would be no breakdown region, which means you could apply as much voltage between the collector-emitter terminals as you like. Second, there would be no collector cutoff current, which means the collector current would be zero for $I_B = 0$. Third, there would be no saturation region, which means the active region would extend all the way down to $V_{CE} = 0$.

Figure 6-11 summarizes the foregoing improvements. That takes care of the collector curves. What about the base curve? The base-emitter part of a transistor is a diode. Ideally, this diode would have no barrier potential and no bulk resistance. It would be the ideal diode we discussed earlier. Therefore, it would act like an intelligent switch that is closed when forward-biased and open when reverse-biased.

Figure 6-12 summarizes the ideal transistor. The input side of an ideal transistor is an ideal diode. The output side is a current source. This

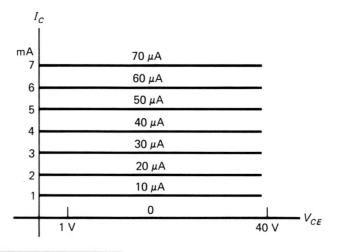

FIGURE 6-11 Ideal collector curves.

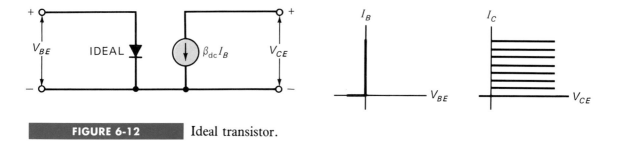

FIGURE 6-12   Ideal transistor.

current source produces a constant current equal to the current gain times the base current:

$$I_C = \beta_{dc}I_B$$

Using this equation is easy. In a transistor circuit, you calculate the base current with an ideal diode. Then you multiply by the current gain to get the collector current. Finally, you use this collector current to calculate the collector-emitter voltage. Examples of this process are given later.

If you have forgotten what a current source is, you should review Secs. 1-3 and 1-5. A current source is the opposite of a voltage source. A voltage source produces a voltage that cannot be changed by the load resistance. A current source is similar because it produces a current that cannot be changed by the load resistance. This will be clearer after you see some examples of transistor-circuit analysis.

The Second Approximation

When the supply voltage for the base circuit is at least 20 times the barrier potential, or about 14 V, the result is an error of less than 5 percent when you use an ideal diode. But the supply voltage is usually less than 14 V, which is why most people use the second approximation of the emitter diode. In other words, they include 0.7 V in the calculation for base current.

Figure 6-13 summarizes the second approximation of a transistor. The only difference between this and the ideal case is that we now approximate the emitter diode by the second approximation of a diode. That is, we assume there is a voltage drop of 0.7 V across the base-emitter part of the transistor. This is important when the supply voltage for the base is only a few volts.

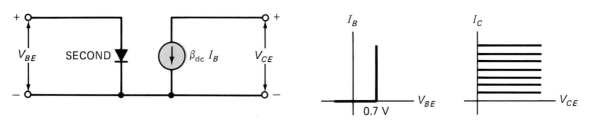

FIGURE 6-13   Second approximation.

The Third Approximation

For troubleshooting and analysis, all you have to know about the third approximation are these two ideas. First, the emitter diode has a bulk resistance which produces a voltage that has to be added to 0.7 V to get the total voltage across the base-emitter terminals. With low-power transistors, this additional voltage is so small that it has almost no effect on the value of V_{BE}. For high-power transistors, the additional voltage may be large enough to produce a V_{BE} greater than 1 V.

Second, the collector diode has a bulk resistance with a few tenths of a volt across it. You don't notice this small voltage unless the transistor is operating in the saturation region. Then, instead of a voltmeter indicating exactly zero, it may indicate 0.1 to 0.2 V with low-power transistors.

Unless you are involved in design work, you don't really need to get into the details of the third approximation. All you need to remember is this: V_{BE} may be larger than 0.7 V because of the emitter bulk resistance, and V_{CE} may be greater than zero when the transistor is saturated. To use the third approximation, you can measure V_{BE} in a built-up circuit and then use this value instead of 0.7 V when calculating the base current.

EXAMPLE 6-7

What is the collector-emitter voltage in Fig. 6-14? Use the ideal transistor.

SOLUTION

An ideal emitter diode means that

$$V_{BE} = 0$$

Therefore, the total voltage across R_B is 15 V. Ohm's law tells us the base current is

$$I_B = \frac{15 \text{ V}}{470 \text{ k}\Omega} = 31.9 \text{ }\mu\text{A}$$

Now we can calculate the collector current. It equals the current gain times the base current:

$$I_C = 100(31.9 \text{ }\mu\text{A}) = 3.19 \text{ mA}$$

Next, we calculate the collector-emitter voltage. It equals the collector supply voltage minus the voltage drop across the collector resistor:

$$V_{CE} = 15 \text{ V} - (3.19 \text{ mA})(3.6 \text{ k}\Omega) = 3.52 \text{ V}$$

In a circuit like Fig. 6-14, the emitter current is not important, so most people would not calculate this quantity. But since this is an example, we will calculate the emitter current. It equals the sum of the collector current and the base current:

$$I_E = 3.19 \text{ mA} + 31.9 \text{ }\mu\text{A} = 3.22 \text{ mA}$$

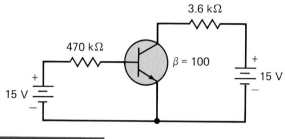

FIGURE 6-14   Example.

This value is extremely close to the value of the collector current, which is another reason for not bothering to calculate it. Most people would say the emitter current is approximately 3.19 mA, the value of the collector current.

In summary, the ideal transistor is an excellent tool for loosening up your thinking. Troubleshooters and designers find it very useful for much of their work. But it is only one of the approximations available for analysis. Sometimes, you get ridiculous answers with an ideal transistor. The next example shows you how to go from the ridiculous to the sublime.

EXAMPLE 6-8

What is the collector-emitter voltage in Fig. 6-14 if you use the second approximation?

SOLUTION

In Fig. 6-14, here is how you would calculate the currents and voltages, using the second approximation. The voltage across the emitter diode is

$$V_{BE} = 0.7 \text{ V}$$

Therefore, the total voltage across R_B is 14.3 V, the difference between 15 and 0.7 V. Ohm's law tells us the base current is

$$I_B = \frac{14.3 \text{ V}}{470 \text{ k}\Omega} = 30.4 \text{ }\mu\text{A}$$

Now we can calculate the collector current. It equals the current gain times the base current:

$$I_C = 100(30.4 \text{ }\mu\text{A}) = 3.04 \text{ mA}$$

Next, we calculate the collector-emitter voltage. It equals the collector supply voltage minus the voltage drop across the collector resistor:

$$V_{CE} = 15 \text{ V} - (3.04 \text{ mA})(3.6 \text{ k}\Omega) = 4.06 \text{ V}$$

The improvement in this answer over the ideal answer is about half a volt: 4.06 versus 3.52 V. Is this half of a volt important? It depends on whether you are troubleshooting, designing, etc. As you gain experience in everyday electronics work, you will begin to develop your own guidelines about which approximation to use. A lot will depend on what kind of work you are doing and the company you work for.

EXAMPLE 6-9

Suppose you measure a V_{BE} of 1 V. What is the collector-emitter voltage in Fig. 6-14 if you use the third approximation?

SOLUTION

The total voltage across R_B is 14 V, the difference between 15 and 1 V. Ohm's law tells us the base current is

$$I_B = \frac{14\text{ V}}{470\text{ k}\Omega} = 29.8\ \mu\text{A}$$

Now, we can calculate the collector current. It equals the current gain times the base current:

$$I_C = 100(29.8\ \mu\text{A}) = 2.98\text{ mA}$$

Next, we calculate the collector-emitter voltage. It equals the collector supply voltage minus the voltage drop across the collector resistor:

$$V_{CE} = 15\text{ V} - (2.98\text{ mA})(3.6\text{ k}\Omega) = 4.27\text{ V}$$

The third approximation is slightly better than the second. And the second was a little bit better than the ideal approximation. This is what you can expect when the base supply voltage is much larger than V_{BE}.

EXAMPLE 6-10

What is the collector-emitter voltage in the three preceding examples if the base supply voltage is 5 V?

SOLUTION

With the ideal diode,

$$I_B = \frac{5\text{ V}}{470\text{ k}\Omega} = 10.6\ \mu\text{A}$$

$$I_C = 100(10.6\ \mu\text{A}) = 1.06\text{ mA}$$

$$V_{CE} = 15\text{ V} - (1.06\text{ mA})(3.6\text{ k}\Omega) = 11.2\text{ V}$$

With the second approximation,

$$I_B = \frac{4.3\,\text{V}}{470\,\text{k}\Omega} = 9.15\,\mu\text{A}$$

$$I_C = 100(9.15\,\mu\text{A}) = 0.915\,\text{mA}$$

$$V_{CE} = 15\,\text{V} - (0.915\,\text{mA})(3.6\,\text{k}\Omega) = 11.7\,\text{V}$$

With the third approximation,

$$I_B = \frac{4\,\text{V}}{470\,\text{k}\Omega} = 8.51\,\mu\text{A}$$

$$I_C = 100(8.51\,\mu\text{A}) = 0.851\,\text{mA}$$

$$V_{CE} = 15\,\text{V} - (0.851\,\text{mA})(3.6\,\text{k}\Omega) = 11.9\,\text{V}$$

This example allows you to compare the three approximations for the case of low base supply voltage. As you can see, all answers are within a volt of each other. This is the first clue as to which approximation to use. If you are troubleshooting this circuit, the ideal analysis will probably be adequate. But if you are designing the circuit, you might want to use the third approximation because of its accuracy. If you are ever in doubt which approximation to use, use the second approximation. This is a good compromise for troubleshooting and design.

6-8 READING DATA SHEETS

Small-signal transistors can dissipate half a watt or less; *power transistors* can dissipate more than half a watt. When you look at a data sheet for either type of transistor, you should start with the maximum ratings because these are the limits on the transistor currents, voltages, and other quantities.

Breakdown Ratings

In the Appendix, the following maximum ratings of a 2N3904 are given:

V_{CB}	60 V
V_{CEO}	40 V
V_{EB}	6 V

These voltage ratings are reverse breakdown voltages, and V_{CB} is the voltage between the collector and the base. The second rating is V_{CEO}, which stands for the voltage from collector to emitter with the base open. Also V_{EB} is the voltage from the emitter to the base. As usual, a conservative design never allows voltages to get even close to the foregoing maximum ratings. If you recall, even getting close to maximum ratings can shorten the lifetime of some devices.

Maximum Current and Power

Also shown in the maximum ratings are these values:

I_C 200 mA dc
P_D 250 mW   (for $T_A = 60°C$)
P_D 350 mW   (for $T_A = 25°C$)
P_D 1 W   (for $T_C = 60°C$)

Here I_C is the maximum dc collector rating. This means a 2N3904 can handle up to 200 mA of direct current. The next three ratings are P_D, the maximum power rating of the device. As you can see, the maximum power that a transistor can handle depends on the temperature. If the temperature of the surrounding air is 60°C, the maximum power rating is 250 mW. This temperature is used because commercial equipment often has to operate over an ambient temperature range of 0 to 60°C. The data sheet is giving you the power dissipation for the worst case of 60°C.

If the ambient temperature is only 25°C, the power rating is 350 mW. The transistor has a higher power rating because the internal temperature is lower. It is the internal or junction temperature that determines when a transistor burns out. If the outside temperature is lower, the internal temperature is lower and further from the burnout point. In this case, the transistor can dissipate more power.

The transistor package or case has a temperature that is usually higher than the ambient temperature. If the designer knows what this case temperature is, she or he may prefer to work with the power rating given for the case temperature.

Derating Factors

As discussed in Chap. 5, the derating factor tells you how much you have to reduce the power rating of a device. The derating factor of the 2N3904 is given as 2.8 mW/°C. This means that you have to reduce the power rating of 350 mW by 2.8 mW for each degree above 25°C. We won't discuss the other derating factor in the Appendix at this time.

Heat Sinks

One way to increase the power rating of a transistor is to get rid of the internal heat faster. This is the purpose of a *heat sink* (a mass of metal). If we increase the surface area of the transistor case, we allow the heat to escape more easily into the surrounding air. For instance, Fig. 6-15a shows one type of heat sink. When this is pushed on to the transistor case, heat radiates more quickly because of the increased surface area of the fins.

Figure 6-15b shows another approach. This is the outline of a power-tab transistor. A metal tab provides a path out of the transistor for heat. This metal tab can be fastened to the chassis of electronic equipment. Because the chassis is a massive heat sink, heat can easily escape from the transistor to the chassis.

(a)

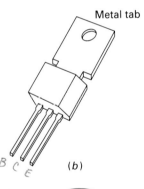

Metal tab

B   C   E   (b)

Collector
connected
to case

⊙ 2

⊙ 1

Pin 1. Base
2. Emitter
Case collector
(c)

FIGURE 6-15

(a) Push-on heat sink; (b) power-tab transistor; (c) power transistor with collector connected to case.

Large power transistors like Fig. 6-15c have the collector connected directly to the case to let heat escape as easily as possible. The transistor case is then fastened to the chassis. To prevent the collector from shorting to chassis ground, a thin mica washer is used between the transistor case and the chassis. The important idea here is that heat can leave the transistor more rapidly, which means that the transistor has a higher power rating at the same ambient temperature. Sometimes, the transistor is fastened to a large heat sink with fins; this is even more efficient in removing heat from the transistor.

No matter what kind of heat sink is used, the purpose is to lower the case temperature because this will lower the internal or junction temperature of the transistor. The data sheet includes other quantities called *thermal resistances*. These allow a designer to work out the case temperature for different heat sinks.

Current Gain

In another system of analysis called the *h parameters*, h_{FE} rather than β_{dc} is used as the symbol for the current gain. The two quantities are equal:

$$\beta_{dc} = h_{FE} \qquad (6\text{-}8)$$

Remember this relation because data sheets use the symbol h_{FE} for the current gain.

In the section labeled "On Characteristics," the data sheet of a 2N3904 lists the values of h_{FE} as follows:

I_C, mA	Min. h_{FE}	Max. h_{FE}
0.1	40	—
1	70	—
10	100	300
50	60	—
100	30	—

The 2N3904 works best when the collector current is in the vicinity of 10 mA. At this level of current, the minimum current gain is 100 and the maximum current gain is 300. What does this mean? It means if you mass-produce a circuit using 2N3904s and a collector current of 10 mA, then some of the transistors will have a current gain as low as 100, and some will have a current gain as high as 300. Most of the transistors will have a current gain somewhere in the middle of this range.

Notice how the minimum current gain decreases for collector currents that are less than or greater than 10 mA. At 0.1 mA, the minimum current gain is 40. At 100 mA, the minimum current gain is 30. These are the sorts of things a designer has to worry about. The data sheet only shows the minimum current gain for currents different from 10 mA because the minimum values represent the worst case. Designers usually do a worst-case design, meaning they figure out how the circuit will work when the transistor characteristics such as current gain are at their worst case.

EXAMPLE 6-11

A 2N3904 has $V_{CE} = 10$ V and $I_C = 10$ mA. What is the power dissipation? How safe is this level of power dissipation if the ambient temperature is 25°C?

SOLUTION

Multiply V_{CE} by I_C to get

$$P_D = (10 \text{ V})(10 \text{ mA}) = 100 \text{ mW}$$

Is this safe? If the ambient temperature is 25°C, the transistor has a power rating of 350 mW. This means the transistor is well within its power rating.

As you know, a good design includes a safety factor to ensure a longer operating life for the transistor. Safety factors of 2 or more are common. A safety factor of 2 means the designer would allow up to half of 350 mW, or 175 mW. Therefore, a power of only 100 mW is very conservative, provided the ambient temperature stays at 25°C.

EXAMPLE 6-12

How safe is the level of power dissipation if the ambient temperature is 100°C in Example 6-11?

SOLUTION

First, work out the number of degrees that the new ambient temperature is above the reference temperature of 25°C. Do this as follows:

$$100°C - 25°C = 75°C$$

Sometimes, you will see this written as

$$\Delta T = 75°C$$

where Δ stands for "difference in." Read the equation as the difference in temperature equals 75°C.

Now, multiply the derating factor by the difference in temperature to get

$$(2.8 \text{ mW/°C})(75°C) = 210 \text{ mW}$$

You often see this written as

$$\Delta P = 210 \text{ mW}$$

where ΔP stands for the difference in power. Finally, you subtract the difference in power from the power rating at 25°C:

$$P_{D(\text{max})} = 350 \text{ mW} - 210 \text{ mW} = 140 \text{ mW}$$

This is the power rating of the transistor when the ambient temperature is 100°C.

How safe is this design? The transistor is still all right because its power is 100 mW compared with the maximum rating of 140 mW. But we no longer have a safety factor of 2. If the ambient temperature were to increase further, or if the power dissipation were to increase, the transistor could get dangerously close to the burnout point. Because of this, a designer might redesign the circuit to restore the safety factor of 2. This means changing circuit values to get a power dissipation of half of 140 mW, or 70 mW. One way would be to reduce the collector-emitter voltage to 7 V. Then the power would decrease to

$$P_D = (7 \text{ V})(10 \text{ mA}) = 70 \text{ mW}$$

6-9   TROUBLESHOOTING

Figure 6-16 shows a common-emitter circuit with grounds. A base supply of 15 V forward-biases the emitter diode through a resistance of 470 kΩ. A collector supply of 15 V reverse-biases the collector diode through a resistance of 1 kΩ. Let us use the ideal approximation to find the collector-emitter voltage. The calculations are as follows:

$$I_B = \frac{15 \text{ V}}{470 \text{ k}\Omega} = 31.9 \text{ } \mu\text{A}$$

$$I_C = 100(31.9 \text{ } \mu\text{A}) = 3.19 \text{ mA}$$

$$V_{CE} = 15 \text{ V} - (3.19 \text{ mA})(1 \text{ k}\Omega) = 11.8 \text{ V}$$

Common Troubles

If you are troubleshooting a circuit like Fig. 6-16, one of the first things to measure is the collector-emitter voltage. It should have a value in the vicinity of 11.8 V. Why don't we use the second or third approximation to get a more accurate answer? Because resistors usually have a tolerance

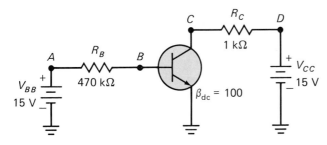

FIGURE 6-16   Troubleshooting a transistor circuit.

of at least ±5 percent, which causes the collector-emitter voltage to differ from your calculations, no matter what approximation you use.

In fact, when troubles come, they are usually big troubles like shorts or opens. Shorts may occur because of damaged devices or solder splashes across resistors. Opens may occur when components burn out. Troubles like these produce big changes in currents and voltages. For instance, one of the most common troubles is no supply voltage reaching the collector. This could happen in a number of ways, such as a trouble in the power supply itself, an open lead between the power supply and the collector resistor, an open collector resistor, etc. In any of these cases, the collector voltage of Fig. 6-16 will be approximately zero because there is no collector supply voltage.

Another possible trouble is an open base resistor, which drops the base current to zero. This forces the collector current to drop to zero and the collector-emitter voltage to rise to 15 V, the value of the collector supply voltage. An open transistor has the same effect.

How Troubleshooters Think

The point is this: Typical troubles cause big deviations in transistor currents and voltages. Troubleshooters are seldom looking for differences in tenths of a volt. They are looking for voltages that are radically different from the ideal values. This is why the ideal transistor is useful as a starting point in troubleshooting. Furthermore, it explains why many trouble-shooters don't even use calculators to find the collector-emitter voltage.

If they don't use calculators, what do they do? They mentally estimate the value of the collector-emitter voltage. Here is the thinking of an experienced troubleshooter while estimating the collector-emitter voltage in Fig. 6-16.

> The voltage across the base resistor is about 15 V. A base resistance of 1 MΩ would produce a base current of about 15 μA. Since 470 kΩ is half of 1 MΩ, the base current is twice as much, approximately 30 μA. A current gain of 100 gives a collector current of about 3 mA. When this flows through 1 kΩ, it produces a voltage drop of 3 V. Subtracting 3 V from 15 V leaves 12 V across the collector-emitter terminals. So, V_{CE} should measure in the vicinity of 12 V, or else there is something wrong in this circuit.

Analyzing without a calculator is something you should practice. If you are always using a calculator, your thinking is too tight, too left-brained. Analyzing without a calculator works well not only for trouble-shooters, but also for designers. When designers start to create a new circuit, the last thing they want is a calculator. A calculator is an anchor in the early stages of a design because designing is done with the right brain. Creative design means flying through a large number of possibilities while looking for anything that will work. Then when you find something that does work, you begin to refine your design by using a calculator and the more advanced approximations.

A Table of Troubles

As discussed in Chap. 5, a shorted component is equivalent to a resistance of zero, while an open component is equivalent to a resistance of infinity. For instance, the base resistor R_B may be shorted or open. Let us designate these states by R_{BS} and R_{BO}, respectively. Similarly, the collector resistor may be shorted or open, symbolized by R_{CS} and R_{CO}, respectively.

Table 6-1 shows a few of the troubles that could occur in a circuit like Fig. 6-16. The voltages were calculated by using the second approximation. When the circuit is operating normally, you should measure a collector voltage of approximately 12 V. If the base resistor were shorted, $+15$ V would appear at the base. This large voltage would destroy the emitter diode. The collector diode would probably open as a result, forcing the collector voltage to go to 15 V. This trouble R_{BS} and its voltages are shown in Table 6-1.

Table 6-1 Troubles and Symptoms

Trouble	V_B, V	V_C, V	Comments
None	0.7	12	No trouble
R_{BS}	15	15	Transistor blown
R_{BO}	0	15	No base or collector current
R_{CS}	0.7	15	
R_{CO}	0.7	0	
No V_{BB}	0	15	Check supply and lead
No V_{CC}	0.7	0	Check supply and lead

If the base resistor were open, there would be no base voltage or current. Furthermore, the collector current would be zero, and the collector voltage would increase to 15 V. This trouble R_{BO} and its voltages are shown in Table 6-1. Continuing like this, we can get the remaining entries of the table.

OPTIONAL TOPICS

The following material continues the earlier discussions at a more advanced and specialized level. All the topics are optional because they are not used in any of the basic discussions in later chapters. This section will be a useful reference when you are in industry because then you will probably want more advanced viewpoints.

6-10 ENERGY VIEWPOINT

Here's another way to visualize transistor action. Figure 6-17 shows the energy levels of an unbiased transistor. The emitter depletion layer is steeper than the collector depletion layer because of the heavier doping

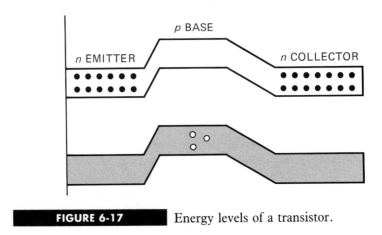

FIGURE 6-17 Energy levels of a transistor.

in the emitter region. The most important thing to notice is that no free electrons in the emitter have enough energy to enter the base region, which is at a higher energy level.

When the transistor is forward-biased, however, the energy levels shift as shown in Fig. 6-18. The emitter bands move up because the emitter diode is forward-biased. The collector bands move down because the collector diode is reverse-biased. Because of this, emitter electrons now have enough energy to diffuse into the base.

Upon entering the base, the free electrons become minority carriers because they are inside a p region. In almost any transistor, more than 95 percent of these free electrons have a long enough lifetime to diffuse into the collector depletion layer and fall down the collector-energy hill. As they fall, they give up energy in the form of heat. The collector must be able to dissipate this heat, and for this reason, it is usually the largest of the three doped regions. Usually, less than 5 percent of the electrons in the base fall along the recombination path shown in Fig. 6-18. Those that do recombine become valence electrons and flow through base holes into the external base lead.

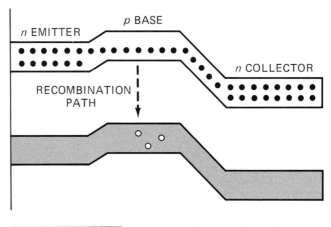

FIGURE 6-18 Energy levels when biased.

6-11 DC ALPHA

Saying that more than 95 percent of injected electrons reach the collector is the same as saying that the collector current almost equals the emitter current. The dc alpha of a transistor indicates how close in value the two currents are; it is defined as

$$\alpha_{dc} = \frac{I_C}{I_E} \tag{6-9}$$

For instance, if we measure an I_C of 4.9 mA and an I_E of 5 mA, then

$$\alpha_{dc} = \frac{4.9\,\text{mA}}{5\,\text{mA}} = 0.98$$

The thinner and more lightly doped the base is, the higher the dc alpha. Ideally, if all injected electrons went on to the collector, the dc alpha would equal unity. Many transistors have dc alphas greater than 0.99, and almost all have dc alphas greater than 0.95. Because of this, we can approximate dc alpha as 1 in most analyses.

6-12 RELATION BETWEEN ALPHA AND BETA

Kirchhoff's current law gives

$$I_E = I_C + I_B \tag{6-10}$$

This tells you that emitter current is the sum of collector current and base currents. Always remember the following: The emitter current is the largest of the three currents, the collector current is almost as large, and the base current is much smaller.

Dividing both sides of Eq. (6-10) by I_C gives

$$\frac{I_E}{I_C} = 1 + \frac{I_B}{I_C}$$

or

$$\frac{1}{\alpha_{dc}} = 1 + \frac{1}{\beta_{dc}}$$

With algebra, we can rearrange to get

$$\beta_{dc} = \frac{\alpha_{dc}}{1 - \alpha_{dc}} \tag{6-11}$$

As an example, if $\alpha_{dc} = 0.98$, the value of β_{dc} is

$$\beta_{dc} = \frac{0.98}{1 - 0.98} = \frac{0.98}{0.02} = 49$$

Occasionally we need a formula for α_{dc} in terms of β_{dc}. With algebra, we can rearrange Eq. (6-11) to get

$$\alpha_{dc} = \frac{\beta_{dc}}{\beta_{dc} + 1} \tag{6-12}$$

For instance, for a β_{dc} of 100,

$$\alpha_{dc} = \frac{100}{100+1} = \frac{100}{101} = 0.99$$

☐ 6-13 BASE CURVES

Figure 6-19 shows a graph of base current versus base-emitter voltage. Since the base-emitter section of a transistor is a diode, we expect to see a graph that resembles a diode curve. And that is what we get, almost. Remember, there are more variables in a transistor than in a diode. At higher collector voltages, the collector gathers in a few more electrons. This reduces the base current. Figure 6-19 echos the idea. The curve with the higher V_{CE} has slightly less base current for a given V_{BE}. This phenomenon, called the *Early effect*, results from internal transistor feedback from the collector diode to the emitter diode. The gap between the curves of Fig. 6-19 is quite small, not even noticeable on an oscilloscope. For this reason, we ignore the Early effect in all preliminary analysis. (The *h* parameters, a high-level method of analysis, include the Early effect.)

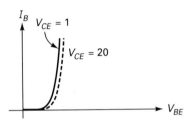

FIGURE 6-19 Base curves: (*a*) ideal; (*b*) Early effect.

☐ 6-14 CUTOFF AND BREAKDOWN

The lowest collector curve is for zero base current. The condition $I_B = 0$ is equivalent to having an open base lead (see Fig. 6-20*a*). The collector current with an open base lead is designated I_{CEO}, where the subscript

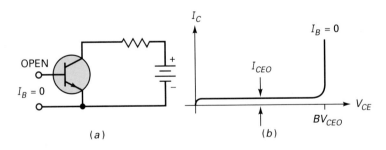

FIGURE 6-20 Cutoff current and breakdown voltage.

CEO stands for collector to emitter with open base and I_{CEO} is caused partly by thermally produced carriers and partly by surface-leakage current.

Figure 6-20b shows the $I_B = 0$ curve. With a large enough collector voltage, we reach a breakdown voltage labeled BV_{CEO}, where the subscript again stands for collector to emitter with an open base. For normal transistor operation, we must keep V_{CE} less than BV_{CEO}. Most transistor data sheets list the value of BV_{CEO} among the maximum ratings. This breakdown voltage may be less than 20 or over 200 V, depending on the transistor type.

As a rule, a good design includes a safety factor to keep V_{CE} well below BV_{CEO}. The transistor lifetime may be shortened by a design that pushes the absolute maximum ratings of a transistor. A safety factor of 2 (V_{CE} less than half BV_{CEO}) is common. Some conservative designs use a safety factor as large as 10 (V_{CE} less than one-tenth of BV_{CEO}).

❏ 6-15 THE THIRD APPROXIMATION

The emitter diode has a bulk resistance. Because it is very small, this bulk resistance usually has only a small IR drop, which means V_{BE} is only slightly larger than 0.7 V.

The collector diode also has a small bulk resistance. This bulk resistance has no effect on the active region. You notice the effect of this bulk resistance only when the transistor is operating in the saturation region. The bulk resistance of the collector diode is what produces the upward slope in the saturation region of the collector curves. As the bulk resistance decreases toward zero, the collector curves move toward the ideal curves discussed earlier. When the bulk resistance is zero, the collector curves are ideal.

The bulk resistances of the two diodes have only minor effects on the currents and voltages of low-power transistors. Low-power transistors, also called *small-signal transistors*, have a power rating of less than half a watt. For instance, a 2N3904 with a collector current of 100 mA has a V_{BE} of 0.85 V instead of 0.7 V. When operating in the saturation region, this transistor has a collector-emitter voltage of only 0.28 V for a collector current of 100 mA.

Power transistors are different. They have power ratings of more than half a watt. They are designed to handle large currents. Because of the large current, the IR drop across the bulk resistances will be important. A designer of transistor power circuits will need to include these bulk resistances in his or her calculations. A troubleshooter of power circuits must at least be aware of larger V_{BE} drops than usual. For instance, a 2N3055 is a transistor with a power rating of 115 W. With a collector current of 10 A, this transistor has a V_{BE} of 1.6 V and a V_{CE} of 0.5 V in the saturation region.

Figure 6-21 shows the third approximation of a transistor. The base curve slopes upward and to the right. Then, as the base current increases, the voltage across the bulk resistance has to be added to the knee voltage to get the total V_{BE}. For instance, the emitter diode of a 2N3904 has a

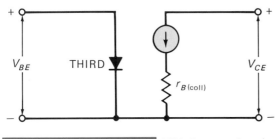

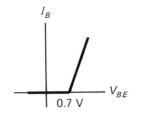

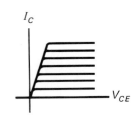

FIGURE 6-21 Third approximation.

bulk resistance of 1.5 Ω. When the emitter current is 100 mA, the additional I_R drop is

$$I_E r_{B(\text{emit})} = (100 \text{ mA})(1.5 \text{ }\Omega) = 0.15 \text{ V}$$

Therefore, the total V_{BE} drop is

$$V_{BE} = 0.7 \text{ V} + 0.15 \text{ V} = 0.85 \text{ V}$$

As another example, the emitter diode of a 2N3055 has a bulk resistance of 0.09 Ω. When the emitter current is 10 A, the additional I_R drop is

$$I_E r_{B(\text{emit})} = (10 \text{ A})(0.09 \text{ }\Omega) = 0.9 \text{ V}$$

In this case, the total V_{BE} drop is

$$V_{BE} = 0.7 \text{ V} + 0.9 \text{ V} = 1.6 \text{ V}$$

The third approximation includes a resistor $r_{B(\text{coll})}$ in series with the collector-current source. This resistor has no effect in the active region. But when the transistor is saturated, the voltage across this resistor prevents V_{CE} from decreasing to zero. The collector-emitter voltage in the saturation region is given by

$$V_{CE(\text{sat})} = I_{C(\text{sat})} r_{B(\text{coll})} \tag{6-13}$$

For instance, the collector diode of 2N3904 has a bulk resistance of 2.8 Ω. If the transistor is saturated and the collector current is 100 mA,

$$V_{CE(\text{sat})} = (100 \text{ mA})(2.8 \text{ }\Omega) = 0.28 \text{ V}$$

The only time this voltage is important is when the transistor is operating in the saturation region.

6-16 BASE-SPREADING RESISTANCE

With two depletion layers penetrating the base, the base holes are confined to the thin channel of p-type semiconductor shown in Fig. 6-22. The resistance of this thin channel is called the *base-spreading resistance* r_b'. Increasing the V_{CB} reverse bias on the collector diode decreases the width of the p channel, which is equivalent to an increase in r_b'.

The recombination current in the base must flow down through r_b'. When it does, it produces a voltage. We discuss the importance of this voltage later. For now, just be aware that r_b' exists and that it depends on the width of the p channel in Fig. 6-22 as well as on the doping of the

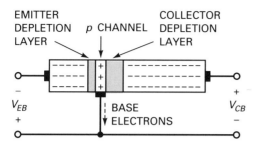

EMITTER DEPLETION LAYER   *p* CHANNEL   COLLECTOR DEPLETION LAYER

BASE ELECTRONS

V_{EB} V_{CB}

| FIGURE 6-22 | Base spreading resistance. |

base. In rare cases, r_b' may be as high as 1000 Ω. Typically, it is in the range of 50 to 150 Ω. The effects of r_b' are important in high-frequency circuits. At low frequencies, r_b' usually has little effect. For this reason, we ignore the effects of r_b' until later chapters.

6-17 THE EBERS-MOLL MODEL

To remember the main ideas of transistor action, look at the equivalent circuit of Fig. 6-23a, shown for conventional current. The voltage V_{BE} is the voltage across the emitter depletion layer. When this voltage is greater than approximately 0.7 V, the emitter injects electrons into the base. As mentioned earlier, the current in the emitter diode controls the collector current. For this reason, the collector-current source forces a current of $\alpha_{dc}I_E$ to flow in the collector circuit. The equivalent circuit of Fig. 6-23a assumes that V_{CE} is greater than a volt or so but less than the breakdown voltage. In other words, the equivalent circuit assumes that the transistor is operating in the active region. The internal voltage V_{BE}' differs from the applied voltage V_{BE} by the drop across r_b':

$$V_{BE} = V_{BE}' + I_B r_b'$$

When the $I_B r_b'$ drop is small, V_{BE} is approximately equal to V_{BE}'.

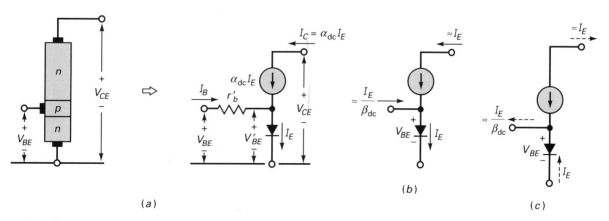

| FIGURE 6-23 | (a) Equivalent circuit for transistor; (b) Ebers-Moll model; (c) electron-flow model. |

Figure 6-23*b* shows a common way to draw the equivalent circuit of a transistor. This is for conventional-current users. If you prefer electron flow, use the equivalent circuit of Fig. 6-23*c*. This equivalent circuit of a transistor, an emitter diode in series with a collector-current source, is called the *Ebers-Moll model*. In using the Ebers-Moll model, we usually approximate as follows:

1. Let V_{BE} equal 0.7 V for silicon transistors (0.3 V for germanium).

2. Disregard the $I_B r_b'$ voltage (this is equivalent to treating the product of I_B and r_b' as negligibly small).

3. Treat I_C as equal to I_E because α_{dc} approaches unity.

4. Approximate I_B as I_E/β_{dc} because I_C almost equals I_E.

6-18 EXAMPLE OF THIRD APPROXIMATION

Bulk resistances are included in the third approximation. To get the values of these bulk resistances, you have to consult the data sheet for the particular transistor. For instance, the data sheet of a 2N3904 shows a graph similar to Fig. 6-24. The graph we want is the one in the middle because the two dashed-line curves are for the saturation region. The curve in the middle is for the active region. To calculate the bulk resistance, read the voltage and current for the highest point shown: 1 V and 200 mA. Now calculate the emitter bulk resistance as follows:

$$r_{B(\text{emit})} = \frac{1\,\text{V} - 0.7\,\text{V}}{200\,\text{mA}} = 1.5\,\Omega$$

The lowest curve of Fig. 6-24 can be used to calculate the collector bulk resistance. As discussed earlier, we need this bulk resistance only

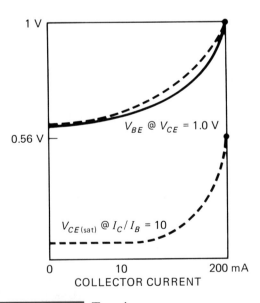

FIGURE 6-24　　Transistor curves.

when the transistor is operating in the saturation region. Therefore, we don't need to calculate the collector bulk resistance for this example, but here is how you do it anyway. Read the voltage and current for the highest point: 0.56 V and 200 mA. Then calcualte as follows:

$$r_{B(\text{coll})} = \frac{0.56 \text{ V}}{200 \text{ mA}} = 2.8 \ \Omega$$

Because of manufacturing tolerances, the curves of Fig. 6-24 apply only to typjcal 2N3904s. Furthermore, the bulk resistances also depend on which particular point on the curves you read. We used the highest points on the curves because these are convenient points well above the knee of the diode. Therefore, the bulk resistances we have are not exact values. They are only estimates. But this is good enough because they have only a slight effect on the currents and voltages in the circuit.

Since the transistor is operating in the active region, only the emitter bulk resistance has an effect on the currents and voltages. The voltage across the emitter bulk resistance has to be added to the barrier potential as follows:

$$V_{BE} = 0.7 \text{ V} + I_E r_{B(\text{emit})} \qquad (6\text{-}14)$$

Because the additional voltage is usually small compared to 0.7 V, an estimate of the emitter bulk resistance is often adequate for this calculation.

Here is how to refine your answers. Calculate V_{BE} with Eq. (6-14). Then use this value in Eq. (6-5) to get a more accurate value of the base current. Then continue with Eqs. (6-3) and (6-6) to get the collector-emitter voltage. The resulting answer is quite accurate.

Often, there is no improvement in using the third approximation to analyze a transistor circuit. Why? Because the collector current is usually small. And a small collector current means that only a small voltage appears across the emitter bulk resistance. As a guideline, unless the additional voltage is at least 0.1 V, it isn't worth using the third approximation. In symbols, use the third approximation only when

$$I_E r_{B(\text{emit})} > 0.1 \text{ V}$$

In summary, when V_{BB} is greater than 14 V, the ideal transistor will be adequate for most jobs. When V_{BB} is less than 14 V, the second approximation is better to use if you need accuracies of at least 5 percent. If the collector current is large, check the value of $I_E r_{B(\text{emit})}$ to see if it is greater than 0.1 V. If so, you may want to use the third approximation.

6-19 THE CB CONNECTION

Forward-bias the emitter diode and reverse-bias the collector diode, and the unexpected happens. In Fig. 6-25a, we expect a large emitter current because the emitter diode is forward-biased. But we do not expect a large collector current because the collector diode is reverse-biased. Nevertheless, the collector current is almost as large as the emitter current.

Here is a brief explanation of why we get a large collector current in Fig. 6-25a. At the instant the forward bias is applied to the emitter diode, electrons in the emitter have not yet entered the base region (see Fig.

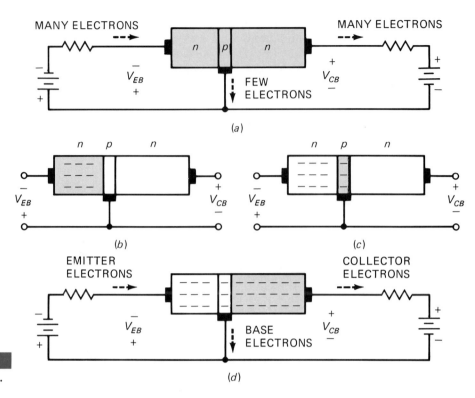

FIGURE 6-25

Common-base connection.

6-25b). If V_{EB} is greater than the barrier potential (0.6 to 0.7 V for silicon transistors), many emitter electrons enter the base region, as shown in Fig. 6-25c. These electrons in the base can flow in either of two directions: down the thin base into the external base lead or across the collector junction into the collector region. The downward component of base current is called the *recombination current*. It is small because the base is lightly doped with only a few holes.

A second crucial idea in transistor action is that the base is very thin. In Fig. 6-25c, the base is teeming with injected conduction-band electrons, causing diffusion into the collector depletion layer. Once inside this layer, the free electrons are pushed by the depletion-layer field into the collector region (see Fig. 6-25d). These collector electrons can then flow into the external collector lead as shown.

Here's our final picture of what's going on. In Fig. 6-25d, we visualize a steady stream of electrons leaving the negative source terminal and entering the emitter region. The V_{EB} forward bias forces these emitter electrons to enter the base region. The thin and lightly doped base gives almost all these electrons enough lifetime to diffuse into the collector depletion layer. The depletion-layer field then pushes a steady stream of electrons into the collector region. These electrons leave the collector, enter the external collector lead, and flow into the positive terminal of the voltage source. In most transistors, more than 95 percent of the emitter-injected electrons flow to the collector; less than 5 percent fall into base holes and flow out the external base lead.

A final point: The circuit of Fig. 6-25 is called a *common-base* circuit because the common or ground side of each supply is connected to the base.

STUDY AIDS

The following study aids will help to reinforce the ideas discussed in this chapter. For best results, use these study aids within 6 hours of reading the earlier material. Then review these study aids a week later and a month later to ensure that the concepts remain in your long-term memory.

SUMMARY

Sec. 6-1 The Unbiased Transistor

A transistor has three doped regions: an emitter, a base and a collector. A *pn* junction exists between the base and the emitter; this part of the transistor is called the emitter diode. Another *pn* junction exists between the base and the collector; this part of the transistor is called the collector diode.

Sec. 6-2 The Biased Transistor

For normal operation, you forward-bias the emitter diode and reverse-bias the collector diode. Under these conditions, the emitter sends free electrons into the base. Most of these free electrons pass through the base to the collector. Because of this, the collector current approximately equals the emitter current. The base current is much smaller, typically less than 5 percent of the emitter current.

Sec. 6-3 Transistor Currents

The ratio of the collector current to the base current is called the current gain, symbolized as β_{dc} or h_{FE}. For low-power transistors, this is typically 100 to 300. The emitter current is the largest of the three currents, the collector current is almost as large, and the base current is much smaller.

Sec. 6-4 The CE Connection

The emitter is grounded or common in a CE circuit. The base-emitter part of a transistor acts approximately like an ordinary diode. The base-collector part acts like a current source that is equal to β_{dc} times the base current. The transistor has an active region, a saturation region, a cutoff region, and a breakdown region. The active region is used in linear amplifiers. Saturation and cutoff are used in digital circuits.

Sec. 6-5 The Base Curve

The graph of base current versus base-emitter voltage looks like the graph of an ordinary diode. Because of this, we can use any of the three diode approximations to calculate the base current. Most of the time, the ideal and the second approximation are all that is necessary.

Sec. 6-6 Collector Curves

The four distinct operating regions of a transistor are the active region, the saturation region, the cutoff region, and the breakdown region. When it is used as an amplifier, the transistor operates in the active region. When it is used in digital circuits, the transistor usually operates in the saturation and cutoff regions. The breakdown region is usually avoided because the risk of transistor destruction is too high.

Sec. 6-7 Transistor Approximations

Exact answers are a waste of time in most electronics work. Almost everybody is using approximations because the answers are adequate for most applications. The ideal transistor is useful for basic troubleshooting. The third approximation is needed for precise design. The second approximation is a good compromise for both troubleshooting and design.

Sec. 6-8 Reading Data Sheets

Transistors have maximum ratings on their voltages, currents, and powers. Small-signal transistors can dissipate half a watt or less. Power transistors can dissipate more than half a watt. Temperature can change the value of the transistor characteristics. Maximum power decreases with a temperature increase. Also, current gain varies a lot with temperature.

Sec. 6-9 Troubleshooting

When troubles come, they usually produce large changes in transistor voltages. This is why ideal analysis is usually adequate for troubleshooters. Furthermore, many troubleshooters spurn the use of calculators because it slows down their thinking. The best troubleshooters learn to mentally estimate the voltages they want to measure.

VOCABULARY

In your own words, explain what each of the following terms mean. Keep your answers short and to the point. If necessary, verify your answer by rereading the appropriate discussion or by looking at the end-of-book Glossary.

active region	emitter
amplifier	emitter diode
base	heat sink
breakdown region	ideal transistor
collector	power transistor
collector cutoff current	saturation region
collector diode	second approximation
common-emitter circuit	small-signal transistor
current gain	third approximation
cutoff region	

IMPORTANT EQUATIONS

Here are some important equations. Say each of the following equations in symbols, then say each in words. Try to explain what the equation means and how it is used. Then read the description that follows.

Eq. 6-1 Transistor Currents

$$I_E = I_C + I_B$$

This is Kirchhoff's current law applied to the three transistor currents. It says the emitter current equals the collector current plus the base current.

Eq. 6-2 Definition of Current Gain

$$\beta_{dc} = \frac{I_C}{I_B}$$

Somebody made this up. In other words, somebody decided that the ratio of collector current to base current was important enough to be called the current gain, symbolized β_{dc}. The equation says the current gain equals the collector current divided by the base current.

Eq. 6-5 Base Current

$$I_B = \frac{V_{BB} - V_{BE}}{R_B}$$

This is Ohm's law applied to the base resistor. The voltage across the resistor equals the difference between the node voltages on the two sides of the resistor. The equation says the base current equals the voltage across the base resistor divided by the resistance. It doesn't get any easier than this.

Eq. 6-6 Collector-Emitter Voltage

$$V_{CE} = V_{CC} - I_C R_C$$

This is absolutely essential for troubleshooting, design, etc. You have to remember this formula or the process behind it. The equation says the collector-emitter voltage equals the collector supply voltage minus the voltage drop across the collector resistor.

Eq. 6-7 Transistor Power Dissipation

$$P_D = V_{CE} I_C$$

The transistor gets hot because of the internal power. You can calculate this power by multiplying the collector-emitter voltage and the collector current.

STUDENT ASSIGNMENTS

QUESTIONS

The following may have more than one right answer. Select the best answer. This is the one that is always true, or covers more situations, or is closest to the truth, etc.

1. A transistor has how many doped regions?
   a. 1
   b. 2
   c. 3
   d. 4

2. What is one important thing transistors do?
   a. Amplify weak signals
   b. Rectify line voltage
   c. Regulate voltage
   d. Emit light

3. Who invented the first junction transistor?
   a. Bell
   b. Faraday
   c. Marconi
   d. Shockley

4. An advantage of a transistor is
   a. Its small size
   b. Lack of heater
   c. It lasts indefinitely
   d. All the above

5. In an *npn* transistor, the majority carriers in the base are
   a. Free electrons
   b. Holes
   c. Neither
   d. Both

6. The barrier potential across each depletion layer is
   a. 0
   b. 0.3 V
   c. 0.7 V
   d. 1 V

7. The emitter diode is usually
   a. Forward-biased
   b. Reverse-biased
   c. Nonconducting
   d. Operating in the breakdown region

8. For normal operation of the transistor, the collector diode has to be
   a. Forward-biased
   b. Reverse-biased
   c. Nonconducting
   d. Operating in the breakdown region

9. The base is thin and
   a. Heavily doped
   b. Lightly doped
   c. Metallic
   d. Doped by a pentavalent material

10. Most of the electrons in the base of an *npn* transistor flow
    a. Out of the base lead
    b. Into the collector
    c. Into the emitter
    d. Into the base supply

11. Most of the electrons in the base of an *npn* transistor do not recombine because they
    a. Have a long lifetime
    b. Have a negative charge
    c. Must flow a long way through the base
    d. Flow out of the base

12. Most of the electrons that flow through the base will
    a. Flow into the collector
    b. Flow out of the base lead
    c. Recombine with base holes
    d. Recombine with collector holes

13. The current gain of a transistor is the ratio of the
    a. Collector current to emitter current
    b. Collector current to base current
    c. Base current to collector current
    d. Emitter current to collector current

14. Increasing the collector supply voltage will increase
    a. Base current
    b. Collector current
    c. Emitter current
    d. None of the above

15. The fact that only a few holes are in the base region means the base is
    a. Lightly doped
    b. Heavily doped
    c. Undoped
    d. None of the above

16. In a normally biased *npn* transistor, the electrons in the emitter have enough energy to overcome the barrier potential of the
    a. Base-emitter junction
    b. Base-collector junction
    c. Collector-base junction
    d. Recombination path

17. When a free electron recombines with a hole in the base region, the free electron becomes
    a. Another free electron
    b. A valence electron
    c. A conduction-band electron
    d. A majority carrier

18. What is the most important fact about the collector current?
    a. It is measured in milliamperes.
    b. It equals the base current divided by the current gain.
    c. It is small.
    d. It approximately equals the emitter current.

19. If the current gain is 200 and the collector current is 100 mA, the base current is
    a. 0.5 mA
    b. 2 mA
    c. 2 A
    d. 20 A

20. The base-emitter voltage is usually
    a. Less than the base supply voltage
    b. Equal to the base supply voltage
    c. More than the base supply voltage
    d. Cannot answer

21. The collector-emitter voltage is usually
    a. Less than the collector supply voltage
    b. Equal to the collector supply voltage
    c. More than the collector supply voltage
    d. Cannot answer

22. The power dissipated by a transistor approximately equals the collector current times
    a. Base-emitter voltage
    b. Collector-emitter voltage
    c. Base supply voltage
    d. 0.7 V

23. A small collector current with zero base current is caused by the leakage current of the
    a. Emitter diode
    b. Collector diode
    c. Base diode
    d. Transistor

24. The collector curves for the second approximation have no
    a. Leakage curent
    b. Breakdown voltage
    c. Saturation region
    d. All the above

25. A transistor acts like a diode and a
    a. Voltage source
    b. Current source
    c. Resistance
    d. Power supply

26. If the base current is 100 mA and the current gain is 30, the collector current is
    a. 300 mA
    b. 3 A
    c. 3.33 A
    d. 10 A

27. The base-emitter voltage of an ideal transistor is
    a. 0
    b. 0.3 V
    c. 0.7 V
    d. 1 V

28. In the saturation region, the collector-emitter voltage of an ideal transistor is
    a. 0
    b. 0.3 V
    c. 0.7 V
    d. 1 V

29. If you recalculate the collector-emitter voltage with the second approximation, the answer will usually be
    a. Smaller than the ideal value
    b. The same as the ideal value
    c. Larger than the ideal value
    d. Inaccurate

30. In the active region, the collector current is not changed significantly by
    a. Base supply voltage
    b. Base current
    c. Current gain
    d. Collector resistance

31. The base-emitter voltage of the second approximation is
    a. 0
    b. 0.3 V
    c. 0.7 V
    d. 1 V

32. If the base resistor is open, what is the collector current?
   a. 0
   b. 1 mA
   c. 2 mA
   d. 10 mA

BASIC PROBLEMS

Sec. 6-3 Transistor Currents

6-1. A transistor has an emitter current of 10 mA and a collector current of 9.95 mA. What is the base current?

6-2. The collector current is 5 mA, and the base current is 0.02 mA. What is the current gain?

6-3. A transistor has a current gain of 125 and a base current of 30 μA. What is the collector current?

6-4. If the collector current is 50 mA and the current gain is 65, what is the base current?

Sec. 6-5 The Base Curve

6-5. What is the base current in Fig. 6-26?

6-6. If the current gain decreases from 200 to 100 in Fig. 6-26, what is the base current?

6-7. If the 330 kΩ of Fig. 6-26 has a tolerance of ±5 percent, what is the maximum base current?

Sec. 6-6 Collector Curves

6-8. A transistor circuit similar to Fig. 6-26 has a collector supply voltage of 20 V, a collector resistance of 1.5 kΩ, and a collector current of 5 mA. What is the collector-emitter voltage?

6-9. If a transistor has a collector current of 100 mA and a collector-emitter voltage of 3.5 V, what is its power dissipation?

Sec. 6-7 Transistor Approximations

6-10. What are the collector-emitter voltage and the transistor power dissipation in Fig. 6-26? (Give answers for the ideal and the second approximation.)

6-11. Figure 6-27a shows a simpler way to draw a transistor circuit. It works the same as the circuits already discussed. What is collector-emitter voltage? The transistor power dissipation? (Give answers for the ideal and the second approximation.)

6-12. When the base and collector supplies are equal, the transistor can be drawn as shown in Fig. 6-27b. What is the collector-emitter voltage in this circuit? The transistor power? (Give answers for the ideal and the second approximation.)

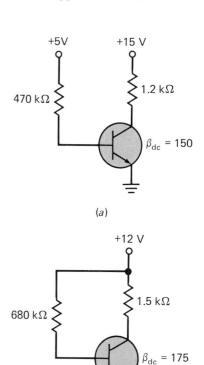

(a)

(b)

FIGURE 6-27

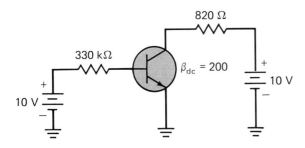

FIGURE 6-26

Sec. 6-8 Reading Data Sheets

6-13. What is the storage temperature range of a 2N3904?

6-14. What is the maximum h_{FE} for a 2N3903 for a collector current of 10 mA and a collector-emitter voltage of 1 V?

6-15. A transistor has a power rating of 1 W. If the collector-emitter voltage is 10 V and the collector current is 120 mA, what happens to the power rating?

6-16. A 2N3904 has a power dissipation of 150 mW without a heat sink. If the ambient temperature is 65°C, what happens to the power rating?

Sec. 6-9 Troubleshooting

6-17. In Fig. 6-26, does the collector-emitter voltage increase, decrease, or remain the same for each of these troubles?
   a. 330 kΩ is shorted
   b. 330 kΩ is open
   c. 820 Ω is shorted
   d. 820 Ω is open
   e. No base supply voltage
   f. No collector supply

ADVANCED PROBLEMS

6-18. What is the dc alpha of a transistor that has a current gain of 200?

6-19. What is the current gain of a transistor with a dc alpha of 0.994?

6-20. Design a CE circuit to meet these specifications: $V_{BB} = 5$ V, $V_{CC} = 15$ V, $h_{FE} = 120$, $I_C = 10$ mA, and $V_{CE} = 7.5$ V.

6-21. A 2N5067 is a power transistor with an r_b' of 10 Ω. How much $I_B r_b'$ drop is there when $I_B = 1$ mA? When $I_B = 10$ mA? When $I_B = 50$ mA?

6-22. A 2N3904 has a power rating of 350 mW at room temperature (25°C). If the collector-emitter voltage is 10 V, what is the maximum current that the transistor can handle for an ambient temperature of 50°C?

6-23. Suppose we connect a LED in series with the 820 Ω of Fig. 6-26. What does the LED current equal?

6-24. The emitter diode of Fig. 6-26 has a bulk resistance of 2 Ω. Use the third approximation to calculate the collector-emitter voltage.

6-25. What is the collector-emitter saturation voltage of a 2N3904 when the collector current is 100 mA? Use the lower curve of Fig. 17 on the data sheet.

SOFTWARE ENGINE PROBLEMS

Use Fig. 6-28 for the remaining problems. This software engine is similar to the one given in Chap. 3. If you haven't already read the introductory material on the software engine, read Sec. 3-10 and Example 3-9 before attempting these problems. Assume increases of approximately 10 percent in the independent variable, and use the second approximation of the transistor. A response should be an N (no change) if the change in a dependent variable is so small that you would have difficulty measuring it. For instance, you probably would find it difficult to measure a change of less than 1 percent. To a troubleshooter, a change like this is usually considered to be no change at all.

6-26. Try to predict the response of each dependent variable in the box labeled V_{BB}. Check your answers. Then answer the following question as simply and directly as possible. What effect does an increase in the base supply voltage have on the dependent variables of the circuit?

6-27. Predict the response of each dependent variable in the box labeled V_{CC}. Check your answers. Then summarize your findings in a one or two sentences.

6-28. Predict the response of each dependent variable in the box labeled R_B. Check your answers. List the dependent variables that decrease. Explain why these variables decrease, using Ohm's law or similar basic ideas.

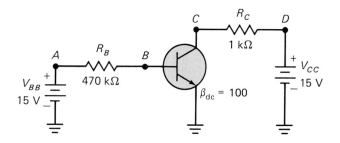

V_{BB}	V_{CC}	R_B	R_C	β_{dc}
V_A : D1	V_A : C2	V_A : B1	V_A : E1	V_A : F1
V_B : E3	V_B : B1	V_B : C2	V_B : F3	V_B : B5
V_C : C6	V_C : A4	V_C : B4	V_C : B2	V_C : E6
V_D : E1	V_D : F5	V_D : F1	V_D : C4	V_D : E5
I_B : F5	I_B : E3	I_B : D3	I_B : E1	I_B : F1
I_C : A2	I_C : B5	I_C : E6	I_C : B1	I_C : D1
P_B : A4	P_B : B1	P_B : A5	P_B : E5	P_B : D6
P_C : D2	P_C : E3	P_C : F6	P_C : C3	P_C : D2
P_D : B4	P_D : A2	P_D : B2	P_D : D5	P_D : A2

	1	2	3	4	5	6
A	U	U	D	U	D	D
B	N	D	D	U	N	N
C	D	N	U	N	U	D
D	U	U	D	U	D	N
E	N	D	N	D	N	D
F	N	U	N	D	U	D

RESPONSES

FIGURE 6-28 Software Engine™. (*Patent pending: Courtesy of Malvino Inc.*)

6-29. Predict the response of each dependent variable in the box labeled R_C. List the dependent variables that show no change. Explain why these variables show no change.

6-30. Predict the response of each dependent variable in the box labeled β_{dc}. List the dependent variables that decrease. Explain why these variables decrease.

7

TRANSISTOR FUNDAMENTALS

The CE circuit discussed throughout Chap. 6 is called the *base bias* because it is designed to set up a specific value of the base current. Although used a lot for digital and switching circuits, it has a fatal flaw that prevents its widespread use as an amplifier. What is the problem? The current gain! As it turns out, the current gain of a transistor has a wide variation with collector current, temperature change, and transistor replacement. The nature of the problem and the solution are the subjects of this chapter.

7-1 VARIATIONS IN CURRENT GAIN

Because of manufacturing tolerances, the current gain of a transistor may vary over as much as a 3:1 range when you change from one transistor to another of the same type. For instance, the data sheet for a 2N3904 lists a minimum h_{FE} of 100 and a maximum of 300 when the collector current is 10 mA. If you mass-produce thousands of circuits with 2N3904s, you will find that some of the transistors have a current gain as low as 100, while others have a current gain as high as 300.

There are two more factors that affect the current gain of a transistor. Figure 7-1 shows the graph of the minimum current gain for a 2N3904. Notice that the value of the current gain depends on the value of the collector current and the junction temperature. As you can see, the variations in the current gain are huge. Assuming a transistor circuit may have to operate anywhere on earth or in the sky above it, you can appreciate the importance of a design that does not rely on a fixed value of the current gain.

Let us inject some reality into the situation. The current gain of a transistor can have more than a 10:1 variation because of the different factors that affect it. Therefore, it is foolish to try to get exact answers when you analyze a base-biased circuit. Approximations are the only sane and sensible approach to take.

Base bias has its uses, especially in digital and switching circuits where the exact value of the current gain does not matter. But when it comes to amplifiers, we need a new circuit, one that is immune to the large variations in current gain. Before we discuss this new circuit, we need to

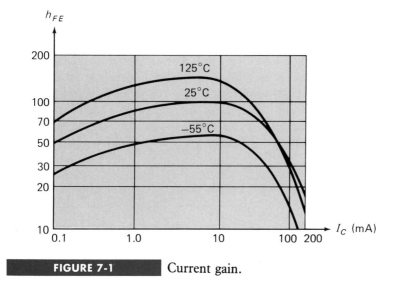

FIGURE 7-1   Current gain.

introduce the load line, a visual concept that will make it easier for you to cope with a large variation in current gain.

EXAMPLE 7-1

What is the current gain of a 2N3904 at 25°C for collector currents of 0.1, 1, 10, and 100 mA?

SOLUTION

Refer to Fig. 7-1. A 2N3904 has a current gain of 50 when the collector current is 0.1 mA, a current gain of 75 at 1 mA, a current gain of 100 at 10 mA, and a current gain of 35 at 100 mA. Over this current range, the maximum h_{FE} is 100 and the minimum is 35, approximately a 3:1 spread. And remember, all these variations take place for a junction temperature of 25°C.

EXAMPLE 7-2

What is the current gain of a 2N3904 at 10 mA for junction temperatures of −55, 25, and 125°C?

SOLUTION

In Fig. 7-1, the current gain is 55 at −55°C, 100 at 25°C, and 150 at 125°C. The spread here is about 3:1.

When you combine the 3:1 spread of Example 7-1 with the 3:1 spread of this example, you are looking at a variation in h_{FE} that is about 9:1. In other words, the current gain is a highly unpredictable characteristic of the transistor. Keep this in mind when you work with transistors.

7-2 THE LOAD LINE

A *load line* is a line drawn over the collector curves to show each and every possible operating point of a transistor. For instance, Fig. 7-2 shows a base-biased circuit with a collector supply voltage of 15 V and a collector resistance of 3 kΩ. To the right of the circuit is a graph of collector curves. The load line is the line drawn between 5 mA on the vertical axis and 15 V on the horizontal axis.

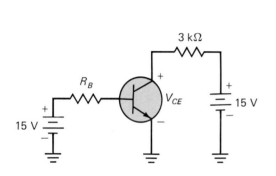

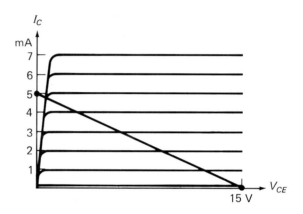

Load line.

Why is the load line useful? Because it contains every possible operating point for the circuit. Stated another way, when the base resistance varies from zero to infinity, the collector current and the collector-emitter voltage vary. If you plot every pair of I_C and V_{CE} values, you will get a string of operating points that fall on the load line. Therefore, the load line is a visual summary of the possible operating points.

The Saturation Point

The *saturation point* is the point where the load line intersects the saturation region of the collector curves. Because the collector-emitter voltage at saturation is very small, the saturation point is almost identical to the upper end of the load line. From now on, we approximate the saturation point as the upper end of the load line, bearing in mind that there is a slight error.

The saturation point tells you the maximum possible collector current for the circuit. For this example, the maximum possible collector current in Fig. 7-2 is approximately 5 mA. If we change the collector supply voltage or the collector resistance, we get a different saturation point.

There is an easy way to find the current at the saturation point. Visualize a short between the collector and emitter of Fig. 7-2. Then V_{CE} drops to zero. All the 15 V from the collector supply will be across the 3 kΩ. Therefore, the current through the collector resistor is

$$I_C = \frac{15\ \text{V}}{3\ \text{k}\Omega} = 5\ \text{mA}$$

You can apply this method to any base-biased circuit. The trick is to visualize a short across the collector-emitter terminals of the transistor. Then calculate the collector current that exists under this condition.

The foregoing method will work for any base-biased circuit. Here is the procedure for finding the saturation current:

1. *Short the collector-emitter terminals.* You can do this mentally or in the laboratory with a built-up circuit.

2. *Find the collector current.* You can calculate this if you are looking at a schematic diagram, or you can measure it if you have a built-up circuit.

These steps are called a *process*, a set of actions performed in a special order. The best troubleshooters and designers use a process, rather than a formula. Why? Because formulas are hard to remember, and they often apply only to one particular circuit. But a process is valid for many different circuits. Furthermore, a process is easier to remember because it makes sense. And a formula is often an inert and meaningless recipe that can lead you astray.

If you like formulas, here is the formula for base-biased circuits:

$$I_{C(\text{sat})} = \frac{V_{CC}}{R_C} \tag{7-1}$$

This says the saturated value of the collector current equals the collector supply voltage divided by the collector resistance. Remember that this formula applies only to the base-biased circuit shown in Fig. 7-2. Don't blindly use this formula for other circuits. It won't work.

The Cutoff Point

The *cutoff point* is the point where the load line intersects the cutoff region of the collector curves. Because the collector current at cutoff is very small, the cutoff point is almost identical to the lower end of the load line. From now on, we approximate the cutoff point as the lower end of the load line.

The cutoff point tells you the maximum possible collector-emitter voltage for the circuit. In Fig. 7-2, the maximum possible collector-emitter voltage is approximately 15 V, the value of the collector supply voltage. If we change the collector supply voltage in Fig. 7-2, we get a different cutoff point.

There is a simple process for finding the cutoff voltage. Visualize the transistor of Fig. 7-2 internally open between the collector and the emitter. Since there is no current through the collector resistor for this open condition, all the 15 V from the collector supply will appear at the collector terminal. Therefore, the voltage between the collector and the ground will equal 15 V. Since the emitter is grounded, the collector-emitter voltage has the same value as the collector-to-ground voltage:

$$V_{CE} = 15 \text{ V}$$

You can apply this process to any circuit you encounter. The trick is to visualize the collector-emitter terminals open. Then figure out the collector-emitter voltage for this condition. Here is the process for finding the collector-emitter voltage for any design:

1. *Open the collector-emitter terminals.*

2. *Find the collector-emitter voltage.*

As usual, you can use this as a mental process or as a laboratory process. This process plus concepts like Ohm's law will allow you to work out the cutoff voltage for a transistor circuit of any base-biased design.

Here is the formula for the cutoff voltage of Fig. 7-2:

$$V_{CE(\text{cut})} = V_{CC} \qquad (7\text{-}2)$$

If you can remember the process for finding the cutoff voltage, you don't have to remember this formula. In general, most formulas are only compact summaries of what you already know or can work out with a process. This is why you are far better off concentrating on the process behind a circuit than on the final formulas for the circuit.

EXAMPLE 7-3

Suppose the collector supply voltage of Fig. 7-2 is increased to 30 V. What happens to the dc load line?

SOLUTION

Visualize a short between the collector and the emitter. Then all the 30 V is across the collector resistance of 3 kΩ. This gives a saturated collector current of 10 mA. Next, visualize an open between the collector and the emitter. Then all the 30 V appears at the collector. Since the emitter is grounded, the collector-emitter voltage is 30 V. The new load line passes through 10 mA on the vertical axis and 30 V on the horizontal axis.

EXAMPLE 7-4

Suppose the collector resistance of Fig. 7-2 is increased to 6 kΩ. What happens to the dc load line?

SOLUTION

Visualize a short between the collector and the emitter. Then all the 15 V is across the collector resistance of 6 kΩ. This gives a saturated collector current of 2.5 mA. Next, visualize an open between the collector and the emitter. Then all the 15 V appears at the collector. Since the emitter is grounded, the collector-emitter voltage is 15 V. The new load line passes through 2.5 mA on the vertical axis and 15 V on the horizontal axis.

7-3 THE OPERATING POINT

Every transistor circuit has a load line. Given any circuit, you work out the saturation current and the cutoff voltage. These values are plotted on the vertical and horizontal axes. Then you draw a line through these two points to get the load line. If the base resistance is given, you can also calculate the current and voltage for the operating point.

Plotting the Q Point

For instance, Fig. 7-3 shows a base-biased circuit with a base resistance of 500 kΩ. We get the saturation current and cutoff voltage by the process given earlier. First, visualize a short across the collector-emitter terminals. Then all the collector supply voltage appears across the collector resistor, which means the saturation current is 5 mA. Second, visualize the collector-emitter terminals open. Then there is no current, and all the supply voltage appears across the collector-emitter terminals, which means the cutoff voltage is 15 V. If we plot the saturation current and cutoff voltage, we can draw the load line shown in Fig. 7-3.

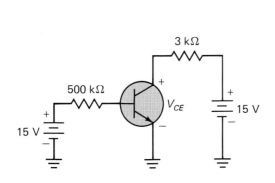
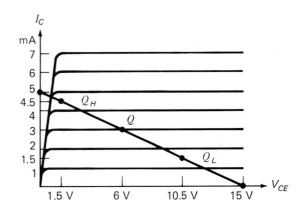

FIGURE 7-3 Plotting the Q point.

Let us keep the discussion simple for now by assuming an ideal transistor. This means that all the base supply voltage will appear across the base resistor. Therefore, the base current is

$$I_B = \frac{15 \text{ V}}{500 \text{ k}\Omega} = 30 \text{ μA}$$

You don't need to use a calculator for this. A troubleshooter would do it mentally as follows: 15 V over 500 kΩ is the same as 30 V over 1 MΩ. So the current is 30 μA.

We cannot proceed until we have a value for the current gain. Suppose the current gain of the transistor is 100. Then the collector current is

$$I_C = 100(30 \text{ μA}) = 3 \text{ mA}$$

This current flowing through 3 kΩ produces a voltage of 9 V across the collector resistor. When you subtract this from the collector supply voltage, you get the voltage across the transistor. Here are the formal calculations:

$$V_{CE} = 15 \text{ V} - (3 \text{ mA})(3 \text{ k}\Omega) = 6 \text{ V}$$

By plotting 3 mA and 6 V (the collector current and voltage), we get the operating point shown on the load line of Fig. 7-3. The operating point is labeled Q because this point is often called the *quiescent point*. (Quiescent means quiet, still, or resting.)

Why the Q Point Varies

We used the ideal transistor for the calculations. What happens with the second approximation? Instead of 15 V across the base resistor, we would have 14.3 V. This means the base current is a bit smaller. The collector current is also smaller, but the collector-emitter voltage is larger. On the load line, this means the Q point moves down a little bit from where it is shown in Fig. 7-3.

What about the current gain? We assumed a current gain of 100. What happens if the current gain is 50? If it is 150? To begin, the base current remains the same because the current gain has no effect on the base current. Ideally, the base current is 30 μA. When the current gain is 50,

$$I_C = 50(30 \text{ μA}) = 1.5 \text{ mA}$$

and the collector-emitter voltage is

$$V_{CE} = 15 \text{ V} - (1.5 \text{ mA})(3 \text{ k}\Omega) = 10.5 \text{ V}$$

Plotting the values gives the low point Q_L shown in Fig. 7-3.
But if the current gain is 150, then

$$I_C = 150(30 \text{ μA}) = 4.5 \text{ mA}$$

and the collector-emitter voltage is

$$V_{CE} = 15 \text{ V} - (4.5 \text{ mA})(3 \text{ k}\Omega) = 1.5 \text{ V}$$

Plotting these values gives the high point Q_H point shown in Fig. 7-3.

The Formulas

The formulas for calculating the Q point are as follows:

$$I_B = \frac{V_{BB} - V_{BE}}{R_B} \tag{7-3}$$

$$I_C = \beta_{dc} I_B \tag{7-4}$$

$$V_{CE} = V_{CC} - I_C R_C \tag{7-5}$$

There is nothing wrong with formulas if they are used to jog your memory about what you already know. That's what they are for. They are compact summaries of what is going on in a circuit. But an experienced technician or engineer normally does not use them. She or he just looks at the circuit and then works out the base current, collector current, and collector-emitter voltage. His or her mental reactions to Eqs. (7-3) through (7-5) go something like this:

> The first equation gives the base current. The second equation gives the collector current. The third equation gives the collector-emitter voltage. So what? I don't use these equations. In other words, I don't plug numbers into these equations to calculate the values. Instead, I find the base current, collector current, and collector-emitter voltage by looking at the circuit and remembering fundamentals like Ohm's law and other basic ideas.

Formulas are not the answer to your problems. The answer is to learn how to apply Ohm's law, Kirchhoff's laws, and the other basic concepts to the analysis of transistor circuits. Most formulas for transistor circuits are Ohm's law in disguise, or a variation of some basic idea you already know. When a truly new formula does appear in this book, you will be told to memorize it if it is important enough.

Conclusion

The exact operating point of the transistor in Fig. 7-3 depends on the base supply voltage, the base resistance, the current gain, the collector resistance, and the collector supply voltage. In other words, it depends on everything. We can easily hold four of the five circuit values constant. For instance, we can use a regulated power supply to produce a rock-solid 15 V. We can also use precision resistors (1 percent tolerance) to get almost exact values of 500 and 3 kΩ. But there is nothing we can do to hold the current gain constant.

As mentioned earlier, the current gain changes randomly from one transistor to the next, it increases with temperature, and it varies with collector current. In mass production, therefore, the circuit of Fig. 7-3 (base bias) has a Q point that will move up or down the load line with temperature changes and transistor replacements. This is not suitable for mass production. What we need is a transistor circuit whose Q point is predictable, one whose Q point is pegged at a particular point on the load line, regardless of temperature changes, transistor replacement, etc.

EXAMPLE 7-5

Suppose the base resistance of Fig. 7-3 is increased to 1 MΩ. What happens to the collector-emitter voltage if β_{dc} is 100?

SOLUTION

Ideally, the base current would decrease to 15 μA, the collector current would decrease to 1.5 mA, and the collector-emitter voltage would increase to

$$V_{CE} = 15 - (1.5 \text{ mA})(3 \text{ k}\Omega) = 10.5 \text{ V}$$

To a second approximation, the base current would decrease to 14.3 μA, and the collector current would decrease to 1.43 mA. The collector-emitter voltage would increase to

$$V_{CE} = 15 - (1.43 \text{ mA})(3 \text{ k}\Omega) = 10.7 \text{ V}$$

7-4 PROOF OF THE LOAD LINE

You may be wondering where the load line came from and why it works. Specifically, you may want to know why the load line represents all possible operating points. What kind of proof will you accept? Experimental? Mathematical? Logical?

Experimental Proof

If you want experimental proof of the load line, build the circuit of Fig. 7-4. Then vary the base resistance from zero to infinity. As you do this, measure the collector current and the collector-emitter voltage. If you plot each operating point, you will get a straight line that passes through these two points:

$$I_C = 5 \text{ mA} \quad \text{and} \quad V_{CE} = 0 \text{ V}$$

and

$$I_C = 0 \quad \text{and} \quad V_{CE} = 15 \text{ V}$$

In other words, the load line is an experimental fact. When you plot each value of collector current and collector-emitter voltage, you will always get a point that falls on the straight line that we have called the load line.

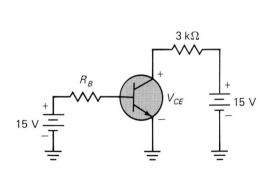

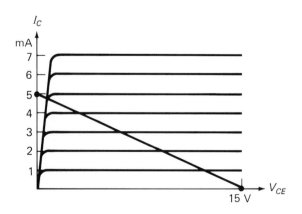

FIGURE 7-4   Load line.

Mathematical Proof

The mathematical proof of the load line is based on the concept of a linear equation and its graph. Here is the basis of the proof. Any equation of the form

$$Ay + Bx = C$$

is called a *linear equation*. It can be shown that the graph of a linear equation is a straight line like Fig. 7-5. If you are interested in more details, see the "Optional Topics."

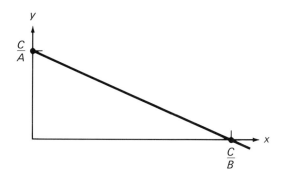

FIGURE 7-5   Vertical and horizontal intercepts.

Logical

The collector voltage of Fig. 7-4 equals the collector supply voltage minus the voltage across the collector resistor. The voltage across the collector resistor is the quantity that is changing. This voltage equals

$$V_{RC} = (3 \text{ k}\Omega)(I_C)$$

This shows that the voltage is directly proportional to the current. Here are some current and voltage values that can exist:

$$V_{RC} = 3 \text{ V} \quad \text{and} \quad I_C = 1 \text{ mA}$$
$$V_{RC} = 6 \text{ V} \quad \text{and} \quad I_C = 2 \text{ mA}$$
$$V_{RC} = 9 \text{ V} \quad \text{and} \quad I_C = 3 \text{ mA}$$
$$V_{RC} = 12 \text{ V} \quad \text{and} \quad I_C = 4 \text{ mA}$$

These values show that the voltage increases 3 V each time the current increases 1 mA. Since this voltage is being subtracted from a collector supply voltage of 15 V to get the collector-emitter voltage, we can conclude that V_{CE} decreases 3 V each time I_C increases 1 mA. Because the changes are directly proportional, the graph of I_C versus V_{CE} will be a straight line. Since the saturation point and the cutoff point are two known points, the load line that represents all operating points must pass through saturation and cutoff.

7-5 RECOGNIZING SATURATION

When you first look at a transistor circuit, you usually cannot tell if it is saturated or operating in the active region. This section discusses two methods for recognizing saturation.

Reductio ad Absurdum

Here is one method for analyzing a new circuit. Assume the transistor operates in the active region, and see if a contradiction arises. Troubleshooters and designers use this powerful technique, known as *reductio ad absurdum,* because it quickly flushes out the operating region of the transistor. Here is the process:

1. Assume the active region.

2. Carry out your calculations.

3. If an absurd answer arises, the assumption is false.

For instance, Fig. 7-6 shows a base-biased circuit. Suppose you want to know what the collector-emitter voltage is. Then you can proceed like this: The base current is ideally 0.1 mA. The current gain of 50 applies only to the active region. Assuming the transistor operates in the active region, the collector current is

$$I_C = 50(0.1 \text{ mA}) = 5 \text{ mA}$$

and the collector-emitter voltage is

$$V_{CE} = 20 \text{ V} - (5 \text{ mA})(10 \text{ k}\Omega) = -30 \text{ V}$$

This answer is impossible. The collector-emitter voltage cannot be negative when the transistor is operating in the active region. We got this absurd result because we assumed the active region. In reality, the transistor has gone into saturation. Therefore, the answer to the original question is this: The collector-emitter voltage is ideally 0 V, or a few tenths of a volt if you take the collector bulk resistance into account.

Another Method

This method is less dramatic, but equally effective as reductio ad absurdum. Start by calculating the collector saturation current of Fig. 7-6:

$$I_{C(\text{sat})} = \frac{20 \text{ V}}{10 \text{ k}\Omega} = 2 \text{ mA}$$

This is the maximum possible value because it occurs at the upper end of the load line. Beyond this point, the transistor goes into saturation.

The base current is ideally 0.1 mA, previously calculated. Assuming a current gain of 50 as shown, the collector current is

$$I_C = 50(0.1 \text{ mA}) = 5 \text{ mA}$$

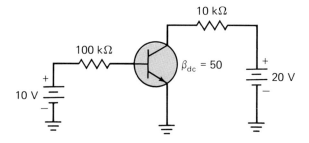

FIGURE 7-6   Recognizing saturation.

But this is greater than the saturation value of 2 mA. Therefore, the transistor must be saturated.

Both methods amount to the same thing. You get an answer that is a contradiction of some sort. In the first case, we got a negative collector-emitter voltage, which is impossible with the *npn* circuit being analyzed. In the second case, we got a collector current that was greater than the saturation current. If either of these cases arises, you automatically know the transistor is saturated.

Current Gain Is Less in Saturation Region

Just because a data sheet tells you the current gain is 50 does not mean this number applies to your circuit under all conditions. When you are given the current gain, it is almost always for the active region. For instance, the current gain of Fig. 7-6 is shown as 50. This means the collector current will be 50 times the base current, provided the transistor is operating in the active region.

When a transistor is saturated, the current gain will be less than the current gain in the active region. You can calculate the saturated current gain as follows:

$$\beta_{dc(sat)} = \frac{I_{C(sat)}}{I_B} \tag{7-6}$$

In Fig. 7-6, the saturated current gain is

$$\beta_{dc(sat)} = \frac{2\ \text{mA}}{0.1\ \text{mA}} = 20$$

Hard Saturation

When a transistor is saturated, the Q point is approximately at the upper end of the load line (see Fig. 7-7). At this operating point, the collector current is maximum. Nothing you do can increase the collector current. For instance, you can increase the base current, but the collector current remains pegged at a value of $I_{C(sat)}$. The only thing that changes with an increase in base current is the current gain; it decreases when the base current increases.

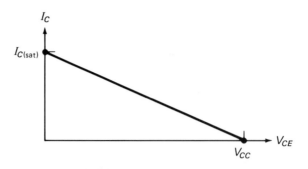

FIGURE 7-7 Load line with endpoints.

For instance, the saturated current gain of Fig. 7-6 is 20, previously calculated. Suppose we change the base resistance to 50 kΩ. Then the base current increases to 0.2 mA, and the saturated current gain decreases to

$$\beta_{dc(sat)} = \frac{2\,mA}{0.2\,mA} = 10$$

A designer who wants a transistor to operate in the saturation region often selects a base resistance that produces a saturated current gain of 10. This is called *hard saturation*, because there is more than enough base current to saturate the transistor.

For the transistor of Fig. 7-6, it takes only

$$I_B = \frac{2\,mA}{50} = 0.04\,mA$$

to saturate the transistor. Therefore, a base current of 0.2 mA drives the transistor deep into saturation.

Why does a designer use hard saturation? Recall that the current gain changes with collector current, temperature variation, and transistor replacement. To make sure that the transistor does not slip out of saturation at low collector currents, low temperatures, etc., the designer uses hard saturation to ensure transistor saturation under all operating conditions.

From now on, *hard saturation* will refer to any design that makes the saturated current gain approximately 10. *Soft saturation* will refer to any design where the transistor is barely saturated, that is, where the saturated current gain is only a little less than the active current gain.

Recognizing Hard Saturation at a Glance

Here is how you can quickly tell if a transistor is in hard saturation. Often, the base supply voltage and the collector supply voltage are equal: $V_{BB} = V_{CC}$. When this is the case, a designer will use the 10:1 rule, which says make the base resistance approximately 10 times as large as the collector resistance.

Figure 7-8 was designed by using the 10:1 rule. Therefore, whenever you see a circuit with a 10:1 ratio (R_B to R_C), you can expect it to be

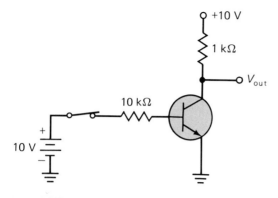

FIGURE 7-8   Hard saturation.

saturated. Even if the supplies are not equal, the chances are good that the transistor is still saturated.

EXAMPLE 7-6

Suppose the base resistance of Fig. 7-6 is increased to 1 MΩ. Is the transistor still saturated?

SOLUTION

Assume the transistor is operating in the active region, and see if a contradiction arises. Ideally, the base current is 10 V divided by 1 MΩ, or 10 μA. The collector current is 50 times 10 μA, or 0.5 mA. This current produces 5 V across the collector resistor. Subtract 5 from 20 V to get

$$V_{CE} = 15 \text{ V}$$

There is no contradiction here. If the transistor were saturated, we would have calculated a negative number or at most 0 V. Because we got 15 V, we know the transistor is operating in the active region.

If the second approximation had been used, the base current would be slightly lower, the collector-emitter voltage would be slightly higher, and the transistor would operate in the active region.

EXAMPLE 7-7

Suppose the collector resistance of Fig. 7-6 is decreased to 5 kΩ. Does the transistor remain in the saturation region?

SOLUTION

Assume the transistor is operating in the active region, and see if a contradiction arises. We can use the same approach as in Example 7-6, but for variety, let us try the second method.

Start by calculating the saturation value of the collector current. Visualize a short between the collector and the emitter. Then you can see that 20 V will be across 5 kΩ. This gives a saturated collector current of

$$I_{C(\text{sat})} = 4 \text{ mA}$$

The base current is ideally 10 V divided by 100 kΩ, or 0.1 mA. The collector current is 50 times 0.1 mA, or 5 mA.

There is a contradiction. The collector current cannot be greater than 4 mA because the transistor hits the saturation point when $I_C = 4$ mA. The only thing that can change at this point is the current gain. The base current is still 0.1 mA, but the current gain decreases to

$$\beta_{\text{dc(sat)}} = \frac{4 \text{ mA}}{0.1 \text{ mA}} = 40$$

This reinforces the idea discussed earlier. A transistor has two current gains, one in the active region and another in the saturation region. The second is equal to or smaller than the first.

7-6 THE TRANSISTOR SWITCH

Base bias is useful in *digital circuits* because these circuits are usually designed to operate at saturation and cutoff. Because of this, they have either low output voltage or high output voltage. In other words, none of the Q points between saturation and cutoff are used. For this reason, variations in the Q point don't matter, because the transistor remains in saturation or cutoff when the current gain changes.

Here is an example of using a base-biased circuit to switch between saturation and cutoff. Figure 7-8, found on page 211, shows an example of a transistor in hard saturation. Therefore, the output voltage is approximately 0 V. This means the Q point is at the upper end of the load line.

When the switch opens, the base current drops to zero. Because of this, the collector current drops to zero. With no current through the 1 kΩ, all the collector supply voltage will appear across the collector-emitter terminals. Therefore, the output voltage rises to +10 V.

The circuit can have only two output voltages: 0 or +10V. This is how you can recognize a digital circuit. It has only two output levels: low or high. The exact values of the two output voltages are not important. All that matters is that you can distinguish the voltages as low or high.

Digital circuits are often called *switching circuits* because their Q point switches between two points on the load line. In most designs, the two points are saturation and cutoff. Another name often used is *two-state circuits*, referring to the low and high outputs.

EXAMPLE 7-8

The collector supply voltage of Fig. 7-8 is decreased to 5 V. What are the two values of the output voltage? If the saturation voltage $V_{CE(\text{sat})}$ is 0.15 V and the collector leakage current I_{CEO} is 50 nA, what are the two values of the output voltage?

SOLUTION

The transistor switches between saturation and cutoff. Ideally, the two values of output voltage are 0 and 5 V. The first voltage is the voltage across the saturated transistor, and the second voltage is the voltage across the cutoff transistor.

If you include the effects of saturation voltage and collector leakage current, the output voltages are 0.15 and 5 V. The first voltage is the voltage across the saturated transistor, which is given as 0.15 V. The second voltage is the collector-emitter voltage with 50 nA flowing through 1 kΩ:

$$V_{CE} = 5\text{ V} - (50\text{ nA})(1\text{ k}\Omega) = 4.99995\text{ V}$$

which rounds to 5 V.

Unless you are a designer, it is waste of time to include the saturation voltage and the leakage current in your calculations of switching circuits. With switching circuits, all you need is two distinct voltages, one low and the other high. It doesn't matter whether the low voltage is 0, 0.1, or 0.15 V, etc. Likewise, it doesn't matter whether the high voltage is 5, 4.9, or 4.5 V. All that usually matters in the analysis of switching circuits is that you can distinguish the low voltage from the high voltage.

7-7 EMITTER BIAS

Digital circuits are the type of circuits used in computers. In this area, base bias and circuits derived from base bias are useful. But when it comes to amplifiers, we need circuits whose Q points are immune to changes in current gain. This calls for a new design.

Figure 7-9 shows *emitter bias*. As you see, the resistor has been moved from the base circuit to the emitter circuit. That's it. That one change makes all the difference in the world. The Q point of this new circuit is now rock-solid. When the current gain changes from 50 to 150, the Q point shows almost no movement along the load line.

Basic Idea

The base supply voltage is now applied directly to the base. Therefore, a troubleshooter will read V_{BB} between the base and ground. The emitter

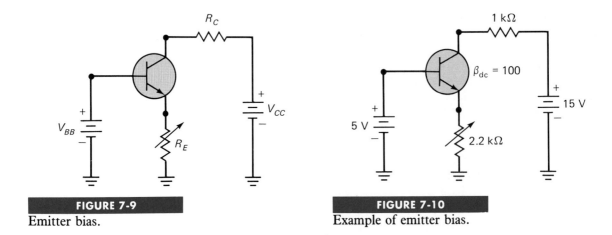

FIGURE 7-9

Emitter bias.

FIGURE 7-10

Example of emitter bias.

is no longer grounded. Now the emitter is above the ground and has a voltage given by

$$V_E = V_{BB} - V_{BE} \qquad (7\text{-}7)$$

You can use any of the three transistor approximations here. If V_{BB} is more than 20 times V_{BE}, the ideal approximation will be accurate. If V_{BB} is less than 20 times V_{BE}, you may want to use the second approximation. Designers may prefer the third approximation if the collector current is large, such as more than 100 mA.

Finding the Q Point

Let us analyze the emitter-biased circuit of Fig. 7-10. The base supply voltage is only 5 V, so we use the second approximation. The voltage between the base and ground is 5 V. From now on, we refer to this base-to-ground voltage simply as the base voltage, or V_B. The voltage across the base-emitter terminals is 0.7 V. We refer to this voltage as the base-emitter voltage, or V_{BE}.

The voltage between the emitter and ground will be called the *emitter voltage*. It equals

$$V_E = 5 \text{ V} - 0.7 \text{ V} = 4.3 \text{ V}$$

This voltage is across the emitter resistance, so we can use Ohm's law to find the emitter current:

$$I_E = \frac{4.3 \text{ V}}{2.2 \text{ k}\Omega} = 1.95 \text{ mA}$$

This means the collector current is 1.95 mA to a close approximation. When this collector current flows through the collector resistor, it produces a voltage drop of 1.95 V. Subtracting this from the collector supply voltage gives the voltage between the collector and ground:

$$V_C = 15 \text{ V} - (1.95 \text{ mA})(1 \text{ k}\Omega) = 13.1 \text{ V}$$

From now on, we refer to this collector-to-ground voltage simply as the collector voltage.

This is the voltage a troubleshooter would measure when testing a transistor circuit. Notice that one lead of the voltmeter would be connected to the collector and the other lead would be connected to ground. If you want the collector-emitter voltage, you have to subtract the emitter voltage from the collector voltage as follows:

$$V_{CE} = 13.1 \text{ V} - 4.3 \text{ V} = 8.8 \text{ V}$$

So, the emitter-biased circuit of Fig. 7-10 has a Q point with these coordinates: $I_C = 1.95$ mA and $V_{CE} = 8.8$ V.

Here is an important idea. The collector-emitter voltage is the voltage used for drawing load lines and for reading transistor data sheets. As a formula,

$$V_{CE} = V_C - V_E \tag{7-8}$$

Normally, a troubleshooter does *not* measure the collector-emitter voltage directly because the common lead of many voltmeters is internally grounded. Why is this a problem? Because connecting the plus lead of the voltmeter to the collector and the common lead to the emitter will *short the emitter to ground*. This gives an incorrect reading.

Therefore, the usual way a troubleshooter gets the collector-emitter voltage is via a three-step process:

1. Measure the collector-to-ground voltage V_C.

2. Measure the emitter-to-ground voltage V_E.

3. Subtract V_E from V_C to get V_{CE}.

Remember this. Otherwise, you may get false readings when trouble-shooting or testing circuits.

Circuit Is Immune to Changes in Current Gain

Here is why emitter bias shines. The Q point of an emitter-biased circuit is immune to changes in current gain. The reason lies in the process used to analyze the circuit. Here are the steps we used earlier:

1. Get the emitter voltage.

2. Calculate the emitter current.

3. Find the collector voltage.

4. Subtract the emitter from the collector voltage to get V_{CE}.

At no time do we need to use the current gain in the foregoing process. Since we don't use it to find the emitter current, collector current, etc., the exact value no longer matters.

Is it a magician's trick? No. Just good design. Here is what happens. By moving the resistor from the base to the emitter circuit, we force the base-to-ground voltage to equal the base supply voltage. Before, almost all this supply voltage was across the base resistor, setting up a fixed base current. Now, all this supply voltage minus 0.7 V is across the emitter resistor, setting up a *fixed emitter current*. Do you see the difference? If not, think about it until the full impact sinks in.

In a base-biased circuit, the base current is fixed even though temperature changes, transistors are replaced, etc. To get the collector current, we have to multiply the fixed base current by the current gain. And that is where we run into trouble. Since the current gain is highly unstable (variable), the collector current is unstable because it is directly proportional to the current gain.

Emitter bias is different. The emitter current is fixed even though temperature changes, transistors are replaced, etc. The current gain still changes, and this forces the base current to change. But that has nothing to do with the collector current. It is approximately equal to the emitter current, which means it is fixed in value.

Minor Effect of Current Gain

The exact truth is this: The current gain has a minor effect on the collector current. (If you blink, you will miss it.) Here is how you can find the exact effect. Under all operating conditions, the three currents are related by

$$I_E = I_C + I_B$$

which can be rearranged as

$$I_E = I_C + \frac{I_C}{\beta_{dc}}$$

Solve this for the collector current, and you get

$$I_C = \frac{\beta_{dc}}{\beta_{dc} + 1} I_E \qquad (7\text{-}9)$$

The quantity that multiplies I_E is called a *correction factor*. It tells you how I_C differs from I_E. When the current gain is 100, the correction factor is

$$\frac{\beta_{dc}}{\beta_{dc} + 1} = \frac{100}{100 + 1} = 0.99$$

This means that the collector current is equal to 99 percent of the emitter current. Therefore, we get only a 1 percent error when we ignore the correction factor and say that the collector current equals the emitter current.

As pointed out in earlier discussions, much everyday work in electronics is done at the 5 percent level. That is, you don't bother with correction factors unless ignoring them introduces more than 5 percent error. In this case, it takes a current gain as low as 20 to produce the critical level:

$$\frac{\beta_{dc}}{\beta_{dc} + 1} = \frac{20}{20 + 1} = 0.952$$

The point is this: Unless you are doing precise design work, you don't need to include the correction factor when you want the value of the collector current. You just use the value of the emitter current as an approximation for the collector current. In the back of your mind, you know that the collector current is slightly less than the emitter current, but the difference is too small to quibble about for most of the work you do. If you ever do need the exact value, you can use Eq. (7-9). From now on, we ignore the correction factor unless otherwise indicated.

Drawing the Load Line

Only one dc load line is possible with base bias. With base bias, you vary the base resistor to create the dc load line. Emitter bias is different; it has two possible dc load lines. The first dc load line can be created by varying the base supply voltage. The second dc load line can be created by varying the emitter resistor. We are going to concentrate on the second dc load line because it is more useful with ac operation of transistors.

In Fig. 7-10, assume we decrease the emitter resistance until the transistor is saturated. Then, the collector-emitter voltage is approximately zero. Since the collector current equals the voltage across the collector resistor, we can write

$$I_{C(\text{sat})} = \frac{15 \text{ V} - 4.3 \text{ V}}{1 \text{ k}\Omega} = 10.7 \text{ mA}$$

Next, assume we increase the emitter resistance toward infinity. Then, the collector current decreases to zero and the collector-to-ground voltage increases to 15 V. In this case,

$$V_{CE(\text{cut})} = 15 \text{ V} - 4.3 \text{ V} = 10.7 \text{ V}$$

Figure 7-11 shows this cutoff voltage at the lower end of the load line. Notice that the saturation current is at the upper end of the load line.

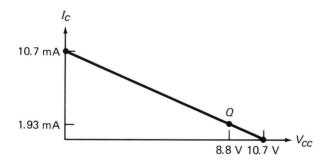

FIGURE 7-11　　Load line for emitter bias.

Effect of Current Gain on the Q Point

Earlier, we found that the Q point of Fig. 7-10 had these coordinates:

$$I_C = 1.95 \text{ mA} \quad \text{and} \quad V_{CE} = 8.8 \text{ V}$$

This Q point ignores the correction factor for the collector current, that is, the effect of the current gain. Now, we want to find out how changes in current gain affect the Q point. We already know the changes will be small. So the question is, How small?

To begin, let us correct the Q point for a current gain of 100. When the current gain is 100, the correction factor is

$$\frac{\beta_{dc}}{\beta_{dc} + 1} = \frac{100}{100 + 1} = 0.99$$

This means the collector current is

$$I_C = 0.99(1.95 \text{ mA}) = 1.93 \text{ mA}$$

and the collector voltage is

$$V_C = 15 \text{ V} - (1.93 \text{ mA})(1 \text{ k}\Omega) = 13.1 \text{ V}$$

and the collector-emitter voltage is

$$V_{CE} = 13.1 \text{ V} - 4.3 \text{ V} = 8.8 \text{ V}$$

Figure 7-11 shows the Q point.

In Sec. 7-3, we analyzed a base-biased circuit for a current gain between 50 and 150. We saw that the Q point varied all over the load line. Let us subject the emitter-biased circuit of Fig. 7-10, on page 214, to the same range in current gain. When the current gain is 50, the calculations look like this:

$$\frac{\beta_{dc}}{\beta_{dc} + 1} = \frac{50}{50 + 1} = 0.98$$

$$I_C = 0.98(1.95 \text{ mA}) = 1.91 \text{ mA}$$

$$V_C = 15 \text{ V} - (1.91 \text{ mA})(1 \text{ k}\Omega) = 13.1 \text{ V}$$

$$V_{CE} = 13.1 \text{ V} - 4.3 \text{ V} = 8.8 \text{ V}$$

When the current gain is 150, the calculations look like this:

$$\frac{\beta_{dc}}{\beta_{dc} + 1} = \frac{150}{150 + 1} = 0.993$$

$$I_C = 0.993(1.95 \text{ mA}) = 1.94 \text{ mA}$$

$$V_C = 15 \text{ V} - (1.94 \text{ mA})(1 \text{ k}\Omega) = 13.1 \text{ V}$$

$$V_{CE} = 13.1 \text{ V} - 4.3 \text{ V} = 8.8 \text{ V}$$

Here is what happened. The changes in collector current are so small that the collector current changes by only the slightest amount when the current gain varies from 50 to 150. Furthermore, when we calculate the collector voltage and round the answer to three significant digits, we get *no change* in the collector voltage. This means the new Q points are so close to the original Q point of Fig. 7-11 that they superimpose it. Simply put, the Q point of the emitter-biased circuit doesn't move despite the current-gain variation from 50 to 150. That's the kind of performance we need to mass-produce circuits. And that is why emitter bias and circuits derived from emitter bias are used in the design of amplifiers.

EXAMPLE 7-9

The emitter resistance of Fig. 7-10 is decreased to 1 kΩ, and the collector resistance is increased to 2 kΩ. What is the voltage between the collector and ground? Between the collector and the emitter?

SOLUTION

The base voltage is 5 V. The emitter voltage is 0.7 V less than this, or

$$V_E = 5 \text{ V} - 0.7 \text{ V} = 4.3 \text{ V}$$

This voltage is across the emitter resistance, which is now 1 kΩ. Therefore, the emitter current is 4.3 V divided by 1 kΩ, or

$$I_E = \frac{4.3 \text{ V}}{1 \text{ k}\Omega} = 4.3 \text{ mA}$$

The collector current is approximately equal to 4.3 mA. When this current flows through the collector resistance (now 2 kΩ), it produces a voltage of

$$I_C R_C = (4.3 \text{ mA})(2 \text{ k}\Omega) = 8.6 \text{ V}$$

When you subtract this voltage from the collector supply voltage, you get

$$V_C = 15 \text{ V} - 8.6 \text{ V} = 6.4 \text{ V}$$

Remember, this is the voltage at the collector node, that is, the voltage between the collector and ground. This is what you would measure when troubleshooting.

As discussed, you should not attempt to connect a voltmeter directly between the collector and the emitter because this may short the emitter to ground. If you wanted to know the value of V_{CE}, you should measure the collector-to-ground voltage, then measure the emitter-to-ground voltage, and subtract the two. In this case,

$$V_{CE} = 6.4 \text{ V} - 4.3 \text{ V} = 2.1 \text{ V}$$

7-8 LED DRIVERS

You have learned that base-biased circuits set up a fixed value of base current, while emitter-biased circuits set up a fixed value of emitter current. Because of the problem with current gain, base-biased circuits are normally designed to switch between saturation and cutoff, whereas emitter-biased circuits are usually designed to operate in the active region.

In this section, we discuss two circuits that can be used as LED drivers. The first circuit uses base bias, and the second circuit uses emitter bias. This will give you a chance to see how each circuit performs in the same application.

Base-Biased LED Driver

The base current is zero in Fig. 7-12, which means the transistor is at cutoff. When the switch of Fig. 7-12 closes, the transistor goes into hard saturation. Visualize a short between the collector-emitter terminals. Then the collector supply voltage (15 V) appears across the series connection of the 1.5 kΩ and the LED. If we ignore the voltage drop across the LED, the collector current is ideally 10 mA. But if we allow 2 V across the LED, then there is 13 V across the 1.5 kΩ, and the collector current is 13 V divided by 1.5 kΩ, or 8.67 mA.

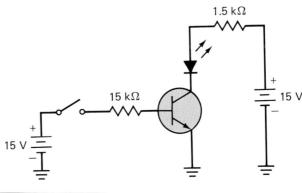

| **FIGURE 7-12** | Base-biased LED driver.

There is nothing wrong with this circuit. It makes a fine LED driver because it is designed for hard saturation, where the current gain doesn't matter. If you want to change the LED current in this circuit, you can change either the collector resistance or the collector supply voltage. The base resistance is made 10 times larger than the collector resistance because we want hard saturation when the switch is closed.

Emitter-Biased LED Driver

The emitter current is zero in Fig. 7-13, which means the transistor is at cutoff. When the switch of Fig. 7-13 closes, the transistor goes into the active region. Ideally, the emitter voltage is 15 V. This means we get an emitter current of 10 mA. This time, the LED voltage drop has no effect. It doesn't matter whether the exact LED voltage is 1.8, 2, or 2.5 V. This is an advantage of the emitter-biased design over the base-biased design. The LED current is independent of the LED voltage. Another advantage is that the circuit doesn't require a collector resistor.

The emitter-biased circuit of Fig. 7-13 operates in the active region when the switch is closed. To change the LED current, you can change the base supply voltage or the emitter resistance. For instance, if you vary the base supply voltage, the LED current varies in direct proportion. This opens up some new possibilities that do not exist with the base-

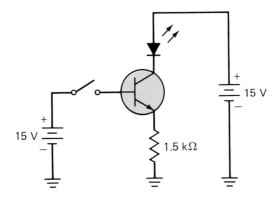

FIGURE 7-13   Emitter-biased LED driver.

biased circuit. In fact, here are some of the thoughts that may run through an inventor's mind when she or he looks at Fig. 7-13:

> With the switch closed, there's approximately 10 mA of LED current. What if I couple a sinusoidal voltage into the base? This will produce a sinusoidal variation in the LED current. Then what if I send the light from this LED down a fiber-optic cable? On the receiving end of the cable, I could use a photodiode to change the varying light to a sinusoidal voltage. What if the original sinusoidal input comes from a microphone? Then . . .

The thoughts running through the inventor's mind are filled with what-if questions. This is one way to start your creative juices flowing. You temporarily suspend all the rules and start asking a lot of what-if questions. You will be surprised at how easy it is to become creative when you start asking, What if? Even more surprising are the answers you sometimes come up with.

EXAMPLE 7-10

We want to get approximately 25 mA of LED current when the switch is closed in Fig. 7-13. How can we do it?

SOLUTION

One solution is to increase the base supply. We want 25 mA to flow through the emitter resistance of 1.5 kΩ. Ohm's law tells us the emitter voltage has to be

$$V_E = (25 \text{ mA})(1.5 \text{ k}\Omega) = 37.5 \text{ V}$$

Ideally, $V_{BB} = 37.5$ V. To a second approximation, $V_{BB} = 38.2$ V. This is a bit high for typical power supplies. But the solution is workable if the particular application allows this high a supply voltage.

A supply voltage of 15 V is common in electronics. Therefore, a better solution in most applications is to decrease the emitter resistance. Ideally, the emitter voltage will be 15 V, and we want 25 mA through the emitter resistor. Again, Ohm's law comes through:

$$R_E = \frac{15 \text{ V}}{25 \text{ mA}} = 600 \; \Omega$$

The nearest standard value with a tolerance of 5 percent is 620 Ω. If we use the second approximation, the resistance is

$$R_E = \frac{14.3 \text{ V}}{25 \text{ mA}} = 572 \; \Omega$$

The nearest standard value is 560 Ω.

7-9 THE EFFECT OF SMALL CHANGES

Understanding how a circuit works means more than being able to calculate the voltages and currents with formulas. It also implies that you can predict how the voltages and currents will react to small changes in the circuit values. Earlier chapters introduced up-down thinking, which is helpful to anyone trying to understand circuits beyond the point of simply plugging numbers into formulas. For the up-down thinking of Fig. 7-14, a small change means a change of approximately 10 percent (the tolerance of many resistors).

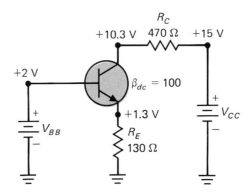

FIGURE 7-14 Up-down thinking.

For instance, Fig. 7-14 shows an emitter-biased circuit with these circuit values:

$$V_{BB} = 2 \text{ V} \quad V_{CC} = 15 \text{ V} \quad R_E = 130 \; \Omega \quad R_C = 470 \; \Omega \quad \beta_{dc} = 100$$

These are the independent variables of the circuit (often called the *circuit values*) because their values are independent of one another. In other words, if any of these variables changes, it has no effect on the other independent variables.

The remaining voltages and currents are as follows:

$$V_E = 1.3 \text{ V} \quad V_C = 10.3 \text{ V} \quad I_B = 99 \text{ μA} \quad I_C = 9.9 \text{ mA} \quad I_E = 10 \text{ mA}$$

Each of these is called a *dependent variable* because its value may change when one of the independent variables changes. If you really understand how a circuit works, you can tell whether a dependent variable will increase, decrease, or remain the same when an independent variable increases.

For instance, suppose V_{BB} increases by about 10 percent in Fig. 7-14. Will V_C increase, decrease, or remain the same? It will decrease. Why? Because an increase in the base supply voltage will increase the emitter current, increase the collector current, increase the voltage across the collector resistor, and decrease the collector voltage.

Table 7-1 summarizes the effects of small increases in the independent variables of Fig. 7-14. We use U for up, D for down, and N for no change (change of less than 1 percent). These results assume the second approximation. By studying this table and asking why the changes occur, you can nail down your understanding of how this circuit works.

Table 7-1 Up-Down Thinking

	V_E	I_E	I_B	I_C	V_C	V_{CE}
V_{BB} increase	U	U	U	U	D	D
V_{CC} increase	N	N	N	N	U	U
R_E increase	N	D	D	D	U	U
R_C increase	N	N	N	N	D	D
β_{dc} increase	N	N	D	N	N	N

The ability to determine the response of dependent variables to small changes in the independent variables depends more on logic than on mathematics. It is a very important ability for troubleshooting, designing, etc. Develop this ability, and your thinking will have the intuitive edge of a professional.

7-10 TROUBLESHOOTING

Many things can go wrong with a transistor. Since it contains two diodes, exceeding any of the breakdown voltages, maximum currents, or power ratings can damage either or both diodes. The troubles may include shorts, opens, high leakage currents, reduced β_{dc}, and other troubles.

Out-of-Circuit Tests

One way to test transistors is with an ohmmeter. You can begin by measuring the resistance between the collector and the emitter. This should be very high in both directions because the collector and emitter diodes are back to back in series. One of the most common troubles is a collector-emitter short, produced by exceeding the power rating. If you read zero to a few thousand ohms in either direction, the transistor is shorted and should be thrown away.

Assuming that the collector-emitter resistance is very high in both directions (in megohms), you can read the reverse and forward resistances of the collector diode (collector-base terminals) and the emitter diode (base-emitter terminals). You should get a high reverse/forward ratio for both diodes, typically more than 1000:1 (silicon). If you do not, the transistor is defective.

Even if the transistor passes the ohmmeter tests, it still may have some faults. After all, the ohmmeter only tests each transistor junction under dc conditions. You can use a curve tracer to look for more subtle faults, such as too much leakage current, low β_{dc}, or insufficient breakdown voltage. Commercial transistor testers are also available; these check the leakage current, β_{dc}, and other quantities.

In-Circuit Tests

The simplest in-circuit tests are to measure transistor voltages with respect to ground. For instance, measuring the collector voltage V_C and the emitter voltage V_E is a good start. The difference $V_C - V_E$ should be more than 1 V but less than V_{CC}. If the reading is less than 1 V, the transistor may be shorted. If the reading equals V_{CC}, the transistor may be open.

The foregoing test usually pins down a dc trouble if one exists. Many people include a test of V_{BE}, done as follows: Measure the base voltage V_B and the emitter voltage V_E. The difference of these readings is V_{BE}, which should be 0.6 to 0.7 V for small-signal transistors operating in the active region. For power transistors, V_{BE} may be 1 V or more because of the bulk resistance of the emitter diode. If the V_{BE} reading is less than 0.6 V, the emitter diode is not being forward-biased. The trouble could be in the transistor or in the biasing components.

Some people include a cutoff test, performed as follows: Short the base-emitter terminals with a jumper wire. This removes the forward-bias on the emitter diode and should force the transistor into cutoff. The collector-to-ground voltage should equal the collector supply voltage. If it does not, something is wrong with the transistor or the circuitry.

A Table of Troubles

As discussed in Chap. 6, a shorted component is equivalent to a resistance of zero, while an open component is equivalent to a resistance of infinity. For instance, the emitter resistor may be shorted or open. Let us designate these states by R_{ES} and R_{EO}, respectively. Similarly, the collector resistor may be shorted or open, symbolized R_{CS} and R_{CO}, respectively.

When a transistor is defective, anything can happen. For instance, one or both diodes may be internally shorted or open. We are going to limit the number of possibilities to the most likely defects as follows: a collector-emitter short (*CES*) will represent all three terminals shorted together (base, collector, and emitter), while a collector-emitter open (*CEO*) stands for all three terminals open. A base-emitter open (*BEO*) means the base-emitter diode is open, and a collector-base open (*CBO*) means the collector-base diode is open.

Table 7-2 shows a few of the troubles that could occur in a circuit like Fig. 7-14. The voltages were calculated by using the second approximation. When the circuit is operating normally, you should measure a base voltage of 2 V, an emitter voltage of 1.3 V, and a collector voltage of approximately 10.3 V. If the emitter resistor were shorted, $+2$ V would appear across the emitter diode. This large voltage would destroy the transistor, probably producing a collector-emitter open. This trouble R_{ES} and its voltages are shown in Table 7-2.

Table 7-2 Troubles and Symptoms

Trouble	V_B, V	V_E, V	V_C, V	Comments
None	2	1.3	10.3	No trouble
R_{ES}	2	0	15	Transistor blown (CEO)
R_{EO}	2	1.3	15	No base or collector current
R_{CS}	2	1.3	15	
R_{CO}	2	1.3	1.3	
No V_{BB}	0	0	15	Check supply and lead
No V_{CC}	2	1.3	1.3	Check supply and lead
CES	2	2	2	All transistor terminals shorted
CEO	2	0	15	All transistor terminals open
BEO	2	0	15	Base-emitter diode open
CBO	2	1.3	15	Collector-base diode open

If the emitter resistor were open, there would be no emitter current. Furthermore, the collector current would be zero, and the collector voltage would increase to 15 V. This trouble R_{EO} and its voltages are shown in Table 7-2. Continuing like this, we can get the remaining entries of the table.

Notice the entry for no V_{CC}. This is worth commenting on. Your initial instinct might be that the collector voltage is zero, because there is no collector supply voltage. But that is not what you will measure with a voltmeter. When you connect a voltmeter between the collector and ground, the base supply will set up a small forward current through the collector diode in series with the voltmeter. Since the base voltage is fixed at 2 V, the collector voltage is 0.7 V less than this. Therefore, the voltmeter will read 1.3 V between the collector and ground. In other words, the voltmeter completes the circuit to ground because the voltmeter looks like a very large resistance in series with the collector diode.

❑

OPTIONAL TOPICS

The following material continues the earlier discussions at a more advanced and specialized level. All the topics are optional because they are not used in any of the basic discussions in later chapters. This section will be a useful reference when you are in industry because then you will probably want more advanced viewpoints.

7-11 MORE ON THE LOAD LINE

Apply Kirchhoff's voltage law to the collector loop of Fig. 7-2, on page 200, to get

$$(3 \text{ k}\Omega)(I_C) + V_{CE} = 15 \text{ V}$$

This is a linear equation in two unknowns, I_C and V_{CE}. It has the basic algebraic form of

$$Ay + Bx = C$$

In basic algebra, you learn that the graph of an equation in two first-order terms is a straight line like Fig. 7-5, on page 207.

The Intercepts

Furthermore, the vertical intercept equals

$$y_i = \frac{C}{A}$$

and the horizontal intercept equals

$$x_i = \frac{C}{B}$$

In Fig. 7-2, $A = 3 \text{ k}\Omega$, $B = 1$, and $C = 15$ V. When we solve for the intercepts, we get a vertical intercept of

$$y_i = \frac{15 \text{ V}}{3 \text{ k}\Omega} = 5 \text{ mA}$$

and a horizontal intercept of

$$x_i = \frac{15 \text{ V}}{1} = 15 \text{ V}$$

In conclusion, applying Kirchhoff's voltage law to the collector loop gives an equation of the form

$$Ay + Bx = C$$

Since the graph of this equation is always a straight line through the intercepts, our collector equation produces a straight line through the saturation current and cutoff voltage. Mathematically speaking, the load line is the *locus* of all operating points and is a straight line because the collector resistor is linear.

Exact Cutoff and Saturation

The exact location of the cutoff and saturation points is slightly different from the horizontal and vertical intercepts. You can see this by looking at Fig. 7-15. The point at which the load line intersects the $I_B = 0$ curve

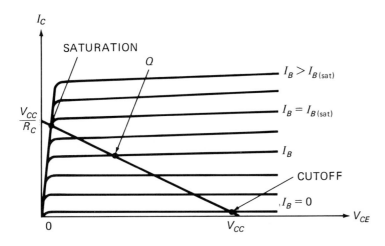

FIGURE 7-15 Exact cutoff and saturation.

is the exact location of the cutoff point. At this point, the base current is zero and the collector current is negligibly small (only leakage current I_{CEO} exists). At cutoff, the emitter diode comes out of forward bias, and normal transistor action is lost. To a close approximation, the collector-emitter voltage with base bias equals the lower end of the load line:

$$V_{CE(\text{cut})} = V_{CC}$$

The intersection of the load line and the $I_B = I_{B(\text{sat})}$ curve is the exact location of the saturation point. At this point, the base current equals $I_{B(\text{sat})}$, and the collector current is maximum. At saturation, the collector diode comes out of reverse bias, and normal transistor action is again lost. To a close approximation, the collector current with base bias equals the upper end of the load line:

$$I_{C(\text{sat})} = \frac{V_{CC}}{R_C}$$

In Fig. 7-15, $I_{B(\text{sat})}$ represents the amount of base current that just produces saturation. If the base current is less than $I_{B(\text{sat})}$, the transistor operates in the active region somewhere between saturation and cutoff. In other words, the operating point is somewhere along the dc load line. On the other hand, if the base current is greater than $I_{B(\text{sat})}$, the collector current approximately equals V_{CC}/R_C, the maximum value. Graphically, this means that the intersection of the load line with any base current greater than $I_{B(\text{sat})}$ produces the saturation point in Fig. 7-15.

Compliance

The *voltage compliance* (or simply *compliance*) of a current source is the range of voltage over which it can operate. With respect to a transistor, the compliance is the voltage range of the active region. The dc load line tells us at a glance what the compliance of a transistor is. In Fig. 7-15, the transistor has a compliance from approxiamtely a few tenths of a volt to approximately V_{CC}. In other words, the transistor acts like a current source anywhere between, but not including, saturation and cutoff.

7-12 MORE ON THE TRANSISTOR SWITCH

The simplest way to use a transistor is as a switch, meaning that we operate it at either saturation or cutoff but nowhere else along the load line. When a transistor is saturated, it is like a closed switch from the collector to the emitter. When a transistor is cut off, it's like an open switch.

Base Current

Figure 7-16*a* shows the circuit we have been analyzing up to now. Summing voltages around the input loop gives

$$I_B R_B + V_{BE} - V_{BB} = 0$$

Solving for I_B, we get

$$I_B = \frac{V_{BB} - V_{BE}}{R_B}$$

If the base current is greater than or equal to $I_{B(\text{sat})}$, the operating point Q is at the upper end of the load line (Fig. 7-16*b*). In this case, the transistor appears like a closed switch. On the other hand, if the base current is zero, the transistor operates at the lower end of the load line and the transistor appears like an open switch.

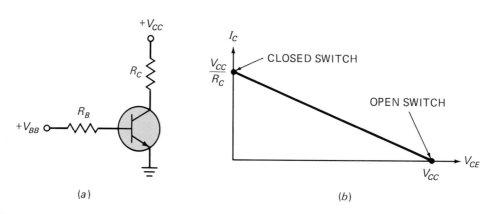

(a)   (b)

FIGURE 7-16   Transistor switching circuit.

Design Rule

Soft saturation means that we barely saturate the transistor; that is, the base current is just enough to operate the transistor at the upper end of the load line. Soft saturation is not reliable in mass production because of the variation in β_{dc} and $I_{B(\text{sat})}$. Don't try to use soft saturation in a transistor switching circuit.

Hard saturation means having sufficient base current to saturate the transistor at all the values of β_{dc} encountered in mass production. For the worst case of temperature and current, almost all small-signal silicon transistors have a β_{dc} greater than 10. Therefore, a design guideline for hard saturation is to have a base current that is approximately one-tenth of the saturated value of the collector current; this guarantees hard saturation under all operating conditions. For instance, if the upper end of the load line has a collector current of 10 mA, then we set up a base current of 1 mA. This guarantees saturation for all transistors, currents, temperatures, etc.

Unless otherwise indicated, we will use the 10:1 rule when designing transistor-switching circuits. Remember, this is only a guideline. If standard resistance values produce an I_C/I_B ratio slightly greater than 10, almost any small-signal transistor will still go into hard saturation.

An Example

Figure 7-17 shows a transistor-switching circuit driven by a voltage step. This is the kind of waveform you get in a digital computer. When the input voltage is zero, the transistor is cut off. In this case, it appears like an open switch. With no current through the collector resistor, the output voltage equals $+15$ V.

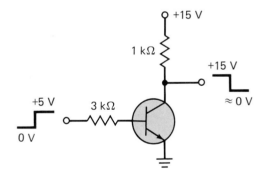

FIGURE 7-17   Transistor used as switch.

When the input voltage is $+5$ V, the base current is

$$I_B = \frac{5\,\text{V} - 0.7\,\text{V}}{3\,\text{k}\Omega} = 1.43\,\text{mA}$$

Visualize the transistor shorted between the collector and the emitter. Then the output voltage ideally drops to zero, and the saturation current is

$$I_{C(\text{sat})} = \frac{15\,\text{V}}{1\,\text{k}\Omega} = 15\,\text{mA}$$

This is approximately 10 times the base current, easily small enough to produce hard saturation in almost any small-signal transistor. This means that the transistor acts like a closed switch and the output voltage is approximately zero.

7-13 TRANSISTOR CURRENT SOURCE

Figure 7-17 shows a transistor switch, one of the basic ways to use a transistor. It is basically the same circuit we called base bias in earlier discussions. Figure 7-18 shows a transistor current source, another basic way to use a transistor. As you see, it is basically the same as the emitter bias in earlier discussions. It is important that you understand that base bias tends to use the transistor as a switch, while emitter bias tends to use the transistor as a current source. This is a critical difference between the two basic circuits, which is why the discussion continues.

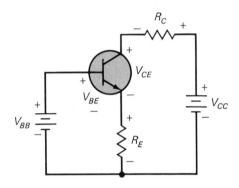

FIGURE 7-18   Transistor used as current source.

Emitter Current

Figure 7-18 has a resistor R_E between the emitter and the common point. Emitter current flows through this resistor, producing a voltage drop $I_E R_E$. Let us sum voltages around the input loop of Fig. 7-18:

$$V_{BE} + I_E R_E - V_{BB} = 0$$

Solving for I_E, we get

$$I_E = \frac{V_{BB} - V_{BE}}{R_E}$$

Here $V_{BB} - V_{BE}$ is the voltage across the emitter resistor. Therefore, this equation is nothing more than Ohm's law applied to the emitter resistor.

Since V_{BB}, V_{BE}, and R_E are all constant (approximately), the emitter current is constant. In other words, given a built-up circuit, only V_{BE} changes slightly with temperature. Disregarding this small change, we can say that the emitter current is approximately constant. Since the collector current is approximately equal to the emitter current, the collector current is approximately constant. But this is identical to saying the collector current is coming out of a current source. This means you can change the collector resistor without changing the collector current.

Emitter Current Is Fixed

Figure 7-19a is the usual way to draw a transistor current source. Given a base voltage V_{BB}, you can set up a fixed emitter current by selecting a value of R_E. This is useful in many applications because the circuit is relatively immune to changes in β_{dc}. Here's why. If β_{dc} changes, the base current changes but the collector current remains essentially the same. This is because the circuit of Fig. 7-19a produces a fixed value of emitter current. The use of an emitter resistor is the key to rock-solid values of collector current. The larger R_E is, the more stable the collector current.

The action is quite different from the transistor switch of Fig. 7-19b where the base current is fixed by V_{BB} and R_B. In a transistor switch, we set up a fixed base current that is large enough to drive the transistor into hard saturation. No attempt is made to operate in the active region because the variations in β_{dc} would cause the operating point to drift all over the load line. In a transistor switch, the emitter is grounded; this is how you can recognize a transistor switch.

Bootstrap Concept

The voltage across the emitter resistor of Fig. 7-19a is

$$V_E = V_{BB} - V_{BE}$$

Because V_{BE} is fixed at approximately 0.7 V, V_E will follow the changes in V_{BB}. For instance, if V_{BB} increases from 2 to 10 V, then V_E will increase from 1.3 to 9.3 V. This kind of follow-the-leader action is called *bootstrapping*. In a transistor current source, the emitter is bootstrapped to the input voltage, which means that it is always within 0.7 V of it.

Again, notice how different this is from the transistor switch of Fig. 7-19b. Since the emitter in a transistor switch is grounded, the emitter is not bootstrapped to the input voltage. Instead, the emitter remains at ground potential no matter what the input voltage does.

Voltage Source Versus Current Source

Another way to distinguish a transistor current source from a transistor switch is by the type of source driving the base. In Fig. 7-19a, the source voltage is applied directly to the base; no base resistor is used. Therefore, a voltage source drives the base. Because of the small V_{BE} drop, most of this source voltage appears across the emitter resistor. This produces a fixed emitter current and a solid Q point in the active region.

On the other hand, the base resistor in the transistor switch of Fig. 7-19b makes the base supply act more like a current source because most of V_{BB} is dropped across the base resistor. Current-sourcing the base is all right with switching circuits because the variations in β_{dc} are swamped out by the hard saturation.

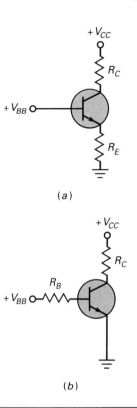

(a)

(b)

FIGURE 7-19

(a) Transistor current source; (b) transistor switch.

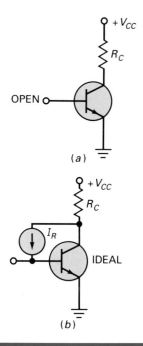

FIGURE 7-20

Transistor with open base.

7-14 MORE OPTOELECTRONIC DEVICES

As mentioned earlier, a transistor with an open base has a small collector current consisting of thermally produced minority carriers and surface leakage. By exposing the collector junction to light, a manufacturer can produce a phototransistor, a transistor that has more sensitivity to light than a photodiode.

Basic Idea of Phototransistors

Figure 7-20a shows a transistor with an open base. As mentioned earlier, a small collector current exists in this circuit. Forget about the surface-leakage component, and concentrate on the thermally produced carriers in the collector diode. Visualize the reverse current produced by these carriers as an ideal current source in parallel with the collector-base junction of an ideal transistor (Fig. 7-20b).

Because the base lead is open, all the reverse current is forced into the base of the transistor. The resulting collector current is

$$I_{CEO} = \beta_{dc} I_R$$

where I_R is the reverse minority-carrier current. This says the collector current is higher than the original reverse current by a factor of β_{dc}. The collector diode is sensitive to light as well as heat. In a phototransistor, light passes through a window and strikes the collector-base junction. As the light increases, I_R increases, and so does I_{CEO}.

Phototransistor Versus Photodiode

The main difference between a phototransistor and a photodiode is the current gain β_{dc}. The same amount of light striking both devices produces β_{dc} times more current in a phototransistor than in a photodiode. The increased sensitivity of a phototransistor is a big advantage over a photodiode.

Figure 7-21a shows the schematic symbol of a phototransistor. Notice the open base. This is the usual way to operate a phototransistor. You can control the sensitivity with a variable base return resistor (Fig. 7-21b), but the base is usually left open to get maximum sensitivity to light.

The price paid for increased sensitivity is reduced speed. A phototransistor is more sensitive than a photodiode, but it cannot turn on and off as fast. On one hand, a photodiode has typical output currents in microamperes and can switch on or off in nanoseconds. On the other hand, the phototransistor has typical output currents in milliamperes but switches on or off in microseconds.

Optocoupler

Figure 7-22 shows a LED driving a phototransistor. This is a much more sensitive optocoupler than the LED-photodiode discussed earlier. The idea is straightforward. Any changes in V_S produce changes in the LED current, which changes the current through the phototransistor. In turn, this produces a changing voltage across the collector-emitter terminals.

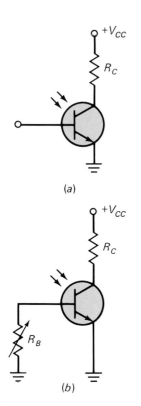

FIGURE 7-21

Phototransistor.

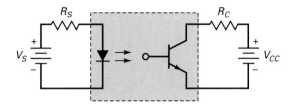

FIGURE 7-22   Optocoupler with LED and phototransistor.

Therefore, a signal voltage is coupled from the input circuit to the output circuit.

Again, the big advantage of an optocoupler is the electrical isolation between the input and output circuits. Stated another way, the common for the input circuit is different from the common for the output circuit. Because of this, no conductive path exists between the two circuits. This means that you can ground one of the circuits and float the other. For instance, the input circuit can be grounded to the chassis of the equipment, while the common of the output side is ungrounded.

An Example

The 4N24 optocoupler of Fig. 7-23*a* provides isolation from the power line and detects zero crossings of line voltage. The graph of Fig. 7-23*b* shows how the collector current is related to the LED current. Here is how you can calculate the peak output voltage from the optocoupler.

The bridge rectifier produces a full-wave current through the LED. Ignoring diode drops, the peak current through the LED is

$$I_{\text{LED}} = \frac{1.414(115 \text{ V})}{16 \text{ k}\Omega} = 10.2 \text{ mA}$$

The saturated value of the phototransistor current is

$$I_{C(\text{sat})} = \frac{20 \text{ V}}{10 \text{ k}\Omega} = 2 \text{ mA}$$

Figure 7-23*b* shows the static curves of phototransistor current versus LED current for three different optocouplers. With a 4N24 (top curve), a LED current of 10.2 mA produces a collector current of approximately 15 mA when the load resistance is zero. In Fig. 7-23*a*, the phototransistor current never reaches 15 mA because the phototransistor saturates a 2 mA. In other words, there is more than enough LED current to produce saturation. Since the peak LED current is 10.2 mA, the transistor is saturated during most of the cycle. At this time, the output voltage is approximately zero, as shown in Fig. 7-23*c*.

The zero crossings occur when the line voltage is changing polarity, from positive to negative, or vice versa. At a zero crossing, the LED current drops to zero. At this instant, the phototransistor becomes an open circuit, and the output voltage increases to approximately 20 V, as indicated in Fig. 7-23*c*. As you see, the output voltage is near zero most of the cycle. At the zero crossings, it increases rapidly to 20 V and then decreases to the baseline.

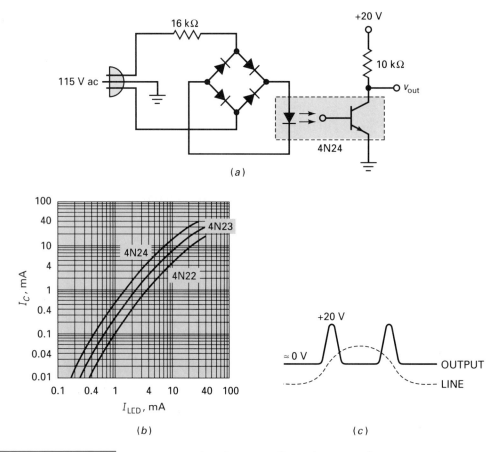

FIGURE 7-23 (*a*) Zero-crossing detector; (*b*) static curves for optocoupler; (*c*) output of detector.

A circuit like Fig. 7-23*a* is useful because it does not require a transformer to provide isolation from the line. The photocoupler takes care of this. Furthermore, the circuit detects zero crossings, desirable in applications where you want to synchronize some other circuit to the frequency of the line voltage.

STUDY AIDS

The following study aids will help to reinforce the ideas discussed in this chapter. For best results, use these study aids within 6 hours of reading the earlier material. Then review these study aids a week later and a month later to ensure that the concepts remain in your long-term memory.

SUMMARY

Sec. 7-1 Variations in Current Gain
The current gain of a transistor is an unpredictable quantity. Because of manufacturing tolerances, the current gain of a transistor may vary over as much

as a 3:1 range when you change from one transistor to another of the same type. Changes in the temperature and the collector current produce additional variations in the dc gain.

Sec. 7-2 The Load Line
The dc load line contains all the possible dc operating points of a transistor circuit. The upper end of the load line is called saturation, and the lower end is called cutoff. The key step in finding the saturation current is to visualize a short between the collector and the emitter. The key step to finding the cutoff voltage is to visualize an open between the collector and emitter.

Sec. 7-3 The Operating Point
The operating point of the transistor is on the dc load line. The exact location of this point is determined by the collector current and the collector-emitter voltage. With base bias, the Q point moves whenever any of the circuit values change.

Sec. 7-4 Proof of the Load Line
The experimental proof is to build the circuit and then measure the collector current and the collector-emitter voltage for different biasing conditions. If you plot the pairs of I_C and V_{CE}, you will get a straight line known as the dc load line. The mathematical proof is to review basic algebra, particularly the theory of linear equations. The logical proof is to realize that changes in I_C and V_{CE} have to be proportional changes because they involve a fixed resistor R_C.

Sec. 7-5 Recognizing Saturation
The idea is to assume the *npn* transistor is operating in the active region. If this leads to a contradiction (such negative collector-emitter voltage, collector current greater than saturation current, etc.), then you know the transistor is operating in the saturation region. Another way to recognize saturation is by comparing the base resistance to the collector resistance. If the ratio is in the vicinity of 10:1, the transistor is probably saturated.

Sec. 7-6 The Transistor Switch
Base bias tends to use the transistor as a switch. The switching action is between cutoff and saturation. This type of operation is useful in digital circuits. Another name for switching circuits is two-state circuits.

Sec. 7-7 Emitter Bias
Emitter bias is virtually immune to changes in current gain. The process for analyzing emitter bias is to find the emitter voltage, emitter current, collector voltage, and collector-emitter voltage. All you need for this process is Ohm's law.

Sec. 7-8 LED Drivers
A base-biased LED driver uses a saturated or cutoff transistor to control the current through a LED. An emitter-biased LED driver uses the active region and cutoff to control the current through the LED.

Sec. 7-9 The Effect of Small Changes
Useful to both troubleshooters and designers is the ability to predict the direction of change for a dependent voltage or current when one of the circuit values changes. When you can do this, you can better understand what happens for different troubles and can more easily design circuits.

Sec. 7-10 Troubleshooting
You can use an ohmmeter to test a transistor. This is best done with the transistor disconnected from the circuit. When the transistor is still in the circuit with the power on, you can measure its voltages. These voltages are clues to what is wrong. A collector voltage equal to the supply voltage means there is no collector current. A collector voltage equal to the emitter voltage means the transistor is saturated.

VOCABULARY

In your own words, explain what each of the following terms means. Keep your answers short and to the point. If necessary, verify your answer by rereading the appropriate discussion or by looking at the end-of-book Glossary.

active current gain	LED driver
base bias	load line
cutoff point	saturated current gain
emitter bias	saturation point
hard saturation	soft saturation

IMPORTANT EQUATIONS

Here are some important equations. Say each of the following equations in symbols, then say each in words. Try to explain what the equation means and how it is used. Then read the description that follows.

Eq. 7-6 Saturated Current Gain

$$\beta_{dc(sat)} = \frac{I_{C(sat)}}{I_B}$$

This is important because the current gain of a transistor decreases rapidly once the transistor enters the saturation region. The equation says the saturated current gain equals the saturated value of the collector current divided by the base current.

Eq. 7-7 Emitter Voltage

$$V_E = V_{BB} - V_{BE}$$

This is a very important equation because it represents the first step in analyzing an emitter-biased circuit. Since there is no base resistor, all the base supply voltage appears at the base. If you subtract the base-emitter voltage, you get the voltage at the emitter node. This voltage is across the emitter resistor.

Eq. 7-8 Collector-Emitter Voltage

$$V_{CE} = V_C - V_E$$

Since the emitter is not grounded in an emitter-biased circuit, you have to subtract the emitter voltage from the collector voltage to get the collector-emitter voltage. This is important in troubleshooting because you cannot normally measure V_{CE} directly. You have to make two separate measurements.

IMPORTANT PROCESSES

Review each of the following processes. The steps are abbreviated. For instance, *base current* means to calculate the base current. All these processes are important. If you can remember the basic ideas, you will be able to solve problems more easily.

Saturated Collector Current

1. Short the collector-emitter terminals.
2. Find the collector current.

Cutoff Voltage

1. Open the collector-emitter terminals.
2. Find the collector-emitter voltage.

Analyzing Base Bias

1. Base current
2. Multiply by current gain to get collector current
3. Collector voltage

Analyzing Emitter Bias

1. Emitter voltage
2. Emitter current
3. Collector voltage
4. Subtract emitter voltage from collector voltage

Measuring Collector-Emitter Voltage

1. Measure the collector-to-ground voltage.
2. Measure the emitter-to-ground voltage.
3. Subtract V_E from V_C to get V_{CE}.

STUDENT ASSIGNMENTS

QUESTIONS

The following questions may refer to the figures you have seen in this chapter. It should be clear from the question which figure is being used. For instance, if the question mentions a base resistor, then the circuit is base-biased. If the question mentions an emitter resistor, the circuit is emitter-biased.

1. The current gain of a transistor is defined as the ratio of the collector current to the
   a. Base current
   b. Emitter current
   c. Supply current
   d. Collector current

2. The graph of current gain versus collector current indicates that the current gain
   a. Is constant
   b. Varies slightly
   c. Varies enormously
   d. Equals the collector current divided by the base current

3. When the collector current increases, what does the current gain do?
   a. Decreases
   b. Stays the same
   c. Increases
   d. Any of the above

4. As the temperature increases, the current gain
   a. Decreases
   b. Remains the same
   c. Increases
   d. Can be any of the above

5. When the base resistor decreases, the collector voltage will probably
   a. Decrease
   b. Stay the same
   c. Increase
   d. Do all the above

6. If the base resistor is very small, the transistor will operate in the
   a. Cutoff region
   b. Active region
   c. Saturation region
   d. Breakdown region

7. Ignoring the bulk resistance of the collector diode, the saturation voltage is
   a. 0
   b. A few tenths of a volt
   c. 1 V
   d. Supply voltage

8. Three different Q points are shown on a load line. The upper Q point represents the
   a. Minimum current gain
   b. Intermediate current gain
   c. Maximum current gain
   d. Cutoff point

9. If a transistor operates at the middle of the load line, an increase in the base resistance will move the Q point
   a. Down
   b. Up
   c. Nowhere
   d. Off the load line

10. If a transistor operates at the middle of the load line, an increase in the current gain will move the Q point
    a. Down
    b. Up
    c. Nowhere
    d. Off the load line

11. If the base supply voltage increases, the Q point moves
    a. Down
    b. Up
    c. Nowhere
    d. Off the load line

12. Suppose the base resistor is open. The Q point will be
    a. In the middle of the load line
    b. At the upper end of the load line
    c. At the lower end of the load line
    d. Off the load line

13. If the base supply voltage is disconnected, the collector-emitter voltage will equal
    a. 0 V
    b. 6 V
    c. 10.5 V
    d. Collector supply voltage

14. If the base resistor is shorted, the transistor will probably be
    a. Saturated
    b. Cut off
    c. Destroyed
    d. None of the above

15. If the collector resistor decreases to zero in a base-biased circuit, the load line will become
    a. Horizontal
    b. Vertical
    c. Useless
    d. Flat

16. The saturation point is approximately the same as
    a. Cutoff point
    b. Lower end of the load line
    c. Upper end of the load line
    d. Infinity

17. When the vertical intercept of the load line increases, the collector current
    a. Decreases
    b. Stays the same
    c. Increases
    d. Does none of the above

18. The collector current is 10 mA. If the current gain is 100, the base current is
    a. 1 μA
    b. 10 μA
    c. 100 μA
    d. 1 mA

19. The base current is 50 μA. If the current gain is 125, the collector current is closest in value to
    a. 40 μA
    b. 500 μA
    c. 1 mA
    d. 6 mA

20. If the current gain increases from 50 to 100, the collector current with base bias will approximately
    a. Drop in half
    b. Double
    c. Stay the same
    d. Destroy the transistor

21. When the collector resistance decreases in a base-biased circuit, the load line becomes
    a. More horizontal
    b. More vertical
    c. Fixed
    d. None of the above

22. When the Q point moves along the load line, the voltage increases when the current
    a. Decreases
    b. Stays the same
    c. Increases
    d. Does none of the above

23. At cutoff, the Q point is at
    a. Upper end of the load line
    b. Middle of the loadline
    c. Lower end of the load line
    d. None of the above

24. When there is no base current in a transistor switch, the output voltage from the transistor is
    a. Low
    b. High
    c. Unchanged
    d. Unknown

25. A circuit with a fixed emitter current is called
    a. Base bias
    b. Emitter bias
    c. Transistor bias
    d. Two-supply bias

26. The first step in analyzing emitter-biased circuits is to find the
    a. Base current
    b. Emitter voltage
    c. Emitter current
    d. Collector current

27. If the current gain is unknown in an emitter-biased circuit, you cannot calculate the
    a. Emitter voltage
    b. Emitter current
    c. Collector current
    d. Base current

28. If the emitter resistor is open, the collector voltage is
    a. Low
    b. High
    c. Unchanged
    d. Unknown

29. If the collector resistor is open, the collector voltage is
    a. Low
    b. High
    c. Unchanged
    d. Unknown

30. When the current gain increases from 50 to 300 in an emitter-biased circuit, the collector current
    a. Remains almost the same
    b. Decreases by a factor of 6
    c. Increases by a factor of 6
    d. Is zero

31. If the emitter resistance decreases, the collector voltage
    a. Decreases
    b. Stays the same
    c. Increases
    d. Breaks down the transistor

32. If the emitter resistance decreases, the
    a. Q point moves up along the load line
    b. Collector current decreases
    c. Q point stays where it is
    d. Current gain increases

BASIC PROBLEMS

Sec. 7-1 Variations in Current Gain

7-1. Refer to Fig. 7-1 on page 199. What is the current gain of a 2N3904 when the collector current is 200 mA and the junction temperature is 25°C?

7-2. Refer to Fig. 7-1. The junction temperature is 125°C, and the collector current is 0.1 mA. What is the current gain?

Sec. 7-2 The Load Line

7-3. Draw the load line for Fig. 7-24a. What is the collector current at the saturation point? The collector-emitter voltage at the cutoff point?

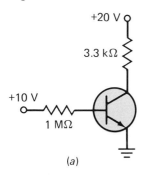

(a)

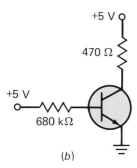

(b)

FIGURE 7-24

7-4. If the collector supply voltage is reduced to 10 V in Fig. 7-24a, what happens to the load line?

7-5. If the collector resistance is reduced to 1 kΩ in Fig. 7-24a, what happens to the load line?

7-6. If the base resistance of Fig. 7-24a is doubled, what happens to the load line?

7-7. Draw the load line for Fig. 7-24b. What is the collector current at the saturation point? The collector-emitter voltage at the cutoff point?

7-8. If the collector supply voltage is doubled in Fig. 7-24b, what happens to the load line?

7-9. If the collector resistance is increased to 1 kΩ in Fig. 7-24b, what happens to the load line?

Sec. 7-3 The Operating Point

7-10. In Fig. 7-24a, what is the voltage between the collector and ground if the current gain is 100?

7-11. The current gain varies from 25 to 300 in Fig. 7-24a. What is the minimum voltage from the collector to ground? The maximum?

7-12. The resistors of Fig. 7-24a have a tolerance of ±5 percent. The supply voltages have a tolerance of ±10 percent. If the current gain can vary from 50 to 150, what is the minimum possible voltage from the collector to ground? The maximum?

7-13. In Fig. 7-24b, what is the voltage between the collector and ground if the current gain is 100?

7-14. The current gain varies from 25 to 300 in Fig. 7-24b. What is the minimum voltage from the collector to ground? The maximum?

7-15. The resistors of Fig. 7-24b have a tolerance of ±5 percent. The supply voltages have a tolerance of ±10 percent. If the current gain can vary from 50 to 150, what is the minimum possible voltage from the collector to ground? The maximum?

Sec. 7-5 Recognizing Saturation

7-16. In Fig. 7-24a, use the circuit values shown unless otherwise indicated. Determine whether the transistor is saturated for each of these changes:
   a. $R_B = 33$ kΩ and $h_{FE} = 100$
   b. $V_{BB} = 5$ V and $h_{FE} = 200$
   c. $R_C = 10$ kΩ and $h_{FE} = 50$
   d. $V_{CC} = 10$ V and $h_{FE} = 100$

7-17. In Fig. 7-24b, use the circuit values shown unless otherwise indicated. Determine whether the transistor is saturated for each of these changes:
   a. $R_B = 47$ kΩ and $h_{FE} = 100$
   b. $V_{BB} = 10$ V and $h_{FE} = 500$
   c. $R_C = 10$ kΩ and $h_{FE} = 100$
   d. $V_{CC} = 10$ V and $h_{FE} = 100$

Sec. 7-6 The Transistor Switch

7-18. The 680 kΩ in Fig. 7-24b is replaced by 4.7 kΩ and a series switch. Assuming an ideal transistor, what is the collector voltage if the switch is open? What is the collector voltage if the switch is closed?

7-19. Repeat Prob. 7-18, except use $V_{CE(\text{sat})}$ = 0.2 V and I_{CEO} = 100 nA.

Sec. 7-7 Emitter Bias

7-20. What is the collector voltage in Fig. 7-25a? The emitter voltage?

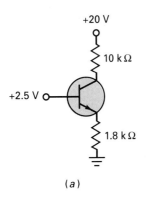

+20 V

10 kΩ

+2.5 V

1.8 kΩ

(a)

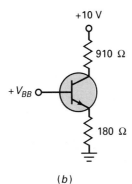

+10 V

910 Ω

+V_{BB}

180 Ω

(b)

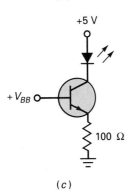

+5 V

+V_{BB}

100 Ω

(c)

FIGURE 7-25

7-21. If the emitter resistor is doubled in Fig. 7-25a, what is the collector-emitter voltage?

7-22. If the collector supply voltage is decreased to 15 V in Fig. 7-25a, what is the collector voltage?

7-23. What is the collector voltage in Fig. 7-25b if V_{BB} = 2 V?

7-24. If the emitter resistor is doubled in Fig. 7-25b, what is the collector-emitter voltage for a base supply voltage of 2.3 V?

7-25. If the collector supply voltage is increased to 15 V in Fig. 7-25b, what is the collector-emitter voltage for V_{BB} = 1.8 V?

Sec. 7-8 LED Drivers

7-26. If the base supply voltage is 2 V in Fig. 7-25c, what is the current through the LED?

7-27. If V_{BB} = 1.8 V in Fig. 7-25c, what is the LED current? The approximate collector voltage?

Sec. 7-9 The Effect of Small Changes

Use the letters U (up), D (down), and N (no change) for your answers in the following problems.

7-28. The base supply voltage of Fig. 7-26a increases by 10 percent. What happens to the base current, collector current, and collector voltage?

7-29. The base resistance of Fig. 7-26a increases by 10 percent. What happens to the base current, collector current, and collector voltage?

7-30. The collector resistance of Fig. 7-26a increases by 10 percent. What happens to the base current, collector current, and collector voltage?

7-31. The collector supply voltage of Fig. 7-26a increases by 10 percent. What happens to the base current, collector current, and collector voltage?

7-32. The base supply voltage of Fig. 7-26b increases by 10 percent. What happens to

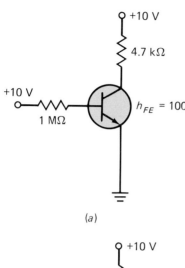

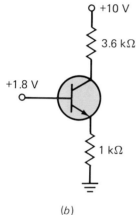

(b)

FIGURE 7-26

the base current, collector current, and collector voltage?

7-33. The emitter resistance of Fig. 7-26b increases by 10 percent. What happens to the emitter current, collector current, and collector voltage?

7-34. The collector resistance of Fig. 7-26b increases by 10 percent. What happens to the emitter current, collector current, and collector voltage?

7-35. The collector supply voltage of Fig. 7-26b increases by 10 percent. What happens to the emitter current, collector current, and collector voltage?

Sec. 7-10 Troubleshooting

7-36. A voltmeter reads 10 V at the collector of Fig. 7-26a. What are some of the troubles that can cause this high reading?

7-37. What if the ground on the emitter is open in Fig. 7-26a? What will a voltmeter read for the base voltage? For the collector voltage?

7-38. A dc voltmeter measures a very low voltage at the collector of Fig. 7-26a. What are some of the possible troubles?

7-39. A voltmeter reads 10 V at the collector of Fig. 7-26b. What are some of the troubles that can cause this high reading?

7-40. What if the emitter resistor is open in Fig. 7-26b? What will a voltmeter read for the base voltage? For the collector voltage?

7-41. A dc voltmeter measures 1.1 V at the collector of Fig. 7-26b. What are some of the possible troubles?

UNUSUAL PROBLEMS

7-42. You have built the circuit of Fig. 7-26a, and it is working normally. Now your job is to destroy the transistor. In other words, you are trying to find the ways in which to ruin the transistor. What will you try?

7-43. A first-year electronics student invents a new circuit. It works quite well when the current gain is between 90 and 110. Outside this range, it fails. The student plans to mass-produce the circuit by hand selecting 2N3904s that have the right current gain. He asks for your advice. What are some of the things you would say?

7-44. Somebody swears that she can build a base-biased circuit with a load line that is not straight. She is willing to bet you $50 that it can be done. Should you take the bet? Explain your answer.

7-45. A student wants to measure the collector-emitter voltage in Fig. 7-26b and so connects a voltmeter between the collector and emitter. What does the voltmeter read? (*Note:* There are many right answers.)

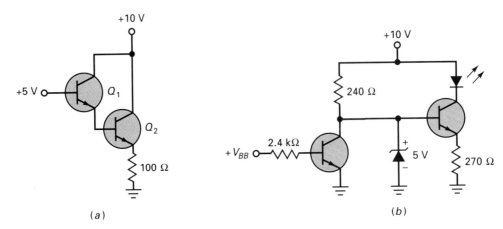

(a)

FIGURE 7-27

ADVANCED PROBLEMS

7-46. What is the collector current in Fig. 7-27a?

7-47. In Fig. 7-27a, the first transistor has a current gain of 100, and the second transistor has a current gain of 50. What is the base current in the first transistor?

7-48. What is the current through the LED of Fig. 7-27b if $V_{BB} = 0$? If $V_{BB} = 10$ V?

7-49. The zener diode of Fig. 7-27b is replaced by a 1N748. What is the LED current when $V_{BB} = 0$?

7-50. What is the maximum possible value of current through the 2 kΩ of Fig. 7-28a?

7-51. Figure 7-28b applies to the 4N33 of Fig. 7-28a. If the voltage across the 2 kΩ is 2 V, what is the value of V_{BB}?

7-52. The LED is open in Fig. 7-28a, and $V_{BB} = 3$ V. A voltmeter is connected between the collector of the 2N3904 and ground. What does the voltmeter read?

7-53. A VOM has a sensitivity of 20,000 Ω/V. The VOM is connected between the collector of Fig. 7-24a and ground. If the 3.3 kΩ is open, what does the VOM read on its 50-V range?

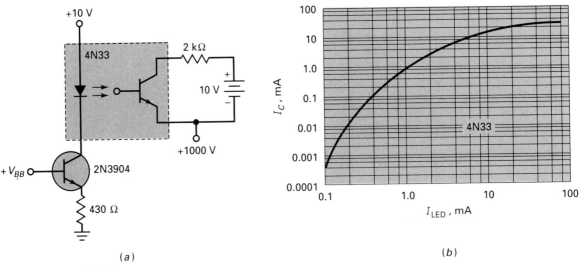

(a)

FIGURE 7-28

SOFTWARE ENGINE PROBLEMS

Use Fig. 7-29 for the remaining problems. Assume increases of approximately 10 percent in the independent variable, and use the second approximation of the transistor. A response should be an N (no change) if the change in a dependent variable is so small that you would have difficulty measuring it. For instance, you probably would find it difficult to measure a change of less than 1 percent. To a troubleshooter, a change like this is usually considered to be no change at all.

7-54. Try to predict the response of each dependent variable in the box labeled V_{BB}. Check your answers. Then answer the following question as simply and directly as possible. What effect does an increase in the base-supply voltage have on the dependent variables of the circuit?

7-55. Predict the response of each dependent variable in the box labeled V_{CC}. Check your answers. Then summarize your findings in one or two sentences.

7-56. Predict the response of each dependent variable in the box labeled R_E. Check your answers. List the dependent variables that decrease. Explain why these variables decrease, using Ohm's law or similar basic ideas.

7-57. Predict the response of each dependent variable in the box labeled R_C. List the dependent variables that show no change. Explain why these variables show no change.

7-58. Predict the response of each dependent variable in the box labeled β_{dc}. List the dependent variables that change. Explain this unusual result.

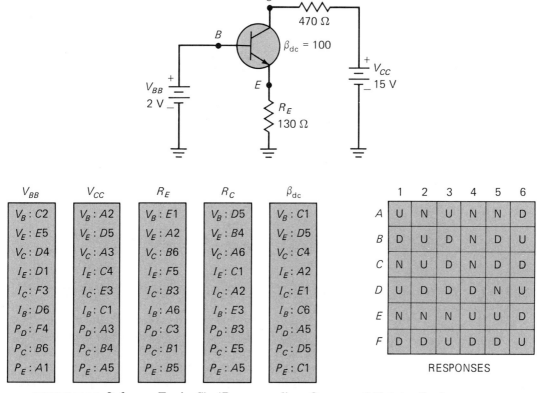

FIGURE 7-29 Software Engine™. (*Patent pending: Courtesy of Malvino Inc.*)

8 TRANSISTOR BIASING

A *prototype* is a basic circuit that a designer can modify to get more advanced circuits. Base bias is a prototype used in the design of digital circuits. Emitter bias is a prototype used in the design of amplifier circuits. In this chapter, we emphasize emitter bias and the important circuits derived from it.

Since you understand emitter bias, it is a simple matter to show you how circuits are derived from it. Furthermore, since you understand the process for analyzing emitter bias, you can easily learn the processes for analyzing advanced circuits based on emitter bias. You don't have to memorize many formulas. All you have to do is to reduce the advanced circuit to emitter bias.

8-1 VOLTAGE-DIVIDER BIAS

The most famous circuit based on the emitter-bias prototype is called the *voltage-divider bias*. Here is how it is derived from the emitter bias. Figure 8-1 shows the emitter-bias prototype. Before we start, recall these abbreviated steps in the analysis of emitter bias:

1. Emitter voltage
2. Emitter current
3. Collector current
4. Voltage drop across collector resistor
5. Collector voltage
6. Collector-emitter voltage

The first step means to find the emitter voltage by subtracting 0.7 V from the base voltage, the second step means to use Ohm's law to calculate the current through the emitter resistor, and so on with the remaining steps. You learned this process thoroughly in Chap. 7.

The six steps just given are more detailed than necessary. You ought to get along fine with these three steps:

1. Emitter current

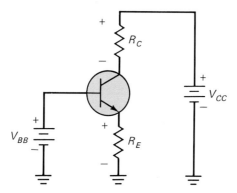

FIGURE 8-1 Emitter-bias prototype.

2. Collector voltage

3. Collector-emitter voltage

These are the crucial steps for analyzing the circuit. There is no need to memorize any formulas here because all you need is Ohm's law.

The Voltage Divider

Sometimes, the voltage from a power supply is too large to apply directly to the base, as shown in Fig. 8-1. How can we reduce this voltage without redesigning the power supply? The simplest way is to insert a voltage divider as shown in Fig. 8-2a. By selecting appropriate values of R_1 and R_2, we can lower the voltage to whatever level is suitable for our design.

The voltage across R_2 is symbolized as V_2. This voltage is applied directly to the base, which means that $V_B = V_2$. The process of analysis is the same as before, except that we start by finding the voltage across R_2. Once we have this voltage, we subtract 0.7 V to find the emitter voltage, and then we are on the way to a solution.

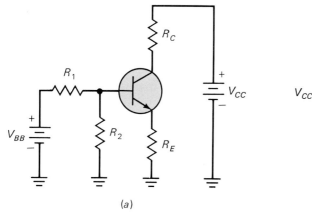
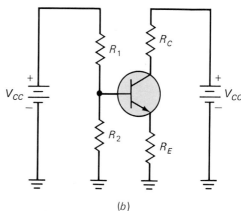

(a)

(b)

FIGURE 8-2 Modifying emitter bias.

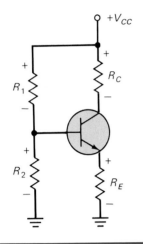

FIGURE 8-3

Voltage-divider bias.

One-Supply Systems

Some electronic systems have only one power-supply voltage. In this case, our circuit has to be designed as shown in Fig. 8-2b. This is all right because we can use whatever values of R_1 and R_2 are necessary to get the desired base voltage. For instance, if the voltage out of the power supply were 10 V, the base supply voltage would equal 10 V. If we wanted to have a base voltage of 2 V, we could use a 4:1 ratio for the voltage divider. This would produce 2 V at the base.

Take a good look at Fig. 8-2b, and answer this question. Can you see any way to draw this circuit more simply? Think about this for a few minutes, and then read on.

Whenever two unconnected nodes are at the same potential, you can connect a wire between them without changing the operation of the circuit. In Fig. 8-2b, the voltage between the top of R_1 and ground equals V_{CC}. Also, the voltage between the top of R_C and ground equals V_{CC}. Since these two node voltages are equal, we can connect a wire between them and redraw the circuit as shown in Fig. 8-3. In this simplified circuit, the voltage at the top of R_1 and R_C is still V_{CC}; therefore, the circuit operates as before.

8-2 VDB ANALYSIS

Figure 8-4 shows a voltage-divider-biased (VDB) circuit with specific values. The approach in analyzing the circuit is to start by finding the base voltage. Then the rest of the process is the same as discussed for emitter bias. But the way you carry out the process depends on your motive, specifically, whether you are troubleshooting, designing, or somewhere in between. What follows is a middle-of-the-road method that uses the second approximation. (Later in Sec. 8-7 we show you the troubleshooter's method, a high-speed way to analyze the circuit.)

The Assumption

The designer of a circuit like Fig. 8-4 needs more accurate answers than a troubleshooter, but even here there is room for sensible compromise. Because design is an open-ended question with many right answers, there are no formal rules to cover all cases. In some designs, calculation errors of up to 20 percent may be all right. In other designs, calculation errors may need to be less than 1 percent.

What we discuss here is the typical design that tolerates errors of 5 percent or less. The designer usually starts with the second approximation. Then, while closing in on the final design, he or she might use the third approximation to get almost perfect answers. Here are how the calculations would look.

The process begins with finding the base voltage. To do this, we make an assumption. We assume the base current is so small that it has no effect on the voltage divider. In this case, we can calculate the current through the voltage divider of Fig. 8-4 as follows:

$$I = \frac{V_{CC}}{R_1 + R_2} \tag{8-1}$$

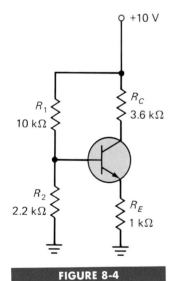

FIGURE 8-4

Example of voltage-divider bias.

This is Ohm's law applied to the total resistance of the voltage divider. In Fig. 8-4, the collector supply voltage is 10 V, and the total divider resistance is 12.2 kΩ. Therefore, the current through the voltage divider is approximately

$$I = \frac{10 \text{ V}}{12.2 \text{ k}\Omega} = 0.82 \text{ mA}$$

Remember: This first calculation is valid only when the base current is small enough. But in any well-designed circuit, the base current will be small enough. Because of this, you can almost always start the process of analysis by using Eq. (8-1).

How Small Is Small?

How small should the base current be? The usual design rule is that it should be at least 20 times smaller than the current through the voltage divider. This limits the calculation error to 5 percent, which is in line with our earlier discussion of error and the 20:1 rule. In this example, the base current should be less than

$$I_B = \frac{0.82 \text{ mA}}{20} = 41 \text{ }\mu\text{A}$$

When this condition is satisfied, we can calculate the base voltage as follows:

$$V_B = IR_2 \qquad (8\text{-}2)$$

This says the base voltage equals the current through the voltage divider times the lower resistance of the voltage divider. In Fig. 8-4,

$$V_B = (0.82 \text{ mA})(2.2 \text{ k}\Omega) = 1.8 \text{ V}$$

This is the voltage across R_2. Since the base is connected to the top of R_2, the base voltage is also 1.8 V.

Emitter Voltage and Current

The next step in the process is to get the emitter voltage with

$$V_E = V_B - V_{BE} \qquad (8\text{-}3)$$

This says that the emitter voltage equals the base voltage minus the voltage drop across the emitter diode. Using the second approximation gives

$$V_E = 1.8 \text{ V} - 0.7 \text{ V} = 1.1 \text{ V}$$

The emitter current is found with Ohm's law:

$$I_E = \frac{V_E}{R_E} \qquad (8\text{-}4)$$

In words, the emitter current equals the emitter voltage divided by the emitter resistance. Since there is 1.1 V across 1 kΩ, the emitter current is

$$I_E = \frac{1.1 \text{ V}}{1 \text{ k}\Omega} = 1.1 \text{ mA}$$

Collector Voltage and Collector-Emitter Voltage

As usual, we calculate the collector voltage with

$$V_C = V_{CC} - I_C R_C \tag{8-5}$$

This says the collector voltage equals the collector supply voltage minus the voltage dropped across the collector resistor. Since collector current is approximately equal to emitter current,

$$V_C = 10 \text{ V} - (1.1 \text{ mA})(3.6 \text{ k}\Omega) = 6.04 \text{ V}$$

Because the emitter is above the ground, we have to use this equation to get the collector-emitter voltage:

$$V_{CE} = V_C - V_E \tag{8-6}$$

This says you subtract the emitter voltage from the collector voltage to get the collector-emitter voltage. Therefore, we can calculate

$$V_{CE} = 6.04 \text{ V} - 1.1 \text{ V} = 4.94 \text{ V}$$

Checking the Assumption

At this point, the designer has to check the original assumption about the base current being 20 times smaller than the current through the voltage divider. Here is an example of how it is done. Suppose the current gain can vary from 36 to 300. The worst case is the lowest current gain because this produces the largest base current. In the worst case, the base current is

$$I_B = \frac{1.1 \text{ mA}}{36} = 30.5 \text{ }\mu\text{A}$$

This base current is less than the critical level of 41 μA that we calculated earlier. Therefore, the calculation error is less than 5 percent when we ignore the base current during the calculation of the base voltage.

Stiff Voltage Divider

There is a connection between voltage-divider bias and the stiff voltage divider discussed in Chap. 1. If you recall, a stiff voltage divider has a load voltage that is within 1 percent of the unloaded output voltage. When voltage-divider bias has been correctly designed, the voltage divider approaches the stiff condition. Because of this, the base voltage is almost constant and equal to the ideal voltage out of an unloaded voltage divider. The almost-constant base voltage is what keeps all other currents and voltages fixed, despite changes in transistors, temperature, etc. (For more discussion, see the "Optional Topics.")

Correction Factor for Current Gain

In Chap. 7, we discussed the correction factor for the collector current:

$$I_C = \frac{\beta_{dc}}{\beta_{dc} + 1}(I_E)$$

Since the worst-case current gain is 36, the correction factor produces this collector current:

$$I_C = \frac{36}{37}(1.1\text{ mA}) = 1.07\text{ mA}$$

As you see, the improvement in accuracy is only minor. In fact, the improvement is less than 5 percent, which means it isn't worth doing if the design accepts errors of 5 percent. The current gain has to hit 20 before the calculation error is 5 percent. (*Note:* 20 divided by 21 is 0.95, which produces a collector current that is 5 percent less than 1.1 mA.)

Q Point Is Immune to Changes in Current Gain

Keeping the base current small simplifies the analysis of the voltage-divider bias. But even more important, it makes the Q point immune to changes in the current gain. Here is why. If the base current is so small that it has no effect on the base voltage, then changes in the current gain have no effect on the base voltage. But a fixed base voltage means a fixed emitter voltage, a fixed emitter current, a fixed collector current, and a fixed collector voltage. In other words, the Q point is immune to changes in the current gain. When the current gain changes, the only other thing that changes is the base current. It may increase or decrease. But this doesn't matter because the base current is effectively out of the picture.

In fact, voltage-divider bias is really emitter bias in disguise. When the voltage divider is stiff, it produces a rock-solid voltage at the base. But this is equivalent to an emitter-biased circuit. In Chap. 7, you saw how perfect (almost) the emitter bias is for holding the Q point fixed. Therefore, a voltage-divider biased circuit approaches the performance level of an emitter-biased circuit.

Summary of Formulas and Process

Here are the formulas we used in our analysis:

Divider Current: $I = \dfrac{V_{CC}}{R_1 + R_2}$

Base Voltage: $V_B = IR_2$

Emitter Voltage: $V_E = V_B - V_{BE}$

Emitter Current: $I_E = \dfrac{V_E}{R_E}$

Collector Voltage: $V_C = V_{CC} - I_C R_C$

Collector-Emitter Voltage: $V_{CE} = V_C - V_E$

Here is the process used to analyze the circuit:

1. Find current I through R_1 and R_2 (ignore base current).

2. Find the voltage across R_2 (I times R_2).

3. Use the emitter-bias process from here on.

or simply

1. I

2. V_B

3. EB process

For a 5 percent analysis, a designer will add these steps:

4. Divide I by 20 to get the critical base current.

5. Divide I_C by the worst-case current gain to get the base current.

6. Make sure the base current is less than the critical level.

or simply

4. $I/20$

5. I_C/β_{dc}

6. $I_C/\beta_{dc} < I/20$

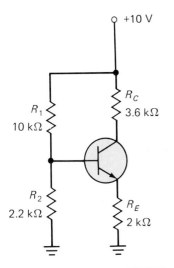

FIGURE 8-5

Example.

EXAMPLE 8-1

In Fig. 8-5, what is the collector-emitter voltage?

SOLUTION

This is the same circuit we analyzed earlier, except that the emitter resistance has doubled. Therefore, the base voltage is unaffected; it remains at 1.8 V. Also, the emitter voltage is still 1.1 V. Since the emitter resistance has doubled, the emitter current decreases to half its original value, which is

$$I_E = \frac{1.1\,\text{mA}}{2} = 0.55\,\text{mA}$$

Another way to get this value is with Ohm's law:

$$I_E = \frac{1.1\,\text{V}}{2\,\text{k}\Omega} = 0.55\,\text{mA}$$

The collector voltage is

$$V_C = 10\,\text{V} - (0.55\,\text{mA})(3.6\,\text{k}\Omega) = 8.02\,\text{V}$$

and the collector-emitter voltage is

$$V_{CE} = 8.02\,\text{V} - 1.1\,\text{V} = 6.92\,\text{V}$$

EXAMPLE 8-2

The collector supply voltage of Fig. 8-5 is tripled. What happens to the collector-emitter voltage?

SOLUTION

You don't have to recalculate the base voltage because you can use a shortcut. Since the supply voltage triples, the current through the divider triples. This means the base voltage triples, from 1.8 to 5.4 V. Therefore, the emitter voltage increases from 1.1 V to

$$V_E = 5.4 \text{ V} - 0.7 \text{ V} = 4.7 \text{ V}$$

The emitter current increases to

$$I_E = \frac{4.7 \text{ V}}{2 \text{ k}\Omega} = 2.35 \text{ mA}$$

The collector voltage is

$$V_C = 30 \text{ V} - (2.35 \text{ mA})(3.6 \text{ k}\Omega) = 21.5 \text{ V}$$

and the collector-emitter voltage is

$$V_{CE} = 21.5 \text{ V} - 4.7 \text{ V} = 16.8 \text{ V}$$

EXAMPLE 8-3

The resistors of Fig. 8-5 have a tolerance of ± 5 percent. What is the collector voltage if all resistances are 5 percent higher than their nominal values?

SOLUTION

Here are the new resistances values:

$$R_1 = 1.05(10 \text{ k}\Omega) = 10.5 \text{ k}\Omega \qquad R_E = 1.05(2 \text{ k}\Omega) = 2.1 \text{ k}\Omega$$

$$R_2 = 1.05(2.2 \text{ k}\Omega) = 2.31 \text{ k}\Omega \qquad R_C = 1.05 \, (3.6 \text{ k}\Omega) = 3.78 \text{ k}\Omega$$

Now, analyze the circuit as previously discussed. Visualize 10 V across the voltage divider. Since the total resistance of the voltage divider is larger, Ohm's law gives

$$I = \frac{10 \text{ V}}{10.5 \text{ k}\Omega + 2.31 \text{ k}\Omega} = 0.781 \text{ mA}$$

When this current flows through R_2, it produces a voltage of

$$V_2 = (0.781 \text{ mA})(2.31 \text{ k}\Omega) = 1.8 \text{ V}$$

Subtract the voltage across the emitter diode to get the emitter voltage:

$$V_E = 1.8 \text{ V} - 0.7 \text{ V} = 1.1 \text{ V}$$

Next, calculate the emitter current:

$$I_E = \frac{1.1 \text{ V}}{2.1 \text{ k}\Omega} = 0.524 \text{ mA}$$

This flows through the collector resistor and produces a collector voltage of

$$V_C = 10 \text{ V} - (0.524 \text{ mA})(3.78 \text{ k}\Omega) = 8 \text{ V}$$

8-3 VDB LOAD LINE AND *Q* POINT

As discussed in Sec. 7-7, emitter bias can have two different dc load lines. We will use the second dc load line, the one you get by varying the emitter resistor. This second dc load line is more useful with transistor amplifiers. Because of the stiff voltage divider in Fig. 8-6, the emitter voltage is held constant at 1.1 V in the following discussion.

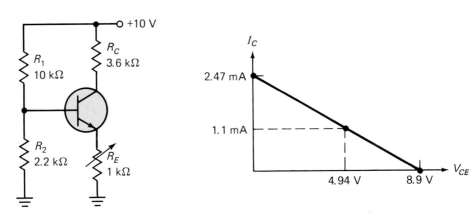

FIGURE 8-6 Voltage-divider bias and load line.

Saturation and Cutoff Points

Here is how to find the saturation current. Assume we decrease the emitter resistor until the transistor is barely saturated. Then, the collector-emitter voltage is approximately zero. This means the voltage across the collector resistor is

$$V_{RC} = 10 \text{ V} - 1.1 \text{ V} = 8.9 \text{ V}$$

Since there is 8.9 V across the collector resistor, the saturation current is

$$I_{C(\text{sat})} = \frac{8.9 \text{ V}}{3.6 \text{ k}\Omega} = 2.47 \text{ mA}$$

Next, assume we increase the emitter resistance toward infinity. Under this condition, there is no current through the collector resistor and the entire supply voltage appears at the collector node. This means the cutoff voltage is

$$V_{CE(\text{cut})} = 10 \text{ V} - 1.1 \text{ V} = 8.9 \text{ V}$$

The *Q* Point

The *Q* point was calculated in Sec. 8-2. It has a collector current of 1.1 mA and a collector-emitter voltage of 4.94 V. These values are plotted to get the *Q* point shown in Fig. 8-6. Since voltage-divider bias is derived from emitter bias, the *Q* point is virtually immune to changes in current gain. The way to move the *Q* point on this load line is by varying the emitter resistor. A designer can fine-tune the value of this resistance to get the *Q* point wherever she or he wants it along the given load line.

For instance, if the emitter resistance is changed to 2.2 kΩ, the collector current decreases to

$$I_E = \frac{1.1\,\text{V}}{2.2\,\text{k}\Omega} = 0.5\,\text{mA}$$

The voltages change as follows:

$$V_C = 10\,\text{V} - (0.5\,\text{mA})(3.6\,\text{k}\Omega) = 8.2\,\text{V}$$

and

$$V_{CE} = 8.2\,\text{V} - 1.1\,\text{V} = 7.1\,\text{V}$$

Therefore, the new Q point will have coordinates of 0.5 mA and 7.1 V, about half way down the load line from the present position.

On the other hand, if we decrease the emitter resistance to 510 Ω, the emitter current increases to

$$I_E = \frac{1.1\,\text{V}}{510\,\Omega} = 2.15\,\text{mA}$$

and the voltages change to

$$V_C = 10\,\text{V} - (2.15\,\text{mA})(3.6\,\text{k}\Omega) = 2.26\,\text{V}$$

and

$$V_{CE} = 2.26\,\text{V} - 1.1\,\text{V} = 1.16\,\text{V}$$

In this case, the Q point shifts up the load line to a new position with coordinates of 2.15 mA and 1.16 V.

Q Point in Middle of Load Line

Now V_{CC}, R_1, R_2, and R_C control the saturation current and the cutoff voltage. A change in any of these quantities will change $I_{C(\text{sat})}$ and/or $V_{CE(\text{cut})}$. Once the designer has established the values of the foregoing variables, he or she uses the emitter resistance to set the Q point at any position along the load line.

If R_E is too large, the Q point moves into the cutoff point. If R_E is too small, the Q point moves into saturation. Some designers set the Q point at the middle of the load line. The reason for this is given in the next chapter. For Fig. 8-6, this means using an emitter resistance that sets up a collector current of half of 2.47 mA, or 1.23 mA. Since the emitter voltage is at 1.1 V, this requires an emitter resistance of

$$R_E = \frac{1.1\,\text{V}}{1.23\,\text{mA}} = 894\,\Omega$$

The nearest standard value is 910 Ω, which is what we would use in a practical circuit.

Equations for Ends of Load Line

Other dc load lines are possible, but the one discussed earlier is the most practical for our purposes. Specifically, the dc load line we are interested in is the one where the emitter voltage is fixed and controlled by the voltage divider in the base circuit. Under this condition, the cutoff voltage is

$$V_{CE(\text{cut})} = V_{CC} - V_E \qquad (8\text{-}7)$$

This says the cutoff voltage equals the collector supply voltage minus the emitter voltage. Given a specific circuit, you work out the dc emitter voltage. Then you use the process described earlier, or you can substitute the values of V_{CC} and V_E into Eq. (8-7) to get the cutoff voltage.

Similarly, the saturation current is

$$I_{C(\text{sat})} = \frac{V_{CC} - V_E}{R_C} \qquad (8\text{-}8)$$

This says the collector saturation current equals the cutoff voltage divided by the collector resistance. Again, you can use the process described earlier, or you can substitute directly into this equation to find the saturation current.

EXAMPLE 8-4

The collector supply voltage doubles in Fig. 8-6. What happens to the load line?

SOLUTION

The base voltage will double from 1.8 to 3.6 V. The emitter voltage will more than double because

$$V_E = 3.6\ \text{V} - 0.7\ \text{V} = 2.94\ \text{V}$$

Visualize this voltage at the emitter node of Fig. 8-6. Hold it fixed in your mind as you short the collector and the emitter. Then you can see 2.94 V at the bottom of the 3.6 kΩ. With Ohm's law, the collector current for this condition is

$$I_{C(\text{sat})} = \frac{20\ \text{V} - 2.94\ \text{V}}{3.6\ \text{k}\Omega} = 4.74\ \text{mA}$$

This represents the upper end of the load line.

The lower end of the load line is obtained by mentally opening the transistor between the collector and the emitter. In this case, all the collector supply voltage appears at the collector node. The emitter node voltage is still 2.94 V because of the voltage divider in the base circuit. Therefore, the cutoff voltage is

$$V_{CE(\text{cut})} = 20\ \text{V} - 2.94\ \text{V} = 17.1\ \text{V}$$

8-4 TWO-SUPPLY EMITTER BIAS

Some electronic equipment has a power supply that produces both positive and negative supply voltages. For instance, Fig. 8-7 shows a transistor circuit with two power supplies: $+10$ and -2 V. The negative supply

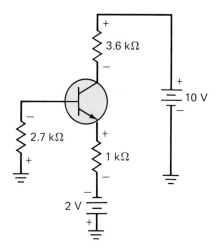

FIGURE 8-7 Two-supply emitter bias.

forward-biases the emitter diode. The positive supply reverse-biases the collector diode. This circuit is derived from emitter bias. For this reason, we refer to it as two-supply emitter bias.

Analysis

The first thing to do is to redraw the circuit as it usually appears on schematic diagrams. This means deleting the battery symbols as shown in Fig. 8-8. This is necessary on schematic diagrams because there usually is no room for battery symbols on complicated diagrams. All the information is still on the diagram, except that it is in condensed form. That is, a negative supply voltage of -2 V is applied to the bottom of the 1 kΩ, and a positive supply voltage of $+10$ V is applied to the top of the 3.6 kΩ.

When this type of circuit is correctly designed, the base current will be small enough to ignore. This is equivalent to saying that the base voltage is 0 V as shown in Fig. 8-9. Later, we reexamine this assumption to see how a designer satisfies the condition. But for now assume the base voltage is so small that it is approximately 0 V.

The voltage across the emitter diode is 0.7 V, which is why -0.7 V is shown on the emitter node. If this is not clear, stop and think about it. There is a plus-to-minus drop of 0.7 V in going from the base to the emitter. If the base voltage is 0 V, the emitter voltage cannot be $+0.7$ V; it must be -0.7 V.

In Fig. 8-9, the emitter resistor again plays the key role in setting up the emitter current. To find this current, you apply Ohm's law to the emitter resistor as follows: The top of the emitter resistor has a voltage of -0.7 V, and the bottom has a voltage of -2 V. Therefore, the voltage across the emitter resistor equals the difference between the two voltages. To get the right answer, you subtract the more negative value from the more positive value. In this case, the more negative value is -2 V, so

$$V_{RE} = -0.7 \text{ V} - (-2 \text{ V}) = 1.3 \text{ V}$$

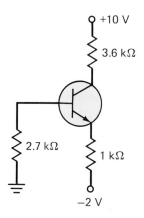

FIGURE 8-8
Two-supply emitter bias redrawn.

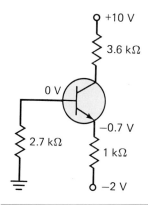

FIGURE 8-9
Two-supply emitter bias with voltages.

If you don't like mathematical tricks with negative signs, you can always use your common sense as follows: The top of the emitter resistor has a node voltage of -0.7 V, and the bottom has a node voltage of -2 V. As we move from -2 to -0.7 V, we pass through a positive change of 1.3. Therefore, the difference of potential across the emitter resistor is 1.3 V.

Once you have found the voltage across the emitter resistor, calculate the emitter current with Ohm's law:

$$I_E = \frac{1.3 \text{ V}}{1 \text{ k}\Omega} = 1.3 \text{ mA}$$

This current flows through the 3.6 kΩ and produces a voltage drop which we subtract from $+10$ V as follows:

$$V_C = 10 \text{ V} - (1.3 \text{ mA})(3.6 \text{ k}\Omega) = 5.32 \text{ V}$$

The collector-emitter voltage is the difference between the collector voltage and the emitter voltage:

$$V_{CE} = 5.32 \text{ V} - (-0.7 \text{ V}) = 6.02 \text{ V}$$

A More Accurate Analysis

Recall that voltage-divider bias requires a small base current. Two-supply emitter bias has the same requirement. Again, we are in the area of an *open-ended* question, one that has many right answers depending on the situation. On the other hand, the *closed-ended* question has only one right answer. In industry, open-ended questions are much more common.

What follows is one of the right answers to this question: How small does the base current have to be for the Q point to be immune to changes in current gain? For this discussion, assume the current gain can vary from 36 to 300 with temperature change, transistor replacement, etc.

Ideally, the base voltage is 0 V. In this case, the voltage across the emitter resistor has a constant value that is unaffected by the changes in current gain. In reality, the base current is greater than zero. Because of the direction of the current through the base resistor, the base voltage is slightly negative. Assume we can tolerate a base voltage as large as -0.1 V. This means the base voltage may vary from 0 to -0.1 V when the current gain varies with temperature, collector current, or transistor replacement.

In the worst case, therefore, the base voltage may be -0.1 V. Then the emitter voltage can be as negative as -0.8 V. In this case, the voltage across the emitter resistor becomes 1.2 instead of 1.3 V, calculated as follows:

$$V_{RE} = -0.8 \text{ V} - (-2 \text{ V}) = 1.2 \text{ V}$$

Therefore, the emitter current will be slightly lower:

$$I_E = \frac{1.2 \text{ V}}{1 \text{ k}\Omega} = 1.2 \text{ mA}$$

The other voltages are slightly higher:

$$V_C = 10 \text{ V} - (1.2 \text{ mA})(3.6 \text{ k}\Omega) = 5.68 \text{ V}$$

and

$$V_{CE} = 5.68\,\text{V} - (-0.8\,\text{V}) = 6.48\,\text{V}$$

Notice that this voltage is larger than the collector voltage because we subtracted a negative quantity.

At this point, the designer has to check the assumption about the base voltage being -0.1 V. In the worst case, the current gain is minimum and the base current is

$$I_B = \frac{1.2\,\text{mA}}{36} = 33.3\,\mu\text{A}$$

When this base current flows through the base resistor, it produces a voltage of

$$V_B = -(33.3\,\mu\text{A})(2.7\,\text{k}\Omega) = -0.0899\,\text{V}$$

This base voltage is less than the limit of -0.1 V.

What it means is this. When the current gain is on the high side (near 300), the base current is very small and the base voltage is near 0 V. When the current gain is on the low side (near 36), the base current has increased to 33.3 μA and the base voltage to -0.0899 V. Since this is less than our worst-case analysis, we know the actual operating point lies somewhere between these two possibilities:

Best case: 1.3 mA and 6.02 V
Worst case: 1.2 mA and 6.48 V

It means the Q point is not exactly constant, but almost. This is how a well-designed circuit should behave. The Q point should vary only slightly under all operating conditions.

From now on, when you see a two-supply emitter-biased circuit like Fig. 8-9, assume that it is well designed. For all calculations, therefore, you can use a base voltage of 0 V. If you are ever in doubt about this assumption, calculate the base current and multiply by the base resistance. This voltage should be small, typically less than -0.1 V. If you are troubleshooting a circuit like this, a voltmeter reading between the base and ground should produce a low reading; otherwise, something is wrong with the circuit.

The Process

Here is the process used to analyze the circuit:

1. Find the voltage across R_E (assume V_B is 0).
2. Find the current through R_E.
3. Find the collector voltage.
4. Find the collector-emitter voltage.

or simply

1. V_{RE}
2. I_E
3. V_C
4. V_{CE}

A designer will add these steps:

5. Divide I_C by the worst-case current gain to get the base current.
6. Multiply the base current and the base resistance to get V_B.
7. Voltage V_B should be small enough to satisfy design specifications.

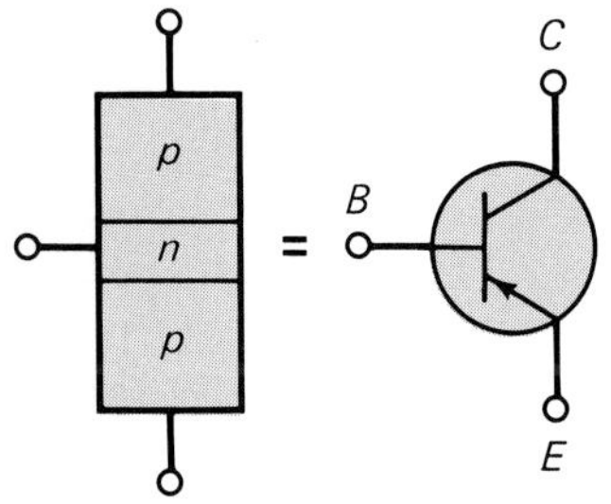

FIGURE 8-10

Pnp transistor.

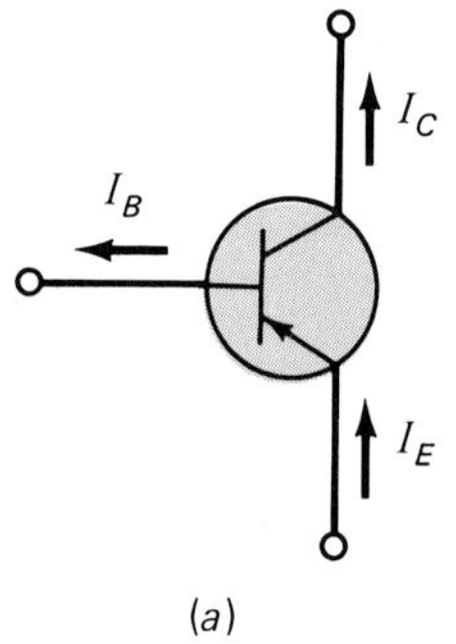

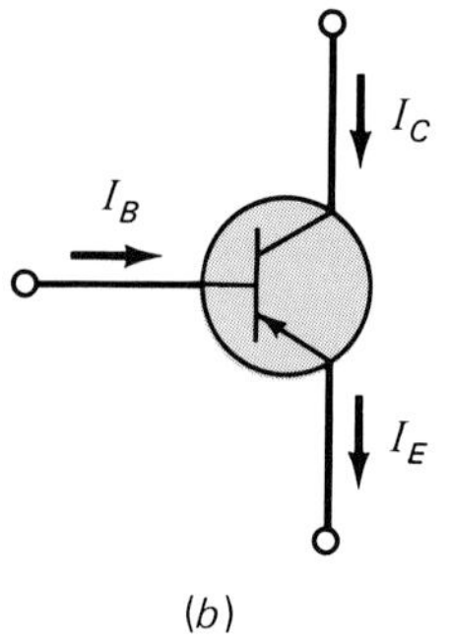

EXAMPLE 8-5

What is the approximate collector voltage in Fig. 8-9, on page 255, if the emitter resistor is increased to 1.8 kΩ?

SOLUTION

The voltage across the emitter resistor is still 1.3 V. The emitter current is

$$I_E = \frac{1.3\ \text{V}}{1.8\ \text{k}\Omega} = 0.722\ \text{mA}$$

The collector voltage is

$$V_C = 10\ \text{V} - (0.722\ \text{mA})(3.6\ \text{k}\Omega) = 7.4\ \text{V}$$

8-5 PNP TRANSISTORS

If you understand how an *npn* transistor works, the road to understanding a *pnp* transistor has only a few small bumps. Figure 8-10 shows the structure of a *pnp* transistor along with its schematic symbol. Because the doped regions are of the opposite type, we have to turn our thinking around. Specifically, it means that holes are the majority carriers in the emitter instead of free electrons.

Basic Ideas

Briefly, here is what happens at the atomic level. The emitter injects holes into the base. The majority of these holes flow on to the collector. For this reason, the collector current is almost equal to the emitter current. The base current is much smaller than either. As before, the current gain of the transistor equals the collector current divided by the base current. As with *npn* transistors, the current gain varies enormously with collector current, temperature, and transistor replacement.

Figure 8-11a shows the three transistor currents with the conventional-flow viewpoint. Figure 8-11b shows the current from the electron-flow viewpoint. As before, these currents are related as follows:

$$I_E = I_C + I_B \tag{8-9}$$

The current gain is expressed as

$$\beta_{dc} = \frac{I_C}{I_B} \tag{8-10}$$

(a)

(b)

FIGURE 8-11

Schematic symbol for *pnp* transistor.

which has these equivalent forms:

$$I_C = \beta_{dc}I_B \quad \text{and} \quad I_B = \frac{I_C}{\beta_{dc}}$$

Negative Supply

If you have a circuit with *npn* transistors, you can often use the same circuit with a negative power supply and *pnp* transistors. For instance, Fig. 8-12 shows voltage-divider bias with a *pnp* transistor and a negative supply voltage of − 10 V. The 2N3906 is the complement of the 2N3904, meaning its characteristics have the same absolute values as those of the 2N3904, but all currents and voltage polarities are reversed. The circuit is similar to the *npn* circuit of Fig. 8-4. There is no need to get tangled up with minus signs when we analyze this circuit. If we use our common sense, the minus signs will not be a problem.

For instance, here is how a troubleshooter will take apart this circuit:

> The current through the voltage divider is about 1 mA. Therefore, I've got about 2 V across the 2.2 kΩ. Because the supply voltage is negative, the base voltage should be around − 2 V if I measure it. There's a voltage drop of 0.7 V across the emitter diode, so only 1.3 V is across the emitter resistor. This produces a collector current of about 1.3 mA, so I get a drop of 4 V across the collector resistor. This means the collector voltage is about 4 less than 10 V, or about 6 V. Again, the supply voltage is negative which means the voltage should be around − 6 V if I measure it.

The only thing that makes this circuit harder to work with is the negative signs. You will get used to this with practice. We will not spend too much time on *pnp* transistors and negative power supplies because they are not as important as *pnp* transistors used with positive power supplies.

Positive Power Supply

Positive power supplies are far more common than negative power supplies. Because of this, you often see *pnp* transistors drawn upside-down as shown in Fig. 8-13. It takes a while to get used to this upside-down circuit, but it's worth the effort because you will see many schematic diagrams drawn like this in industry.

How do you analyze this circuit? The process is very much the same as before. To get the basic idea quickly, here is the troubleshooter's method of analysis:

> The current through the voltage divider is around 0.8 mA. This produces about +8 V at the base node. Because of the 0.7 V across the emitter diode, the emitter node is 0.7 V higher than the base node, or +8.7 V. This means the voltage across the emitter resistor is about 10 minus 8.7, or 1.3 V. The current through the 1 kΩ is 1.3 V divided by 1 kΩ, or 1.3 mA. This sets up a collector current of approximately 1.3 mA. When this flows through the 3.6 kΩ, it produces a voltage of about 4 V. Therefore, I should measure a collector-to-ground voltage in the vicinity of 4 V. Otherwise, something is wrong with the circuit.

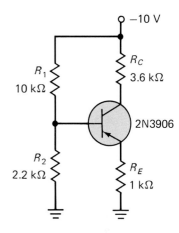

FIGURE 8-12

Voltage-divider bias with *pnp* transistor.

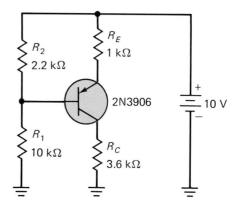

Upside-down *pnp* transistor.

The designer goes through the same process as the troubleshooter, but calculates more accurately by using a calculator or a computer. You don't need any new equations here. If you just apply Ohm's law and the basic ideas discussed earlier, you can analyze this circuit quickly and easily.

For instance, here is a more accurate analysis of Fig. 8-13. If the base current is small enough to ignore, we can calculate the current through the voltage divider of Fig. 8-13 like this:

$$I = \frac{10\,\text{V}}{12.2\,\text{k}\Omega} = 0.82\,\text{mA}$$

In line with our 20:1 rule, the base current should be less than

$$I_B = \frac{0.82\,\text{mA}}{20} = 41\,\mu\text{A}$$

If the base current is less than 41 μA, we can calculate the voltage across R_2 as follows:

$$V_2 = (0.82\,\text{mA})(2.2\,\text{k}\Omega) = 1.8\,\text{V}$$

Now, we subtract 0.7 V from this voltage to get the voltage across the emitter resistor:

$$V_{RE} = 1.8\,\text{V} - 0.7\,\text{V} = 1.1\,\text{V}$$

Next, we calculate the emitter current as

$$I_E = \frac{1.1\,\text{V}}{1\,\text{k}\Omega} = 1.1\,\text{mA}$$

and the collector voltage as

$$V_C = (1.1\,\text{mA})(3.6\,\text{k}\Omega) = 3.96\,\text{V}$$

This is the voltage between the collector and ground. It is the voltage you would measure when you connect the active lead of a voltmeter to the collector and the common lead to ground.

The emitter voltage equals the supply voltage minus the voltage drop across the emitter resistor:

$$V_E = 10\,\text{V} - 1.1\,\text{V} = 8.9\,\text{V}$$

This is the voltage between the emitter and ground. It is the voltage you would measure with the active lead of a voltmeter on the emitter and the common lead on ground.

The collector-emitter voltage is the algebraic difference of the collector and emitter voltages:

$$V_{CE} = 3.96 \text{ V} - 8.9 \text{ V} = -4.94 \text{ V}$$

Notice that this voltage is negative because it is the voltage between the collector and emitter. Stated another way, it is the voltage of the collector with the emitter as a starting point. If you start at the emitter, you have to move from 8.94 to 3.96 V, which is a negative change.

At this point, the designer can check the assumption about the base current being small. If the worst-case current gain is as low as 36, then the base current is

$$I_B = \frac{1.1 \text{ mA}}{36} = 30.6 \text{ μA}$$

This base current is less than the critical level of 41 μA calculated earlier. Therefore, the calculation error is less than 5 percent when we ignore the base current.

If you remember the process, you will not get confused by the minus signs. Here is the process for analyzing upside-down *pnp* voltage-divider bias:

1. Find current I through R_1 and R_2.

2. Find the voltage across R_2.

3. Subtract 0.7 V to get the voltage across the emitter resistor.

4. Find the emitter current.

5. Find the collector voltage.

6. Find the emitter voltage.

7. Find the collector-emitter voltage (it's negative).

Each step requires at most Ohm's law, so you don't have to clutter your mind with a lot of formulas. If you forget how to carry out any step, reread the discussion. Everything makes sense, so it should be easy to remember.

Negative Collector-Emitter Voltage

When you look at some data sheets for *pnp* transistors, they will show negative values for the collector-emitter voltage. For instance, the Texas Instruments data sheet for a 2N3906 shows a collector breakdown voltage of

$$V_{CEO} = -40 \text{ V}$$

Similarly, the minimum current gain is listed as

$$h_{FE} = 100 \qquad \text{when } I_C = 10 \text{ mA}, V_{CE} = -1 \text{ V}$$

But not all data sheets use negative values. For instance, the Motorola data sheet for a 2N3906 shows

$$V_{CEO} = 40 \text{ V}$$

and a minimum current gain of

$$h_{FE} = 100 \quad \text{when } I_C = 10 \text{ mA}, V_{CE} = 1 \text{ V}$$

What's going on here? Is the collector-emitter voltage positive or negative? The answer is simple. The Motorola data sheets use the absolute values, while the Texas Instruments data sheets include the negative sign.

Who's right? Motorola or Texas Instruments? They're both right. If you want to ignore the minus sign, you can, as long as you remember that you are dealing with absolute values. If you want to be mathematically precise, you can include the minus sign. The minus sign removes all doubt, but it does clutter the data sheets. After you have worked with *pnp* transistors for a while, you automatically know that the collector-emitter voltage is negative whether or not a data sheet shows it as such.

How are you going to survive in such a messy world, a world where different people are using different rules? By learning to use your right brain as well as your left. Remember the left brain wants everything done the same way: only one right answer and only one right way to get the answer, a formula for everything and no loose ends. However, the real world is not a left-brain world. To succeed, you must learn to use both sides of your brain.

EXAMPLE 8-6

In Fig. 8-14, what is the collector-emitter voltage?

SOLUTION

Begin by calculating the current through the voltage divider. Visualize 20 V across the voltage divider. Since the total resistance of the voltage divider is 12.2 kΩ, Ohm's law gives

$$I = \frac{20 \text{ V}}{12.2 \text{ k}\Omega} = 1.64 \text{ mA}$$

When this current flows through R_2, it produces a voltage of

$$V_2 = (1.64 \text{ mA})(2.2 \text{ k}\Omega) = 3.61 \text{ V}$$

In Fig. 8-14, this voltage is across the emitter diode in series with the emitter resistor. If we subtract the voltage across the emitter diode, we get the voltage across the emitter resistor:

$$V_{RE} = 3.61 \text{ V} - 0.7 \text{ V} = 2.91 \text{ V}$$

Next, we calculate the emitter current:

$$I_E = \frac{2.91 \text{ V}}{1.5 \text{ k}\Omega} = 1.94 \text{ mA}$$

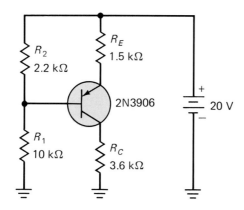

FIGURE 8-14   Example.

This flows through the collector resistor and produces a collector voltage of

$$V_C = (1.94 \text{ mA})(3.6 \text{ k}\Omega) = 6.98 \text{ V}$$

How are we going to find the collector-emitter voltage? Here is what we already know:

$$V_C = 6.98 \text{ V} \qquad V_{RE} = 2.91 \text{ V}$$

Look at Fig. 8-14 and visualize 2.91 V across the emitter resistor and 6.98 V at the collector node. Can you see what has to be done? Think about this for a few minutes before you read what comes next.

The supply voltage puts +20 V at the top of the emitter resistor. Subtract 2.91 V to get the voltage at the emitter node:

$$V_E = 20 \text{ V} - 2.91 \text{ V} = 17.1 \text{ V}$$

The collector-emitter voltage is

$$V_{CE} = 6.98 \text{ V} - 17.1 \text{ V} = -10.1 \text{ V}$$

The minus sign occurs because the transistor is upside-down.

Including the minus sign is the mathematical approach to circuit analysis. Many people prefer to drop the minus sign and use absolute values. This is perfectly all right as long as you know what you are doing. When you look at Fig. 8-14, you can see that the emitter is more positive than the collector. That's all that matters. You don't have to include a minus sign to remind you of this reversal. If you ask 9 troubleshooters out of 10 what the collector-emitter voltage is in Fig. 8-14, they probably would answer like this:

I hate minus signs. They prevent me from working with the true polarities of the circuit. As far as I'm concerned, they are a mathematical excuse for not knowing what is going on. When I look at a circuit, I prefer to use my head instead of a minus sign.

In Fig. 8-14, V_{CE} is 10.1 V with the emitter more positive than the collector.

When you say that V_{CE} is 10.1 V with the emitter more positive than the collector, you have said it all. It is equivalent to saying $V_{CE} = -10.1$ V. Both approaches are valid, and the one you use often depends on what you are trying to do. If you are flexible, you will be able to use either approach, without the minus sign for troubleshooting, with the minus sign when programming a computer, etc.

8-6 OTHER TYPES OF BIAS

In this final section, we discuss some other types of bias. A detailed analysis of these types of bias is not necessary because they are rarely used. But you should at least be aware of their existence in case you see them on a schematic diagram. The main reason for including this section is that it allows us to introduce some ideas used in other areas of electronics, troubleshooting, design, etc., ideas like negative feedback, trial-and-error solutions, etc.

Emitter-Feedback Bias

Figure 8-15a shows base bias. This circuit is a disaster when it comes to setting up a fixed Q point. Why? Because in the active region the collector current is hypersensitive to the current gain. This means the changes in the Q point are directly proportional to the changes in the current gain.

Historically, the first attempt at stabilizing the Q point was to introduce an emitter resistor as shown in Fig. 8-15b. How can this emitter resistor stabilize the Q point? The inventor's thinking process for the circuit went like this:

> Assume the current gain increases because of a temperature increase. This will increase the collector current and emitter current. An increase in the emitter current will increase the emitter voltage, which in turn increases the base voltage. An increase in the base voltage will produce a decrease in the voltage across the base resistor, which means less base current.

At this point, the inventor is close to a new discovery. She or he has seen that an increase in current gain produces a decrease in base current. Here is what comes next:

> I've got it! Less base current means less collector current and less emitter current. These decreasing changes oppose the original increasing changes. So, the emitter resistor produces a voltage that opposes the change in current gain.

The inventor has rediscovered negative feedback, a concept known by earlier workers in electronics.

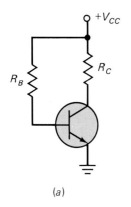

(a)

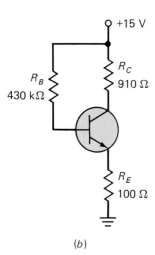

(b)

FIGURE 8-15

(a) Base bias; (b) emitter-feedback bias.

What is negative feedback? The base current is an input to the transistor, while the collector and emitter currents are outputs. When the current gain increases, it increases an output quantity (emitter current). This output quantity flows through the emitter resistor, which decreases an input quantity (base current). This is very important: *The output has changed the input.* This is the first time you have encountered this phenomenon in this book. It's called *feedback*, and it refers to the output controlling the input, at least partially. Because an increase in emitter current produces a decrease in base current, we refer to this type of feedback as *negative feedback*.

Negative feedback is very useful in stabilizing electronic circuits. You will see more examples in later chapters. This is your introduction to it. All you are supposed to get from this nonmathematical discussion is the basic idea: Negative feedback exists when an increase in an output quantity produces a decrease in an input quantity. The converse is also true: Negative feedback exists when a decrease in an output quantity produces an increase in an input quantity.

Analysis of Emitter-Feedback Bias

Although the idea sounds good, it doesn't work that well. That is why emitter-feedback bias never became popular. The negative feedback does reduce the shift in Q point with changes in current gain, but it does not eliminate it the way the voltage-divider bias does. The movement of the Q point is still too large for most applications.

An exact analysis of this circuit is a waste of time. But a troubleshooter's analysis is not because it will introduce another way to find right answers. What you are about to see is a method that pure mathematicians hate. And yet, the method can produce usable answers in many situations where you don't know how to find the exact mathematical answers.

The method is called *trial and error*. If you feel an aversion to this method, it is because of those early years of mathematical training where you were taught there's only one right way to get an answer. Some of the best troubleshooters and designers use trial and error when all else fails, or when it takes too long to derive the exact formulas.

Here is how a troubleshooter might analyze the circuit of Fig. 8-15*b* to find the collector voltage:

> **I'm going to assume the collector voltage is 7.5 V, half the supply voltage. This means there's 7.5 V across the 910 Ω. And 7.5 V is almost 8 V, and 910 Ω is almost 1 kΩ, so the collector current is roughly 8 mA. The emitter current is about 8 mA and produces an emitter voltage of 0.8 V. The base voltage is 0.7 V higher than this, or 1.5 V. This means there is 13.5 V across the base resistor. Since 13.5 is almost 15 V and 430 is almost 500 kΩ, I can double 15 to get 30 V and can double 500 kΩ to get 1 MΩ. So, the base current is roughly 30 μA.**

At this point, the troubleshooter has these rough answers: $I_C = 8$ mA and $I_B = 30$ μA. The original assumption was a collector voltage of 7.5 V.

Working with these rough answers this is how the troubleshooter's thinking continues:

> The current gain is 8 mA divided by 30 μA. That's roughly the same as around 10 mA divided by 40 μA, or a current gain of 250. But 250 sounds too high. Suppose the current gain is only 100. I am going to assume the base current stays around 30 μA, which means the collector current is around 3 mA. When this flows through the 910 Ω, it produces a voltage of about 3 V. Subtracting this from 15 gives 12 V, the estimated value of collector voltage.

This method is called *trial-and-error* because you assume an answer (7.5 V) and then work backward from this answer to the other currents and voltages. When you have the collector current and base current, you can calculate the current gain. If this current gain is too high, you can change it to whatever value seems reasonable and recalculate the collector voltage (12 V).

The important thing about this method is that you don't need the exact formulas to analyze the circuit. Incidentally, here are the exact formulas:

$$I_C = \frac{V_{CC} - V_{BE}}{R_E + R_B/(\beta_{dc} + 1)} \tag{8-11}$$

and

$$V_C = V_{CC} - I_C R_C \tag{8-12}$$

And here are the exact calculations for a current gain of 100:

$$I_C = \frac{15\,V - 0.7\,V}{100\,\Omega + 430\,k\Omega/101} = 3.28\,mA$$

and

$$V_C = 15\,V - (3.28\,mA)(910\,\Omega) = 12\,V$$

This exact answer for the collector voltage is the same as the estimated answer found by trial and error. This is just a coincidence. Usually, the trial-and-error answer is only approximately correct.

The important idea to learn is that there is more than one way to find the collector voltage. First, there is the troubleshooter's way of using trial-and-error. This approach is acceptable because troubleshooters do not need exact answers. Furthermore, troubleshooters rarely use exact formulas like Eqs. (8-11) and (8-12). They don't have the time, and the job doesn't require it.

Second, there is the exact approach using the mathematical equations. Although some designers may use trial and error and a calculator to get exact answers, other designers may prefer using Eqs. (8-11) and (8-12).

A final point. If you are wondering where Eqs. (8-11) and (8-12) came from, here is the answer. By writing two simultaneous equations (one for the base circuit and the other for the collector circuit), we can derive Eqs. (8-11) and (8-12). We do not show this derivation, however, because the circuits are no longer in the mainstream of electronics.

Collector-Feedback Bias

Figure 8-16 shows collector-feedback bias. Historically, this was another attempt at stabilizing the Q point with negative feedback. How can the circuit stabilize the Q point? The inventor's thinking process for the circuit went like this:

> Assume the current gain increases because of a temperature increase. This will increase the collector current, which decreases the collector voltage. A decrease in collector voltage decreases the voltage across the base resistor, which means less base current. But less base current means less collector current. So, the resistor between the collector and the base is producing negative feedback.

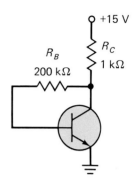

FIGURE 8-16
Collector-feedback bias.

Again, the idea is like a new-born baby. It has great potential but is not in its final form. In other words, the collector-feedback bias will reduce the shift in the Q point, but the improvement still falls short of the rock-solid Q point needed for mass production.

As before, an exact analysis is a waste of time. But a troubleshooter's analysis is worthwhile because it will give you additional insight into how to use your mind when you troubleshoot:

> I'm going to assume a collector voltage of 7.5 V because circuits are often designed with a collector voltage of half the supply voltage. In this case, the collector current is 7.5 mA. The voltage across the base resistor is 7.5 minus 0.7, which is approximately 7 V. Then 7 V divided by 200 kΩ is equivalent to 3.5 V divided by 100 kΩ, or 35 μA. The current gain is 7.5 mA divided by 35 μA, or a little bit more than 200.
>
> Suppose the current gain is only 100. I'm going to assume the base current stays in the vicinity of 35 μA, which means the collector current is around 3.5 mA. When this flows through the 1 kΩ, it produces a voltage of about 3.5 V. Subtracting this from 15 gives 11.5 V, the estimated value of the collector voltage.

Again, the important thing about this trial-and-error method is that you are liberated from the dependence on formulas. All you need is Ohm's law and your whole brain.

When exact answers are needed, you do need to work with formulas. By writing two loop equations, we can derive these exact formulas:

$$I_C = \frac{V_{CC} - V_{BE}}{R_C + R_B/(\beta_{dc} + 1)} \qquad (8\text{-}13)$$

and

$$V_C = V_{CC} - I_C R_C \qquad (8\text{-}14)$$

For a current gain of 100, the collector current and voltage are

$$I_C = \frac{15\,\text{V} - 0.7\,\text{V}}{1\,\text{k}\Omega + 200\,\text{k}\Omega/101} = 4.79\,\text{mA}$$

and

$$V_C = 15\,\text{V} - (4.79\,\text{mA})(1\,\text{k}\Omega) = 10.2\,\text{V}$$

This collector voltage is different from the estimated answer found by trial and error (10.2 versus 11.5 V). This difference is not great enough to concern a troubleshooter. He or she knows the estimate of 11.5 V is only a ballpark value. If a troubleshooter does measure 10.2 instead of 11.5 V, here is how she or he would probably react:

> I estimated 11.5 V for the collector voltage and measured 10.2 V. That's close enough. After all, I was rounding heavily when I calculated the collector voltage. The transistor is probably all right.

If there is any doubt at point, the troubleshooter can always refine the calculations as follows:

> Since I measured 10.2 V at the collector, the voltage across the base resistor is actually 9.5 V. Dividing this by 200 kΩ gives 47.5 µA. Multiplying by a current gain of 100 gives a collector current of 4.75 mA. When this flows through 1 kΩ, the voltage drop is 4.75 V. Subtracting this from the supply voltage leaves 10.25 V. That's better. That's almost exactly what I measured. So there is no question about it: The transistor is all right.

Collector- and Emitter-Feedback Bias

Emitter-feedback bias and collector-feedback bias were the first steps toward a more stable bias for transistor circuits. Even though the idea of negative feedback is a good one, these circuits fall short because there is not enough negative feedback to do the job. This is why the next step in biasing was the circuit shown in Fig. 8-17. Here you can see what the inventor had in mind:

> If I need more negative feedback, why not use a combination of an emitter resistor and a collector resistor? Then I ought to get still better results. It's worth a try.

As it turns out, more is not always better. But you don't know until you try. In this case, combining both types of feedback in one circuit helps but still falls short of the performance needed for mass production.

We do not analyze this circuit because we have already covered negative feedback and trial-and-error methods with the earlier discussion. The main reason for showing you this circuit is that it was the third step in the evolution toward voltage-divider bias. If you should come across this circuit, here are the equations for analyzing it:

$$I_C = \frac{V_{CC} - V_{BE}}{R_C + R_E + R_B/(\beta_{dc} + 1)} \tag{8-15}$$

and

$$V_C = V_{CC} - I_C R_C \tag{8-16}$$

In Eq. (8-15), the combined effect of the collector and emitter resistances is to reduce the impact of changes in the current gain. No matter how

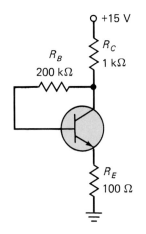

FIGURE 8-17

Collector- and emitter-feedback bias.

you select the circuit values, however, you cannot achieve the stability achieved with voltage-divider bias.

A final point about this third failure. Failure means little to an inventor. Some of the best inventions have come after repeated failures. Why? Because failure often gives the inventor new ideas that never would have occurred to her or him without the failure. The best inventors look upon failure as an opportunity because within every failure is the seed of an eventual success.

When Thomas Edison was asked how he managed to come up with over 1000 inventions, he gave this answer:

> I just fail my way to success. I try something and it fails. Then I ask why did it fail? When I find an answer, I try again and I keep trying until I succeed. Did you know that I failed over 10,000 times before I finally got the incandescent lamp to work?

The idea that you have to succeed the first time is a left-brain trap. The greatest achievers in all fields agree that you learn more from your failures than from your successes. It is almost as though you have to make mistakes to grow. The most important thing is to keep trying.

Voltage-Divider Bias

When you learn to use your right brain as well as your left, anything can happen. You can become a great poet, painter, entrepreneur, technician, engineer, teacher, etc. You might even invent a circuit like the voltage-divider bias. Figure 8-18 shows how transistor bias evolved from its most primitive form until it culminated in voltage-divider bias. How do you think the inventor finally hit upon the ultimate feedback solution of voltage-divider bias?

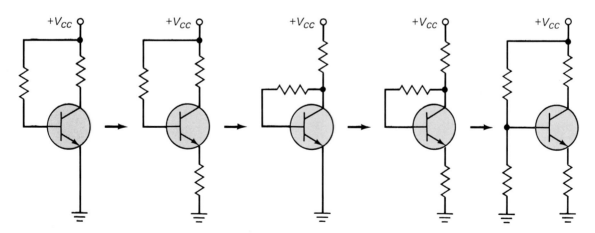

FIGURE 8-18 Evolution of transistor bias.

It's possible the inventor reasoned (left brain) to arrive at this design, but not likely. Most inventions originate in the right brain. Because this side of the brain is nonverbal, the answers it comes up with are transmitted to the left brain through something called *intuition*. Some people claim there is no such thing as intuition. But you will never hear an inventor say that. Inventors have always felt the best ideas seem to come out of nowhere. Recent brain research tells us the source of those great ideas is nowhere else but the right brain.

Voltage-divider bias is the ultimate form of negative feedback, what the inventor was looking for in the first place. When the current gain increases, the emitter current increases. This increases the emitter voltage. Since the base voltage is fixed by the voltage divider, the base-emitter voltage decreases, causing the base current to decrease. This time, the negative feedback works extremely well because almost all the increase in current gain produces a proportional decrease in base current. The collector and emitter currents show only the slightest increase. As you saw in the earlier analysis of this circuit, shifting of the Q point with changes in current gain is almost nil.

8-7   TROUBLESHOOTING

Let us discuss the troubleshooting of voltage-divider bias because this biasing method is the most widely used. In this section, we will see how a troubleshooter thinks while analyzing the voltage-divider bias.

Analysis

Time is of the essence when you are troubleshooting, especially if you are in business for yourself. Even if you work for a company, you cannot sit back on your calculator and take all day to find the exact values of the voltages you will be measuring. For this reason, the best troubleshooters use their heads instead of their calculators to analyze a circuit. This requires the courage and the sense to settle for rough approximations.

Here is how a troubleshooter would attack the circuit of Fig. 8-19:

> I know the base current is very small in this circuit, so small that I can ignore it. This means the current through the voltage divider is 10 V divided by about 12 kΩ, which is roughly 1 mA. When this 1 mA flows through the 2.2 kΩ, it produces a voltage of around 2 V. This is the approximate voltage at the base.

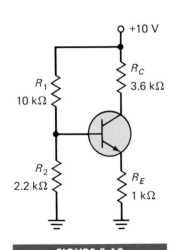

FIGURE 8-19

Troubleshooting voltage-divider bias.

Notice how the usual left-brain caution is thrown to wind. Brutal approximations are used to allow mental calculations. The exact analysis of this circuit shows that the base voltage is 1.8 V. This means our hypothetical troubleshooter is not far off the mark in using 2 instead of 1.8 V. The analysis of Fig. 8-19 continues like this:

> Now 2 minus 0.7 is 1.3 V. This sets up 1.3 mA of emitter current. When this flows through the collector resistor, it produces a drop of roughly 4 V. Subtracting 4 from 10 leaves 6 V at the collector. So, if this circuit is working correctly, I will measure about 6 V between the collector and ground. If the measured voltage is not somewhere in the vicinity of 6 V, I will know there is some kind of trouble in the circuit.

The troubleshooter knows that defective components, shorts, opens, etc. usually produce large changes in the voltages she or he measures. And the troubleshooter is smart enough to know that rough mental calculations will produce answers that are adequate 9 times out of 10. It is very rare to see a troubleshooter pick up a calculator to work out some voltage he or she is about to measure.

The strange thing about rough mental calculations is that they often produce fairly accurate answers. For this circuit, an exact analysis using the third approximation of a transistor shows that the collector voltage is 6.14 V, which is quite close to the troubleshooter's answer of 6 V. Rough answers are usually not this close, but they are often closer than you would expect.

Troubleshooting is an art. Because of this, it cannot be reduced to a set of rules for your left brain to memorize. So how do you learn troubleshooting? Mostly from experience. Remember, it is a skill, something you get better at it but never totally master. You have already learned a lot about troubleshooting just by reading the foregoing passages of what goes through a troubleshooter's mind. Reading about how a troubleshooter thinks will point you in the right direction. It will turn on your right brain as well as your left.

A final point. Your mind was made for flying, not for crawling. A troubleshooter knows this. So, too, does an experienced designer, and anyone else who has worked in everyday electronics. You never calculate anything to a greater degree of accuracy than is needed for the immediate job at hand. So when you are troubleshooting, throw away your calculator and start rounding the numbers generously.

Common Troubles

After making some rough approximations of the voltages in a circuit, the troubleshooter is ready to troubleshoot. For instance, suppose a troubleshooter measures 10 V between the base and ground in Fig. 8-19. Here is how she or he might react:

> Something has to be wrong with R_1. Either it is open, or it is shorted. An open resistor always has zero current and an unknown voltage. That description doesn't fit this situation at all. On the other hand, a shorted resistor always has zero voltage and an unknown current. Perfect. That has to be it. A shorted R_1 would place 10 V directly on the base of the transistor.

As discussed earlier, when resistors burn out, they open up rather than short like some semiconductor devices. Therefore, a shorted R_1 means an indirect short caused by a solder splash or other condition that is shorting R_1. So, the troubleshooter will look for some sort of mechanical short across R_1.

Most of the time, an open or shorted component produces unique voltages. For instance, the only way to get 10 V at the base of the transistor in Fig. 8-19 is with a shorted R_1. No other shorted or open component can produce the same result. A trouble like this is unique. A competent troubleshooter can find it in a flash.

Some troubles do not produce unique voltages. In other words, either of two troubles can produce the same voltages. In a case like this, the troubleshooter may have to disconnect one of the suspected components and use an ohmmeter or other instrument to test it. Here is an example. Suppose a troubleshooter measures these voltages in Fig. 8-19: $V_B = 1.8$ V, $V_E = 1.1$ V, and $V_C = 10$ V. If you think about this for a while, you will realize that two possible troubles can produce these voltages. The troubles are either a shorted collector resistor or an open emitter resistor. The first trouble places 10 V directly on the collector. The second trouble does the same thing because there is no collector current. You have 1.1 V on the emitter with an open emitter resistor because that is what you will measure when you connect a voltmeter between the emitter and ground. *Remember:* The voltmeter acts like a large resistance in series with the emitter diode, so 1.1 V will appear across the voltmeter. Since either of these troubles may be the culprit, the troubleshooter can disconnect the emitter resistor and measure its resistance. The results of this measurement will immediately indicate the trouble by process of elimination.

There is a T-shooter at the end of this chapter that allows you to continue your troubleshooting practice. If you do use it, give it your best effort. Like all the arts, you have to keep practicing troubleshooting to get the hang of it. The T-shooter of this chapter has 12 different troubles. If you work your way through all these troubles, you will know how to troubleshoot any voltage-divider biased circuit. This is your chance to learn. Don't be discouraged if you make mistakes at first. *Remember:* You usually learn a lot more from your mistakes than from your successes.

❑ OPTIONAL TOPICS

The following material continues the earlier discussions at a more advanced and specialized level. All the topics are optional because they are not used in any of the basic discussions in later chapters. This section will be a useful reference when you are in industry because then you will probably want more advanced viewpoints.

❑ 8-8 MORE VOLTAGE-DIVIDER BIAS

This section will give you a more advanced mathematical discussion of voltage-divider bias. This may be useful to designers and others who want more discussion of the mathematics behind this method of transistor bias.

Emitter Current

Figure 8-20a shows voltage-divider bias. Mentally open the base lead in Fig. 8-20a. Then you are looking at an unloaded voltage divider whose Thevenin voltage is

$$V_{TH} = \frac{R_2}{R_1 + R_2} V_{CC} \qquad (8\text{-}17)$$

Now mentally reconnect the base lead. If the voltage divider is stiff, more than 99 percent of the Thevenin voltage drives the base. In other words, the circuit simplifies to Fig. 8-20b. In this equivalent circuit, the emitter current is

$$I_E = \frac{V_{TH} - V_{BE}}{R_E} \qquad (8\text{-}18)$$

The collector current approximately equals this value.

Notice that β_{dc} does not appear in this formula. This means that the circuit is immune to variations in β_{dc}, which implies a fixed Q point. Because of this, voltage-divider bias is the preferred form of bias in linear transistor circuits. You see it used almost universally, which is why it is also called *universal bias*.

Stiff Voltage Divider

The key to a well-designed circuit is the *stiffness* of the voltage divider. Here is how to get a design guideline for stiffness. If we thevenize the circuit of Fig. 8-20a, we get the equivalent circuit of Fig. 8-21, in which

$$R_{TH} = \frac{R_1 R_2}{R_1 + R_2} \qquad (8\text{-}19)$$

For simplicity, this is often written as

$$R_{TH} = R_1 \parallel R_2 \qquad (8\text{-}20)$$

where the vertical bars stand for "is parallel with." You read Eq. (8-20) as "R_{TH} equals R_1 in parallel with R_2." Summing voltages around the base loop of Fig. 8-21 gives

$$V_{BE} + I_E R_E - V_{TH} + I_B R_{TH} = 0$$

Since $I_B = I_E/\beta_{dc}$, the foregoing equation reduces to

$$I_E = \frac{V_{TH} - V_{BE}}{R_E + R_{TH}/\beta_{dc}} \qquad (8\text{-}21)$$

If R_E is 100 times greater than R_{TH}/β_{dc}, the second term is swamped out and the equation simplifies to

$$I_E = \frac{V_{TH} - V_{BE}}{R_E}$$

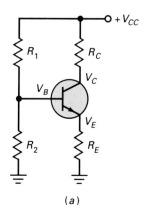

(a)

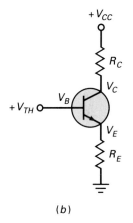

(b)

FIGURE 8-20

Voltage-divider bias.

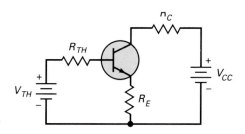

FIGURE 8-21 Equivalent circuit for voltage-divider bias.

In this book, a *stiff* voltage-divider-biased circuit is one that satisfies this condition:

$$R_{TH} < 0.01\beta_{dc}R_E \qquad (8\text{-}22)$$

This 100:1 rule must be satisfied for the minimum β_{dc} encountered under all conditions. For instance, if a transistor has a β_{dc} that varies from 80 to 400, use the lower value (80).

Usually, R_2 is smaller than R_1, and Eq. (8-22) is simplified to

$$R_2 < 0.01\beta_{dc}R_E \qquad (8\text{-}23)$$

This is conservative because satisfying Eq. (8-23) automatically satisfies Eq. (8-22). For convenience, we use Eq. (8-23) when designing stiff voltage dividers.

Firm Voltage Divider

Sometimes a stiff design results in such small values of R_1 and R_2 that other problems arise (discussed later). In this case, many designers compromise by using this rule:

$$R_{TH} < 0.1\beta_{dc}R_E \qquad (8\text{-}24)$$

Once more, it is convenient to work with this design rule:

$$R_2 < 0.1\beta_{dc}R_E \qquad (8\text{-}25)$$

In the worst case, satisfying this rule means that the collector current will be approximately 10 percent lower than the ideal value given by Eq. (8-18).

From now on, we refer to a voltage divider as *firm* when it satisfies Eq. (8-25). As a guideline, we usually try to make the voltage divider stiff. For reasons given later (input impedance), we sometimes compromise by using a firm voltage divider because this may give us a better all-around circuit design.

Design Guidelines

Figure 8-22 shows an amplifier. The capacitors couple an ac signal into and out of the amplifier. As far as the direct current is concerned, the capacitors appear like open circuits. So you may ignore the capacitors in the following discussion.

Unless otherwise indicated, we use the one-tenth rule, which makes the emitter voltage approximately one-tenth of the supply voltage:

$$V_E = 0.1V_{CC} \qquad (8\text{-}26)$$

This design rule is suitable for most circuits, but remember, it is only a guideline. Not everyone uses this rule, so don't be surprised to find emitter voltages at values different from one-tenth of the supply voltage.

Start by calculating the R_E needed to set up the specified collector:

$$R_E = \frac{V_E}{I_E} \qquad (8\text{-}27)$$

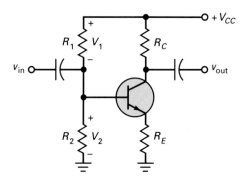

FIGURE 8-22 An amplifier.

Also, locate the Q point at approximately the middle of the dc load line. This means that about $0.5V_{CC}$ appears across the collector-emitter terminals. The remaining $0.4V_{CC}$ appears across the collector resistor; therefore,

$$R_C = 4R_E \qquad (8\text{-}28)$$

Next, you can design a stiff voltage divider, using the 100:1 rule:

$$R_2 < 0.01\beta_{dc}R_E$$

If you prefer a firm voltage divider, then apply the 10:1 rule:

$$R_2 < 0.1\beta_{dc}R_E$$

Finally, calculate R_1 by using proportion:

$$R_1 = \frac{V_1}{V_2}R_2 \qquad (8\text{-}29)$$

8-9 EMITTER-FEEDBACK BIAS

Figure 8-23 shows emitter-feedback bias. If we sum the voltages around the collector loop, we get

$$V_{CE} + I_ER_E - V_{CC} + I_CR_C = 0$$

Since I_E approximately equals I_C, this equation can be rearranged as

$$I_C = \frac{V_{CC} - V_{CE}}{R_C + R_E} \qquad (8\text{-}30)$$

By setting V_{CE} equal to zero, we get a saturation current of $V_{CC}/(R_C + R_E)$. By setting I_C equal to zero, we get a cutoff voltage of V_{CC}.

Next, we can sum voltages around the base loop to get

$$V_{BE} + I_ER_E - V_{CC} + I_BR_B = 0$$

Since $I_E = I_C$ and $I_B = I_C/\beta_{dc}$, we can rewrite the equation as

$$I_C = \frac{V_{CC} - V_{BE}}{R_E + R_B/\beta_{dc}} \qquad (8\text{-}31)$$

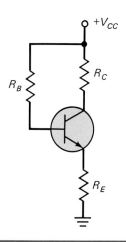

FIGURE 8-23

Emitter-feedback bias.

The collector voltage is given by

$$V_C = V_{CC} - I_C R_C \tag{8-32}$$

and the emitter voltage by

$$V_E = I_E R_E \tag{8-33}$$

and the collector-emitter voltage by

$$V_{CE} = V_C - V_E \tag{8-34}$$

The intent of emitter-feedback bias is to swamp out the variations in β_{dc}; this is equivalent to R_E being much larger than R_B/β_{dc}. In practical circuits, however, you cannot make R_E large enough to swamp out the effects of β_{dc} without saturating the transistor. For typical designs, it turns out that emitter-feedback bias is almost as sensitive to changes in β_{dc} as is the base bias. For instance, Fig. 8-24a shows an emitter-feedback-biased circuit. Figure 8-24b shows its dc load line and the operating points for two different current gains. As you can see, a 3:1 variation in current gain produces a large variation in collector current.

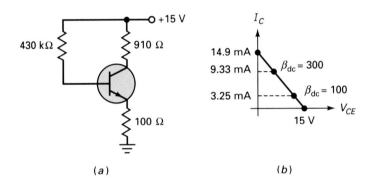

(a)   (b)

FIGURE 8-24   Emitter-feedback bias and load line.

8-10 COLLECTOR-FEEDBACK BIAS

Figure 8-25 shows collector-feedback bias (also called *self-bias*). Summing voltages around the collector loop gives

$$V_{CE} - V_{CC} + (I_C + I_B)R_C = 0$$

Since I_B is much smaller than I_C in the active region, we can ignore I_B and rearrange the equation as

$$I_C = \frac{V_{CC} - V_{CE}}{R_C} \tag{8-35}$$

By setting V_{CE} equal to zero, we get a saturation current of V_{CC}/R_C. By setting I_C equal to zero, we get a cutoff voltage of V_{CC}.

If we sum the voltages around the base loop, then

$$V_{BE} - V_{CC} + (I_C + I_B)R_C + I_B R_B = 0$$

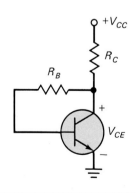

FIGURE 8-25
Collector-feedback bias.

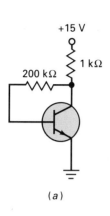

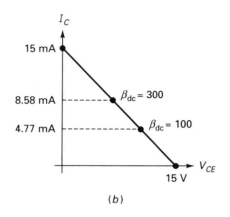

(a)

(b)

FIGURE 8-26   Collector-feedback bias and load line.

or

$$V_{BE} - V_{CC} + I_C R_C + I_B R_B = 0$$

Since $I_B = I_C/\beta_{dc}$, the foregoing equation can be solved for I_C:

$$I_C = \frac{V_{CC} - V_{BE}}{R_C + R_B/\beta_{dc}} \qquad (8\text{-}36)$$

Collector-feedback bias has another advantage over emitter-feedback bias: You cannot saturate the transistor. As you decrease the base resistance, the operating point moves toward the saturation point on the dc load line. But it can never reach saturation, no matter how low the base resistance is.

The Q point is usually set up near the middle of the dc load line. With collector-feedback bias, this requires

$$R_B = \beta_{dc} R_C \qquad (8\text{-}37)$$

The easiest way to see this is by substituting this value into Eq. (8-36). This gives a collector current that is approximately half of the saturation value.

Collector-feedback bias is more effective than emitter-feedback bias. Although the circuit is still sensitive to changes in β_{dc}, it is used in practice. It has the advantage of simplicity and improved frequency response (discussed later). Figure 8-26a shows a collector-feedback-biased circuit. Figure 8-24b shows its dc load line and the operating points for two different current gains. As you can see, a 3:1 variation in current gain produces less variation in collector current than emitter-feedback bias (Fig. 8-24b).

8-11 TWO SUPPLY EMITTER BIAS

Figure 8-27 shows emitter bias, which is sometimes used when a split supply is available (positive and negative voltages). If R_B is small enough, the base voltage is approximately zero. The emitter voltage is one V_{BE}

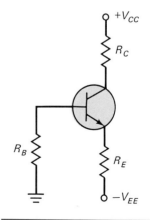

FIGURE 8-27

Two-supply emitter bias.

drop below this. Therefore, the voltage across the emitter-feedback resistor is $V_{EE} - V_{BE}$, and the emitter current is

$$I_E = \frac{V_{EE} - V_{BE}}{R_E} \tag{8-38}$$

Since β_{dc} does not appear in this formula, the Q point is fixed. Whenever a split supply is available, two-supply emitter bias may be used, because it provides a rock-solid Q point just as voltage-divider bias does.

The key to a well-designed circuit is the size of R_B. It must be small. But how small? By a derivation similar to that given for voltage-divider bias, the exact formula for emitter current is

$$I_E = \frac{V_{EE} - V_{BE}}{R_E + R_B/\beta_{dc}}$$

In a stiff design, R_E is at least 100 times greater than R_B/β_{dc}. This is equivalent to

$$R_B < 0.01\beta_{dc}R_E \tag{8-39}$$

When troubleshooting this circuit, you need to estimate the transistor voltages with respect to ground. The collector voltage is

$$V_C = V_{CC} - I_C R_C \tag{8-40}$$

In a stiff design, the base voltage is approximately 0 V, and the emitter voltage is approximately -0.7 V.

STUDY AIDS

The following study aids will help to reinforce the ideas discussed in this chapter. For best results, use these study aids within 6 hours of reading the earlier material. Then review these study aids a week later and a month later to ensure that the concepts remain in your long-term memory.

SUMMARY

Sec. 8-1 Voltage-Divider Bias
The most famous circuit based on the emitter-bias prototype is called voltage-divider bias. You can recognize it by the voltage divider in the base circuit.

Sec. 8-2 VDB Analysis
The key idea is for the base current to be much smaller than the current through the voltage divider. When this condition is satisfied, the voltage divider holds the base voltage almost constant and equal to the unloaded voltage out of the voltage

divider. This produces a solid Q point under all operating conditions.

Sec. 8-3 VDB Load Line and Q Point
The load line is drawn through saturation and cutoff. The Q point lies on the load line with the exact location determined by the biasing. Large variations in current gain have almost no effect on the Q point because this type of bias sets up a constant value of emitter current.

Sec. 8-4 Two-Supply Emitter Bias
This design uses two power supplies, one positive and the other negative. The idea is to set up a constant value of emitter current. The circuit is a variation of the emitter-bias prototype discussed earlier.

Sec. 8-5 PNP Transistors
These pnp devices have all currents and voltages reversed from their npn counterparts. They may be used with negative power supplies; more com-

monly, they are used with positive power supplies in an upside-down configuration.

Sec. 8-6 Other Types of Bias

This section introduced negative feedback, a phenomenon that exists when an increase in an output quantity produces a decrease in an input quantity. It is a brilliant idea that led to voltage-divider bias. The other types of bias cannot use enough negative feedback, so they fail to attain the performance level of voltage-divider bias.

Sec. 8-7 Troubleshooting

Troubleshooting is an art. Because of this, it cannot be reduced to a set of rules for your left brain to memorize. You learn troubleshooting mostly from experience.

VOCABULARY

In your own words, explain what each of the following terms means. Keep your answers short and to the point. If necessary, verify your answer by rereading the appropriate discussion or by looking at the end-of-book Glossary.

absolute value	prototype
failure	stiff voltage divider
intuition	trial and error
negative feedback	upside-down *pnp* bias
pnp transistor	

IMPORTANT EQUATIONS

Here are some important equations. Say each of the following equations in symbols, then say each in words. Try to explain what the equation means and how it is used. Then read the description that follows.

Eq. 8-1 Current Through Voltage Divider

$$I = \frac{V_{CC}}{R_1 + R_2}$$

This is the first step in analyzing voltage-divider bias. This is the current through an unloaded voltage divider, or one where the base current is small enough to ignore. It says the current equals the collector supply voltage divided by the total resistance. It's nothing more than Ohm's law.

Eq. 8-2 Base Voltage

$$V_B = IR_2$$

This is the second step in analyzing voltage-divider bias. It says the base voltage equals the current through the voltage divider times the divider resistance R_2. Again, this is Ohm's law.

Eq. 8-3 Emitter Voltage

$$V_E = V_B - V_{BE}$$

This is the third step in analyzing voltage-divider bias. You can get fancy here and use the exact value of V_{BE} if it is known. Most people use 0.7 or even 0 V. The equation says the emitter voltage equals the base voltage minus the voltage across the emitter diode.

Eq. 8-4 Emitter Current

$$I_E = \frac{V_E}{R_E}$$

This is the fourth step in analyzing voltage-divider bias. You apply Ohm's law to the emitter resistor.

Eq. 8-5 Collector Voltage

$$V_C = V_{CC} - I_C R_C$$

This is very familiar by now.

Eq. 8-6 Collector-Emitter Voltage

$$V_{CE} = V_C - V_E$$

Because the emitter is no longer at ground, you have to subtract the emitter voltage from the collector voltage when you want to know the value of collector-emitter voltage.

IMPORTANT PROCESSES

Review each of the following processes. The steps are abbreviated, because all you need is a memory jog. If you can remember the basic ideas, you will be able to solve problems more easily.

VDB Analysis

1. Divider current
2. Base voltage

3. Emitter voltage
4. Emitter current
5. Collector voltage
6. Collector-emitter voltage

Is Base Current Small Enough?

1. Divide I by 20.
2. Divide I_C by the worst-case current gain.
3. Check that the base current is less than the critical level.

VDB Saturation Current

1. Get emitter voltage.
2. Short the collector and emitter.
3. Work out the collector current.

VDB Cutoff Voltage

1. Get the emitter voltage.
2. Open the collector and emitter.
3. Work out the collector-emitter voltage.

Trial and Error

1. Guess a value for the answer.
2. Calculate other quantities including the answer.
3. Compare calculated answer to guessed answer.
4. Repeat process until two answers are approximately equal.

STUDENT ASSIGNMENTS

QUESTIONS

The following questions may refer to the figures you have seen in this chapter. It should be clear from the question which figure is being used. For instance, if the question mentions a base resistor, then the circuit is base-biased. If the question mentions as emitter resistor, the circuit is emitter-biased.

1. For the emitter bias, the voltage across the emitter resistor is the same as the voltage between the emitter and the
   a. Base
   b. Collector
   c. Emitter
   d. Ground

2. For emitter bias, the voltage at the emitter is 0.7 V less than the
   a. Base voltage
   b. Emitter voltage
   c. Collector voltage
   d. Ground voltage

3. With voltage-divider bias, the base voltage is
   a. Less than the base supply voltage
   b. Equal to the base supply voltage
   c. Greater than the base supply voltage
   d. Greater than the collector supply voltage

4. VDB is noted for its
   a. Unstable collector voltage
   b. Varying emitter current
   c. Large base current
   d. Stable Q point

5. With VDB, an increase in emitter resistance will
   a. Decrease the emitter voltage
   b. Decrease the collector voltage
   c. Increase the emitter voltage
   d. Decrease the emitter current

6. VDB has a stable Q point like
   a. Base bias
   b. Emitter bias
   c. Collector-feedback bias
   d. Emitter-feedback bias

7. VDB needs
   a. Only three resistors
   b. Only one supply
   c. Precision resistors
   d. More resistors to work better

8. VDB normally operates in the
   a. Active region
   b. Cutoff region
   c. Saturation region
   d. Breakdown region

9. The collector voltage of a VDB circuit is not sensitive to changes in the
   a. Supply voltage
   b. Emitter resistance
   c. Current gain
   d. Collector resistance

10. If the emitter resistance increases in a VDB circuit, the collector voltage
    a. Decreases
    b. Stays the same
    c. Increases
    d. Doubles

11. Base bias is associated with
    a. Amplifiers
    b. Digital circuits
    c. Stable Q point
    d. Fixed emitter current

12. If the emitter resistance doubles in a VDB circuit, the collector current will
    a. Double
    b. Drop in half
    c. Remain the same
    d. Increase

13. If the collector resistance increases in a VDB circuit, the collector voltage will
    a. Decrease
    b. Stay the same
    c. Increase
    d. Double

14. The Q point of a VDB circuit is
    a. Hypersensitive to changes in current gain
    b. Somewhat sensitive to changes in current gain
    c. Almost totally insensitive to changes in current gain
    d. Greatly affected by temperature changes

15. The base voltage of *two-supply emitter bias* (TSEB) is
    a. 0.7 V
    b. Very large
    c. Near 0 V
    d. 1.3 V

16. If the emitter resistance doubles with the TSEB, the collector current will
    a. Drop in half
    b. Stay the same
    c. Double
    d. Increase

17. If a splash of solder shorts the collector resistor of the TSEB, the collector voltage will
    a. Drop to zero
    b. Equal the collector supply voltage
    c. Stay the same
    d. Double

18. If the emitter resistance increases with the TSEB, the collector voltage will
    a. Decrease
    b. Stay the same
    c. Increase
    d. Equal the collector supply voltage

19. If the emitter resistor opens with the TSEB, the collector voltage will
    a. Decrease
    b. Stay the same
    c. Increase slightly
    d. Equal the collector supply voltage

20. In the TSEB, the base current must be very
    a. Small
    b. Large
    c. Unstable
    d. Stable

21. The Q point of the TSEB does not depend on the
    a. Emitter resistance
    b. Collector resistance
    c. Current gain
    d. Emitter voltage

22. The majority carriers in the emitter of a *pnp* transistor are
    a. Holes
    b. Free electrons
    c. Trivalent atoms
    d. Pentavalent atoms

23. The current gain of a *pnp* transistor is
    a. The negative of the *npn* current gain
    b. The collector current divided by the emitter current
    c. Near zero
    d. The ratio of collector current to base current

24. Which is the largest current in a *pnp* transistor?
    a. Base current
    b. Emitter current
    c. Collector current
    d. None of these

25. The currents of a *pnp* transistor are
    a. Usually smaller than *npn* currents
    b. Opposite *npn* currents
    c. Usually larger than *npn* currents
    d. Negative

26. With *pnp* voltage-divider bias, you must use
    a. Negative power supplies
    b. Positive power supplies
    c. Resistors
    d. Grounds

BASIC PROBLEMS

Sec. 8-2 VDB Analysis

8-1. What is the emitter voltage in Fig. 8-28? The collector voltage?

8-2. What is the emitter voltage in Fig. 8-29? The collector voltage?

8-3. What is the emitter voltage in Fig. 8-30? The collector voltage?

8-4. What is the emitter voltage in Fig. 8-31? The collector voltage?

8-5. All resistors in Fig. 8-30 have a tolerance of ± 5 percent. What is the lowest possible value of the collector voltage? The highest?

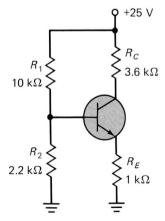

FIGURE 8-28

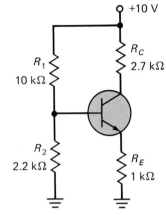

FIGURE 8-29

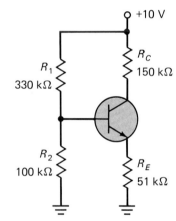

FIGURE 8-30

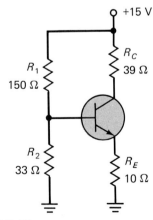

FIGURE 8-31

8-6. The power supply of Fig. 8-31 has a tolerance of ± 10 percent. What is the lowest possible value of the collector voltage? The highest? (Use nominal values for the resistors.)

Sec. 8-3 VDB Load Line and Q Point

8-7. Draw the dc load line and Q point for Fig. 8-28.

8-8. Draw the dc load line and Q point for Fig. 8-29.

8-9. Draw the dc load line and Q point for Fig. 8-30.

8-10. Draw the dc load line and Q point for Fig. 8-31.

8-11. All resistors in Fig. 8-30 have a tolerance of ± 5 percent. What is the lowest value of the saturation current? The highest?

8-12. The power supply of Fig. 8-31 has a tolerance of ± 10 percent. What is the lowest possible value of the saturation current? The highest? (Use nominal values for the resistors.)

Sec. 8-4 Two-Supply Emitter Bias

8-13. What is the emitter current in Fig. 8-32? The collector voltage?

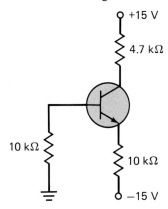

FIGURE 8-32

8-14. If all resistances are doubled in Fig. 8-32, what is the emitter current? The collector voltage?

8-15. All resistors in Fig. 8-32 have a tolerance of ± 5 percent. What is the lowest possible value of the collector voltage? The highest?

Sec. 8-5 PNP Transistors

8-16. What is the collector voltage in Fig. 8-33?

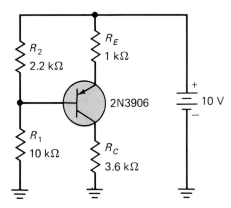

FIGURE 8-33

8-17. What is the collector-emitter voltage in Fig. 8-33?

8-18. What is the collector saturation current in Fig. 8-33? The collector-emitter cutoff voltage?

8-19. What is the emitter voltage in Fig. 8-34? The collector voltage?

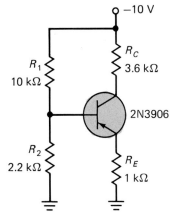

FIGURE 8-34

Sec. 8-6 Other Types of Bias

8-20. Does the collector voltage increase, decrease, or remain the same in Fig. 8-31 for small changes in each of the following?
a. R_1 increases d. R_C decreases
b. R_2 decreases e. V_{CC} increases
c. R_E increases f. β_{dc} decreases

8-21. Does the collector voltage increase, decrease, or remain the same in Fig. 8-33 for small increases in each of the following circuit values?
a. R_1 d. R_C
b. R_2 e. V_{CC}
c. R_E f. β_{dc}

Sec. 8-7 Troubleshooting

8-22. What is approximate value of the collector voltage in Fig. 8-31 for each of these troubles?

   a. R_1 open   **d.** R_C open
   b. R_2 open   **e.** Collector-emitter open
   c. R_E open

8-23. What is approximate value of the collector voltage in Fig. 8-33 for each of these troubles?

   a. R_1 open   **d.** R_C open
   b. R_2 open   **e.** Collector-emitter open
   c. R_E open

UNUSUAL PROBLEMS

8-24. Somebody has built the circuit of Fig. 8-31, except for changing the voltage divider as follows: $R_1 = 150$ kΩ and $R_2 = 33$ kΩ. The builder cannot understand why the base voltage is only 0.77 instead of 2.7 V (the ideal output of the voltage divider). Can you explain what is happening?

8-25. Somebody builds the circuit of Fig. 8-31 with a 2N3904. What do you have to say about that?

8-26. A student wants to measure the collector-emitter voltage in Fig. 8-31, so she connects a voltmeter between the collector and the emitter. What does it read?

8-27. You can vary any circuit value in Fig. 8-31. Name all the ways you can think of to destroy the transistor.

8-28. The power supply of Fig. 8-31 has to supply current to the transistor circuit. Name all the ways you can think of to find this current.

ADVANCED PROBLEMS

8-29. Calculate the collector voltage for each transistor of Fig. 8-35. (*Hint:* Capacitors are open to direct current.)

8-30. The circuit of Fig. 8-36*a* uses silicon diodes. What is the emitter current? The collector voltage?

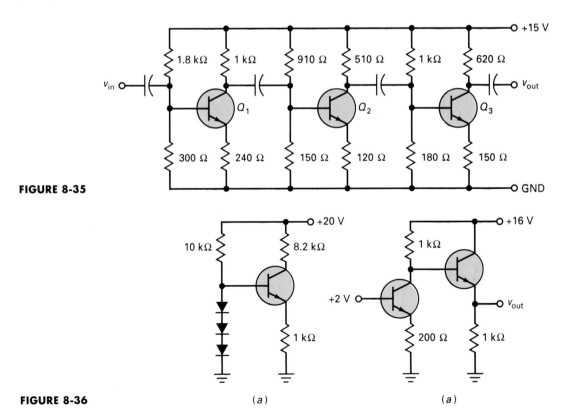

FIGURE 8-35

FIGURE 8-36   (*a*)   (*a*)

8-31. What is the output voltage in Fig. 8-36*b*?

8-32. How much current is there through the LED of Fig. 8-37*a*?

8-33. What is the LED current in Fig. 8-37*b*?

8-34. We want the voltage divider of Fig. 8-30 to be stiff. Change R_1 and R_2 as needed without changing the Q point.

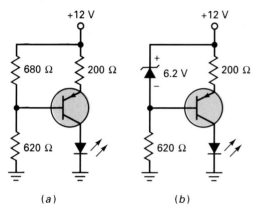

(a)   (b)

FIGURE 8-37

T-SHOOTER PROBLEMS

Use Fig. 8-38 for the remaining problems.

8-35. Find trouble 1.

8-36. Find trouble 2.

8-37. Find troubles 3 and 4.

8-38. Find troubles 5 and 6

8-39. Find troubles 7 and 8.

8-40. Find troubles 9 and 10.

8-41. Find troubles 11 and 12.

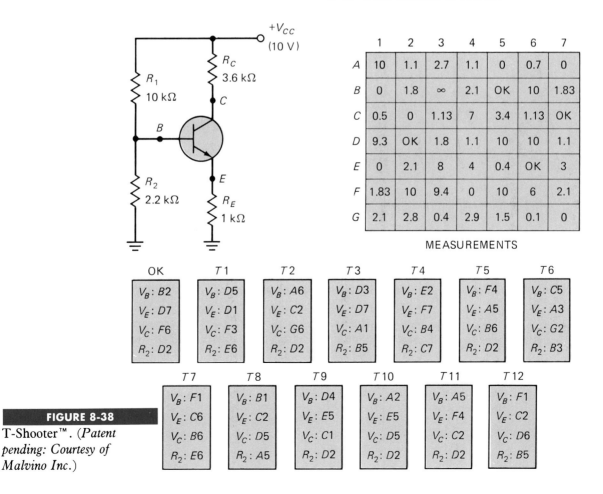

FIGURE 8-38

T-Shooter™. (*Patent pending: Courtesy of Malvino Inc.*)

9

AC MODELS

Now the fun begins. After a transistor has been biased with the Q point near the middle of the load line, we can put a small ac voltage on the base. This produces a large ac voltage at the collector. This ac collector voltage looks like the original signal, except it's a lot bigger. In other words, if we look at the collector voltage with an oscilloscope, we will see a magnified version of the input voltage. This increase in the signal is called *amplification*.

Without amplification, nothing would be possible in electronics. There would be no radio, no television, no computers. In this chapter you will learn more about amplification. In particular, you will learn how amplifiers work, what their limitations are, and how to analyze them.

9-1 COUPLING CAPACITOR

Capacitors are more interesting than resistors. When the frequency increases, the opposition of a resistor does not change. A capacitor is different. When the frequency increases, the opposition to current decreases. This is the kind of action that inventors and designers love. It means they can create circuits that are frequency sensitive.

DC Open and AC Short

When it comes to amplifiers, there are two fundamental ways in which capacitors are used. First, they are used to couple or transmit ac signals from one circuit to another. Second, they are used to bypass or short ac signals to ground. Both of these uses depend on a relation that you learned in the study of ac circuits:

$$X_C = \frac{1}{2\pi f C} \tag{9-1}$$

This formula says the capacitive reactance is inversely proportional to frequency and to capacitance. If you double the frequency, the reactance drops in half. When the frequency is high enough, the reactance approaches zero. This means a capacitor is an *ac short* at high frequencies. The opposite is also true. When the frequency decreases to zero, the reactance becomes infinite. This means a capacitor is a *dc open* at low frequencies.

When a capacitor is in a circuit, it can act like two different things at the same time. At low frequencies, it will act like an open circuit. At high frequencies, it will act like a short circuit. In other words, a capacitor is like an intelligent switch that has the sense to open at low frequencies and to close at high frequencies.

The foregoing capacitor action is analogous to an ideal diode that opens when reverse-biased and closes when forward-biased. But it's different because frequency is what controls the switching action. One more time, this is what you must remember:

A capacitor is open at low frequencies and shorted at high frequencies.

This basic idea is one of the keys to understanding amplifier circuits.

What It Does

A coupling capacitor transmits an ac voltage from one node or point to another. Figure 9-1 shows a coupling capacitor. An ac generator produces an alternating current through the series components. How much current? That depends on the frequency of the generator voltage. At low frequencies, the capacitor appears open and the current is approximately zero. At high frequencies, the capacitor appears shorted and the current equals

$$I_{\text{max}} = \frac{V_G}{R} \qquad (9\text{-}2)$$

where R is the total resistance, the sum of R_G and R_L. The current is labeled I_{max} in Eq. (9-2) because it is the maximum current that can exist in the circuit.

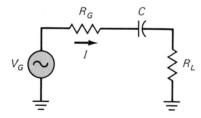

| **FIGURE 9-1** | Coupling capacitor. |

For a coupling capacitor to work properly, it has to act like an ac short at the lowest frequency that the generator can have. For instance, if the generator frequency can vary from 20 Hz to 20 kHz, the lowest frequency is 20 Hz. This is the worst-case frequency the designer has to worry about when selecting the size of the capacitor. The rule he/she uses is simple:

Make the reactance at least 10 times smaller than the total resistance in series with the capacitor.

Mathematically, this is written as

$$X_C < 0.1R$$

where $R = R_G + R_L$

Critical Frequency

Over and over again in electronics, the concept of *critical frequency* keeps reappearing. So, you may as well learn about it now if you are not already familiar with it. The critical frequency of Fig. 9-1 is the frequency where the reactance of the capacitor equals the total resistance. In symbols, it is the frequency where

$$X_C = R \tag{9-3}$$

For this condition

$$I = 0.707 I_{max}$$

In other words, at the critical frequency the rms current, in Fig. 9-1, decreases to 70.7 percent of the maximum value.

We can rewrite Eq. (9-3) as

$$\frac{1}{2\pi f C} = R$$

Next, we can solve this equation for frequency to get

$$f = \frac{1}{2\pi R C}$$

Because this frequency is the critical frequency, a subscript of c is usually attached to get the final formula:

$$f_c = \frac{1}{2\pi R C} \tag{9-4}$$

where R is the sum of R_G and R_L. This is a formula worth memorizing because it is the key to coupling circuits and other *RC* circuits.

Critical Frequency and High Frequency Border

We know a coupling capacitor acts like a short at high frequencies. But what does "high" mean? High means 10 times as high as the critical frequency. Where does the 10 come from? It comes from the 10:1 rule for reactance. When we say the reactance has to be at least 10 times smaller than the total resistance, we are saying that the frequency has to be at least 10 times higher than the critical frequency. As a formula,

$$f_h > 10 f_c \tag{9-5}$$

Given an *RC* circuit, you can find its critical frequency with Eq. (9-4). Then you can multiply by 10 to get the frequency where high frequencies begin. For any frequency greater than this, the coupling capacitor acts like an ac short. We will call the frequency given by Eq. (9-5) the *high-frequency border* because this is where high frequencies begin for the coupling capacitor. Above the high-frequency border, the load current is within 1 percent of its maximum value (see "Optional Topics," later in the chapter for the proof).

EXAMPLE 9-1

Figure 9-2 shows a capacitor coupling a signal from a generator to a load. What is the maximum current in this circuit? What is the high-frequency border?

SOLUTION

At high frequencies the coupling capacitor is a short, so that the maximum current is

$$I_{max} = \frac{1 \text{ V}}{5 \text{ k}\Omega} = 200 \text{ mA}$$

The critical frequency is

$$f_c = \frac{1}{2\pi(5 \text{ k}\Omega)(100 \text{ }\mu\text{F})} = 0.318 \text{ Hz}$$

Ten times this critical frequency gives the high-frequency border:

$$f_h = 3.18 \text{ Hz}$$

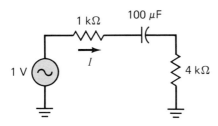

FIGURE 9-2   Example of coupling capacitor.

As long as the generator has a frequency that is greater than 3.18 Hz, the coupling capacitor acts like an ac short. For instance, if the generator is an audio generator (20 Hz to 20 kHz), the coupling capacitor looks like an ac short for all frequencies because the lowest frequency (20 Hz) is greater than the high-frequency border (3.18 Hz).

That's it. What could be simpler? If you are a troubleshooter working with a circuit that is in production, you can assume that all coupling capacitors act like ac shorts. You don't have to calculate the critical frequency or the high-frequency border because the designer has already done that.

9-2 BYPASS CAPACITOR

Figure 9-3a shows a bypass capacitor. To begin with, it is not connected in series like a coupling capacitor. Instead, it is connected in parallel across a resistor. The reason for doing this is to bypass or shunt ac current away from the resistor. In other words, when the frequency is high enough, the capacitor looks like a short. This provides a very low impedance path for alternating current. As a result, the alternating current will flow into the 100 μF rather than the 4 kΩ.

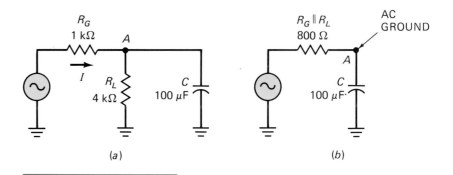

FIGURE 9-3 Bypass capacitor.

Since the capacitor is an ac short at high frequencies, point A is shorted to ground as far as ac signals are concerned. If we look at point A with an oscilloscope, we will see nothing at high frequencies because it is at ground potential.

High-Frequency Border

Again, we have to ask the question, "What is high frequency?" To get the answer, we can use the famous Thevenin theorem that is discussed in basic electricity courses. The rest of the analysis is almost the same as for a coupling capacitor. The critical frequency is given by the same formula as before:

$$f_c = \frac{1}{2\pi RC} \tag{9-6}$$

In this formula, R is the Thevenin resistance facing the capacitor:

$$R = \frac{R_G R_L}{R_G + R_L}$$

This is the equivalent parallel resistance of R_G and R_L. Sometimes, you will see this written as

$$R = R_G \parallel R_L$$

where the two vertical lines stand for "in parallel with."

The high-frequency border is the same as before:

$$f_h = 10 f_c \tag{9-7}$$

When the generator frequency is equal to or greater than this value, the bypass capacitor acts like a short and node A is grounded to ac signals. In other words, point A is an ac ground.

AC Ground

You may never have heard of ac ground before, but it really does exist as a separate kind of ground that is different from a mechanical ground. A mechanical ground is what you get when you connect a wire between node A and ground. This kind of ground shorts node A to ground for all frequencies.

Ac ground is different. Because it is being produced by a bypass capacitor, this kind of ground exists only at high frequencies. A designer uses ac ground when he wants node A to be frequency sensitive. That is, when he wants a node A to be normal at low frequencies but grounded at high frequencies.

EXAMPLE 9-2

Calculate the high-frequency border in Fig. 9-3a.

SOLUTION

First, get the Thevenin resistance facing the capacitor. This resistance is the equivalent resistance of 1 kΩ in parallel with 4 kΩ. Using the product-over-sum rule,

$$R = \frac{(1 \text{ k}\Omega)(4 \text{ k}\Omega)}{1 \text{ k}\Omega + 4 \text{ k}\Omega} = 800 \ \Omega$$

Figure 9-3b shows the Thevenized circuit. The voltage at node A of this circuit will be exactly the same as the voltage at node A of the original circuit.

Second, calculate the critical frequency:

$$f_c = \frac{1}{2\pi(800 \ \Omega)(100 \ \mu\text{F})} = 1.99 \text{ Hz}$$

The high-frequency border is 10 times the critical frequency:

$$f_c = 10(1.99 \text{ Hz}) = 19.9 \text{ Hz}$$

When the generator frequency is equal to or greater than 19.9 Hz, point A is an ac ground point. This means the ac voltage appearing at this point is much smaller than the generator voltage.

9-3 SUPERPOSITION IN AMPLIFIERS

Figure 9-4 shows a transistor amplifier. V_{CC} is the dc supply voltage that sets up the Q point. V_G is the ac generator voltage. C_1 couples the generator signal into the base, while C_2 couples the amplified signal into

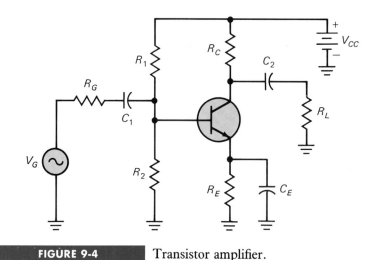

FIGURE 9-4 Transistor amplifier.

the load resistor. C_E bypasses the emitter node to ground. Because an ac signal is coupled into the base, it produces ac variations in the base current. These ac variations are multiplied by the current gain to produce large ac variations in the collector current. Because the collector current flows through the collector resistor, the collector voltage has large ac variations.

DC and AC Equivalent Circuits

The simplest way to analyze the circuit is to split the analysis into two parts: a dc analysis and an ac analysis. In other words, we can use the *superposition theorem*. If you recall, the superposition theorem is used when a circuit has more than one source. The theorem says you can find the effect produced by each source acting alone, and then add the individual effects to get the total effect.

To isolate each source, we have to transform the circuit into something simpler and easier to work with. Here is a trick that really helps. The capacitors are open to dc and shorted to ac. Because of this, we can change the original circuit into two new circuits: one for dc and another for ac. Here is the process:

1. Reduce the ac source to zero.

2. Open all capacitors.

3. Analyze the dc equivalent circuit.

In Fig. 9-4, reducing the ac source to zero is equivalent to replacing it by a short circuit. Opening all capacitors is the same as disconnecting them. The circuit that remains after these transformations is called the *dc equivalent circuit*. We can analyze this equivalent circuit to find the dc currents and voltages. Then we can go on to the ac analysis.

The process continues like this:

4. In the original circuit, reduce all dc sources to zero.

5. Short all capacitors.

6. Analyze the ac equivalent circuit.

Reducing the dc source to zero is equivalent to replacing it by a short circuit. Shorting all capacitors means replacing them by shorts. The circuit that remains is called the *ac equivalent circuit*. We can analyze this circuit to find the ac currents and voltages.

The process ends like this:

7. Add the dc current and the ac current to get the total current in any branch.

8. Add the dc voltage and the ac voltage to get the total voltage at any node or across any resistor.

All these steps are straightforward, but you will need to study some examples and work out some problems until the process feels natural.

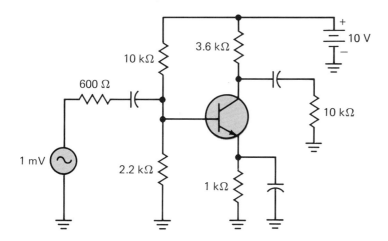

| FIGURE 9-5 | Transistor amplifier with circuit values

Basic Idea

Figure 9-5 shows a transistor amplifier. If you have a feeling of *deja vu* (having been here before), you are right. The inner part of this amplifier is the voltage-divider bias analyzed in the preceding chapter. What we have is voltage-divider bias with coupling and bypass capacitors.

Here is how the amplifier works. The ac generator has a voltage of 1 mV and a resistance of 600 Ω. The input capacitor couples some of the generator voltage into the base. This ac voltage produces an ac variation in base current of the same frequency as the generator. Because of the current gain, the ac variation in the collector current is a magnified version of the ac base current. Since the collector current flows through the collector resistor, there is a varying ac voltage across the collector resistor. The collector voltage equals the supply voltage minus the voltage across the collector resistor. Therefore, the collector voltage will now be a voltage that has an ac variation. In other words, if we look at the collector voltage with an oscilloscope, we will see an amplified version of the ac base voltage.

The output capacitor couples the ac collector voltage to the load resistor. This load resistor can be anything. It might be another discrete resistor or it might be the equivalent resistance of a device like a loudspeaker. If it's a loudspeaker, we will hear the amplified signal as a sound. For example, a radio receiver has several transistor stages that amplify a weak signal from an antenna until it is large enough to drive a loudspeaker.

The purpose of the emitter capacitor cannot be fully explained right now, except to say that without it, the amplifier has less amplification. If we are looking at the ac voltage at the collector, removing the emitter capacitor will make the signal much smaller.

DC Analysis

The process for analyzing Fig. 9-5 starts with dc analysis. The dc equivalent circuit is the simplified circuit you can use when you want to calculate only dc currents and voltages. Recall that the first three steps are:

1. Reduce the ac source to zero.

2. Open all capacitors.

3. Analyze the dc equivalent circuit.

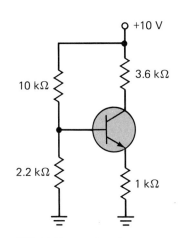

FIGURE 9-6

Dc equivalent circuit.

The first step in this process removes the ac signal. The second step disconnects the generator and the load from the circuit, as well as the emitter capacitor. In other words, the original circuit is transformed to the simpler circuit shown in Fig. 9-6. This is the only part of the circuit that matters when you are calculating dc currents and voltages.

This circuit is the voltage-divider bias analyzed in Chap. 8. The supply voltage and resistors are identical, so the results for step 3 are the same as before:

$$V_B = 1.8 \text{ V}$$
$$V_E = 1.1 \text{ V}$$
$$I_E = 1.1 \text{ mA}$$
$$V_C = 6.04 \text{ V}$$
$$V_{CE} = 4.94 \text{ V}$$

These are the currents and voltages when the dc source is acting alone. To get these results, we temporarily removed the ac source and associated components like capacitors and resistors not affecting the dc operation of the circuit. This completes the dc analysis of the circuit.

AC Analysis

The process for analyzing Fig. 9-5 continues with ac analysis. The ac equivalent circuit is the simplified circuit you can use to calculate ac currents and voltages. The next three steps are:

4. In the original circuit, reduce all dc sources to zero.

5. Short all capacitors.

6. Analyze the ac equivalent circuit.

The fourth step in this process grounds the top of the 10 kΩ and 3.6 kΩ in Fig. 9-5. The fifth step connects the generator and the load to the transistor and grounds the emitter. In other words, the original circuit is transformed to the simpler circuit shown in Fig. 9-7*a*. The circuit has been redrawn with the grounded ends on the bottom instead of the top. This is the circuit that determines ac currents and voltages.

We are not yet ready to calculate ac currents and voltages, but we are close. There is something we can do right now, however. On the input side of the transistor, 10 kΩ is in parallel with 2.2 kΩ. This parallel connection always happens when you are analyzing the ac equivalent circuit, and you can always handle it as follows. Use the product-over-the-sum rule to find the equivalent resistance:

$$R = \frac{(10\,\text{k}\Omega)(2.2\,\text{k}\Omega)}{12.2\,\text{k}\Omega} = 1.8\,\text{k}\Omega$$

The two resistors in parallel are equivalent to a single resistance of 1.8 kΩ as shown in Fig. 9-7*b*. This simplification is important for a troubleshooter to remember because he often wants to measure the ac voltage between the base and ground.

In Fig. 9-7*b*, we can now see that the ac base voltage has to be less than 1 mV. Why? Because the 600 Ω and the 1.8 kΩ form a voltage divider. Therefore, there will be less than 1 mV of ac voltage appearing at the base. How much less? You will have to wait to find out because

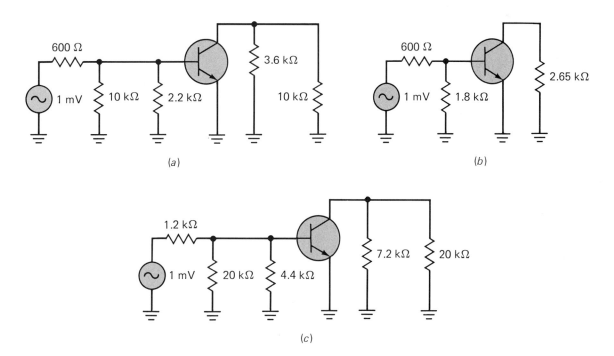

(a)

(b)

(c)

FIGURE 9-7   Ac equivalent circuit.

the ac base current has to be taken into account when calculating the ac base voltage.

The final simplification we can make in Fig. 9-7a is in the collector circuit. As far as the ac signal is concerned, 3.6 kΩ is in parallel with 10 kΩ. This produces an equivalent resistance of

$$R = \frac{(3.6 \text{ k}\Omega)(10 \text{ k}\Omega)}{13.6 \text{ k}\Omega} = 2.65 \text{ k}\Omega$$

This equivalent resistance is called the ac collector resistance because it is the resistance or opposition to the ac collector current. In other words, seen through the eyes of the ac signal, there are two paths for current: the path through the 10 kΩ and the path through the 3.6 kΩ.

The ac collector resistance is different from the dc collector resistance. The dc collector resistance is 3.6 kΩ because this is the opposition to dc current in Fig. 9-5. The 10 kΩ cannot affect the dc current since the capacitor is open at low frequencies.

Recap

All of this may seem quite different from the way you have analyzed circuits in the past. The analysis is different because we are moving from one-dimensional thinking (circuits with one source) to two-dimensional thinking (circuits with dc and ac sources). Now, we have to separate the analysis into two problems, one for low frequencies (dc) and one for high frequencies (ac).

An amplifier like the one in Fig. 9-5 is really two circuits rolled into one because it acts one way to dc and another way to ac. The superposition theorem allows us to transform this circuit into its dc equivalent circuit and its ac equivalent circuit. You already know how to analyze the dc equivalent circuit, and you are on the verge of analyzing the ac equivalent circuit.

EXAMPLE 9-3

All resistances are doubled in Fig. 9-5, on page 293. What happens to the dc equivalent circuit? The ac equivalent circuit?

SOLUTION

In the dc equivalent circuit, the base voltage and emitter voltage are still 1.8 and 1.1 V. But the emitter current decreases to

$$I_E = \frac{1.1 \text{ V}}{2 \text{ k}\Omega} = 0.55 \text{ mA}$$

The collector voltage is

$$V_C = 10 \text{ V} - (0.55 \text{ mA})(7.2 \text{ k}\Omega) = 6.04 \text{ V}$$

This is the same as before. In other words, doubling all resistances decreases all currents by a factor of 2, but the voltages remain the same.

Figure 9-7c shows the ac equivalent circuit. As you can see, all resistances have been doubled. In the base circuit, we can calculate the equivalent parallel resistance with the product-over-sum rule:

$$R = \frac{(20 \text{ k}\Omega)(4.4 \text{ k}\Omega)}{24.4 \text{ k}\Omega} = 3.6 \text{ k}\Omega$$

Another way to find this resistance is the following. If you double two parallel resistances, the equivalent resistance is twice as much. Originally,

$$R_1 \parallel R_2 = 10 \text{ k}\Omega \parallel 2.2 \text{ k}\Omega = 1.8 \text{ k}\Omega$$

Since the individual resistances are doubled,

$$R_1 \parallel R_2 = 20 \text{ k}\Omega \parallel 4.4 \text{ k}\Omega = 3.6 \text{ k}\Omega$$

If you don't understand this shortcut, you should think about it until you do.

A similar idea applies to the collector circuit. The two resistances changed from 3.6 kΩ and 10 kΩ to 7.2 kΩ and 20 kΩ. Either with the product-over-the-sum rule or with the shortcut,

$$R_C \parallel R_L = 7.2 \text{ k}\Omega \parallel 20 \text{ k}\Omega = 5.3 \text{ k}\Omega$$

9-4 SMALL-SIGNAL OPERATION

Figure 9-8 shows the graph of current versus voltage for the emitter diode. The point, labeled Q, represents the quiescent (at rest) operating point. The quiescent point is the same as the dc operating point. If you did not bias a transistor, the Q point would be at the origin. As you increase the dc emitter current, the Q point moves higher up the curve.

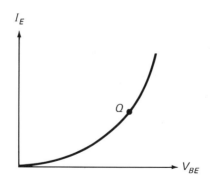

FIGURE 9-8 Graph of current versus voltage.

Instantaneous Operating Point Moves

The curve has been distorted for this discussion. In reality, the portion of the graph below the knee of the curve hugs the horizontal axis until V_{BE} is approximately 0.7 V. We have shown the curve rounded because we now want to talk about the effect of an ac signal.

When we couple an ac voltage into the base of a biased transistor, we force the instantaneous operating point to move up and down about the

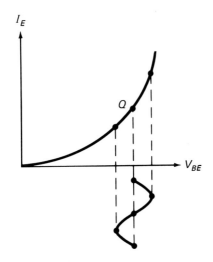

FIGURE 9-9

AC voltage applied to base-emitter diode.

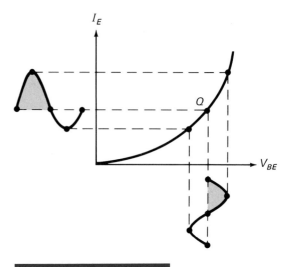

FIGURE 9-10

Corresponding emitter current.

quiescent point. This instantaneous operating point is different from the quiescent operating point (Q point). The Q point is what you get with dc voltages and currents. When an ac voltage is added to the dc voltage, the actual or *instantaneous operating point* moves away from the quiescent point.

For instance, Fig. 9-9 shows an ac voltage across the base-emitter terminals. When the sine wave increases to its positive peak, the instantaneous operating point moves from Q to the upper point. On the other hand, when the sine wave decreases to its negative peak, the instantaneous operating point moves from Q to the lower point.

The total base-emitter voltage of Fig. 9-9 is a dc voltage plus an ac voltage. The size of the ac voltage determines how far the instantaneous point moves away from the Q point. Large ac voltages produce large variations, while small ac voltages produce only small variations.

The motion of the instantaneous operating point around the Q point reminds us of a swing. The point moves to the left of the Q point, then to the right of the Q point, then to left, and so on. The Q point is the center of the swing, the point where the action begins and ends.

Distortion

The ac voltage on the base produces an ac emitter current, as shown in Fig. 9-10. This ac emitter current has the same frequency as the ac base voltage. For instance, if the ac generator driving the base has a frequency of 1 kHz, the ac emitter current has a frequency of 1 kHz. The ac emitter current also has approximately the same shape as the ac base voltage. If the ac base voltage is sinusoidal, then the ac emitter current is approximately sinusoidal.

The reason the ac emitter current is not a perfect replica of the ac base voltage is because of the curvature of the graph. Since the graph is concave upward, the positive half-cycle of ac emitter current is elongated (stretched) while the negative half-cycle is compressed. This stretching and compressing of alternate half-cycles is called *distortion*. It is undesirable in high-fidelity amplifiers because it changes the sound of voice and music.

Reducing Distortion

One way to reduce distortion in Fig. 9-10 is by keeping the ac base voltage small. When you reduce the peak value of the base voltage, you reduce the movement of the instantaneous operating point. The smaller this swing or variation, the less concave the graph appears. If the signal is small enough, the graph appears linear.

The idea is similar to the curvature of the earth. High above the earth, you can see the curvature, but as you move down to earth, the earth begins to look flat. The same concept applies to Fig. 9-10. To a large signal the graph looks curved, but to a small signal it appears flat or linear.

Why is this important? Because there is no distortion for a small signal. When the signal is small, the changes in ac emitter current are directly proportional to the changes in ac base voltage because the graph appears linear. In other words, if the ac base voltage is a small enough sine wave, the ac emitter current will also be a small sine wave with no stretching or compression of half-cycles.

The 10 Percent Rule

But now we have a question to answer. How small is small? The rule we will use in this book is the 10 percent rule which says

 The ac signal is small when the peak-to-peak ac emitter current is less than 10 percent of the dc emitter current.

The 10 percent rule does not totally eliminate distortion but it reduces it to a small enough level for most applications.

From now on, we will refer to amplifiers that satisfy the 10 percent rule as small-signal amplifiers. This type of amplifier is used at the front end of radio and television receivers. Why? Because the signal coming in from the antenna is a very weak signal. When coupled into a transistor amplifier, a weak signal produces very small variations in emitter current, much less than the 10 percent rule requires.

As the signal passes through successive amplifier stages, it becomes larger and larger until the operation is no longer small signal. At this point, distortion appears. How we solve this problem will be discussed later.

EXAMPLE 9-4

A transistor amplifier has a dc emitter current of 10 mA. If the amplifier is operating as a small-signal amplifier, what is the ac emitter current?

SOLUTION

Figure 9-11 shows the dc operating point of 10 mA. The peak-to-peak ac emitter current has to be less than 1 mA. If this condition is satisfied, the operation is small signal. This makes sense visually because you can see that a small arc has to be very close to linear.

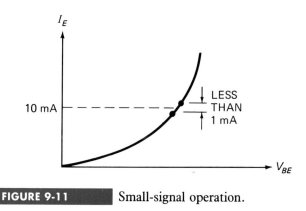

FIGURE 9-11 Small-signal operation.

9-5 AC RESISTANCE OF THE EMITTER DIODE

We have the basic idea of how an amplifier works. We know a transistor is biased to set up a quiescent operating point. We know an ac voltage is coupled into the base to produce ac variations in emitter current. We know that these variations produce an ac collector voltage, which is an amplified version of the ac base voltage. But we don't know how to calculate the value of the amplification. For instance, if the ac base voltage has a peak-to-peak value of 1 mV, what is the peak-to-peak value of the ac collector voltage? Until we can answer this question, we cannot troubleshoot or design amplifiers.

DC Resistance

We need a new idea, something you probably never heard of before. It is called *ac resistance*. Before you can understand what this is, you have to understand what dc resistance is. Recall how resistance was defined when you first studied basic electricity:

$$R = \frac{V}{I}$$

In this equation, the V and the I are dc voltage and current. Whenever you divide dc voltage across a component by dc current through the component, you get the dc resistance of the component. To emphasize that this resistance is valid only for dc, the formula may be written as

$$R_{dc} = \frac{V}{I}$$

Whenever you divide total voltage by total current, you always get dc resistance.

For instance, in Fig. 9-12 the emitter diode has a Q point with a dc voltage of 0.7 V and a dc current of 1 mA. Therefore, the emitter diode has a dc resistance of

$$R_{dc} = \frac{0.7 \text{ V}}{1 \text{ mA}} = 700 \text{ }\Omega$$

This value is useless except to show you what the dc resistance is and to prepare you for ac resistance.

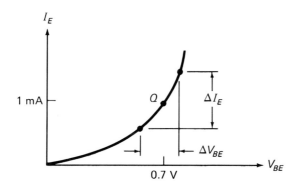

| **FIGURE 9-12** | Ac resistance of emitter diode. |

AC Resistance

What is ac resistance? It is the resistance to a small ac signal. For a linear resistor like an ordinary carbon-composition resistor, the ac resistance is the same as the dc resistance. But when you have a nonlinear device like a diode or transistor, this is no longer true. The ac signal sees or uses only a small part of the nonlinear graph. Because of this, the ac resistance is different from the dc resistance, which uses the entire graph.

Mathematically, ac resistance is defined as ac voltage across a component divided by ac current through the component. The formula is

$$R_{ac} = \frac{\Delta V_{BE}}{\Delta I_E} \tag{9-8}$$

where the symbol Δ stands for the words "the change in." You read Eq. (9-8) as

 The ac resistance equals the change in base-emitter voltage divided by the change in emitter current.

Figure 9-12 shows the visual meaning of the Δ quantities. The Δ quantities are the peak-to-peak values of the ac voltage and current.

For instance, suppose an ac base voltage of 1 mV produces an ac emitter current of 40 μA. Then the emitter diode has an ac resistance of

$$R_{ac} = \frac{1 \text{ mV}}{40 \text{ μA}} = 25 \text{ }\Omega$$

This is the resistance that you would use if you were calculating ac quantities. Stated another way, it is the resistance you would use in an ac equivalent circuit.

Compare the ac resistance (25 Ω) to the dc resistance calculated earlier (700 Ω). They are quite different. And why not? In the first case, we are talking about a small ac voltage and current that use only a small part of a nonlinear graph. In the second case, we are talking about a large dc voltage and current that use all of the nonlinear graph.

Formula for AC Emitter Resistance

The graph of the emitter current versus the base-emitter voltage has a special mathematical property (known as *exponential*). Using calculus, it is possible to derive this formula for the ac resistance of the emitter diode:

$$R_{ac} = \frac{25\,\text{mV}}{I_E}$$

In the derivation of this formula, an important assumption is made: the operation must be small-signal. When this condition is satisfied, the formula is quite accurate. You do not need to know how to prove the formula. All you need to know is that the formula gives us the ac resistance of the emitter diode.

Because the emitter diode is so special in transistor analysis, the foregoing formula is usually written like this:

$$r'_e = \frac{25\,\text{mV}}{I_E} \tag{9-9}$$

The subscript e reminds us of the emitter. The lowercase r is used to indicate an ac resistance. And the prime, ′, indicates an internal resistance; something that is inside the transistor. In words, the formula says

The ac resistance of the emitter diode equals 25 mV divided by the dc current through the emitter diode.

This relation applies to all transistors. In other words, Eq. (9-9) is a universal formula that you can use with any transistor. It is based on a perfect base-emitter junction, so there will be deviations in commercially produced transistors. But almost all commercial transistors have an ac emitter resistance that is between 25 mV/I_E and 50 mV/I_E.

Another important idea about Eq. (9-9) is this. The ac emitter resistance is inversely proportional to the dc emitter current. When the dc emitter current is small, the ac emitter resistance is large. If you change the biasing to get more dc emitter current, the ac emitter resistance decreases. With Eq. (9-9), you can calculate the ideal value of ac emitter resistance for any value of dc emitter current.

Equation (9-9) must be memorized. It is a new concept like Ohm's law, something you memorize because you expect to use it a few thousand times when troubleshooting, analyzing, or designing transistor circuits. This equation is the link between dc and ac. The denominator is a dc quantity, but the resistance is an ac quantity. This is one of those beautiful concepts that inventors dream about because it produces an incredible simplicity in the design of new circuits.

EXAMPLE 9-5

A voltage-divider biased circuit has an emitter voltage of 2 V and an emitter resistor of 4.7 kΩ. What is the ac resistance of the emitter diode?

SOLUTION

First, get the dc emitter current:

$$I_E = \frac{2\text{ V}}{4.7\text{ k}\Omega} = 0.426\text{ mA}$$

Second, calculate the ac resistance of the emitter diode:

$$r'_e = \frac{25\text{ mV}}{0.426\text{ mA}} = 58.7\ \Omega$$

9-6 AC BETA

Figure 9-13 shows an amplifier. We know the transistor is biased to set up a quiescent operating point. Our earlier analysis of this circuit gave us these dc quantities:

$V_B = 1.8$ V
$V_E = 1.1$ V
$I_E = 1.1$ mA
$V_C = 6.04$ V
$V_{CE} = 4.94$ V

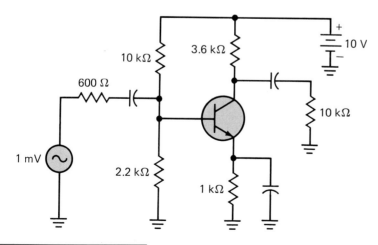

FIGURE 9-13   Transistor amplifier.

The dc emitter current is 1.1 mA. What follows shows you how to find the ac resistance of the emitter diode.

Calculating the AC Emitter Resistance

After you have finished your dc analysis, you can proceed with the ac analysis. The first thing to do is to calculate the ac resistance of the emitter diode as follows:

$$r'_e = \frac{25\,\text{mV}}{1.1\,\text{mA}} = 22.7\,\Omega$$

This says the emitter diode has an ac resistance of 22.7 Ω. This is the resistance that you would use when calculating ac quantities. In other words, this is the resistance that relates the peak-to-peak voltage and current:

$$\frac{\Delta V_{BE}}{\Delta I_E} = 22.7\,\Omega$$

DC Current Gain

At this point we have to bring the current gain of the transistor into our analysis. But this raises a question. What is the current gain in an ac equivalent circuit? As you may suspect, the current gain of a transistor is different in an ac equivalent circuit than in a dc equivalent circuit.

Earlier, you saw that there are two kinds of resistance: dc resistance and ac resistance. The dc resistance is the total voltage divided by the total current. The ac resistance is the change in total voltage divided by the change in total current. The basic concept can be applied to other quantities. The current gain in all discussions up to this point has been dc current gain. This was defined as follows:

$$\beta_{\text{dc}} = \frac{I_C}{I_B} \tag{9-10}$$

This says that the dc current gain is the ratio of collector current to base current.

The currents in this formula are total or dc quantities. Put another way, they are the currents at the quiescent operating point of the transistor. One way to measure the dc current gain is to use a dc ammeter to measure the collector current and then the base current. After you divide the measured collector current by the measured base current, you have the dc current gain.

AC Current Gain

Figure 9-14 shows the typical graph of collector current versus base current. The Q point is the point in the middle. The collector current and base current for this Q point are the dc collector current and the dc base current that you would use in Eq. (9-10).

What is the ac current gain? It is the current gain for the ac signal. The ac signal sees or uses only a small part of the graph on both sides of

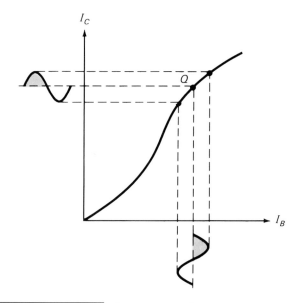

FIGURE 9-14   Ac current gain.

the Q point. Because of this, the ac current gain is different from the dc current gain, which uses the entire graph.

Mathematically, the ac current gain is defined as the ac collector current divided by the ac base current. The formula is

$$\beta = \frac{\Delta I_C}{\Delta I_B} \tag{9-11}$$

You read Eq. (9-11) as

 The ac current gain equals the change in collector current divided by the change in base current.

Figure 9-14 shows the visual meaning of the Δ quantities. The Δ quantities are the peak-to-peak values of the ac collector current and the ac base current. Because of this, you often see the formula written like this:

$$\beta = \frac{i_c}{i_b} \tag{9-12}$$

where the lower-case letters symbolize ac. Graphically, β is the slope of the curve at point Q. For this reason, it has different values at different Q locations.

On data sheets, β is listed as h_{fe}. The subscripts on h_{fe} are lower-case letters, whereas the subscripts on h_{FE} are capital letters. When reading data sheets, do not confuse the two current gains. The quantity h_{FE} is the dc current gain, identical to β_{dc}. This is the current gain to use for the dc equivalent circuit. On the other hand, h_{fe} is the ac current gain, identical to β. This is the current gain that we will use in the ac equivalent circuit.

9-7 CE AMPLIFIER

Figure 9-15 shows an amplifier. Because the emitter is at ac ground, this is called a *common-emitter* (CE) amplifier. The ac generator voltage has a value of 1 mV. Unless otherwise indicated, this refers to the peak-to-peak value of the sinusoidal voltage rather than the rms value. The 600 Ω is the internal resistance of the ac generator. (If you studied Thevenin's theorem in basic electricity, the 600 Ω is the Thevenin impedance of the ac generator.)

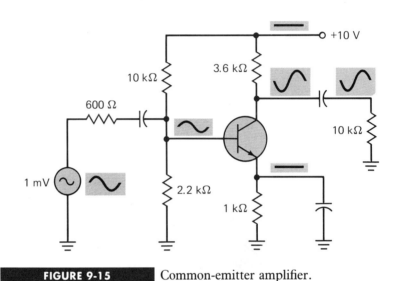

FIGURE 9-15   Common-emitter amplifier.

Input Coupling

The generator voltage is coupled through the input capacitor into the base of the transistor. The ac base voltage of Fig. 9-15 is smaller than the generator voltage because there is some loss of voltage across the 600 Ω. Since the emitter is at ac ground, all of the ac base voltage appears across the base-emitter terminals, or what amounts to the same thing, the emitter diode. Because of this ac voltage, the emitter diode has a corresponding ac current. In other words, the ac emitter current will have the same frequency and phase as the ac base voltage. The ac base voltage is often referred to as the input voltage because this is the voltage that drives the amplifier.

Phase Inversion

Because the total collector current is approximately equal to the total emitter current, the ac collector current is approximately equal to the ac emitter current. When the ac collector current flows through the ac collector resistance, it produces an ac collector voltage. On the positive half-cycle of input voltage, the total collector current is increasing, which means there is more voltage drop across the collector resistor. In turn, this means there is less total voltage at the collector node. Stated another way, the amplified ac collector voltage is inverted as shown in Fig. 9-15, equivalent to being 180° out of phase with the input voltage.

Output Capacitor Blocks DC Voltage

The total collector voltage is the superposition of a dc voltage and an ac voltage. In Fig. 9-15, the dc collector voltage is approximately 6 V. Centered on this dc level is a sinusoidal voltage that swings above and below 6 V.

The output capacitor couples the amplified and inverted ac collector voltage to the load resistor. Because a capacitor is open to dc and shorted to ac, it will block the dc collector voltage but pass the ac collector voltage. For this reason, the final load voltage is a pure ac voltage.

No AC Voltage at Emitter Node

If you look at the emitter voltage with an oscilloscope, you will see a horizontal line as shown in Fig. 9-15. This line represents a dc voltage of approximately 1.1 V. There is no sine wave appearing on the pattern because there is no ac voltage between the emitter and ground. You already know why. The emitter is at ac ground because the bypass capacitor is an ac short (ideally).

No AC Voltage on Supply Line

Another point. An oscilloscope connected between the supply line and ground would also show nothing but a horizontal line at the +10 V level. There would be no ac voltage on the screen because the power supply has a large filter capacitor, equivalent to a bypass capacitor. In other words, a complete schematic diagram of the amplifier and the power supply would reveal a large filter capacitor between the supply line and ground. This large filter capacitor doubles as a bypass capacitor. Therefore, the entire length of the supply line is an ac ground point.

9-8 AC MODEL OF A CE AMPLIFIER

Whenever you look at an amplifier like the one shown in Fig. 9-15, remember that it acts one way to dc and another way to ac. To dc, all the capacitors are open, so all you have left to analyze is the voltage-divider biased circuit in the center. But to ac, all the capacitors appear shorted, so you have a slightly more difficult circuit to work with. To analyze a CE amplifier, you need to reduce it to an ac equivalent circuit that you can apply Ohm's law to. Earlier, we reduced the CE amplifier of Fig. 9-15 to an ac equivalent circuit like Fig. 9-16. We stopped at this point because we had not yet discussed the concept of ac beta. Now we are ready to go on.

Input Impedance of the Base

With the circuit in the form of Fig. 9-16, we can see a voltage divider exists on the input side of the transistor. This means the ac base voltage

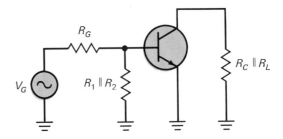

FIGURE 9-16 . Ac equivalent circuit.

will be less than the ac generator voltage. In other words, when generator current flows through R_G, it produces an ac voltage across R_G. This voltage has to be subtracted from the generator voltage to get the ac base voltage.

The amount of the voltage drop across the generator resistance depends on the value of R_1 in parallel with R_2. But there is another factor to include in the calculations. The base lead draws current from the junction of R_G and R_1 parallel R_2. Therefore, the base acts like an equivalent resistance of

$$R_{ac} = \frac{v_b}{i_b} \qquad (9\text{-}13)$$

where v_b and i_b are the peak-to-peak ac base voltage and current. For instance, if the ac base voltage is 1 mV and the ac base current is 0.4 μA, then the ac resistance looking into the base is

$$R_{ac} = \frac{1\,\text{mV}}{0.4\,\mu\text{A}} = 2.5\,\text{k}\Omega$$

This is another new concept to get used to. You may never have seen anything like this before. Nevertheless, it is valid. Whenever you have a voltage between a node and ground, and a current flowing into the node, you can replace the node by an equivalent resistance that equals the voltage divided by the current. This is important enough to repeat again as follows:

 The input resistance looking into a node equals the node voltage divided by the node current.

Remember the node voltage is the voltage between the node and ground. It is what you would measure by connecting the active lead of a measuring instrument to the node and the common lead to ground.

Equation (9-13) is usually written like this:

$$z_{\text{in(base)}} = \frac{v_b}{i_b} \qquad (9\text{-}14)$$

In this form, we are reminded of two things. First, the ac resistance looking into the base is also known as the input impedance. Recall that impedance is the opposition to alternating current. Because of this, it may include reactance. Below 100 kHz all reactances can be ignored because they have negligible effects. In other words, below 100 kHz the impedance looking into the base is purely resistive.

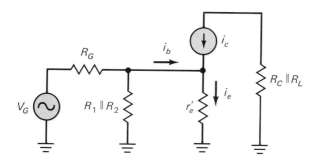

FIGURE 9-17   Transistor replaced by model T.

Model T

Beauty is in the eye of the beholder. This means the same thing can look different to different people. When the ac signal looks at the transistor of Fig. 9-16, it sees a T junction with a current source on top and an ac emitter resistance on the bottom as shown in Fig. 9-17. In this equivalent circuit, the ac base voltage appears across R_1 parallel R_2. Because r_e' is in parallel with R_1 parallel R_2, the ac base voltage is directly across r_e'. Therefore, we can calculate the ac emitter current like this:

$$i_e = \frac{v_b}{r_e'} \tag{9-15}$$

As long as the transistor is operating in the active region, the ac collector current is approximately equal to the ac emitter current. When the ac collector current flows through the ac collector resistance, it produces an ac collector voltage of

$$v_c = i_c r_c \tag{9-16}$$

Model II

As already observed, the same thing can look different to different people at different times in different places. In other words, get over the idea that everything is always the same. You can look at things from different viewpoints and still get the same answers. The model T of Fig. 9-17 is one way to visualize the inside of a transistor. This ac model works because it is based on the way the transistor voltages and currents are known to act.

Figure 9-18 shows model II, another ac model that gives the same answers as model T. This second ac model of a transistor will be called model II because it looks somewhat like a II. Where does this new ac model come from? In the model T of Fig. 9-17, the input impedance looking into the base is

$$z_{\text{in(base)}} = \frac{v_b}{i_b}$$

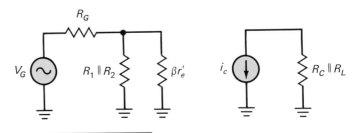

FIGURE 9-18   Transistor replaced by model II.

We already know that $v_b = i_e r'_e$, so we can substitute to get

$$z_{\text{in(base)}} = \frac{i_e r'_e}{i_b}$$

But i_e/i_b is approximately equal to β. Therefore,

$$z_{\text{in(base)}} = \beta r'_e \tag{9-17}$$

This equation implies the following. The transistor still acts like a current source on the collector side; but on the base side, it appears to be a resistance of β times r'_e. We can summarize this by redrawing the circuit as shown in Fig. 9-18. Looking into the base, the ac signal sees $\beta r'_e$. Looking into the collector, the ac load sees a current source. Therefore, we have derived another ac model for the transistor.

Which ac model of the transistor do you have to use? Use whichever one works, whichever one makes your life easier, whichever one gets the job done. Always remember that there will usually be many right ways to do things in industry. Get over the false idea that you must do things in a certain way, or that there is only one right way to do things or to solve problems. To be successful in industry, you have to find more than one right answer, because the second or third right answer is often the best solution for the particular job.

In the present discussion, we have discussed two ac models of the transistor in Figs. 9-17 and 9-18. Both give the same answers for transistor currents and voltages. Like the two sides of a coin, they are two views of the same thing. From now on, we will use whichever one is better suited to the job at hand.

Input Impedance of the Stage

In a complicated circuit, a *stage* is a transistor with its biasing resistors and capacitors. For instance, Fig. 9-15 on page 306 is a stage. If you were to connect two circuits like this together, you would have an amplifier with two stages. More will be said about this in the next chapter.

The input impedance of a stage is the combined effect of the biasing resistors and the input impedance of the base (Fig. 9-18). Since the resistors appear in parallel to the ac signal, the input impedance is the equivalent parallel resistance:

$$z_{\text{in}} = R_1 \| R_2 \| \beta r'_e \tag{9-18}$$

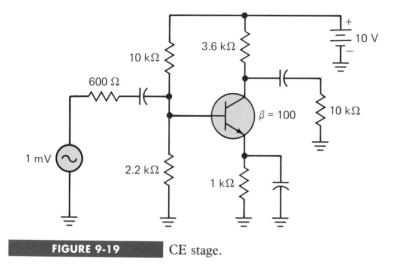

FIGURE 9-19   CE stage.

This input impedance is always less than the input impedance of the base.

EXAMPLE 9-6

What is the ac equivalent circuit of Figure 9-19?

SOLUTION

We already have calculated the following dc and ac quantities for the circuit:

$$V_B = 1.8 \text{ V}$$
$$V_E = 1.1 \text{ V}$$
$$I_E = 1.1 \text{ mA}$$
$$V_C = 6.04 \text{ V}$$
$$V_{CE} = 4.94 \text{ V}$$
$$r_c = 2.65 \text{ k}\Omega$$
$$r'_e = 22.7 \text{ }\Omega$$

Now, we can proceed to calculate the ac quantities. The input impedance of the base is

$$z_{\text{in(base)}} = 100(22.7 \text{ }\Omega) = 2.27 \text{ k}\Omega$$

This resistance appears in parallel with the two biasing resistors. The input impedance of the stage is

$$z_{\text{in}} = 10 \text{ k}\Omega \parallel 2.2 \text{ k}\Omega \parallel 2.27 \text{ k}\Omega = 1 \text{ k}\Omega$$

Now, we can draw the ac equivalent circuit of the stage as shown in Fig. 9-20. This is a simple circuit to analyze. It has a voltage divider in the base circuit, and a current source in the collector

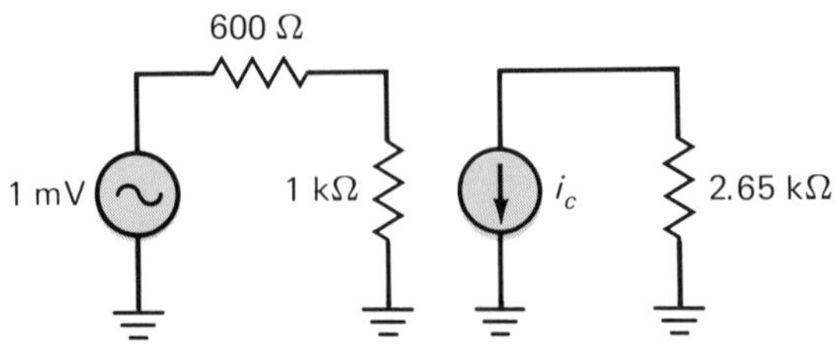

FIGURE 9-20   Ac equivalent circuit.

current. With more practice, you will be able to look at the complete amplifier of Fig. 9-19 and see Fig. 9-20 in your mind's eye. Troubleshooters and designers do it all the time. They mentally fuse the three resistors in the base circuit into a single equivalent resistance of 1 kΩ. Similarly, they mentally combine the 3.6 and 10 kΩ into a single equivalent ac collector resistance of 2.65 kΩ.

Because amplifiers contain so many new concepts, our discussion of these important circuits continues in the next chapter. There, you will learn how to calculate additional quantities of interest. The main objective of this chapter has been to arrive at the ac equivalent circuit shown in Fig. 9-20. Once you have reduced an amplifier to this equivalent circuit, you have an easy job of analysis. In fact, Ohm's law and a few other basic ideas are all you need to analyze this circuit.

9-9 AC QUANTITIES ON THE DATA SHEET

Refer to the data sheet of a 2N3904 in the Appendix during the following discussion. The ac quantities appear in the section labeled "Small-signal Characteristics." In this section, you will find four new quantities labeled h_{fe}, h_{ie}, h_{re}, and h_{oe}. These are called h parameters. What are they?

When the transistor was first invented, people knew very little about its internal operation. Because of this, an approach known as the h parameters was initially used to analyze and design transistor circuits. This is a mathematical approach that models the transistor on what is happening at its terminals without regard for the physical processes taking place inside the transistor. The h parameters are too mathematical for most people. In fact, most troubleshooters and designers hate the h parameters because they prevent them from understanding what is happening inside the transistor.

A more practical approach is the one we are using. It is called the r' parameter method, and it uses quantities like β, r'_e. With this approach, you can use Ohm's law and other basic ideas in the analysis and design of transistor circuits. This is why the r' parameters are better suited to most people. With the r' parameters, you can use your right brain as well as your left. With the h parameters, you get to use only half your brain.

This does not mean the h parameters are useless. Not quite. They have survived on data sheets because they are easier to measure than r' parameters. When you read data sheets, therefore, don't look for β, r'_e, and other r' parameters. You won't find them. Instead, you will find h_{fe}, h_{ie}, h_{re}, and h_{oe}. These four h parameters give useful information when translated into r' parameters.

For instance, h_{fe} given in the "Small-signal Characteristics" section of the Appendix is identical to the ac current gain. In symbols this is represented as:

$$\beta = h_{fe} \qquad (9\text{-}19)$$

The data sheet lists a minimum h_{fe} of 100 and a maximum of 400. Therefore, β may be as low as 100 or as high as 400. These values are for a collector current of 1 mA and a collector-emitter voltage of 10 V.

Another h parameter is the quantity h_{ie}. The data sheets give a minimum h_{ie} of 1 kΩ and a maximum of 10 kΩ. It is related to r' parameters like this:

$$r'_e = \frac{h_{ie}}{h_{fe}} \qquad (9\text{-}20)$$

For instance, the maximum values of h_{ie} and h_{fe} are 10 kΩ and 400. Therefore,

$$r'_e = \frac{10 \text{ k}\Omega}{400} = 25 \ \Omega$$

The last two h parameters, h_{re} and h_{oe}, are not needed for troubleshooting and basic design. (If you want to know more about them, see "Optional Topics," on the next page.)

Other quantities listed under "Small-signal Characteristics" include f_T, C_i, C_{ob}, and NF. The first, f_T, gives information about the high-frequency limitations on a 2N3904. The second and third quantities, C_{ib} and C_{ob}, are the input and output capacitances of the device. The final quantity NF is the noise figure; it indicates how much noise the 2N3904 produces. All of these quantities are of interest to a designer. You will learn more about them in later chapters.

The data sheet of a 2N3904 includes a lot of graphs, which are worth looking at. For instance, Fig. 11 on the data sheet gives the *current gain*. It show that h_{fe} increases from approximately 70 to 160 when the collector current increases from 0.1 mA to 10 mA. Notice that h_{fe} is approximately 125 when the collector current is 1 mA. This graph is for a typical 2N3904 at room temperature. If you recall that the minimum and maximum h_{fe} values were given as 100 and 400, then you can see that h_{fe} will have a large variation in mass production. Also worth remembering is that it changes with temperature. Unless you are a designer, all you have to remember is that h_{fe} can have very large variations due to temperature, transistor replacement, and collector current.

Take a look at Fig. 13 on the data sheet of the 2N3904. Notice how h_{ie} decreases from approximately 20 kΩ to 500 Ω when the collector current increases from 0.1 mA to 10 mA. Equation (9-20) tells us how to calculate r'_e. It says to divide h_{ie} by h_{fe} to get r'_e. Let's try it. If you read the value of h_{fe} and h_{ie} at a collector current of 1 mA from Figs. 11 and

13 on the data sheet, you get these approximate values: $h_{fe} = 125$ and $h_{ie} = 3.6$ kΩ. With Eq. (9-20),

$$r'_e = \frac{3.6 \text{ k}\Omega}{125} = 28.8 \text{ }\Omega$$

The ideal value of r'_e is

$$r'_e = \frac{25 \text{ mV}}{1 \text{ mA}} = 25 \text{ }\Omega$$

Similar results occur at other collector currents. This means you don't need data sheets if an approximate value of r'_e is adequate.

In conclusion, the data sheet of the 2N3904 contains much useful information. If you are a designer, most of the information on the data sheet is indispensable because you have to include temperature effects, replacement effects, and tolerances in your design. The data sheet tells what the minimums and maximums are for the different characteristics, which allows you to work out a worst-case design. On the other hand, if you are a troubleshooter, the data sheet is less useful, but it still has a few pieces of valuable information like maximum power ratings and current ratings.

OPTIONAL TOPICS

The following material continues the earlier discussions at a more advanced and specialized level. All the topics are optional because they are not used in any of the basic discussions in later chapters. This section will be a useful reference when you are in industry because then you will probably want more advanced viewpoints.

9-10 MORE ON CAPACITORS

For a capacitor to couple a signal, the capacitive reactance should be at least 10 times smaller than the total resistance. In symbols,

$$X_C < 0.1R$$

where $R = R_G + R_L$. When a designer satisfies this condition, the capacitor acts approximately like an ac short. This condition is equivalent to the high-frequency border discussed earlier. In other words, $X_C = 0.1R$ when $f = 10f_c$.

Where does this 10:1 rule come from? The rms current in a series RC circuit equals the generator voltage divided by the impedance:

$$I = \frac{V_G}{\sqrt{R^2 + X_C^2}}$$

When $X_C = 0.1R$, the current is

$$I = \frac{V_G}{\sqrt{R^2 + (0.1R)^2}} = \frac{V_G}{\sqrt{1.01R^2}} = \frac{0.995V_G}{R}$$

This current is only half a percent less than the maximum current, which equals V_G/R. Therefore, the coupling capacitor is having a negligible effect on the current. This is why we can treat a coupling capacitor as an ac short when its reactance is at least 10 times smaller than the total resistance.

9-11 MORE ON AC EMITTER RESISTANCE

One of the key formulas in transistor circuit analysis is

$$r_e' = \frac{25 \text{ mV}}{I_E}$$

Where does this formula come from? Shockley, the inventor of the junction transistor, derived this formula for the current through the emitter diode:

$$I_E = I_S(e^{Vq/kT} - 1)$$

At approximately 25°C, the equation simplifies to

$$I_E = I_S(e^{40V} - 1)$$

where I_S is the reverse saturation current and V is the voltage across the diode. With calculus, you can take the derivative of I_E with respect to V. Then you can rearrange the equation to get

$$r_e' = \frac{25 \text{ mV}}{I_E + I_S}$$

In a practical circuit, I_E is much greater than I_S, so the equation reduces to

$$r_e' = \frac{25 \text{ mV}}{I_E}$$

This is for a junction temperature of approximately 25°C.

When the junction temperature is different from 25°C, you can use this approximation to get the ac resistance:

$$r_e' = \frac{25 \text{ mV}}{I_E} \frac{T + 273}{298} \qquad (9\text{-}21)$$

where T is the junction temperature in Celsius degrees. For instance, if $T = 100°C$, the foregoing gives

$$r_e' = \frac{31.3 \text{ mV}}{I_E}$$

Equation (9-21) tells us that r_e' increases when temperature increases.

9-12 MEANING OF H PARAMETERS

Figure 9-21a shows the circuit model used with *hybrid* (*h*) parameters. Voltages are considered positive when they have the plus-minus polarity

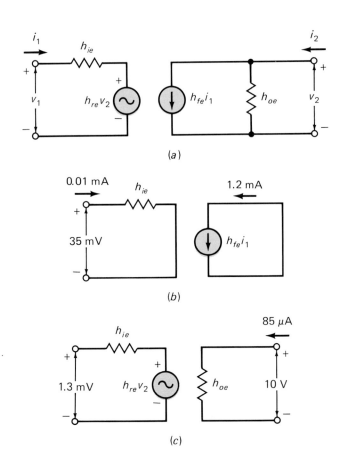

FIGURE 9-21   *H* parameters.

shown. Also, conventional currents are considered positive when they enter the circuit as shown. The Kirchhoff equations for the circuit are represented as follows:

$$v_1 = h_{ie}i_1 + h_{re}v_2 \tag{9-22}$$

$$i_2 = h_{fe}i_1 + h_{oe}v_2 \tag{9-23}$$

Input Impedance, h_{ie}

To discover the meaning of h_{ie} and h_{fe}, we proceed as follows. Suppose there is an ac short across the output terminals. Then $v_2 = 0$ and the hybrid equations reduce to

$$v_1 = h_{ie}i_1 \tag{9-24}$$

$$i_2 = h_{fe}i_1 \tag{9-25}$$

Solving the first equation, we get

$$h_{ie} = \frac{v_1}{i_1} \quad \text{(output shorted)} \tag{9-26}$$

Since it has the dimensions of volts divided by amperes, h_{ie} is an impedance. For instance, if $v_1 = 35$ mV and $i_1 = 0.01$ mA,

$$h_{ie} = \frac{35 \text{ mV}}{0.01 \text{ mA}} = 3.5 \text{ k}\Omega$$

So, h_{ie} is the input impedance of a network when the output is shorted, as shown in Fig. 9-21b.

Current Gain, h_{fe}

Next, we can solve Eq. (9-25) to get

$$h_{fe} = \frac{i_2}{i_1} \quad \text{(output shorted)} \tag{9-27}$$

Since it is the ratio of output to input current, h_{fe} is called the current gain with the output shorted. If $i_2 = 1.2$ mA and $i_1 = 0.01$ mA, then

$$h_{fe} = \frac{1.2 \text{ mA}}{0.01 \text{ mA}} = 120$$

Therefore, h_{fe} is the current gain of a network when the output is shorted (see Fig. 9-21b).

Reverse Voltage Gain, h_{re}

What is the meaning of h_{re} and h_{oe}? If the input terminals are open, then $i_1 = 0$ and Eqs. (9-22) and (9-23) simplify to

$$v_1 = h_{re}v_2 \tag{9-28}$$
$$i_2 = h_{oe}v_2 \tag{9-29}$$

Solving the first equation gives

$$h_{re} = \frac{v_1}{v_2} \quad \text{(input open)} \tag{9-30}$$

Because it is the ratio of input voltage to output voltage, h_{re} is called the reverse voltage gain with the input open. In other words, if we deliberately drive the output with a signal generator and measure the signal feeding back to the input side, then we can calculate the reverse voltage gain with the input open.

For instance, suppose the voltage driving the output is 10 V and the voltage appearing at the input is only 1.3 mV. Then

$$h_{re} = \frac{1.3 \text{ mV}}{10 \text{ V}} = 1.3(10^{-4})$$

As you can see, the reverse voltage gain is very small. This means that the circuit does not work too well in the reverse direction. Remember, h_{re} is the reverse voltage gain with the input open as shown in Fig. 9-21c.

Output Admittance, h_{oe}

Finally, we can solve Eq. (9-29) to get

$$h_{oe} = \frac{i_2}{v_2} \quad \text{(input open)} \qquad (9\text{-}31)$$

This is the ratio of output current to output voltage; therefore, h_{oe} is the output admittance with the input open. As an example, if $i_2 = 85$ μA and $v_2 = 10$ V, then

$$h_{oe} = \frac{85 \text{ μA}}{10 \text{ V}} = 8.5 \text{ μS}$$

(*Note:* S stands for siemens or mhos, the reciprocal of ohms.) In Fig. 9-21c, h_{oe} is an admittance.

Measuring H Parameters

Because it is easy to put an ac short across the output of a transistor amplifier or to put an ac open at the input, manufacturers usually measure and specify the small-signal characteristics of a transistor with h parameters. For instance, the typical h parameters of a 2N3904 in a CE connection with 1 mA of collector current are as follows:

$h_{ie} = 3.5 \text{ k}\Omega$
$h_{re} = 1.3(10^{-4})$
$h_{fe} = 120$
$h_{oe} = 8.5 \text{ μS}$

Incidentally, h_{oe} equals the slope of the collector curves seen on a curve tracer. Pick any two points on the almost horizontal part of a collector curve; the ratio of the change in current to the change in voltage equals h_{oe}. The more horizontal the collector curves are, the smaller the value of h_{oe}, which is equivalent to saying that the collector current source has a higher impedance.

STUDY AIDS

The following study aids will help to reinforce the ideas discussed in this chapter. For best results, use these study aids within 6 hours of reading the earlier material. Then review these study aids a week later and a month later to ensure that the concepts remain in your long-term memory.

SUMMARY

Sec. 9-1 Coupling Capacitor

When a capacitor is in a circuit, it can act like two different things at the same time. At low frequen- cies, it will act like an open circuit. At high frequencies, it will act like a short circuit. A coupling capacitor transmits a signal from the generator to the load at high frequencies. The basic rule is this: make the reactance of the capacitor at least 10 times smaller than the total resistance in series with the capacitor.

Sec. 9-2 Bypass Capacitor

An ac ground is different from a mechanical ground. Ac ground is produced by a bypass capacitor. This kind of ground exists only at high frequencies. A designer uses ac ground when he wants a node to be normal at low frequencies but grounded at high frequencies. The basic rule is this: make the reactance of the capacitor at least 10 times smaller than the Thevenin resistance facing the capacitor.

Sec. 9-3 Superposition in Amplifiers

An amplifier is really two circuits rolled into one because it acts one way to dc and another way to ac. The superposition theorem allows us to transform the amplifier into its dc equivalent circuit and its ac equivalent circuit.

Sec. 9-4 Small-Signal Operation

The ac signal is small when the peak-to-peak ac emitter current is less than 10 percent of the dc emitter current. When the signal is small, the changes in ac emitter current are directly proportional to the changes in ac base voltage because the graph appears linear. Because of this, there is almost no distortion with small-signal operation.

Sec. 9-5 AC Resistance of the Emitter Diode

This is the resistance to a small ac signal. The ac resistance of the emitter diode equals 25 mV divided by the dc current through the emitter diode. The ac emitter resistance is inversely proportional to the dc emitter current.

Sec. 9-6 AC Beta

The ac current gain, β, equals the change in collector current divided by the change in base current. On data sheets, β is listed as h_{fe}.

Sec. 9-7 CE Amplifier

Because the emitter is at ac ground, this is called a common-emitter (CE) amplifier. The ac signal is coupled through a capacitor to the base. The ac output signal is coupled through a capacitor to the load. The amplified ac output voltage is inverted, equivalent to being 180° out of phase with the input voltage.

Sec. 9-8 AC Model of a CE Amplifier

Whenever you look at an amplifier, remember that it acts one way to dc and another way to ac. To dc, all the capacitors are open. To ac, all the capacitors are shorted. The transistor may be visualized either as a model T or a model II. The input impedance of the stage is the equivalent parallel resistance of the biasing resistors and the input impedance of the base. The ac collector resistance is the equivalent parallel resistance of the collector resistance and the load resistance.

VOCABULARY

In your own words, explain what each of the following terms means. Keep your answers short and to the point. If necessary, verify your answer by rereading the appropriate discussion or by looking at the end-of-book Glossary.

ac current gain	dc equivalent circuit
ac equivalent circuit	h parameters
ac ground	high-frequency border
ac resistance	r' parameters
bypass capacitor	small-signal operation
coupling capacitor	superposition
critical frequency	

IMPORTANT EQUATIONS

Here are some of the most important equations you will find in transistor circuit analysis. If you don't know these, you don't know transistor circuits. Say each of the following equations in words. Then read the description that follows.

Eq. 9-1 Capacitive Reactance

$$X_C = \frac{1}{2\pi f C}$$

If you don't have this equation memorized yet, then now is the time to do so. No technician or engineer can survive without this formula.

Eqs. 9-4 and 9-6 Critical Frequency

$$f_c = \frac{1}{2\pi RC}$$

This relation keeps reappearing throughout electronics because the critical frequency exists in all kinds of circuits. Memorize it. For a coupling circuit, R is the total resistance in series with the coupling capacitor. For a bypass circuit, R is the Thevenin resistance facing the bypass capacitor.

Eq. 9-9 AC Resistance of Emitter Diode

$$r'_e = \frac{25\text{ mV}}{I_E}$$

Another bread-and-butter equation. Troubleshooters and designers love it because it cuts through layers of difficulty and gets right to the point. It applies to an ideal junction transistor at room temperature. Commercially available transistors may have a somewhat different value of r'_e. Nevertheless, this equation is an excellent starting point for all troubleshooting and design.

Eq. 9-12 AC Current Gain

$$\beta = \frac{i_c}{i_b}$$

Analogous to the dc current gain, except that you use ac quantities. It says that the ac current gain equals the ac collector current divided by the ac base current.

Eq. 9-17 Input Impedance of Base

$$z_{\text{in(base)}} = \beta r'_e$$

This is the input impedance that an ac signal sees looking into the base of a CE amplifier. In other words, an ac voltage v_b between the base and ground will produce an ac base current i_b. The value of v_b/i_b equals $\beta r'_e$.

Eq. 9-18 Input Impedance of Stage

$$z_{\text{in}} = R_1 \parallel R_2 \parallel \beta r'_e$$

This is the input impedance of a voltage-divider biased CE stage. As far as the ac signal is concerned,

the input impedance is the parallel of the biasing resistors and the input impedance of the base.

Eq. 9-19 AC Current Gain

$$\beta = h_{fe}$$

The only time you need this is when you look at a data sheet.

IMPORTANT PROCESSES

Each of the following processes is like a map that guides you through the analysis of an amplifier. If you can remember the basic ideas in each step, you will be able to solve problems more easily.

DC Equivalent Circuit

1. Reduce all ac sources to zero.
2. Open all capacitors.
3. Analyze the dc equivalent circuit.

AC Equivalent Circuit

4. Reduce all dc sources to zero.
5. Short all capacitors.
6. Analyze the ac equivalent circuit.

Final Step in Superposition

7. Add the dc and ac currents to get the total current in any branch.
8. Add the dc and the ac voltages to get the total voltage across any resistor.

STUDENT ASSIGNMENTS

QUESTIONS

The following questions refer to the figures you have seen in this chapter. You may be able to find more than one right answer in some of the questions. You are to select the best answer, the one that is in step with the approximations of this chapter or that most accurately describes the situation.

1. For dc, the current in a coupling circuit is
   a. Zero
   b. Maximum
   c. Minimum
   d. Average

2. The current in a coupling circuit for high frequencies is
   a. Zero
   b. Maximum
   c. Minimum
   d. Average

3. A capacitor is
   a. A dc open
   b. An ac short
   c. A dc short and an ac open
   d. A dc open and an ac short

4. In a bypass circuit, the top of the capacitor is
   a. An open
   b. A short
   c. An ac ground at high frequencies
   d. A mechanical ground

5. The capacitor that produces an ac ground is called a
   a. Bypass capacitor
   b. Coupling capacitor
   c. Condenser
   d. DC open

6. The amplifier discussed in this chapter is
   a. A CB amplifier
   b. An emitter follower
   c. A CC amplifier
   d. Voltage-divider biased

7. The ac base voltage of a CE amplifier is
   a. Less than the ac emitter voltage
   b. Equal to the generator voltage
   c. Greater than the ac emitter voltage
   d. Greater than the generator voltage

8. The capacitors of a CE amplifier appear
   a. Open to ac
   b. Shorted to dc
   c. Open to supply voltage
   d. Shorted to ac

9. Reducing all dc sources to zero is one of the steps in getting the
   a. DC equivalent circuit
   b. AC equivalent circuit
   c. Complete amplifier circuit
   d. Voltage-divider biased circuit

10. The ac base voltage of an amplifier is usually
    a. Less than the generator voltage
    b. Equal to the generator voltage
    c. Greater than the generator voltage
    d. Equal to the supply voltage

11. The ac equivalent circuit is derived from the original circuit by shorting all
    a. Resistors
    b. Capacitors
    c. Inductors
    d. Transistors

12. When the ac base voltage is too large, the ac emitter current is
    a. Sinusoidal
    b. Constant
    c. Distorted
    d. Alternating

13. In a CE amplifier with a large input signal, the positive half-cycle of the ac emitter current is
    a. Equal to the negative half-cycle
    b. Smaller than the negative half-cycle
    c. Larger than the negative half-cycle
    d. Equal to the negative half-cycle

14. Ac emitter resistance equals 25 mV divided by the
    a. Quiescent base current
    b. DC emitter current
    c. AC emitter current
    d. Change in collector current

15. To reduce the distortion in a CE amplifier, reduce the
    a. DC emitter current
    b. Base-emitter voltage
    c. Collector current
    d. AC base voltage

16. If the ac voltage across the emitter diode is 1 mV and the ac emitter current is 100 µA, the ac resistance of the emitter diode is
    a. 1 Ω   c. 100 Ω
    b. 10 Ω   d. 1 kΩ

17. A graph of ac emitter current versus ac base-emitter voltage applies to the
    a. Transistor
    b. Emitter diode
    c. Collector diode
    d. Power supply

18. The output voltage of a CE amplifier is
    a. Amplified
    b. Inverted
    c. 180° out of phase with the input
    d. All of the above

19. The emitter of a CE amplifier has no ac voltage because of the
    a. DC voltage on it
    b. Bypass capacitor
    c. Coupling capacitor
    d. Load resistor

20. The voltage across the load resistor of a CE amplifier is
    a. DC and ac
    b. DC only
    c. AC only
    d. Neither dc nor ac

21. In a CE amplifier, the generator voltage is usually
    a. Smaller than the ac base voltage
    b. Equal to the ac base voltage
    c. Larger than the ac base voltage
    d. Alternating

22. The ac collector resistance of a CE amplifier is usually
    a. Smaller than the dc collector resistance
    b. Larger than the dc collector resistance
    c. Smaller than the generator resistance
    d. Larger than the generator resistance

23. The ac collector current is approximately equal to the
    a. AC base current
    b. AC emitter current
    c. AC supply current
    d. AC generator current

24. The ac base voltage is usually less than the ac generator voltage because of the voltage loss across the
    a. AC collector resistance
    b. Voltage-divider resistors
    c. Generator resistance
    d. Emitter resistance

25. The ac emitter current times the ac emitter resistance equals the
    a. Generator voltage
    b. AC base voltage
    c. AC collector voltage
    d. Supply voltage

26. The output voltage of a CE amplifier equals the ac collector current times the
    a. AC collector resistance
    b. AC emitter resistance
    c. Generator resistance
    d. AC base current

27. The ac collector current equals the ac base current times the
    a. AC collector resistance
    b. DC current gain
    c. AC current gain
    d. Generator voltage

28. The input impedance of a CE amplifier is
    a. Less than the generator resistance
    b. More than the generator resistance
    c. Equal to the generator resistance
    d. Measured in ohms

29. The input impedance of a CE stage equals the ac base voltage divided by the ac
    a. Emitter current
    b. Base current
    c. Generator current
    d. Collector current

BASIC PROBLEMS

Sec. 9-1 Coupling Capacitor

9-1. What is the maximum current in Fig. 9-22? The maximum load voltage?

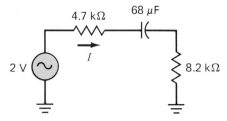

FIGURE 9-22

9-2. In Fig. 9-22, what is the critical frequency? The high-frequency border?

9-3. The generator voltage is doubled in Fig. 9-22. What happens to the high-frequency border?

9-4. All resistances are doubled in Fig. 9-22. What happens to the maximum current? Maximum load voltage? Critical frequency? High-frequency border?

9-5. The capacitance is reduced by a factor of 2 in Fig. 9-22. What happens to the high-frequency border?

Sec. 9-2 Bypass Capacitor

9-6. What is the maximum generator current in Fig. 9-23? The maximum voltage across the capacitor?

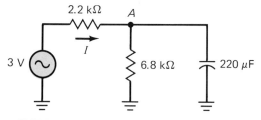

FIGURE 9-23

9-7. In Fig. 9-23, what is the critical frequency? The high-frequency border?

9-8. The generator voltage is doubled in Fig. 9-23. What happens to the high-frequency border?

9-9. All resistances are doubled in Fig. 9-23. What happens to the maximum current? Maximum load voltage? Critical frequency? High-frequency border?

9-10. The capacitance is reduced by a factor of 2 in Fig. 9-23. What happens to the high-frequency border?

Sec. 9-3 Superposition in Amplifiers

9-11. The supply voltage is doubled in Fig. 9-24. What is the dc voltage between the collector and ground? The ac collector resistance?

9-12. What is the ac collector resistance in Fig. 9-25?

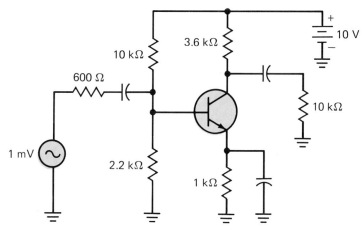

FIGURE 9-24

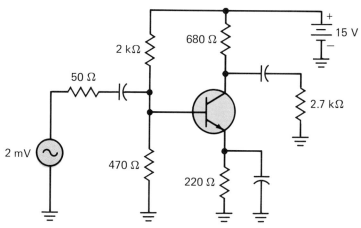

FIGURE 9-25

9-13. Do the following for Fig. 9-25. First, draw the ac equivalent circuit in unsimplified form (similar to Fig. 9-7a). Second, draw the ac equivalent circuit in simplified form (Fig. 9-7b).

9-14. All resistances are doubled in Fig. 9-25. Draw the simplified ac equivalent circuit (similar to Fig. 9-7b).

Sec. 9-4 Small-Signal Operation

9-15. A transistor amplifier has a dc collector current of 5 mA. What is the maximum allowable ac emitter curent if the amplifier is operating small-signal?

9-16. The supply voltage is doubled in Fig. 9-24. What is the maximum ac emitter current allowed if the amplifier is to operate in the small-signal mode?

9-17. What is the maximum ac emitter current allowed in Fig. 9-25 if the operation is to remain small-signal?

Sec. 9-5 AC Resistance of the Emiter Diode

9-18. A transistor amplifier has a dc collector current of 5 mA. What is the ac resistance of the emitter diode?

9-19. The supply voltage is doubled in Fig. 9-24. What is the ac resistance of the emitter diode?

9-20. What is the ac resistance of the emitter diode in Fig. 9-25?

Sec. 9-8 AC Model of a CE Amplifier

9-21. A transistor amplifier has a dc collector current of 5 mA. What is the ac resistance of the base if $\beta = 200$?

9-22. The supply voltage is doubled in Fig. 9-24. What is the ac resistance of the base if $\beta = 250$? The input impedance of the stage?

9-23. What is the ac resistance of the base in Fig. 9-25 if $\beta = 125$? The input impedance of the stage?

Sec. 9-9 AC Quantities on the Data Sheet

9-24. What are the minimum and maximum values listed under "Small-signal Characteristics" in the Appendix for the h_{fe} of a 2N3903? For what collector current are these values given? For what temperature are these values given?

9-25. Refer to the data sheet of a 2N3904 for the following. What is the typical value of r'_e that you calculate from the h parameters if the transistor operates at a collector current of 5 mA? Is this smaller or larger than the ideal value of r'_e calculated with $25 \text{ mV}/I_E$?

UNUSUAL PROBLEMS

9-26. Somebody has built the circuit of Fig. 9-22. The builder cannot understand why he measures a very small dc voltage across the 8.2 kΩ when the source is 2 V at zero frequency. Can you explain what is happening?

9-27. Assume you are in the laboratory testing the circuit of Fig. 9-23. As you increase the frequency of the generator, the voltage at node A decreases until it becomes too small to measure. If you continue to increase the frequency well into the megahertz region, the voltage at node A begins to increase. Can you explain why this happens? (*Hint:* it will happen with any bypass circuit if the frequency is well into the megahertz region.)

9-28. If one says that the dc resistance and ac resistance have the same values for a resistor, one is both right and wrong. Under what conditions is the statement right? Under what conditions is it wrong?

ADVANCED PROBLEMS

9-29. A transistor has a junction temperature of 75°C and a collector current of 2 mA. What is the ac resistance of the emitter diode?

9-30. What is the critical frequency in Fig. 9-26a? Does the critical frequency decrease or increase when you remove the 20 kΩ?

9-31. What is the critical frequency in Fig. 9-26b? What happens to the critical frequency if you double the supply voltage?

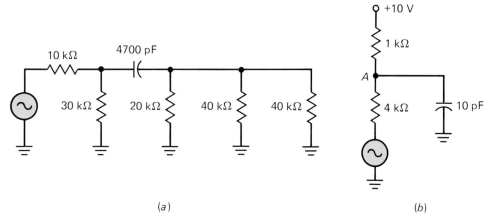

(a)   (b)

FIGURE 9-26

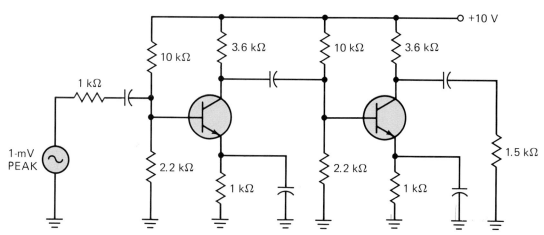

FIGURE 9-27

9-32. What is the ac collector resistance of the second stage in Fig. 9-27? The input impedance of the second stage for $\beta = 100$?

9-33. Draw the simplified ac equivalent circuit for Fig. 9-27.

9-34. What is the ac collector resistance in Fig. 9-28? The input impedance of the stage for $\beta = 200$?

9-35. A generator has an impedance of 600 Ω and an amplifier has an input impedance of 2 kΩ. Select a capacitor that will couple the ac signal over a frequency range of 20 Hz to 20 kHz.

9-36. Look at Fig. 9-24. An engineer tells you that the impedance facing the emitter bypass capacitor is 30 Ω. If the emitter is supposed to be ac ground over a frequency range of 20 Hz to 20 kHz, what size should the bypass capacitor be?

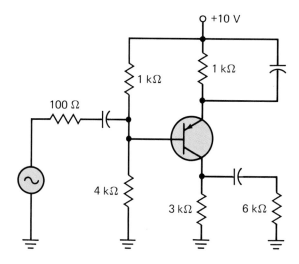

FIGURE 9-28

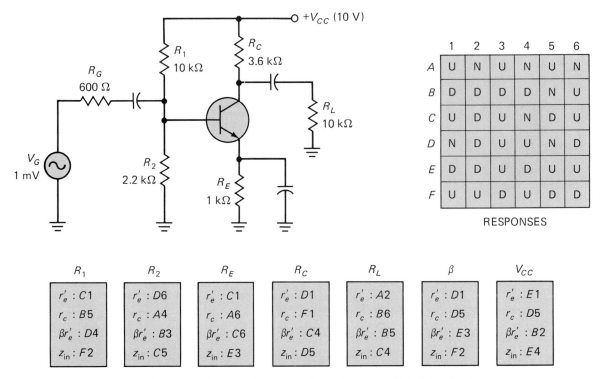

FIGURE 9-29 Software Engine™. (*Patent pending: Courtesy of Malvino Inc.*)

SOFTWARE ENGINE PROBLEMS

Use Fig. 9-29 for the remaining problems. Assume increases of approximately 10 percent in the independent variable and use the second approximation of the transistor. A response should be an N (no change) if the change in a dependent variable is so small that you would have difficulty measuring it.

9-37. Try to predict the response of each dependent variable in the box labeled R_1. Check your answers. Then answer the following question as simply and directly as possible. What effect does an increase in R_1 have on the dependent variables of the circuit?

9-38. Predict the response of each dependent variable in the box labeled R_2. Check your answers. Then summarize your findings in one or two sentences.

9-39. Predict the response of each dependent variable in the box labeled R_E. Check your

answers. List the dependent variables that decrease. Explain why these variables decrease, using Ohm's law or similar basic ideas.

9-40. Predict the response of each dependent variable in the box labeled R_C. List the dependent variables that show no change. Explain why these variables show no change.

9-41. Predict the response of each dependent variable in the box labeled R_L. List the dependent variables that increase. Explain why these variables show an increase.

9-42. Predict the response of each dependent variable in the box labeled β. List the dependent variables that increase. Explain why these variables show an increase.

9-43. Predict the response of each dependent variable in the box labeled V_{CC}. Check your answers. Then summarize your findings in one or two sentences.

10

VOLTAGE AMPLIFIERS

Now that you know how to reduce an amplifier to an ac equivalent circuit, you are in a position to analyze it mathematically. For troubleshooting and design, the most important quantity you need to find is the *voltage gain*, defined as the output voltage divided by the input voltage. When an amplifier is optimized for its voltage gain, it is called a *voltage amplifier*.

The previous chapter was about the CE amplifier. This chapter continues the discussion of CE amplifiers and shows you how a troubleshooter and a designer can calculate the voltage gain. Because of variations in β and r'_e, the CE amplifier has an unpredictable voltage gain. For this reason, we will also discuss the swamped amplifier, a CE amplifier whose voltage gain is stabilized by negative feedback.

10-1 HIGHLIGHTS OF A CE AMPLIFIER

Figure 10-1 shows a CE amplifier. Recall what CE stands for. It means common-emitter, and it refers to the fact that the emitter is common or grounded in the ac equivalent circuit. In Chap. 9, we introduced the superposition theorem for this CE amplifier. The key idea was this. An amplifier acts like two different circuits, one for dc and another for ac. At first, this seemed to make things more complicated. But in reality, it made two simple problems out of one big one.

Basic Operation

In Fig. 10-1, the generator voltage is the source of the ac signal. Because every voltage source has some internal resistance, we have to include R_G in the analysis. When R_G is very small compared to the input impedance, all of the generator voltage appears across the emitter diode. But when R_G is large compared to the input impedance, the ac base voltage is noticeably less than the generator voltage. Unless otherwise indicated, all ac voltages are peak-to-peak values because these are easy to measure on an oscilloscope.

The generator voltage is coupled through the input capacitor into the base of the transistor. Since the emitter is at ac ground, all of the ac base

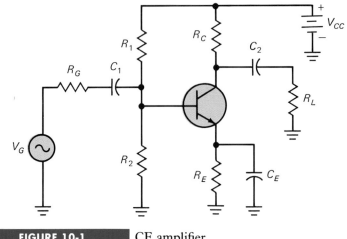

FIGURE 10-1   CE amplifier.

voltage appears across the emitter diode. Because of this, the ac emitter current will have the same frequency and phase as the ac base voltage.

The ac collector current is approximately equal to the ac emitter current. When the ac collector current flows through the ac collector resistance, it produces an ac collector voltage. The amplified ac collector voltage is inverted, equivalent to being 180° out of phase with the input voltage.

In Fig. 10-1, the total collector voltage is a dc voltage and an ac voltage. The output capacitor couples the amplified and inverted ac collector voltage to the load resistor. Because a capacitor is open to dc and shorted to ac, it will block the dc collector voltage but pass the ac collector voltage. For this reason, the final load voltage is a pure ac voltage.

Important DC Quantities

Whenever you look at an amplifier like the one shown in Fig. 10-2, remember that it acts one way to dc and another way to ac. Since all capacitors are open to dc, all that remains in the dc equivalent circuit is a voltage-divider biased circuit. We have analyzed this circuit many times with the following approximate results:

$V_B = 1.8\,\text{V}$
$V_E = 1.1\,\text{V}$
$V_C = 6.04\,\text{V}$
$I_E = 1.1\,\text{mA}$

These are the most important dc quantities. The dc voltages are valuable for troubleshooting, while the dc emitter current is the link to ac operation.

Important AC Quantities

The connection between dc and ac operation is given by

$$r'_e = \frac{25\,\text{mV}}{I_E} \tag{10-1}$$

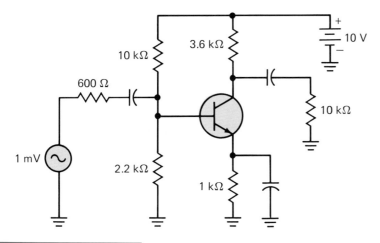

FIGURE 10-2 CE amplifier with circuit values.

For a dc emitter current of 1.1 mA, this gives an ac emitter resistance of 22.7 Ω. This resistance is one of the key quantities in the ac equivalent circuit. Another important quantity is the ac collector resistance. Since R_C (3.6 kΩ) and R_L (10 kΩ) are in ac parallel, the equivalent resistance seen by the ac signal is

$$r_c = 2.65 \text{ k}\Omega$$

Also important is the input impedance of the base. Because of the current gain, the ac emitter resistance is transformed into a larger impedance when seen from the base terminal. The transformation factor is β, which means

$$z_{\text{in(base)}} = \beta r'_e \qquad (10\text{-}2)$$

This impedance appears in ac parallel with the two biasing resistors. Therefore, the amplifier or stage has an input impedance of

$$z_{\text{in}} = R_1 \| R_2 \| \beta r'_e \qquad (10\text{-}3)$$

The Achilles Heel

Equation (10-3) reveals one of the shortcomings of a CE amplifier. It says that the input impedance is β sensitive. Since β has large variations due to quiescent current, temperature change, and transistor replacement, the ac base voltage depends on the input impedance of the amplifier. This means the amplifier's performance is β sensitive. For $\beta = 50$ in Fig. 10-2, the input impedance is approximately

$$z_{\text{in}} = 698 \ \Omega$$

For $\beta = 300$, the input impedance increases to

$$z_{\text{in}} = 1.42 \text{ k}\Omega$$

Because of the generator resistance, less ac base voltage exists when $\beta = 50$ than when $\beta = 300$. Calculations show that the ac base voltages for the two cases are 0.538 mV and 0.703 mV.

The main thing to be aware of is the effect that β has on the input voltage. When β increases, the ac base voltage increases. This will increase the output voltage. Whether or not the β variations are important will depend on the particular application. More will be said about this later in the chapter.

Lowercase Notation for AC Quantities

In case it was not crystal clear in the preceding chapter, Δ quantities are always ac quantities, so you can tell at a glance when ac voltages and currents are involved. Another system that you will see used in industry is the lowercase letters approach. In this system, lowercase letters are used to represent ac voltages and currents according to these equivalences:

$$v_b = \Delta V_B$$
$$v_e = \Delta V_E$$
$$v_c = \Delta V_C$$
$$i_b = \Delta I_B$$
$$i_e = \Delta I_E$$
$$i_c = \Delta I_C$$

Six of one, half dozen of another. They stand for the same thing. Whether you use v_b or ΔV_B, you are talking about the ac base voltage. A similar statement applies to the other quantities. Any of them may be used to represent ac quantities.

Our discussion has emphasized peak-to-peak voltages and currents. But this is not absolutely necessary. If it is more convenient, you can use rms values in the place of peak-to-peak values. But whichever you use, you must be consistent. For instance, if you use rms voltage for v_b, then all calculations for i_e, i_c, and v_c are rms values. Unless otherwise indicated, we will always use peak-to-peak values.

RMS and Peak-to-Peak Values

If you ever need them, here are two formulas that relate rms and peak-to-peak:

$$v_{\text{rms}} = \frac{0.707 v_{\text{pp}}}{2}$$

This makes sense. We divide the peak-to-peak value by 2 to get the peak value, then we multiply by 0.707 to get rms. Use this formula to convert peak-to-peak values to their rms equivalent values. If you want to convert the other way, use

$$v_{\text{pp}} = 2(1.414) v_{\text{rms}}$$

If you leave the formula unsimplified like this, you are more likely to remember it. It says to multiply the rms value by 1.414 to get the peak value, then multiply by 2 to get the peak-to-peak value.

EXAMPLE 10-1

Calculate the input impedance of the stage in Fig. 10-2 for β range of 50 to 300.

SOLUTION

We previously calculated 22.7 Ω for the ac resistance of the emitter diode. Therefore, the minimum input impedance of the base is

$$z_{in(base)} = 50(22.7 \ \Omega) = 1.14 \ k\Omega$$

and the maximum is

$$z_{in(base)} = 300(22.7 \ \Omega) = 6.81 \ k\Omega$$

The input impedance of the stage includes the effect of the biasing resistors as follows: The minimum input impedance of the stage is

$$z_{in} = 10 \ k\Omega \parallel 2.2 \ k\Omega \parallel 1.14 \ k\Omega = 698 \ \Omega$$

and the maximum input impedance of the stage is

$$z_{in} = 10 \ k\Omega \parallel 2.2 \ k\Omega \parallel 6.81 \ k\Omega = 1.42 \ k\Omega$$

EXAMPLE 10-2

Convert 1 V rms to its peak-to-peak value.

SOLUTION

The process looks like this:

1. Multiply rms by 1.414 to get peak

2. Double peak to get peak-to-peak

Because this process makes sense, you are more likely to remember it than a formula. If you know the process, the rest is easy:

$$v_{pp} = 2(1.414)(1 \ V) = 2.828 \ V$$

10-2   VOLTAGE GAIN

The preceding chapter showed you how to convert the amplifier of Fig. 10-2 into an ac equivalent circuit. For $\beta = 100$ and $r'_e = 22.7 \ \Omega$, the input impedance of the base is

$$z_{in(base)} = 2.27 \ k\Omega$$

This is in parallel with the biasing resistors, so that the input impedance of the stage is

$$z_{in} = 1 \ k\Omega$$

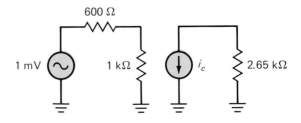

FIGURE 10-3 Ac equivalent circuit.

After replacing the transistor by its II model, we arrived at the final ac equivalent circuit shown in Fig. 10-3. Now, we are ready to continue our mathematical analysis of the CE amplifier.

Finding the Input Voltage

The input voltage to the amplifier is the same as the ac base voltage. Because the voltage divider is formed by the generator resistance and the input impedance of Fig. 10-3, the input voltage has to be less than the generator voltage. One way to find the input voltage is by multiplying the generator current and the input impedance. The generator current is

$$i_g = \frac{1\,\text{mV}}{1.6\,\text{k}\Omega} = 0.625\,\mu\text{A}$$

This produces an input voltage of

$$v_b = (0.625\,\mu\text{A})(1\,\text{k}\Omega) = 0.625\,\text{mV}$$

You can almost see the foregoing answer mentally when you look at Fig. 10-3, if you recall the voltage-divider theorem discussed in basic electricity. It says the voltage across the input impedance equals the resistance ratio times the generator voltage. In the ac equivalent circuit, this means

$$v_b = \frac{z_{\text{in}}}{R_G + z_{\text{in}}}\, v_g \tag{10-4}$$

The first factor is the resistance ratio; it equals $1\,\text{k}\Omega/1.6\,\text{k}\Omega$ or approximately 0.6. This resistance ratio times 1 mV gives 0.6 mV, which is approximately the same as the more accurate answer we just calculated. This approach allows a troubleshooter to quickly estimate the ac base voltage.

If you don't want to use Eq. (10-4), you don't have to. Just remember to calculate the generator current with

$$i_g = \frac{v_g}{R_G + z_{\text{in}}} \tag{10-5}$$

and the ac base voltage with

$$v_b = i_g z_{\text{in}} \tag{10-6}$$

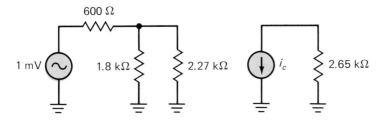

FIGURE 10-4 Ac equivalent circuit with model II.

If you stop and think about it, Eq. (10-4) comes from Eqs. (10-5) and (10-6). So, it should be fairly easy to remember Eq. (10-4).

Calculating the AC Collector Voltage

Here is one way to calculate the ac emitter current. Figure 10-4 shows the ac equivalent circuit with the model II of a transistor. In this form, it is clear that the ac base voltage is across the input impedance of the base, which means the ac base current is

$$i_b = \frac{0.625 \text{ mV}}{2.27 \text{ k}\Omega} = 0.275 \text{ }\mu\text{A}$$

Since $\beta = 100$, the ac collector current is

$$i_c = 100(0.275 \text{ }\mu\text{A}) = 27.5 \text{ }\mu\text{A}$$

When this flows through the ac collector resistance, it produces an ac collector voltage of

$$v_c = (27.5 \text{ }\mu\text{A})(2.65 \text{ k}\Omega) = 72.9 \text{ mV}$$

This is the same as the output voltage because the output capacitor couples the ac collector voltage to the load resistor.

Another Way to Find AC Collector Voltage

You usually don't know the value of β, so the foregoing method forces you to guess a value of β and work out an approximate answer. There is a way around this. Instead of using model II, use model T. Remember, we can use either model because they both produce the same final answers.

Figure 10-5 shows the ac equivalent circuit with model T of a transistor. Sometimes, you will have the ac base voltage directly from a measurement. In this case, you can get an answer without needing to know the value of β. For instance, suppose you measure an ac base voltage of 0.625 mV. In Fig. 10-5, this ac base voltage is directly across the r'_e of the transistor. Therefore, the ac emitter current must equal

$$i_e = \frac{0.625 \text{ mV}}{22.7 \text{ }\Omega} = 27.5 \text{ }\mu\text{A}$$

The ac collector current is approximately equal to this. As before, the ac collector voltage is the product of 27.5 μA and 2.65 kΩ, so the final output voltage is still 72.9 mV.

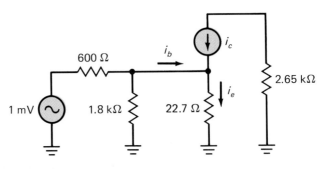

FIGURE 10-5 Ac equivalent circuit with model T.

Do you see the difference? Model II is useful when you have the value of β and want to calculate the ac base voltage. On the other hand, if you are given the ac base voltage or if you can measure it, you are better off using model T because it does not require knowing the value of β.

Calculating Voltage Gain

After you have measured or calculated the output voltage, you can calculate the *voltage gain* of the amplifier. The voltage gain tells you how much amplification the circuit has. It is defined as the output voltage divided by the input voltage:

$$A = \frac{v_{\text{out}}}{v_{\text{in}}} \tag{10-7}$$

The symbol A reminds us of amplification. You will also see the symbol A_V used in some books because the type of amplification being produced is voltage gain.

Troubleshooters use this formula when they test amplifiers. With an oscilloscope, a troubleshooter can measure the input voltage and the output voltage. If the amplifier is working correctly, the output voltage will be much larger than the input voltage. For instance, if a troubleshooter measures an input voltage of 0.625 mV and an output voltage of 72.9 mV, he can calculate a voltage gain of

$$A = \frac{72.9\,\text{mV}}{0.625\,\text{mV}} = 117$$

This is the voltage gain of the amplifier we have been discussing. If you recall, we calculated an ac base voltage of 0.625 mV and an ac collector voltage of 72.9 mV.

By cross-multiplying, Eq. (10-7) can be rearranged to get

$$v_{\text{out}} = A v_{\text{in}} \tag{10-8}$$

In this form, we can see that the output voltage is A times the input voltage. For example, a designer may indicate that his circuit should have a voltage gain of 180. This tells the troubleshooter that the ac output voltage should be 180 times larger than the ac input voltage. If the troubleshooter measures an ac base voltage of 2 mV, the output voltage should measure

$$v_{out} = 180(2 \text{ mV}) = 360 \text{ mV}$$

The Process

Let's pull the various ideas of our discussion together. Here is a process you can use to calculate the voltage gain, given the ac base voltage:

1. Find the ac resistance of the emitter diode.

2. Find the ac emitter current.

3. Find the ac collector voltage.

4. Divide the ac output voltage by the ac input voltage.

The first step uses the 25 mV divided by I_E. The second step uses an ac version of Ohm's law applied to the emitter diode. The third step uses an ac version of Ohm's law applied to the collector circuit. The fourth step uses a new formula, $A = v_{out}/v_{in}$.

EXAMPLE 10-3

The output voltage of an amplifier is 5 V rms and the input voltage is 0.1 V rms. What is the voltage gain?

SOLUTION

Since voltage gain is the ratio of output voltage to input voltage, it doesn't matter whether you use rms or peak-to-peak values. But whichever you use, be consistent by using one or the other in both the numerator and denominator. In this case,

$$A = \frac{5 \text{ V}}{0.1 \text{ V}} = 50$$

EXAMPLE 10-4

What is the ac base voltage in Fig. 10-6 if $\beta = 100$?

SOLUTION

This is the same circuit discussed earlier, except that the supply voltage has doubled. Therefore, the dc base voltage increases from 1.8 V to 3.6 V. The dc emitter voltage increases from 1.1 V to 2.9 V. The dc emitter current increases to

$$I_E = \frac{2.9 \text{ V}}{1 \text{ k}\Omega} = 2.9 \text{ mA}$$

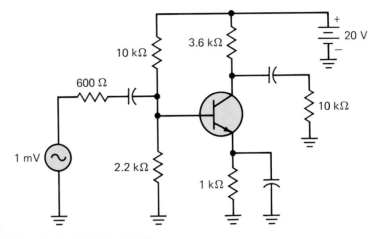

FIGURE 10-6   Example.

and the ac resistance of the emitter diode decreases to

$$r'_e = \frac{25\,\text{mV}}{2.9\,\text{mA}} = 8.62\,\Omega$$

The input impedance of the base decreases to

$$z_{\text{in(base)}} = 100(8.62\,\Omega) = 862\,\Omega$$

and the input impedance of the stage decreases to

$$z_{\text{in}} = 10\,\text{k}\Omega \parallel 2.2\,\text{k}\Omega \parallel 862\,\Omega = 583\,\Omega$$

The ac generator current increases to

$$i_g = \frac{1\,\text{mV}}{600\,\Omega + 583\,\Omega} = 0.845\,\mu\text{A}$$

Finally, the ac base voltage decreases to

$$v_b = (0.845\,\mu\text{A})(583\,\Omega) = 0.493\,\text{mV}$$

Notice the effect of increasing the supply voltage. The emitter current is increased, and the input impedance of the base is decreased. This lowered the input impedance of the stage, which resulted in less ac base voltage. Now, we have only 0.493 mV instead of 0.625 mV.

EXAMPLE 10-5

What is the voltage gain in Fig. 10-6 if $\beta = 100$?

SOLUTION

In the previous example, we calculated an ac base voltage of 0.493 mV. Once you have the ac base voltage, you can continue in either of two ways. You can calculate the ac base current and multiply by β to get the ac collector current, or you can divide the ac base voltage by r'_e to get the ac emitter current. The first approach looks like this:

$$i_b = \frac{0.493 \text{ mV}}{862 \ \Omega} = 0.572 \ \mu\text{A}$$

and

$$i_c = 100(0.572 \ \mu\text{A}) = 57.2 \ \mu\text{A}$$

The second method looks like this:

$$i_e = \frac{0.493 \text{ mV}}{8.62 \ \Omega} = 57.2 \ \mu\text{A}$$

Either way, you get 57.2 μA for the approximate value of ac collector current.

Since the ac collector resistance is 2.65 kΩ, the ac collector voltage is approximately

$$v_c = (57.2 \ \mu\text{A})(2.65 \text{ k}\Omega) = 152 \text{ mV}$$

Now we can divide this by the ac base voltage to get the voltage gain of the amplifier:

$$A = \frac{152 \text{ mV}}{0.493 \text{ mV}} = 308$$

The voltage gain is 308 for a supply of 20 V compared to 117 for a supply of 10 V. Whenever you keep the resistances constant and increase the supply voltage, the same thing will happen. You will get more voltage gain. The next section tells you why.

10-3 PREDICTING VOLTAGE GAIN

Recall how a troubleshooter tests an amplifier. He measures the output voltage (same as ac collector voltage), measures the input voltage (same as ac base voltage), and calculates the ratio of v_out to v_in. As a formula,

$$A = \frac{v_\text{out}}{v_\text{in}}$$

This is the true voltage gain of the amplifier. If the amplifier is not working correctly, this voltage gain may be much lower than it should be. For instance, if the transistor is open, a troubleshooter may measure 2 mV for the ac base voltage and 0 mV for ac collector voltage, in which case the voltage gain is zero.

There is another formula that a troubleshooter uses to predict the voltage gain before measuring it. The formula is

$$A = \frac{r_c}{r'_e} \tag{10-9}$$

Look at how simple this is. It says the voltage gain of an amplifier equals the ac collector resistance divided by the ac emitter resistance. Given the circuit values, you can work out the values of r_c and r'_e, then divide to get the voltage gain. This is the voltage gain you should get when the amplifier is working normally.

Where Does It Come From

How do we know Eq. (10-9) is true? Where did it come from? Start with the defining formula for voltage gain, which is

$$A = \frac{v_{\text{out}}}{v_{\text{in}}}$$

We call this the defining formula because this is where the idea of voltage gain starts. The defining formula cannot be proved because it was made up by someone. Defining formulas are rock bottom. You can't look back any further than the defining formula because there is nothing that precedes it.

In other words, the first time we talked about voltage gain, we said that it equaled the output voltage divided by the input voltage. Therefore, whenever anyone uses the phrase "voltage gain," he or she means the number you get when you divide v_{out} by v_{in}.

Using the defining formula

$$A = \frac{v_{\text{out}}}{v_{\text{in}}}$$

as a starting point, we can mathematically derive a new formula for voltage gain as follows. First, realize that the output voltage is given by

$$v_{\text{out}} = i_c r_c$$

and that the input voltage is given by

$$v_{\text{in}} = i_e r'_e$$

Substitute these two expressions into the defining formula to get

$$A = \frac{i_c r_c}{i_e r'_e}$$

Since electronics is not an exact science, we don't hesitate saying i_c approximately equals i_e. This means the voltage gain is approximately equal to

$$A = \frac{r_c}{r'_e} \tag{10-10}$$

Equation (10-10) is quite accurate for small-signal operation. If you recall, this means the peak-to-peak collector current is less than 10 percent of the dc emitter current. But even for large-signal operation, you can use the formula to estimate the voltage gain. With large-signal operation, the value of r'_e is no longer ideally equal to 25 mV/I_E because of the distortion that appears. Nevertheless, don't hesitate to use Eq. (10-10) for large signals as well as small signals. Just bear in mind that the value you calculate for voltage gain is only an approximation of the truth.

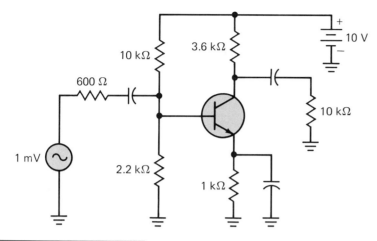

FIGURE 10-7   Example of voltage gain.

Equation (10-10) is useful not only for a troubleshooter but for a designer as well. It gives an accurate approximation of the voltage gain for a CE stage. Because of this, it is adequate for most preliminary design. To refine the answer, a designer would include temperature effects, variations in r'_e calculated with the h parameters given on the data sheet.

Measured versus Predicted Voltage Gain

With Eq. (10-10), you can calculate the voltage gain that an amplifier should have, given its circuit values. Then you can build the amplifier and measure its voltage gain with

$$A = \frac{v_{\text{out}}}{v_{\text{in}}}$$

This formula gives you the measured voltage gain of the amplifier, whereas Eq. (10-10) gives you the predicted voltage gain. If the amplifier is working correctly, the measured voltage gain will approximately equal the predicted voltage gain.

EXAMPLE 10-6

What is the predicted voltage gain for the circuit of Fig. 10-7?

SOLUTION

You have to divide the ac collector resistance by the ac resistance of the emitter diode. Earlier we found both r_c and r'_e:

$$r_c = 2.65 \text{ k}\Omega$$

and

$$r'_e = 22.7 \ \Omega$$

The predicted voltage gain of the circuit is

$$A = \frac{2.65 \text{ k}\Omega}{22.7 \text{ }\Omega} = 117$$

EXAMPLE 10-7

What is the predicted voltage gain for the circuit of Fig. 10-7 if the supply voltage is doubled?

SOLUTION

We analyzed the circuit in Example 10-4 and found that doubling the supply voltage decreases the ac resistance of the emitter diode to

$$r_e' = 8.62 \text{ }\Omega$$

Therefore, the predicted voltage gain is

$$A = \frac{2.65 \text{ k}\Omega}{8.62 \text{ }\Omega} = 307$$

10-4 SIMPLIFIED ANALYSIS

When is an exact answer wrong? When you are doing a routine analysis, a preliminary design, and especially when you're troubleshooting. You won't last five minutes in the real world if you waste your time trying to get exact answers. The bulk of everyday work uses approximations of different accuracies. For instance, if a troubleshooter were testing the amplifier of Fig. 10-7, the output voltage would need to be estimated. We have already analyzed this circuit with fairly accurate methods. Now, we want to discuss a method used by a troubleshooter who's in a hurry.

Result of the Accurate Method

The earlier analysis of Fig. 10-7 was accurate to within 5 percent and gave these results for $\beta = 100$:

$V_B = 1.8 \text{ V}$	$z_{\text{in}} = 1 \text{ k}\Omega$
$V_E = 1.1 \text{ V}$	$i_g = 0.625 \text{ }\mu\text{A}$
$V_C = 6.04 \text{ V}$	$v_b = 0.625 \text{ mV}$
$I_E = 1.1 \text{ mA}$	$i_b = 0.275 \text{ }\mu\text{A}$
$r_c = 2.65 \text{ k}\Omega$	$i_e = 27.5 \text{ }\mu\text{A}$
$r_e' = 22.7 \text{ }\Omega$	$i_c = 27.5 \text{ }\mu\text{A}$
$\beta r_e' = 2.27 \text{ k}\Omega$	$v_c = 72.9 \text{ mV}$

To get answers like these, you need to use a calculator. If you are designing circuits, calculators are no problem. In fact, a designer may have a computer with programs that automatically calculate the foregoing quantities.

Troubleshooting Method

Time is money to troubleshooters. They want to find the trouble in the least amount of time. This produces the greatest amount of money for them in the long run. To troubleshooters, exact answers are an overkill. They don't shoot ants with a cannon, nor do they find accurate answers with a calculator. There may be exceptions. They might occasionally need an accurate answer. But usually, the goal is to get a ball-park estimate of the quantity to be measured.

Here is how a typical troubleshooter might go about estimating the ac output voltage for Fig. 10-7. First, a quick dc analysis is done:

> The voltage divider formed by the biasing resistors has a total resistance of about 12 kΩ, so the divider current is around 10 V divided by 12 kΩ, or about 0.8 mA. When this flows through the 2.2 kΩ, it produces a dc voltage of roughly 2 V at the base. Subtract 0.7 V to get 1.3 V at the emitter. This sets up a dc emitter current of approximately 1.3 mA.

At this point, an error has already crept into the analysis because the actual dc emitter current is closer to 1.1 mA. The thinking continues like this:

> 25 mV divided by 1.3 mA gives an r'_e of about 20 Ω. Typical ac current gain is 100, so the input impedance of the base is around 2 kΩ.

The accurate answer found earlier was 2.27 kΩ, which is not that much better than 2 kΩ. The troubleshooting of Fig. 10-7 continues like this:

> Now, the ac signal sees three resistors in parallel: 10 kΩ, 2.2 kΩ, and 2 kΩ. The first two have an equivalent resistance of about 2 kΩ. This is in parallel with another 2 kΩ to give me 1 kΩ.

Is it luck? Troubleshooter has arrived at the same answer we found earlier using more accurate methods. What has happened is this. The troubleshooter rounds off generously. Sometimes, this produces an answer on the low side; other times, the answer comes out on the high side. After rounding off several times, the errors may cancel as you have just seen. It does not always happen, but the troubleshooter's mental calculations often produce answers within 10 percent of the exact answer. In troubleshooting, answers within 10 to 20 percent are usually adequate. If not, the troubleshooter can always reach for a calculator (but only as a last resort).

If you have ever watched troubleshooters, you have seen whole-brain artists in action. The troubleshooters fly; they are not bogged down by formulas. They know the basic formulas like Ohm's law, input impedance of the base, and so on. But they do not "plug and chug." That is, they do not substitute numbers into one formula after another and work out exact values with a calculator. Neither should you, at least not when you are troubleshooting.

The troubleshooter's thinking process for Fig. 10-7 continues something like this:

> In my mind's eye, I can see a generator resistance of 600 Ω in series with an input impedance of 1 kΩ, which gives a total resistance of 1.6 kΩ. The generator voltage is 1 mV, so about 0.6 μA flows in the circuit. This 0.6 μA flows through the 1 kΩ to produce an ac base voltage of 0.6 mV. That makes sense because the voltage divider has a resistance ratio of about 0.6.

Remember this thinking is taking place at high speed, unhampered by the use of a calculator. The troubleshooter probably gets to this value in less than a minute and now knows the ac base voltage is approximately 0.6 mV. Compare this to the accurate answer of 0.625 mV and you can see how well the thinking process works. The rest of the analysis goes like this:

> The ac collector resistance is 3.6 kΩ in parallel with 10 kΩ. This gives an equivalent resistance somewhat less than 3.6 kΩ, say 2.5 kΩ. The r'_e is around 20, so when I divide 2.5 kΩ by 20 Ω, I get a voltage gain of 125. The final ac output voltage is 125 times 0.6 mV, or about 70 mV. That's the ball-park value I should measure if the amplifier is working correctly.

The accurate method gave us an answer of 72.9 mV, so that the troubleshooter has come in with an impressively close answer. Solutions may not always be this close, but then again, they don't have to be. When an amplifier is in trouble, its output voltage is usually far different from its normal value.

Don't Plug and Chug

There is another factor that dooms the person who plugs and chugs: nobody knows the exact value of ac current gain. As mentioned earlier, β will have large variations comparable to β_{dc}. We are talking about 3:1, 5:1, and 9:1, depending on the transistor, temperature, and current. So, accurate calculations while troubleshooting are virtually impossible because one big piece of the puzzle is missing (the value of β).

Learn from troubleshooters. Exact answers are useless in many environments. In electronics, you usually are dealing with component tolerances of 5 percent or more. Furthermore, you often don't have all of the transistor characteristics. Even if you have data sheets, you still are only getting the typical characteristics for the transistor. And finally, you are being paid to find troubles, not exact answers.

If that's not enough, then read again what Aristotle had to say about exact answers:

> It is the mark of an instructed mind to rest satisfied with that degree of precision which the nature of the subject admits, and not to seek exactness where only an approximation of the truth is possible.

He's telling us that exactness doesn't work in the real world. It's human to want the comfort and certainty of exact answers, but the only place you get these answers is in pure mathematics, accounting, and subjects where exactness is possible. It is not possible in electronics. The sooner you accept this reality, the sooner you'll become a whole-brain thinker.

Recap

There is a time for exact answers, a time for calculator approximations, and a time for mental estimates. Don't be fooled into thinking you must do things in a certain way. Don't accept the idea that there is only one right answer to every problem and only one way to get it. Every top professional agrees that there are many right answers and many ways to find them. The best strategy is to stay flexible and open-minded. Visualize what you are trying to do, then listen to whatever comes into your mind. Let the mind go where it will. You will be surprised at how many right answers it comes up with.

10-5 SWAMPED AMPLIFIER

A CE amplifier has a few shortcomings. Its output voltage changes with quiescent current, temperature variations, and transistor replacement. Why does the output change? When the dc emitter current increases, the ac emitter resistance, r_e', decreases. This changes the voltage gain because it equals r_c divided by r_e'. Similarly, when the temperature varies or when a transistor is replaced, the ac current gain, β, varies. This is to be expected since β is derived from the graph of β_{dc}. Since $\beta r_e'$ affects the ac base voltage, the ac output voltage is affected.

Variations May Be Acceptable

The point is this. The ac collector voltage of a CE amplifier depends on the exact characteristics of the transistor, which means the output voltage will change with temperature variation, transistor replacement, etc. This is acceptable in many applications. For instance, when you change the volume control of a radio, you can adjust the voltage gain of an amplifier to compensate for changes in voltage gain caused by temperature variation and transistor replacement.

AC Emitter Feedback

But not all electronics equipment has a volume control. Furthermore, some equipment needs a voltage gain that is constant under all operating conditions. How can we make the voltage gain constant? One way to stabilize the voltage gain is by leaving some of the emitter resistance unbypassed as shown in Fig. 10-8. This 180 Ω is called a feedback resistor because it produces negative feedback, first described in Sec. 8-6.

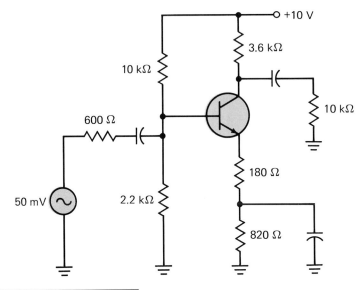

FIGURE 10-8   Ac emitter feedback.

How does it work? The ac emitter current has to flow through the 180 Ω before it reaches the bypass capacitor and the ac ground point. Because of this, an ac voltage appears across the 180 Ω. This ac emitter voltage is the feedback voltage. In symbols,

$$v_f = v_e$$

For instance, if the input voltage at the base increases, more ac voltage will appear across the 180 Ω. This means that there is more feedback voltage.

Without the feedback resistor, the emitter diode has all of the ac input voltage across it. But with the feedback resistor, the ac input voltage appears across the emitter diode and the 180 Ω. In symbols,

$$v_{\text{in}} = v_{be} + v_e$$

Therefore, the ac voltage across the emitter diode equals the ac base voltage minus the feedback voltage:

$$v_{be} = v_{\text{in}} - v_e$$

When the input voltage increases, the emitter voltage increases. This means the feedback voltage is in phase with ac input voltage. As a result, there is less voltage across the emitter diode than before. The feedback is negative because the feedback voltage reduces the ac voltage across the emitter diode.

Formula for Voltage Gain

In the dc equivalent circuit of Fig. 10-8, the total emitter resistance is 1 kΩ, the sum of 180 Ω and 820 Ω. Therefore, the dc voltages and currents are the same as before. The dc emitter current is still 1.1 mA, which gives an r_e' of 22.7 Ω.

Next, we can draw the ac equivalent circuit shown in Fig. 10-9. This is almost the same as before, except that 180 Ω is in series with 22.7 Ω. Now, the ac signal sees an ac emitter resistance equal to the sum of 22.7 Ω and 180 Ω, approximately 203 Ω. This is the total ac emitter resistance. This is what one uses in the formula for voltage gain.

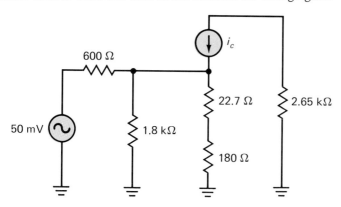

FIGURE 10-9   Ac equivalent circuit with model T.

By a mathematical proof similar to what went before, the voltage gain is given by

$$A = \frac{r_c}{r_e + r_e'} \tag{10-11}$$

where r_e equals the feedback resistance. In Eq. (10-11), this gives a voltage gain of

$$A = \frac{2.65 \text{ k}\Omega}{203 \text{ }\Omega} = 13.1$$

You can remember Eq. (10-11) more easily if you say it in words:

> The voltage gain equals the ac collector resistance divided by the total ac resistance of the emitter branch.

Tradeoff Gain for Stability

Design is a game of compromise. You find that by sacrificing in one area, you can gain in another. Here, we have an example of trading off voltage gain for stability. With no feedback resistor the voltage gain is

$$A = \frac{2.65 \text{ k}\Omega}{22.7 \text{ }\Omega} = 117$$

With a feedback resistor of 180 Ω, the voltage gain is

$$A = \frac{2.65 \text{ k}\Omega}{203 \text{ }\Omega} = 13.1$$

We get a larger output signal, but not as large as we do without the feedback resistor.

Do we get anything else? Yes, we get a voltage gain that varies less with quiescent current, temperature variations, and transistor replacement. Why? Because the changes in r_e' are being reduced by the feedback resistance. The fastest way to see this is to look at the formula for voltage gain:

$$A = \frac{r_c}{r_e + r_e'}$$

If r_e is much larger than r_e', changes in r_e' are *swamped* (greatly reduced in effect). It's like immersing a hyperactive child in a tub of molasses; the variations are inhibited.

For instance, the voltage gain is 117 when there is no feedback resistance. If r_e' doubles from 22.7 Ω to 45.4 Ω, the voltage gain drops to

$$A = \frac{2.65 \text{ k}\Omega}{45.4 \text{ }\Omega} = 59$$

This is a decrease of 50 percent (117 down to 59). But with a feedback resistance of 180 Ω, the voltage gain only decreases to

$$A = \frac{2.65 \text{ k}\Omega}{180 \text{ }\Omega + 45.4 \text{ }\Omega} = 11.8$$

This is a decrease of only 10 percent (13.1 to 11.8).

In conclusion, by giving up some voltage gain, we get a more stable or constant voltage gain. We are not at the mercy of the transistor characteristics which vary all over the place. We do not have to accept changes of 50 percent or more in voltage gain if we can settle for less voltage gain. After all, we can always use another stage to amplify the signal out of the first stage. In this way, we can boost the voltage gain back up to a higher value.

Troubleshooting Shortcut

Using local feedback (an unbypassed emitter resistor) is not the only way to stabilize voltage gain nor the best, but it is one of the basic tricks used by designers. For a troubleshooter the circuit offers a new shortcut for estimating the voltage gain. Since the whole point of using the feedback resistor is to swamp or eliminate the effect of r_e', it means the troubleshooter can ignore r_e' in his estimates for voltage gain. For quick gain estimates, the following formula is used:

$$A = \frac{r_c}{r_e} \tag{10-12}$$

This is derived from Eq. (10-11) by assuming r_e is much larger than r_e'. It says the voltage gain equals the ac collector resistance divided by the feedback resistance. In Fig. 10-8, it gives

$$A = \frac{2.65 \text{ k}\Omega}{180 \text{ }\Omega} = 14.7$$

This is a good estimate for troubleshooting purposes, because the accurate answer is only slightly different (13.1).

Input Impedance of Base

The feedback resistor has another bonus effect. It increases the input impedance of the base. We can prove this mathematically, but it also makes sense logically. Since the ac emitter voltage is in phase with the ac base voltage, there is less ac voltage across the emitter diode. Less ac voltage produces less ac current. Less ac current means more opposition, equivalent to increasing the input impedance.

Here is the mathematical proof. Recall how we defined the input impedance of the base. It equals the ac base voltage divided by the ac base current:

$$z_{\text{in(base)}} = \frac{v_b}{i_b}$$

Since $v_b = i_e(r_e + r_e')$, we can substitute to get

$$z_{\text{in(base)}} = \frac{i_e(r_e + r_e')}{i_b}$$

But i_e/i_b is approximately equal to β. Therefore,

$$z_{\text{in(base)}} = \beta(r_e + r_e') \tag{10-13}$$

For troubleshooting purposes, you can usually ignore r_e' and use βr_e as an estimate for the input impedance of the base.

Think of it this way. The total ac emitter resistance is the sum of r_e and r_e'. The ac current through this resistance is i_e. To see things from the base side of the transistor means to use base current instead of emitter current. But base current is approximately β times smaller than emitter current. This is like saying the input impedance is β times larger than r_e and r_e'. In conclusion, the impedance is stepped up by a factor of β when you change your point of view from the emitter to the base.

Less Distortion with Large Signals

There is a third bonus effect in using negative feedback. Since the feedback resistor swamps the emitter diode, it indirectly reduces the distortion that occurs for large-signal operation. The nonlinearity of the emitter-diode curve is the source of the large-signal distortion. So, it follows that reducing the effect of the emitter diode on voltage gain also reduces the distortion that occurs for large-signal operation.

Without the feedback resistor, the voltage gain is

$$A = \frac{r_c}{r_e'}$$

Since the r_e' is current-sensitive, its value changes for large-signal operation. This means the voltage gain changes during the cycle of a large signal, which distorts the signal.

With the feedback resistor, the voltage gain is

$$A = \frac{r_c}{r_e + r_e'}$$

When r_e is much larger than r_e', the voltage gain no longer changes with large-signal operation because of changes in r_e'. In other words, the voltage gain remains constant throughout the cycle; therefore, the distortion is eliminated.

In conclusion, using negative feedback in the emitter of an amplifier has three beneficial effects. First, it reduces the effects of changing transistor characteristics, which means the voltage gain is more stable. Second, it increases the input impedance of the base, which results in more ac base voltage. This second effect partially makes up for the loss in gain when using negative feedback. Third, it permits undistorted large-signal operation. When you are troubleshooting a swamped amplifier, you can estimate the voltage gain as r_c/r_e, and the input impedance of the base as βr_e.

EXAMPLE 10-8

What is the ac input voltage in Fig. 10-9, on page 345, if $\beta = 100$? The ac output voltage?

SOLUTION

In Fig. 10-9, the input impedance of the base is

$$z_{in(base)} = 100(180\ \Omega + 22.7\ \Omega) = 20.3\ k\Omega$$

Figure 10-10 shows the model II equivalent circuit. Does your intuition tell you anything about the ac input voltage? Will it be lower or higher than it is without the feedback resistor? The answer is higher. The reason is because there is less loading of the voltage divider. When we combine the two parallel resistances, we get

$$z_{in} = 1.8\ k\Omega \parallel 20.3\ k\Omega = 1.65\ k\Omega$$

For variety, let us use the voltage-divider method to get v_{in}:

$$v_{in} = \frac{1.65\ k\Omega}{600\ \Omega + 1.65\ k\Omega}\ 50\ mV = 36.7\ mV$$

Earlier, we calculated a voltage gain of 13.1 for the circuit. Since the ac base voltage is 36.7 mV, the ac collector voltage is

$$v_{out} = 13.1(36.7\ mV) = 481\ mV$$

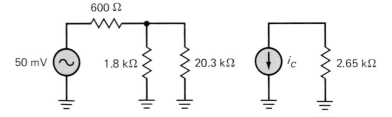

FIGURE 10-10 Ac equivalent circuit with model II.

Incidentally, this represents large-signal operation because the ac emitter current is greater than 10 percent of the dc emitter current. Here is the proof. The peak-to-peak ac emitter current equals the ac emitter voltage divided by the total ac emitter resistance.

$$i_e = \frac{36.7\,\text{mV}}{180\,\Omega + 22.7\,\Omega} = 0.181\,\text{mA}$$

Since the dc emitter current is 1.1 mA, the ac emiter current is more than 10 percent of the dc emitter current and the operation is no longer small signal. Nevertheless, the signal has almost no distortion because of the negative feedback.

10-6 CASCADED STAGES

To get more voltage gain out of an amplifier, we can connect two stages as shown in Fig. 10-11. This is called *cascading* the stages; it means the amplified voltage out of the first transistor is coupled into the base of the second transistor. The second transistor then amplifies the signal, so that the final signal is much larger than when it started.

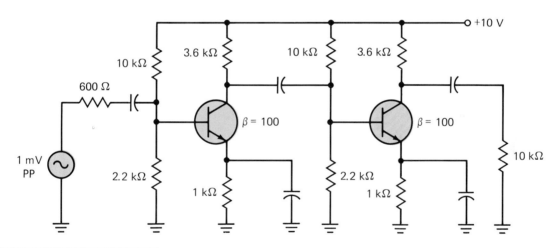

FIGURE 10-11 Two-stage amplifier.

Loading Effect of Second Stage

The ac current at the first collector sees all kinds of paths that it can flow through. It sees the following paths:

Through the 3.6 kΩ of the first stage to the supply line
Through the 10 kΩ of the second stage to the supply line
Through the 2.2 kΩ of the second stage to ground
Into the base of the second transistor

Which of these paths does it flow through? All of them. The ac collector current divides among the resistances because they are all in parallel in

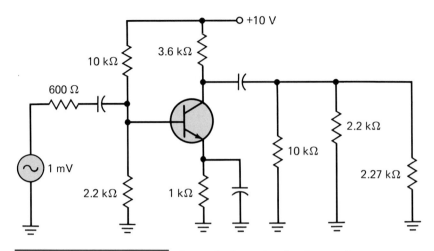

FIGURE 10-12   Ac equivalent circuit of second stage.

the ac equivalent circuit. To make this clear, we can reduce the original circuit to the equivalent circuit of Fig. 10-12. In this ac equivalent circuit, we see how the base circuit of the second stage appears to an ac signal at the collector of the first stage. For a β of 100, the input impedance of the second base is

$$z_{\text{in(base)}} = 100(22.7 \ \Omega) = 2.27 \ \text{k}\Omega$$

The input impedance of the second stage is the combined effect of all three resistors in parallel:

$$z_{\text{in}} = 10 \ \text{k}\Omega \ \| \ 2.2 \ \text{k}\Omega \ \| \ 2.27 \ \text{k}\Omega = 1 \ \text{k}\Omega.$$

Analysis of the First Stage

The original circuit acts the same as a single stage with a load resistance of 1 kΩ. We already know how to analyze a single stage. In Fig. 10-12, the ac collector resistance is the parallel of the collector resistance (3.6 kΩ) and the input impedance of the second stage (1 kΩ):

$$r_c = 3.6 \ \text{k}\Omega \ \| \ 1 \ \text{k}\Omega = 783 \ \Omega$$

Therefore, the voltage gain of the first stage is

$$A = \frac{783 \ \Omega}{22.7 \ \Omega} = 34.5$$

The first stage has the same input impedance as the second stage. The generator voltage is 1 mV. Since 600 Ω is in series with 1 kΩ, the ac base voltage is

$$v_b = \frac{1 \ \text{mV}}{1.6 \ \text{k}\Omega} 1 \ \text{k}\Omega = 0.625 \ \text{mV}$$

and the ac collector voltage of the first stage is

$$v_c = 34.5(0.625 \ \text{mV}) = 21.6 \ \text{mV}$$

Analysis of the Second Stage

Because of the coupling capacitor between the stages, the ac base voltage of the second stage equals 21.6 mV. All we have to do now is find the final ac output voltage. The second stage has a voltage gain of

$$A = \frac{2.65 \text{ k}\Omega}{22.7 \text{ }\Omega} = 117$$

So, the final output voltage is

$$v_{\text{out}} = 117(21.6 \text{ mV}) = 2.53 \text{ V}$$

Unknown β

That's all there is to the analysis of a two-stage amplifier. The same approach works with any multistage amplifier (two or more stages). You work out the input impedance of each stage. This input impedance acts like the load resistance for the preceding stage. Since you know how to analyze a single stage, you calculate the ac collector voltage of each successive stage until you arrive at the final output. Then you are finished.

A few more pointers. If you don't have a value for the β of each transistor, assume it's 100. The answer you get will be an approximation of the truth, but that's better than no truth at all. If you have data sheets for the transistors, you can use typical values for β and work out a typical answer. This answer will still be approximate because you don't have the exact value of β. If you are a designer, you can analyze the circuit twice: once for the minimum possible β and again for the maximum possible β. This will give you two extreme answers. These two answers will bracket the exact voltage gain. Then you will know the true answer lies somewhere between the two extremes.

Total Voltage Gain

In the foregoing analysis, we discovered that the first stage had a voltage gain of

$$A_1 = 34.5$$

and the second stage had a voltage gain of

$$A_2 = 117$$

Maybe your intuition is telling you that the total voltage gain of the amplifier is somehow related to the voltage gain of each stage. If so, your intuition is right. Does it feel right to say the total voltage gain equals the product of the individual voltage gains? Because that is exactly what the total voltage gain equals.

In other words, the voltage gain from the first base to the second collector is given by

$$A = (34.5)(117) = 4037$$

You can check this by dividing the output voltage by the input voltage:

$$\frac{v_{\text{out}}}{v_{\text{in}}} = \frac{2.53 \text{ V}}{0.625 \text{ mV}} = 4048$$

which is approximately the same as 4037. (*Note:* The discrepancy between 4037 and 4048 occurs because of round-off errors. Exact calculations show that the two values are identical.)

As a formula, the total voltage gain for a two-stage amplifier is given by

$$A = A_1 A_2 \qquad (10\text{-}14)$$

The mathematical proof is simple. Just substitute into the foregoing equation to check if both sides are identical. Here's how you do it:

$$\frac{v_{\text{out}}}{v_{\text{in}}} = \frac{v_{c1}}{v_{b1}} \frac{v_{c2}}{v_{b2}} = \frac{v_{c2}}{v_{b1}} = \frac{v_{\text{out}}}{v_{\text{in}}}$$

The intermediate steps are

1. v_{c1} equals and cancels v_{b2}

2. v_{c2} equals v_{out} and v_{b1} equals v_{in}

Since both ends of the equation are identical, Eq. (10-14) has been proved.

The same principle applies to any number of stages. The total voltage gain equals the product of the individual voltage gains:

$$A = A_1 A_2 A_3 \ldots$$

If we cascaded three swamped amplifiers each with a voltage gain of 10, the total voltage gain would be

$$A = (10)(10)(10) = 1000$$

The Process

Here are the important quantities to find when analyzing a two-stage amplifier:

1. Input impedance of the second stage

2. Ac collector resistance of the first stage

3. Voltage gain of the first stage

4. Voltage gain of the second stage

5. Total voltage gain

6. Input voltage to the first stage

7. Output voltage from the second stage

These steps will guide you through the analysis of a two-stage amplifier. Study the following examples to see the process in action.

EXAMPLE 10-9

What is the ac output voltage in Fig. 10-11, on page 349, if the supply voltage is increased to 15 V?

SOLUTION

A supply voltage of 15 V produces a dc base voltage of 2.7 V, a dc emitter voltage of 2 V, a dc emitter current of 2 mA, and an r'_e of 12.5 Ω. This means the second stage has an input impedance of

$$z_{\text{in}} = 10 \text{ k}\Omega \parallel 2.2 \text{ k}\Omega \parallel 1.25 \text{ k}\Omega = 738 \, \Omega$$

The first stage has an ac collector resistance of

$$r_c = 3.6 \text{ k}\Omega \parallel 738 \, \Omega = 612 \, \Omega$$

and a voltage gain of

$$A_1 = \frac{612 \, \Omega}{12.5 \, \Omega} = 49$$

The second stage has an ac collector resistance of

$$r_c = 3.6 \text{ k}\Omega \parallel 10 \text{ k}\Omega = 2.65 \text{ k}\Omega$$

and a voltage gain of

$$A_2 = \frac{2.65 \text{ k}\Omega}{12.5 \, \Omega} = 212$$

The total voltage gain is the product of the stage gains:

$$A = (49)(212) = 10,400$$

The ac input voltage to the first stage is

$$v_b = \frac{1 \text{ mV}}{600 \, \Omega + 738 \, \Omega} 738 \, \Omega = 0.552 \text{ mV}$$

The ac output voltage from the second stage is

$$v_{\text{out}} = 10,400(0.552 \text{ mV}) = 5.74 \text{ V}$$

As you see, increasing the supply voltage produces more ac output voltage. With a supply of 10 V, we got 2.53 V for the output voltage. With a supply of 15 V, we get 5.74 V. Why does it happen? Because increasing the supply voltage decreases the r'_e of each stage. This increases the voltage gain of each stage. Even though the input impedance of each stage decreases and produces more loading effect, the reduction in each input voltage is more than offset by the increase in stage gain.

EXAMPLE 10-10

What is the ac output voltage in Fig. 10-11 if $\beta = 50$ for both transistors?

SOLUTION

A supply voltage of 10 V produces a dc base voltage of 1.8 V, a dc emitter voltage of 1.1 V, a dc emitter current of 1.1 mA, and an r'_e of 22.7 Ω. For a β of 50, the input impedance of the second base is

$$z_{\text{in(base)}} = 50(22.7 \ \Omega) = 1.14 \text{ k}\Omega$$

The second stage has an input impedance of

$$z_{\text{in}} = 10 \text{ k}\Omega \parallel 2.2 \text{ k}\Omega \parallel 1.14 \text{ k}\Omega = 698 \ \Omega$$

The first stage has an ac collector resistance of

$$r_c = 3.6 \text{ k}\Omega \parallel 698 \ \Omega = 585 \ \Omega$$

and a voltage gain of

$$A_1 = \frac{585 \ \Omega}{22.7 \ \Omega} = 25.8$$

The second stage has an ac collector resistance of

$$r_c = 3.6 \text{ k}\Omega \parallel 10 \text{ k}\Omega = 2.65 \text{ k}\Omega$$

and a voltage gain of

$$A_2 = \frac{2.65 \text{ k}\Omega}{22.7 \ \Omega} = 117$$

The total voltage gain is the product of the stage gains:

$$A = (25.8)(117) = 3020$$

The ac input voltage to the first stage is

$$v_b = \frac{1 \text{ mV}}{600 \ \Omega + 698 \ \Omega} \, 698 \ \Omega = 0.538 \text{ mV}$$

The ac output voltage from the second stage is

$$v_{\text{out}} = 3020(0.538 \text{ mV}) = 1.62 \text{ V}$$

With a β of 100, we got 2.53 V for the output voltage. With a β of 50, we get 1.62 V. Why does it happen? Decreasing β will decrease the input impedance of each stage and increase the loading effect. Although the second stage has the same gain, the first stage has less voltage gain. The overall result is less output voltage.

10-7 TROUBLESHOOTING

Most of the time, a troubleshooter needs only rough estimates of the dc and ac voltages in a transistor circuit. This is because troubles usually produce large changes from the normal values. This reality frees the troubleshooter from the need for exact answers. In fact, it usually frees the troubleshooter from relying on a calculator. Instead, one can mentally estimate the approximate values of the dc and ac voltages in the circuit. This section continues the discussion of troubleshooting.

When an amplifier is not working, a troubleshooter can start by measuring dc voltages. These voltages are estimated mentally as discussed earlier, then the voltages are measured to see if they are approximately correct. If the dc voltages are distinctly different from the estimated voltages, the possible troubles include open resistors (burned out), shorted resistors (solder bridges across them), incorrect wiring, and shorted capacitors. A short across a coupling or bypass capacitor will change the dc equivalent circuit, which means radically different dc voltages.

If all dc voltages measure okay, the troubleshooting is continued by considering what can go wrong in the ac equivalent circuit. If there is generator voltage but there is no ac base voltage, something may be open between the generator and the base. Perhaps a connecting wire is not in place, or maybe the input coupling capacitor is open. Similarly, if there is no final output voltage but there is an ac collector voltage, the output coupling capacitor may be open, or a connection may be missing.

Normally, there is no ac voltage between the emitter and ground because the emitter is at ac ground. When an amplifier is not working properly, one of the things a troubleshooter checks with an oscilloscope is the emitter voltage. If there is any ac voltage at the emitter, it means the bypass capacitor is not working. The capacitor may be defective, or not connected, or otherwise inactive.

For instance, an open bypass capacitor means the emitter is no longer at ac ground. Because of this, the ac emitter current flows through R_E instead of through the bypass capacitor. This produces an ac emitter voltage which you can see with an oscilloscope. So, if you see an ac emitter voltage comparable in size to the ac base voltage, check the emitter bypass capacitor. It may be defective or not properly connected.

Under normal conditions, the supply line is an ac ground point because of the filter capacitor in the power supply. If the filter capacitor is defective, the ripple becomes huge. This unwanted ripple gets to the base through the voltage divider. Then it is amplified the same as the generator signal. This amplified ripple will produce a hum of 120 Hz when the amplifier is connected to a loudspeaker. So, if you ever hear excessive hum coming out of a loudspeaker, one of the prime suspects is an open filter capacitor in the power supply.

EXAMPLE 10-11

The CE amplifier of Fig. 10-13 has an ac load voltage of zero. If the dc collector voltage is 6 V and the ac collector voltage is 70 mV, what is the trouble?

SOLUTION

Too bad all troubles aren't this easy. Since the dc and ac collector voltages are normal, there are only two components that can be the

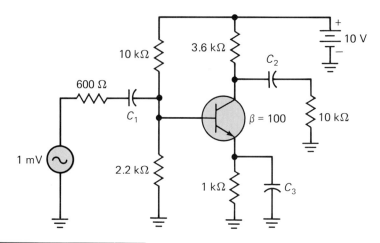

FIGURE 10-13 Troubleshooting a CE amplifier.

trouble: C_2 or R_L. If you ask four what-if questions about these components, you can flush out the trouble. The four what-ifs are

What if C_2 is shorted?
What if C_2 is open?
What if R_L is shorted?
What if R_L is open?

An experienced troubleshooter has the answers to all four questions in about 2 microseconds. The answers are

A shorted C_2 decreases the dc collector voltage significantly.
An open C_2 breaks the ac path but does not change the dc or ac collector voltages.
A shorted R_L kills the ac collector voltage.
An open R_L increases the ac collector voltage significantly.

The trouble is now obvious: C_2 is open. When you first learn how to troubleshoot, you may have to ask yourself what-if questions to isolate the trouble. After you gain experience, the whole process becomes automatic and approaches the speed of light. An experienced troubleshooter would have found this trouble almost instantaneously.

EXAMPLE 10-12

The CE amplifier of Fig. 10-13 has an ac emitter voltage of 0.75 mV and an ac collector voltage of 2 mV. What is the trouble?

SOLUTION

Because troubleshooting is an art, you have to ask what-if questions that make sense to *you* and in any order that helps you to find the trouble. If you haven't figured out this trouble yet, start to ask what-if questions about each component and see if you can find the trouble. Then read what comes next.

No matter which component you select, your what-if questions will not produce the symptoms given here until you start asking *these* what-if questions:

What if C_3 is shorted?
What if C_3 is open?

A shorted C_3 cannot produce the symptoms, but an open C_3 does. Why? Because with an open C_3, the input impedance of the base is much higher as the ac base voltage increases from 0.625 mV to 0.75 mV. Since the emitter is no longer ac grounded, almost all of this 0.75 mV appears at the emitter. Since the amplifier has a swamped voltage gain of 2.65, the ac collector voltage is approximately 2 mV. (*Note:* There is a T-shooter at the end of this chapter for additional troubleshooting practice.)

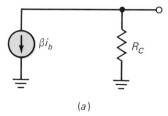

(a)

OPTIONAL TOPICS

The following material continues the earlier discussions at a more advanced and specialized level. All the topics are optional because they are not used in any of the basic discussions in later chapters. This section will be a useful reference when you are in industry because then you will probably want more advanced viewpoints.

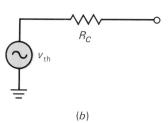

(b)

10-8 OUTPUT IMPEDANCE

The input impedance of a CE amplifier is the parallel equivalent resistance of the biasing resistors and the input impedance of the base. The input impedance is important because it interacts with the generator resistance to determine the amount of ac voltage at the base of the CE amplifier.

There is also an *output impedance* for the CE amplifier. It is important because it interacts with the load resistor to determine the ac voltage across the load. The output impedance is also called the Thevenin impedance. You can see why, when you look at Fig. 10-14a. This is the equivalent circuit for the output side of the CE amplifier. The load resistor has been disconnected because we are going to Thevenize the circuit.

The Thevenin voltage across the output is given by

$$v_{th} = \beta i_b R_C$$

But this is the same as

$$v_{th} = i_c R_C = \frac{v_{in}}{r'_e} R_C$$

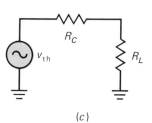

(c)

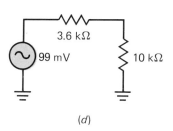

(d)

FIGURE 10-14
Output impedance.

or

$$v_{th} = A_{th}v_{in} \qquad (10\text{-}15)$$

where

$$A_{th} = \frac{R_C}{r_e'} \qquad (10\text{-}16)$$

This is the voltage gain when the load resistor has been disconnected. It will be higher than the voltage gain when the load resistor is reconnected.

To get the Thevenin impedance, you have to reduce the current source to zero. In this case, the current source is like an open circuit. Therefore, the Thevenin impedance is

$$r_{th} = R_C$$

As mentioned earlier, this Thevenin impedance is also known as the output impedance of the CE amplifier.

Figure 10-14*b* shows the Thevenin equivalent circuit for the output side of a CE amplifier. When the load resistor is reconnected as shown in Fig. 10-14*c*, the ac load voltage will be less than the Thevenin voltage. In other words, some of the generator voltage is lost across the output impedance of the amplifier. This idea is similar to what happens on the input side of the amplifier where R_G and z_{in} interact.

EXAMPLE 10-13

What is the Thevenin voltage in Fig. 10-13? The load voltage?

SOLUTION

Figure 10-13 is the CE amplifier we have been discussing throughout this chapter. We already know the input voltage is 0.625 mV and r_e' is 22.7 Ω. The Thevenin voltage gain is

$$A_{th} = \frac{3.6\,k\Omega}{22.7\,\Omega} = 159$$

and the Thevenin voltage is

$$v_{th} = 159(0.625\,\text{mV}) = 99\,\text{mV}$$

Since $v_{th} = 99$ mV and $r_{th} = 3.6$ kΩ, we can draw the Thevenin equivalent circuit shown in Fig. 10-14*d*. The load resistor has been reconnected because we want to calculate the load voltage. The circuit is nothing more than a series circuit, which is easy to analyze. To get the load voltage, calculate the current and multiply by the load resistance. Alternatively, use the voltage-divider theorem. Either way, you get

$$v_{out} = \frac{99\,\text{mV}}{13.6\,k\Omega}\,10\,k\Omega = 72.8\,\text{mV}$$

Ignoring the round-off error, this answer is identical to the 72.9 mV calculated earlier with a voltage gain of 117.

Here is a summary:

Method 1: $v_{in} = 0.625\,\text{mV}, A = 117, v_{out} = 72.9\,\text{mV}$

Method 2: $v_{in} = 0.625\,\text{mV}, A_{th} = 159, v_{th} = 99\,\text{mV}, v_{out} = 72.8\,\text{mV}$

You can use either method: the one discussed earlier or this Thevenin method. Both methods give the same load voltage. The first method is easier to work with, so we will emphasize it in later discussions. The second method gives us an important design clue worth remembering: In Fig. 10-14c, the larger R_L is compared to R_C, the greater the output voltage.

▢ 10-9 MORE ON NEGATIVE FEEDBACK

As you saw earlier, a feedback resistor in the emitter produces three beneficial effects. First, it stabilizes the voltage gain. Second, it increases the input impedance. Third, it reduces the distortion. In all cases, the improvement is given by

$$D = \frac{r_e + r'_e}{r'_e} \qquad (10\text{-}17)$$

This value is sometimes called the *desensitivity* factor or *sacrifice* factor because it represents how much voltage gain you have sacrificed. If you reduce the voltage gain from 80 to 10, you have reduced the voltage gain by a factor of 8. The sacrifice factor is also the value of how much you improve matters.

The proof of Eq. (10-17) is given in a later chapter on negative feedback. But it makes sense. What you give up in one area, you get back in another. By giving up some voltage gain, we improve other characteristics of the amplifier. Equation (10-17) pins down the amount of improvement.

EXAMPLE 10-14

In the earlier discussions we used $r_e = 180\,\Omega$ and $r'_e = 22.7\,\Omega$. What is the sacrifice factor? What effect does this have on the amplifier?

SOLUTION

The design produces a sacrifice factor of

$$D = \frac{180\,\Omega + 22.7\,\Omega}{22.7\,\Omega} = 8.93$$

This sacrifice factor tells us how much the voltage gain is reduced by the feedback resistor. But it also tells us how much improvement occurs in the gain stability, input impedance, and distortion. It means the changes in voltage gain are reduced by a factor of 8.93, the input impedance is increased by a factor of 8.93, and the distortion is reduced by a factor of 8.93.

10-10 CASCADED STAGES: THEVENIN METHOD

Figure 10-15a shows a two-stage amplifier using cascaded CE stages. An ac source with a source resistance of R_G drives the input of the amplifier. The grounded-emitter stage amplifies the signal which is then coupled into the next CE stage. Then the signal is amplified once more to get a final output that is considerably larger than the source signal.

As discussed in Sec. 10-8, we can use the Thevenin theorem to analyze a CE stage. Therefore, we can draw the ac equivalent circuit shown in Fig. 10-15b. This is very slick because now we have three series circuits. Only the numbers differ from one to the next. The equivalent circuit helps us to understand how the stages interact. In Fig. 10-15b, we can see what has to be done to get more ac output voltage. We have to make z_{in} and R_L as large as possible. Then, minimum signal voltage is lost across the generator impedance and the output impedance of each stage.

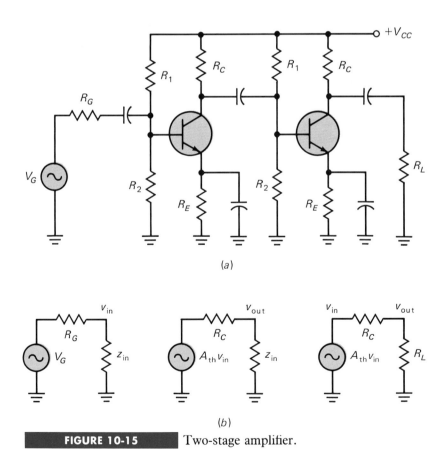

(a)

(b)

FIGURE 10-15   Two-stage amplifier.

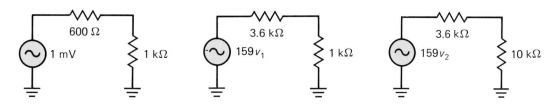

FIGURE 10-16 Ac equivalent circuit.

EXAMPLE 10-15

Reanalyze the two-stage amplifier of Fig. 10-11, on page 349, using the Thevenin method.

SOLUTION

We already know that each stage has an input impedance of 1 kΩ and an output impedance of 3.6 kΩ. The Thevenin voltage gain of each stage is

$$A_{th} = \frac{3.6\text{ k}\Omega}{22.7\ \Omega} = 159$$

Figure 10-16 shows the ac equivalent circuit for the two-stage amplifier where v_1 is the input voltage to the first stage and v_2 is the input voltage to the second stage.

The first series circuit is the familiar one that we have analyzed before. It governs the interaction between the generator resistance and the input impedance of the first stage. The input voltage to the first stage is

$$v_1 = \frac{1\text{ mV}}{1.6\text{ k}\Omega}\,1\text{ k}\Omega = 0.625\text{ mV}$$

The second series circuit has a Thevenin voltage of

$$v_{th} = 159(0.625\text{ mV}) = 99\text{ mV}$$

This voltage is across the series connection of 3.6 kΩ and 1 kΩ. Therefore, the input voltage to the second stage is

$$v_2 = \frac{99\text{ mV}}{4.6\text{ k}\Omega}\,1\text{ k}\Omega = 21.5\text{ mV}$$

The third series circuit has a Thevenin voltage of

$$v_{th} = 159(21.5\text{ mV}) = 3.42\text{ V}$$

This voltage is across 3.6 kΩ in series with 10 kΩ. Therefore, the output voltage is

$$v_2 = \frac{3.42\text{ V}}{13.6\text{ k}\Omega}\,10\text{ k}\Omega = 2.51\text{ V}$$

If you ignore round-off errors, this is the same output voltage found by the direct method of Sec. 10-6.

The direct method discussed earlier is better for most troubleshooting and design, but the Thevenin method discussed here gives us more insight into how the stages interact with each other. When we look at Fig. 10-16, we can see at a glance that the output impedance of the first stage is larger than the input impedance of the second stage. This tells us that we will lose a lot of the ac signal across the output impedance of the first stage. If a designer wanted to get more voltage gain, he would redesign the amplifier to make the output impedance of the first stage smaller than the input impedance of the second stage.

❑ 10-11 *H* PARAMETERS

The hybrid (h) parameters are an advanced mathematical approach to linear transistor circuit analysis. They represent the ultimate tool for finding the exact voltage gain, input impedance, and output impedance of a transistor amplifier. Because using the h parameters requires a lot of time-consuming calculations, this approach is sensible only if two conditions are satisfied: First, you are doing design work that requires the most accurate answers you can get and second, you have access to a computer.

Formulas

For a CE amplifier, the formulas that describe the amplifier are

$$A = \frac{-h_{fe}r_c}{h_{ie} + (h_{ie}h_{oe} - h_{re}h_{fe})r_c} \tag{10-18}$$

$$z_{\text{in}} = h_{ie} - \frac{h_{re}h_{fe}r_c}{1 + h_{oe}r_c} \tag{10-19}$$

$$z_{\text{out}} = \frac{r_s + h_{ie}}{(r_s + h_{ie})h_{oe} - h_{re}h_{fe}} \tag{10-20}$$

where $r_s = R_G \parallel R_1 \parallel R_2$

As you can see, these formulas are as left-brained as you can get. Formulas like these are perfect for computers but not for human beings. It is too easy to make a calculation error with these complicated formulas. Even worse, you lose touch with the physical action of the amplifier.

Nevertheless, the formulas will allow a designer to get exact answers, provided he has exact values for the h parameters. But that's the rub. It is next to impossible to get the exact values for the h parameters. So, the value of these exact formulas is thrown into doubt. In special circumstances, Eqs. (10-18) through (10-20) may be helpful. But most people shudder at the sight of these formulas. They have a way of smothering creativity, of making you wish you worked in another field. Fortunately, they are no longer in the mainstream of everyday electronics.

Variations in *H* Parameters

If you have a computer at your disposal, there is another problem. Data sheets do not always supply the *h* parameters of a transistor. If the data sheets do supply parameter graphs, you must remember that these are only typical *h* parameters, values that are somewhere between the minimum and maximum parameters of the transistor type. For instance, besides the typical values given earlier, a 2N3904 has these minimum and maximum values listed on its data sheet for a quiescent collector current of 1 mA:

Parameter	Min	Max
h_{ie}	1 kΩ	10 kΩ
h_{re}	$0.5(10^{-4})$	$8(10^{-4})$
h_{fe}	100	400
h_{oe}	1 μS	40 μS

These parameter spreads are very large. For this reason, it's impossible to know the exact *h* parameters of a transistor used in mass production. In other words, we have exact formulas, but they are useless without an exact set of *h* parameters. For mass production, some designers use the typical *h* parameters listed on the data sheet and include negative feedback to stabilize the transistor amplifier.

STUDY AIDS

The following study aids will help to reinforce the ideas discussed in this chapter. For best results, use these study aids within 6 hours of reading the earlier material. Then review these study aids a week later and a month later to ensure that the concepts remain in your long-term memory.

SUMMARY

Sec. 10-1 Highlights of a CE Amplifier

The generator voltage is coupled through the input capacitor into the base of the transistor. Since the emitter is at ac ground, all of the ac base voltage appears across the emitter diode. When the ac collector current flows through the ac collector resistance, it produces an amplified ac voltage that is 180° out of phase with the input voltage. The output capacitor couples the amplified and inverted ac collector voltage to the load resistor. Because a capacitor is open to dc and shorted to ac, it blocks the dc collector voltage but passes the ac collector voltage.

Sec. 10-2 Voltage Gain

The input voltage to an amplifier is the same as the ac base voltage. The output voltage of an amplifier is the same as the ac load voltage. The voltage gain equals the output voltage divided by the input voltage.

Sec. 10-3 Predicting Voltage Gain

Before you build an amplifier, you can predict its voltage gain for normal operation. The voltage gain of a CE amplifier should equal the ac collector resistance divided by the ac resistance of the emitter diode. Although this is only an approximation, the predicted voltage gain is usually close to what you will find after building and testing the circuit.

Sec. 10-4 Simplified Analysis

Exact answers may be wrong in some situations. When troubleshooting, you usually need only rough estimates of the dc and ac voltages in a circuit. This is why the best troubleshooters learn how to analyze circuits without using calculators. Your mind is capable of extraordinary things, given the freedom to work in its own individual way. With any real-world problem, there are usually many right answers and many ways to find them.

Sec. 10-5 Swamped Amplifier

Sometimes the voltage gain of a circuit has to be constant, despite the changes in temperature and other quantities. One method of stabilizing the voltage gain is to use a feedback resistor in the emitter circuit. The voltage across this feedback resistor opposes the input voltage, so that negative feedback is present. This negative feedback reduces the voltage gain, but improves other characteristics of the amplifier including its gain stability, input impedance, and distortion.

Sec. 10-6 Cascaded Stages

In a two-stage amplifier, each stage amplifies the ac signal and the final output voltage is much larger than with only one stage. The second stage loads down the first stage. This means that the input impedance of the second stage becomes the load resistance seen by the first stage. The overall voltage gain of cascaded stages is equal to the product of the individual stage gains.

Sec. 10-7 Troubleshooting

Shorted capacitors will change the biasing of the transistor, which usually means radically different dc voltages than normal. Open capacitors virtually eliminate the ac output voltage. Similarly, shorted and open resistors may affect the dc voltages and ac voltages. The supply line should be an ac ground. If you detect a large signal of 120 Hz at the output of an amplifier, then check the filter capacitor in the power supply.

VOCABULARY

In your own words, explain what each of the following terms mean. Keep your answers short and to the point. If necessary, verify your answer by rereading the appropriate discussion or by looking at the end-of-book Glossary.

cascaded stages

measured voltage gain

plug and chug

predicted voltage gain

swamped amplifier

voltage amplifier

voltage gain

IMPORTANT EQUATIONS

Here are some of the most important equations you will find in transistor circuit analysis. Say each of the following equations in words. Then read the description that follows.

Eq. 10-7 Definition of Voltage Gain

$$A = \frac{v_{out}}{v_{in}}$$

This is the defining formula for voltage gain. A defining formula is the starting point in a discussion. There is nothing that precedes it, so you don't have to ask where it came from. When something is defined for the first time, all you have to do is memorize it just as you would memorize a code. Somebody made up this formula because he realized that the ratio of output voltage to input voltage would be an excellent way to determine how well an amplifier was working.

Eq. 10-8 Output Voltage

$$v_{out} = Av_{in}$$

If someone asks where this formula came from, the answer is that it is derived mathematically from the defining formula for voltage gain. Simply put, you multiply both sides of the defining formula by v_{in} to get the new derived equation. It says the output voltage equals the voltage gain times the input voltage.

Eq. 10-9 Predicted Voltage Gain

$$A = \frac{r_c}{r_e'}$$

This is an equation that is mathematically derived from the defining formula for voltage gain. It says that a CE stage should have a voltage gain equal to the ac collector resistance divided by the ac resistance of the emitter diode.

Eq. 10-11 Swamped Amplifier

$$A = \frac{r_c}{r_e + r_e'}$$

This is the voltage gain for a CE amplifier that has part of its emitter resistance unbypassed. The feedback resistance, r_e, should be much larger than r_e'. This will eliminate the effects of changes in r_e' on the voltage gain. The equation says that the voltage gain equals the ac collector resistance divided by the total ac resistance in the emitter branch.

Eq. 10-14 Cascaded Stages

$$A = A_1 A_2$$

The voltage gain of a two-stage amplifier equals the product of the stage gains. There is a loading effect of the second stage on the first stage, which means A_1 is usually smaller than A_2.

IMPORTANT PROCESSES

Each of the following processes is like a map that guides you through the analysis of an amplifier. If you can remember the basic ideas in each step, you will be able to solve problems more easily.

CE Voltage Gain: Direct Method

1. Find the ac resistance of the emitter diode
2. Find the ac emitter current
3. Divide the ac output voltage by the ac input voltage.

CE Voltage Gain: Measured

1. Measure the ac output voltage
2. Measure the ac input voltage
3. Divide the ac output voltage by the ac input voltage.

CE Voltage Gain: Predicted

1. Calculate the ac collector resistance
2. Calculate the ac resistance of the emitter diode
3. Divide the ac collector resistance by the ac emitter resistance

Two-Stage Amplifier

1. Input impedance of the second stage
2. Ac collector resistance of the first stage
3. Voltage gain of the first stage
4. Voltage gain of the second stage
5. Total voltage gain
6. Input voltage to the first stage
7. Output voltage from the second stage

STUDENT ASSIGNMENTS

QUESTIONS

The following questions refer to the figures you have seen in this chapter. You may be able to find more than one right answer in some of the questions. You are to select the best answer, the one that is in step with the approximations of this chapter or that most accurately describes the situation.

1. The emitter is at ac ground in a
   a. CB stage
   b. CC stage
   c. CE stage
   d. None of these

2. The output voltage of a CE stage is usually
   a. Constant
   b. Dependent on β
   c. Small
   d. Less than one

3. The input voltage is usually
   a. Equal to the generator voltage
   b. Less than the generator voltage
   c. More than the generator voltage
   d. Zero

4. The voltage gain equals the output voltage divided by the
   a. Input voltage
   b. AC emitter resistance
   c. AC collector resistance
   d. Generator voltage

5. The input impedance of the base increases when
   a. β increases
   b. Suppy voltage increases
   c. β decreases
   d. AC collector resistance increases

6. Voltage gain is directly proportional to
   a. β
   b. r_e'
   c. DC collector voltage
   d. AC collector resistance

7. The ac emitter voltage of a swamped amplifier
   a. Opposes the input voltage
   b. Aids the input voltage
   c. Increases with a decrease in β
   d. Is zero

8. Compared to the ac resistance of the emitter diode, the feedback resistance of a swamped amplifier should be
   a. Small
   b. Equal
   c. Large
   d. Zero

9. Compared to a CE stage, a swamped amplifier has an input impedance that is
   a. Smaller
   b. Equal
   c. Larger
   d. Zero

10. To reduce the distortion of an amplified signal, you can increase the
    a. Collector resistance
    b. Emitter feedback resistance
    c. Generator resistance
    d. Load resistance

11. The emitter of a swamped amplifier
    a. Is grounded
    b. Has no dc voltage
    c. Has an ac voltage
    d. Has no ac voltage

12. A swamped amplifier uses
    a. Feedback
    b. Positive feedback
    c. Negative feedback
    d. A grounded emitter

13. In a swamped amplifier, the effects of the emitter diode become
    a. Important to voltage gain
    b. Critical to input impedance
    c. Significant to the analysis
    d. Unimportant

14. The feedback resistor
    a. Increases voltage gain
    b. Reduces distortion
    c. Decreases collector resistance
    d. Decreases input impedance

15. The feedback resistor
    a. Stabilizes voltage gain
    b. Increases distortion
    c. Increases collector resistance
    d. Decreases input impedance

16. The ac collector resistance of the first stage includes the
    a. Load resistance
    b. Input impedance of first stage
    c. Emitter resistance of first stage
    d. Input impedance of second stage

17. The ac collector resistance of the first stage in a two-stage amplifier is the equivalent parallel resistance of how many resistances?
    a. 1
    b. 2
    c. 3
    d. 4

18. If the emitter bypass capacitor opens, the ac output voltage will
    a. Decrease
    b. Increase
    c. Remain the same
    d. Equal zero

19. If the collector resistor is shorted, the ac output voltage will
    a. Decrease
    b. Increase
    c. Remain the same
    d. Equal zero

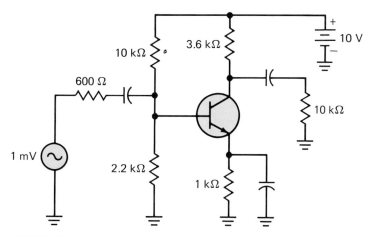

FIGURE 10-17

20. If the load resistance is open, the ac output voltage will
  a. Decrease
  b. Increase
  c. Remain the same
  d. Equal zero

21. If any capacitor is open, the ac output voltage will
  a. Decrease
  b. Increase
  c. Remain the same
  d. Equal zero

22. If the input coupling capacitor is open, the ac input voltage will
  a. Decrease
  b. Increase
  c. Remain the same
  d. Equal zero

23. If the bypass capacitor is open, the ac input voltage will
  a. Decrease
  b. Increase
  c. Remain the same
  d. Equal zero

24. If the output coupling capacitor is open, the ac input voltage will
  a. Decrease
  b. Increase
  c. Remain the same
  d. Equal zero

25. If the emitter resistor is open, the ac input voltage will
  a. Decrease
  b. Increase
  c. Remain the same
  d. Equal zero

26. If the collector resistor is open, the ac input voltage will
  a. Decrease
  b. Increase
  c. Remain the same
  d. Equal approximately zero

27. If the emitter bypass capacitor is shorted, the ac input voltage will
  a. Decrease
  b. Increase
  c. Remain the same
  d. Equal zero

BASIC PROBLEMS
(*Note:* Assume $\beta = 200$ in all problems unless otherwise indicated.)

Sec. 10-1 Highlights of a CE Amplifier

10-1. The generator voltage of Fig. 10-17 doubles. What is the input impedance?

10-2. The generator resistance of Fig. 10-17 doubles. What is the input impedance?

10-3. The load resistance of Fig. 10-17 is reduced to 4.7 kΩ. What is the input impedance?

10-4. The supply voltage of Fig. 10-17 is tripled. What is the input impedance?

10-5. All resistances are doubled in Fig. 10-17. What is the input impedance?

Sec. 10-2 Voltage Gain

10-6. The generator voltage of Fig. 10-17 doubles. What is the ac output voltage?

10-7. The generator resistance of Fig. 10-17 doubles. What is the voltage gain?

10-8. The load resistance of Fig. 10-17 is reduced to 4.7 kΩ. What is the ac output voltage?

10-9. The supply voltage of Fig. 10-17 is tripled. What is the voltage gain?

10-10. All resistances are doubled in Fig. 10-17. What is the ac output voltage?

Sec. 10-3 Predicting Voltage Gain

10-11. The generator voltage of Fig. 10-17 doubles. What is the ac output voltage?

10-12. The generator resistance of Fig. 10-17 doubles. What is the ac output voltage?

10-13. The load resistance of Fig. 10-17 is reduced to 4.7 kΩ. What is the voltage gain?

10-14. The supply voltage of Fig. 10-17 is tripled. What is the voltage gain?

10-15. All resistances are doubled in Fig. 10-17. What is the voltage gain?

Sec. 10-5 Swamped Amplifier

10-16. The generator voltage of Fig. 10-18 doubles. What is the ac output voltage?

10-17. The generator resistance of Fig. 10-18 doubles. What is the ac output voltage?

10-18. The load resistance of Fig. 10-18 is reduced to 4.7 kΩ. What is the voltage gain?

10-19. All resistances are doubled in Fig. 10-18. What is the voltage gain?

10-20. All resistances in Fig. 10-18 have a tolerance of ±5 percent. What is the minimum voltage gain? The maximum?

Sec. 10-6 Cascaded Stages

10-21. The supply voltage is increased to 12 V in Fig. 10-19. What is the final ac output voltage?

10-22. If β = 300 for the transistors of Fig. 10-19, what is the total voltage gain?

10-23. The load resistor of Fig. 10-19 has a tolerance of ±5 percent. What is the minimum ac output voltage? The maximum?

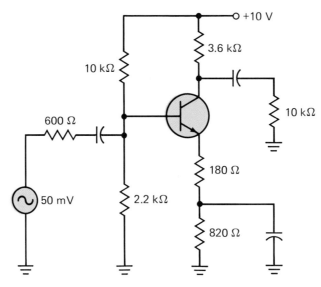

FIGURE 10-18

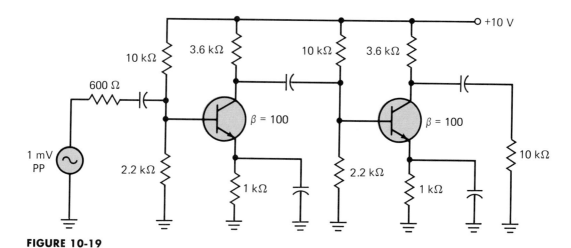

FIGURE 10-19

10-24. Does the output voltage of Fig. 10-19 increase, decrease, or remain the same for each of these changes:
  a. V_{CC} increases
  b. β of first stage decreases
  c. Generator resistance decreases
  d. Load resistance increases

10-25. In Fig. 10-19, does v_{out} increase, decrease, or remain the same for each of these changes:
  a. R_1 in first stage is 10 percent low
  b. R_2 in first stage is 10 percent high
  c. R_E in second stage is 10 percent low
  d. R_C in second stage is 10 percent high

Sec. 10-7 Troubleshooting

10-26. In Fig. 10-19, the emitter bypass capacitor is open in the first stage. What happens to the dc voltages of the first stage? To the ac input voltage to the second stage? To the final output voltage?

10-27. There is no ac load voltage in Fig. 10-19. The ac input voltage to the second stage is approximately 20 mV. Name some of the troubles.

ADVANCED PROBLEMS

10-28. All resistors are doubled in Fig. 10-17. What is the output impedance? What happens to this output impedance if β doubles?

10-29. The generator voltage of Fig. 10-17 is increased to 3 mV. What is the Thevenin voltage? Is this Thevenin voltage lower, higher, or equal to the load voltage when the load resistor is connected?

10-30. The feedback resistor of Fig. 10-18 is increased from 180 Ω to 270 Ω. What is the desensitivity? What happens to the voltage gain, input impedance and distortion?

10-31. What is the output impedance of each stage in Fig. 10-19? What happens to this output impedance of β increases? If the all resistors are 10 percent low?

10-32. What is the Thevenin voltage at each collector in Fig. 10-19? Is this Thevenin voltage lower, higher, or equal to the actual voltage at each collector?

10-33. The transistor of Fig. 10-17 has these h parameters:

Parameter	Min	Max
h_{ie}	1 kΩ	10 kΩ
h_{re}	$0.5(10^{-4})$	$8(10^{-4})$
h_{fe}	100	400
h_{oe}	1 μS	40 μS

What is the voltage gain if all parameters are minimum? If they are all maximum?

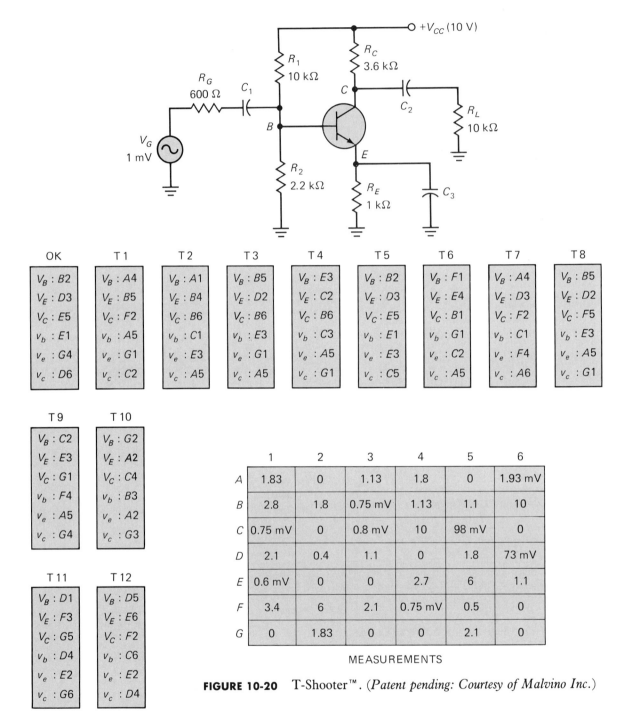

FIGURE 10-20 T-Shooter™. (*Patent pending: Courtesy of Malvino Inc.*)

T-SHOOTER PROBLEMS

Use Fig. 10-20 for the remaining problems.

10-34. Find Trouble 1.

10-35. Find Trouble 2.

10-36. Find Troubles 3 and 4.

10-37. Find Troubles 5 and 6.

10-38. Find Troubles 7 and 8.

10-39. Find Troubles 9 and 10.

10-40. Find Troubles 11 and 12.

POWER AMPLIFIERS

After several stages of voltage gain, the signal swing uses the entire load line. Any further gain has to be power gain rather than voltage gain. In these later stages the collector currents are much larger because the load impedances are much smaller. In a typical AM radio, for instance, the final load impedance is 3.2 Ω, the impedance of a small loudspeaker. The final stage of amplification has to produce enough current to drive this low impedance.

As mentioned in Chap. 6, small-signal transistors have a power rating of less than half a watt and power transistors have a power rating of more than half a watt. Small-signal transistors are typically used near the front end of systems where the signal power is low, and power transistors are used near the end of systems because the signal power is high.

11-1 THE AC LOAD LINE

Every amplifier sees two loads: a dc load and an ac load. Because of this, every amplifier has two load lines: a dc load line and an ac load line. In earlier chapters, we used the dc load line to analyze biasing circuits. In this chapter, we will use the ac load line to analyze large-signal operations.

Same AC and DC Collector Resistances

Figure 11-1a shows the amplifier we have been analyzing in earlier chapters, except that it has no source or load resistors. The dc quantities are still the same as before:

$$V_B = 1.8\,\text{V}$$
$$V_E = 1.1\,\text{V}$$
$$I_E = 1.1\,\text{mA}$$
$$V_C = 6.04\,\text{V}$$
$$V_{CE} = 4.94\,\text{V}$$

The ac input signal produces variations in two transistor voltages (base and collector), as well as all three transistor currents. The only quantity

371

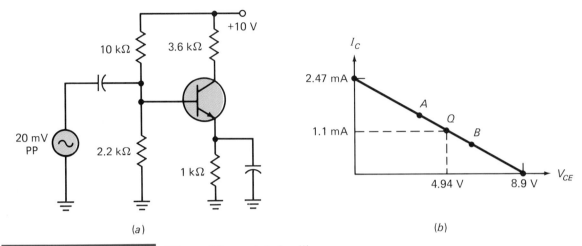

(a)   (b)

FIGURE 11-1   CE amplifier and dc loadline.

that does not vary is the emitter voltage because the emitter is at ac ground.

Figure 11-1*b* shows the dc load line for the amplifier with its *Q* point at 1.1 mA and 4.94 V. When an ac signal is present, it forces the instantaneous operating point to swing above and below the *Q* point. Since Fig. 11-1*a* has no load resistor, the ac load resistance is the same as the dc load resistance. In symbols,

$$r_c = R_C$$

For this condition, the signal swing is along the dc load line. Stated another way, the ac load line is the same as the dc load line.

Since there is no generator resistance, all of the 20 mV appear directly across the emitter diode. We already know this circuit has $r'_e = 22.7\ \Omega$. Therefore, the ac collector current is approximately

$$i_c = \frac{20\ \text{mV}}{22.7\ \Omega} = 0.881\ \text{mA}$$

This flows through the 3.6 kΩ which produces an ac collector voltage of

$$v_c = (0.881\ \text{mA})(3.6\ \text{k}\Omega) = 3.17\ \text{V}$$

As usual, these are peak-to-peak current and peak-to-peak voltage.

In Fig. 11-1*b*, we can visualize the instantaneous operating point starting at the *Q* point. During the positive half-cycle of input voltage, the instantaneous operating point swings up to point *A*, then it swings back through *Q* to point *B*. The peak-to-peak current swing is 0.881 mA and the peak-to-peak voltage swing is 3.17 V.

In conclusion, the foregoing results apply only to the special case of $r_c = R_C$, that is, the ac collector resistance is equal to the dc collector resistance. When the ac collector resistance is different from the dc collector resistance, something new happens.

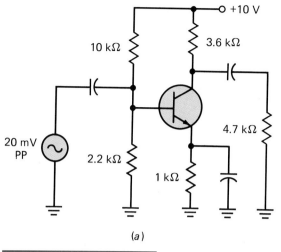
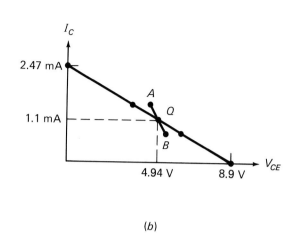

(a)   (b)

FIGURE 11-2   Dc and ac loadline.

Different AC and DC Collector Resistances

Figure 11-2*a* shows the same amplifier, except that it now has a load resistance of 4.7 kΩ. The ac collector resistance is the parallel resistance of 3.6 kΩ and 4.7 kΩ, which is

$$r_c = 2.04 \text{ k}\Omega$$

Because the ac collector resistance (2.04 kΩ) is different from the dc collector resistance (3.6 kΩ), the operating point no longer moves along the dc load line. Instead, it moves along the new line shown in Fig. 11-2*b*. In other words, the ac load line is different from the dc load line.

Here is why the instantaneous operating point moves along a different load line. The peak-to-peak current swing is the same as before:

$$i_c = \frac{20 \text{ mV}}{22.7 \ \Omega} = 0.881 \text{ mA}$$

But the peak-to-peak voltage swing is less because this current flows through a smaller ac resistance:

$$v_c = (0.881 \text{ mA})(2.04 \text{ k}\Omega) = 1.8 \text{ V}$$

In Fig. 11-2*b*, we can visualize the instantaneous operating point starting at the Q point. During the positive half-cycle of input voltage, the instantaneous operating point swings up to point A, then it swings back through Q to point B. The peak-to-peak current swing is still 0.881 mA, but the peak-to-peak voltage swing is only 1.8 V. As you see, the smaller peak-to-peak voltage swing means the ac load line has to be different from the dc load line. Nothing else would make sense.

AC Saturation and Cutoff

Here is what happens when the ac collector resistance is different from the dc load collector resistance. As shown in Fig. 11-3, the saturation and cutoff points on the ac load line are different from those on the dc

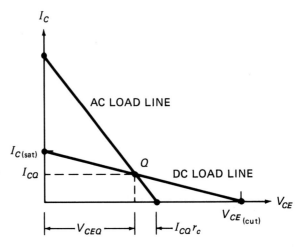

FIGURE 11-3 Maximum swing with ac loadline.

load line. Also, the ac load line is steeper (has more slope) than the dc load line because the ac collector resistance is smaller than the dc collector resistance. The mathematical derivation of Fig. 11-3 is given in the "Optional Topics" section.

Although the two load lines are different, they both include the Q point. We know why the Q point is on the dc load line. But why is it also on the ac load line? Because the signal swing starts and ends at the Q point. Therefore, all changes on the ac load line have to be measured from the Q point. To symbolize the current and voltage at the Q point, we will use these notations in subsequent discussions:

$$I_{CQ} = \text{quiescent collector current}$$

$$V_{CEQ} = \text{quiescent collector-emitter voltage}$$

These values are identical to the dc collector current and the dc collector-emitter voltage of earlier discussions.

Equations

Equations are useful mathematical summaries of what is going on in a circuit. They help to jog your memory about the key points of a discussion. They help to remind you of the processes behind a circuit. With this caution in mind, here are the equations describing the discussion to this point. In Fig. 11-3, the lower end of the dc load line is given by

$$V_{CE(cut)} = V_{CC} - V_E$$

The upper end of the dc load line is given by

$$I_{C(sat)} = \frac{V_{CC} - V_E}{R_C}$$

The dc collector current is approximately

$$I_{CQ} = \frac{V_E}{R_E}$$

and the dc collector-emitter voltage is

$$V_{CEQ} = V_{CC} - I_C R_C - V_E$$

EXAMPLE 11-1

In Fig. 11-2, the emitter resistor is changed from 1 kΩ to 820 Ω. What effect does this have on the dc load line and the Q point?

SOLUTION

The dc emitter voltage is still equal to 1.1 V. Recall how to find the cutoff voltage between the collector and the emitter. You visualize the transistor open between the collector and the emitter. Therefore, the cutoff voltage across the collector-emitter terminals is still

$$V_{CE(\text{cut})} = 10 \text{ V} - 1.1 \text{ V} = 8.9 \text{ V}$$

This is the cutoff voltage, the lower end of the dc load line.

To get the upper end of the dc loadline, visualize a short between the collector and the emitter. Then, calculate the saturation current as follows:

$$I_{C(\text{sat})} = \frac{8.9 \text{ V}}{3.6 \text{ k}\Omega} = 2.47 \text{ mA}$$

A glance at Fig. 11-2b shows that the dc load line is identical, even though the emitter resistor has decreased from 1 kΩ to 820 Ω.

What does change? The Q point. The dc emitter current now equals

$$I_E = \frac{1.1 \text{ V}}{820 \text{ }\Omega} = 1.34 \text{ mA}$$

and the dc collector voltage is approximately

$$V_C = 10 \text{ V} - (1.34 \text{ mA})(3.6 \text{ k}\Omega) = 5.18 \text{ V}$$

and the dc collector-emitter voltage is

$$V_{CE} = 5.18 \text{ V} - 1.1 \text{ V} = 4.08 \text{ V}$$

In Fig. 11-2b, this means the Q point is slightly higher on the dc load line. The new position is given by the coordinates 1.34 mA and 4.08 V.

11-2 ■ LIMITS ON SIGNAL SWING

For the amplifier we have been analyzing, the quiescent or dc values are

$$I_{CQ} = 1.1 \text{ mA}$$

$$V_{CEQ} = 4.94 \text{ V}$$

These values are very important because they are the key to how large the ac signal swing can become before it hits the ends of the ac load line. In Fig. 11-3, the maximum voltage swings from the Q point are

$$\text{Left swing} = V_{CEQ}$$

$$\text{Right swing} = I_{CQ}r_c$$

The proof of these two formulas is too complicated to show here. Refer to the "Optional Topics" if you want to see the mathematical derivation. For the amplifier we have been discussing, this means

$$\text{Left swing} = 4.94 \text{ V}$$

$$\text{Right swing} = (1.1 \text{ mA})(2.04 \text{ k}\Omega) = 2.24 \text{ V}$$

$I_{CQ}r_c$ Clipping

Because the right swing is smaller than the left, the signal will be clipped on the right swing before it is clipped on the left swing. Here is the idea. Fig. 11-4a shows the dc and ac load lines. Assume the input voltage is just large enough to produce the signal swing shown here. Notice that the ac collector voltage is sinusoidal and the right peak is just reaching the lower end of the ac load line.

If the input voltage is increased, the ac collector voltage will be clipped on the right side as shown in Fig. 11-4b. This is undesirable because it results in excessive distortion of the signal. This would be disastrous in a high-fidelity system because the clipped signal would sound terrible.

The maximum peak-to-peak (MPP) unclipped ac voltage that we can get with the amplifier is therefore limited to two times the value of $I_{CQ}r_c$. As a formula,

$$MPP = 2I_{CQ}r_c \tag{11-1}$$

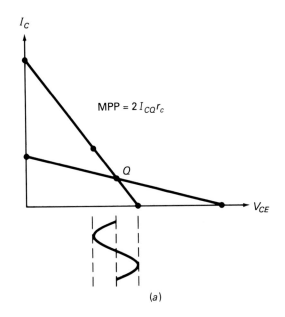

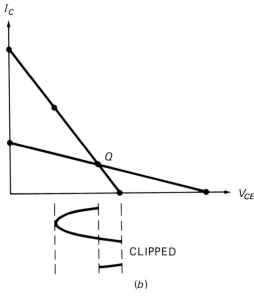

FIGURE 11-4 (a) Maximum peak-to-peak signal; (b) clipping.

Whether you are a troubleshooter or a designer, you need to be aware of this equation. It places an upper limit on how large the peak-to-peak ac voltage can be without clipping.

Optimum *Q* Point

Is there any way we can get more signal swing? Yes. At this point, a designer would probably think along these lines:

> The *Q* point of Fig. 11-3 is near the middle of the dc load line. If I move the *Q* point toward saturation on the dc load line, I ought to get a larger $I_{CQ}r_c$ swing before clipping occurs.

In other words, we can get a larger signal swing if we increase the quiescent collector current. In fact, the best thing to do is to slide the *Q* point upward until it is at the center of the ac load line as shown in Fig. 11-5.

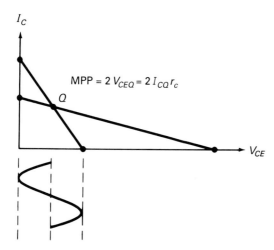

FIGURE 11-5   Optimizing the *Q* point.

When the *Q* point is at the center of the ac load line, the signal can swing equally in both directions before clipping occurs. In other words, the left swing equals the right swing. In symbols,

$$V_{CEQ} = I_{CQ}r_c$$

This means the maximum peak-to-peak swing is given by either

$$\text{MPP} = 2V_{CEQ} \qquad (11\text{-}2)$$

or

$$\text{MPP} = 2I_{CQ}r_c \qquad (11\text{-}3)$$

Both equations are valid when the *Q* point is at the center of the ac load line. You should memorize these equations, or at least remember that the left swing is limited to V_{CEQ} and the right swing to $I_{CQ}r_c$.

As long as the input signal does not produce an output voltage that exceeds the limits given by either of these two equations, you will not get clipping on the peaks of the signal. If the input signal is too large, however, you will get clipping on both peaks. The output voltage will then look more like a square wave than a sine wave.

How to Locate the Optimum Q Point

It is possible to derive a mathematical formula for the optimum Q point, but many designers prefer trial and error. The idea is to decrease the dc emitter resistance until you find a Q point that produces equal values for V_{CEQ} and $I_{CQ}r_c$. For instance, in the amplifier of Fig. 11-6, the left and right swings are

$$V_{CEQ} = 4.94 \text{ V}$$

$$I_{CQ}r_c = (1.1 \text{ mA})(2.04 \text{ k}\Omega) = 2.24 \text{ V}$$

Right now, the two swings are unequal. We want to make them equal by using more dc collector current.

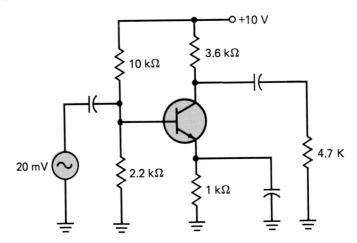

FIGURE 11-6 Locating the optimum Q point.

Suppose we try approximately half the emitter resistance, $R_E = 510 \ \Omega$. This will increase the dc collector current, equivalent to moving the Q point toward saturation. The new value of dc collector current is

$$I_{CQ} = \frac{1.1 \text{ V}}{510 \ \Omega} = 2.16 \text{ mA}$$

and the new dc collector-emitter voltage is

$$V_{CEQ} = 10 \text{ V} - (2.16 \text{ mA})(3.6 \text{ k}\Omega) - 1.1 \text{ V} = 1.12 \text{ V}$$

This means the two swings are

$$V_{CEQ} = 1.12 \text{ V}$$

$$I_{CQ}r_c = (2.16 \text{ mA})(2.04 \text{ k}\Omega) = 4.41 \text{ V}$$

We went too far. The first clipping takes place on the V_{CEQ} side. But now we know that the dc emitter resistance has to be between 510 Ω and 1 kΩ.

Another trial or two will lead to these values:

$$R_E = 680 \, \Omega$$
$$I_{CQ} = 1.61 \, \text{mA}$$
$$V_{CEQ} = 10 \, \text{V} - (1.61 \, \text{mA})(3.6 \, \text{k}\Omega) - 1.1 \, \text{V} = 3.1 \, \text{V}$$
$$I_{CQ}r_c = (1.61 \, \text{mA})(2.04 \, \text{k}\Omega) = 3.28 \, \text{V}$$

This is as close as we can get by using a standard value for R_E. Notice that the two swings are almost equal. Since the V_{CEQ} limit is smaller, the maximum peak-to-peak ac collector voltage is

$$\text{MPP} = 2(3.1 \, \text{V}) = 6.2 \, \text{V}$$

Points to Remember

When the ac load line is different from the dc load line, you get clipping on either or both swings of the signal, if the input voltage is too large. If the Q point is near the middle of the dc load line, you will get $I_{CQ}r_c$ clipping first. In this case, the maximum peak-to-peak output voltage is $2I_{CQ}r_c$.

If the Q point is above the center of the dc load line, the amplifier can produce a larger unclipped output. Regardless of where the Q point is located, the maximum peak-to-peak output voltage is the smaller of $2V_{CEQ}$ or $2I_{CQ}r_c$.

EXAMPLE 11-2

Figure 11-7 shows the amplifier discussed in the preceding chapter. What is its MPP?

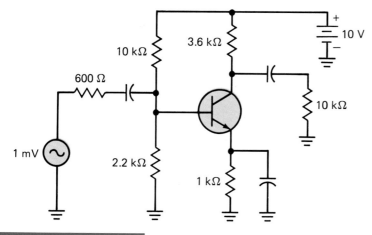

FIGURE 11-7   Example.

SOLUTION

The three key quantities are I_{CQ}, V_{CEQ}, and r_c. Once you have these three quantities, you can work out the maximum swings on both sides of the Q point. For Fig. 11-7, we have the following values calculated in the preceding chapter:

$$I_{CQ} = 1.1 \text{ mA}$$
$$V_{CEQ} = 4.94 \text{ V}$$
$$r_c = 2.65 \text{ k}\Omega$$

The maximum left swing is
$$V_{CEQ} = 4.94 \text{ V}$$

The maximum right swing is
$$I_{CQ}r_c = (1.1 \text{ ma})(2.65 \text{ k}\Omega) = 2.92 \text{ V}$$

Therefore, the maximum peak-to-peak signal is
$$\text{MPP} = 2(2.92 \text{ V}) = 5.84 \text{ V}$$

EXAMPLE 11-3

The supply voltage is increased to 25 V in Fig. 11-7. What is the new value of MPP?

SOLUTION

The supply voltage increases from 10 to 25 V, a factor of 2.5. The dc base voltage was 1.8 V before the increase. Therefore, the new dc base voltage is

$$V_B = 2.5(1.8 \text{ V}) = 4.5 \text{ V}$$

In other words, the dc base voltage is directly proportional to the supply voltage. Since the supply voltage increased by a factor of 2.5, the dc base voltage must increase by the same factor.

If you don't like the foregoing method, you can use the direct method. Get the current through the voltage divider, then multiply by 2.2 kΩ as follows:

$$V_B = \frac{25 \text{ V}}{12.2 \text{ k}\Omega} 2.2 \text{ k}\Omega = 4.5 \text{ V}$$

This means the dc emitter voltage is

$$V_E = 4.5 \text{ V} - 0.7 \text{ V} = 3.8 \text{ V}$$

The dc emitter current is

$$I_E = \frac{3.8 \text{ V}}{1 \text{ k}\Omega} = 3.8 \text{ mA}$$

The dc collector voltage is

$$V_C = 25\text{ V} - (3.8\text{ mA})(3.6\text{ k}\Omega) = 11.3\text{ V}$$

The values needed to solve this problem are

$$V_{CEQ} = 11.3\text{ V} - 3.8\text{ V} = 7.5\text{ V}$$
$$I_{CQ} = 3.8\text{ mA}$$
$$r_c = 2.65\text{ k}\Omega$$

The left swing is

$$V_{CEQ} = 7.5\text{ V}$$

The right swing is

$$I_{CQ}r_c = (3.8\text{ mA})(2.65\text{ k}\Omega) = 10.1\text{ V}$$

The left swing is smaller, so it produces clipping first. This means the maximum peak-to-peak unclipped signal is

$$\text{MPP} = 2(7.5\text{ V}) = 15\text{ V}$$

11-3 CLASS A OPERATION

Class A operation means that the transistor operates in the active region at all times. This is equivalent to saying that collector current flows for 360° of the ac cycle. In this section, we will discuss some properties of a class A amplifier needed for troubleshooting and design.

Power Gain

Besides voltage gain A, an amplifier has power gain, defined as

$$A_p = \frac{P_{\text{out}}}{P_{\text{in}}} \tag{11-4}$$

For instance, if the amplifier has an output power of 10 mW and an input power of 10 μW, it has a power gain of

$$A_p = \frac{10\text{ mW}}{10\text{ }\mu\text{W}} = 1000$$

Load Power

Figure 11-8 shows the amplifier of earlier discussions, except that we have decreased the emitter resistor to 680 Ω. This design locates the Q point at the center of the ac load line. As shown earlier, an amplifier like this can produce a maximum unclipped peak-to-peak output voltage of approximately 6.2 V.

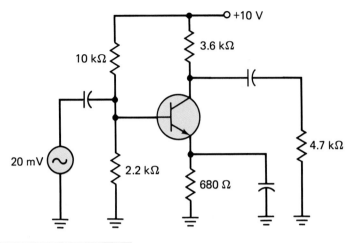

FIGURE 11-8 Load power.

The dc quantities for the circuit are

$$V_B = 1.8\,\text{V}$$
$$V_E = 1.1\,\text{V}$$
$$I_{CQ} = 1.61\,\text{mA}$$
$$V_C = 4.2\,\text{V}$$
$$V_E = 1.1\,\text{V}$$
$$V_{CEQ} = 3.1\,\text{V}$$

The load resistance seen by an amplifier may be another device such as a loudspeaker or a motor. The dc collector resistor, on the other hand, is usually an ordinary resistor that is part of the voltage-divider biasing. We are interested in the power that gets to the load resistor because this can do useful work. On the other hand, any power in the dc collector resistor is wasted power because it leaves the resistor as heat.

In other words, when we say output power, we usually mean useful load power. This load power is given by

$$P_L = V_L I_L$$

where V_L = rms load voltage

I_L = rms load current

An alternative equation for load power is

$$P_L = \frac{V_L^2}{R_L} \tag{11-5}$$

This is convenient to use when you measure the ac load voltage with an rms voltmeter.

On the other hand, if you are using an oscilloscope, it is easier to measure peak-to-peak voltage. In this case, you can use this equivalent formula:

$$P_L = \frac{V_{\text{out}}^2}{8R_L} \tag{11-6}$$

where V_{out} is the peak-to-peak voltage out of the amplifier. For instance, when the amplifier of Fig. 11-8 is producing its maximum unclipped output, the load power is

$$P_L = \frac{(6.2 \text{ V})^2}{8(4.7 \text{ k}\Omega)} = 1.02 \text{ mW}$$

Incidentally, the proof of Eq. (11-6) is given in the "Optional Topics."

Transistor Power Dissipation

When no signal drives an amplifier, the power dissipation of the transistor equals the product of dc voltage and current:

$$P_D = V_{CEQ}I_{CQ} \tag{11-7}$$

This power dissipation must not exceed the power rating of the transistor. In Fig. 11-8, the transistor power is

$$P_D = (3.1 \text{ V})(1.61 \text{ mA}) = 4.99 \text{ mW}$$

This is the approximate power in the transistor when there is no ac signal. It represents the worst case, because this power decreases when a signal is present.

Current Drain

The current drain is the dc current that a circuit draws from the power supply. Whoever designs the power supply has to know the current drain of each stage. The total current drain is the sum of the stage current drains.

In an amplifier like Fig. 11-8, the dc voltage source must supply direct current to the voltage divider and to the collector circuit. The voltage divider has a dc current of approximately

$$I_1 = \frac{V_{CC}}{R_1 + R_2} \tag{11-8}$$

The total supply current or current drain is the sum of the divider current and the collector current:

$$I_S = I_1 + I_{CQ} \tag{11-9}$$

This is the current drain of the stage.

Efficiency

The total dc power supplied to an amplifier is

$$P_S = V_{CC}I_S \tag{11-10}$$

where V_{CC} is the supply voltage and I_S is the current drain of the stage. To compare the efficiency of one design with that of another, we can use the efficiency, given by

$$\eta = \frac{P_L}{P_S} \times 100\% \tag{11-11}$$

This says that the efficiency equals the ac load power divided by the dc power from the supply times 100 percent.

The efficiency of any amplifier is a value between 0 and 100 percent. Why? Because the amplifier converts dc power to ac power. If it were 100 percent efficient, all of the dc input power would be converted to ac output power. But this never happens because of power losses in the resistors and the transistor. In other words, the ac power delivered to the load resistor has to come from somewhere. The only place it can come from is the power supply. Efficiency is a way of seeing how well an amplifier uses the dc power from the supply to produce useful load power. This is especially important in battery-operated equipment because high efficiency means the batteries last longer.

Incidentally, class A amplifiers have poor efficiency, typically well under 25 percent (the theoretical limit). This is because of power losses in the biasing resistors, the collector resistor, the emitter resistor, and the transistor. The next chapter will discuss class B operation, a design that eliminates some of the resistors and uses the transistor more efficiently. With class B operation, the efficiency can approach a theoretical limit of 78.5 percent.

EXAMPLE 11-4

What is the current drain for the amplifier of Fig. 11-8?

SOLUTION

The amplifier of Fig. 11-8 has a divider current of

$$I_1 = \frac{10\,\text{V}}{12.2\,\text{k}\Omega} = 0.82\,\text{mA}$$

We already know the amplifier has a dc collector current of

$$I_{CQ} = 1.61\,\text{mA}$$

Therefore, the total dc current drain of the stage is

$$I_S = 0.82\,\text{mA} + 1.61\,\text{mA} = 2.43\,\text{mA}$$

This is the direct current that the power supply is delivering to the entire stage.

EXAMPLE 11-5

What is the total dc power supplied to the amplifier of Fig. 11-8?

SOLUTION

The total dc power supplied to the circuit is the product of the dc supply voltage and the dc current drain:

$$P_S = (10\,\text{V})(2.43\,\text{mA}) = 24.3\,\text{mW}$$

EXAMPLE 11-6

What is the maximum efficiency of the amplifier in Fig. 11-8?

SOLUTION

For the maximum unclipped output signal, the earlier discussion calculated a load power of

$$P_L = \frac{(6.2 \text{ V})^2}{8(4.7 \text{ k}\Omega)} = 1.02 \text{ mW}$$

Therefore, the maximum efficiency we can get with this design is

$$\eta = \frac{1.02 \text{ mW}}{24.3 \text{ mW}} \times 100\% = 4.2\%$$

EXAMPLE 11-7

Example 11-5 calculated a total dc power of 24.3 mW being supplied to the circuit of Fig. 11-8. The amplifier converts 1.02 mW to useful ac load power. What happens to the rest of the power supplied to the circuit?

SOLUTION

Here are all the power losses in the circuit without the ac signal:

$$P_1 = (0.82 \text{ mA})^2(10 \text{ k}\Omega) = 6.72 \text{ mW}$$
$$P_2 = (0.82 \text{ mA})^2(2.2 \text{ k}\Omega) = 1.48 \text{ mW}$$
$$P_E = (1.61 \text{ mA})^2(680 \text{ }\Omega) = 1.76 \text{ mW}$$
$$P_C = (1.61 \text{ mA})^2(3.6 \text{ k}\Omega) = 9.33 \text{ mW}$$
$$P_D = (3.1 \text{ V})(1.61 \text{ mA}) = 4.99 \text{ mW}$$

The total of these is 24.3 mW, which equals the dc power supplied to the amplifier.

When the ac signal is present, all of the foregoing powers remain the same, except for P_D. This power decreases by 1.02 mW, which represents the power converted to ac load power. In other words, the transistor power decreases from 4.99 mW to 3.97 mW when the ac signal is present, and the load power increases from 0 to 1.02 mW.

The dc power supplied to the amplifier equals the sum of all unwanted power losses plus the ac load power:

$$\text{DC power} = \text{Unwanted losses} + \text{Ac load power}$$

or

$$24.3 \text{ mW} = 23.3 \text{ mW} + 1.02 \text{ mW}$$

The key to building a more efficient amplifier is to reduce the unwanted power losses that occur in the biasing resistors and the transistor. More is said about this in the next chapter when we discuss class B operation.

11-4 TRANSISTOR POWER RATING

The temperature at the collector junction places a limit on the allowable power dissipation P_D. Depending on the transistor type, a junction temperature in the range of 150 to 200°C will destroy the transistor. Data sheets specify this maximum junction temperature as $T_{J(\text{max})}$. For instance, the data sheet of a 2N3904 gives a $T_{J(\text{max})}$ of 150°C; the data sheet of a 2N3719 specifies a $T_{J(\text{max})}$ of 200°C.

Ambient Temperature

The heat produced at the junction passes through the transistor case (metal or plastic housing) and radiates to the surrounding air. The temperature of this air, known as the ambient temperature, is around 25°C, but it can get much higher on hot days. Also, the ambient temperature may be much higher inside a piece of electronic equipment.

Derating Factor

Data sheets often specify the $P_{D(\text{max})}$ of a transistor at an ambient temperature of 25°C. For instance, the 2N1936 has a $P_{D(\text{max})}$ of 4 W for an ambient temperature of 25°C. This means a 2N1936 used in a class A amplifier can have a quiescent power dissipation as high as 4 W. As long as the ambient temperature is 25°C or less, the transistor is within its specified power rating.

What do you do if the ambient temperature is greater than 25°C? You have to derate (reduce) the power rating. Data sheets sometimes include a derating curve like the one in Fig. 11-9. As you can see, the power rating decreases when the ambient temperature increases. For instance,

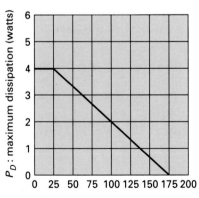

FIGURE 11-9   Power derating curve for ambient temperature.

at an ambient temperature of 100°C, the power rating is 2 W. Notice that the power rating decreases linearly with temperature.

Some data sheets do not give a derating curve like the one in Fig. 11-9. Instead, they list a derating factor, D. For instance, the derating factor of a 2N1936 is 26.7 mW/°C. This means that you have to subtract 26.7 mW for each degree the ambient temperature is above 25°C. In symbols,

$$\Delta P = D(T_A - 25°C) \qquad (11\text{-}12)$$

where ΔP = decrease in power rating
$\quad\ D$ = derating factor
$\quad\ T_A$ = ambient temperature

As an example, if the ambient temperature rises to 75°C, you have to reduce the power rating by

$$\Delta P = 26.7 \text{ mW}(75 - 25) = 1.34 \text{ W}$$

Since the power rating is 4 W at 25°C, the new power rating is

$$P_{D(\max)} = 4 \text{ W} - 1.34 \text{ W} = 2.66 \text{ W}$$

This agrees with the derating curve of Fig. 11-9.

Whether you get the reduced power rating from a derating curve like the one in Fig. 11-9 or from a formula like the one in Eq. (11-12), the important thing to be aware of is the reduction in power rating as the ambient temperature increases. Just because a circuit works well at 25°C doesn't mean it will perform well over a large temperature range. When you design circuits, therefore, you must take the operating temperature range into account by derating all transistors for the highest expected ambient temperature.

Heat Sinks

As discussed in Chap. 6, one way to increase the power rating of a transistor is to get rid of the heat faster. This is why heat sinks are used. If we increase the surface area of the transistor case, we allow the heat to escape more easily into the surrounding air. As a review, look at Fig. 11-10a. When this type of heat sink is pushed on to the transistor case, heat radiates more quickly because of the increased surface area of the fins.

Figure 11-10b shows the power-tab transistor. The metal tab provides a path out of the transistor for heat. This metal tab can be fastened to the chassis of electronics equipment. Because the chassis is a massive heat sink, heat can easily escape from the transistor to the chassis.

Large power transistors like Fig. 11-10c have the collector connected directly to the case to let heat escape as easily as possible. The transistor case is then fastened to the chassis. To prevent the collector from shorting to the chassis ground, a thin mica washer is used between the transistor case and the chassis. The important idea here is that heat can leave the transistor more rapidly, which means that the transistor has a higher power rating at the same ambient temperature. Sometimes, the transistor is fastened to a large heat sink with fins; this is even more efficient in removing heat from the transistor.

(a)

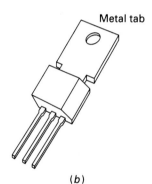

Metal tab

(b)

Collector
connected
to case

Pin 1. Base
2. Emitter
Case collector
(c)

FIGURE 11-10

(a) Push-on heat sink; (b) power-tab transistor; (c) power transistor with collector connected to case.

Case Temperature

When heat flows out of a transistor, it passes through the case of the transistor and into the heat sink, which then radiates the heat into the surrounding air. The temperature of the transistor case, T_C, will be slightly higher than the temperature of the heat sink, T_S, which in turn is slightly higher than the ambient temperature, T_A.

The data sheets of large power transistors give derating curves for the case temperature rather than the ambient temperature. For instance, Fig. 11-11 shows the derating curve of a 2N5877. The power rating is 150 W at a case temperature of 25°C; then it decreases linearly with temperature until it reaches zero for a case temperature of 200°C.

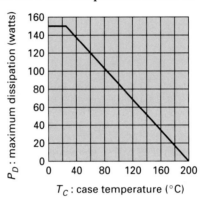

FIGURE 11-11   Power derating curve for case temperature.

Sometimes you get a derating factor instead of a derating curve. In this case, you can use the following equation to calculate the reduction in power rating:

$$\Delta P = D(T_C - 25°\text{C}) \tag{11-13}$$

where ΔP = decrease in power rating
$\quad D$ = derating factor
$\quad T_C$ = case temperature

To use the derating curve of a large power transistor, you need to know what the case temperature will be in the worst case. Then you can derate the transistor to arrive at its maximum power rating. To calculate the case temperature, you need to know something about thermodynamics: the study of heat flow. If interested in this, see "Optional Topics," on the next page.

EXAMPLE 11-8

The circuit of Fig. 11-12 is to operate over an ambient temperature range of 0 to 50°C. What is the maximum power rating of the transistor for the worst-case temperature?

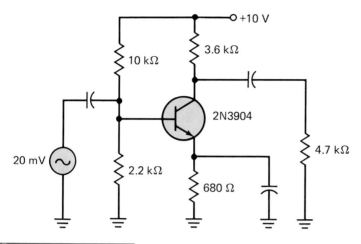

FIGURE 11-12 Example.

SOLUTION

The worst-case temperature is the highest one because you have to derate the power rating given on a data sheet. If you look at the data sheet of a 2N3904 in the Appendix, you will see the maximum power rating is listed as

$$P_D = 350 \text{ mW at } 25°C \text{ ambient}$$

and the derating factor is given as

$$D = 2.8 \text{ mW/°C}$$

With Eq. (11-12), we can calculate

$$\Delta P = 2.8 \text{ mW}(50 - 25) = 70 \text{ mW}$$

Therefore, the maximum power rating at 50°C is

$$P_{D(\text{max})} = 350 \text{ mW} - 70 \text{ mW} = 280 \text{ mW}$$

Recall that the transistor power dissipation with no ac signal is 4.99 mW (calculated in Example 11-7). This means the transistor is operating well within its maximum power rating at the worst-case ambient temperature.

OPTIONAL TOPICS

The following material continues the earlier discussions at a more advanced and specialized level. All the topics are optional because they are not used in any of the basic discussions in later chapters. This section will be a useful reference when you are in industry because then you will probably want more advanced viewpoints.

☐ 11-5 AC SATURATION AND CUTOFF

The saturation and cutoff points on the ac load line are different from those of the dc load line when the ac load resistance is different from the dc load resistance. Here is how to derive equations for the intercepts of the ac load line. We can sum ac voltages around the collector loop to get

$$v_{ce} + i_c r_c = 0$$

or

$$i_c = \frac{-v_{ce}}{r_c} \tag{11-14}$$

The ac collector current is given by

$$i_c = \Delta I_C = I_C - I_{CQ}$$

and the ac collector voltage is

$$v_{ce} = \Delta V_{CE} = V_{CE} - V_{CEQ}$$

Substituting these expressions into Eq. (11-14) and rearranging them gives

$$I_C = I_{CQ} + \frac{V_{CEQ}}{r_c} - \frac{V_{CE}}{r_c} \tag{11-15}$$

This is the equation of the ac load line. We can find the intercepts in the usual way. When the transistor goes into saturation, V_{CE} is zero and Eq. (11-15) gives

$$I_{C(\text{sat})} = I_{CQ} + \frac{V_{CEQ}}{r_c} \quad \text{(upper end)} \tag{11-16}$$

where $I_{C(\text{sat})}$ = ac saturation current
$\quad I_{CQ}$ = dc collector current
$\quad V_{CEQ}$ = dc collector-emitter voltage
$\quad r_c$ = ac resistance seen by collector

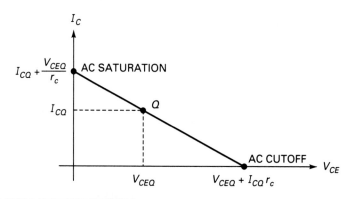

FIGURE 11-13   Ac loadline for a CE amplifier.

When the transistor goes into cutoff, I_C equals zero and we get an ac cutoff voltage of

$$V_{CE(\text{cut})} = V_{CEQ} + I_{CQ}r_c \quad \text{(lower end)} \qquad (11\text{-}17)$$

Figure 11-13 shows the ac load line with its saturation current and cutoff voltage. This is the ac load line because it represents all possible ac operating points. At any instant during the ac cycle, the operating point of the transistor is somewhere along the ac load line, the exact point determined by the amount of change from the Q point.

11-6 AC OUTPUT COMPLIANCE

The ac load line is a visual aid for understanding large-signal operation. During the positive half cycle of ac source voltage, the collector voltage swings from the Q point toward saturation. On the negative half cycle, the collector voltage swings from the Q point toward cutoff. For a large enough ac signal, clipping can occur on either or both signal peaks.

The maximum unclipped peak-to-peak signal out of an amplifier is also referred to as the *ac output compliance*. In other words, a more formal way to discuss the value of MPP is to talk about the ac output compliance. It means the same thing, but it sounds a lot more impressive. For instance, in Fig. 11-14, the ac output compliance is 2 V. If we try to get more than 2 V peak to peak, the output signal will be clipped.

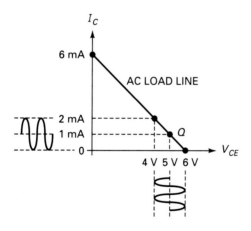

FIGURE 11-14 Maximum peak-to-peak unclipped output.

Once we know the ac output compliance of an amplifier, we know its large-signal limit. As before, we symbolize the ac output compliance of an amplifier as MPP, a reminder that it is the maximum unclipped peak-to-peak voltage that an amplifier can produce. In Fig. 11-14, the amplifier has an MPP of 2 V.

Since the ac cutoff voltage is $V_{CEQ} + I_{CQ}r_c$, the maximum positive swing from the Q point is

$$V_{CEQ} + I_{CQ}r_c - V_{CEQ} = I_{CQ}r_c$$

Since the ac saturation voltage is ideally zero, the maximum negative swing from the Q point is

$$0 - V_{CEQ} = -V_{CEQ}$$

The ac output compliance of a CE amplifier, therefore, is given by the smaller of these two approximate values:

$$\text{MPP} = 2I_{CQ}r_c \qquad (11\text{-}18)$$

or

$$\text{MPP} = 2V_{CEQ} \qquad (11\text{-}19)$$

Maximum AC Output Compliance

In earlier chapters, we set the Q point near the middle of the dc load line. This was done to keep things simple. We can increase the ac output compliance if we set the Q point higher than the center of the dc load line. Figure 11-15 illustrates why this is so. Q_1 is the Q point at the center of the dc load line. Notice how large the corresponding ac output voltage is. Q_2 is a Q point further up the dc load line. As you can see, the higher Q point results in a larger unclipped ac output voltage. Therefore, if you are designing a large-signal amplifier and want to get maximum ac output compliance, locate the Q point above the center of the dc load line.

As discussed earlier, what you are trying to do in your design is to get equal voltage swing in both directions as shown in Fig. 11-16. This allows maximum swing along the ac load line for each half cycle and produces maximum ac output compliance. To get equal voltage swing in either direction, you must satisfy the following relation.

$$I_{CQ}r_c = V_{CEQ} \quad \text{(CE stage)} \qquad (11\text{-}20)$$

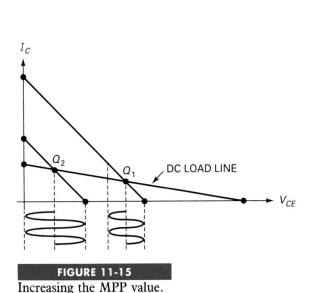

FIGURE 11-15
Increasing the MPP value.

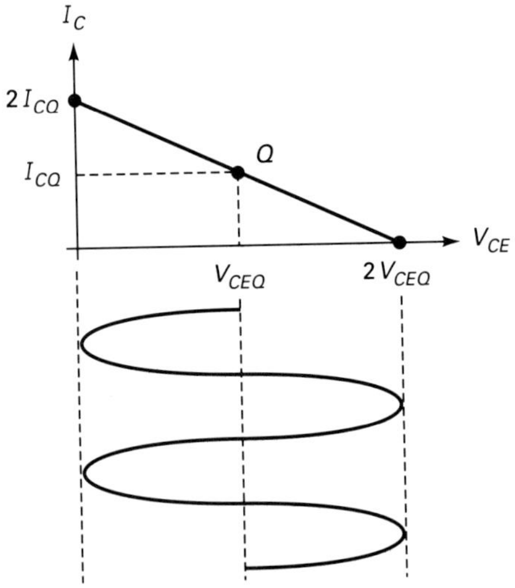

FIGURE 11-16
Optimum Q point for maximum output.

Most designers use a trial-and-error approach here. Try a collector current and see if the equation is approximately satisfied; then try again until the answer is close enough. With trial-and-error (also known as successive approximations), you can sneak up on the optimum Q point. (Other design alternatives include a graphical solution and a computer solution.)

11-7 MORE ON CLASS A OPERATION

Earlier, we discussed the basic idea behind class A operation. As you saw, this is a design where collector current flows during the entire cycle. At no time during the cycle does the transistor go into saturation or cutoff. Now, we will discuss additional properties of a class A amplifier in more depth.

Voltage Gain

In the CE amplifier of Fig. 11-17a, an ac voltage, v_{in}, drives the base, producing an ac output voltage, v_{out}. The voltage gain is sometimes written like this:

$$A_v = \frac{r_c}{r'_e} \tag{11-21}$$

For instance, if $R_C = 10 \text{ k}\Omega$, $R_L = 30 \text{ k}\Omega$, and $r'_e = 50 \text{ }\Omega$, then

$$A_v = \frac{10 \text{ k}\Omega \parallel 30 \text{ k}\Omega}{50 \text{ }\Omega} = 150$$

The symbol, A_v, is used here because we are about to discuss two other types of gain.

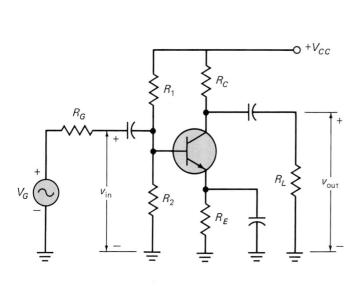

(a)

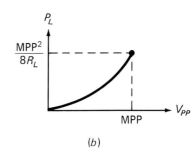

(b)

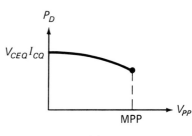

(c)

FIGURE 11-17   (a) CE amplifier; (b) load power; (c) transistor power dissipation.

Current Gain

In Fig. 11-17a, the current gain of the transistor is the ratio of ac collector current to ac base current. In symbols,

$$A_i = \frac{i_c}{i_b} \qquad (11\text{-}22)$$

This time, the subscript is i instead of v because the gain is for two currents.

Power Gain

In Fig. 11-17a, the ac input power to the base is

$$p_{in} = v_{in}i_b$$

The ac output power from the collector is

$$p_{out} = v_{out}i_c$$

The ratio p_{out}/p_{in} is called the power gain and is symbolized by A_p. Taking the ratio of p_{out} to p_{in}, we get

$$A_p = \frac{p_{out}}{p_{in}} = \frac{v_{out}i_c}{v_{in}i_b} \qquad (11\text{-}23)$$

Since $A_v = v_{out}/v_{in}$ and $A_i = i_c/i_b$,

$$A_p = A_v A_i \qquad (11\text{-}24)$$

The equation makes sense. It says that power gain equals the product of voltage gain and current gain.

For instance, if a CE amplifier has $r_c = 7500\ \Omega$, $r'_e = 50\ \Omega$, and $\beta = 125$, the voltage gain is

$$A_v = \frac{7500\ \Omega}{50\ \Omega} = 150$$

The current gain is

$$A_i = 125$$

The power gain is

$$A_p = (150)(125) = 18{,}750$$

This means that an ac input power of 1 μW results in an ac output power of 18,750 μW or 18.75 mW.

Load Power

As discussed earlier, the load power is given by

$$P_L = \frac{V_L{}^2}{R_L} \qquad (11\text{-}25)$$

This is a convenient equation to use when you measure the ac load voltage with a voltmeter because the typical voltmeter is calibrated in rms values.

Often you look at the ac output voltage with an oscilloscope. In this case, it is convenient to have a formula that uses peak-to-peak voltage instead of rms voltage. Since

$$V_L = 0.707V_p$$

and

$$V_p = \frac{V_{pp}}{2}$$

we can write

$$V_L = 0.707V_p = \frac{0.707V_{pp}}{2}$$

Substitute this into Eq. (11-25) and you get

$$P_L = \frac{V_{pp}^2}{8R_L} \qquad (11\text{-}26)$$

You will find this useful when you measure the peak-to-peak voltage with an oscilloscope.

Maximum AC Load Power

What is the maximum ac load power you can get from a CE amplifier operated as a class A amplifier? The ac output compliance, MPP, equals the maximum unclipped output voltage. Therefore, we can rewrite Eq. (11-26) as

$$P_{L(max)} = \frac{MPP^2}{8R_L} \qquad (11\text{-}27)$$

This is the maximum ac load power that a class A amplifier can produce without clipping.

Figure 11-17b shows how load power varies with the peak-to-peak load voltage. This is a parabolic curve because power is directly proportional to the square of voltage. As you can see, the maximum load power occurs when the peak-to-peak load voltage equals the ac output compliance.

Transistor Power Dissipation

When no signal drives an amplifier, the power dissipation of the transistor equals the product of dc voltage and current:

$$P_{DQ} = V_{CEQ}I_{CQ} \qquad (11\text{-}28)$$

where P_{DQ} = quiescent power dissipation
V_{CEQ} = quiescent collector-emitter voltage
I_{CQ} = quiescent collector current

This power dissipation must not exceed the power rating of the transistor. If it does, you run the risk of damaging the transistor. For instance, if $V_{CEQ} = 10$ V and $I_{CQ} = 5$ mA, then

$$P_{DQ} = (10 \text{ V})(5 \text{ mA}) = 50 \text{ mW}$$

A 2N3904 has a power rating of 350 mW for an ambient temperature of 25°C. Therefore, a 2N3904 would have no problem dissipating 50 mW of quiescent power when the ambient temperature is 25°C.

Figure 11-17c shows how transistor power dissipation varies with the peak-to-peak load voltage. P_D is maximum when there is no input signal. It decreases when the peak-to-peak load voltage increases. In the worst case, the transistor must have a power rating that is greater than P_{DQ}, the quiescent dissipation. In symbols,

$$P_{D(\text{max})} = P_{DQ} \tag{11-29}$$

Therefore, a designer must make sure that P_{DQ} is less than the power rating of the transistor being used because P_{DQ} represents the worst case.

Equation (11-29) is true only for a class A operation. That is, it is only in class A operation that the worst-case dissipation of the transistor occurs under no-signal conditions. For the other classes of operation studied later, more transistor power dissipation occurs when the signal is present.

☐ 11-8 THERMAL RESISTANCE

With power transistors, a designer often uses a heat sink to get a higher power rating for the transistor. If you recall, the heat sink allows the internally generated heat to escape more easily from the transistor. This reduces the junction temperature, equivalent to increasing the maximum power rating. Thermal resistance, θ, is the resistance to heat flow between two temperature points. For instance, Fig. 11-18a shows three temperature points: the case temperature, the sink temperature, and the ambient temperature. Heat flows from the case of the transistor to the heat sink and on to the surrounding air. As this heat flows from the case to the heat sink, it encounters the thermal resistance θ_{CS}. When the heat passes from the sink to the surrounding air, it passes through a thermal resistance θ_{SA}. As a guide, θ_{CS} is from 0.2 to 1°C/W and θ_{SA} is from 1 to 100°C/W, depending on the size of the heat sink, number of fins, finish, and other factors. For instance, if the data sheet of a heat sink lists $\theta_{CS} = 0.5$°C/W and $\theta_{SA} = 1.5$°C/W, then the thermal resistances are as shown in Fig. 11-18b.

The transistor power dissipation P_D is the same as the rate at which heat flows out of the transistor. In thermodynamics, the rate of heat flow is analogous to current, thermal resistance to resistance, and temperature difference to voltage:

$$P_D \longrightarrow \text{current}$$
$$\theta \longrightarrow \text{resistance}$$
$$T_1 - T_2 \longrightarrow \text{voltage}$$

where T_1 and T_2 are the temperatures of any two points. Using this analogy, Ohm's law for thermodynamics can be written as

$$P_D = \frac{T_1 - T_2}{\theta} \tag{11-30}$$

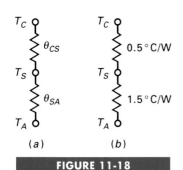

FIGURE 11-18

Thermal resistances.

The thermal resistances of Fig. 11-18*a* are in series and can be added to get the total thermal resistance between the case and the surrounding air:

$$\theta_{CA} = \theta_{CS} + \theta_{SA}$$

So, we can rewrite Eq. (11-30) as

$$P_D = \frac{T_C - T_A}{\theta_{CS} + \theta_{SA}}$$

Solving for case temperature gives

$$T_C = T_A + P_D(\theta_{CS} + \theta_{SA}) \qquad (11\text{-}31)$$

where T_C = case temperature

T_A = ambient temperature

P_D = transistor power dissipation

θ_{CS} = thermal resistance between case and sink

θ_{SA} = thermal resistance between sink and surrounding air

This is the formula needed to calculate the case temperature of a power transistor. A designer can use this formula with power transistors.

EXAMPLE 11-9

A circuit must operate over an ambient temperature range of 0 to 70°C. A 2N5877 and a heat sink have the following thermal resistances: $\theta_{CS} = 0.5°C/W$ and $\theta_{SA} = 1.5°C/W$. If the transistor has a power dissipation of 30 W, what is the maximum case temperature of the transistor? With the derating curve of Fig. 11-11 on page 388, what is the power rating of the 2N5877 at the maximum case temperature?

SOLUTION

The highest case temperature occurs when the ambient temperature is 70°C. With Eq. (11-31),

$$T_C = 70°C + (30 \text{ W})(0.5°C/W + 1.5°C/W) = 130°C$$

This tells us that the case temperature is 130°C when the transistor is dissipating 30 W. With the derating curve of Fig. 11-11, the power rating of the 2N5877 is

$$P_{D(\text{max})} = 60 \text{ W}$$

In summary, the highest ambient temperature is 70°C and the highest case temperature is 130°C. The transistor power dissipation of 30 W is still well within the maximum power rating of 60 W at the highest temperature.

STUDY AIDS

The following study aids will help to reinforce the ideas discussed in this chapter. For best results, use these study aids within 6 hours of reading the earlier material. Then review these study aids a week later and a month later to ensure that the concepts remain in your long-term memory.

SUMMARY

Sec. 11-1 The AC Load Line

A CE amplifier has two load lines: a dc load line and an ac load line. The load lines are different whenever the ac collector resistance is different from the dc collector resistance. The load lines pass through the dc or quiescent operating point. Because of this, I_{CQ} and V_{CEQ} are very important in large-signal ac operation.

Sec. 11-2 Limits on Signal Swing

When the ac signal is large, clipping may occur on either or both half-cycles. When the Q point is at the center of the dc load line, $I_{CQ}r_c$ clipping occurs first. When the Q point is above the center of the dc load line, either V_{CEQ} clipping or $I_{CQ}r_c$ clipping may occur first. It depends on which value (V_{CEQ} or $I_{CQ}r_c$) is smaller.

Sec. 11-3 Class A Operation

Class A operation means the transistor is conducting throughout the ac cycle without going into saturation or cutoff. The efficiency of the amplifier is defined as the ac load power divided by the dc supply power times 100 percent. The efficiency of a class A amplifier is low, usually well under 25 percent.

Sec. 11-4 Transistor Power Rating

The temperature at the collector junction limits the power a transistor can dissipate without being destroyed. The case temperature is between the junction temperature and the ambient temperature. Heat sinks allow the heat to escape more easily from a transistor, which lowers the junction temperature.

VOCABULARY

In your own words, explain what each of the following terms mean. Keep your answers short and to the point. If necessary, verify your answer by rereading the appropriate discussion or by looking at the end-of-book Glossary.

ac cutoff	efficiency
ac load line	heat sink
ac saturation	load power
case temperature	optimum Q point
class A operation	power gain
derating factor	

IMPORTANT EQUATIONS

Here are some of the most important equations you will find in transistor circuit analysis. Say each of the following equations in words. Then read the description that follows.

Eq. 11-1 Maximum Unclipped Output

$$MPP = 2I_{CQ}r_c$$

This is worth memorizing. And if you do forget it, at least remember the concept: there is a limit to the peak-to-peak ac output voltage from an amplifier because clipping can occur on either end of the ac load line. This equation is for clipping on the cutoff side of the load line. It says the maximum peak-to-peak output voltage without clipping is two times the dc collector current times the ac collector resistance.

Eq. 11-2 Maximum Unclipped Output

$$MPP = 2V_{CEQ}$$

Whenever the Q point is above the middle of the dc load line, clipping may occur on the saturation side of the load line. If this is the case, then the maximum peak-to-peak output voltage without clipping is equal to two times the dc collector-emitter voltage.

Eq. 11-4 Power Gain

$$A_p = \frac{p_{\text{out}}}{p_{\text{in}}}$$

Power gain is defined as output power divided by input power. This is a definition, a starting point, so it cannot be derived from other equations.

Eq. 11-5 Load Power

$$P_L = \frac{V_L{}^2}{R_L}$$

If you measure ac voltage with an rms voltmeter, this is the way to calculate the load power. The equation says that the ac load power equals the rms voltage squared and divided by the load resistance.

Eq. 11-6 Load Power

$$P_L = \frac{V_{\text{out}}{}^2}{8R_L}$$

If you measure ac voltage with an oscilloscope, this is the way to calculate the load power. The equation says that the ac load power equals the peak-to-peak output voltage squared and divided by eight times the load resistance.

Eq. 11-7 Transistor Power Dissipation

$$P_D = V_{CEQ}I_{CQ}$$

This is the dc power dissipation of a transistor. For a class A amplifier, it represents the worst case. This is the equation a designer uses. This power dissipation has to be less than the maximum power rating of the transistor at the highest temperature

to be encountered. The equation says that the power dissipation in a transistor equals the collector-emitter voltage times the collector current. (Note: The foregoing value is a close approximation. There is also some base power dissipation given by $V_{BE}I_B$. This base power is much smaller and most designers ignore it.)

Eq. 11-11 Efficiency

$$\eta = \frac{P_L}{P_S} \times 100\%$$

One of the most important characteristics of a power amplifier is its efficiency. A designer tries to get as high an efficiency as possible, especially in battery-powered equipment. The efficiency equals the ac load power divided by the dc power supplied to the amplifier times 100 percent.

IMPORTANT PROCESSES

Each of the following processes is important. If you can remember the basic ideas in each step, you will be able to solve problems more easily.

Limits on Signal Swing

1. Left swing is V_{CEQ}
2. Right swing is $I_{CQ}r_c$
3. Maximum unclipped is two times smaller

Optimum Q Point

1. Move Q point up dc load line
2. Stop when $V_{CEQ} = I_{CQ}r_c$

STUDENT ASSIGNMENTS

QUESTIONS

The following questions refer to the figures you have seen in this chapter. You may be able to find more than one right answer in some of the questions. Select the best answer, the one that is in step with the approximations of this chapter or that most accurately describes the situation.

1. The ac load line is the same as the dc load line when the ac collector resistance equals the
    a. DC emitter resistance
    b. AC emitter resistance
    c. DC collector resistance
    d. Supply voltage divided by collector current

2. If $R_C = 3.6$ kΩ and $R_L = 10$ kΩ, the ac load resistance equals
   a. 10 kΩ
   b. 2.65 kΩ
   c. 1 kΩ
   d. 3.6 kΩ

3. The quiescent collector current is the same as the
   a. DC collector current
   b. AC collector current
   c. Total collector current
   d. Voltage-divider current

4. The lower side of the ac load line corresponds to the
   a. Positive peak of input voltage
   b. Negative peak of input voltage
   c. Negative peak of output voltage
   d. Supply voltage

5. The upper side of the ac load line corresponds to the
   a. Positive peak of input voltage
   b. Negative peak of input voltage
   c. Positive peak of output voltage
   d. Supply voltage

6. The ac load line usually
   a. Equals the dc load line
   b. Has less slope than the dc load line
   c. Is steeper than the dc load line
   d. Is horizontal

7. For maximum unclipped output, the Q point is at the center of the
   a. AC load line
   b. DC load line
   c. Collector voltage
   d. Emitter voltage

8. For a Q point near the center of the dc load line, clipping is more likely to occur on the
   a. Positive peak of input voltage
   b. Negative peak of output voltage
   c. Positive peak of output voltage
   d. Negative peak of emitter voltage

9. If clipping occurs in an amplifier, it makes music or voice
   a. Sound better
   b. Louder
   c. Sound terrible
   d. One octave lower

10. In a class A amplifier, the collector current flows for
    a. Less than half the cycle
    b. Half the cycle
    c. Less than the whole cycle
    d. The entire cycle

11. When a voltage-divider biased amplifier has its Q point near the middle of the dc load line, the maximum unclipped peak-to-peak output voltage equals
    a. V_{CEQ}
    b. $2V_{CEQ}$
    c. $I_{CQ}r_c$
    d. $2I_{CQ}r_c$

12. Normally, the output signal should be
    a. Unclipped
    b. Clipped on positive voltage peak
    c. Clipped on negative voltage peak
    d. Clipped on negative current peak

13. The instantaneous operating point swings along the
    a. AC load line
    b. DC load line
    c. Both load lines
    d. Neither load line

14. With $I_{CQ}r_c$ clipping, the output signal is
    a. Unclipped
    b. Clipped on positive load voltage peak
    c. Clipped on negative load voltage peak
    d. Clipped on negative load current peak

15. Clipping is more likely to occur on the
    a. Positive peak of input voltage
    b. Negative peak of output voltage
    c. Positive peak of output voltage
    d. Both peaks simultaneously

16. The maximum unclipped peak-to-peak output voltage may equal
    a. Twice the dc collector-emitter voltage
    b. Two times the dc collector current
    c. Twice the load voltage
    d. The quiescent collector voltage times current

17. The current drain of an amplifier is the
    a. Total ac current from the generator
    b. Total dc current from the supply
    c. Current gain from base to collector
    d. Current gain from collector to base

18. The power gain of an amplifier
    a. Is the same as the voltage gain
    b. Is smaller than the voltage gain
    c. Equals output power divided by input power
    d. Equals load power

19. Clipping produces excessive
    a. Noise
    b. Distortion
    c. Current
    d. Voltage

20. To improve the efficiency of an amplifier, you have to
    a. Reduce load power
    b. Increase supply current
    c. Reduce supply voltage
    d. Decrease unwanted power losses

21. Heat sinks reduce the
    a. Transistor power
    b. Ambient temperature
    c. Junction temperature
    d. Collector current

22. When the ambient temperature increases, the maximum transistor power rating
    a. Decreases
    b. Increases
    c. Remains the same
    d. None of the above

23. The ac load resistance is different from the dc load resistance:
    a. Always
    b. Usually
    c. Seldom
    d. Never

24. Increasing the supply voltage will increase
    a. Input impedance
    b. r_e'
    c. MPP
    d. Load power

25. If the supply voltage doubles, the transistor power will
    a. Decrease
    b. Double
    c. Increase
    d. Stay the same

26. If the load power is 3 mW and the dc power is 150 mW, the efficiency is
    a. 0
    b. 2 percent
    c. 3 percent
    d. 20 percent

BASIC PROBLEMS

Sec. 11-2 Limits on Signal Swing

11-1. The emitter resistor of Fig. 11-19 is changed to 560 Ω. What is the value of MPP?

11-2. In Fig. 11-19, what is the value of the generator voltage that produces clipping of the amplified signal?

11-3. The supply voltage of Fig. 11-19 is doubled. What is the new value of MPP? What is the generator voltage that produces output clipping?

11-4. What is the MPP of Fig. 11-20?

11-5. Select a new value of emitter resistance in Fig. 11-20 to get the maximum value of MPP.

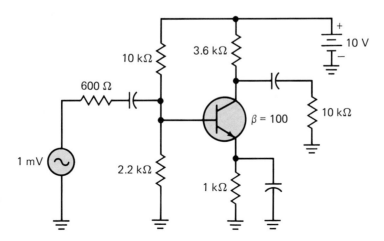

FIGURE 11-19

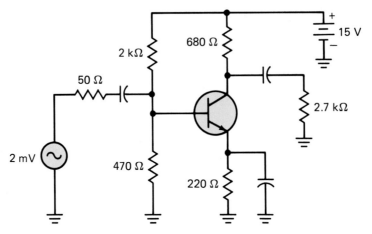

FIGURE 11-20

Sec. 11-3 Class A Operation

11-6. What is the current drain in Fig. 11-19?

11-7. What is the current drain in Fig. 11-20?

11-8. What is the total dc power supplied in Fig. 11-19?

11-9. What is the total dc power supplied in Fig. 11-20?

11-10. What is the maximum efficiency in Fig. 11-19?

11-11. What is the maximum efficiency in Fig. 11-20?

11-12. What is the transistor power dissipation in Fig. 11-19?

11-13. What is the transistor power dissipation in Fig. 11-20?

Sec. 11-4 Transistor Power Rating

11-14. A 2N3904 is used in Fig. 11-20. If the circuit has to operate over an ambient temperature range of 0 to 100°C, what is the maximum power rating of the transistor in the worst case?

11-15. A transistor has the derating curve shown in Fig. 11-9. What is the maximum power rating for an ambient temperature of 100°C?

11-16. The data sheet of a 2N3055 lists a power rating of 115 W for a case temperature of 25°C. If the derating factor is 0.657 W/°C,

what is the $P_{D(\text{max})}$ when the case temperature is 90°C?

11-17. The emitter resistor decreases in Fig. 11-20. Do each of the following increase, decrease, or remain the same:
  a. load power
  b. MPP
  c. efficiency
  d. transistor power

11-18. The supply voltage decreases in Fig. 11-20. Do each of the following increase, decrease, or remain the same:
  a. load power
  b. MPP
  c. efficiency
  d. transistor power

UNUSUAL PROBLEMS

11-19. The output of an amplifier is a square wave output even though the input is a sine wave. What is the explanation?

11-20. A power transistor like the one in Fig. 11-10c, on page 387, is used in an amplifier. Somebody tells you that since the case is grounded, you can safely touch the case. What do you think about this?

11-21. You are in a bookstore and you read the following in an electronics book: "Some power amplifiers can have an efficiency of 125 percent." Would you buy the book? Explain your answer.

11-22. Normally, the ac load line is more vertical than the dc load line. Someone is willing to bet that they can draw a circuit whose ac load line is less vertical than the dc load line. Would you take the bet? Explain.

ADVANCED PROBLEMS

11-23. Draw the dc and ac load lines for Fig. 11-20.

11-24. What is the MPP in Fig. 11-21? To increase the MPP to its maximum value, what value should the emitter resistor be? If β = 100, what is the maximum allowable generator voltage that can be used without getting output clipping?

11-25. What is the MPP in Fig. 11-22?

11-26. What is maximum unclipped load power in Fig. 11-22?

11-27. What is the MPP of each stage in Fig. 11-23?

11-28. A circuit operates over an ambient temperature range of 0 to 80°C. A transistor

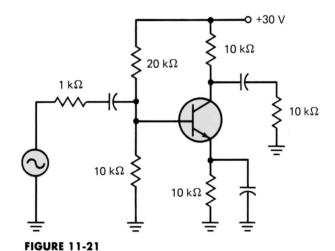

FIGURE 11-21

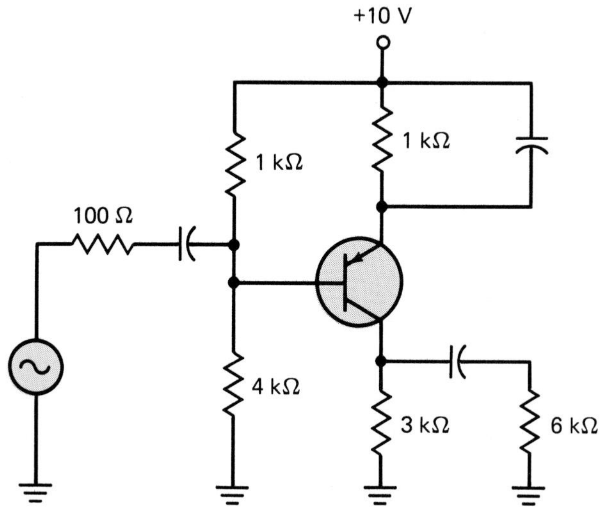

FIGURE 11-22

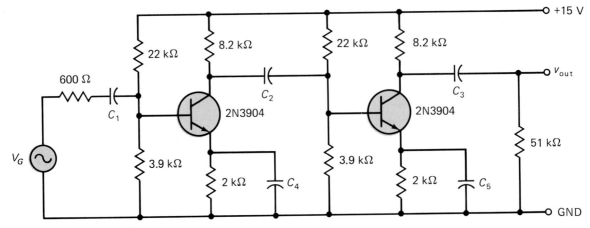

FIGURE 11-23

and heat sink have these thermal resistances: $\theta_{CS} = 0.3°C/W$ and $\theta_{SA} = 2.3°C/W$. If the transistor power dissipation is 40 W, what is the maximum case temperature?

SOFTWARE ENGINE PROBLEMS

Use Fig. 11-24 for the remaining problems. Assume increases of approximately 10 percent in the independent variable and use the second approximation

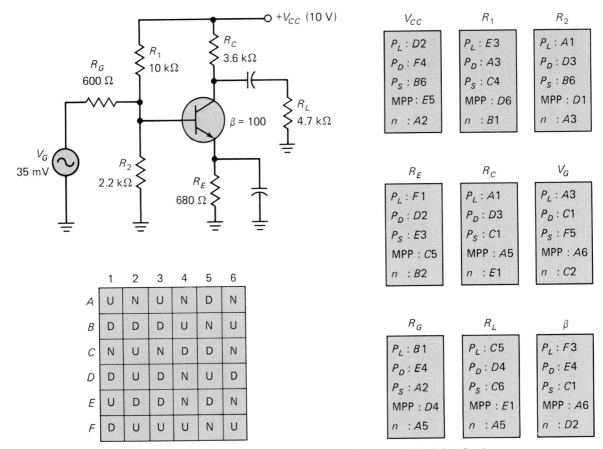

FIGURE 11-24 Software Engine™. (*Patent pending: Courtesy of Malvino Inc.*)

of the transistor. A response should be an N (no change) if the change in a dependent variable is so small that you would have difficulty measuring it.

11-29. Try to predict the response of each dependent variable in the box labeled V_{CC}. Check your answers. Then, answer the following question as simply and directly as possible. What effect does an increase in V_{CC} have on the dependent variables of the circuit?

11-30. Predict the response of each dependent variable in the box labeled R_1. Check your answers. Then summarize your findings in one or two sentences.

11-31. Predict the response of each dependent variable in the box labeled R_2. Check your answers. List the dependent variables that decrease. Explain why these variables decrease, using Ohm's law or similar basic ideas.

11-32. Predict the response of each dependent variable in the box labeled R_E. List the dependent variables that show no change. Explain why these variables show no change.

11-33. Predict the response of each dependent variable in the box labeled R_C. List the dependent variables that increase. Explain why these variables show an increase.

11-34. Predict the response of each dependent variable in the box labeled V_G. List the dependent variables that increase. Explain why these variables show an increase.

11-35. Predict the response of each dependent variable in the box labeled R_G. Check your answers. Then summarize your findings in one or two sentences.

11-36. Predict the response of each dependent variable in the box labeled R_L. List the dependent variables that increase. Explain why these variables show an increase.

11-37. Predict the response of each dependent variable in the box labeled β. List the dependent variables that increase. Explain why these variables show an increase.

EMITTER FOLLOWER

When you connect a load resistor to a CE amplifier, you decrease the ac collector resistance, which decreases the voltage gain. If the load resistance is very small, the voltage gain drops to a very low value. This problem is called *overloading* and is a direct result of using a load resistance that is much smaller than the dc collector resistance.

One way to prevent overloading is to use an *emitter follower*, also known as a CC amplifier. What you do is connect an emitter follower between the CE amplifier and the small load resistance. Because of its properties, the emitter follower can drive the small load resistance with almost no loss of voltage gain. Furthermore, the emitter follower has such a high input impedance that it does not overload the CE amplifier.

12-1 CC AMPLIFIER

Figure 12-1 shows an emitter follower. Because the collector is at ac ground, this is also known as a *common-collector (CC) amplifier*. The generator signal is coupled through the input capacitor to the base of the transistor. The ac base voltage is almost equal to the generator voltage because there is only a small loss of voltage across the 600 Ω.

The ac base voltage produces an ac emitter current. This ac emitter current sees 4.3 kΩ in parallel with 10 kΩ, which is 3 kΩ. When the ac emitter current flows through the 3 kΩ, an ac voltage appears at the emitter. Because of the output capacitor, the ac emitter voltage is coupled to the load resistor. This output voltage is approximately equal to the input voltage at the base.

Negative Feedback

Like a swamped amplifier, the emitter follower uses negative feedback. But with the emitter follower, the feedback resistance equals all of the emitter resistance. Because of this, the negative feedback is very pronounced. All of the benefits of the swamped amplifier are optimized in the emitter follower because the negative feedback is as heavy as it can become. In other words, the voltage gain is ultrastable, the distortion is almost nonexistent, and the input impedance of the base is very high.

Since the emitter is no longer at ac ground, the ac base voltage equals the ac voltage across the emitter diode plus the ac emitter voltage. In symbols,

$$v_b = v_{be} + v_e \qquad (12\text{-}1)$$

or

$$v_{\text{in}} = v_{be} + v_{\text{out}}$$

The reason the circuit is called an emitter follower is because the output voltage follows the input voltage. This means v_e is in phase with v_b and it has approximately the same peak value.

Output Capacitor Blocks DC Voltage

The total emitter voltage is the superposition of a dc voltage and an ac voltage. In Fig. 12-1, the dc emitter voltage is approximately 4.3 V. Centered on this dc level is a sinusoidal voltage that swings above and below 4.3 V.

The output capacitor couples the in-phase ac emitter voltage to the load resistor. Because the capacitor is open to dc and shorted to ac, it blocks the dc emitter voltage but passes the ac emitter voltage. For this reason, the final load voltage is a pure ac voltage.

No AC Voltage at Collector

If you look at the collector voltage with an oscilloscope, you will see a horizontal line as shown in Fig. 12-1. This line represents a dc voltage of 10 V. There is no sine wave appearing on the collector because there is no ac voltage between the collector and ground. The collector is at ac ground because the entire supply line is an ac ground point (ideally). If you recall, the supply line is ac grounded because the power supply has a large filter capacitor, equivalent to a bypass capacitor.

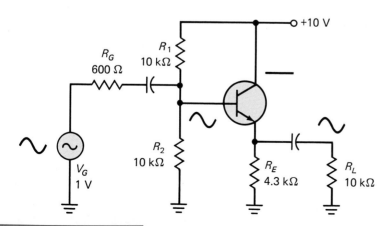

FIGURE 12-1   Emitter follower.

12-2 AC MODEL OF A CC AMPLIFIER

Whenever you look at an emitter follower like the one shown in Fig. 12-1, remember that it acts one way to dc and another way to ac. To dc, all the capacitors are open, so all you have left to analyze is the voltage-divider biased circuit in the center. But to ac, all the capacitors appear shorted. This means the collector is at ac ground and R_1 is in parallel with R_2, which is equivalent to

$$R_1 \parallel R_2 = 10 \text{ k}\Omega \parallel 10 \text{ k}\Omega = 5 \text{ k}\Omega$$

Also, R_E is in parallel with R_L, which is equivalent to

$$R_E \parallel R_L = 4.3 \text{ k}\Omega \parallel 10 \text{ k}\Omega = 3 \text{ k}\Omega$$

When we redraw the circuit as it appears to the ac signal, we get Fig. 12-2.

Input Impedance of the Base

As before, the base acts like an equivalent resistance of

$$z_{\text{in(base)}} = \frac{v_b}{i_b} \tag{12-2}$$

We can replace the transistor with its T model to get Fig. 12-3. Now, we can calculate the ac emitter current like this:

$$i_e = \frac{v_b}{r_e + r_e'} \tag{12-3}$$

By cross-multiplying the equation, we get another useful relation:

$$v_b = i_e(r_e + r_e') \tag{12-4}$$

When the ac emitter current flows through the external ac emitter resistance, it produces an ac emitter voltage of

$$v_e = i_e r_e \tag{12-5}$$

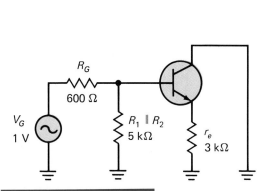

FIGURE 12-2

Ac equivalent circuit.

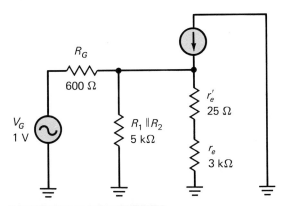

FIGURE 12-3

Ac equivalent circuit with model T.

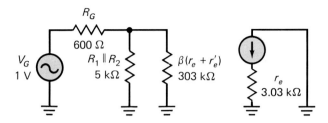

FIGURE 12-4   Ac equivalent circuit with model II.

Another AC Equivalent Circuit

Figure 12-4 shows model II assuming a β of 100. We already know that $v_b = i_e(r_e + r'_e)$, so we can substitute into Eq. (12-2) to get

$$z_{in(base)} = \frac{i_e(r_e + r'_e)}{i_b}$$

But i_e/i_b is approximately equal to β. Therefore,

$$z_{in(base)} = \beta(r_e + r'_e) \qquad (12\text{-}6)$$

On the base side, the transistor appears to be a resistance of $\beta(r_e + r'_e)$. For a β of 100,

$$z_{in(base)} = 100(3\ k\Omega + 25\ \Omega) = 303\ k\Omega$$

Input Impedance of the Stage

The input impedance of the stage is the combined effect of the biasing resistors and the input impedance of the base. Since these three resistors appear in parallel to the ac signal, the input impedance is the equivalent parallel resistance:

$$z_{in} = R_1 \| R_2 \| \beta(r_e + r'_e) \qquad (12\text{-}7)$$

With an emitter follower, the input impedance of the base is usually large enough to ignore, so that the approximate input impedance of the stage is

$$z_{in} = R_1 \| R_2 \qquad (12\text{-}8)$$

EXAMPLE 12-1

What is the impedance of the base in Fig. 12-5 if β = 200? What is the input impedance of the stage?

SOLUTION

Because each resistance in the voltage divider is 10 kΩ, the dc base voltage is half the supply voltage, or 5 V. The dc emitter voltage is

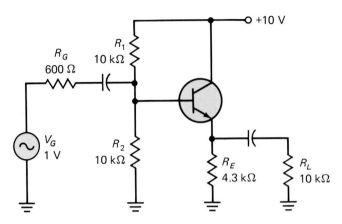

FIGURE 12-5   Voltage gain of emitter follower.

0.7 V less, or 4.3 V. The dc emitter current is 4.3 V divided by 4.3 kΩ, or 1 mA. Therefore, the ac resistance of the emitter diode is

$$r_e' = \frac{25\ \text{mV}}{1\ \text{mA}} = 25\ \Omega$$

The external ac emitter resistance is the parallel equivalent of R_E and R_L, which is

$$r_e = 4.3\ \text{k}\Omega \parallel 10\ \text{k}\Omega = 3\ \text{k}\Omega$$

Since the transistor has an ac current gain of 200, the input impedance of the base is

$$z_{\text{in(base)}} = 200(3\ \text{k}\Omega + 25\ \Omega) = 605\ \text{k}\Omega$$

Notice how high this impedance is. This is typical of an emitter follower. Because the load resistance is in the emitter circuit, the total ac emitter resistance is higher than in a CE amplifier. After multiplying by the ac current gain, therefore, the input impedance of the base is much higher than with a CE amplifier.

The input impedance of the base appears in parallel with the two biasing resistors. The input impedance of the stage is

$$z_{\text{in}} = 10\ \text{k}\Omega \parallel 10\ \text{k}\Omega \parallel 605\ \text{k}\Omega = 4.96\ \text{k}\Omega$$

Because the 605 kΩ is much larger than 5 kΩ, troubleshooters usually approximate the input impedance of the stage as the parallel of the biasing resistors only:

$$z_{\text{in}} = 10\ \text{k}\Omega \parallel 10\ \text{k}\Omega = 5\ \text{k}\Omega$$

EXAMPLE 12-2

Assuming a β of 200, what is the ac input voltage to the emitter follower of Fig. 12-5?

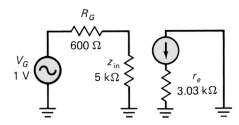

FIGURE 12-6   Ac equivalent circuit.

SOLUTION

Figure 12-6 shows the ac equivalent circuit. The ac base voltage appears across z_{in}. This is a simple circuit to analyze, nothing more than a voltage divider. Because the input impedance of the stage is large compared to the generator resistance, most of the generator voltage appears at the base. A troubleshooter would look at this voltage divider and probably think along these lines:

> The input impedance of the stage is roughly 10 times the generator resistance, so I will lose about 10 percent of the voltage across the 600 Ω. 10 percent of 1 V is 0.1 V. So, the ac base voltage should be in the vicinity of 0.9 V.

If the troubleshooter did want a more accurate answer, he would use Ohm's law twice. Once to calculate the generator current. And a second time to calculate the ac base voltage. Both operations can be combined into one equation like this:

$$v_b = \frac{1\,\text{V}}{5.6\,\text{k}\Omega} 5\,\text{k}\Omega = 0.893\,\text{V}$$

As before, the ac base voltage of the amplifier is the same as the input voltage to the amplifier, so we may write

$$v_{in} = 0.893\,\text{V}$$

The emitter follower tends to have more input voltage than a CE amplifier because it can use larger biasing resistors. Why? Since the output is taken from the emitter instead of the collector, we can use a larger dc emitter voltage and a larger emitter resistor. Typically, the emitter follower is biased with a dc emitter voltage of at least half the supply voltage. You will see why when we discuss the ac load line.

12-3 VOLTAGE GAIN

The voltage gain of an emitter follower is defined as the output voltage divided by the input voltage. The voltage gain is approximately equal to one. The following discussion tells you why.

For $\beta = 100$ in Fig. 12-5, the input impedance of the base is

$$z_{\text{in(base)}} = 303 \text{ k}\Omega$$

This is in parallel with the biasing resistors, so that the input impedance of the stage is approximately

$$z_{\text{in}} = 5 \text{ k}\Omega$$

After replacing the transistor by its II model and simplifying, we arrived at the final ac equivalent circuit shown in Fig. 12-6.

The input voltage to the amplifier is the same as the ac base voltage. Because the voltage divider is formed by the generator resistance and the input impedance of the stage, the ac base voltage is slightly less than the generator voltage. We calculated an ac base voltage of

$$v_b = 0.893 \text{ V}$$

The formula we used to find this ac base voltage was

$$v_b = \frac{v_g}{R_G + z_{\text{in}}} z_{\text{in}} \tag{12-9}$$

Calculating the AC Emitter Voltage

Figure 12-7 shows the transistor in its model II form. This assumes a β of 100. The ac base current equals the ac base voltage divided by the input impedance of the base:

$$i_b = \frac{0.893 \text{ V}}{303 \text{ k}\Omega} = 2.95 \text{ } \mu\text{A}$$

Since $\beta = 100$, the ac emitter current is approximately

$$i_e = 100(2.95 \text{ } \mu\text{A}) = 0.295 \text{ mA}$$

When the ac emitter current flows through the external ac emitter resistance, it produces an ac emitter voltage of

$$v_e = (0.295 \text{ mA})(3 \text{ k}\Omega) = 0.885 \text{ V}$$

This is the same as the output voltage because the output capacitor couples the ac emitter voltage to the load resistor.

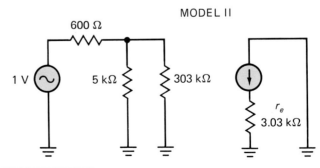

FIGURE 12-7   Ac equivalent circuit with model II.

Another Way to Find AC Emitter Voltage

You usually don't know the value of β, so the foregoing method forces you to guess a value of β and work out an approximate answer. There is a way around this. Instead of using the model II, use the model T. Remember, we can use either model because they both produce the same final answers.

Figure 12-8 shows the ac equivalent circuit with the model T of the transistor. The ac base voltage is directly across r'_e and r_e. Therefore, the ac emitter current equals

$$i_e = \frac{0.893 \text{ mV}}{3.03 \text{ k}\Omega} = 0.295 \text{ mA}$$

We found this ac emitter current *without* using the value of β. As before, the ac emitter voltage is the product of 0.295 mA and 3 kΩ, so the final output voltage is 0.885 V. This is the same answer as found with the model II.

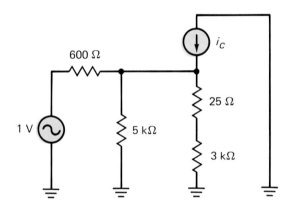

FIGURE 12-8   Ac equivalent circuit with model T.

Calculating Voltage Gain

After you have the output voltage, you can calculate the voltage gain of the emitter follower. As before, it is defined as the output voltage divided by the input voltage:

$$A = \frac{v_{\text{out}}}{v_{\text{in}}} \tag{12-10}$$

For the emitter follower we have been discussing, the voltage gain is

$$A = \frac{0.885 \text{ V}}{0.893 \text{ V}} = 0.99$$

As you can see, the voltage gain is approximately equal to one. This is typical of an emitter follower. It usually has a voltage gain that is approximately equal to one because almost all of the ac base voltage appears at the emitter. This is a direct consequence of the heavy negative feedback.

Predicting Voltage Gain

There is a formula that a troubleshooter or designer can use to predict the voltage gain of an emitter follower before measuring it. The formula is

$$A = \frac{r_e}{r_e + r_e'}$$

Where does it come from? Start with the defining formula for voltage gain, which is

$$A = \frac{v_{\text{out}}}{v_{\text{in}}}$$

First, realize that the output voltage is given by

$$v_{\text{out}} = i_e r_e$$

and that the input voltage is given by

$$v_{\text{in}} = i_e(r_e + r_e')$$

Substitute these two expressions into the defining formula to get

$$A = \frac{i_e r_e}{i_e(r_e + r_e')}$$

which simplifies to

$$A = \frac{r_e}{r_e + r_e'} \tag{12-11}$$

Usually, r_e is much greater than r_e', so that the voltage gain is approximately one in the typical emitter follower.

The emitter follower represents swamping at its very best. In our example, $r_e = 3\ \text{k}\Omega$ versus $r_e' = 25\ \Omega$, the ratio is more than 100:1. Any changes in r_e' are swamped out of sight. For instance, r_e' can change from 25 to 50 Ω and the voltage gain will still be very close to one. Because r_e' is effectively eliminated in the calculation of voltage gain, the output voltage is almost a perfect replica of the input voltage. This is the same as saying there is no distortion of the signal.

Highlights of Emitter Follower

With an emitter follower, the generator voltage is the source of the ac signal. Because every voltage source has some internal resistance, we have to include R_G in the our analysis. Usually, R_G is small compared to the input impedance of the stage. Therefore, almost all of the generator voltage appears at the base.

The ac generator voltage is coupled through the input capacitor into the base of the transistor. Almost all of the ac base voltage appears at the emitter. Because of the output capacitor, the ac emitter voltage is coupled to the load resistor. The output voltage is in phase with the input voltage and it has approximately the same peak value.

The heavy swamping of an emitter follower virtually eliminates the effect of r_e' on the voltage gain. This is why the voltage gain is ultrastable

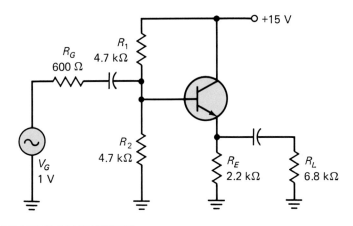

FIGURE 12-9   •   Example.

and approximately equal to one. The other benefits of the heavy swamping are a very high input impedance looking into the base and almost no distortion of the signal.

The main use of the emitter follower is in isolating a low-impedance load from a CE amplifier. If a CE amplifier tries to drive a low-impedance load directly, it will be overloaded because the ac resistance it sees is too low. By using an emitter follower between the CE amplifier and the low-impedance load, a designer can prevent the overloading. You can think of an emitter follower as a circuit that steps up the impedance. The low impedance of the load is stepped up by a factor of β. This prevents the overloading that would occur without the emitter follower.

EXAMPLE 12-3

What is the voltage gain of the emitter follower in Fig. 12-9? If β = 150, what is the ac load voltage?

SOLUTION

The dc base voltage is half the supply voltage:

$$V_B = 7.5 \text{ V}$$

The dc emitter current is

$$I_E = \frac{6.8 \text{ V}}{2.2 \text{ k}\Omega} = 3.09 \text{ mA}$$

and the ac resistance of the emitter diode is

$$r'_e = \frac{25 \text{ mV}}{3.09 \text{ mA}} = 8.09 \text{ }\Omega$$

The external ac emitter resistance is

$$r_e = 2.2 \text{ k}\Omega \parallel 6.8 \text{ k}\Omega = 1.66 \text{ k}\Omega$$

The voltage gain equals

$$A = \frac{1.66 \text{ k}\Omega}{1.66 \text{ k}\Omega + 8.09 \text{ }\Omega} = 0.995$$

The input impedance of the base is

$$z_{\text{in(base)}} = 150(1.66 \text{ k}\Omega + 8.09 \text{ }\Omega) = 250 \text{ k}\Omega$$

This is much larger than the biasing resistors. Therefore, to a close approximation, the input impedance of the emitter follower is

$$z_{\text{in}} = 4.7 \text{ k}\Omega \parallel 4.7 \text{ k}\Omega = 2.35 \text{ k}\Omega$$

The ac input voltage is

$$v_{\text{in}} = \frac{1 \text{ V}}{600 \text{ }\Omega + 2.35 \text{ k}\Omega} 2.35 \text{ k}\Omega = 0.797 \text{ V}$$

The ac output voltage is

$$v_{\text{out}} = 0.995(0.797 \text{ V}) = 0.793 \text{ V}$$

Again, you have seen how the voltage gain of the emitter follower is approximately equal to one. Also, the input impedance of the emitter follower is approximately equal to the parallel equivalent resistance of the two biasing resistors in the base. Because the input impedance of the emitter follower is 2.35 kΩ, about 20 percent of the ac generator voltage is dropped across the 600 Ω. In other words, the ac input voltage is 0.797 V instead of 1 V. Since the voltage gain is close to one, the ac output voltage is 0.793 V.

12-4 MAXIMUM UNCLIPPED OUTPUT

Figure 12-10 shows an emitter follower. When the generator voltage increases, the output becomes larger until clipping occurs on either or both peaks. The question now is how can we find the maximum unclipped peak-to-peak output? In Fig. 12-10, the dc emitter current and the dc emitter voltage are

$$I_E = 1 \text{ mA}$$

$$V_E = 4.3 \text{ V}$$

The collector current is approximately equal to the emitter current. Furthermore, the voltage between the collector and the emitter is

$$V_{CE} = 10 \text{ V} - 4.3 \text{ V} = 5.7 \text{ V}$$

The external ac emitter resistance is

$$r_e = 4.3 \text{ k}\Omega \parallel 10 \text{ k}\Omega = 3 \text{ k}\Omega$$

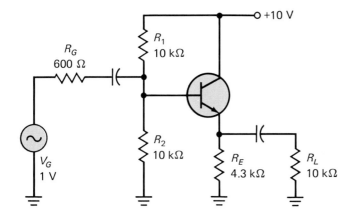

FIGURE 12-10   Maximum unclipped output.

Therefore, we can use these key values in the discussion that follows:

$$I_{CQ} = 1\,\text{mA}$$
$$V_{CEQ} = 5.7\,\text{V}$$
$$r_e = 3\,\text{k}\Omega$$

The Limits

The emitter follower has two load lines: a dc load line and an ac load line. As before, both load lines pass through the Q point. Furthermore, the ac load line determines the limits of the signal swing. These limits are similar to those of a CE amplifier, except that we need to use r_e instead of r_c.

In Fig. 12-11, the maximum voltage swings from the Q point are

$$\text{Left swing} = V_{CEQ}$$
$$\text{Right swing} = I_{CQ}r_e$$

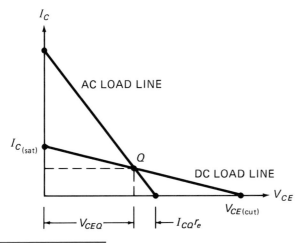

FIGURE 12-11   Maximum voltage swings.

For the emitter follower of Fig. 12-11, this means

$$\text{Left swing} = 5.7\text{ V}$$

$$\text{Right swing} = (1\text{ mA})(3\text{ k}\Omega) = 3\text{ V}$$

$I_{CQ}r_e$ Clipping

Because the right swing is smaller than the left, the signal will be clipped on the right swing before it is clipped on the left swing. The maximum peak-to-peak unclipped ac voltage that we can get with the emitter follower is therefore limited to two times the value of $I_{CQ}r_e$. As a formula,

$$\text{MPP} = 2I_{CQ}r_e \qquad (12\text{-}12)$$

Optimum Q Point

As before, we can get a larger signal swing if we increase the dc collector current. The easiest way to do this is to slide the Q point upward until it is at the center of the ac load line as shown in Fig. 12-12. When the Q point is at the center of the ac load line, the signal can swing equally in both directions before clipping occurs. In other words, the left swing equals the right swing. In symbols,

$$V_{CEQ} = I_{CQ}r_e$$

This means the maximum peak-to-peak swing is given by either

$$\text{MPP} = 2V_{CEQ} \qquad (12\text{-}13)$$

or

$$\text{MPP} = 2I_{CQ}r_e \qquad (12\text{-}14)$$

Both equations are valid when the Q point is at the center of the ac load line.

As long as the input signal does not produce an output voltage that exceeds the limits given by either of these two equations, you will not get clipping on the peaks of the signal. If the input signal is too large,

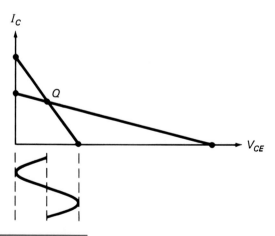

FIGURE 12-12   Optimum Q point.

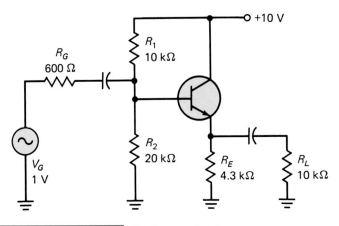

FIGURE 12-13 Optimum circuit.

however, you will get clipping on both peaks. The output voltage will then look more like a square wave than a sine wave.

How to Locate the Optimum Q Point

We can increase the dc base voltage until we find a Q point that produces equal values for V_{CEQ} and $I_{CQ}r_e$. For instance, in the amplifier of Fig. 12-10, the left and right swings are

$$V_{CEQ} = 5.7\,\text{V}$$

$$I_{CQ}r_e = (1\,\text{mA})(3\,\text{k}\Omega) = 3\,\text{V}$$

Right now, the two swings are unequal. In this case, the MPP is limited to two times the smaller of these, or 6 V. We want to increase MPP by rebiasing the transistor with more dc collector current.

Suppose we increase the base voltage by using a larger value for R_2. Let's try $R_2 = 20\,\text{k}\Omega$ as shown in Fig. 12-13. Then, the dc base voltage increases to approximately

$$V_B = \frac{10\,\text{V}}{30\,\text{k}\Omega}\,20\,\text{k}\Omega = 6.67\,\text{V}$$

and the dc emitter voltage increases to

$$V_E = 6.67\,\text{V} - 0.7\,\text{V} = 5.97\,\text{V}$$

and the dc emitter current increases to

$$I_E = \frac{5.97\,\text{V}}{4.3\,\text{k}\Omega} = 1.39\,\text{mA}$$

Now, the two limits are

$$V_{CEQ} = 10\,\text{V} - 5.97\,\text{V} = 4.03\,\text{V}$$

and

$$I_{CQ}r_e = (1.39\,\text{mA})(3\,\text{k}\Omega) = 4.17\,\text{V}$$

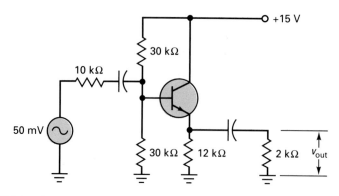

FIGURE 12-14   Example.

V_{CEQ} and $I_{CQ}r_e$ are almost equal. This is as close as we can get with standard resistance values. Therefore, the MPP value for the new design is

$$\text{MPP} = 2(4.03 \text{ V}) = 8.06 \text{ V}$$

or approximately 8 V peak-to-peak.

EXAMPLE 12-4

What is the MPP in Fig. 12-14?

SOLUTION

The dc base voltage is half the supply voltage:

$$V_B = 7.5 \text{ V}$$

The dc emitter voltage is 0.7 V less than this, or 6.8 V. The dc emitter current is

$$I_E = \frac{6.8 \text{ V}}{12 \text{ k}\Omega} = 0.567 \text{ mA}$$

The dc collector-emitter voltage is

$$V_{CEQ} = 15 \text{ V} - 6.8 \text{ V} = 8.2 \text{ V}$$

and the external ac emitter resistance is

$$r_e = 12 \text{ k}\Omega \parallel 2 \text{ k}\Omega = 1.71 \text{ k}\Omega$$

The two limits on the signal swing are

$$V_{CEQ} = 8.2 \text{ V}$$

and

$$I_{CQ}r_e = (0.567 \text{ mA})(1.71 \text{ k}\Omega) = 0.97 \text{ V}$$

Clipping occurs first on the $I_{CQ}r_e$ side of the Q point. Therefore, the maximum unclipped peak-to-peak output is

$$MPP = 2(0.97 \text{ V}) = 1.94 \text{ V}$$

EXAMPLE 12-5

What is the MPP in Fig. 12-14 if R_2 is increased to 68 kΩ?

SOLUTION

The dc base voltage is increased to

$$V_B = \frac{15 \text{ V}}{30 \text{ k}\Omega + 68 \text{ k}\Omega} 68 \text{ k}\Omega = 10.4 \text{ V}$$

The dc emitter voltage is

$$V_E = 10.4 \text{ V} - 0.7 \text{ V} = 9.7 \text{ V}$$

The dc emitter current is

$$I_E = \frac{9.7 \text{ V}}{12 \text{ k}\Omega} = 0.808 \text{ mA}$$

The dc collector-emitter voltage is

$$V_{CEQ} = 15 \text{ V} - 9.7 \text{ V} = 5.3 \text{ V}$$

The external ac emitter resistance is still

$$r_e = 12 \text{ k}\Omega \parallel 2 \text{ k}\Omega = 1.71 \text{ k}\Omega$$

The two limits on the signal swing are

$$V_{CEQ} = 5.3 \text{ V}$$

and

$$I_{CQ}r_e = (0.808 \text{ mA})(1.71 \text{ k}\Omega) = 1.38 \text{ V}$$

The MPP is

$$MPP = 2(1.38 \text{ V}) = 2.76 \text{ V}$$

12-5 CASCADING CE AND CC

Suppose we have a load resistance of 270 Ω. If we try to couple the output of a CE amplifier directly into this load resistance, we may overload the amplifier. One way to avoid this overload is by using an emitter follower between the CE amplifier and the load resistance. The signal can be coupled *capacitively* (this means through coupling capacitors) or it may be *direct coupled* as shown in Fig. 12-15.

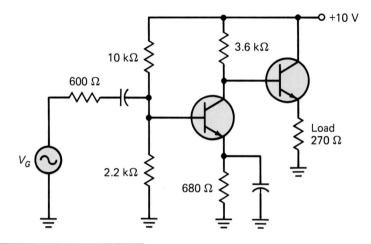

Direct coupled output stage.

This design is elegant because it eliminates the biasing resistors and the input coupling capacitor of the emitter follower. As you can see, the base of the second transistor is connected directly to the collector of the first transistor. Because of this, the dc collector voltage of the first transistor is used to bias the second transistor. If the dc current gain of the second transistor is 100, the dc resistance looking into the base of the second transistor is

$$R_{\text{in}} = 100(270 \ \Omega) = 27 \ \text{k}\Omega$$

Because 27 kΩ is large compared to the 3.6 kΩ, the dc collector voltage of the first stage is only slightly disturbed.

In Fig. 12-15, the amplified voltage out of the first stage drives the emitter follower and appears across the final load resistance of 270 Ω. Without the emitter follower, the 270 Ω would overload the first stage. But with the emitter follower, its impedance effect is increased by a factor of β. Instead of appearing like 270 Ω, it now looks like 27 kΩ in both the dc and the ac equivalent circuits. This is one application of the emitter follower. You also will see the circuit used at the front end of measuring instruments because it has a very high input impedance looking into the base.

EXAMPLE 12-6

What is the voltage gain of the CE stage in Fig. 12-15 for a β of 100?

SOLUTION

The dc base voltage of the CE stage is 1.8 V and the dc emitter voltage is 1.1 V. The dc emitter current is

$$I_E = \frac{1.1 \ \text{V}}{680 \ \Omega} = 1.61 \ \text{mA}$$

and the ac resistance of the emitter diode is

$$r_e' = \frac{25 \text{ mV}}{1.61 \text{ mA}} = 15.5 \ \Omega$$

Next, we need to calculate the input impedance of the emitter follower. Since there are no biasing resistors, the input impedance equals the input impedance looking into the base:

$$z_{\text{in}} = (100)(270 \ \Omega) = 27 \text{ k}\Omega$$

The ac collector resistance of the CE amplifier is

$$r_c = 3.6 \text{ k}\Omega \parallel 27 \text{ k}\Omega = 3.18 \text{ k}\Omega$$

and the voltage gain of this stage is

$$A = \frac{3.18 \text{ k}\Omega}{15.5 \ \Omega} = 205$$

EXAMPLE 12-7

Suppose the emitter follower is removed in Fig. 12-15 and a capacitor is used to couple the ac signal to the 270 Ω. What happens to the voltage gain of the CE amplifier?

SOLUTION

The value of r_e' remains the same for the CE stage: 15.5 Ω. But the ac collector resistance is much lower. To begin with, the ac collector resistance is the parallel resistance of 3.6 kΩ and 270 Ω:

$$r_c = 3.6 \text{ k}\Omega \parallel 270 \ \Omega = 251 \ \Omega$$

Because this is much lower, the voltage gain decreases to

$$A = \frac{251 \ \Omega}{15.5 \ \Omega} = 16.2$$

This shows you the effects of overloading a CE amplifier. The load resistance should be much greater than the dc collector resistance to get maximum voltage gain. We have just the opposite; the load resistance (270 Ω) is much smaller than the dc collector resistance (3.6 kΩ). This is undesirable because it knocks the voltage gain way down.

12-6 DARLINGTON TRANSISTOR

A *Darlington connection* consists of cascaded emitter followers, typically a pair like the ones in Fig. 12-16. The base current of the second transistor comes from the emitter of the first transistor. Therefore, the current gain between the first base and the second emitter is

$$\beta = \beta_1 \beta_2 \tag{12-15}$$

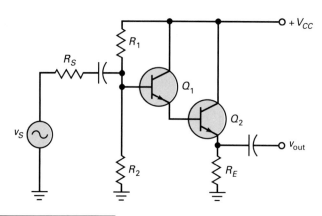

FIGURE 12-16 Darlington amplifier.

In other words, the two transistors have a total current gain equal to the product of the individual current gains. The main advantage of a Darlington connection is the high input impedance looking into the base of the first transistor. In Fig. 12-16, the impedance is approximately

$$z_{\text{in(base)}} = \beta R_E \qquad (12\text{-}16)$$

Transistor manufacturers can put two transistors inside a single transistor housing as shown in Fig. 12-17. This three-terminal device, known as a *Darlington transistor*, acts like a single transistor with an extremely high β. For instance, the TP101 is a Darlington transistor with a minimum β of 1000 and a maximum β of 20,000. The analysis of a circuit using a Darlington transistor is almost identical to the process discussed earlier, except for one thing. Since there are two transistors, there are two V_{BE} drops. For instance, if the dc base voltage is 5 V, the dc emitter voltage is

$$V_E = 5\text{ V} - 1.4\text{ V} = 3.6\text{ V}$$

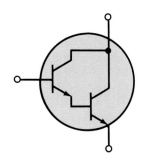

FIGURE 12-17
Darlington transistor.

EXAMPLE 12-8

In Fig. 12-16, $R_E = 100\ \Omega$. If $\beta_1 = 100$ and $\beta_2 = 75$, what does the input impedance of the first base equal?

SOLUTION

Since $\beta_1 = 100$ and $\beta_2 = 75$, the overall current gain will be given by

$$\beta = (100)(75) = 7500$$

The input impedance of the first base is

$$z_{\text{in(base)}} = 7500(100\ \Omega) = 750\text{ k}\Omega$$

12-7 CLASS B OPERATION

Class A is the common way to run a transistor in linear circuits because it leads to the simplest and most stable biasing circuits. But class A is not the most efficient way to operate a transistor. In some applications, like battery-powered systems, current drain and stage efficiency become important considerations in the design. Class B operation of a transistor means that collector current flows for only 180° of the ac cycle. This implies that the Q point is located approximately at cutoff on both the dc and ac load lines. The advantage of class B operation is lower transistor power dissipation and reduced current drain.

Push-Pull Circuit

When a transistor operates class B, it clips off half a cycle because it conducts for only 180° of the cycle. To avoid distortion, a designer uses two transistors in a *push-pull* arrangement. This means that one transistor conducts during one half-cycle, and the other transistor conducts during the other half-cycle. With push-pull circuits, it is possible to build class B amplifiers that have low distortion, large load power, and high efficiency.

Figure 12-18 shows one way to connect a class B push-pull emitter follower. The transistor on top is an *npn* emitter follower, and the one on the bottom is a *pnp* emitter follower. The designer selects biasing resistors to set the Q point at cutoff. This biases the emitter diode of each transistor between 0.6 and 0.7 V, depending on the transistor. When either transistor is conducting, that transistor's operating point swings upward along the ac load line; the operating point of the other transistor remains at cutoff. The voltage swing of the conducting transistor can go all the way from cutoff to saturation as shown in Fig. 12-19. On the

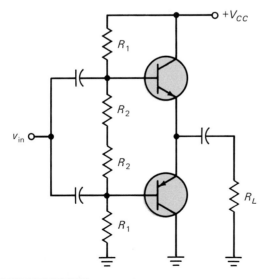

FIGURE 12-18   Class B push-pull emitter follower.

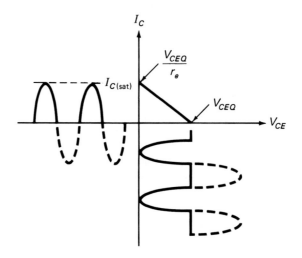

Class B current and voltage.

alternate half cycle, the other transistor does the same thing. This means that the maximum unclipped peak-to-peak output of a class B push-pull amplifier is higher than class A because it now equals

$$\text{MPP} = 2V_{CEQ} \qquad (12\text{-}17)$$

Circuit Action

On the positive half-cycle of input voltage, the upper transistor of Fig. 12-18 conducts and the lower one cuts off. The upper transistor acts like an ordinary emitter follower, so that the output voltage approximately equals the input voltage. On the negative half-cycle of input voltage, the upper transistor cuts off and the lower transistor conducts. The lower transistor acts like an ordinary emitter follower and produces a load voltage approximately equal to the input voltage. The upper transistor handles the positive half-cycle of input voltage, and the lower transistor takes care of the negative half-cycle. During either half-cycle, the source sees a high input impedance looking into either base.

As discussed earlier, a large-signal class A amplifier elongates one half-cycle and squashes the other. One cure for this is swamping, which reduces the distortion to acceptable levels. The class B push-pull emitter follower reduces the distortion even further because both half-cycles are identical in shape. Although some distortion still remains, it is much less than with class A.

Voltage-Divider Bias

The hardest thing about designing a class B amplifier is setting up a stable Q point near cutoff. Figure 12-20a shows voltage-divider bias for a class B push-pull circuit. The two transistors have to be *complementary*, meaning similar V_{BE} curves, maximum ratings, etc. For instance, the 2N3904 and 2N3906 are complementary, the first being an *npn* transistor and the second a *pnp*; they have similar V_{BE} curves, maximum ratings, and so on. Complementary pairs like these are available for almost any class B push-pull design.

In Fig. 12-20a, the collector and emitter currents are approximately equal. Because of the series connection of complementary transistors, each transistor drops half the supply voltage. To avoid distortion, a designer will set the Q point slightly above cutoff, with the correct V_{BE} somewhere between 0.6 and 0.7, depending on the transistor type, the temperature, and other factors. Data sheets indicate that an increase of 60 mV in V_{BE} produces 10 times as much emitter current. Because of this, it is extremely difficult to find standard resistors that can produce the correct V_{BE}. Almost always, an adjustable resistor is needed to set the correct Q point.

But an adjustable resistor does not solve the temperature problem. For a given collector current, V_{BE} decreases approximately 2 mV per degree rise. In other words, the V_{BE} required to set up a particular collector current decreases as the temperature increases. In Fig. 12-20a, the voltage dividers produce a stiff drive for each emitter diode. Therefore, as the temperature increases, the fixed voltage on each emitter diode forces the collector current to increase.

The ultimate danger is a condition called *thermal runaway*. When the temperature increases, the collector current increases. As the Q point moves toward higher collector currents, the temperature of the transistor increases, further reducing the correct V_{BE}. This escalating situation means that the Q point may "run away" by increasing until excessive power destroys the transistor. Whether or not thermal runaway takes place depends on the thermal properties of the transistor, how it is cooled, and the type of heat sink used.

Diode Bias

One way to avoid thermal runaway is with diode bias, shown in Fig. 12-20b. The idea is to use diodes to provide the biasing voltage to the emitter diodes. For this to work, the diode curves must match the V_{BE} curves of the transistors. Then, any increase in temperature reduces the biasing voltage developed by the diodes. For instance, assume a biasing voltage of 0.65 V sets up 1 mA of quiescent collector current. If the temperature rises 30°C, the voltage across each diode decreases about 60 mV. Since the required V_{BE} of each transistor also decreases by approximately 60 mV, the quiescent collector current remains at approximately 1 mA. Because the external diodes compensate for changes in temperature, they are called *compensating diodes*.

When you see diode bias as in Fig. 12-20b, you may assume the current through the emitter diodes is equal to the current through the compensating diodes. The reason is because the compensating diodes are matched to the emitter diodes. For instance, if you know that 10 mA flows through the compensating diodes; then you will automatically know that 10 mA flows through each emitter diode.

Class B Driver

In the initial discussion of the class B push-pull emitter follower, capacitors were used to couple the ac signal into the amplifier. This is not the best

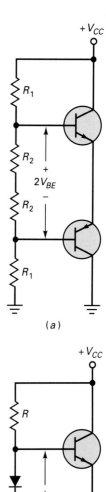

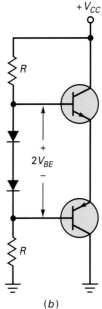

FIGURE 12-20

(*a*) Voltage-divider bias for class B; (*b*) diode bias.

way to drive a class B stage. It is easier to use a direct coupled CE driver, as shown in Fig. 12-21. The amplifier has three stages: a small-signal amplifier (Q_1), a large-signal class A amplifier (Q_2), and a class B push-pull emitter follower (Q_3 and Q_4).

The three-stage amplifier of Fig. 12-21 is fairly complicated. Nevertheless, it can be analyzed without too much trouble if you make sensible approximations. The following examples will show you how it is done. These examples are valuable because they will pull together a lot of ideas discussed in this and earlier chapters.

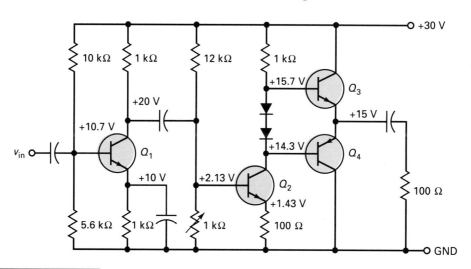

FIGURE 12-21   Complete amplifier.

EXAMPLE 12-9

What is the voltage gain of the first stage in Fig. 12-21? Assume all β's are equal to 100.

SOLUTION

First, get the r_e' of Q_1 as follows:

$$r_e' = \frac{25\text{ mV}}{10\text{ mA}} = 2.5\ \Omega$$

Second, let us work out the input impedance of the second stage. Looking into the base of the second stage, the ac signal sees an impedance of approximately

$$z_{\text{in(base)}} = 100(100\ \Omega) = 10\text{ k}\Omega$$

Notice that we are ignoring the r_e' of the second stage. (Note: it is only 1.75 Ω, which is much smaller than 100 Ω.) The input impedance of the second stage includes the biasing resistors:

$$z_{\text{in}} = 12\text{ k}\Omega \parallel 1\text{ k}\Omega \parallel 10\text{ k}\Omega = 844\ \Omega$$

Third, work out the ac collector resistance of the first stage and the voltage gain. The ac collector resistance is

$$r_c = 1 \text{ k}\Omega \parallel 844 \text{ }\Omega = 458 \text{ }\Omega$$

and the voltage gain is

$$A = \frac{458 \text{ }\Omega}{2.5 \text{ }\Omega} = 183$$

EXAMPLE 12-10

If $\beta = 120$, what is the voltage gain of the second stage in Fig. 12-21?

SOLUTION

First, get the r_e' of Q_2 as follows:

$$r_e' = \frac{25 \text{ mV}}{14.3 \text{ mA}} = 1.75 \text{ }\Omega$$

The second stage is a swamped amplifier because there is no emitter bypass capacitor. Therefore, the total ac resistance of the emitter branch is

$$r_e + r_e' = 100 \text{ }\Omega + 1.75 \text{ }\Omega = 102 \text{ }\Omega$$

This shows you how little effect r_e' has in the second stage. If we ignore it, an error of less than 2 percent will occur.

Second, let us work out the input impedance of the third stage. The third stage operates class B. This means that only one of the transistors is conducting at a time. No matter which is conducting, the input impedance looking into its base is

$$z_{\text{in(base)}} = 100(100 \text{ }\Omega) = 10 \text{ k}\Omega$$

Third, work out the ac collector resistance of the second stage. The collector of Q_2 sees two conducting diodes in series with 1 kΩ. The dc current through these diodes is approximately equal to the dc current through the emitter resistor of the second stage, which is 14.3 mA. Therefore, each diode has an ac resistance of

$$r_{\text{ac}} = \frac{25 \text{ mV}}{14.3 \text{ mA}} = 1.75 \text{ }\Omega$$

This ac resistance is much smaller than the 1 kΩ collector resistor of the second stage. Therefore, in the ac equivalent circuit we can ignore the ac resistance of the two diodes. This is why it is valid to say that the ac collector resistance of the second stage is approximately

$$r_c = 1 \text{ k}\Omega \parallel 10 \text{ k}\Omega = 909 \text{ }\Omega$$

This is the parallel equivalent of the 1 kΩ in the second stage and the 10 kΩ of input impedance for the third stage.

Now, we can calculate the voltage gain of the second stage:

$$A = \frac{909\ \Omega}{102\ \Omega} = 8.91$$

Because of the swamping resistor in the second stage, most people would ignore the r_e' and calculate a voltage gain of

$$A = \frac{909\ \Omega}{100\ \Omega} = 9.09$$

In fact, a troubleshooter looking at the second stage of Fig. 12-21 would think along these lines:

> The input impedance of the third stage is much larger than 1 kΩ, so I can ignore it. The ac resistance of the two diodes is much smaller than 1 kΩ, so I can ignore the diodes. That means that the collector of the second stage sees only 1 kΩ. Since the second stage has a swamping resistor of 100 Ω, the voltage gain is approximately
>
> $$A = \frac{1\ k\Omega}{100\ \Omega} = 10$$

A designer probably would prefer the first or second answer, because the additional accuracy improves one's confidence in the design. But a troubleshooter doesn't need much accuracy because troubles cause large changes from normal values.

Remember, there is never only one right answer to a real-life situation. There is only a best answer selected from several right answers. The best answer depends on your goal and the reason for analyzing the circuit. Don't accept the idea that there is only one correct way to do anything. There are many ways to solve problems. The goal determines which way is the best way. If your goal is to design a circuit, the goal will shape your thinking and control the level of approximation you use. If your goal is to find out why a circuit is not working, this goal will change your approach and desire for accurate answers. Therefore, stay flexible in every situation, use your head as much as possible, and do it your way at all times.

EXAMPLE 12-11

What is the voltage gain of the three-stage amplifier in Fig. 12-21? What is the value of MPP for the output stage? What is the approximate input voltage that produces the MPP?

SOLUTION

The third stage is an emitter follower. Each compensating diode has 14.3 mA flowing through it. Therefore, each emitter diode in the final stage has 14.3 mA, which is equivalent to an r'_e of 1.75 Ω. This means the third stage has a voltage gain of

$$A = \frac{100\ \Omega}{100\ \Omega + 1.75\ \Omega} = 0.983$$

In previous examples, we found that the first stage had a gain of 183 and the second stage had a gain of 8.91. The total gain of the amplifier is the product of the three stage gains:

$$A = (183)(8.91)(0.983) = 1603$$

Here is how you get the MPP of the third stage. The supply voltage is 30 V, so each transistor in the third stage has a quiescent collector-emitter voltage of

$$V_{CEQ} = 15\ \text{V}$$

Equation (12-17) tells us that MPP is two times the value of V_{CEQ}:

$$\text{MPP} = 2(15\ \text{V}) = 30\ \text{V}$$

This is approximately the largest peak-to-peak output voltage without clipping.

The approximate input voltage that produces the MPP is

$$v_{\text{in}} = \frac{30\ \text{V}}{1600} = 18.8\ \text{mV}$$

In other words, when the input voltage to the three-stage amplifier is approximately 18.8 mV peak-to-peak, the output voltage is approximately 30 V peak-to-peak. Any further increase in the input voltage will produce clipping of the output signal.

❑ OPTIONAL TOPICS

The following material continues the earlier discussions at a more advanced and specialized level. All the topics are optional because they are not used in any of the basic discussions in later chapters. This section will be a useful reference when you are in industry because then you will probably want more advanced viewpoints.

❑ 12-8 OUTPUT IMPEDANCE

Figure 12-22a shows an emitter follower with the load resistance disconnected. Looking back into the emitter follower, the load resistance sees an output impedance equivalent to the Thevenin impedance of the emitter follower. This output impedance is very low, which is one of the big advantages of the emitter follower. The following discussion will show you how to find the output impedance of an emitter follower.

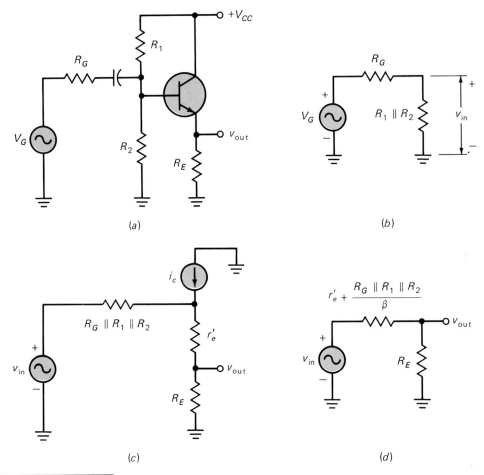

FIGURE 12-22 Deriving an ac model for the emitter follower.

To get an ac model of the emitter follower, we need to start by applying Thevenin's theorem to the base circuit (Fig. 12-22b). The Thevenin resistance is

$$R_{TH} = R_G \parallel R_1 \parallel R_2$$

Since this is the Thevenin resistance seen from the base, we can replace the transistor by its T model to get Figure 12-22c. Summing ac voltages around the loop gives

$$i_e r'_e + i_e R_E - v_{in} + i_b(R_G \parallel R_1 \parallel R_2) = 0$$

Since $i_b = i_c/\beta = i_e/\beta$, we can solve for i_e, to get

$$i_e = \frac{v_{in}}{R_E + r'_e + (R_G \parallel R_1 \parallel R_2)/\beta}$$

Figure 12-22d shows the equivalent output circuit for this emitter current. The emitter resistor, R_E, is driven by an ac source with a Thevenin impedance of

$$r_{th} = r'_e + \frac{R_G \parallel R_1 \parallel R_2}{\beta}$$

In any practical circuit, R_E is large enough to ignore. For this reason, the output impedance of the emitter follower is approximately equal to the output impedance looking back into the emitter:

$$r_{\text{out}} = r'_e + \frac{R_G \| R_1 \| R_2}{\beta} \tag{12-18}$$

Notice how the ac resistance of the base is decreased by a factor of β when seen from the emitter. This makes sense. When we move from the emitter to the base, the impedance is stepped up a factor of β. When we move from the base to the emitter, the impedance is stepped down by a factor of β.

EXAMPLE 12-12

The emitter follower of Fig. 12-23 has these circuit values: $R_G = 600\ \Omega$, $R_1 = 10\ \text{k}\Omega$, $R_2 = 20\ \text{k}\Omega$, $r'_e = 18\ \Omega$, and $\beta = 100$. What is the output impedance of the emitter follower?

SOLUTION

Start with this calculation:

$$R_G \| R_1 \| R_2 = 600\ \Omega \| 10\ \text{k}\Omega \| 20\ \text{k}\Omega = 550\ \Omega$$

The output impedance is

$$r_{\text{out}} = 18\ \Omega + \frac{550\ \Omega}{100} = 23.5\ \Omega$$

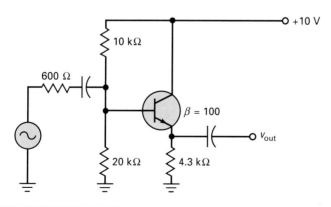

FIGURE 12-23 Darlington emitter follower.

As you can see, the output impedance of the emitter is very low. This is one of the advantages of an emitter follower. A low output impedance implies a stiff voltage source. In this example, the emitter follower will be stiff to all load resistances greater than 2.35 kΩ ($100 \times r_{\text{out}}$).

12-9 IMPROVED VOLTAGE REGULATION

An emitter follower can improve the performance of a zener regulator. Figure 12-24a shows a *zener follower*, a circuit that combines a zener regulator and an emitter follower. Here is how it works. The zener voltage is the input to the base. Therefore, the dc output voltage is

$$V_{\text{out}} = V_Z - V_{BE} \qquad (12\text{-}19)$$

This output voltage is fixed, equal to the zener voltage minus the V_{BE} drop of the transistor. If the supply voltage changes, the zener voltage remains approximately constant, and so does the output voltage. In other words, the circuit acts like a voltage regulator.

The zener follower has two advantages over an ordinary zener regulator. First, the direct current through R_S is the sum of the zener current and the base current, which equals

$$I_B = \frac{I_L}{\beta_{\text{dc}}} \qquad (12\text{-}20)$$

Since this base current is much smaller than the load current, a smaller zener diode can be used. For instance, if you are trying to supply amperes to a load resistor, an ordinary zener regulator requires a zener diode capable of handling amperes. On the other hand, with the improved regulator of Fig. 12-24a, the zener diode needs to handle only tens of milliamperes because of the reduction by β_{dc}.

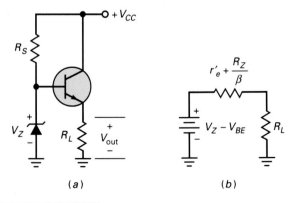

(a) (b)

FIGURE 12-24 Zener follower. (a) Circuit; (b) equivalent.

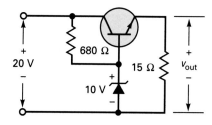

FIGURE 12-25 Example.

In an ordinary zener regulator, the load resistor sees an output impedance of approximately R_Z, the zener impedance. But in the zener follower, the output impedance is

$$r_{\text{out}} = r_e' + \frac{R_Z}{\beta} \qquad (12\text{-}21)$$

This means that the equivalent output circuit looks like Fig. 12-24b. A circuit like this can hold the load voltage almost constant because the source looks stiff.

The two advantages of a zener follower, less load on the zener diode and lower output impedance, allow us to design stiff voltage regulators. The key idea is that the emitter follower increases the current-handling capability of a zener regulator. The zener follower increases the load current by a factor of β_{dc}.

EXAMPLE 12-13

Figure 12-25 shows the zener regulator as it is usually drawn on a schematic diagram. This emphasizes the input and output voltages. If $\beta_{\text{dc}} = 80$, what is the current through the zener diode.

SOLUTION

The current through the series-limiting resistor is

$$I_S = \frac{20\,\text{V} - 10\,\text{V}}{680\,\Omega} = 14.7\,\text{mA}$$

Next, find the base current. The load voltage is

$$V_{\text{out}} = 10\,\text{V} - 0.7\,\text{V} = 9.3\,\text{V}$$

The load current is

$$I_L = \frac{9.3\,\text{V}}{15\,\Omega} = 0.62\,\text{A}$$

The base current is

$$I_B = \frac{0.62\,\text{A}}{80} = 7.75\,\text{mA}$$

The zener current is

$$I_Z = 14.7 \text{ mA} - 7.75 \text{ mA} = 6.95 \text{ mA}$$

Notice how much smaller the zener current is than the load current. This is the whole point of the circuit. A simple zener regulator capable of handling milliamperes is cascaded to an emitter follower to get a regulated load current in hundreds of milliamperes.

STUDY AIDS

The following study aids will help to reinforce the ideas discussed in this chapter. For best results, use these study aids within 6 hours of reading the earlier material. Then review these study aids a week later and a month later to ensure that the concepts remain in your long-term memory.

SUMMARY

Sec. 12-1 CC Amplifier
A CC amplifier has its collector at ac ground. The input signal drives the base and the output signal comes from the emitter. The circuit is better known as an emitter follower. Because it is heavily swamped by the emitter resistor, it has a stable voltage gain, a high input impedance, and little distortion.

Sec. 12-2 AC Model of a CC Amplifier
The input impedance of the base is β times the emitter resistance. Because this is usually much larger than the biasing resistors, the input impedance of an emitter follower is typically equal to the parallel equivalent resistance of the biasing resistors.

Sec. 12-3 Voltage Gain
Typically, the voltage gain is close to one. The main use for an emitter follower is to step up the impedance. It prevents small load resistances from overloading CE stages.

Sec. 12-4 Maximum Unclipped Output
The left swing is given by V_{CEQ} and the right swing by $I_{CQ}r_e$. The value of MPP is twice the smaller swing. A designer can locate the Q point at the middle of the ac load line to get maximum MPP.

Sec. 12-5 Cascading CE and CC
An emitter follower is often used between a low-impedance load and a CE stage. Often, the CE stage is direct-coupled to the emitter follower. This eliminates biasing resistors and a coupling capacitor.

Sec. 12-6 Darlington Transistor
Two transistors may be connected as a Darlington pair. The emitter of the first is connected to the base of the second. This connection produces a current gain equal to the product of individual current gains.

Sec. 12-7 Class B Operation
Class B is more efficient than class A. One popular circuit is the class B push-pull emitter follower. In this circuit, an *npn* transistor and a *pnp* transistor are used. The *npn* transistor conducts during one half-cycle, and the *pnp* transistor conducts during the other half-cycle.

VOCABULARY

In your own words, explain what each of the following terms means. Keep your answers short and to the point. If necessary, verify your answer by rereading the appropriate discussion or by looking at the end-of-book Glossary.

class B operation

common-collector
   amplifier

compensating diodes

complementary
   transistors

Darlington transistor

direct coupling

emitter follower

overloading

push-pull connection

thermal runaway

IMPORTANT EQUATIONS

Here are some important equations. Say each of the following equations in symbols, then say each of them in words. Try to explain what the equation means and how it is used. Then, read the description that follows.

Eq. 12-8 Input Impedance of an Emitter Follower

$$z_{in} = R_1 \parallel R_2$$

This is for an emitter follower that is voltage-divider biased. It is an approximation because it ignores the input impedance of the base. Typically, the input impedance of the base is much larger than the combined effect of the biasing resistors. The equation says that the input impedance of an emitter follower equals the parallel equivalent resistance of the two biasing resistors.

Eq. 12-11 Voltage Gain of an Emitter Follower

$$A = \frac{r_e}{r_e + r_e'}$$

Typically, r_e is much larger than r_e'. Therefore, the voltage gain is approximately equal to one. This means the voltage gain is very stable and the distortion is almost nonexistent.

Eqs. 12-13 and 12-14 MPP of an Emitter Follower

$$MPP = 2V_{CEQ}$$

or

$$MPP = 2I_{CQ}r_e$$

The instantaneous operating point on the ac load line swings to the left and to the right of the Q point. The maximum peak-to-peak unclipped output from an emitter follower equals two times the smaller swing.

Eq. 12-17 MPP for a Class B

$$MPP = 2V_{CEQ}$$

Because of the symmetry of a class B push-pull emitter follower, each transistor has the same dc collector-emitter voltage. Therefore, each has the same swing from the Q point to saturation. Because of this, the MPP is two times V_{CEQ}.

IMPORTANT PROCESSES

Each of the following processes is important for class A emitter followers. If you can remember the basic ideas in each step, you will be able to solve problems more easily.

Output Voltage of Emitter Follower

1. Get input impedance
2. Calculate input voltage
3. Multiply by voltage gain to get output

Limits on Signal Swing of Class A

1. Left swing is V_{CEQ}
2. Right swing is $I_{CQ}r_e$
3. Maximum unclipped is two times smaller

Optimum Q Point of Class A

1. Move Q point up dc load line
2. Stop when $V_{CEQ} = I_{CQ}r_e$

QUESTIONS

The following questions refer to the figures you have seen in this chapter. Select the best answer, the one that is in step with the approximations of this chapter or that most accurately describes the situation.

1. An emitter follower has a voltage gain that is
   a. Much less than one
   b. Approximately equal to one
   c. Greater than one
   d. Zero

2. The total ac emitter resistance of an emitter follower equals
   a. r_e
   b. r_e
   c. $r_e + r_e'$
   d. R_E

3. The input impedance of the base of an emitter follower is usually
   a. Low
   b. High
   c. Shorted to ground
   d. Open

4. The dc emitter current for class A emitter followers is
   a. The same as the ac emitter current
   b. V_E divided by R_E
   c. V_C divided by R_C
   d. The same as the load current

5. The ac base voltage of an emitter follower is across the
   a. Emitter diode
   b. DC emitter resistor
   c. Load resistor
   d. Emitter diode and external ac emitter resistance

6. The output voltage of an emitter follower is across the
   a. Emitter diode
   b. DC collector resistor
   c. Load resistor
   d. Emitter diode and external ac emitter resistance

7. If $\beta = 200$ and $r_e = 150\ \Omega$, the input impedance of the base is
   a. 30 kΩ
   b. 600 Ω
   c. 3 kΩ
   d. 5 kΩ

8. The input voltage to an emitter follower is
   a. Less than the generator voltage
   b. Equal to the generator voltage
   c. Greater than the generator voltage
   d. Equal to the supply voltage

9. The ac emitter current is closest to
   a. V_G divided by r_e
   b. v_{in} divided by r_e'
   c. V_G divided by r_e'
   d. v_{in} divided by r_e

10. The output voltage of an emitter follower is approximately
    a. 0
    b. V_G
    c. v_{in}
    d. V_{CC}

11. The ac load line of an emitter follower is usually
    a. The same as the dc load line
    b. More horizontal than the dc load line
    c. Steeper than the dc load line
    d. Vertical

12. If the input voltage to an emitter follower is too large, the output voltage will be
    a. Smaller
    b. Larger
    c. Equal
    d. Clipped

13. If the Q point is at the middle of the dc load line, clipping will first occur on the
    a. Left voltage swing
    b. Upward current swing
    c. Positive half-cycle of input
    d. Negative half-cycle of input

14. If an emitter follower has $V_{CEQ} = 5$ V, $I_{CQ} = 1$ mA, and $r_e = 1\ \text{k}\Omega$, the maximum peak-to-peak unclipped output is
    a. 1 V
    b. 2 V
    c. 5 V
    d. 10 V

15. If the load resistance of an emitter follower is very large, the external ac emitter resistance equals
    a. Generator resistance
    b. Impedance of the base
    c. DC emitter resistance
    d. DC collector resistance

16. If an emitter follower has $r_e' = 10\ \Omega$ and $r_e = 90\ \Omega$, the voltage gain is approximately
    a. 0
    b. 0.5
    c. 0.9
    d. 1

17. A square wave out of an emitter follower implies
    a. No clipping
    b. Clipping at saturation
    c. Clipping at cutoff
    d. Clipping on both peaks

18. A Darlington transistor has
    a. A very low input impedance
    b. Three transistors
    c. A very high current gain
    d. One V_{BE} drop

19. The ac load line of the emitter follower is
    a. The same as the dc load line
    b. Different from the dc load line
    c. Horizontal
    d. Vertical

20. If the generator voltage is 5 mV in an emitter follower, the output voltage across the load is closest to
    a. 5 mV
    b. 150 mV
    c. 0.25 V
    d. 0.5 V

21. If the load resistor of Fig. 12-1 is shorted, which of the following are different from their normal values:
    a. Only ac voltages
    b. Only dc voltages
    c. Both dc and ac voltages
    d. Neither dc nor ac voltages

22. If R_1 is open in an emitter follower, which of these is true?
    a. DC base voltage is V_{CC}
    b. DC collector voltage is zero
    c. Output voltage is normal
    d. DC base voltage is zero

23. Usually, the distortion in an emitter follower is
    a. Very low
    b. Very high
    c. Large
    d. Not acceptable

24. The distortion in an emitter follower is
    a. Seldom low
    b. Often high
    c. Always low
    d. High when clipping occurs

25. If a CE stage is direct-coupled to an emitter follower, how many coupling capacitors are there between the two stages?
    a. 0
    b. 1
    c. 2
    d. 3

26. A Darlington transistor has a β of 8000. If R_E = 1 kΩ and R_L = 100 Ω, the input impedance of the base is closest to
    a. 8 kΩ
    b. 80 kΩ
    c. 800 kΩ
    d. 8 MΩ

27. The transistors of a class B push-pull emitter follower are biased at or near
    a. Cutoff
    b. The center of the dc load line
    c. Saturation
    d. The center of the ac load line

28. Thermal runaway is
    a. Good for transistors
    b. Always desirable
    c. Useful at times
    d. Usually destructive

29. The ac resistance of compensating diodes
    a. Must be included
    b. Is usually small enough to ignore
    c. Compensates for temperature changes
    d. Is very high

BASIC PROBLEMS

Sec. 12-2 AC Model of a CC Amplifier
12-1. In Fig. 12-26, what is the input impedance of the base?

12-2. What is the input voltage in Fig. 12-26 if β varies over a range of 50 to 300?

12-3. The supply voltage of Fig. 12-26 is doubled. If β = 200, what is the input voltage?

12-4. All resistors are doubled in Fig. 12-26. What happens to the input impedance of the stage? To the input voltage?

Sec. 12-3 Voltage Gain
12-5. What is the voltage gain in Fig. 12-26? The output voltage?

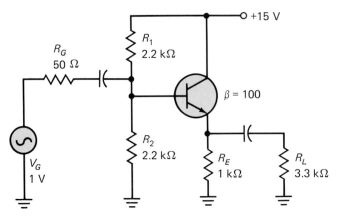

FIGURE 12-26

12-6. What is the output voltage in Fig. 12-26 if β varies over a range of 50 to 300?

12-7. The supply voltage of Fig. 12-26 is doubled. If β = 200, what is the output voltage?

12-8. All resistors are doubled in Fig. 12-26. What happens to the voltage gain? To the output voltage?

Sec. 12-4 Maximum Unclipped Output

12-9. What is the MPP in Fig. 12-26?

12-10. What is the MPP in Fig. 12-26 if R_2 is increased to 4.7 kΩ?

12-11. What happens to MPP in Fig. 12-26 if the supply voltage is doubled?

12-12. The load resistance of Fig. 12-26 is changed to 470 Ω. What is the MPP?

Sec. 12-5 Cascading CE and CC

12-13. What is the voltage gain of the CE stage in Fig. 12-27 if the second transistor has a dc and ac current gain of 150?

12-14. If both transistors have dc and ac current gains of 120, what is the output voltage in Fig. 12-27 if the generator voltage is 2 mV?

12-15. What is the MPP of the first stage in Fig. 12-27 if all β's are 100?

12-16. What is the MPP of the second stage in Fig. 12-27?

12-17. If both transistors have dc and ac current gains of 100, what is the output voltage in Fig. 12-28?

12-18. What is the MPP of each stage in Fig. 12-28?

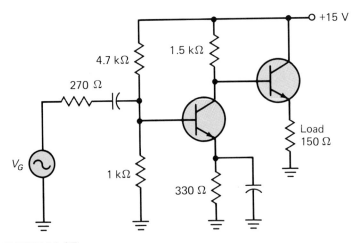

FIGURE 12-27

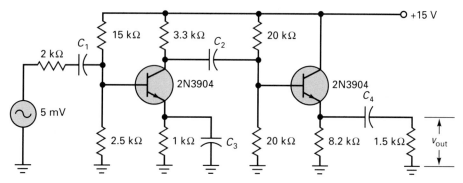

FIGURE 12-28

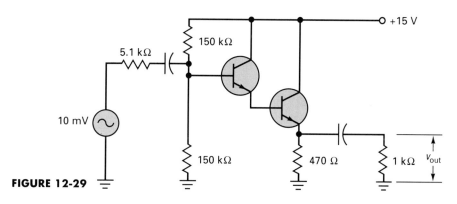

FIGURE 12-29

Sec. 12-6 Darlington Transistor

12-19. Both transistors have a β of 150 in Fig. 12-29. What is the input impedance of the second base? Of the first base?

12-20. What is the dc emitter current in Fig. 12-29 of the second transistor?

12-21. What is the output voltage in Fig. 12-29 if β is 100 for both transistors?

Sec. 12-7 Class B Operation

12-22. All resistors in Fig. 12-30 have a tolerance of ±10 percent. What is the minimum voltage gain of the first stage? The maximum?

12-23. All resistors in Fig. 12-30 have a tolerance of ±10 percent. What is the minimum dc collector voltage in the first stage?

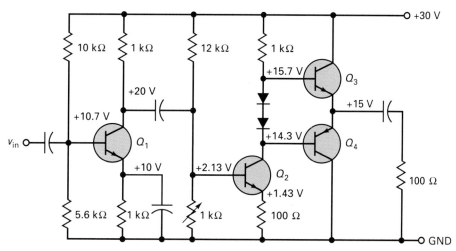

FIGURE 12-30

12-24. The supply voltage of Fig. 12-30 is decreased to 25 V. What is the MPP of the first stage?

ADVANCED PROBLEMS

12-25. What is the output impedance of the emitter follower of Fig. 12-26 if $\beta = 300$?

12-26. If each transistor of Fig. 12-29 has a β of 100, what is the output impedance of the stage?

12-27. If all β's are 100 in Fig. 12-30, what is the output impedance of the first stage?

12-28. The transistor of Fig. 12-31 has a β of 80. What is the output voltage? The zener current? The power dissipation of the transistor?

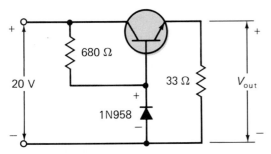

FIGURE 12-31

12-29. In Fig. 12-30, the output stage has an ideal MPP of 30 V. If you build and test

the circuit, however, you will find that the MPP is approximately 24 V peak-to-peak. Why is this?

12-30. Design an emitter follower to meet these specifications: $V_{CC} = 15$ V, $I_C = 2$ mA, $\beta = 100$, $z_{in} > 10$ kΩ, and $z_{out} < 112$ Ω.

12-31. Design a zener follower to meet these specifications: $V_{CC} = 15$ V, $V_{out} = 5.5$ V, and $I_L = 0.5$ A.

12-32. In Fig. 12-32a, the transistor has a β_{dc} of 150. Calculate the following dc quantities: V_B, V_E, V_C, I_E, I_C, and I_B.

12-33. If an input signal with a peak-to-peak value of 5 mV drives the circuit of Fig. 12-32a, what are the two ac output voltages? What do you think is the purpose of this circuit?

12-34. Figure 12-32b shows a circuit where the control voltage can be 0 V or $+5$ V. If the audio input voltage is 10 mV peak-to-peak, what is the audio output voltage when the control voltage is 0 V? When the control voltage is $+5$ V? What do you think this circuit is supposed to do?

T-SHOOTER PROBLEMS

Use Fig. 12-33 for the remaining problems. This T-shooter measures ac voltages only. The big box labeled "Millivolts" contains the measurements of the ac voltages expressed in millivolts. For instance,

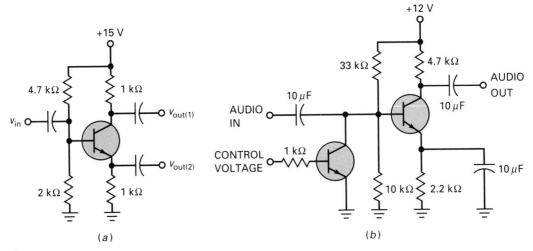

(a)   (b)

FIGURE 12-32

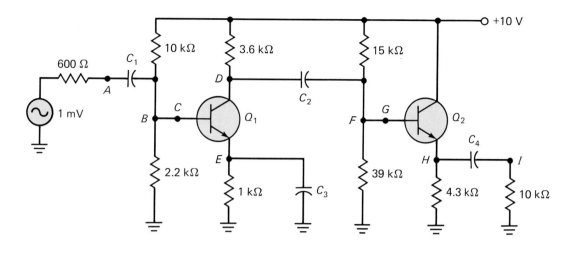

FIGURE 12-33 T-Shooter™. (*Patent pending: Courtesy of Malvino Inc.*)

	1	2	3	4	5	6	7
A	0.75	0.6	1	95	2	70	0
B	2	0.6	70	0	0.75	0	0.6
C	0.6	0.75	2	50	0.6	70	0
D	95	70	0	0.6	0	32	70
E	0.75	0	14	2	0	70	0.75
F	95	95	70	0	70	19	3
G	70	0	39	2	28	0	0.6

MILLIVOLTS

token $C4$ has a value of 50, equivalent to 50 mV. For this exercise, all resistors are normal. The troubles are limited to open capacitors, open connecting wires, and open transistors.

12-35. Find Trouble 1.

12-36. Find Trouble 2.

12-37. Find Trouble 3.

12-38. Find Trouble 4.

12-39. Find Trouble 5.

12-40. Find Trouble 6.

12-41. Find Trouble 7.

FIELD-EFFECT TRANSISTORS

The *bipolar transistor* is the backbone of linear electronics. Its operation relies on two types of charge, electrons and holes. This is why it is called bipolar; the prefix *bi* stands for "two." For many linear applications, the bipolar transistor is the best choice. But there are some applications where a *unipolar transistor* is better suited. The operation of a unipolar transistor depends on only one type of charge, either electrons or holes. This is why it is called unipolar; the prefix *uni* stands for "one."

The *field-effect transistor* (FET) is an example of a unipolar transistor. This chapter discusses the three basic FETs, their structure, and how they work. The FET is more similar to a bipolar transistor than it is different. Because of this, almost everything you learned earlier about bipolar transistors applies to FETs with certain qualifications. In other words, almost everything you learned about voltage gain, load lines, etc. can be interpreted for the new device about to be discussed. Wherever possible, we use the similarities between bipolars and FETs to simplify the discussion. After all, learning is supposed to connect all things, not isolate them.

13-1 THE JFET

The first kind of FET that we discuss is the *junction FET*, abbreviated JFET. Here is the basic idea behind a JFET. Figure 13-1*a* shows a piece of *n*-type semiconductor. This is not a JFET, but it is the first step in making a JFET. The lower end is called the *source*, and the upper end is called the *drain*. The supply voltage V_{DD} forces free electrons to flow from the source to the drain. The source and drain of a JFET are analogous to the emitter and collector of a bipolar transistor.

To produce a JFET, a manufacturer diffuses two areas of *p*-type semiconductor into the *n*-type semiconductor, as shown in Fig. 13-1*b*. Each of these *p* regions is called a *gate*. When a manufacturer connects a separate lead to each gate, the device is called a *dual-gate JFET*. The main use of a dual-gate JFET is with a *mixer*, a special circuit used in communications equipment.

Most JFETs have the two gates connected internally to get a single external gate lead as shown in Fig. 13-1*c*. Because the two gates are always at the same potential, the device acts as though it has only a single gate. We concentrate on the single-gate JFET for the remainder of this chapter because it is more widely used than the dual-gate JFET. Incidentally, the gate of a JFET is analogous to the base of a bipolar transistor.

In Fig. 13-1c, the gate is a *p* region, while the source and the drain are *n* regions. Because of this, a JFET is similar to two diodes. The gate and the source form one of the diodes, and the gate and the drain form the other diode. From now on, we refer to these diodes as the *gate-source diode* and the *gate-drain diode*. Since JFETs are silicon devices, it takes only 0.7 V of forward bias to get significant current in either diode. These diodes conduct the same as any silicon diode. For instance, if the gate is more positive than 0.7 V with respect to the source, the gate-source diode conducts heavily.

Look at how similar a JFET is to a bipolar transistor. Both devices have three external leads, both have two built-in diodes with a barrier potential of 0.7 V, and both have three regions of interest. An *analogy* is defined as a likeness in some ways between things that are otherwise unlike. Analogy is the stuff that inventions are made of. The ability to see the likeness between dissimilar things is extremely important. Not only does it simplify learning and remembering, but also it is at the core of all creativity.

A strong analogy exists between the bipolar transistor and the JFET. Because of this analogy, many of the formulas describing JFET circuits are nothing more than bipolar formulas written for JFET quantities. For this reason, we stress the similarities between bipolars and JFETs whenever possible. For instance, the first analogy to note is the three regions:

Bipolar	JFET
Emitter	Source
Base	Gate
Collector	Drain

Because of these similar regions, many of the JFET formulas are bipolar formulas in disguise. The key idea is to change the subscripts as follows:

Bipolar		JFET
E	$\longrightarrow$	S
B	$\longrightarrow$	G
C	$\longrightarrow$	D

For instance, instead of a dc emitter current I_E, a JFET has a dc source current I_S. Instead of a dc base current I_B, it has a dc gate current I_G. Instead of a dc collector current I_C, it has a dc drain current I_D.

One caution, however. The foregoing similarities might lead you to think that you can substitute a JFET directly into a bipolar circuit. Not so! Even though the devices are more similar than they are different, the differences require some redesign if a JFET is used instead of the bipolar transistor. The discussion that follows will bring out the first major difference between a JFET and a bipolar transistor. This difference is both the strength and the weakness of a JFET.

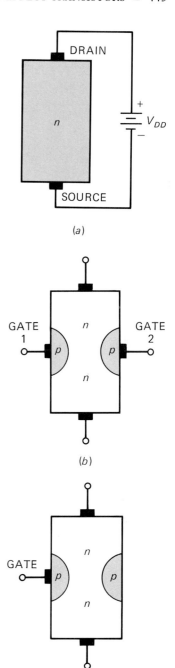

FIGURE 13-1

(*a*) Part of JFET; (*b*) dual-gate JFET; (*c*) single-gate JFET.

13-2 THE BIASED JFET

Figure 13-2a shows the normal way to bias a JFET. Look carefully and notice that this is distinctly different from the way we bias a bipolar transistor. See if you can figure out what the specific difference is before you continue reading.

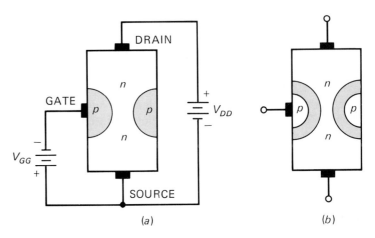

(a)   (b)

FIGURE 13-2   (a) Normal biasing of JFET; (b) depletion layers.

Gate Current

The big difference is this: In a bipolar transistor, we forward-bias the base-emitter diode, but in a JFET, we always *reverse-bias* the gate-source diode. Because of the reverse bias, only a very small reverse current can exist in the gate lead. As an approximation, the gate current is zero. In symbols,

$$I_G = 0$$

Let that sink in. If a device has no input current, what does that tell you about its input resistance? It should tell you that the device has an infinite input resistance.

For instance, if $V_{GG} = 2$ V and $I_G = 0$, the input resistance is

$$R_{IN} = \frac{2\,\text{V}}{0} = \infty$$

The reality of the situation is that I_G is not quite zero, so the input resistance is not quite infinite. But it's close. A typical JFET has an input resistance in the hundreds of megohms. This is the big advantage that a JFET has over a bipolar transistor. And it is the reason that JFETs excel in applications where a high input impedance is required. One of the most important applications of the JFET is the *source follower*, a circuit that is analogous to the emitter follower, except that the input impedance is in the hundreds of megohms for lower frequencies.

Field Effect

The term *field effect* is related to the depletion layers around each p region as shown in Fig. 13-2b. The junctions between each p region and the n regions have depletion layers because free electrons diffuse from the n regions into the p regions. The recombination of free electrons and holes then creates the depletion layers shown by the shaded areas of Fig. 13-2b.

Notice the area between the two depletion layers. When electrons flow from the source to the drain, they must pass through the narrow *channel* between the two depletion layers. The more negative the gate voltage is, the tighter the channel becomes. In other words, the gate voltage can control the current through the channel. The more negative the gate voltage, the smaller the current between the source and the drain.

Since the gate of a JFET is reverse-biased rather than forward-biased, the JFET acts as a *voltage-controlled device* rather than a current-controlled device. In a JFET, the controlling input quantity is the gate-to-source voltage V_{GS}. Changes in V_{GS} determine how much current can flow from source to drain. This is distinctly different from the bipolar transistor where the controlling input quantity is the base current I_B. Voltage control goes hand in hand with high input impedance, whereas current control implies a lower input impedance.

In Fig. 13-2a, the drain supply voltage is positive, and the gate supply voltage is negative. Because of this, the voltage between the gate and the drain is negative. Therefore, the gate-drain diode is reverse-biased. As you see, both diodes in a JFET are reverse-biased for normal operation. There are no exceptions. Keep this in mind when we discuss biasing a JFET later.

How It Works

At the instant the drain supply voltage is applied to the circuit, free electrons start to flow from the source to the drain. These free electrons have to pass through the narrow channel between the depletion layers. The gate voltage controls the width of this channel. The more negative the gate voltage, the narrower the channel and the smaller the drain current.

Almost all the free electrons passing through the channel flow to the drain. Because of this,

$$I_D = I_S$$

This is a highly accurate approximation. The only error is the extremely small reverse current to the gate. But we are talking about more than 99.99 percent of the electrons from the source going on to the drain. For this reason, everyone treats the drain current as equal to the source current.

The Price

Sometimes the strength of a device is also its weakness. The JFET has almost infinite input impedance, but the price paid for this is a loss of control over the output current. In other words, a JFET is less sensitive to changes in the input voltage than a bipolar transistor. In almost any JFET a change in V_{GS} of 0.1 V produces a change in the drain current of less than 10 mA. But in a bipolar transistor the same change in V_{BE} produces a change in the output current of much greater than 10 mA.

What does this mean? It means a JFET amplifier has much less voltage gain than a bipolar amplifier. For this reason, the first design rule governing the two devices is this: Use bipolars for large voltage gain, and use JFETs for high input impedance. Often, a designer combines a JFET and a bipolar transistor to get the best of all worlds. For instance, the first stage may be a JFET source follower, and the second stage may be a bipolar *CE* amplifier. This gives a multistage amplifier a high input impedance and a large voltage gain. More will be said about this later.

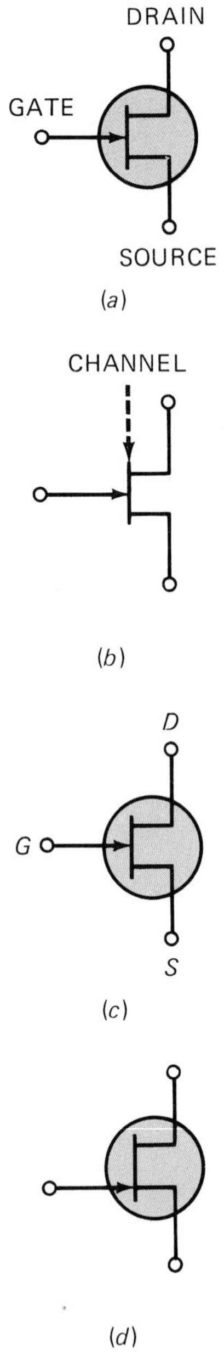

Schematic Symbol

The JFET we have been discussing is called an *n-channel JFET* because the channel between the depletion layers is made of *n*-type semiconductor. Figure 13-3a shows the schematic symbol for an *n*-channel JFET. As a memory aid, visualize the thin vertical line (see Fig. 13-3b) as the *n* channel; the source and drain connect to this line. Since the gate is a *p* region and the channel is an *n* region, the *p*-type gate points toward the *n* channel.

In many low-frequency applications, the source and the drain are interchangeable because you can use either end as the source and the other end as the drain. For this reason, the JFET symbol of Fig. 13-3a is drawn symmetrically with the arrow pointing to the center of the device. When you use the symmetric JFET symbol, you can label the terminals as in Fig. 13-3a or use abbreviations as in Fig. 13-3c. On many schematic diagrams, labels or abbreviations are not used. In this case, you have to figure out which end is the source from your understanding of the circuit operation. Occasionally, you will see the nonsymmetric symbol of Fig. 13-3d where the arrow is drawn closer to the source.

Although either end of most JFETs may be used as the source at low frequencies, this is not true at high frequencies. Almost always, the manufacturer minimizes the internal capacitance on the drain side of the JFET. You will learn more about internal capacitances later. All you need to know is this: The capacitance between the gate and the drain is smaller than the capacitance between the gate and the source. These internal capacitances degrade the high-frequency performance of a JFET circuit. For this reason, there is a preferred source terminal and a preferred drain terminal. The data sheet shows you exactly which end is which. All high-frequency circuits should use the ends shown on the data sheet. (For instance, see the data sheet of the MPF102 in the Appendix.)

A final point. There is also a *p*-channel JFET. It consists of a *p*-type material with diffused islands of *n*-type material. The schematic symbol for a *p*-channel JFET is similar to that for the *n*-channel JFET, except that the gate arrow points from the channel to the gate. The action of a *p*-channel JFET is complementary, which means that all voltages and currents are reversed. To reduce the confusion while you are learning about FETs, we emphasize the *n*-channel devices in the remainder of this chapter. Once you understand how *n*-channel devices work, a little bit of

DRAIN

GATE

SOURCE

(a)

CHANNEL

(b)

D

G

S

(c)

(d)

FIGURE 13-3

Schematic symbol for *n*-channel JFET.

reverse thinking should allow you to understand how complementary
p-channel circuits work.

EXAMPLE 13-1

The Appendix shows the data sheet of an MPF102, a widely used
n-channel JFET. What is the dc input resistance of the device?

SOLUTION

If you examine the data sheet, you will find this specification for
the gate reverse current:

$$I_{GSS} = -2 \text{ nA}$$

This value is given for $V_{GS} = -15$ V and $V_{DS} = 0$. The dc input
resistance of the device is

$$R_{in} = \frac{15 \text{ V}}{2 \text{ nA}} = 7.5(10^9) \ \Omega = 7500 \text{ M}\Omega$$

By way of comparison, a bipolar transistor with 10 μA of base
current has a dc input resistance of

$$R_{in} = \frac{0.7 \text{ V}}{10 \text{ μA}} = 70 \text{ k}\Omega$$

These numbers show the overwhelming superiority of a JFET when
it comes to dc input resistance.

Incidentally, don't let the minus signs on a data sheet confuse
you. They are only a reminder of the direction of current and are
not necessary for calculations. You can ignore the minus signs in
your calculations and use the magnitudes if you prefer. For instance,
the data sheet gives $I_{GSS} = -2$ nA and $V_{GS} = -15$ V. A pure
mathematician would calculate the dc input resistance as follows:

$$R_{in} = \frac{-15 \text{ V}}{-2 \text{ nA}} = 7.5 \,(10^9) \ \Omega = 7500 \text{ M}\Omega$$

But we are not dealing with pure mathematics, so we don't have to
use minus signs. In fact, many data sheets don't include the minus
signs. They specify the data as follows:

$$I_{GSS} = 2 \text{ nA at } V_{GS} = 15 \text{ V}$$

They leave out the minus signs because anyone who knows how
JFETs work already knows that the gate-source diode is reverse-
biased. When you know what you are doing, you can use or delete
the minus signs.

A final point. The data sheet specifies an I_{GSS} of 2 nA maximum
at $V_{DS} = 0$. This is the worst case. Normally, V_{DS} is greater than
zero, and this increases the reverse bias between the gate and the
drain. The effect of increasing V_{DS} is to decrease I_{GSS}. In other
words, the dc input resistance will be even higher than 7500 MΩ
for V_{DS} greater than zero.

13-3 DRAIN CURVES

Figure 13-4a shows a JFET with normal biasing voltages. In this simple circuit, the gate-source voltage V_{GS} equals the gate supply voltage V_{GG}, and the drain-source voltage V_{DS} equals the drain supply voltage V_{DD}. Although the circuit is not practical, it will be useful for our discussion of characteristic curves. Because of all that you already know about bipolar transistors, the discussion of JFET curves does not have start from ground zero. In fact, we can move a lot faster because of the bipolar-JFET analogy.

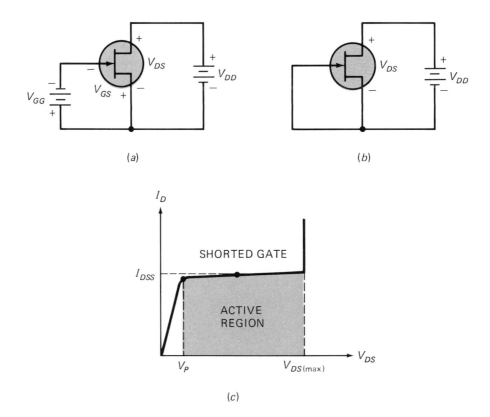

(a)   (b)

(c)

FIGURE 13-4   (a) Normal bias for JFET; (b) zero gate voltage; (c) shorted gate drain current.

Maximum Drain Current

The maximum drain current out of a JFET occurs when the gate-source voltage is zero as shown in Fig. 13-4b. Here you see the gate supply voltage replaced by a short circuit, which guarantees that

$$V_{GS} = 0$$

Figure 13-4c shows the corresponding graph of drain current I_D versus drain-source voltage V_{DS}. Notice the similarity to a collector curve. The drain current increases rapidly at first, then levels off and becomes almost horizontal. In the region between V_P and $V_{DS(\max)}$, the drain current is almost constant. If the drain voltage is too large, the JFET breaks down as shown.

Similar to a bipolar transistor, a JFET acts like a current source when it is operating along the almost-horizontal part of the drain curve. This almost-horizontal part of the drain curve is between a minimum voltage of V_P and a maximum voltage of $V_{DS(\max)}$. The minimum voltage V_P is called the *pinchoff voltage*, and maximum voltage $V_{DS(\max)}$ is called the *breakdown voltage*. Between pinchoff and breakdown, the JFET acts approximately like a current source with a value of I_{DSS}.

Now I_{DSS} stands for the current from drain to source with a shorted gate, and I_{DSS} is the maximum drain current a JFET can produce. All data sheets for JFETs list the value of I_{DSS}. This is one of the most important JFET quantities, and you should always look for it first because it gives the limitation on the JFET current. For instance, the MPF102 has a typical I_{DSS} of 6 mA. This tells you that no matter what the circuit design is, the drain current will be between 0 and 6 mA for a typical MPF102. (This is different from a bipolar transistor which has no upper limit aside from its burnout value.)

Gate Cutoff and Pinchoff

Figure 13-5 shows a set of drain curves for a JFET with an I_{DSS} of 10 mA. The top curve is for $V_{GS} = 0$. The pinchoff voltage is 4 V, and the breakdown voltage is 30 V. The next curve down is for $V_{GS} = -1$ V, the next for $V_{GS} = -2$ V, and so on. As you see, the more negative the gate-source voltage, the smaller the drain current.

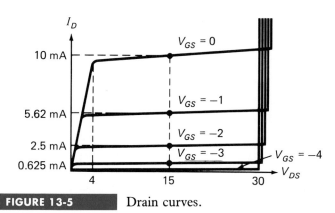

FIGURE 13-5　Drain curves.

The bottom curve is especially important. Notice that a V_{GS} of -4 V reduces the drain current to almost zero. This voltage is called the *gate-source cutoff voltage*. On data sheets, it is symbolized as $V_{GS(\text{off})}$. In Fig. 13-5, notice that

$$V_{GS(\text{off})} = -4 \text{ V} \qquad \text{and} \qquad V_P = 4 \text{ V}$$

Is this a coincidence? Not at all. For advanced reasons that we won't go into, the magnitudes of these two voltages are always equal. This is worth remembering because many data sheets will list one value but not the other. They do this because everyone is supposed to know that the two voltages are equal in magnitude. Giving you the value of one is equivalent to giving you the other. For instance, the data sheet of an MPF102 gives

$$V_{GS(\text{off})} = -8 \text{ V}$$

for the gate-source cutoff voltage. Although the pinchoff value is not given, we know automatically the $V_P = 8$ V.

Here is a formal reminder of how the gate-source cutoff voltage is related to the pinchoff voltage:

$$V_{GS(\text{off})} = -V_P \tag{13-1}$$

This says the gate-source voltage equals the negative of the pinchoff voltage.

The Ohmic Region

In Fig. 13-5, the pinchoff voltage is the voltage where the highest drain curve changes from almost vertical to almost horizontal. It is a very important voltage because it separates two major operating regions of the JFET. The almost-vertical part of the drain curve is called the *ohmic region,* equivalent to the saturation region of a bipolar transistor. When operated in the ohmic region, a JFET acts as a small resistor with a value of approximately

$$R_{DS} = \frac{V_P}{I_D} \tag{13-2}$$

The next chapter is about JFET applications. At that time, we will have a lot more to say about the ohmic region and using the JFET as a voltage-controlled resistor.

EXAMPLE 13-2

A data sheet gives these JFET values: $I_{DSS} = 20$ mA and $V_P = 5$ V. What is the maximum drain current? What is the gate-source cutoff voltage?

SOLUTION

For any gate voltage, the drain current has to be in this range:

$$0 < I_{DSS} < 20 \text{ mA}$$

When the gate voltage is zero, the drain current has its maximum value of

$$I_D = 20 \text{ mA}$$

The gate-source voltage has the same magnitude as the pinchoff voltage but the opposite sign. Since the pinchoff voltage is 5 V,

$$V_{GS(\text{off})} = -5 \text{ V}$$

EXAMPLE 13-3

In Example 13-2, what is the dc resistance of the JFET in the ohmic region?

SOLUTION

The dc resistance is equal to the pinchoff voltage divided by the maximum drain current:

$$R_{DS} = \frac{5\text{ V}}{20\text{ mA}} = 250\ \Omega$$

13-4 THE TRANSCONDUCTANCE CURVE

The *transconductance curve* of a JFET is a graph of drain current versus gate-source voltage, or I_D versus V_{GS}. By reading the values of I_D and V_{GS} in Fig. 13-5, we can plot the transconductance curve shown in Fig. 13-6a. In general, the transconductance curve of any JFET will have the same shape as Fig. 13-6a; only the numbers will be different.

Figure 13-6b shows how the transconductance curve of any JFET will appear. Why is this? The physics behind JFET operation is the same for all JFETs. Only the size of the doped regions, the level of doping, etc. change from one JFET to the next. Because of this, all JFETs have a transconductance curve that is the graph of the following equation:

$$I_D = I_{DSS}\left(1 - \frac{V_{GS}}{V_{GS(\text{off})}}\right)^2 \tag{13-3}$$

This equation can be derived with advanced physics and mathematics. We do not show the derivation because it is too complicated.

With Eq. (13-3), we can calculate the drain current once given the maximum drain current, the gate-source cutoff voltage, and the gate voltage. This is the algebraic way to find the drain current. On the other

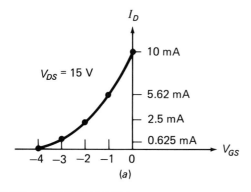

(a)

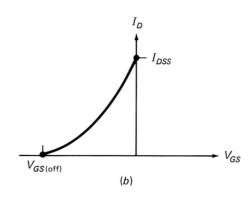

(b)

FIGURE 13-6   Transconductance curves.

hand, some data sheets include graphs like Fig. 13-6a. In this case, you don't have to use Eq. (13-3). You can read the values of drain current directly from the graphs. This is the graphical way to find the drain current.

For instance, Fig. 13-6a is good for quick and approximate answers. You can see at a glance that the maximum drain current is 10 mA and that the gate-source cutoff voltage is −4 V. In between these extreme points on the graph, you can see that the graph is *nonlinear*. In fact, the shape of this graph is part of a *parabola*, a curve that exists when quantities are squared. The quantity that multiplies I_{DSS} in the foregoing equation is the K factor, given by

$$K = \left(1 - \frac{V_{GS}}{V_{GS(\text{off})}}\right)^2 \tag{13-4}$$

We are going to use the K factor in later discussions. For now, notice that we can rewrite Eq. (13-3) as

$$I_D = KI_{DSS} \tag{13-5}$$

If we have the value of K for any circuit, we can quickly calculate the value of drain current, given the maximum drain current.

Incidentally, the *square law* is another name for *parabolic*. This is why JFETs are often called *square-law devices*. And this is another big difference between a bipolar transistor and a JFET. The square-law property gives JFETs a major advantage over bipolar transistors when it comes to *mixers*, circuits used in communications equipment.

EXAMPLE 13-4

Suppose a JFET has $I_{DSS} = 7$ mA and $V_{GS(\text{off})} = -3$ V. Calculate the drain current for a gate-source voltage of −1 V.

SOLUTION

With Eq. (13-4), you can work out the K factor as follows:

$$K = \left(1 - \frac{1 \text{ V}}{3 \text{ V}}\right)^2 = 0.667^2 = 0.445$$

Now, multiply the K factor by I_{DSS} to get the drain current:

$$I_D = 0.445(7 \text{ mA}) = 3.12 \text{ mA}$$

13-5 JFET APPROXIMATIONS

As with bipolar transistors, exact analysis of JFET circuits is a waste of time. The manufacturing spreads of JFETs are even worse than they are for bipolar transistors. For instance, a 2N3904 has minimum and maximum β values of 100 to 300, a 3:1 spread. An MPF102 has minimum and maximum I_{DSS} values of 2 and 20 mA, a 10:1 spread. When you have a 10:1 spread like this, the only sensible approach is to use reasonable approximations. Anything beyond this is a flight from reality, at least for the bulk of the everyday work done with JFETs.

The Ideal JFET

At this time, we are going to discuss two dc approximations for any JFET. Both approximations are derived as follows: If a manufacturer could produce an ideal JFET, here is what would happen to the curves of Fig. 13-5 on page 451. First, there would be no breakdown region. Second, all drain curves would superimpose in the ohmic region. Third, all drain curves would be horizontal in the current-source region.

Figure 13-7 shows the drain curves of an ideal JFET and a typical dc load line. The ideal JFET has two major regions of operation: the ohmic region (saturation) and the current-source region (active). The ohmic region of the JFET is highly desirable because it can be used in all kinds of analog-switching applications. This is why we have included the almost-vertical part of the drain curves in Fig. 13-7. When we want a JFET to act like a resistor, we have to make sure that the JFET is saturated, that its operating point is on the almost-vertical part of the drain curves. But when we want a JFET to act as a current source, we have to make sure that the operating point is on the horizontal part of the drain curves.

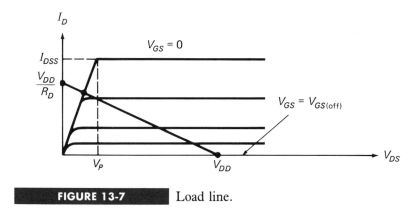

FIGURE 13-7   Load line.

Since there are two major regions of operation, we need two basic models or equivalent circuits to describe dc operation. First, we approximate a JFET by the dc model shown in Fig. 13-8a. As you see, the input side of the JFET has a dc input resistance of R_{GS}. If necessary, you can estimate its value by taking the ratio of the V_{GS} and I_{GS} values

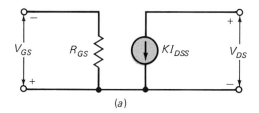
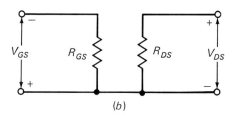

FIGURE 13-8   (a) Current-source model; (b) ohmic model.

given on the data sheet. But most of the time, you can ignore R_{GS} because it is almost infinite.

On the output side of Fig. 13-8a, the JFET acts like a current source of KI_{DSS}. This is the dc model that we can use when the JFET operates in the active region. Recall that the pinchoff voltage is the clue here. When V_{DS} is greater than V_P, the JFET will act like a current source for any gate voltage. Given I_{DSS} and $V_{GS(off)}$, we can calculate the value of K for any V_{GS} input voltage.

Figure 13-8b shows a second model for a JFET. This is the ohmic model because it is valid whenever the JFET is operating on the almost-vertical part of the drain curves. Notice that the JFET is no longer a current source on the output side. Rather, it acts like a resistance of R_{DS}. You can estimate the value of R_{DS} by the ratio of V_P to I_{DSS}.

Proportional Pinchoff

The pinchoff voltage of Fig. 13-7 separates the ohmic region from the active region when V_{GS} is zero. When V_{GS} does not equal zero, we can use the *proportional pinchoff voltage* as our guide. Symbolized V'_P, this voltage is the border between the ohmic region and the current-source region for any value of V_{GS}. This quantity is given by

$$V'_P = I_D R_{DS} \tag{13-6}$$

Here is how you use this equation. First, you calculate R_{DS} by dividing V_P by I_{DSS}. Then you multiply R_{DS} by the actual drain current to find the value of V'_P. This value is the border between the two operating regions.

Figure 13-9 shows you why Eq. (13-6) is valid. Here you see the ohmic region of an ideal JFET. The highest point in the ohmic region has coordinates of I_{DSS} and V_P. The other point represents any point in the ohmic region. The coordinates of any point in the ohmic region are I_D and V'_P. With basic geometry, you can see this proportional relation:

$$\frac{V'_P}{I_D} = \frac{V_P}{I_{DSS}}$$

But this is the equivalent to

$$\frac{V'_P}{I_D} = R_{DS}$$

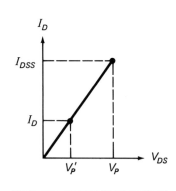

FIGURE 13-9

Proportional pinchoff.

If you solve this for V'_P, you get Eq. (13-6).

Memorize Eq. (13-6). Designers use the JFET in two basic ways: as a resistor and as a current source. When you analyze JFET circuits, you have to figure out which way the JFET is being used. Then you will know whether to use the current-source model (Fig. 13-8a) or the ohmic model (Fig. 13-8b). Here is the process for deciding which model to use:

1. Divide V_P by I_{DSS} to get R_{DS}.

2. Multiply I_D by R_{DS} to get V'_P.

3. If $V_{DS} > V'_P$, use the current-source model.

4. If $V_{DS} < V'_P$, use the ohmic model.

Analyzing JFET Circuits

We are about to look at several examples of analyzing a JFET. Before we do, let us summarize the important quantities and equations that we need. To begin, we must have I_{DSS} and $V_{GS(off)}$. These quantities are rock bottom. Without them, you don't have enough information to analyze the circuit. Depending on how the analysis goes, you will need some of or all the following useful formulas:

$$V_P = -V_{GS(off)} \qquad\qquad (13\text{-}7)$$

$$R_{DS} = \frac{V_P}{I_{DSS}} \qquad\qquad (13\text{-}8)$$

$$K = \left(1 - \frac{V_{GS}}{V_{GS(off)}}\right)^2 \qquad\qquad (13\text{-}9)$$

$$I_D = KI_{DSS} \qquad\qquad (13\text{-}10)$$

$$V'_P = I_D R_{DS} \qquad\qquad (13\text{-}11)$$

Reductio ad Absurdum

You already know about reductio ad absurdum, which was introduced with bipolar transistors. Recall the basic idea. When you are not sure which region a device is operating in, you assume an operating region and see if your calculations produce an absurd or a contradictory result. If so, then you know the device cannot operate in the assumed region.

If you are analyzing a JFET circuit and you are not sure of the operating region, then proceed as follows:

1. Assume the current-source region.

2. Carry out your calculations.

3. If an absurd answer arises, the assumption is false.

4. Change to the ohmic model.

The following examples illustrate these steps and help to pull together all the ideas of this section.

EXAMPLE 13-5

In Fig. 13-10, what is the drain-source voltage when V_{GS} is zero?

SOLUTION

Assume the JFET acts as a current source. Since the gate voltage is zero, the drain current is at its maximum value of 10 mA. Figure 13-11a shows the equivalent circuit for the drain circuit. Therefore, the drain-source voltage is

$$V_{DS} = 10 \text{ V} - (10 \text{ mA})(360 \text{ }\Omega) = 6.4 \text{ V}$$

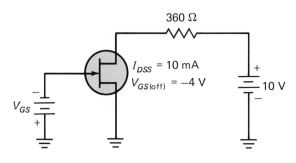

FIGURE 13-10 Example.

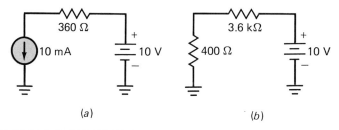

(a)   (b)

FIGURE 13-11 Equivalent circuits.

Since $V_{GS(\text{off})} = -4$ V, the pinchoff voltage V_P is 4 V. Because V_{DS} is greater than 4 V, the assumption about a current source is correct.

The calculation for V_{DS} is identical to the calculation for V_{CE} in a bipolar transistor, except for a change in the subscripts. Here is how the calculation looks as a JFET formula:

$$V_{DS} = V_{DD} - I_D R_D \qquad (13\text{-}12)$$

The corresponding bipolar formula is

$$V_{CE} = V_{CC} - I_C R_C$$

The two equations have the same format; they differ only in their subscripts.

This is an example of what we mean by analogy. When an old system and a new system are governed by the same fundamental laws, their final equations are the same in appearance. If you already know a lot about the old system, you don't have to rediscover everything for the new system. You can take advantage of the similarities in the old system to understand the new system.

The analogy between bipolar and JFET circuits gives us all kinds of powerful shortcuts for solving new JFET circuits with old bipolar methods. Since Ohm's and Kirchhoff's laws are the fundamental laws behind bipolar and JFET circuits, many JFET equations are nothing more than bipolar equations with their subscripts changed as follows:

Bipolar		JFET
E	$\longrightarrow$	S
B	$\longrightarrow$	G
C	$\longrightarrow$	D

Because of the analogy between bipolars and JFETs, many of the new JFET formulas you see will be a lot easier to remember.

EXAMPLE 13-6

In Fig. 13-10, the resistor is changed to 3.6 kΩ. If $V_{GS} = 0$, what is the drain-source voltage?

SOLUTION

Assume the JFET acts as a current source. Since the gate voltage is zero, the drain current is at its maximum value of 10 mA. Therefore, the drain-source voltage is

$$V_{DS} = 10 \text{ V} - (10 \text{ mA})(3.6 \text{ k}\Omega) = -26 \text{ V}$$

Impossible! The drain voltage cannot be negative. We have an absurd result, which means the JFET cannot be operating in the current-source region. It must be operating in the ohmic region.

Here is what to do next. Since the JFET is operating in the ohmic region, we need to calculate the value of R_{DS}. It equals the pinchoff voltage divided by the maximum drain current:

$$R_{DS} = \frac{4 \text{ V}}{10 \text{ mA}} = 400 \text{ }\Omega$$

Figure 13-11b shows the equivalent circuit for the drain circuit. The drain-source voltage is across 400 Ω. This is an Ohm's law problem, which is easily solved as follows:

$$V_{DS} = \frac{10 \text{ V}}{4 \text{ k}\Omega}(400 \text{ }\Omega) = 1 \text{ V}$$

EXAMPLE 13-7

What is the drain-source voltage in Example 13-6 for $V_{GS} = -2.2 \text{ V}$?

SOLUTION

Since V_{GS} has changed from 0 to -2.2 V, there is less drain current and it is possible that the JFET no longer operates in the ohmic region. Here is how to proceed. Assume the JFET is operating as a current source. First, get the K factor and the drain current as follows:

$$K = \left(1 - \frac{2.2\,\text{V}}{4\,\text{V}}\right)^2 = 0.45^2 = 0.203$$

and

$$I_D = 0.203(10\,\text{mA}) = 2.03\,\text{mA}$$

Second, the drain-source voltage is

$$V_{DS} = 10\,\text{V} - (2.03\,\text{mA})(3.6\,\text{k}\Omega) = 2.69\,\text{V}$$

Third, calculate the proportional pinchoff voltage:

$$V'_P = (2.03\,\text{mA})(400\,\Omega) = 0.812\,\text{V}$$

This voltage separates the ohmic region and the active region when $V_{GS} = -2.2$ V. Since a V_{DS} of 2.69 V is greater than a V'_P of 0.812 V, the JFET is operating as a current source. This agrees with the original assumption. Therefore, the final answer is

$$V_{DS} = 2.69\,\text{V}$$

13-6 THE DEPLETION-MODE MOSFET

The *metal-oxide semiconductor FET* or MOSFET, has a source, gate, and drain. Unlike a JFET, however, the gate is electrically insulated from the channel. Because of this, the gate current is extremely small whether the gate is positive or negative. The MOSFET is often referred to as an *IGFET*, which stands for insulated-gate FET. There are two kinds of MOSFETs. This section discusses the depletion-mode MOSFET, a device that is quite similar to the JFET. The next section discusses the enhancement-mode MOSFET, a device that is quite different from the JFET.

The Basic Idea

Figure 13-12 shows an n-channel depletion-mode MOSFET. It is a piece of n material with a p region on the right and an insulated gate on the left. Free electrons can flow from the source to the drain through the n material. The p region is called the *substrate* (or *body*). Electrons flowing from source to drain must pass through the narrow channel between the gate and the p region.

A thin layer of silicon dioxide (SiO_2) is deposited on the left side of the channel. Silicon dioxide is the same as glass, which is an insulator. In a MOSFET, the gate is metallic. Because the metallic gate is insulated from the channel, negligible gate current flows even when the gate voltage is positive. In other words, the gate-source diode and the gate-drain diode of the JFET have been eliminated in the MOSFET.

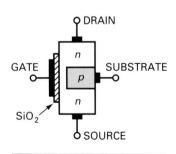

FIGURE 13-12

Depletion-mode MOSFET.

Figure 13-13*a* shows a depletion-mode MOSFET with a negative gate. The V_{DD} supply forces free electrons to flow from source to drain. These electrons flow through the narrow channel on the left of the *p* substrate. As with a JFET, the gate voltage controls the width of the channel. The more negative the gate voltage, the smaller the drain current. When the gate voltage is negative enough, the drain current is cut off. Therefore, the operation of a MOSFET is similar to that of a JFET when V_{GS} is negative.

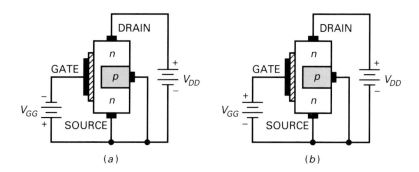

FIGURE 13-13   (*a*) Negative gate voltage; (*b*) positive gate voltage.

Because the gate of a MOSFET is electrically insulated from the channel, we can apply a positive voltage to the gate, as shown in Fig. 13-13*b*. The positive gate voltage increases the number of free electrons flowing through the channel. The more positive the gate voltage, the greater the conduction from source to drain. Being able to use a positive gate voltage is what distinguishes the depletion-mode MOSFET from the JFET. This is useful in some applications that are discussed in the next chapter.

Graphs

Figure 13-14*a* shows typical drain curves for an *n*-channel MOSFET. Notice that the upper curves have a positive V_{GS} and the lower curves have a negative V_{GS}. The bottom drain curve is for $V_{GS} = V_{GS(\text{off})}$.

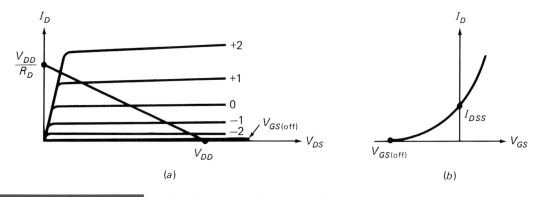

FIGURE 13-14   (*a*) Drain curves; (*b*) transconductance curve.

Along this cutoff curve, the drain current is approximately zero. When V_{GS} is between $V_{GS(\text{off})}$ and zero, we get *depletion-mode* operation. And V_{GS} greater than zero gives *enhancement-mode* operation. These drain curves again display an ohmic region, a current-source region, and a cutoff region. Like the JFET, the depletion-mode MOSFET has two major applications: a current source or a resistance.

Figure 13-14*b* is the transconductance curve of a depletion-mode MOSFET, and I_{DSS} is the drain current with a shorted gate. Since the curve extends to the right of the origin, I_{DSS} is no longer the maximum possible drain current. Mathematically, this curve is still part of a parabola, and the same square-law relation exists as with a JFET. In fact, the depletion-mode MOSFET has a drain current given by the same transconductance equation as before, Eq. (13-3). Furthermore, it has the same equivalent circuits as a JFET. Because of this, the analysis of a depletion-mode MOSFET circuit is almost identical to that of a JFET circuit. The only difference is the analysis for a positive gate, but even here the same basic formulas are used to find the drain current, gate-source voltage, etc. The examples that follow will show you how to analyze a depletion-mode MOSFET circuit.

Schematic Symbol

Figure 13-15*a* shows the schematic symbol for a depletion-mode MOSFET. Just to the right of the gate is the thin vertical line representing the channel. The drain lead comes out the top of the channel, and the source lead connects to the bottom. The arrow on the p substrate points to the n material. In some applications, a voltage can be applied to the substrate for added control of the drain current. For this reason, some depletion-mode MOSFETs have four external leads. But in most applications, the substrate is connected to the source. Usually, the manufacturer internally connects the substrate to the source. This results in a three-terminal device whose schematic symbol is shown in Fig. 13-15*b*.

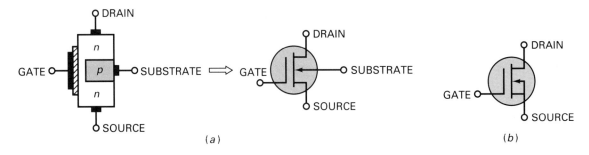

FIGURE 13-15 Schematic symbols.

There is also a p-channel depletion-type MOSFET. It consists of a piece of p material with an n region on the right and an insulated gate on the left. The schematic symbol of a p-channel MOSFET is similar to that of an n-channel MOSFET, except that the arrow points outward. In the remainder of this chapter, we emphasize the n-channel MOSFET. The action of a p-channel MOSFET is complementary, meaning that all voltages and currents are reversed.

EXAMPLE 13-8

In Fig. 13-16a, what is the drain-source voltage when V_{GS} is zero?

SOLUTION

Assume the MOSFET acts like a current source. Since the gate voltage is zero, the drain current is 10 mA as shown in Fig. 13-16b. Therefore, the drain-source voltage is

$$V_{DS} = 20 \text{ V} - (10 \text{ mA})(470 \text{ }\Omega) = 15.3 \text{ V}$$

Since $V_{GS(off)} = -4$ V, the pinchoff voltage V_P is 4 V. Because V_{DS} is greater than 4 V, the assumption about a current source is correct.

EXAMPLE 13-9

In Fig. 13-16a, the resistor is changed to 4.7 kΩ. If $V_{GS} = 0$, what is the drain-source voltage?

SOLUTION

Assume the MOSFET acts like a current source. Since the gate voltage is zero, the drain current is 10 mA. Therefore, the drain-source voltage is

$$V_{DS} = 20 \text{ V} - (10 \text{ mA})(4.7 \text{ k}\Omega) = -27 \text{ V}$$

Impossible! The drain voltage cannot be negative. We have an absurd result, which means the MOSFET cannot be operating in the active region. It must be operating in the ohmic region.

Here is what to do next. Since the MOSFET is operating in the ohmic region, we need to calculate the value of R_{DS}. It equals the pinchoff voltage divided by the maximum drain current:

$$R_{DS} = \frac{4 \text{ V}}{10 \text{ mA}} = 400 \text{ }\Omega$$

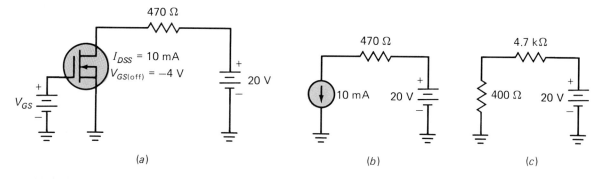

(a)   (b)   (c)

FIGURE 13-16   Example.

In Fig. 13-16c, the MOSFET acts like a resistance of 400 Ω. The total resistance in the drain circuit is the sum of 400 Ω and 4.7 kΩ. Therefore, the drain-source voltage is

$$V_{DS} = \frac{20\text{ V}}{5.1\text{ k}\Omega}(400\text{ }\Omega) = 1.57\text{ V}$$

EXAMPLE 13-10

In Fig. 13-16a, what is the drain-source voltage when $V_{GS} = +1$ V?

SOLUTION

Assume the MOSFET is operating as a current source. First, get the K factor and the drain current as follows:

$$K = \left(1 - \frac{+1\text{ V}}{-4\text{ V}}\right)^2 = 1.25^2 = 1.56$$

and

$$I_D = 1.56(10\text{ mA}) = 15.6\text{ mA}$$

Second, the drain-source voltage is

$$V_{DS} = 20\text{ V} - (15.6\text{ mA})(470\text{ }\Omega) = 12.7\text{ V}$$

Third, calculate the proportional pinchoff voltage:

$$V_P' = (15.6\text{ mA})(400\text{ }\Omega) = 6.24\text{ V}$$

Since V_{DS} is greater than V_P', the MOSFET is operating as a current source.

When you compare the foregoing MOSFET examples with the JFET examples given earlier, you can see that the analysis is identical in its approach. Each device acts as a current source or a resistor. You start the analysis by assuming current-source operation. If you get a contradiction, you know the device is acting as a resistor instead of as a current source. Then you use Ohm's law to calculate V_{DS}. When V_{GS} is zero, the pinchoff voltage separates the ohmic and the current-source regions. When V_{GS} does not equal zero, the proportional pinchoff voltage separates the two regions.

13-7 THE ENHANCEMENT-MODE MOSFET

Although the depletion-mode MOSFET is useful in special situations, it does not enjoy widespread use for reasons given in the next chapter. But it played an important role in history because it was part of the evolution toward the *enhancement-mode MOSFET*, a device that has revolutionized the electronics industry. This second type of MOSFET has become enormously important in digital electronics and computers. Without it, the personal computers that are now so widespread would not exist.

The Basic Idea

Figure 13-17*a* shows an *n*-channel enhancement-type MOSFET. The substrate extends all the way to the silicon dioxide. As you see, there no longer is an *n* channel between the source and the drain.

How does it work? Figure 13-17*b* shows normal biasing polarities. When the gate voltage is zero, the V_{DD} supply tries to force free electrons from source to drain, but the *p* substrate has only a few thermally produced free electrons. Aside from these minority carriers and some surface leakage, the current between source and drain is zero. For this reason, an enhancement-mode MOSFET is normally off when the gate voltage is zero. This is completely different from depletion-mode devices like the JFET or the depletion-mode MOSFET.

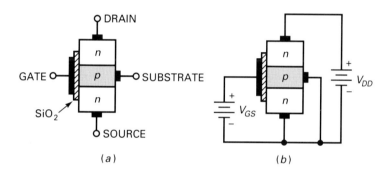

(a)   (b)

FIGURE 13-17   Enhancement-mode MOSFET.

When the gate is positive enough, it attracts free electrons into the *p* region. The free electrons recombine with the holes next to the silicon dioxide. When the gate voltage is positive enough, all the holes touching the silicon dioxide are filled and free electrons begin to flow from the source to the drain. The effect is the same as creating a thin layer of *n*-type material next to the silicon dioxide. This conducting layer is called the *n-type inversion layer*. When it exists, the normally off device suddenly turns on and free electrons flow easily from the source to the drain.

The minimum V_{GS} that creates the *n*-type inversion layer is called the *threshold voltage*, symbolized $V_{GS(th)}$. When V_{GS} is less than $V_{GS(th)}$, the drain current is zero. But when V_{GS} is greater than $V_{GS(th)}$, an *n*-type inversion layer connects the source to the drain and the drain current is large. Depending on the particular device being used, $V_{GS(th)}$ can vary from less than 1 to more than 5 V.

JFETs and depletion-mode MOSFETs are classified as depletion-mode devices because their conductivity depends on the action of depletion layers. The enhancement-mode MOSFET is classified as an enhancement-mode device because its conductivity depends on the action of the *n*-type inversion layer. Depletion-mode devices are normally on when the gate voltage is zero, whereas enhancement-mode devices are normally off when the gate voltage is zero.

Graphs and Formulas

Figure 13-18*a* shows a set of drain curves for an enhancement-mode MOSFET and a typical load line. The lowest curve is the $V_{GS(th)}$ curve. When V_{GS} is less than $V_{GS(th)}$, the drain current is approximately zero. When V_{GS} is greater than $V_{GS(th)}$, the device turns on and the drain current is controlled by the gate voltage. Again, notice the almost-vertical and almost-horizontal parts of the curves. The almost vertical part corresponds to the ohmic region, and the almost horizontal parts correspond to the current-source region. The enhancement-mode MOSFET can operate in either of these regions. In other words, it can act as a current source or as a resistor.

Figure 13-18*b* shows a typical transconductance curve. Again, the curve is parabolic or square-law. The vertex (starting point) of the parabola is at $V_{GS(th)}$. Because of this, the equation for the parabola is different from before. It now equals

$$I_D = k(V_{GS} - V_{GS(th)})^2 \tag{13-13}$$

where k is a constant that depends on the particular MOSFET. Any data sheet for an enhancement-mode MOSFET will include the current, $I_{D(on)}$, and the voltage, $V_{GS(on)}$, for one point well above the threshold as shown in Fig. 13-18*b*.

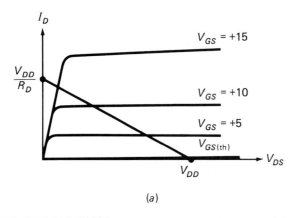

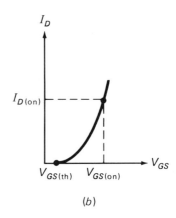

(a)   (b)

FIGURE 13-18   (*a*) Drain curves; (*b*) transconductance curves.

With JFETs and depletion-mode MOSFETs, the values of I_{DSS} and $V_{GS(off)}$ are the key quantities needed for analysis. With enhancement-mode MOSFETs the key quantities are $I_{D(on)}$, $V_{GS(th)}$, and $V_{GS(on)}$, shown in Fig. 13-18*b*. These three quantities are the first items to look for on a data sheet. By substituting these quantities into Eq. (13-13), we can rearrange the equation in a more useful form:

$$I_D = KI_{D(on)} \tag{13-14}$$

where

$$K = \left(\frac{V_{GS} - V_{GS(th)}}{V_{GS(on)} - V_{GS(th)}}\right)^2 \tag{13-15}$$

This expression appears formidable at first, but it is easy to work with after you get used to it.

Schematic Symbol

When $V_{GS} = 0$, the enhancement-mode MOSFET is off because there is no conducting channel between source and drain. The schematic symbol of Fig. 13-19a has a broken channel line to indicate this normally off condition. As you know, a gate voltage greater than the threshold voltage creates an n-type inversion layer that connects the source to the drain. The arrow points to this inversion layer, which acts like an n channel when the device is conducting. There is also a p-channel enhancement-mode MOSFET. The schematic symbol is similar, except that the arrow points outward, as shown in Fig. 13-19b.

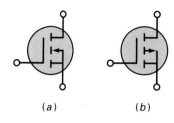

FIGURE 13-19

Schematic symbols; (a) n-channel; (b) p-channel.

Maximum Gate-Source Voltage

MOSFETs have a thin layer of silicon dioxide, an insulator that prevents gate current for positive as well as negative gate voltages. This insulating layer is kept as thin as possible to give the gate more control over the drain current. Because the insulating layer is so thin, it is easily destroyed by excessive gate-source voltage. For instance, a 2N3796 has a $V_{GS(\text{max})}$ rating of ± 30 V. If the gate-source voltage becomes more positive than $+30$ V or more negative than -30 V, the thin insulating layer will be destroyed.

Aside from directly applying an excessive V_{GS}, you can destroy the thin insulating layer in more subtle ways. If you remove or insert a MOSFET into a circuit while the power is on, transient voltages caused by inductive kickback and other effects may exceed $V_{GS(\text{max})}$ rating. This will wipe out the MOSFET. Even picking up a MOSFET may deposit enough static charge to exceed the $V_{GS(\text{max})}$ rating. This is the reason why MOSFETs are often shipped with a wire ring around the leads. You remove the ring after the MOSFET is connected in the circuit.

Some MOSFETs are protected by built-in zener diodes in parallel with the gate and the source. The zener voltage is less than the $V_{GS(\text{max})}$ rating. Therefore, the zener diode breaks down before any damage to the thin insulating layer occurs. The disadvantage of these internal zener diodes is that they reduce the MOSFET's high input resistance. The tradeoff is worth it in some applications because expensive MOSFETs are easily destroyed without zener protection.

Remember this idea: MOSFET devices are delicate and can be easily destroyed. You have to handle them carefully. Furthermore, you should never connect or disconnect them while the power is on. Finally, before you pick up a MOSFET device, you should ground your body by touching the chassis of equipment you are working on.

Equivalent Circuits

Figure 13-20 shows ideal drain curves for an enhancement-mode MOSFET. First, there is no breakdown region. Second, all drain curves superimpose in the ohmic region to produce a single, almost vertical line. Third, all drain curves are horizontal in the current-source region. These ideal drain curves are similar to depletion-mode curves, except for the proportional knee voltage V'_K. This voltage is given by

$$V'_K = I_D R_{DS} \qquad (13\text{-}16)$$

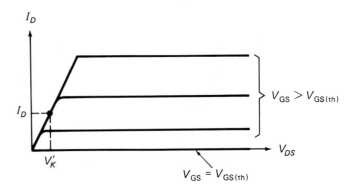

FIGURE 13-20 Ideal drain curves.

This voltage is the border between the ohmic region and the current-source region in an ideal enhancement-mode device. The border concept is identical to V_P'. The reason for not using V_P' is because enhancement-mode MOSFETs don't have a pinchoff voltage where depletion layers come together. Instead, they have an inversion layer. Because a different physical mechanism is involved, we use the symbol V_K' for the border between the two regions.

Figure 13-21 shows the two ideal equivalent circuits. As you see, these equivalent circuits are the same as for a JFET, except for $I_{D(on)}$ and the positive gate voltage. In other words, the enhancement-mode MOSFET can act like a current source or like a resistor. Which of these you use depends on where the operating point is. The proportional knee voltage is your guide. When V_{DS} is greater than V_K', the device is a current source. When V_{DS} is less than V_K', the device is a resistor. Here is the process for deciding which model to use:

1. Calculate V_K'.
2. If $V_{DS} > V_K'$, use a current-source equivalent circuit.
3. If $V_{DS} < V_K'$, use an ohmic equivalent circuit.

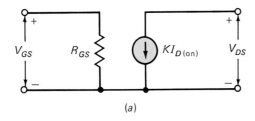

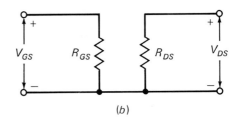

(a) (b)

FIGURE 13-21 Ideal equivalent circuits.

The following examples will show you how to use the different ideas presented in this section.

EXAMPLE 13-11

In Fig. 13-22, what is the drain-source voltage when V_{GS} is zero?

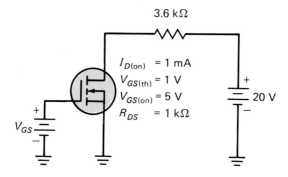

$I_{D(on)} = 1$ mA
$V_{GS(th)} = 1$ V
$V_{GS(on)} = 5$ V
$R_{DS} = 1$ kΩ

3.6 kΩ

20 V

V_{GS}

FIGURE 13-22   Example.

SOLUTION

The gate voltage is less than the threshold voltage. Therefore, the MOSFET is not conducting. In this case, the drain current is zero, and the drain-source voltage rises to the drain supply voltage:

$$V_{DS} = 20 \text{ V}$$

EXAMPLE 13-12

In Fig. 13-22, what is the drain-source voltage when V_{GS} is 5 V?

SOLUTION

Assume the MOSFET acts as a current source. Since the gate voltage equals the on voltage, the drain current is 1 mA. Therefore, the drain-source voltage is

$$V_{DS} = 20 \text{ V} - (1 \text{ mA})(3.6 \text{ k}\Omega) = 16.4 \text{ V}$$

The proportional knee voltage is

$$V'_K = (1 \text{ mA})(1 \text{ k}\Omega) = 1 \text{ V}$$

Since V_{DS} is greater than V'_K, the assumption about current-source operation is correct.

EXAMPLE 13-13

In Fig. 13-22, the drain resistor increases to 36 kΩ. What is the drain-source voltage when V_{GS} is 5 V?

SOLUTION

Assume the MOSFET acts as a current source. Since the gate voltage is +5 V, the drain current is 1 mA. Therefore, the drain-source voltage is

$$V_{DS} = 20 \text{ V} - (1 \text{ mA})(36 \text{ k}\Omega) = -16 \text{ V}$$

Impossible! The drain voltage cannot be negative. We have an absurd result, which means the assumption about the current source is incorrect. The MOSFET cannot be operating in the current-source region. It must be operating in the ohmic region.

The MOSFET acts as a resistance of 1 kΩ. The total resistance in the drain circuit is the sum of 1 and 36 kΩ. Therefore, we can calculate the drain-source voltage like this:

$$V_{DS} = \frac{20\ \text{V}}{1\ \text{k}\Omega + 36\ \text{k}\Omega}(1\ \text{k}\Omega) = 0.54\ \text{V}$$

EXAMPLE 13-14

In Fig. 13-22, what is the drain-source voltage when $V_{GS} = 3$ V?

SOLUTION

Assume the MOSFET is operating as a current source. First, get the K factor by substituting the given quantities into Eq. (13-15):

$$K = \left(\frac{3\ \text{V} - 1\ \text{V}}{5\ \text{V} - 1\ \text{V}}\right)^2 = 0.5^2 = 0.25$$

and

$$I_D = 0.25(1\ \text{mA}) = 0.25\ \text{mA}$$

Second, the drain-source voltage is

$$V_{DS} = 20\ \text{V} - (0.25\ \text{mA})(3.6\ \text{k}\Omega) = 19.1\ \text{V}$$

EXAMPLE 13-15

Repeat Example 13-14 for a gate voltage of 8 V.

SOLUTION

Assume the MOSFET is operating as a current source. First, get the K factor by substituting the given quantities into Eq. (13-15):

$$K = \left(\frac{8\ \text{V} - 1\ \text{V}}{5\ \text{V} - 1\ \text{V}}\right)^2 = 1.75^2 = 3.06$$

and

$$I_D = 3.06(1\ \text{mA}) = 3.06\ \text{mA}$$

Second, the drain-source voltage is

$$V_{DS} = 20\ \text{V} - (3.06\ \text{mA})(3.6\ \text{k}\Omega) = 8.98\ \text{V}$$

The proportional knee voltage is

$$V'_K = (3.06\ \text{mA})(1\ \text{k}\Omega) = 3.06\ \text{V}$$

Since V_{DS} is greater than V'_K, the assumption about current-source operation is correct.

13-8 READING DATA SHEETS

FET data sheets are similar to bipolar data sheets. You will find maximum ratings, dc characteristics, ac characteristics, mechanical data, etc. As usual, a good place to start is with the maximum ratings because these are the limits on the FET currents, voltages, and other quantities.

Breakdown Ratings

The data sheet of the MPF102 describes the device as an n-channel JFET with these maximum ratings:

V_{DS}	25 V
V_{DG}	25 V
V_{GS}	-25 V
I_G	10 mA
P_D	200 mW
T_J	125°C

The voltage ratings are reverse breakdown voltages, and V_{DS} is the voltage between the drain and the source. The second rating is V_{DG}, which is the voltage from the drain to the gate, and V_{GS} is the voltage from the gate to the source. As usual, a conservative design includes a safety factor. Notice that the maximum forward gate current is rated at 10 mA. Normally, the gate is reverse-biased. The data sheet includes this forward rating in case the gate is forward-biased for any reason. There is no reason for forward-biasing the gate, unless it is a rare and unusual application.

As discussed in Chap. 5, the derating factor tells you how much to reduce the power rating of a device. The derating factor of an MPF102 is given as 2 mW/°C. This means that you have to reduce the power rating of 200 mW by 2 mW for each degree above 25°C.

I_{DSS} and $V_{GS(off)}$

Two of the most important pieces of information on the data sheet of a depletion-mode device are the maximum drain current and the gate-source cutoff voltage. These values are given on the data sheet of an MPF102:

Symbol	Minimum	Maximum
$V_{GS(off)}$	—	-8 V
I_{DSS}	2 mA	20 mA

We have already discussed the 10:1 spread in I_{DSS}. This large spread was one of the reasons for using JFET approximations. Another reason for approximating is this: Data sheets often omit values, so you really have no idea what some values may be. In the case of the MPF102, the minimum value of $V_{GS(off)}$ is not given. This makes the case for approximation even stronger.

As you know, the value of I_{DSS} represents the maximum drain current with a JFET, and $V_{GS(off)}$ represents the gate-source voltage needed to turn off the drain current. Also important, the pinchoff voltage V_P has the same magnitude as $V_{GS(\text{off})}$. And finally the ratio of V_P to I_{DSS} gives the R_{DS} of the JFET in the ohmic region.

Enhancement-Mode Data Sheets

Depletion-mode devices have two key quantities: I_{DSS} and $V_{GS(\text{off})}$. Enhancement-mode devices have three key quantities: $I_{D(\text{on})}$, $V_{GS(\text{th})}$, and $V_{GS(\text{on})}$. Data sheets vary from one manufacturer to the next, so you sometimes cannot find all the quantities you would like to see. But almost all data sheets will give $I_{D(\text{on})}$ and $V_{GS(\text{th})}$, and most will include $V_{GS(\text{on})}$. Some also include information on the drain resistance R_{DS}. This resistance may be symbolized R_{DS}, or it may appear as $r_{DS(\text{on})}$, $r_{ds(\text{on})}$, etc. with different conditions attached to the listed value.

For instance, here is how the data sheet of an M116 lists the drain resistance:

$$r_{DS(\text{on})} = 100 \ \Omega \qquad \text{for } V_{GS} = 20 \text{ V and } I_D = 100 \ \mu\text{A}$$
$$r_{DS(\text{on})} = 200 \ \Omega \qquad \text{for } V_{GS} = 10 \text{ V and } I_D = 100 \ \mu\text{A}$$

Sometimes, the data sheet does not specify values of drain resistance. For instance, the data sheet of a 2N4351 gives

$$V_{DS(\text{on})} = 1 \text{ V} \qquad \text{for } V_{GS} = 10 \text{ V and } I_D = 2 \text{ mA}$$

In this case, you can calculate the drain resistance as follows:

$$R_{DS} = \frac{1 \text{ V}}{2 \text{ mA}} = 500 \ \Omega$$

The lack of standardization in the data sheets means different symbols are often used for the same quantity. Also, data sheets may specify a particular quantity measured under different conditions. The world is a messy place, and data sheets prove it. But there is some relief as follows. You rarely need exact answers, so exact data are unnecessary. All you need most of the time are typical and worst characteristics.

❑ OPTIONAL TOPICS

The following material continues the earlier discussions at a more advanced and specialized level. All the topics are optional because they are not used in any of the basic discussions in later chapters. This section will be a useful reference when you are in industry because then you will probably want more advanced viewpoints.

❑ 13-9 ANOTHER KIND OF SATURATION

For whatever reason, some of the terms for bipolar devices are used in a different way for FET devices. For instance, the four major operating regions of a bipolar transistor are the saturation region, active region, cutoff region, and breakdown region. It would have been nice if the same terms had been carried over to FET devices, but unfortunately this did not happen. Instead, here are the four regions and the terms used to describe them:

Bipolar	FET
Saturation	Ohmic
Active	Saturation or pinchoff
Cutoff	Cutoff
Breakdown	Breakdown

In the basic discussion for FETs, the second operating region was called the *current-source region* to avoid the confusion shown here. As you see, the second JFET operating region is formally called the *saturation region*. This name is used because the channel is physically conducting all the current it can. This region is also called the *pinchoff region* because the two depletion layers are trying to pinch off the current, which holds the current at a constant value.

It's unfortunate that the term *saturation* is used in different ways, regardless of how it came about. This material is hard enough without language adding to the confusion. It makes you want to throw your hands in the air and ask, Why did they do it that way? But it is a good example of what you are up against in the real world. It's messy out there, far from ideal. Contradictions are everywhere. Identical words are used in different ways, information is left off the data sheets, documentation has big gaps in it, etc. It is a very confusing, but very real world. This lack of consistency and exactness is too much for the left brain to handle. This half of the brain wants everything to be neat and in its proper place. It wants everything to make sense. It wants exact answers. But you can see how unrealistic such expectations are.

The only hope for success in such a messy world is to use your right brain as much as your left. A good way to do this is to accept the idea that there are many right answers and many ways to find them. This will open the channel to your right brain and let you think in ways you never thought possible. Then when you run into information gaps, omitted data, confusing terms, etc., you will have a much better chance to figure out what's going on. In fact, once you know how to use your right brain, you will find the real world full of opportunities precisely because it is such a mess. Creators, inventors, and whole-brain people love it.

❑ 13-10 MATHEMATICAL DERIVATION

In the discussion of the enhancement-mode MOSFET, the basic equation of drain current was given as

$$I_D = k(V_{GS} - V_{GS(\text{th})})^2 \qquad (13\text{-}17)$$

The derivation of this basic formula is given in engineering textbooks on FETs. Here we want to show how this equation is rearranged into the more useful form

$$I_D = KI_{D(\text{on})} \qquad (13\text{-}18)$$

where

$$K = \left(\frac{V_{GS} - V_{GS\text{(th)}}}{V_{GS\text{(on)}} - V_{GS\text{(th)}}}\right)^2 \qquad (13\text{-}19)$$

To begin, substitute $I_{D\text{(on)}}$ and $V_{GS\text{(on)}}$ into Eq. (13-17) to get

$$I_{D\text{(on)}} = k(V_{GS\text{(on)}} - V_{GS\text{(th)}})^2$$

Solve for k to get

$$k = \frac{I_{D\text{(on)}}}{(V_{GS\text{(on)}} - V_{GS\text{(th)}})^2}$$

Substitute this k into Eq. (13-17) to get

$$I_D = \frac{I_{D\text{(on)}}}{(V_{GS\text{(on)}} - V_{GS\text{(th)}})^2}(V_{GS} - V_{GS\text{(th)}})^2$$

Now, define

$$K = \left(\frac{V_{GS} - V_{GS\text{(th)}}}{V_{GS\text{(on)}} - V_{GS\text{(th)}}}\right)^2$$

which means

$$I_D = KI_{D\text{(on)}}$$

❏ 13-11 ANOTHER LOOK AT THE DRAIN CURVES

A word of caution. To get simple and useful equivalent circuits, we idealized the drain curves of JFETs as shown in Fig. 13-7 on page 455. Above proportional pinchoff, each drain curve is horizontal. Also, the ohmic region is a single, almost vertical line. This ideal set of drain curves allowed us to show the two equivalent circuits of Fig. 13-8 on page 455. We continued to use the ideal drain curves and similar equivalent circuits for the depletion-mode MOSFET and the enhancement-mode MOSFET. The ideal curves and equivalent circuits are fine for troubleshooting and most preliminary analysis. But they will be inadequate in some situations.

The real truth is this: The drain curves look more like Fig. 13-23a. The higher the curve, the less horizontal it appears. Furthermore, the

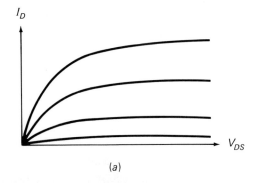

(a)

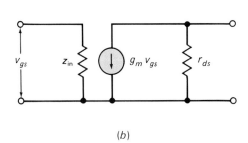

(b)

FIGURE 13-23   Drain curves.

ohmic region is not a single, almost vertical line, but rather many vertical lines of different slopes. So the analysis of a FET can be a lot more complicated if we want to make it so. In most applications, we don't have to take these higher-order effects into account because of the huge tolerances in FET devices. But sometimes a designer does want to look more closely at the real drain curves of a FET.

One of the quantities implied in the more accurate drain curves is the ac drain resistance, which is defined as

$$r_{ds} = \frac{\Delta V_{DS}}{\Delta I_D} \qquad (13\text{-}20)$$

for a constant V_{GS}. This is also written as

$$r_{ds} = \frac{dV_{DS}}{dI_D}$$

where r_{ds} is the derivative of the drain curve at the operating point. Look at the drain curves of Fig. 13-23a and notice that r_{ds} has small values near the origin and large values on the almost horizontal part of the curves.

The drain curves are more horizontal above the knee, which implies a larger value of r_{ds}. This is equivalent to saying the FET acts more as a current source above the knee. To bring out this point, Fig. 13-23b shows the ac equivalent circuit of a FET. When the drain curves are horizontal, r_{ds} approaches infinity and the output side of the circuit becomes a stiff current source. Everything depends on the size of the load resistor. If it is small compared to r_{ds}, then the circuit acts as a stiff current source. But if the load resistance is comparable to r_{ds}, then an accurate analysis will need to include the effects of r_{ds} in parallel with the load resistor.

Below the well-rounded knee of each drain curve, the FET acts more as a resistor than a current source. But notice that r_{ds} depends on which drain curve is being used. This implies that the gate voltage determines the r_{ds} in the ohmic region. For this reason, the drain resistance of a FET can be controlled by the gate voltage. This leads to several interesting applications.

STUDY AIDS

The following study aids will help to reinforce the ideas discussed in this chapter. For best results, use these study aids within 6 hours of reading the earlier material. Then review these study aids a week later and a month later to ensure that the concepts remain in your long-term memory.

SUMMARY

Sec. 13-1 The JFET

The junction FET, abbreviated JFET, has a source, gate, and drain analogous to the emitter, base, and collector of a bipolar transistor. The JFET has two built-in diodes: the gate-source diode and the gate-drain diode. These diodes will conduct if they are forward-biased with more than 0.7 V.

Sec. 13-2 The Biased JFET

Unlike a bipolar transistor where the base-emitter diode is forward-biased, the gate-source diode of a JFET is always reverse-biased. Similarly, the gate-drain diode is reverse-biased for normal operation. The JFET has a dc input resistance that approaches infinity, but it has less voltage gain than a bipolar transistor.

Sec. 13-3 Drain Curves

The drain curves of a JFET are similar to those of a bipolar transistor, except that V_{GS} is the controlling input rather than I_B. The JFET acts as a current source along the horizontal part of the drain curves and as a resistor along the almost-vertical parts of the drain curve. The maximum drain current is symbolized I_{DSS}, and the gate-source cutoff voltage is symbolized $V_{GS(off)}$. The pinchoff voltage has the same magnitude as $V_{GS(off)}$.

Sec. 13-4 The Transconductance Curve

This is the graph of drain current versus gate voltage. The curve is nonlinear, part of a parabola, and also called a square-law curve. The square-law property gives the JFET a major advantage in communications circuits called mixers.

Sec. 13-5 JFET Approximations

The ideal JFET has drain curves that consist of an ohmic region and several horizontal regions. When the operating point is along one of the horizontal lines, the JFET acts as a current source. When the operating point is in the ohmic region, the JFET acts as a resistor. The proportional pinchoff voltage is the border between the ohmic region and the current-source region.

Sec. 13-6 The Depletion-Mode MOSFET

The metal-oxide semiconductor FET, abbreviated MOSFET, has a source, gate, and drain. The gate is electrically insulated from the channel. Because of this, the dc input resistance is even higher than that of a JFET. The depletion-mode MOSFET is normally on when V_{GS} is zero. It has drain curves and equivalent circuits similar to the JFET. The only difference is that it can operate with positive as well as negative gate voltages.

Sec. 13-7 The Enhancement-Mode MOSFET

The enhancement-mode MOSFET has revolutionized the electronics industry, specifically in the area of computers. This device is normally off when gate voltage is zero. To turn it on, you have to apply a positive enough gate voltage. The gate voltage that does this is called the threshold voltage. The device acts as a current source or as a resistor.

Sec. 13-8 Reading Data Sheets

Data sheets list the maximum ratings for voltages, currents, and powers. Besides these absolute maximum ratings, the important quantities for depletion-mode devices are I_{DSS} and $V_{GS(off)}$. The important quantities for enhancement-mode devices are $I_{D(on)}$, $V_{GS(th)}$, and $V_{GS(on)}$.

VOCABULARY

In your own words, explain what each of the following means. Keep your answers short and to the point. If necessary, verify your answer by rereading the appropriate discussion or by looking at the end-of-book Glossary.

analogy	ohmic region
bipolar transistor	pinchoff voltage
depletion-mode MOSFET	proportional pinchoff voltage
enhancement-mode MOSFET	reductio ad absurdum
field-effect transistor	threshold voltage
gate-source cutoff voltage	voltage-controlled device

IMPORTANT EQUATIONS

Here are some important equations. Say each of the following equations in symbols, then say each in words. Try to explain what the equation means and how it is used. Then read the description that follows.

Eq. 13-1 Gate Cutoff and Pinchoff

$$V_{GS(off)} = -V_P$$

This relates two important JFET voltages. It says the two voltages have the same magnitude. They differ only in sign.

Eq. 13-2 Drain-Source Resistance

$$R_{DS} = \frac{V_P}{I_{DSS}}$$

This is a very useful approximation. It says a JFET has a resistance of approximately V_P/I_{DSS} when it is operating in the ohmic region.

Eqs. 13-4 and 13-5 Drain Current as a Function of Gate Voltage

$$I_D = KI_{DSS}$$

where

$$K = \left(1 - \frac{V_{GS}}{V_{GS(\text{off})}}\right)^2$$

The drain current is a fraction of the maximum drain current. The K factor tells you what this fraction is. The K factor is always is a number between 0 and 1.

Eq. 13-6 Proportional Pinchoff

$$V_P' = I_D R_{DS}$$

After you have calculated R_{DS}, you can multiply by the drain current to get the proportional pinchoff voltage. Then V_P' is your clue as to which region the JFET is operating in. If V_{DS} is greater than V_P', the JFET acts as a current source; if V_{DS} is less than V_P', the JFET acts as a resistor.

Eqs. 13-14 and 13-15 Enhancement-Mode Drain Current

$$I_D = KI_{D(\text{on})}$$

where

$$K = \left(\frac{V_{GS} - V_{GS(\text{th})}}{V_{GS(\text{on})} - V_{GS(\text{th})}}\right)^2$$

To calculate the drain current of an enhancement-mode MOSFET, you need three quantities from the data sheet: $I_{D(\text{on})}$, $V_{GS(\text{th})}$, and $V_{GS(\text{on})}$. With these quantities, you can calculate the K factor for any value of V_{GS}. Then you can multiply the K factor by $I_{D(\text{on})}$ to get the drain current.

Incidentally, these formulas are valid only when V_{GS} is greater than $V_{GS(\text{th})}$.

IMPORTANT PROCESSES

Each of the following processes is important for FET circuit analysis. If you can remember the basic ideas in each step, you will be able to solve problems more easily.

Which JFET Model to Use

1. Calculate V_P'.
2. If $V_{DS} > V_P'$, use a current-source equivalent circuit.
3. If $V_{DS} < V_P'$, use an ohmic equivalent circuit.

Reductio ad Absurdum

1. Assume the current-source region.
2. Carry out your calculations.
3. If an absurd answer results, the assumption is false.
4. Change to the ohmic model.

Which Depletion-Mode MOSFET Model to Use

1. Calculate V_P'.
2. If $V_{DS} > V_P'$, use a current-source equivalent circuit.
3. If $V_{DS} < V_P'$, use an ohmic equivalent circuit.

Which Enhancement-Mode MOSFET Model to Use

1. Calculate V_K'.
2. If $V_{DS} > V_K'$, use a current-source equivalent circuit.
3. If $V_{DS} < V_K'$, use an ohmic equivalent circuit.

STUDENT ASSIGNMENTS

QUESTIONS

The following questions may refer to the figure you have seen in this chapter. Select the best answer the one that is in step with the approximations o this chapter, or that most accurately describes the situation, etc.

1. A JFET
   a. Is a voltage-controlled device
   b. Is a current-controlled device
   c. Has a low input resistance
   d. Has a very large voltage gain

2. A unipolar transistor uses
   a. Both free electrons and holes
   b. Only free electrons
   c. Only holes
   d. Either one or the other, but not both

3. The input impedance of a JFET
   a. Approaches zero
   b. Approaches one
   c. Approaches infinity
   d. Is impossible to predict

4. The gate controls
   a. The width of the channel
   b. The drain current
   c. The proportional pinchoff voltage
   d. All the above

5. The gate-source diode of a JFET should be
   a. Forward-biased
   b. Reverse-biased
   c. Either forward- or reverse-biased
   d. None of the above

6. Compared to a bipolar transistor, the JFET has much more
   a. Voltage gain
   b. Input resistance
   c. Supply voltage
   d. Current

7. Compared to a bipolar transistor, the JFET has much less
   a. Voltage gain
   b. Input resistance
   c. Supply voltage
   d. Current

8. The carriers in a p-channel JFET are
   a. Free electrons
   b. Holes
   c. Either free electrons or holes
   d. Both free electrons and holes

9. The pinchoff voltage has the same magnitude as the
   a. Gate voltage
   b. Drain-source voltage
   c. Gate-source voltage
   d. Gate-source cutoff voltage

10. When the drain-source voltage is less than the proportional pinchoff voltage, a depletion-mode device acts as a
    a. Bipolar transistor
    b. Current source
    c. Resistor
    d. Battery

11. The ideal drain resistance in the ohmic region of a depletion-mode device equals the pinchoff voltage divided by the
    a. Drain current
    b. Gate current
    c. Ideal drain current
    d. Drain current for zero gate voltage

12. The transconductance curve is
    a. Linear
    b. Similar to the graph of a resistor
    c. Nonlinear
    d. Like a single drain curve

13. If an absurd answer occurs after the current-source region is assumed, a conducting FET must be operating in the
    a. Breakdown region
    b. Cutoff region
    c. Ohmic region
    d. Enhancement region

14. The depletion-mode MOSFET acts mostly as
    a. A JFET
    b. A current source
    c. A resistor
    d. An enhancement-mode MOSFET

15. Which of the following devices revolutionized the computer industry?
    a. JFET
    b. Depletion-mode MOSFET
    c. Enhancement-mode MOSFET
    d. Zener diode

16. The voltage that turns on an enhancement-mode device is the
    a. Gate-source cutoff voltage
    b. Pinchoff voltage
    c. Threshold voltage
    d. Knee voltage

17. Which of these will you find on the data sheet of a JFET?
    a. $V_{GS(\text{th})}$ c. $V_{GS(\text{on})}$
    b. I_{DSS} d. $V_{S(\text{on})}$

18. Which of these may appear on the data sheet of an enhancement-mode MOSFET?
- **a.** $V_{GS(th)}$
- **b.** $I_{D(on)}$
- **c.** $V_{GS(on)}$
- **d.** All the above

19. When a JFET is cut off, the depletion layers are
- **a.** Far apart
- **b.** Close together
- **c.** Touching
- **d.** Conducting

20. When the gate voltage becomes more negative in an n-channel JFET, the channel between the depletion layers becomes
- **a.** Narrower
- **b.** Wider
- **c.** Conducting
- **d.** Nonconducting

21. If a JFET has $I_{DSS} = 10$ mA and $V_P = 2$ V, then R_{DS} equals
- **a.** 200 Ω
- **b.** 400 Ω
- **c.** 1 kΩ
- **d.** 5 kΩ

22. The $V_{GS(on)}$ of an n-channel enhancement-mode MOSFET is
- **a.** Less than the threshold voltage
- **b.** Equal to the gate-source cutoff voltage
- **c.** Greater than $V_{DS(on)}$
- **d.** Greater than $V_{GS(th)}$

23. The V'_K plays the same role for enhancement-mode devices as which of the following does for depletion-mode devices?
- **a.** V'_{GS}
- **b.** V'_P
- **c.** V'_{DS}
- **d.** $V_{GS(th)}$

24. The real world is a place where
- **a.** There are no contradictions
- **b.** A formula can be found to solve any problem
- **c.** Every problem has only one right answer
- **d.** Many right answers can be found to make it less messy

BASIC PROBLEMS

Sec. 13-2 The Biased JFET

13-1. The data sheet of a 2N5902 lists the following quantities at room temperature: $I_{GSS} = 5$ pA at $V_{GS} = 20$ V. What is the dc input resistance of the gate at room temperature?

13-2. The data sheet of a 2N5902 lists the following quantities at 125°C: $I_{GSS} = 10$ nA at $V_{GS} = 20$ V. What is the dc input resistance of the gate at 125°C?

Sec. 13-3 Drain Curves

13-3. A JFET has $I_{DSS} = 16$ mA and $V_P = 3$ V. What is the minimum drain current? The maximum drain current? The gate-source cutoff voltage?

13-4. A 2N5902 has $I_{DSS} = 500$ μA and $V_{GS(off)} = -2$ V. What is the pinchoff voltage for this JFET? What is the drain resistance R_{DS} in the ohmic region?

13-5. The data sheet of a U311 lists these minimum values: $I_{DSS} = 20$ mA and $V_{GS(off)} = -1$ V. What does R_{DS} equal?

13-6. The data sheet of a U311 lists these maximum values: $I_{DSS} = 60$ mA and $V_{GS(off)} = -6$ V. What does R_{DS} equal?

Sec. 13-4 The Transconductance Curve

13-7. A 2N5457 has these minimum values listed on its data sheet: $I_{DSS} = 1$ mA and $V_{GS(off)} = -0.5$ V. Calculate the K factor for V_{GS} values of 0, -0.1, -0.2, -0.3, -0.4, and -0.5 V. Then calculate the drain currents for each gate voltage.

13-8. A 2N5457 has these maximum values listed on its data sheet: $I_{DSS} = 5$ mA and $V_{GS(off)} = -6$ V. Calculate the K factor for V_{GS} values of 0, -1, -2, -3, -4, -5, and -6 V. Then calculate the drain currents for each gate voltage.

Sec. 13-5 JFET Approximations

13-9. What is the drain-source voltage in Fig. 13-24a?

13-10. If the gate voltage is changed to -1 V in Fig. 13-24a, what is the drain-source voltage?

13-11. The gate voltage is increased to 0 V, and the drain supply voltage is increased to 20 V in Fig. 13-24a. What is the drain-source voltage?

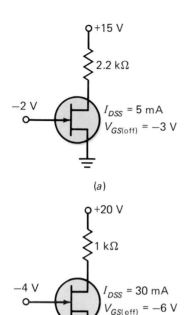

(a)

(b)

FIGURE 13-24

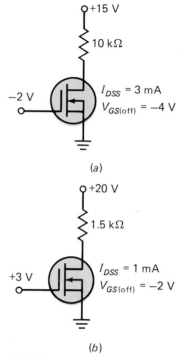

(a)

(b)

FIGURE 13-25

13-12. If the drain resistor is changed to 22 kΩ in Fig. 13-24a, what is the drain-source voltage?

13-13. What is the drain-source voltage in Fig. 13-24b?

13-14. If the gate voltage is changed to −1 V in Fig. 13-24b, what is the drain-source voltage?

13-15. The gate voltage is increased to 0 V, and the drain supply voltage is decreased to 15 V in Fig. 13-24b. What is the drain-source voltage?

13-16. If the drain resistor is changed to 10 kΩ in Fig. 13-24b, what is the drain-source voltage?

Sec. 13-6 The Depletion-Mode MOSFET

13-17. What is the drain-source voltage in Fig. 13-25a?

13-18. If the gate voltage is changed to −1 V in Fig. 13-25a, what is the drain-source voltage?

13-19. The gate voltage is increased to 0 V, and the drain supply voltage is increased to

20 V in Fig. 13-25a. What is the drain-source voltage?

13-20. If the drain resistor is changed to 22 kΩ in Fig. 13-25a, what is the drain-source voltage?

13-21. What is the drain-source voltage in Fig. 13-25b?

13-22. If the gate voltage is changed to +10 V in Fig. 13-25b, what is the drain-source voltage?

13-23. The gate voltage is decreased to 0 V, and the drain supply voltage is decreased to 15 V in Fig. 13-25b. What is the drain-source voltage?

13-24. If the drain resistor is changed to 10 kΩ in Fig. 13-25b, what is the drain-source voltage?

Sec. 13-7 The Enhancement-Mode MOSFET

13-25. What is the drain-source voltage in Fig. 13-26a?

13-26. If the gate voltage is changed to +5 V in Fig. 13-26a, what is the drain-source voltage?

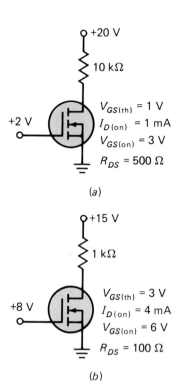

$V_{GS(th)} = 1$ V
$I_{D(on)} = 1$ mA
$V_{GS(on)} = 3$ V

$R_{DS} = 500\ \Omega$

(a)

+15 V

1 kΩ

+8 V

$V_{GS(th)} = 3$ V
$I_{D(on)} = 4$ mA
$V_{GS(on)} = 6$ V

$R_{DS} = 100\ \Omega$

(b)

FIGURE 13-26

13-27. The gate voltage is decreased to 0 V, and the drain supply voltage is increased to 30 V in Fig. 13-26a. What is the drain-source voltage?

13-28. If the drain resistor is changed to 22 kΩ in Fig. 13-26a, what is the drain-source voltage?

13-29. What is the drain-source voltage in Fig. 13-26b?

13-30. If the gate voltage is changed to +15 V in Fig. 13-26b, what is the drain-source voltage?

13-31. The gate voltage is decreased to 0 V, and the drain supply voltage is decreased to 10 V in Fig. 13-26b. What is the drain-source voltage?

13-32. If the drain resistor is changed to 10 kΩ in Fig. 13-26b, what is the drain-source voltage?

Sec. 13-8 Reading Data Sheets

13-33. What is the dc input gate resistance of an MPF102 when the ambient tempera-

ture is 100°C and the gate-source voltage is −15 V?

13-34. What is the drain-source resistance of an MPF102 in the ohmic region? Since the data sheet has some missing values, do the best you can with this and state the conditions for your answer.

ADVANCED PROBLEMS

13-35. If a JFET has the drain curves of Fig. 13-27a, what does I_{DSS} equal? What is the maximum V_{DS} in the ohmic region? Over what voltage range of V_{DS} does the JFET act as a current source?

13-36. Write the transconductance equation for the JFET whose curve is shown in Fig. 13-27b. How much drain current is there when $V_{GS} = -4$ V? When $V_{GS} = -2$ V?

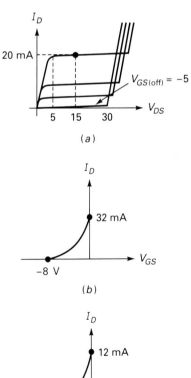

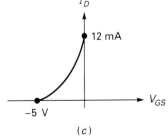

FIGURE 13-27

13-37. If a JFET has a square-law curve like Fig. 13-27c, how much drain current is there when $V_{GS} = -1$ V?

13-38. A MOSFET has the transconductance curve shown in Fig. 13-28a. What is the drain current when the gate voltage is -2 V?

13-39. The MOSFET of Fig. 13-28b has the transconductance curve shown in Fig. 13-28a. What is the dc voltage from the drain to ground?

13-40. Someone mistakenly wires Fig. 13-28b with a drain resistor of 10 kΩ. Given the transconductance curve of Fig. 13-28a, what is the drain-source voltage?

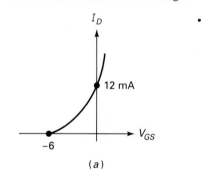

(a)

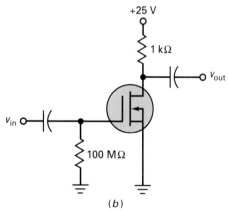

(b)

FIGURE 13-28

13-41. Figure 13-29a shows the transconductance curve for Fig. 13-29b. What is the gate voltage?

13-42. The drain resistor is increased to 2.2 kΩ in Fig. 13-29b. Given the transconductance curve of Fig. 13-29a, what is the gate voltage?

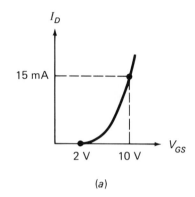

(a)

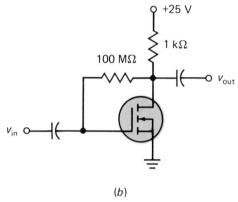

(b)

FIGURE 13-29

SOFTWARE ENGINE PROBLEMS

Use Fig. 13-30 for the remaining problems. Assume increases of approximately 10 percent in the independent variable, and assume current-source operation. A response should be an N (no change) if the change in a dependent variable is so small that you would have difficulty measuring it.

13-43. Try to predict the response of each dependent variable in the box labeled V_{DD}. Check your answers. Then answer the following as simply and directly as possible: What effect does an increase in V_{DD} have on the dependent variables of the circuit?

13-44. Predict the response of each dependent variable in the box labeled V_{GG}. Check your answers. Then summarize your findings in one or two sentences.

13-45. Predict the response of each dependent variable in the box labeled $V_{GS(\text{off})}$. Check your answers. List the dependent varia-

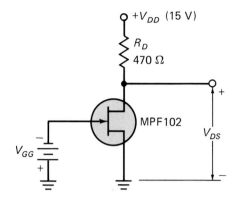

	1	2	3	4	5	6
A	U	N	D	U	D	U
B	N	D	U	U	D	D
C	D	N	D	D	N	U
D	N	U	D	U	D	N
E	U	D	U	U	U	D
F	N	N	U	D	N	N

V_{DD}	V_{GG}	$V_{GS(off)}$	I_{DSS}	I_{GSS}	R_D
I_D : F1	I_D : B2	I_D : A4	I_D : A4	I_D : F1	I_D : F2
V_{DS}: C5	V_{DS}: A6	V_{DS}: D5	V_{DS}: E6	V_{DS}: C5	V_{DS}: B5
R_{DS}: F2	R_{DS}: C5	R_{DS}: B3	R_{DS}: C1	R_{DS}: A2	R_{DS}: D1
R_{GS}: D6	R_{GS}: D1	R_{GS}: F1	R_{GS}: F2	R_{GS}: C3	R_{GS}: C5

FIGURE 13-30 Software Engine™. (*Patent pending: Courtesy of Malvino Inc.*)

bles that increase. Explain why these variables increase, using Ohm's law or similar basic ideas.

13-46. Predict the response of each dependent variable in the box labeled I_{DSS}. List the dependent variables that decrease. Explain why these variables decrease.

13-47. Predict the response of each dependent variable in the box labeled I_{GSS}. List

the dependent variables that decrease. Explain why these variables show a decrease.

13-48. Predict the response of each dependent variable in the box labeled R_D. List the dependent variables that decrease. Explain why these variables show a decrease.

14

FET CIRCUITS

In this chapter we discuss uses for JFETs, depletion-mode MOSFETs, and enhancement-mode MOSFETs. Here you will find out how to bias these devices and how to use them as amplifiers, analog switches, and digital switches. The main uses of a JFET are as a source follower (analogous to the emitter follower) and an analog switch (a circuit that transmits or blocks ac signals). The depletion-mode MOSFET is useful as a very high frequency (VHF) amplifier. The enhancement-mode MOSFET is primarily used as a digital switch, the backbone of computers. In this chapter we discuss some of the key ideas you need to understand how these FET devices are used in their most practical applications.

14-1 SELF-BIAS OF JFETs

You can bias a JFET in all kinds of ways. We will discuss them shortly. The important thing to remember is that the gate-source diode has to be reverse-biased. If you accidentally forward-bias the gate-source diode, the device loses its high input resistance. When this happens, the device loses its strength, its major advantage over the bipolar transistor. Since the gate-source diode has to be reverse-biased, some of the biasing methods used with bipolar transistors will work and others won't.

Bipolar and JFET Biasing

In the following list, those forms of bias that won't work are listed as "no equivalent." When a type of bias does work in both systems, it is renamed in terms of the JFET terminals if necessary. Here are the analogous biasing methods for both devices:

Bipolar transistor	JFET
Voltage-divider bias	Voltage-divider bias
Two-supply emitter bias	Two-supply source bias
Emitter-feedback bias	No equivalent
Collector-feedback bias	No equivalent
Base bias	Gate bias
No equivalent	Self-bias

484

Voltage-divider bias is the preferred biasing method for bipolar transistors used as linear amplifiers, but not with JFETs. Although a few JFET circuits have been designed with voltage-divider bias, it has not emerged as the leading type of JFET bias.

The next type of bipolar bias is two-supply emitter bias. When positive and negative supply voltages are available, most designers will use this type of bipolar bias because it produces a larger undistorted output from the final stage. The analogous form of JFET biasing, two-supply source bias, has been used successfully with JFETs. But again, it is not the bias you see used most often with JFET amplifiers.

Emitter-feedback bias and collector-feedback bias have no equivalent JFET bias because these types of bias can only forward-bias the gate-source diode. Since this is never done in JFET amplifiers, these two forms of unpopular bipolar bias are even more unpopular with JFETs. As indicated in the above list, no equivalent forms of emitter-feedback and collector-feedback bias exist for JFET amplifiers.

Recall that base bias is the worst way to bias a bipolar amplifier. But if the bipolar transistor is used in a switching circuit rather than a linear amplifier, base bias is a useful way to switch between cutoff and saturation. For switching applications of a bipolar transistor, we rely on hard saturation to overcome the large variations in β_{dc}. A similar idea applies to JFETs. Gate bias is the worst way to bias a JFET if you are using it as an amplifier because the Q point varies with the values of I_{DSS} and $V_{GS(off)}$. Since these quantities have huge variations with temperature and JFET replacement, it's impossible to get a stable Q point with gate bias. But if the JFET is used in a switching applications, then gate bias is useful because the Q point will be either at cutoff or in the ohmic region.

Finally, we come to *self-bias*. This type of bias has no equivalent in bipolar circuits. This is the preferred form of biasing a JFET amplifier. You will see this type of bias used more than any other JFET bias. Because of its simplicity, it offers an elegant and effective method of biasing a JFET amplifier. Although the Q point is not rock-solid, it is stable enough for most of the amplifier applications that use JFETs.

The Basic Idea

Figure 14-1a shows self-bias. Only a single supply voltage is needed, the drain supply. There is no gate supply. Since the gate is returned to ground through a resistor, the gate voltage is zero. Because current flows through the source resistor, there is a voltage across this resistor. The basic idea behind self-bias is to use the voltage across R_S to produce the required gate-source biasing voltage.

Let's use reductio ad absurdum to get started in the analysis of self-bias. Assume there is drain current. Then a source current flows through the source resistor, producing a voltage across R_S with the polarity shown in Fig. 14-1a. Because of this voltage, the source terminal is positive with respect to ground. Since the gate is grounded through the gate resistor, the source is positive with respect to the gate. This means the gate is negative with respect to the source.

For instance, suppose the voltage across R_S equals 2 V with the polarity shown in Fig. 14-1a. Then the source has a voltage of $+2$ V with respect

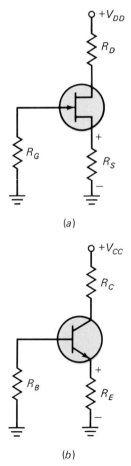

(a)

(b)

FIGURE 14-1

Self-bias.

to the gate. Conversely, the gate has a voltage of -2 V with respect to the source. In symbols,

$$V_G = 0 \text{ V}$$

$$V_S = +2 \text{ V}$$

Therefore

$$V_{GS} = V_G - V_S = -2 \text{ V}$$

Since the gate-source diode is reverse-biased, the JFET is correctly biased. In other words, we have no contradictions. Therefore, the original assumption about drain current is valid. The drain current can and does flow.

If you still don't get it, you will eventually. It takes a little time for the idea to sink in because self-bias is very clever. It is a slick invention, an example of what can be done when you start believing in many right answers. The inventor did not find this biasing method in any bipolar book because it does not exist in bipolar circuits. Self-bias cannot be used with bipolar transistors. Why not? Because it won't work. If you use a bipolar transistor as shown in Fig. 14-1b, here is how the analysis goes: Assume that collector current exists. Then the emitter and base current must also exist. The emitter current produces a voltage across the emitter resistor with the polarity shown. This makes the emitter positive with respect to the base. Conversely, it makes the base negative with respect to the emitter. But a negative V_{BE} would reverse-bias the base-emitter diode. In turn, it would mean the transistor is cut off. This is a contradiction because we originally assumed there was collector current. Therefore, the original assumption is incorrect. There can be no collector current, which means the transistor is cut off. This is why the self-bias will not work with bipolar transistors.

The self-bias of Fig. 14-1a is a form of negative feedback, similar to that used with bipolar transistors. Recall how negative feedback works. If the drain current increases, the voltage across R_S increases. This increases the gate-source reverse voltage, which decreases the drain current. Therefore, the negative feedback prevents the drain current from changing as much as it would without the feedback.

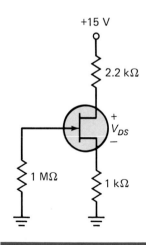

+15 V

2.2 kΩ

$+$
V_{DS}
$-$

1 MΩ   1 kΩ

FIGURE 14-2

Self-bias with circuit values.

EXAMPLE 14-1

If the drain current is 3 mA in Fig. 14-2, what is V_{GS}? What is V_D? What is V_{DS}?

SOLUTION

With 3 mA flowing through the source resistance, the voltage across the source resistor is

$$V_S = (3 \text{ mA})(1 \text{ k}\Omega) = 3 \text{ V}$$

The gate-source voltage is the negative of this, which equals

$$V_{GS} = -3 \text{ V}$$

In Fig. 14-2, the source terminal is no longer ground. Therefore, the drain-source voltage V_{DS} is different from the drain-ground voltage V_D. The drain-ground voltage is

$$V_D = 15 \text{ V} - (3 \text{ mA})(2.2 \text{ k}\Omega) = 8.4 \text{ V}$$

The drain-source voltage is

$$V_{DS} = 8.4 \text{ V} - 3 \text{ V} = 5.4 \text{ V}$$

Note the similarity of these calculations to bipolar calculations.

14-2 GRAPHICAL SOLUTION FOR SELF-BIAS

With a self-biased JFET, the source voltage equals the product of the drain current and the source resistance:

$$V_S = I_D R_S$$

The gate-source voltage is the negative of this, which equals

$$V_{GS} = -I_D R_S \qquad (14\text{-}1)$$

This equation can be used with a transconductance curve to find the Q point of a self-biased JFET.

Drawing the Self-Bias Line

One way to analyze a self-biased circuit is by the graphical approach. Here is how it works. Suppose a self-biased JFET has the transconductance curve shown in Fig. 14-3a. The maximum drain current is 4 mA, and the gate-source cutoff voltage is -2 V. This means the gate voltage has to be between 0 and -2 V. To find out what it is, we can graph Eq. (14-1) and find out where it intersects the transconductance curve. Since Eq. (14-1) is a linear equation, all we have to do is plot two points and draw a line through them.

Suppose the source resistance is 500 Ω. Then Eq. (14-1) becomes

$$V_{GS} = -I_D(500 \ \Omega)$$

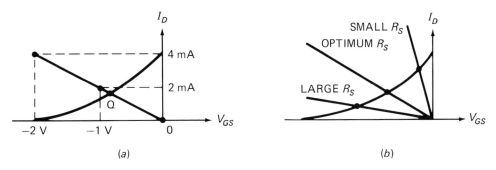

(a)   (b)

FIGURE 14-3   The Q point depends on R_S.

Since any two points can be used, we choose the two convenient points corresponding to $I_D = 0$ and $I_D = I_{DSS}$. At the first point, $I_D = 0$ and

$$V_{GS} = 0(500 \ \Omega) = 0$$

Therefore, the coordinates of the first point are at $(0, 0)$, which is the origin. To get the second point, find V_{GS} for $I_D = I_{DSS}$. In this case, $I_D = 4$ mA and

$$V_{GS} = -(4 \text{ mA})(500 \ \Omega) = -2 \text{ V}$$

Therefore, the coordinates of the second point are at $(4 \text{ mA}, -2 \text{ V})$.

We now have two points on the graph of Eq. (14-1). The two points are $(0, 0)$ and $(4 \text{ mA}, -2 \text{ V})$. By plotting these two points as shown in Fig. 14-3a, we can draw the straight line through the two points as shown. This line will intersect the transconductance curve. This intersection point is the operating point of the self-biased JFET. As you can see, the drain current is slightly less than 2 mA, and the gate-source voltage is slightly less than -1 V.

Designers often use the graphical method when they have the transconductance curve of a JFET. Because many data sheets include these transconductance curves, a designer can plot the two points and draw the straight line directly on the transconductance curves. By reading the coordinates of the point of intersection, he or she will have the quiescent values of I_D and V_{GS}.

Figure 14-3b shows how the Q point changes when the source resistance changes. When R_S is large, the Q point is far down the transconductance curve and the drain current is small. When R_S is small, the Q point is far up the transconductance curve and the drain current is large. In between, there is an optimum value of R_S that sets up a Q point near the middle of the transconductance curve.

The origin is always one of the points on the graph of Eq. (14-1), as shown in Fig. 14-3b. Only the second point is different. Its location depends on the value of R_S. Here is the process for finding the Q point of any self-biased JFET, provided you have the transconductance curve:

1. Multiply I_{DSS} by R_S to get V_{GS} for the second point.

2. Plot the second point (I_{DSS}, V_{GS}).

3. Draw a line through the origin and the second point.

4. Read the coordinates of the intersection point.

Selecting the Source Resistor

If you ever have to select a source resistance for a self-biased JFET, here is the simplest design method known. In Fig. 14-4, a self-bias line is drawn through the point with coordinates I_{DSS} and $V_{GS(off)}$. The point of intersection is not at the exact middle, but it is relatively close to the middle of the transconductance curve. This Q point is acceptable in most self-biased circuits. The source resistance that produces this self-bias line is given by

$$R_S = \frac{-V_{GS(off)}}{I_{DSS}}$$

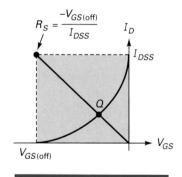

FIGURE 14-4
One way to calculate R_S.

An easy way to remember this is to substitute V_P for $-V_{GS(off)}$. Then you get

$$R_S = \frac{V_P}{I_{DSS}} \qquad (14\text{-}2)$$

This is the same as R_{DS}, the drain resistance in the ohmic region of an ideal JFET. In other words, a quick design rule for self-bias is to use a source resistance that equals the drain resistance in the ohmic region.

EXAMPLE 14-2

A self-biased JFET has the transconductance curve shown in Fig. 14-5. Use the graphical solution to find the Q point for an R_S of 470 Ω.

SOLUTION

Since I_{DSS} = 10 mA, the voltage for the second point is

$$V_{GS} = -(10 \text{ mA})(470 \text{ }\Omega) = -4.7 \text{ V}$$

Plot the point with coordinates of 10 mA and -4.7 V. Then draw a line from the origin through the plotted point to get the line shown in Fig. 14-5. Finally, read the following coordinates at the intersection point:

$$I_D = 4.5 \text{ mA} \qquad V_{GS} = -2 \text{ V}$$

These are the coordinates of the Q point, which is the same as the point of intersection.

EXAMPLE 14-3

A JFET has I_{DSS} = 16 mA and $V_{GS(off)}$ = -5 V. Select a source resistance to produce self-bias for the JFET. Use the simple design rule given earlier.

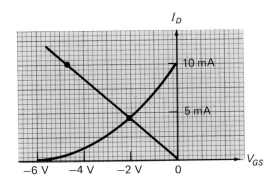

FIGURE 14-5 Example.

14-3 SOLUTION WITH UNIVERSAL JFET CURVE

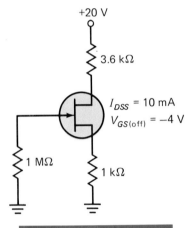

+20 V

3.6 kΩ

I_{DSS} = 10 mA
$V_{GS(\text{off})}$ = −4 V

1 MΩ

1 kΩ

FIGURE 14-6
Self-bias.

How do you find the drain current of a self-biased JFET when you have the values of I_{DSS} and $V_{GS(\text{off})}$, but not the transconductance curve? For instance, how can you find the drain current in Fig. 14-6?

Here is the algebraic way to solve the problem. The drain current is given by

$$I_D = I_{DSS}\left(1 - \frac{V_{GS}}{V_{GS(\text{off})}}\right)^2 \qquad (14\text{-}3)$$

This says the drain current depends on V_{GS}, which equals

$$V_{GS} = -I_D R_S \qquad (14\text{-}4)$$

The quantities I_{DSS}, $V_{GS(\text{off})}$, and R_S are given for a particular JFET circuit. In Fig. 14-6, I_{DSS} = 10 mA, $V_{GS(\text{off})}$ = −4 V, and R_S = 1 kΩ. After the given quantities are substituted, Eqs. (14-3) and (14-4) become two simultaneous equations in two unknowns I_D and V_{GS}, respectively:

$$I_D = (10\,\text{mA})\left(1 + \frac{V_{GS}}{4\,\text{V}}\right)^2 \qquad V_{GS} = -I_D(1\,\text{k}\Omega)$$

By the method of substitution, we can solve for I_D. This requires substituting $-I_D(1\,\text{k}\Omega)$ for V_{GS} in the first equation and solving for I_D. To solve for I_D, you have to use the quadratic formula:

$$x = \frac{-b \pm \sqrt{b^2 - 4ac}}{2a}$$

Because of the quadratic formula, the final equation for I_D is very complicated. In other words, an algebraic solution might be pleasing to a mathematician, but it is unrealistic for a troubleshooter, designer, etc.

Another, much better solution for everyday work is to use the universal curve shown in Fig. 14-7. This graph is called a *universal curve* because you can use it for any self-biased JFET. Given I_{DSS}, $V_{GS(\text{off})}$, and R_S, you can calculate R_S/R_{DS}. Then you can read the value of I_D/I_{DSS} from Fig. 14-7. With this ratio, you can calculate the I_D of the circuit. The rest of the analysis uses Ohm's law as with bipolar transistors. The following example shows you how to analyze a self-biased JFET.

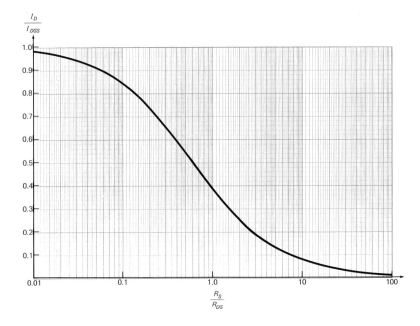

FIGURE 14-7

Universal curve.

Incidentally, the universal curve of Fig. 14-7 was created as follows: A computer was programmed to solve Eqs. (14-3) and (14-4) for different values of I_{DSS}, $V_{GS(\text{off})}$, and R_S. The results were then plotted and graphed to get the universal curve you see in Fig. 14-7.

EXAMPLE 14-4

What are the drain current and the gate-source voltage in Fig. 14-6?

SOLUTION

Calculate the drain resistance in the ohmic region:

$$R_{DS} = \frac{4\text{ V}}{10\text{ mA}} = 400\ \Omega$$

Next, calculate the ratio R_S/R_{DS}:

$$\frac{R_S}{R_{DS}} = \frac{1\text{ k}\Omega}{400\ \Omega} = 2.5$$

In Fig. 14-7, read the corresponding value of I_D/I_{DSS}:

$$\frac{I_D}{I_{DSS}} = 0.215$$

Now calculate the drain current and the gate-source voltage:

$$I_D = 0.215(10\text{ mA}) = 2.15\text{ mA}$$

With this drain current, you can calculate the drain voltage:

$$V_D = 20\text{ V} - (2.15\text{ mA})(3.6\text{ k}\Omega) = 12.2\text{ V}$$

The voltage across the source resistor is

$$V_S = (2.15 \text{ mA})(1 \text{ k}\Omega) = 2.15 \text{ V}$$

The voltage between the drain and the source is

$$V_{DS} = 12.2 \text{ V} - 2.15 \text{ V} = 10.1 \text{ V}$$

and the gate-source voltage is

$$V_{GS} = -2.15 \text{ V}$$

Incidentally, the quantities R_S/R_{DS}, I_D/I_{DSS}, and $V_{GS}/V_{GS(\text{off})}$ are called *normalized variables*. The word *normalized* means a variable is divided by another variable with the same units or dimensions (ohms, milliamperes, volts, etc.). The normalized variable, therefore, is dimensionless. Furthermore, normalizing a variable allows us to produce universal graphs like Fig. 14-7. With normalized variables, a computer can produce a single solution that applies to many different cases. You often see the computer solution of complicated simultaneous equations in the form of universal curves with normalized variables.

14-4 TRANSCONDUCTANCE

To analyze JFET amplifiers, we need to discuss an ac quantity called the *transconductance*, designated g_m. In symbols, transconductance is given by

$$g_m = \frac{\Delta I_D}{\Delta V_{GS}} \qquad (14\text{-}5)$$

Because the changes in I_D and V_{GS} are equivalent to ac current and voltage, Eq. (14-5) can be written as

$$g_m = \frac{i_d}{v_{gs}} \qquad (14\text{-}6)$$

This is a small-signal formula, exact for infinitesimal changes. We use Eq. (14-6) as an approximation for g_m whenever the peak-to-peak value of i_d is less than 10 percent of the quiescent drain current.

If the peak-to-peak values are $i_d = 0.2$ mA and $v_{gs} = 0.1$ V, then

$$g_m = \frac{0.2 \text{ mA}}{0.1 \text{ V}} = 2(10^{-3}) \text{ mho} = 2000 \text{ } \mu\text{mho}$$

The unit "mho" is the ratio of current to voltage. The formal equivalent for the mho is the *siemen* (S). Most data sheets continue to use the mho instead of the siemen. They also use the symbol g_{fs} for g_m. As an example, the data sheet of a 2N5451 lists a typical g_{fs} of 2000 μmho for a drain current of 1 mA. This is identical to saying that the 2N5451 has a typical g_m of 2000 μmho at 1 mA.

Ideal AC JFET Model

Figure 14-8a brings out the meaning of g_m in terms of the transconductance curve. Between points A and B, a change in V_{GS} produces a change in I_D. The ratio of the change in I_D to the change in V_{GS} equals the value of g_m between A and B. If we select another pair of points farther up the curve at C and D, we get more of a change in I_D for a given change in V_{GS}. Therefore, g_m has a larger value farther up the curve. In a nutshell, g_m tells us how much control the gate voltage has over the drain current. The higher g_m is, the more effective the gate voltage is in controlling the drain current. Data sheets for JFETs usually include a graph that shows how g_m varies with the quiescent drain current. Therefore, once we figure out how much dc drain current a JFET amplifier has, we can look up the value of g_m for this value of quiescent drain current.

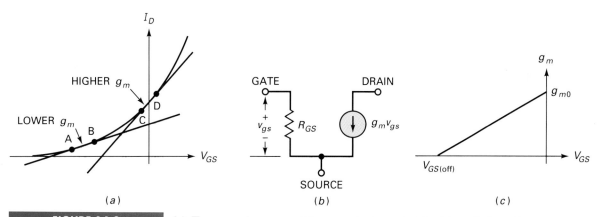

FIGURE 14-8 (a) Transconductance; (b) ac equivalent circuit; (c) variation of transconductance.

Figure 14-8b shows a simple ac equivalent circuit for a JFET. A very high resistance R_{GS} is between the gate and the source. This is well into the tens or hundreds of megohms. The drain of a JFET acts as a current source with a value of $g_m v_{gs}$. If we know g_m and v_{gs}, we can calculate the ac drain current. This model is a first approximation because it does not include the internal resistance of the current source, capacitances inside the JFET, and so on. At low frequencies, we can use this ideal ac model for troubleshooting and preliminary analysis.

Transconductance and Gate-Source Cutoff Voltage

Here is a very useful formula:

$$V_{GS(\text{off})} = \frac{-2I_{DSS}}{g_{m0}} \tag{14-7}$$

This is useful because $V_{GS(\text{off})}$ is difficult to measure accurately. But I_{DSS} and g_{m0} are easy to measure with high accuracy. Therefore, the standard approach is to measure I_{DSS} and g_{m0} and then to calculate $V_{GS(\text{off})}$. This is what is done by manufacturers on data sheets.

When $V_{GS} = 0$, g_m has its maximum value. This maximum value is designated g_{m0}, or g_{fs0} on data sheets. When V_{GS} is negative, g_m decreases in value. Here is the equation of g_m for any value of V_{GS}:

$$g_m = g_{m0}\left(1 - \frac{V_{GS}}{V_{GS(\text{off})}}\right) \tag{14-8}$$

Notice that g_m decreases linearly when V_{GS} becomes more negative, as shown in Fig. 14-8c. This property is useful in automatic gain control, which is discussed later.

If you are wondering where Eqs. (14-7) and (14-8) come from, here is the answer. Calculus is needed to derive them. If you know calculus, take the derivative of Eq. (14-3) and Eq. (14-7) will almost hit you in the face, with Eq. (14-8) only a step beyond. If you don't know calculus, accept the equations as products of higher mathematics.

Bipolar Transconductance

The concept of transconductance can be used with bipolar transistors. It is defined as for JFETs, except for a change in subscripts. If we rewrite Eq. (14-6) in terms of bipolar current and voltage, we get

$$g_m = \frac{i_c}{v_{be}}$$

Recall that $r_e' = v_{be}/i_c$. Therefore, we can write

$$g_m = \frac{1}{r_e'} \tag{14-9}$$

This is useful because it allows us to compare bipolar transistors and JFETs.

EXAMPLE 14-5

A 2N5457 has $I_{DSS} = 5$ mA and $g_{m0} = 5000$ μmho. What is the value of I_D for $V_{GS} = -1$ V? What is the g_m for this drain current?

SOLUTION

Start with Eq. (14-7) to get an accurate value of $V_{GS(\text{off})}$:

$$V_{GS(\text{off})} = \frac{-2(5 \text{ mA})}{5000 \text{ μmho}} = -2 \text{ V}$$

To get the drain current, first calculate the K factor with Eq. (13-4):

$$K = (1 - \tfrac{1}{2})^2 = 0.5^2 = 0.25$$

Then the drain current is

$$I_D = 0.25(5 \text{ mA}) = 1.25 \text{ mA}$$

Next, use Eq. (14-8) to calculate g_m at $V_{GS} = -1$ V:

$$g_m = (5000 \ \mu\text{mho})\left(1 - \frac{1 \text{ V}}{2 \text{ V}}\right) = 2500 \ \mu\text{mho}$$

As you see, g_m is 2500 μmho when I_D is 1.25 mA.

EXAMPLE 14-6

What is the transconductance of a bipolar transistor when the emitter current is 1.25 mA?

SOLUTION

You calculate r_e' in the usual way:

$$r_e' = \frac{25 \text{ mV}}{1.25 \text{ mA}} = 20 \ \Omega$$

Next, calculate the transconductance by taking the reciprocal of r_e':

$$g_m = \frac{1}{20 \ \Omega} = 50,000 \ \mu\text{mho}$$

In Example 14-5, we saw that the JFET has a g_m of 2500 μmho. A bipolar transistor at the same current has a g_m of 50,000 μmho, about 20 times as large. As you will see in the next section, this means the voltage gain of a typical JFET amplifier whose g_m is 2500 μmho will be about 20 times smaller than a comparable bipolar amplifier. As discussed earlier, JFETs are great for a high input resistance, but the price paid is a much lower transconductance, which translates to a much lower voltage gain. For this reason, designers tend to use bipolar transistors when they want a high voltage gain and JFETs when they want a high input resistance. Often, the two devices are combined in a multistage amplifier with a JFET input stage to get the high input resistance and a bipolar second stage to get a high voltage gain.

EXAMPLE 14-7

If a JFET has $g_m = 2500$ μmho, what is the ac drain current for $v_{gs} = 1$ mV? Compare this to a bipolar transistor with a g_m of 50,000 μmho.

SOLUTION

In Fig. 14-8b, the current source has a value of $g_m v_{gs}$. Therefore, an ac input of 1 mV produces

$$i_d = (2500 \ \mu\text{mho})(1 \text{ mV}) = 2.5 \ \mu\text{A}$$

The same ac input voltage to a bipolar transistor would produce an ac collector current of

$$i_c = (50{,}000 \; \mu mho)(1 \; mV) = 50 \; \mu A$$

This bipolar output current is 20 times greater than the JFET output current. Given the same load resistances, a bipolar amplifier would produce 20 times more output voltage than the JFET.

14-5 JFET AMPLIFIERS

Figure 14-9a shows a common-source (CS) amplifier. It is similar to a bipolar CE amplifier. Therefore, many of the ideas that you learned earlier about bipolar transistors apply here. For instance, the coupling and bypass capacitors act as ac shorts. Because of this, the ac input voltage is coupled directly into the gate. Since the source is bypassed to ground, all the ac input voltage appears between the gate and the source. This produces an ac drain current. Since this ac current flows through the drain resistor, we get an amplified and inverted ac output voltage. This output signal is then coupled to the load resistor.

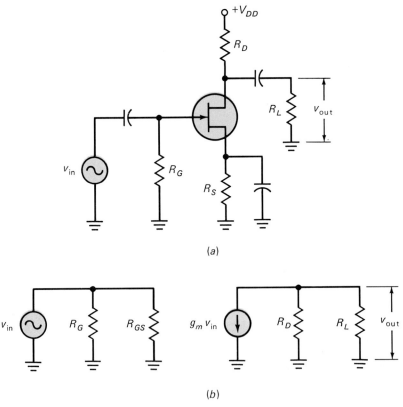

(a)

(b)

FIGURE 14-9 (a) Common-source amplifier; (b) equivalent circuit.

Figure 14-9*b* shows the ac equivalent circuit, drawn by using the rules you learned in bipolar analysis. Here the gate resistance is in parallel with the gate-source resistance of the JFET. Since the ac input voltage appears across the gate-source terminals, the current source has a value of $g_m v_{in}$. This ac drain current flows through the ac drain resistance, the parallel resistance of R_D and R_L.

Voltage Gain

In Fig. 14-9*a*, the ac drain resistance is

$$r_d = R_D \parallel R_L$$

When the output current $g_m v_{in}$ flows through r_d, it produces an output voltage of

$$v_{out} = g_m v_{in} r_d$$

Divide both sides by v_{in} and you get

$$\frac{v_{out}}{v_{in}} = g_m r_d$$

Recall that voltage gain is defined as the output voltage divided by the input voltage. Therefore, the preceding equation may be written as

$$A = g_m r_d \qquad (14\text{-}10)$$

This says the voltage gain of a *CS* amplifier equals the transconductance times the ac drain resistance.

A *CS* amplifier like Fig. 14-9*a* is not popular with designers because it has low voltage gain and high distortion. When designers want voltage gain, they normally use a bipolar transistor rather than a JFET. One way to get around the excessive distortion is to use plenty of negative feedback, but this reduces the voltage gain even further. In other words, a *CS* amplifier has little to offer designers that they can't get more easily with bipolar transistors.

Bipolar-to-JFET Shortcuts

Recall that the g_m of a bipolar transistor is given by

$$g_m = \frac{1}{r_e'}$$

Now, look at Eq. (14-10). Does your intuition tell you anything at all? Do you sense some sort of connection between the voltage gain of a JFET and a bipolar transistor? The ability to see connections, however darkly and vaguely, is often the beginning of a new idea. You should learn to listen to your hunches because sooner or later they will lead you to something valuable and worthwhile.

As it turns out, there is a connection between the voltage gain of a JFET and a bipolar transistor. This connection enables us to use a shortcut for deriving and remembering JFET formulas. Because bipolar transistors and JFETs have similar equivalent circuits, all formulas for voltage gain are analogous to each other. This means we can rewrite any

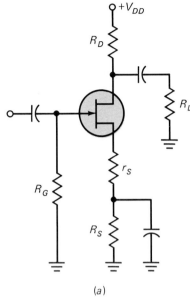

(a)

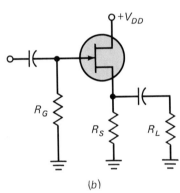

(b)

FIGURE 14-10

Examples.

bipolar formula for a comparable JFET circuit by changing subscripts and replacing r_e' by g_m. For instance, the bipolar *CE* amplifier has a voltage gain of

$$A = \frac{r_c}{r_e'}$$

A JFET *CS* amplifier has the same design configuration. Therefore, we can replace r_c by r_d and r_e' by $1/g_m$ to get

$$A = \frac{r_d}{1/g_m} = g_m r_d$$

This is identical to Eq. (14-10), which we derived earlier.

A swamped JFET amplifier like Fig. 14-10*a* has part of the source resistance unbypassed. This is similar to a swamped bipolar amplifier whose voltage gain is

$$A = \frac{r_c}{r_e + r_e'}$$

If you remember this bipolar formula, you can easily get the equivalent JFET formula. Replace r_c by r_d, r_e by r_s, and r_e' by $1/g_m$ to get

$$A = \frac{r_d}{r_s + 1/g_m} = \frac{g_m r_d}{1 + g_m r_s} \qquad (14\text{-}11)$$

As a final example, the source follower of Fig. 14-10*b* is similar to an emitter follower whose voltage gain is

$$A = \frac{r_e}{r_e + r_e'}$$

When you replace r_e by r_s and r_e' by $1/g_m$, you get the voltage gain of a source follower:

$$A = \frac{r_s}{r_s + 1/g_m} = \frac{g_m r_s}{1 + g_m r_s} \qquad (14\text{-}12)$$

The last equation is worth remembering because the source follower is one of the most widely used JFET circuits. Like the emitter follower, its voltage gain is less than 1. But it has the major advantage of a very high input impedance. You will see the source follower used in all kinds of applications.

Although the *CS* amplifier is not popular, the source follower is. A source follower like Fig. 14-10*b* has maximum negative feedback and minimum distortion. This circuit has emerged as one of the most important JFET applications for two reasons. First, it has an extremely high input resistance. Second, its heavy negative feedback eliminates most of the square-law distortion that appears in *CS* amplifiers. These two reasons make the source follower the designer's choice when it comes to an input stage for many types of equipment. For normal operation, the JFET should act as a current source, not as a resistor. Incidentally, the source follower is formally called a *common-drain* (CD) *amplifier* because the drain is at ac ground.

Similarities exist for dc quantities. For instance, the dc collector voltage of a *CE* amplifier is

$$V_C = V_{CC} - I_C R_C$$

The equivalent JFET relation is

$$V_D = V_{DD} - I_D R_D$$

The dc emitter voltage of a *CE* amplifier is

$$V_E = I_E R_E$$

and the dc source voltage of a *CS* amplifier is

$$V_S = I_S R_S$$

Other shortcuts and memory aids can be applied because of the analogy between bipolar transistors and JFETs. We will point these out later.

The bipolar-JFET analogy makes it easier to remember JFET formulas. It also makes it easier to understand how JFET circuits work. The only danger here is carrying the analogy too far by forgetting the key differences between bipolar transistors and JFETs. One fundamental difference is that you forward-bias the emitter diode and reverse-bias the gate-source diode. This difference means you cannot translate the input impedance of the base to an equivalent JFET input impedance. In short, the devices are cousins, not twins.

EXAMPLE 14-8

If $g_m = 2500$ μmho for the JFET of Fig. 14-11, what is the ac output voltage?

SOLUTION

The ac drain resistance is

$$r_d = 3.6 \text{ k}\Omega \parallel 10 \text{ k}\Omega = 2.65 \text{ k}\Omega$$

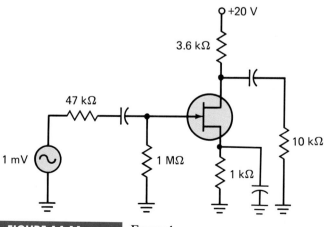

FIGURE 14-11   Example.

The voltage gain is

$$A = (2500 \text{ } \mu\text{mho})(2.65 \text{ k}\Omega) = 6.63$$

The input impedance of the amplifier is

$$z_{\text{in}} = 1 \text{ M}\Omega$$

We are ignoring the R_{GS} of the JFET because it is usually in the hundreds of megohms.

The generator has an internal resistance of 47 kΩ. Therefore, some of the signal voltage is dropped across this 47 kΩ—but not much. The ac voltage at the gate is found with Ohm's law:

$$v_{\text{in}} = \frac{1 \text{ mV}}{47 \text{ k}\Omega + 1 \text{ M}\Omega}(1 \text{ M}\Omega) = 0.955 \text{ mV}$$

The ac output voltage equals the voltage gain times the input voltage:

$$v_{\text{out}} = 6.63(0.955 \text{ mV}) = 6.33 \text{ mV}$$

EXAMPLE 14-9

If $g_m = 2500$ μmho for the source follower of Fig. 14-12, what is the ac output voltage?

SOLUTION

The input voltage drives the gate, and the output voltage appears at the source. The ac source resistance is

$$r_s = 1 \text{ k}\Omega \parallel 1 \text{ k}\Omega = 500 \text{ }\Omega$$

With Eq. (14-12), the voltage gain is

$$A = \frac{(2500 \text{ }\mu\text{mho})(500 \text{ }\Omega)}{1 + (2500 \text{ }\mu\text{mho})(500 \text{ }\Omega)} = 0.556$$

The input impedance of the source follower is

$$z_{\text{in}} = 10 \text{ M}\Omega$$

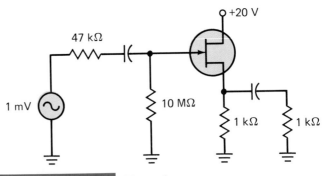

FIGURE 14-12　　Example.

Again, we are ignoring the R_{GS} of the JFET because it is usually in the hundreds of megohms.

The generator has an internal resistance of 47 kΩ. Therefore, almost none of the ac voltage is dropped across the generator resistance:

$$v_{in} = \frac{1\,mV}{47\,k\Omega + 10\,M\Omega}(10\,M\Omega) = 0.995\,mV$$

The ac output voltage equals the voltage gain times the input voltage:

$$v_{out} = (0.556)(0.995\,mV) = 0.553\,mV$$

14-6 THE JFET ANALOG SWITCH

The source follower is one of the major applications of a JFET. In this application, a JFET acts as a current source. Another big application of JFETs is in analog switching. In this application, the JFET acts as a switch. To get this type of operation, the gate-source voltage V_{GS} is restricted to two values: 0 V or a large negative voltage. The negative voltage must be equal to or more negative than $V_{GS(off)}$. The exact value of the negative voltage does not matter. All that is needed is a voltage negative enough to cut off the JFET. For instance, if $V_{GS(off)} = -4$ V, then $V_{GS} = 0$ V and $V_{GS} = -5$ V would be acceptable input voltages for analog switching.

Figure 14-13a is called a JFET *shunt switch*. The JFET is turned on and off by V_{GS}. Because of this, we can use Fig. 14-13b as an equivalent

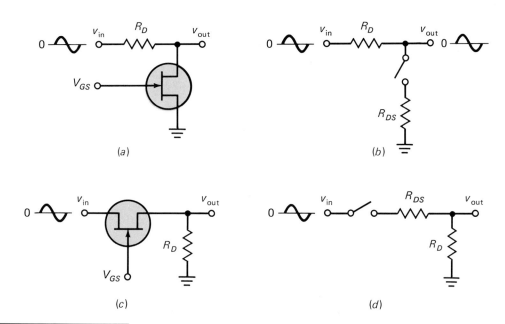

(a) (b)

(c) (d)

FIGURE 14-13   (a) Shunt switch; (b) ac equivalent circuit; (c) series switch; (d) ac equivalent circuit.

circuit. For normal operation, the ac voltage v_{in} is small, typically less than 100 mV. Also, R_D is much greater than R_{DS}. When V_{GS} is zero, the JFET operates in the ohmic region and the switch of Fig. 14-13b is closed. In this case, v_{out} is much smaller than v_{in} because of the voltage divider. When V_{GS} is more negative than $V_{GS(off)}$, the JFET is cut off and the switch of Fig. 14-13b is open. This means that v_{out} equals v_{in}.

Figure 14-13c shows a JFET *series switch*, and Fig. 14-13d is its equivalent circuit. When V_{GS} is zero, the switch is closed and the JFET is equivalent to a resistance of R_{DS}. In this case, the output approximately equals the input. When V_{GS} is equal to or more negative than $V_{GS(off)}$, the JFET is open and v_{out} is approximately zero.

You will see both kinds of JFET switches used in practice. The series switch is used more often because it has a better on-off ratio.

EXAMPLE 14-10

A JFET shunt switch like Fig. 14-13a has $R_D = 10$ kΩ, $I_{DSS} = 10$ mA, and $V_{GS(off)} = -2$ V. If $v_{in} = 10$ mV peak to peak, what does v_{out} equal when $V_{GS} = 0$? When $V_{GS} = -5$ V?

SOLUTION

Calculate the ideal value of R_{DS} as follows:

$$R_{DS} = \frac{2\,V}{10\,mA} = 200\ \Omega$$

When $V_{GS} = 0$, the circuit acts like the equivalent circuit of Fig. 14-14a. With Ohm's law,

$$v_{out} = \frac{10\,mV}{10\,k\Omega + 200\,\Omega}(200\ \Omega) = 0.196\,mV$$

When $V_{GS} = -5$ V, the JFET is like an open circuit. In Fig. 14-14a, visualize the 200 Ω increasing to infinity. You can see that

$$v_{out} = v_{in} = 10\,mV$$

EXAMPLE 14-11

Repeat Example 14-10 for a JFET series switch. Use the same data.

SOLUTION

When $V_{GS} = 0$, the circuit acts like the equivalent circuit of Fig. 14-14b. With Ohm's law,

$$v_{out} = \frac{10\,mV}{10\,k\Omega + 200\,\Omega}(10\,k\Omega) = 9.8\,mV$$

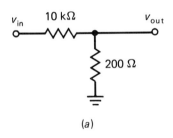

v_{in} 10 kΩ v_{out}

200 Ω

(a)

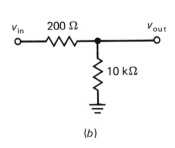

v_{in} 200 Ω v_{out}

10 kΩ

(b)

FIGURE 14-14

Example.

When $V_{GS} = -5$ V, the JFET is like an open circuit. In Fig. 14-14b, visualize the 200 Ω increasing to infinity. You can see that

$$v_{\text{out}} = 0 \text{ mV}$$

The JFET series switch is usually preferred because its on-off ratio is much higher. The on-off ratio of the shunt switch in Example 14-10 is

$$\frac{10 \text{ mV}}{0.196 \text{ mV}} = 51$$

The on-off ratio of the series switch with the same R_D and JFET is

$$\frac{9.8 \text{ mV}}{0} = \infty$$

These are ideal results. In reality, an open JFET may have several hundred megohms of resistance. Therefore, the on-off ratio of the series switch will not be infinite, but it will still be much greater than 51. This is why you will see the series switch used more often than the shunt switch.

14-7 DEPLETION-MODE MOSFET AMPLIFIERS

Because a depletion-mode MOSFET can operate with a positive or a negative gate, we can set its Q point at $V_{GS} = 0$, as shown in Fig. 14-15a. Then an ac input signal to the gate can produce variations above and below the Q point. Being able to use zero V_{GS} allows a designer to use the unique biasing circuit of Fig. 14-15b. This simple circuit has $V_{GS} = 0$ and $I_D = I_{DSS}$. The dc drain voltage is

$$V_{DS} = V_{DD} - I_{DSS}R_D \qquad (14\text{-}13)$$

The zero bias of Fig. 14-15a is unique with depletion-mode MOSFETs. This type of bias will not work with a bipolar transistor, a JFET, or an enhancement-mode MOSFET. Although any of the JFET biasing methods will work with a depletion-mode MOSFET, the zero-bias method of Fig. 14-15b is the preferred form of bias because it does the job simply and adequately.

After the depletion-type MOSFET is biased to a Q point, it can amplify small signals. The JFET formulas for voltage gain apply directly to a MOSFET amplifier. But like the JFET, the depletion-mode MOSFET has a relatively low voltage gain. MOSFETs have excellent low-noise properties, a definite advantage for any stage near the front end of a system where the signal is weak such as a television receiver. As with a JFET, the g_m of a MOSFET can be controlled by changing the gate-source voltage. Because of this, MOSFETs have been used for *automatic gain control* (see "Optional Topics" if interested).

Some MOSFETs are dual-gate devices. This means they have two separate gates, like the dual-gate MOSFET shown in Fig. 14-16a. One

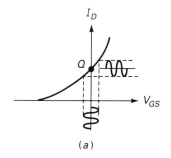

(a)

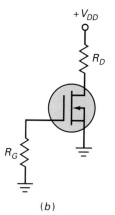

(b)

FIGURE 14-15

Zero bias.

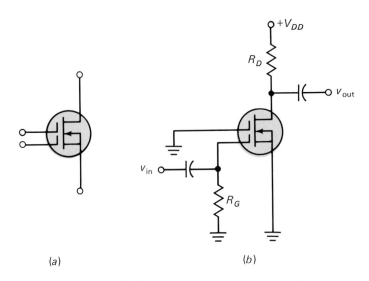

FIGURE 14-16 (*a*) Dual-gate MOSFET; (*b*) cascode amplifier using dual-gate MOSFET.

use for this device is to build a *cascode amplifier* like that in Fig. 14-16*b*. The ac input signal drives the lower gate. The upper gate is grounded. Because of its internal structure, the dual-gate MOSFET is equivalent to one MOSFET driving another MOSFET. As can be shown, the cascode amplifier of Fig. 14-16*b* has a voltage gain of

$$A = g_m R_D$$

This is the same as an ordinary *CS* amplifier. But the input impedance at higher frequencies is larger because the input capacitance is lower.

In the next chapter, we discuss the effects of capacitance and frequency. There you will see how capacitance decreases the voltage gain at higher frequencies. One of the preferred uses for depletion-mode MOSFETs is a cascode amplifier like Fig. 14-16*b* because it continues to provide useful voltage gain at higher frequencies. By higher frequencies, we mean VHF and UHF frequencies. (*Note:* VHF includes all frequencies from 30 to 300 MHz, and UHF is all frequencies from 300 to 3000 MHz.)

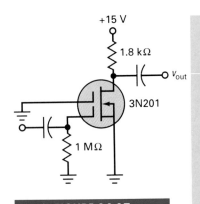

FIGURE 14-17

Cascode amplifier.

EXAMPLE 14-12

The 3N201 of Fig. 14-17 has $I_{DSS} = 4$ mA, $V_{GS(off)} = -2$ V, and $g_m = 10,000$ μmho. What is the dc voltage between the drain and ground? What is the voltage gain?

SOLUTION

The dc voltage at the drain equals the supply voltage minus the voltage drop across the drain resistor:

$$V_D = 15 \text{ V} - (4 \text{ mA})(1.8 \text{ k}\Omega) = 7.8 \text{ V}$$

The voltage gain is

$$A = (10{,}000 \ \mu\text{mho})(1.8 \ \text{k}\Omega) = 18$$

14-8 ENHANCEMENT-MODE MOSFET APPLICATIONS

Although the enhancement-mode MOSFET can be biased and used as an amplifier, this application is not its strength. Again, the bipolar transistor is the preferred device for voltage gain. Unless very high frequencies are involved, the MOSFET has little advantage when it comes to general-purpose amplifiers.

So why has the enhancement-mode MOSFET revolutionized the computer industry? Because of its threshold voltage, the enhancement-mode MOSFET is a dream come true for digital designers. It is ideal for use as a switching device because it is nonconducting when the gate voltage is zero. When the gate voltage is greater than the threshold voltage, the device conducts. This off-on action is the key to building computers. If you study computers, you will see how a typical computer uses millions of enhancement-mode MOSFETs like off-on switches to process data. (Data mean numbers and facts.)

Passive-Load Switching

Figure 14-18 shows an enhancement-mode MOSFET and a *passive load*. The word *passive* means an ordinary resistor. In this circuit, v_{in} is either low or high. By *low*, we mean any voltage that is less than the threshold voltage. By *high*, we mean any voltage that is greater than the threshold voltage. For instance, if $V_{GS(\text{th})} = +2 \ \text{V}$, then $v_{\text{in}} = 0 \ \text{V}$ and $v_{\text{in}} = +5 \ \text{V}$ would be acceptable low and high voltages, respectively.

When v_{in} is low, the MOSFET is off and v_{out} equals the supply voltage. When v_{in} is high, the MOSFET conducts heavily and v_{out} drops to a low value. For the circuit to work properly, the drain resistance R_{DS} in the ohmic region has to be much smaller than the passive drain resistance. In symbols,

$$R_{DS} \ll R_D$$

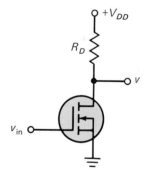

FIGURE 14-18
Passive load.

As a guide, R_{DS} should be at least 10 times smaller than R_D.

A circuit like Fig. 14-18 is the simplest computer circuit that can be built. It is referred to as an *inverter* because the output voltage is the opposite of the input voltage. When the input voltage is low, the output voltage is high. When the input voltage is high, the output voltage is low. If approximate answers were valid in the earlier discussions of amplifiers, they are even more valid now because nobody cares about the exact voltages in switching circuits. Switching circuits are much less demanding than amplifiers. All that matters in a switching circuit is that the input and output voltages can be easily recognized as either low or high.

In a switching circuit like Fig. 14-18, the drain resistance R_{DS} of the MOSFET is much smaller than the passive load resistance R_D. Since all that matters in switching circuits is being able to recognize a voltage as low or high, we can approximate the JFET as simple off-on switch. Troubleshooters who look at a circuit like Fig. 14-18 don't see a MOSFET. They see an off-on switch in series with R_D. When the switch is open, v_{out} is high. When the switch is closed, v_{out} is low.

Active-Load Switching

Integrated circuits (ICs) consist of thousands of microscopically small transistors, either bipolar or MOS. The earliest integrated circuits also included passive load resistances like the one of Fig. 14-18. But a passive load resistance has a major disadvantage: It is physically much larger than a MOSFET. Because of this, integrated circuits with passive-load resistors were much bigger than the integrated circuits now being used. Somebody found a solution to the problem by inventing *active-load resistors*. This greatly reduced the size of integrated circuits and led to the personal computers that we have today.

The key idea was to get rid of passive-load resistors. But how? The inventor's thinking process may have gone something like this:

> The passive resistors are too big compared to the MOSFETs. I've got to get rid of them. Let's see. Is there any way I can use the MOSFET as a resistor?

At this point, the inventor is warm. Let's see what comes next:

> When a MOSFET operates in the ohmic region, it acts as a resistor. Suppose I replace the passive-load resistor of Fig. 14-18 with a MOSFET that is operating in the ohmic region. If I can make the R_{DS} of this MOSFET high enough, the circuit should continue to work as a switching circuit.

Now, the inventor is on the verge of a breakthrough. She is going to replace the passive-load resistor by another MOSFET. This will get rid of the large passive load and replace it with a physically smaller MOSFET. The final piece of the puzzle is almost in place. Here is the breakthrough:

> What if I connect the gate terminal to the drain terminal? Then $V_{GS} = V_{DS}$ at all times. Furthermore, the MOSFET changes from a three-terminal device to a two-terminal device. This two-terminal device will have a resistance. If this resistance is high enough, the switching circuit will work as it did with a passive resistor. I think I've got it.

Figure 14-19 shows the invention. It is called *active-load switching*. The lower MOSFET still acts as a switch, but the upper MOSFET acts as a large resistance. Notice that the upper MOSFET has its gate connected to its drain. Because of this, it becomes a two-terminal device with a resistance of

$$R_{DS} = \frac{V_{DS}}{I_D} \tag{14-14}$$

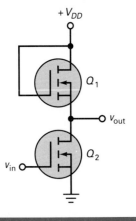

FIGURE 14-19

Active load.

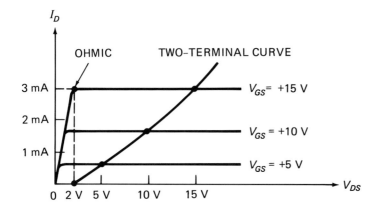

FIGURE 14-20   Two-terminal curve.

For the new circuit to work properly, the R_{DS} of the upper MOSFET has to be large compared to the R_{DS} of the lower MOSFET.

Is the upper R_{DS} much larger than the lower R_{DS}? Yes. You can see this by looking at Fig. 14-20. Connecting the gate to the drain results in $V_{GS} = V_{DS}$ at all times. Therefore, we can plot the two-terminal curve shown here. The ratio of voltage to current at any point on this two-terminal curve is the value of R_{DS} for the upper MOSFET. For instance, when $V_{GS} = V_{DS} = 15$ V, the drain current is approximately 3 mA and the drain resistance of the two-terminal MOSFET is

$$R_{DS} = \frac{15 \text{ V}}{3 \text{ mA}} = 5 \text{ k}\Omega$$

If the lower MOSFET has the same drain curves, then its drain resistance in the ohmic region is approximately

$$R_{DS} = \frac{2 \text{ V}}{3 \text{ mA}} = 667 \ \Omega$$

As you see, the upper R_{DS} is much larger than the lower R_{DS}.

To keep the two drain resistances distinct, some data sheets use $R_{DS(\text{on})}$ for the two-terminal drain resistance and $r_{DS(\text{on})}$ for the ohmic region. For instance, the data sheet of an M116 gives $I_{D(\text{on})} = 2$ mA for $V_{GS} = V_{DS} = +10$ V. With this, you can calculate

$$R_{DS(\text{on})} = \frac{10 \text{ V}}{2 \text{ mA}} = 5 \text{ k}\Omega$$

This is the drain resistance when the device is used as a two-terminal resistor. The same data sheet also gives an $r_{DS(\text{on})}$ of 200 Ω. This is the drain resistance when the device is used as a three-terminal MOSFET in the ohmic region.

In conclusion, the switching circuit of Fig. 14-19 has an input voltage that is either low or high. The output voltage is inverted, being either high or low. The lower MOSFET is ideally equivalent to an off-on switch, while the upper MOSFET is ideally equivalent to a large resistor. The most important feature of this circuit is the active load resistance.

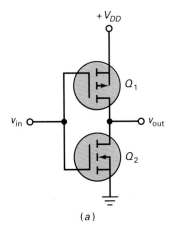

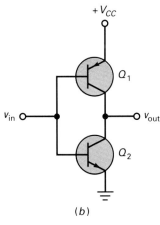

FIGURE 14-21
Complementary circuits.

CMOS Inverter

We can build *complementary MOS* (CMOS) circuits with *p*-channel and *n*-channel MOSFETs. The most important of all is the CMOS inverter shown in Fig. 14-21*a*. Note that Q_1 is a *p*-channel device and Q_2 is an *n*-channel device. This circuit is analogous to the class B push-pull bipolar amplifier of Fig. 14-21*b*. When one device is on, the other is off, and vice versa.

For instance, when v_{in} is low in Fig. 14-21*a*, the lower MOSFET is off but the upper one is on. Therefore, the output voltage is high. But when v_{in} is high, the lower MOSFET is on and the upper one is off. In this case, the output voltage is low. Since the output voltage is always opposite to the phase of the input, the circuit is called an *inverter*.

The CMOS inverter can be modified to build other complementary-type circuits. The key advantage in using CMOS design is its extremely low power consumption. Because both devices are in series, the current is determined by the leakage in the off device, which is typically in nanoamperes. This means that the total power dissipation of the circuit is in nanowatts. The low power consumption is the main reason CMOS circuits are popular in calculators, digital watches, and satellites.

VMOS Transistors

Conventional MOS devices have small drain currents, which implies low power ratings (typically less than 1 W). The VMOS transistor is an enhancement-mode MOSFET modified to handle much larger currents and voltages than a conventional MOSFET. Prior to the invention of the VMOS transistor, MOSFETs could not compete with the power ratings of large bipolar transistors. But now the VMOS offers a new type of MOSFET that is better than the bipolar transistor in many applications requiring high load power.

One major advantage VMOS transistors have over bipolar transistors is the lack of thermal runaway. As you recall, an increase in the device temperature lowers the V_{BE} of a bipolar transistor. This increases the collector current, which raises the temperature further. If the heat sinking is inadequate, the bipolar transistor can go into a thermal runaway and be destroyed by the excessive power dissipation.

A VMOS transistor, on the other hand, has a negative thermal coefficient. As the device's temperature increases, the drain current decreases, which reduces power dissipation. Because of this, the VMOS transistor cannot go into thermal runaway, which is a big advantage in any power amplifier.

Bipolar transistors cannot be connected in parallel to increase the load power because their V_{BE} drops do not match closely enough. If you try to connect them in parallel, current hogging occurs (the one with the lower V_{BE} drop has more collector current). Because of their negative temperature coefficients, two VMOS transistors can be connected in parallel to increase the load power. If one of the parallel VMOS transistors tries to hog the current, its negative temperature coefficient reduces the current through it, so that approximately equal currents flow through the parallel VMOS transistors.

When a small-signal bipolar transistor is used as a saturated switch, conservative design calls for a base current that is approximately one-tenth of the saturated collector current. Since most transistors have a β_{dc} greater than 10, the excess base current guarantees saturation from one transistor to the next. But the excess base current also does something else that we have not mentioned until now. Extra carriers are stored in the base region of a saturated transistor. When the transistor tries to come out of saturation, there is a small delay called the *saturation delay time* (also called the *storage time*). For instance, the storage time of a 2N3713 is 0.3 μs. This means that it takes approximately 0.3 μs for a 2N3713 to come out of saturation after the base drive is removed.

Another advantage the VMOS transistor has over the bipolar transistor is the lack of storage time. Because no extra charges are stored in VMOS when it is conducting, it can come out of saturation almost immediately. Typically, a VMOS transistor can shut off amperes of current in tens of nanoseconds. This is from 10 to 100 times faster than a comparable bipolar transistor. Therefore, the VMOS transistor finds numerous applications in high-speed switching circuits, switching regulators, and so on.

Digital ICs are low-power devices because they can supply only small load currents. *Interfacing* means using some kind of buffer between a low-power device (often a digital IC) and a high-power load (such as a relay, motor, or incandescent lamp). The VMOS transistor is an excellent device for interfacing digital ICs to high-power loads. Interfacing digital ICs (such as CMOS, MOS, or TTL) to high-power loads is one of the most important applications for the VMOS transistor.

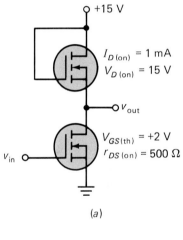

(a)

EXAMPLE 14-13

The input voltage of Fig. 14-22a can have either of two values: 0 or +5 V. What is the output voltage for each input value?

SOLUTION

The upper MOSFET has the following approximate value of R_{DS} in the two-terminal region:

$$R_{DS} = \frac{15\,V}{1\,mA} = 15\,k\Omega$$

The lower MOSFET has the following R_{DS} in the ohmic region:

$$r_{DS(on)} = 500\,\Omega$$

When v_{in} is +5 V, we can visualize the circuit as shown in Fig. 14-22b. In this case, the output voltage is

$$v_{out} = \frac{15\,V}{15\,k\Omega + 500\,\Omega}(500\,\Omega) = 0.484\,V$$

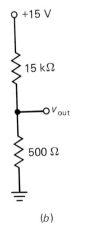

(b)

FIGURE 14-22
Example.

When $v_{in} = 0$, the lower MOSFET is open. If you visualize the 500 Ω changed to infinity in Fig. 14-22b, then you can see the output voltage will increase to

$$v_{out} = 15 \text{ V}$$

14-9 OTHER JFET BIASING

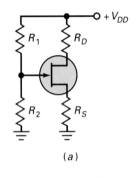

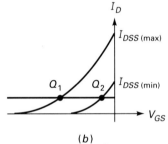

FIGURE 14-23

Voltage-divider bias.

Self-bias is the most popular way to stabilize the Q point of a JFET because small-signal operation is common with JFETs. This type of operation does not require a rock-solid Q point. Sometimes JFETs are used in large-signal applications. In this case, a designer may decide to use a more stable JFET biasing method.

Voltage-Divider Bias

Figure 14-23a shows voltage-divider bias for a JFET. The Thevenin voltage V_{TH} applied to the gate is

$$V_{TH} = \frac{R_2}{R_1 + R_2}(V_{DD}) \qquad (14\text{-}15)$$

This is the dc voltage from the gate to ground. Because of V_{GS}, the voltage from the source to ground is

$$V_S = V_{TH} - V_{GS} \qquad (14\text{-}16)$$

Therefore, the drain current equals

$$I_D = \frac{V_{TH} - V_{GS}}{R_S} \qquad (14\text{-}17)$$

If V_{TH} is large enough to swamp out V_{GS} in Eq. (14-17), the drain current is approximately constant for any JFET, as shown in Fig. 14-23b.

But there is a design problem. The V_{GS} can vary several volts from one JFET to another. With low supply voltages, it is difficult to make V_{TH} large enough to swamp out the variations in V_{GS}. For this reason, voltage-divider bias is more stable than self-bias, but not as stable as it is with bipolar transistors. The designer has to compromise by trying to make V_{TH} as large as possible while keeping the Q point near the middle of the ac load line. For maximum bias stability, the design goals are as high a V_{TH} as possible and as low a V_{GS} as possible.

Source Bias

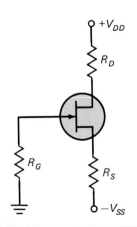

FIGURE 14-24

Source bias.

Figure 14-24 shows *source bias* (similar to two-supply emitter bias). The idea is to swamp out the variations in V_{GS}. Since most of V_{SS} appears across R_S, the drain current is about equal to V_{SS}/R_S. The exact value is given by

$$I_D = \frac{V_{SS} - V_{GS}}{R_S} \qquad (14\text{-}18)$$

For source bias to work well, V_{SS} must be much greater than V_{GS}. However, a typical range for V_{GS} is from -1 to -5 V, so you can see that perfect swamping is not possible with typical supply voltages. Again, maximum bias stability requires as large a V_{SS} as possible and as small a V_{GS} as possible.

Current-Source Bias

When positive and negative supplies are available, you can use *current-source bias*, shown in Fig. 14-25a. Since the bipolar transistor is emitter-biased, its collector current is given by

$$I_C = \frac{V_{EE} - V_{BE}}{R_E} \qquad (14\text{-}19)$$

Because the bipolar transistor acts like a dc current source, it forces the JFET drain current to equal the bipolar collector current:

$$I_D = I_C$$

Figure 14-25b illustrates how effective the current-source bias is. Since I_C is constant, both Q points have the same value of drain current. The current source effectively wipes out the influence of V_{GS}. Although V_{GS} is different for each Q point, it no longer influences the value of the drain current. This is the ultimate bias stability for a JFET.

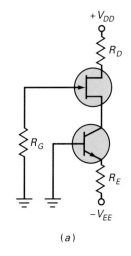

(a)

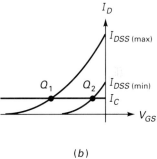

(b)

FIGURE 14-25
Current-source bias.

□ ### OPTIONAL TOPICS

The following material continues the earlier discussions at a more advanced and specialized level. All the topics are optional because they are not used in any of the basic discussions in later chapters. This section will be a useful reference when you are in industry because then you will probably want more advanced viewpoints.

□ ### 14-10 OUTPUT IMPEDANCE OF THE SOURCE FOLLOWER

As discussed in Sec. 12-8, the emitter follower has an output impedance of

$$r_{\text{th}} = r_e' + \frac{R_G \| R_1 \| R_2}{\beta}$$

The way to remember this is as follows: Looking back into the emitter, you see r_e' plus the stepped-down impedance of the base circuit. This stepdown occurs because of the current gain between the base and the emitter. Since impedances are stepped up when they are viewed from the base, they are stepped down when viewed from the emitter.

The source follower is analogous to the emitter follower, so the formula for its output impedance will bear some resemblance to the foregoing

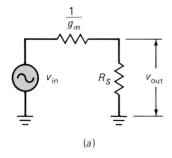

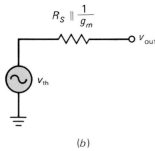

(a)

(b)

FIGURE 14-26

AC equivalent circuit of source follower.

equation. Here is how to derive the output impedance of a source follower: The voltage gain of a source follower is

$$\frac{v_{\text{out}}}{v_{\text{in}}} = \frac{R_S}{R_S + 1/g_m}$$

Therefore, the output voltage is

$$v_{\text{out}} = \frac{R_S}{R_S + 1/g_m}(v_{\text{in}})$$

Figure 14-26a shows the equivalent circuit for this equation. The Thevenin resistance looking back into this circuit is

$$r_{\text{th}} = R_S \parallel \frac{1}{g_m} \qquad (14\text{-}20)$$

Figure 14-26b shows the Thevenin output circuit. The Thevenin voltage is v_{in}, and the output impedance is the equivalent parallel resistance of R_S and $1/g_m$. When R_S is much greater than $1/g_m$, the output impedance of the source follower approximately equals $1/g_m$.

14-11 OTHER FET APPLICATIONS

In this section, we discuss some of the applications in which the FET's properties give it a clear-cut advantage over the bipolar transistor.

Multiplexing

Multiplex means "many into one." Figure 14-27 shows an *analog multiplexer*, a circuit that steers one of the input signals to the output line. Each JFET acts as a single-pole single-throw switch. When the control signals (V_1, V_2, and V_3) are more negative than $V_{GS(\text{off})}$, all input signals are blocked. By making any control voltage equal to zero, we can transmit

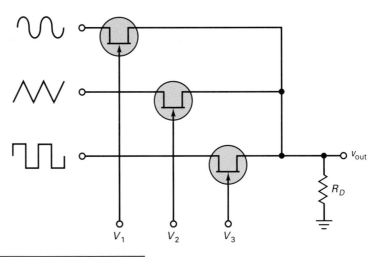

FIGURE 14-27   Analog multiplexer.

one of the inputs to the output. For instance, if V_1 is zero, we get a sinusoidal output. If V_2 is zero, we get a triangular output. And if V_3 is zero, we get a square-wave output. Normally, only one of the control signals is zero.

JFET Choppers

We can build a direct-coupled amplifier by leaving out the coupling and bypass capacitors and connecting the output of each stage directly to the input of the next stage. In this way, direct current is coupled, as well as alternating current. Circuits that can amplify dc signals are called *dc amplifiers*. The major disadvantage of a direct coupling is *drift*, a slow shift in the final output voltage produced by supply, transistor, and temperature variations.

Figure 14-28a shows a way to overcome the drift problem produced by direct coupling. It is based on the *chopper method* of building a dc amplifier. The input dc voltage is chopped by a switching circuit. This results in the square wave shown at the chopper output. The peak value of this square wave equals V_{DC}. Because the square wave is an ac signal, we can use a conventional ac amplifier, one with coupling capacitors between the stages. These capacitors eliminate the drift problem because they do not couple the drift. The amplified output can then be "peak-detected" to recover the amplified dc signal.

If we apply a square wave to the gate of a JFET analog switch, it becomes a chopper (see Fig. 14-28b). The gate square wave is negative-going, swinging from 0 V to at least $V_{GS(off)}$. This alternately saturates and cuts off the JFET. Therefore, the output voltage is a square wave with a peak value of V_{DC}.

If the input is a low-frequency ac signal, it gets chopped into the ac waveform of Fig. 14-28c. This chopped signal can now be amplified by an ac amplifier that is drift-free. The amplified signal can then be peak-

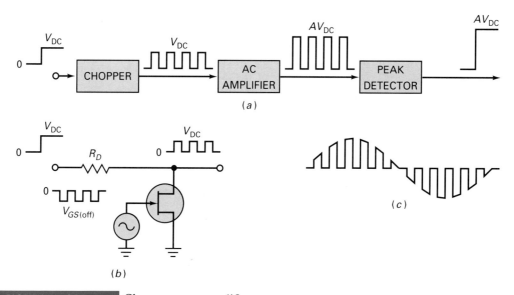

FIGURE 14-28 Chopper-type amplifier.

detected to recover the original input signal. In other words, a chopper amplifier will amplify both dc and ac signals. The only restriction is keeping the frequency of the amplified signal much smaller than the chopping frequency. This is necessary to get a waveform like that in Fig. 14-28c.

Buffer Amplifier

Figure 14-29 shows a buffer amplifier, a stage that isolates the preceding stage from the following stage. Ideally, a buffer should have a high input impedance. If it does, almost all the Thevenin voltage from stage A appears at the buffer input. The buffer should also have a low output impedance. This ensures that all its output voltage reaches the input of stage B.

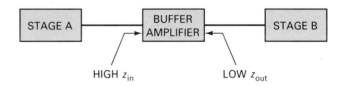

FIGURE 14-29   Buffer amplifier isolates stages *a* and *b*.

The source follower is an excellent buffer amplifier because of its high input impedance (well into the megohms at low frequencies) and its low output impedance (typically a few hundred ohms). The high impedance means light loading of the preceding stage. The low output impedance means that the buffer can drive heavy loads (small load resistances).

Low-Noise Amplifier

Noise is any unwanted disturbance superimposed upon a useful signal. Noise interferes with the information contained in the signal; the greater the noise, the less the information. For instance, the noise in television receivers produces small white or black spots on the picture; severe noise can wipe out the picture. Similarly, the noise in radio receivers produces crackling and hissing, which sometimes completely masks the voice or music. Noise is independent of the signal because it exists even when the signal is off.

The FET is an outstanding low-noise device because it produces very little noise. This is especially important near the front end of receivers and other electronic equipment because the subsequent stages amplify front-end noise along with the signal. If we use a FET amplifier at the front end, we get less amplified noise at the final output.

Voltage-Variable Resistance

In our ideal JFET approximation, $R_{DS} = V_P/I_{DSS}$. This is useful for quick analysis of JFETs operating in the ohmic region. But there is a second approximation of a JFET that more closely models the drain

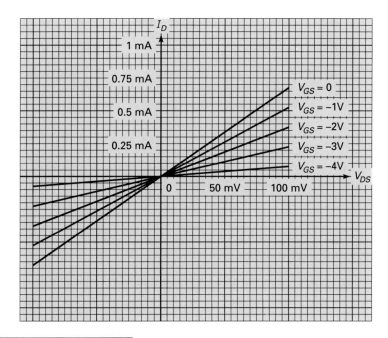

FIGURE 14-30   Drain curves are linear near origin.

curves in the ohmic region. The real truth is that R_{DS} depends on the value of V_{GS} as well as on V_P and I_{DSS}.

Figure 14-30 shows the drain curves of a 2N5951 in the ohmic region. The small-signal resistance R_{DS} depends on the value of V_{GS}. You can calculate R_{DS} by taking the ratio of drain voltage to drain current. For instance, when $V_{GS} = 0$,

$$R_{DS} = \frac{100\,\text{mV}}{0.75\,\text{mA}} = 133\,\Omega$$

When $V_{GS} = -2$ V,

$$R_{DS} = \frac{100\,\text{mV}}{0.4\,\text{mA}} = 250\,\Omega$$

You can see that V_{GS} controls the value of R_{DS}. Because of this a JFET operating in the ohmic region with small ac signals acts as a voltage-controlled resistance. Let that sink in. It opens the door to all kinds of inventions.

Notice that the drain curves of Fig. 14-30 extend on both sides of the origin. This means that a JFET can be used as a voltage-variable resistance for small ac signals, typically those less than 100 mV. When it is used in this way, the JFET does not need a dc drain voltage from the supply. All that is required is an ac input signal. You will see many applications for a voltage-variable resistance in later chapters.

Incidentally, data sheets list the value of the voltage-controlled resistance not as R_{DS} but as $r_{ds(\text{on})}$. Lowercase letters are used because it is a small-signal ac quantity. It is valid only for small signals like 100 mV peak to peak (see Fig. 14-30). We continue to use R_{DS} in this book because it unifies our discussion of drain resistance under different conditions. Here are the different R_{DS} values you will find on data sheets:

$R_{DS(\text{on})}$ — Drain resistance of a two-terminal enhancement-mode MOSFET in the current-source region. This is quite a bit larger than the drain resistance in the ohmic region because $V_{DS(\text{on})}$ is typically 10 to 15 V.

$r_{DS(\text{on})}$ — The dc drain resistance in the ohmic region. It is smaller than $R_{DS(\text{on})}$ because $r_{DS(\text{on})}$ is measured for a V_{DS} that is less than the pinchoff voltage.

$r_{ds(\text{on})}$ — The ac drain resistance in the ohmic region. It is valid only for small signals where the FET is operating in the vicinity of the origin.

Automatic Gain Control

When a receiver is tuned from a weak to a strong station, the loudspeaker will blare unless the volume is immediately decreased. Or the volume may change because of fading, a variation in signal strength caused by an electrical change in the path between the transmitting and receiving antennas. To prevent unwanted changes in volume, most receivers use automatic gain control (AGC).

This is where the JFET comes in. As shown earlier,

$$g_m = g_{m0}\left(1 - \frac{V_{GS}}{V_{GS(\text{off})}}\right)$$

This is a linear equation. When it is graphed, it results in Fig. 14-31a. For a JFET, g_m reaches a maximum value when $V_{GS} = 0$. As V_{GS} becomes more negative, the value of g_m decreases. Since a common-source amplifier has a voltage gain of

$$A = g_m r_d$$

we can control the voltage gain by controlling the value of g_m.

Figure 14-31b shows how it's done. A JFET amplifier is near the front end of a receiver. It has a voltage gain of $g_m r_d$. Subsequent stages amplify the JFET output. This amplified output goes into a negative peak detector that produces voltage V_{AGC}. This negative voltage returns to the JFET amplifier, where it is applied to the gate through a 10-kΩ resistor. When the receiver is tuned from a weak to a strong station, a larger signal is peak-detected and V_{AGC} is more negative; this reduces the gain of the JFET amplifier.

The overall effect of AGC is this: The final signal increases, but not nearly as much as it would without AGC. For instance, in some AGC systems an increase of 100 percent in the input signal results in an increase of less than 1 percent in the final output signal.

Cascode Amplifier

Figure 14-32a is an example of a cascode amplifier, a common-source amplifier driving a common-gate amplifier. Here's how it works: For

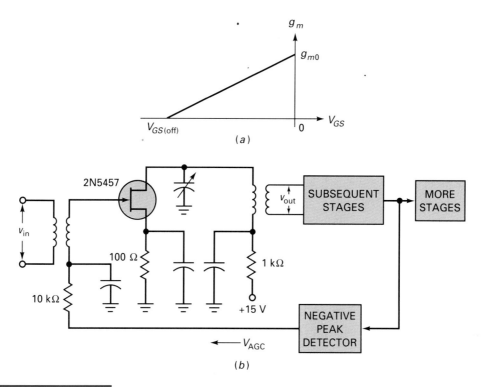

FIGURE 14-31 Automatic gain control.

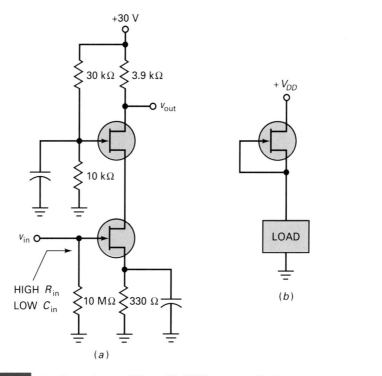

FIGURE 14-32 (a) Cascode amplifier; (b) FET current limiter.

simplicity, assume matched JFETs so that both have the same g_m. The *CS* amplifier has a gain of

$$A_1 = g_m R_D$$

The input impedance of the *CG* amplifier is $1/g_m$; this is the drain resistance seen by the *CS* amplifier. Therefore,

$$A_1 = g_m R_D = g_m \frac{1}{g_m} = 1$$

The upper JFET has a gain of

$$A_2 = g_m R_D$$

So the overall gain of the two JFETs is

$$A = A_1 A_2 = g_m R_D$$

This says that a cascode connection has the same voltage gain as a CS amplifier.

The main advantage of a cascode connection is its low input capacitance, which is considerably less than the input capacitance of a *CS* amplifier. Chapter 16 will explain why a cascode amplifier has a low input capacitance (it is related to the Miller effect). Until then, all you have to remember is that a cascode amplifier has the same voltage gain as a *CS* amplifier, but its input capacitance is very low.

Current Limiting

The JFET of Fig. 14-32*b* can protect a load against excessive current. For instance, suppose the normal load current is 1 mA. If $I_{DSS} = 10$ mA and $r_{DS(\text{on})} = 200\ \Omega$, then a normal load current of 1 mA means the JFET is operating in the ohmic region with a voltage drop of only

$$V_{DS} = (1\ \text{mA})(200\ \Omega) = 0.2\ \text{V}$$

Almost all the supply voltage therefore appears across the load. Now suppose the load shorts. Then the load current tries to increase to an excessive level. The increased load current forces the JFET into the active region, where it limits the current to 10 mA. The JFET now acts as a current source and prevents excessive load current.

A manufacturer can tie the gate to the source and package the JFET as a two-terminal device. This is how constant-current diodes are made. Constant-current diodes are also called *current-regulator diodes*.

Sample-and-Hold Amplifier

Like the JFET, the MOSFET can act as a switch, either in shunt or in series with the load. The enhancement-mode MOSFET is particularly convenient in switching applications because it is normally off. Figure 14-33*a* shows a useful circuit called a *sample-and-hold amplifier*. When V_{GS} is high, the MOSFET turns on and the capacitor charges to the value of input voltage. The charging time constant is very short because $r_{ds(\text{on})}$ is small. When V_{GS} goes low, the MOSFET opens and the capacitor

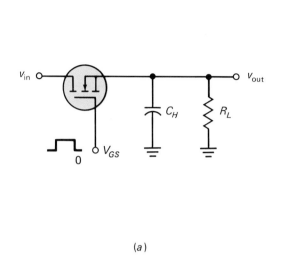

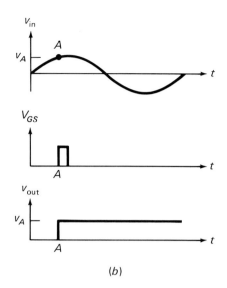

(a)   (b)

FIGURE 14-33   (a) Sample-and-hold amplifier; (b) waveforms.

begins to discharge through the load resistor. If the discharging time constant is very large, the capacitor can hold its charge for a long time.

In many applications we need a dc output equal to the input voltage at a particular instant. For example, suppose we want the value of input voltage at point A in Fig. 14-33b. If we apply a narrow V_{GS} at point A, the v_{out} of the sample-and-hold amplifier can charge to approximately v_A as shown. When V_{GS} returns to zero, the MOSFET opens and the input voltage can no longer affect the value of the output voltage. Given a long time constant, the output voltage holds at v_A for an indefinite period (see Fig. 14-33b).

When V_{GS} is high in Fig. 14-33a, the circuit is sampling the input and the capacitor charges to the approximate value of input voltage. When V_{GS} returns to low, the circuit goes into a hold condition because the capacitor stores the sampled value of the input voltage. Remember the basic idea of a sample-and-hold amplifier; you will see it used a lot with analog-to-digital converters (computer circuits).

Enhancement-Mode Power MOSFETs

As discussed earlier, the VMOS transistor is an enhancement-mode MOSFET optimized for its high power dissipation. One important application for power MOSFETs lies in interfacing low-power digital devices to high-power loads. Interfacing means using some kind of buffer between a low-power device (often a digital IC) and a high-power load (such as a relay, motor, or incandescent lamp). The enhancement-mode power MOSFET is an excellent device for interfacing digital ICs to high-power loads. As shown in Fig. 14-34, a digital IC drives the gate of a power MOSFET. When the digital output is low, the MOSFET is off. When the digital output is high, the MOSFET acts as a closed switch, and maximum current flows through the load. Interfacing digital ICs

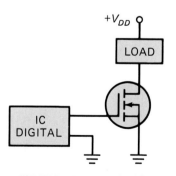

FIGURE 14-34

VMOS transistor interfaces a low-power digital IC with a high-power load.

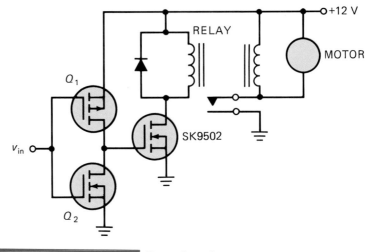

FIGURE 14-35 Part of a robot.

(such as CMOS, MOS, or TTL) to high-power loads is one of the most important applications for the power MOSFET.

Controlling motors is another important application of power MOSFETs. Figure 14-35 shows part of a robot. The SK9502 is a power MOSFET. It has a maximum current rating of 3 A and a breakdown voltage of 20 V. The circuit interfaces a CMOS inverter with a relay. Typically, a relay takes hundred of milliamperes, too heavy a load for a digital IC like a CMOS inverter. The SK9502 has a maximum current rating of 3 A, more than enough to supply the relay current.

When v_{in} is low, the CMOS inverter (Q_1 and Q_2) has a high output, which turns on the MOSFET. Since the MOSFET has a very small $r_{DS(on)}$, it effectively shorts the lower end of the relay to ground. The relay points then close, and the motor starts turning. The motor continues to run as long as v_{in} is low. When v_{in} goes high, the CMOS inverter output goes low. The power MOSFET then shuts off, the relay opens, and the motor stops running.

STUDY AIDS

The following study aids will help to reinforce the ideas discussed in this chapter. For best results, use these study aids within 6 hours of reading the earlier material. Then review these study aids a week later and a month later to ensure that the concepts remain in your long-term memory.

SUMMARY

Sec. 14-1 Self-Bias of JFETs

The drain current produces a voltage across the source resistor. This reverse-biases the JFET. The larger the source resistance, the more negative the gate-source voltage. Self-bias is the most widely used JFET bias. Because self-bias produces negative feedback, it stabilizes the Q point against changes in JFET quantities.

Sec. 14-2 Graphical Solution for Self-Bias

When a transconductance curve is available for a JFET, you can find the Q point of a self-biased circuit by drawing the self-bias line. This is a line that starts at the origin and passes through the point with coordinates of I_{DSS} and $-I_{DSS}R_S$. The intersection of the self-bias line and the transconductance curve is the Q point.

Sec. 14-3 Solution with Universal JFET Curve

A neat way to analyze any self-biased circuit is with a universal curve. For self-bias, this is a curve that shows I_D/I_{DSS} versus R_S/R_{DS}. These quantities are normalized variables and are the key to creating the universal curve. Given any self-biased circuit, you calculate R_{DS} with V_P/I_{DSS}. Next you calculate R_S/R_{DS}. Then you read I_D/I_{DSS} from the universal curve. Finally you calculate the value of I_D.

Sec. 14-4 Transconductance

Transconductance indicates how effectively the input voltage controls the output current. It is an ac quantity because it equals the ratio of a small change in drain current to a small change in gate-source voltage. The transconductance is maximum at $V_{GS} = 0$. It decreases linearly as V_{GS} changes from 0 to $V_{GS(\text{off})}$. The g_m of a bipolar transistor equals $1/r'_e$.

Sec. 14-5 JFET Amplifiers

The voltage gain of a CS amplifier is $g_m r_d$. The voltage gain of other JFET amplifiers is analogous to the equivalent bipolar amplifiers. All you have to do is convert r_c, r_e, and r'_e to r_d, r_s, and $1/g_m$. The CS amplifier is not very popular because it has low voltage gain and high distortion. On the other hand, the source follower is immensely popular because of its high input impedance and low distortion.

Sec. 14-6 The JFET Analog Switch

One of the major applications of a JFET is analog switching. This refers to transmitting or blocking small ac signals. The two common JFET analog switches are the shunt switch and the series switch. In either circuit, the JFET is ideally equivalent to an off-on switch. For normal operation, the ac signal has to be kept small, much less than the pinchoff voltage. In most applications, the series switch is preferred because it has a better off-on ratio.

Sec. 14-7 Depletion-Mode MOSFET Amplifiers

Because of their low noise, depletion-mode MOSFETs are often used in the early stages of high-frequency equipment. Since g_m is controllable by V_{GS}, depletion-mode MOSFET amplifiers can have their voltage gain automatically controlled. Another application is the dual-gate device which can be used as a cascode amplifier. This type of amplifier provides useful voltage gain at higher frequencies than a comparable bipolar amplifier.

Sec. 14-8 Enhancement-Mode MOSFET Applications

The main application for the enhancement-mode MOSFET lies in switching circuits. These are circuits whose output voltage is two-state: either low or high. In the switching circuits, you often will see the enhancement-mode MOSFET used as either a switch or a resistor. Active-load switching means one MOSFET acts as a driver or switch, while the other acts as a resistor.

VOCABULARY

In your own words, explain what each of the following terms means. Keep your answers short and to the point. If necessary, verify by rereading the appropriate discussion or by looking at the end-of-book Glossary.

active-load resistor	shunt switch
CMOS inverter	source follower
normalized variable	transconductance
self-bias	universal curve
series switch	

IMPORTANT EQUATIONS

Here are some important equations. Say each of the following equations in symbols, then say each in words. Try to explain what the equation means and how it is used. Then read the description that follows.

Eq. 14-1 Self-Bias Voltage

$$V_{GS} = -I_D R_S$$

This says the gate-source voltage equals the negative of the source voltage. It makes sense because the gate-source voltage equals the gate voltage minus the source voltage. Since the gate voltage is zero, the gate-source voltage . . . (you know the rest).

Eq. 14-2 Selecting a Source Resistance for Self-Bias

$$R_S = \frac{V_P}{I_{DSS}}$$

This is a quick design rule for self-biased JFETs. It says to use a source resistance that equals the pinchoff voltage divided by the maximum drain current. An easy way to remember this is to use a source resistance equal to the ideal R_{DS} in the ohmic region.

Eqs. 14-5 and 14-6 Transconductance

$$g_m = \frac{\Delta I_D}{\Delta V_{GS}} \quad \text{and} \quad g_m = \frac{i_d}{v_{gs}}$$

Both of these amount to the same thing. The first says to divide the change in drain current by the change in gate-source cutoff voltage. The second says to divide the ac drain current by the ac gate-source voltage. Each equation gives the value of the transconductance.

Eq. 14-7 Gate-Source Cutoff Voltage

$$V_{GS(\text{off})} = \frac{-2I_{DSS}}{g_{m0}}$$

Sometimes data sheets give you I_{DSS} and g_{m0}, but not $V_{GS(\text{off})}$. In this case, you can use this equation to calculate an accurate value of $V_{GS(\text{off})}$. It's not the kind of equation that you have to memorize, but you should remember that it is here if you need it. The equation says to double I_{DSS} and divide by g_{m0}. Add a minus sign, and you have the gate-source cutoff voltage.

Eq. 14-8 Operating Transconductance

$$g_m = g_{m0}\left(1 - \frac{V_{GS}}{V_{GS(\text{off})}}\right)$$

The transconductance at the Q point is less than the maximum transconductance, which occurs when $V_{GS} = 0$. This equation says the operating transconductance decreases linearly when V_{GS} changes from 0 V to $V_{GS(\text{off})}$.

Eq. 14-9 Bipolar Transconductance

$$g_m = \frac{1}{r'_e}$$

This is useful in two ways. First, it allows you to compare a bipolar transistor to a FET. Second, it is part of the bipolar-to-FET shortcut for remembering the voltage gain of different FET amplifiers.

IMPORTANT PROCESSES

Each of the following processes is important for FET circuit analysis. If you can remember the basic ideas in each step, you will be able to solve problems more easily.

Graphical Solution for Self-Bias

1. Multiply I_{DSS} by R_S to get V_{GS} for the second point.
2. Draw a line through the origin and the second point.
3. Read the coordinates of the intersection point.

Self-Bias Solution with Universal Curve

1. Calculate R_{DS} with V_P/I_{DSS}.
2. Calculate R_S/R_{DS}.
3. Read I_D/I_{DSS} from the universal curve.
4. Calculate I_D.

Bipolar-to-FET Conversions

1. Change r_c and r_e to r_d and r_s, respectively.
2. Change r'_e to $1/g_m$.
3. Rearrange the equation for the JFET voltage gain.

Active-Load Switching

1. Calculate R_{DS} of the two-terminal MOSFET with $V_{D(\text{on})}/I_{D(\text{on})}$.
2. Look up $r_{DS(\text{on})}$ of the driver MOSFET on the data sheet.
3. Calculate the output voltage for the voltage divider formed by R_{DS} and $r_{DS(\text{on})}$.

STUDENT ASSIGNMENTS

QUESTIONS

Some questions have more than one right answer. Select the best answer, the one that is always true, or that most accurately describes the situation, etc.

1. The preferred form of biasing a JFET amplifier is through the
   a. Voltage-divider bias
   b. Self-bias
   c. Gate bias
   d. Source bias

2. Self-bias produces
   a. Positive feedback
   b. Negative feedback
   c. Forward feedback
   d. Reverse feedback

3. To get a negative gate-source voltage in a self-biased JFET circuit, you must have a
   a. Voltage divider
   b. Source resistor
   c. Ground
   d. Negative gate supply voltage

4. The quantity R_S/R_{DS} is measured in
   a. Ohms
   b. Amperes
   c. Volts
   d. Dimensionless units

5. Transconductance is measured in
   a. Ohms
   b. Amperes
   c. Volts
   d. Mhos

6. Transconductance indicates how effectively the input voltage controls the
   a. Voltage gain
   b. Input resistance
   c. Supply voltage
   d. Output current

7. The transconductance increases when the drain current approaches
   a. 0
   b. $V_{GS(off)}$
   c. I_{DSS}
   d. V_P/R_S

8. A *CS* amplifier has a voltage gain of
   a. $g_m r_d$
   b. $g_m r_s$
   c. $g_m r_d/(1 + g_m r_s)$
   d. $g_m r_s/(1 + g_m r_s)$

9. A source follower has a voltage gain of
   a. $g_m r_d$
   b. $g_m r_s$
   c. $g_m r_d/(1 + g_m r_s)$
   d. $g_m r_s/(1 + g_m r_s)$

10. A swamped *CS* amplifier has a voltage of
    a. $g_m r_d$
    b. $g_m r_s$
    c. $g_m r_d/(1 + g_m r_s)$
    d. $g_m r_s/(1 + g_m r_s)$

11. For the same output current, the g_m of a JFET is
    a. Much smaller than that of a bipolar transistor
    b. About the same
    c. Much larger
    d. Impossible to predict

12. When the input signal is large, a *CS* amplifier has
    a. A large voltage gain
    b. A small distortion
    c. A low input resistance
    d. None of these

13. When the input signal is large, a source follower has
    a. A voltage gain of less than 1
    b. A small distortion
    c. A high input resistance
    d. All these

14. The input signal used with a JFET analog switch should be
    a. Small
    b. Large
    c. A square wave
    d. Chopped

15. A depletion-mode MOSFET is often used in what part of high-frequency equipment?
    a. Front end
    b. Power supply
    c. Class B push-pull output stage
    d. Middle stages

16. A cascode amplifier has the advantage of
  a. Large voltage gain
  b. Low input capacitance
  c. Low input impedance
  d. Higher g_m

17. VHF stands for frequencies from
  a. 300 kHz to 3 MHz
  b. 3 to 30 MHz
  c. 30 to 300 MHz
  d. 300 MHz to 3 GHz

18. An ordinary resistor is an example of
  a. A three-terminal device
  b. An active load
  c. A passive load
  d. A switching device

19. An enhancement-mode MOSFET with its gate connected to its drain is an example of
  a. A three-terminal device
  b. An active load
  c. A passive load
  d. A switching device

20. An enhancement-mode MOSFET that operates at cutoff or in the ohmic region is an example of
  a. A three-terminal device
  b. An active load
  c. A passive load
  d. A switching device

21. If an enhancement-mode MOSFET has $I_{D(on)} = 10$ mA and $V_{DS(on)} = 2$ V, then its $R_{DS(on)}$ is
  a. 200 Ω
  b. 2 V
  c. 1 kΩ
  d. -2 V

22. CMOS stands for
  a. Common MOS
  b. Active-load switching
  c. p-channel and n-channel devices
  d. Complementary MOS

23. The VMOS transistor is a
  a. Bipolar transistor
  b. Small-signal MOSFET
  c. Medium-power JFET
  d. High-power enhancement-mode MOSFET

24. Self-bias will not work with
  a. JFETs
  b. Depletion-mode MOSFETs
  c. Enhancement-mode MOSFETs
  d. Any depletion-mode device

BASIC PROBLEMS

Sec. 14-1 Self-Bias of JFETs

14-1. Suppose the dc drain current of Fig. 14-36a is 2.5 mA. What are the values of V_D, V_S, and V_{GS}?

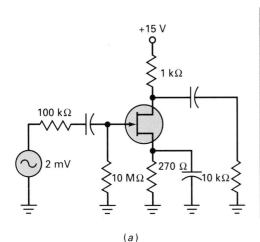

(a)

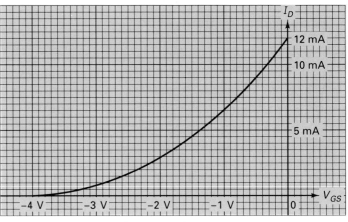

(b)

FIGURE 14-36

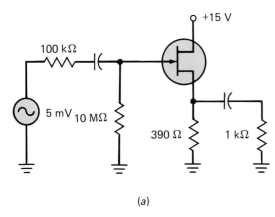

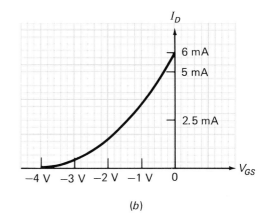

(a)

(b)

FIGURE 14-37

14-2. Somebody builds the circuit of Fig. 14-36a and measures a dc drain-to-ground voltage of 7.5 V. What is the value of V_{GS}?

14-3. When the circuit of Fig. 14-36a is built, the dc voltage between the source and ground is 1 V. What are the values of V_{GS} and V_{DS}?

Sec. 14-2 Graphical Solution for Self-Bias

14-4. The JFET of Fig. 14-36a has the transconductance curve shown in Fig. 14-36b. Use the graphical solution to find the Q point.

14-5. Repeat Prob. 14-4 for $R_S = 100\ \Omega$.

14-6. Repeat Prob. 14-4 for $R_S = 800\ \Omega$.

14-7. A JFET has $I_{DSS} = 2$ mA and $V_{GS(\text{off})} = -5$ V. Use Eq. (14-2) to select a source resistance to produce self-bias for the JFET.

14-8. Repeat Prob. 14-7 for an MPF102 using the maximum I_{DSS} and $V_{GS(\text{off})}$ on the data sheet.

Sec. 14-3 Solution with Universal JFET Curve

14-9. What are the dc drain current and the gate-source voltage in Fig. 14-36a if $I_{DSS} = 10$ mA and $V_{GS(\text{off})} = -5$ V?

14-10. What is the dc drain-to-ground voltage in Fig. 14-36a if $I_{DSS} = 8$ mA and $V_{GS(\text{off})} = -2$ V?

14-11. If $I_{DSS} = 6$ mA and $V_P = 4$ V in Fig. 14-37a, what is the dc voltage between the source and ground?

Sec. 14-4 Transconductance

14-12. A 2N4416 has $I_{DSS} = 10$ mA and $g_{m0} = 4000$ μmho. What is its gate-source cutoff voltage? What is the value of g_m for $V_{GS} = -2$ V?

14-13. A 2N3370 has $I_{DSS} = 2.5$ mA and $g_{m0} = 1500$ μmho. What is the value of g_m for $V_{GS} = -1$ V?

14-14. The JFET of Fig. 14-36a has a g_{m0} of 6000 μmho. If $I_{DSS} = 12$ mA, what is the value of I_D for $V_{GS} = -2$ V? Find the g_m for this I_D.

Sec. 14-5 JFET Amplifiers

14-15. If $g_m = 3000$ μmho in Fig. 14-36a, what is the ac output voltage?

14-16. The JFET amplifier of Fig. 14-36a has the transconductance curve of Fig. 14-36b. In this case, what is the ac output voltage?

14-17. If the source follower of Fig. 14-37a has $g_m = 2000$ μmho, what is the ac output voltage?

14-18. The source follower of Fig. 14-37a has the transconductance curve of Fig. 14-37b. What is the ac output voltage?

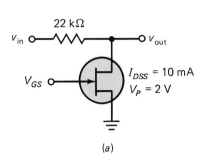

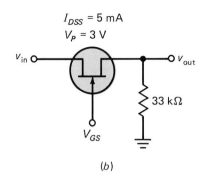

FIGURE 14-38

Sec. 14-6 The JFET Analog Switch

14-19. The input voltage of Fig. 14-38a is 50 mV peak to peak. What is the output voltage when $V_{GS} = 0$ V? When $V_{GS} = -3$ V?

14-20. The input voltage of Fig. 14-38b is 25 mV peak to peak. What is the output voltage when $V_{GS} = 0$ V? When $V_{GS} = -5$ V?

Sec. 14-8 Enhancement-Mode MOSFET Applications

14-21. In Fig. 14-39a, what is v_{out} when $v_{in} = 0$? When $v_{in} = +5$ V?

14-22. If the voltage across the upper MOSFET of Fig. 14-39b is 15 V, what is the drain resistance of this two-terminal device?

14-23. In Fig. 14-39b, what is v_{out} when $v_{in} = 0$? When $v_{in} = +5$ V?

TROUBLESHOOTING PROBLEMS

14-24. In Fig. 14-41, you measure a dc drain-to-ground voltage of $+15$ V. Name some of the troubles that can produce this.

14-25. All dc voltages are normal in Fig. 14-41, but the ac output voltage is zero. Name some of the troubles that can produce this condition.

14-26. In Fig. 14-41, the emitter bypass capacitor is open in the second stage. What happens to the dc voltage at the output? What happens to the ac voltage?

ADVANCED PROBLEMS

14-27. The JFET of Fig. 14-36a has the transconductance curve of Fig. 14-36b. Keep

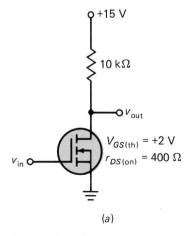

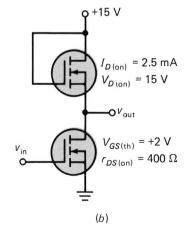

FIGURE 14-39

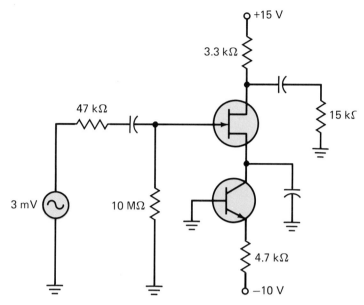

FIGURE 14-40

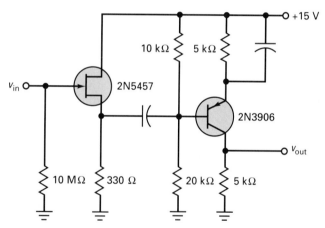

FIGURE 14-41

all resistances the same, except the source resistance R_S.

a. What is the value of R_S that produces an I_D of 5 mA?

b. What is the value of R_S that produces a V_{GS} of -2 V?

14-28. In any way you can think of, get a formula for R_S in terms of R_{DS} that self-biases a JFET at $I_D = 0.5I_{DSS}$.

14-29. What is the dc drain voltage in Fig. 14-40? The ac output voltage if $g_m = 2000$ μmho?

14-30. The 2N5457 has a g_m of 2500 μmho in Fig. 14-41. The 2N3906 has a β of 150. If v_{in} is 1 mV, what is v_{out}?

14-31. If the 2N5457 of Fig. 14-41 has a g_m of 2500 μmho, what is the output impedance of the first stage?

14-32. Figure 14-42 shows a FET dc voltmeter. The zero adjust is set just before a reading is taken. The calibrate adjust is set periodically to give full-scale deflection when $V_{in} = -2.5$ V. A calibrate adjustment

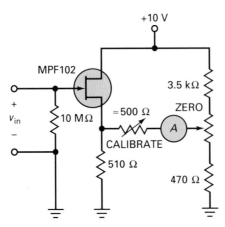

FIGURE 14-42

like this takes care of variations from one FET to another and FET aging effects.

a. The current through the 510 Ω equals 4 mA. How much dc voltage is there from the source to ground?

b. If no current flows through the ammeter, what voltage does the wiper tap off the zero adjust?

c. If an input voltage of 2.5 V produces a deflection of 1 mA, how much deflection does 1.25 V produce?

d. The MPF102 has an I_{GSS} of 2 nA for a V_{GS} of 15 V. What is the input impedance of the voltmeter?

14-33. A JFET can be used as a current limiter to protect a load from excessive current. In Fig. 14-43a, the JFET has an I_{DSS} of 16 mA and an $r_{DS(on)}$ of 200 Ω. If the load accidentally shorts, what are the load current and the voltage across the JFET? If the load has a resistance of 10 kΩ, what are the load current and the voltage across the JFET?

14-34. Figure 14-43b shows part of an AGC amplifier. A dc voltage is fed back from an output stage to an earlier stage such as the one shown here. Figure 14-36b is the transconductance curve. What is the unloaded voltage gain for each of these?

a. $V_{AGC} = 0$
b. $V_{AGC} = -1$ V
c. $V_{AGC} = -2$ V
d. $V_{AGC} = -3$ V
e. $V_{AGC} = -3.5$ V

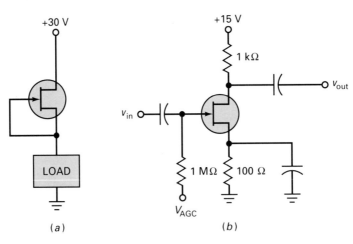

(a)   (b)

FIGURE 14-43

T-SHOOTER PROBLEMS

Use Fig. 14-44 for the remaining problems. The big box labeled "Millivolts unless otherwise indicated" contains measurements of the ac voltages expressed in millivolts. For this exercise, the voltage sources and the JFET are normal.

14-35. Find Trouble 1.

14-36. Find Trouble 2.

14-37. Find Trouble 3.

14-38. Find Trouble 4.

14-39. Find Troubles 5 and 6.

14-40. Find Troubles 7 and 8.

14-41. Find Troubles 9 and 10.

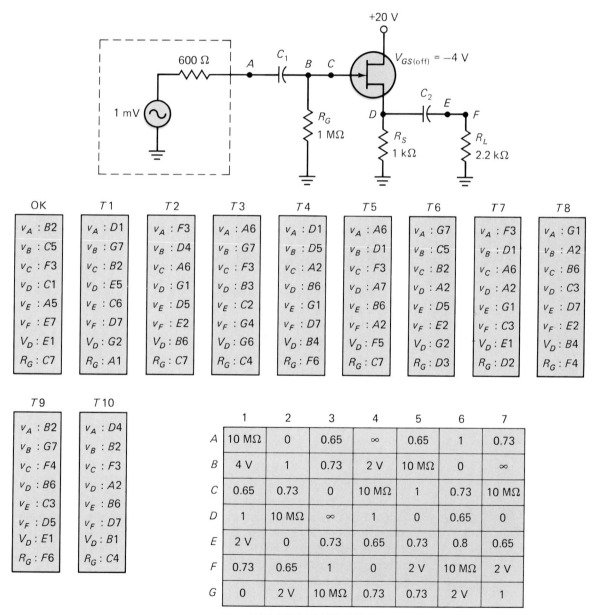

OK

v_A	: B2
v_B	: C5
v_C	: F3
v_D	: C1
v_E	: A5
v_F	: E7
V_D	: E1
R_G	: C7

T1

v_A	: D1
v_B	: G7
v_C	: B2
v_D	: E5
v_E	: C6
v_F	: D7
V_D	: G2
R_G	: A1

T2

v_A	: F3
v_B	: D4
v_C	: A6
v_D	: G1
v_E	: D5
v_F	: E2
V_D	: B6
R_G	: C7

T3

v_A	: A6
v_B	: G7
v_C	: F3
v_D	: B3
v_E	: C2
v_F	: G4
V_D	: G6
R_G	: C4

T4

v_A	: D1
v_B	: D5
v_C	: A2
v_D	: B6
v_E	: G1
v_F	: D7
V_D	: B4
R_G	: F6

T5

v_A	: A6
v_B	: D1
v_C	: F3
v_D	: A7
v_E	: B6
v_F	: A2
V_D	: F5
R_G	: C7

T6

v_A	: G7
v_B	: C5
v_C	: B2
v_D	: A2
v_E	: D5
v_F	: E2
V_D	: G2
R_G	: D3

T7

v_A	: F3
v_B	: D1
v_C	: A6
v_D	: A2
v_E	: G1
v_F	: C3
V_D	: E1
R_G	: D2

T8

v_A	: G1
v_B	: A2
v_C	: B6
v_D	: C3
v_E	: D7
v_F	: E2
V_D	: B4
R_G	: F4

T9

v_A	: B2
v_B	: G7
v_C	: F4
v_D	: B6
v_E	: C3
v_F	: D5
V_D	: E1
R_G	: F6

T10

v_A	: D4
v_B	: B2
v_C	: F3
v_D	: A2
v_E	: B6
v_F	: D7
V_D	: B1
R_G	: C4

	1	2	3	4	5	6	7
A	10 MΩ	0	0.65	∞	0.65	1	0.73
B	4 V	1	0.73	2 V	10 MΩ	0	∞
C	0.65	0.73	0	10 MΩ	1	0.73	10 MΩ
D	1	10 MΩ	∞	1	0	0.65	0
E	2 V	0	0.73	0.65	0.73	0.8	0.65
F	0.73	0.65	1	0	2 V	10 MΩ	2 V
G	0	2 V	10 MΩ	0.73	0.73	2 V	1

MILLIVOLTS UNLESS OTHERWISE INDICATED

FIGURE 14-44 T-Shooter™. (*Patent pending: Courtesy of Malvino Inc.*)

15

THYRISTORS

A *thyristor* is a semiconductor device that uses internal feedback to produce a new kind of switching action. Unlike bipolar transistors and FETs, which can operate either as linear amplifiers or as switches, thyristors can only operate as switches. The main application of this new device is in controlling large amounts of load current for motors, heaters, lighting systems, and other such devices.

15-1 THE FOUR-LAYER DIODE

The word *thyristor* comes from the Greek and means "door," as in opening a door and letting something pass through it. As a start, you can think of a thyristor as a new kind of switch. All thyristors can be explained in terms of the circuit shown in Fig. 15-1a. Notice that the upper transistor Q_1 is a *pnp* device and the lower transistor Q_2 is an *npn* device. The collector of Q_1 drives the base of Q_2, and the collector of Q_2 drives the base of Q_1. This rather unusual connection is what makes the thyristor behave as it does. It is the reason the thyristor can function only as a switch.

Positive Feedback

Because of the unusual connection of Fig. 15-1a, we have *positive feedback*. A change in current at any point in the feedback loop is amplified and returned to the starting point with the same phase. For instance, if the Q_2 base current increases, the Q_2 collector current increases. This forces more base current through Q_1. In turn, this produces a larger Q_1 collector current, which drives the Q_2 base harder. This buildup in currents will continue until both transistors are driven into saturation. In this case, the circuit acts as a closed switch (Fig. 15-1b).

But if something causes the Q_2 base current to decrease, the Q_2 collector current will decrease. This reduces the Q_1 base current. In turn, there will be less Q_1 collector current, which reduces the Q_2 base current even

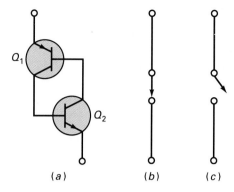

FIGURE 15-1   Transistor latch.

more. This positive feedback continues until both transistors are driven into cutoff. At this time, the circuit acts as an open switch (Fig. 15-1c).

The switching circuit can be in either of two states, closed or open. It will remain in either state indefinitely. If the switching circuit is closed, it stays closed until something causes the currents to decrease. If it is open, it stays open until something else forces the currents to increase. Because this kind of switching action is based on positive feedback, the circuit has been called a *latch*.

Closing a Latch

Assume the latch of Fig. 15-2a is open. Then the equivalent circuit is an open switch as shown in Fig. 15-2b. Because there is no current through the load resistor, the output voltage equals the supply voltage. This means the operating point is at the lower end of the load line (Fig. 15-2d).

How can we close the latch? One way is by *triggering*. The idea is to apply a trigger (sharp pulse) to forward-bias the Q_2 base-emitter diode in

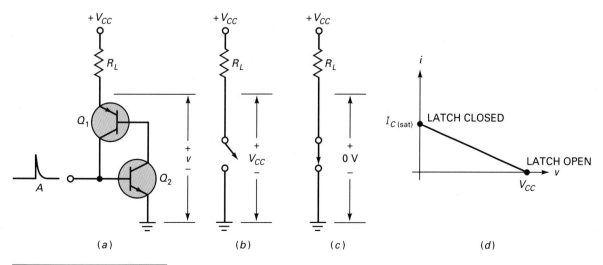

FIGURE 15-2   Transistor latch with trigger input.

Fig. 15-2a. At point A in time, the trigger momentarily turns on the Q_2 base current. The Q_2 collector current suddenly comes on and forces the base current through Q_1. In turn, the Q_1 collector current comes on and drives the Q_2 base harder. Since the Q_1 collector now supplies the Q_2 base current, the trigger pulse is no longer needed. Once the positive feedback starts, it will sustain itself and drive both transistors into saturation. The minimum input current needed to start the switching action is called the *trigger current*.

When saturated, both transistors ideally look like short circuits, and the latch is closed (Fig. 15-2c). Ideally, the latch has zero voltage across it when it is closed, and the operating point is at the upper end of the load line (Fig. 15-2d).

Another way to close a latch is by *breakover*. This means using a large enough supply voltage V_{CC} to break down either collector diode. Once the breakdown begins, current comes out of one of the collectors and drives the other base. The effect is the same as if the base had received a trigger. Although breakover starts with a breakdown of one of the collector diodes, it ends with both transistors in the saturated state. This is why the term *breakover* is used instead of *breakdown* to describe this kind of latch closing.

Opening a Latch

How do we open an ideal latch? One way is to reduce the load current to zero. This forces the transistors to come out of saturation and return to the open state. For instance, in Fig. 15-2a we can open the load resistor. Alternatively, we can reduce the V_{CC} supply to zero. In either case, a closed latch will be forced to open. We call this type of opening a *low-current dropout* because it depends on reducing the latch current to a low value.

Another way to open the latch is to apply a *reverse-bias trigger* in Fig. 15-2a. When a negative trigger is used instead of a positive one, the Q_2 base current decreases. This forces the Q_1 base current to decrease. Since the Q_1 collector current also decreases, the positive feedback will rapidly drive both transistors into cutoff, which opens the latch.

Here are the basic ways to close and open a latch:

1. We can close a latch by forward-bias triggering or by breakover.

2. We can open a latch by reverse-bias triggering or by low-current dropout.

There are many different kinds of thyristors. The simplest type can only be closed with breakover and be opened with low-current dropout. The most popular types are closed by triggering and are opened by low-current dropout. Some rare types can be closed and opened in any of the ways described here.

The Shockley Diode

Figure 15-3a is called a *four-layer diode* (also known as a *Shockley diode*). It is classified as a diode because it has only two external leads. Because

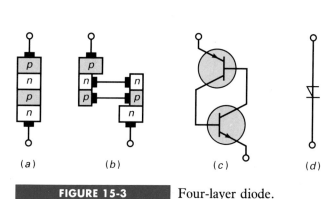

FIGURE 15-3 Four-layer diode.

of its four doped regions, it's often called a *pnpn* diode. The easiest way to understand how it works is to visualize it separated into two halves, as shown in Fig. 15-3b. The left half is a *pnp* transistor, and the right half is an *npn* transistor. Therefore, the four-layer diode is equivalent to the latch shown in Fig. 15-3c.

Because there are no trigger inputs, the only way to close a four-layer diode is by breakover, and the only way to open it is by low-current dropout. With a four-layer diode it is not necessary to reduce the current all the way to zero to open the latch. The internal transistors of the four-layer diode will come out of saturation when the current is reduced to a low value called the *holding current*. Figure 15-3d shows the schematic symbol for a four-layer diode.

After a four-layer diode breaks over, the voltage across it drops to a low value, depending on how much current there is. For instance, Fig. 15-3e shows the current versus voltage for a 1N5158. Notice that the voltage increases with the current through the device: 1 V at 0.2 A, 1.5 V at 0.95 A, 2 V at 1.8 A, and so forth. This means the switching is not perfect. An ideal switch has no voltage across it when closed. The four-layer diode is an imperfect switch because it has a volt or so across it when closed. This is not a serious disadvantage. Read the following examples and you will see why.

Breakover Characteristic

Figure 15-4 shows the graph of current versus voltage for a breakover diode. The device has two operating regions: nonconducting and conducting. When it is nonconducting, it operates on the lower line with no current and a voltage less than V_B. If the voltage tries to exceed V_B, the four-layer diode breaks over and switches along the dashed line to the conducting region. The dashed line in this graph indicates an unstable or a temporary condition. The device can have current and voltage values on this dashed line only briefly as it switches between the two stable operating regions. When the four-layer diode is conducting, it is operating on the upper line. As long as the current through it is greater than the

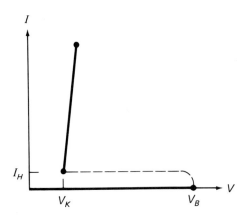

FIGURE 15-4 Breakover characteristic.

holding current I_H, the voltage across it is slightly larger than V_K. If the current tries to decrease to less than I_H, the device switches back along the dashed line to the nonconducting region.

The ideal approximation of a four-layer diode is an open switch when nonconducting and a closed switch when conducting. The second approximation includes the knee voltage V_K shown in Fig. 15-4. This knee voltage depends on the particular device. Often, it is near 0.7 V. To keep things simple and easy to remember, we will use 0.7 V unless a more accurate knee voltage is available from a data sheet. For the third approximation, we can include the bulk resistance of the diode in calculating the total diode voltage:

$$V_D = V_K + I_D R_B \qquad (15\text{-}1)$$

The answer to the question "Which approximation should I use?" depends on the situation. If you are troubleshooting, the ideal approximation is usually adequate. If you are designing, you probably will have a data sheet and may want to use Eq. (15-1) with an accurate value for the knee voltage and bulk resistance. Most of the time, the second approximation is the best compromise. Unless otherwise indicated, we use the second approximation.

EXAMPLE 15-1

In Fig. 15-5a, the input voltage is +15 V. What is the diode current? What is the input voltage at the dropout point?

SOLUTION

We use the second approximation with a knee voltage of 0.7 V. Since the four-layer diode has 0.7 V across it when conducting, the current is

$$I = \frac{15\,\text{V} - 0.7\,\text{V}}{100\,\Omega} = 143\,\text{mA}$$

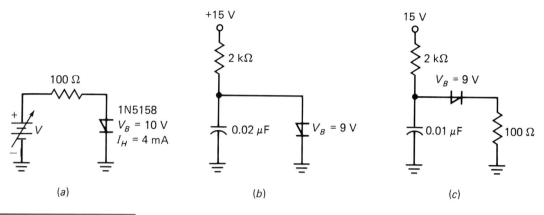

FIGURE 15-5 Example.

To open the four-layer diode, we have to reduce the current below the holding current of 4 mA. This means reducing the input voltage to slightly less than

$$V = 0.7\ \text{V} + (4\ \text{mA})(100\ \Omega) = 1.1\ \text{V}$$

EXAMPLE 15-2

The capacitor of Fig. 15-5b charges from 0.7 to 9 V, causing the four-layer diode to break over. In this example, what is the current through the 2 kΩ just before the diode breaks over and when it is conducting?

SOLUTION

Just before breakover, the voltage across the capacitor and diode is slightly less than 9 V. Therefore, the current through the 2 kΩ is

$$I = \frac{15\ \text{V} - 9\ \text{V}}{2\ \text{k}\Omega} = 3\ \text{mA}$$

Since the diode is off, all the current is charging the capacitor.

When the diode is conducting, it has a voltage of 0.7 V (second approximation). Therefore, the current through the 2 kΩ is

$$I = \frac{15\ \text{V} - 0.7\ \text{V}}{2\ \text{k}\Omega} = 7.15\ \text{mA}$$

EXAMPLE 15-3

In Fig. 15-5c, the capacitor voltage is slightly more than 9 V, and the diode voltage is 0.7 V. In this example, what is the current through the 100 Ω?

SOLUTION

The voltage across the 100 Ω is the difference between the capacitor voltage and the diode voltage:

$$V = 9\text{ V} - 0.7\text{ V} = 8.3\text{ V}$$

Therefore, Ohm's law gives

$$I = \frac{8.3\text{ V}}{100\ \Omega} = 83\text{ mA}$$

If you are wondering what the circuit does, here is a brief explanation. When power is first applied to the circuit, the capacitor voltage is zero. The capacitor charges, and its voltage increases from 0 toward 15 V. When the capacitor voltage is slightly greater than 9 V, the four-layer diode breaks over and becomes equivalent to a closed switch. This forces the capacitor to discharge through the 100 Ω. As the capacitor discharges, its voltage decreases, and this causes the diode current to decrease. When the diode current is slightly less than the holding current, the diode opens. The capacitor then starts to charge all over again. The cycle repeats with the diode breaking over, discharging the capacitor, and so on. The output of the circuit is a sharp voltage pulse across the load resistor. This sharp voltage pulse appears when the diode breaks over. The pulse is very short in duration because the capacitor quickly discharges.

15-2 THE SILICON CONTROLLED RECTIFIER

The *silicon controlled rectifier* (SCR) is more useful than a four-layer diode because it has an extra lead connected to the base of the *npn* section, as shown in Fig. 15-6a. Again, you can visualize the four doped regions separated into two transistors, as shown in Fig. 15-6b. Therefore, the SCR is equivalent to a latch with a trigger input (Fig. 15-6c). Schematic diagrams use the symbol of Fig. 15-6d. Whenever you see this, remember that it is equivalent to a latch with a trigger input.

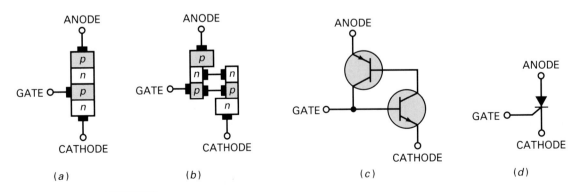

FIGURE 15-6   Silicon controlled rectifier.

Gate Triggering

The gate of an SCR is approximately equivalent to a diode (see Fig. 15-6c). For this reason, it takes at least 0.7 V to trigger an SCR. Furthermore, to get the positive feedback started, a minimum input current is required. Data sheets list the *trigger voltage* and *trigger current* for SCRs. For example, the data sheet of a 2N4441 gives a typical trigger voltage of 0.75 V and a trigger current of 10 mA. The source driving the gate of a 2N4441 has to be able to supply at least 10 mA at 0.75 V; otherwise, the SCR will not latch shut.

When an SCR is conducting, it has a low voltage across it. For instance, a 2N4441 has 1 V across it when it is turned on. To open an SCR, the action is similar to that for a four-layer diode. You have to decrease the SCR current to less than its holding current. For example, the 2N4441 has a holding current of 6 mA. To shut off this SCR, the anode current must decrease to less than 6 mA. Then it will suddenly stop conducting and become an open circuit.

We can again use three approximations to describe the action of an SCR. These are the same as for the four-layer diode. Ideally, an SCR has zero voltage when conducting. To a second approximation, it has V_K across it when conducting. To a third approximation, it has V_K plus the drop across its bulk resistance. If better data are not available, you can use 0.7 V for V_K.

Blocking Voltage

SCRs are not intended for breakover operation. Breakover voltages range from around 50 to more than 2500 V, depending on the SCR type number. Most SCRs are designed for trigger closing and low-current opening. In other words, an SCR stays open until a trigger drives its gate (Fig. 15-6d). Then the SCR latches and remains closed, even though the trigger disappears. The only way to open an SCR is with low-current dropout.

Most people think of an SCR as a device that blocks voltage until the trigger closes it. For this reason, the breakover voltage is often called the *forward blocking voltage* on data sheets. For instance, the 2N4441 has a forward blocking voltage of 50 V. As long as the supply voltage is less than 50 V, the SCR cannot break over. The only way to close it is with a gate trigger.

High Currents

Almost all SCRs are industrial devices that can handle large currents ranging from less than 1 to more than 2500 A, depending on the type number. Because they are high-current devices, SCRs have relatively large trigger and holding currents. The 2N4441 can conduct up to 8 A continuously; its trigger current is 10 mA, and so is its holding current. This means that you have to supply the gate with at least 10 mA to control up to 8 A of anode current. (The anode and cathode are shown in Fig. 15-6d.) As another example, the C701 is an SCR that can conduct up to 1250 A with a trigger current of 500 mA and a holding current of 500 mA.

Critical Rate of Rise

In many applications, an ac supply voltage is used with the SCR. By triggering the gate at a certain point in the cycle, we can control large amounts of ac power to a load such as a motor, a heater, or some other load. Because of junction capacitances inside the SCR, it is possible for a rapidly changing supply voltage to trigger the SCR. Put another way, if the rate of rise of forward voltage is high enough, the capacitive charging current can initiate the positive feedback.

To avoid false triggering of an SCR, the anode rate of voltage change must not exceed the *critical rate of voltage rise* listed on the data sheet. For instance, a 2N4441 has a critical rate of voltage rise of 50 V/μs. To avoid a false triggering, the anode voltage must not rise faster than 50 V/μs. As another example, the C701 has a critical rate of voltage rise of 200 V/μs. To avoid a false closure, the anode voltage must not increase faster than 200 V/μs.

Switching transients are the main cause of exceeding the critical rate of voltage rise. One way to reduce the effects of switching transients is with an *RC snubber*, shown in Fig. 15-7a. If a high-speed switching transient does appear on the supply voltage, its rate of rise is reduced at the anode because of the *RC* circuit. The rate of the anode voltage rise depends on the load resistance as well as on the *R* and *C* values.

Larger SCRs also have a *critical rate of current rise*. For instance, the C701 has a critical rate of current rise of 150 A/μs. If the anode current tries to rise faster than this, the SCR may be destroyed. Including an inductor in series, as shown in Fig. 15-7b, reduces the rate of current rise as well as helps the *RC* snubber decrease the rate of voltage rise.

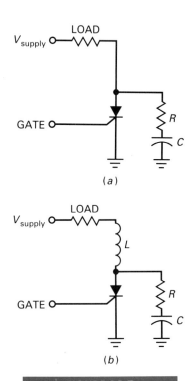

FIGURE 15-7

(*a*) *RC* snubber; (*b*) inductor protects SCR.

Trigger Current and Trigger Voltage

An SCR like the one shown in Fig. 15-8 has a gate voltage V_G. When this gate voltage is in the vicinity of 0.7 V, the SCR will turn on and the output voltage will drop from $+V_{CC}$ to a low value. When a gate resistor is used as shown here, you can calculate the input voltage needed to trigger an SCR by using this equation:

$$V_{in} = V_T + I_T R_G \qquad (15\text{-}2)$$

In this equation, V_T and I_T are the trigger voltage and trigger current, respectively, needed for the gate of the device. You will find this information on data sheets. For instance, the data sheet of a 2N4441 gives $V_T = 0.75$ V and $I_T = 10$ mA. When you have the value of R_G, the calculation of V_{in} is straightforward. Sometimes a gate resistor is not used. In this case, R_G is the Thevenin resistance of the circuit driving the gate. Unless Eq. (15-2) is satisfied, the SCR cannot turn on.

After the SCR has turned on, it stays on even though you reduce V_{in} to zero. In this case, the output voltage remains low indefinitely. The only way to reset the SCR is to reduce its current to less than the holding current. One way to do this is by opening R_C. Another way to do this is by reducing V_{CC} to a low value. There are other ways to turn off the SCR, which are discussed later.

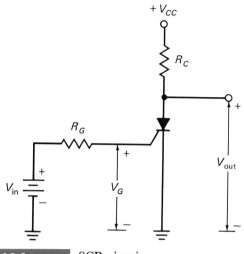

FIGURE 15-8   SCR circuit.

SCR Crowbar

If anything happens inside a power supply to cause its output voltage to go high, the results can be devastating. Why? Because some loads such as expensive ICs cannot withstand excessive supply voltage without being destroyed. One of the most important applications of the SCR is to protect delicate and expensive loads against overvoltages from a power supply.

Figure 15-9 shows a positive supply of V_{CC} applied to a protected load. The load is protected by the zener diode, resistor, and SCR. Under normal conditions, V_{CC} is less than the breakdown voltage of the zener diode. In this case, there is no voltage across R, and the SCR remains open. The load receives a voltage of V_{CC}, and all is well.

Now, assume the supply voltage increases for any reason whatever. When V_{CC} is too large, the zener diode conducts and a voltage appears across R. If this voltage is greater than the trigger voltage of the SCR (typically 0.7 V), the SCR turns on and conducts heavily. The action is

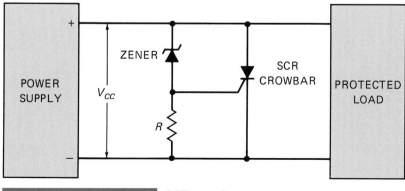

FIGURE 15-9   SCR crowbar.

similar to throwing a *crowbar* across the load terminals. Because the SCR turn-on is very fast (1 μs for a 2N4441), the load is quickly protected against the damaging effects of a large overvoltage.

Crowbarring, though a drastic form of protection, is necessary with many digital ICs; they can't take much overvoltage. Rather than destroy expensive ICs, therefore, we can use an SCR crowbar to short the load terminals at the first sign of overvoltage. Power supplies with an SCR crowbar need a fuse or current limiting to prevent excessive current when the SCR closes.

The crowbar of Fig. 15-9 is a popular design. It is adequate for many applications, provided the components have low tolerances. More advanced crowbar circuits include transistors to improve the turn-on action. In fact, special ICs such as the RCA SK9345 series are off-the-shelf, ready-to-use crowbars. These IC crowbars contain a zener diode, a couple of transistors, and an SCR. If you want more information on practical crowbar circuits, see "Optional Topics."

Incidentally, there is no separate gate resistor in Fig. 15-9 as shown in Fig. 15-8. In this case, V_{in} and R_G are interpreted as the V_{TH} and R_{TH}, respectively, of the circuit facing the gate of the SCR. What is this Thevenin resistance in Fig. 15-9? Looking back from the gate, you see the zener resistance in parallel with R. In a typical design, the Thevenin resistance is small. This means the equivalent input voltage needed to trigger the SCR is only slightly more than 0.7 V.

EXAMPLE 15-4

The SCR of Fig. 15-10 has $V_T = 0.75$ V, $I_T = 7$ mA, and $I_H = 6$ mA. What is the output voltage when the SCR is off? What is the input voltage that triggers the SCR? If V_{CC} is decreased until the SCR opens, what is the value of V_{CC}?

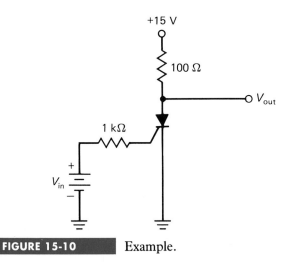

FIGURE 15-10   Example.

SOLUTION

When the SCR is not conducting, there is no current through the 100 Ω and the output voltage equals the supply voltage:

$$V_{\text{out}} = 15 \text{ V}$$

Since the trigger voltage is 0.75 V and the trigger current is 7 mA, the required input voltage for triggering is

$$V_{\text{in}} = 0.75 \text{ V} + (7 \text{ mA})(1 \text{ k}\Omega) = 7.75 \text{ V}$$

Because the holding current is 6 mA, the voltage across the 100 Ω at low-current dropout is

$$V = (6 \text{ mA})(100 \text{ }\Omega) = 0.6 \text{ V}$$

With the ideal approximation, this means we have to decrease V_{CC} to slightly less than 0.6 V to turn off the SCR. With the second approximation, we have to decrease V_{CC} to slightly less than

$$V_{CC} = 0.7 \text{ V} + (6 \text{ mA})(100 \text{ }\Omega) = 1.3 \text{ V}$$

EXAMPLE 15-5

In Fig. 15-11, the 2N4441 has a trigger voltage of 0.75 V. Calculate the supply voltage that turns on the crowbar. Ignore the zener resistance in this calculation.

SOLUTION

Ideally, the 1N752 has a breakdown voltage of 5.6 V. To turn on the SCR, the voltage across the 68 Ω has to be 0.75 V. Therefore, the supply voltage that just turns on the SCR is

$$V_{CC} = 5.6 \text{ V} + 0.75 \text{ V} = 6.35 \text{ V}$$

When the supply voltage reaches this level, the SCR closes like a switch. This immediately forces the supply voltage to drop toward

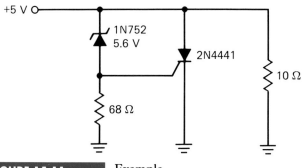

FIGURE 15-11　Example.

zero. The supply voltage drops because it has a fuse that burns out or a feature called *current limiting* (discussed later). In either case, the supply voltage drops to zero or a very low level.

The circuit is all right if the application is not too fussy about the exact supply voltage where the SCR turns on. For instance, the 1N752 has a tolerance of ±10 percent, which means the breakdown voltage can vary from 5.04 to 6.16 V. Furthermore, the trigger voltage of a 2N4441 has a maximum value of 1.5 V, which means the highest overvoltage detected can be as much as

$$6.16 \text{ V} + 1.5 \text{ V} = 7.66 \text{ V}$$

As it stands, we have a circuit that may not crowbar until the supply voltage reaches 7.66 V. If the application can withstand this much voltage, then the circuit is useful.

Many digital ICs require a supply voltage of +5 V, with an absolute maximum of +7 V. Beyond +7 V, they are destroyed. A circuit like Fig. 15-11 therefore falls short of the kind of protection needed by digital ICs. Nevertheless, it is all right for less stringent requirements, and you see it from time to time.

EXAMPLE 15-6

Repeat Example 15-5 with $V_T = 0.75$ V, $I_T = 30$ mA, and a zener resistance of 11 Ω.

SOLUTION

First, refer to Fig. 15-8 on page 539. Realize that V_{in} and R_G are the Thevenin equivalent circuit driving the 2N4441 shown in Fig. 15-11. With this in mind, we can proceed like this: The 1N752 has a zener resistance of 11 Ω. Therefore, the Thevenin resistance facing the gate of the 2N4441 is

$$R_{TH} = 11 \text{ Ω} \parallel 68 \text{ Ω} = 9.47 \text{ Ω}$$

The SCR trigger current has to flow through this Thevenin resistance and produces a voltage of

$$V = (30 \text{ mA})(9.47 \text{ Ω}) = 0.284 \text{ V}$$

This voltage has to be added to the trigger voltage at the gate of the SCR to get the Thevenin voltage:

$$V_{TH} = 0.75 \text{ V} + 0.284 \text{ V} = 1.03 \text{ V}$$

The supply voltage that turns on the crowbar is 5.6 V above this, or

$$V_{CC} = 1.03 \text{ V} + 5.6 \text{ V} = 6.63 \text{ V}$$

You can see from these calculations that the answer is only slightly more accurate than before. In Example 15-5, we got 6.35 V. So, we are in an area where everything depends on what you are trying to do. (Doesn't it always?) If you were troubleshooting, you would ignore the zener resistance and find the supply voltage as shown in Example 15-5. If you were designing, you would include the zener resistance as shown in this example.

15-3 VARIATIONS OF THE SCR

There are other *pnpn* devices whose action is similar to that of the SCR. What follows is a brief description of these SCR variations. The devices discussed are for low-power applications.

Photo-SCR

Figure 15-12*a* shows a *photo-SCR*, also known as a *light-activated SCR* (LASCR). The arrows represent incoming light that passes through a window and hits the depletion layers. When the light is strong enough, valence electrons are dislodged from their orbits and become free electrons. When these free electrons flow out of the collector of one transistor into the base of the other, the positive feedback starts and the photo-SCR closes.

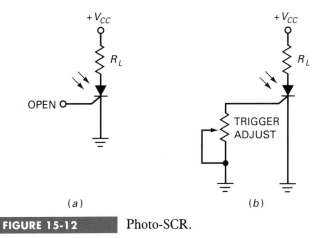

(*a*)   (*b*)

FIGURE 15-12 Photo-SCR.

After a light trigger has closed the photo-SCR, it remains closed, even though the light disappears. For maximum sensitivity to light, the gate is left open, as shown in Fig. 15-12*a*. If you want an adjustable trip point, you can include the trigger adjust shown in Fig. 15-12*b*. The gate resistor diverts some of the light-produced electrons and changes the sensitivity of the circuit to the incoming light.

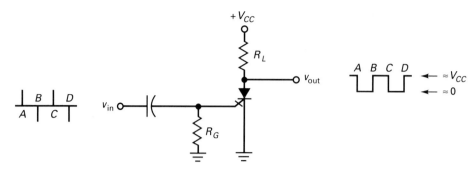

FIGURE 15-13 GCS circuit.

Gate-Controlled Switch

As mentioned earlier, low-current dropout is the normal way to open an SCR. But the *gate-controlled switch* (GCS) is designed for easy opening with a reverse-biased trigger. A GCS is closed by a positive trigger and opened by a negative trigger (or by low-current dropout). Figure 15-13 shows a GCS circuit. Each positive trigger closes the GCS, and each negative trigger opens it. Because of this, we get the square-wave output shown. The GCS is useful in counter, digital circuits, and other applications in which a negative trigger is available for turnoff.

Silicon Controlled Switch

Figure 15-14a shows the doped regions of a *silicon controlled switch* (SCS). Now an external lead is connected to each doped region. Visualize the device separated into two halves (Fig. 15-14b). Therefore, it's equivalent to a latch with access to both bases (Fig. 15-14c). A forward-bias trigger on either base will close the SCS. Likewise, a reverse-bias trigger on either base will open the device.

Figure 15-14d shows the schematic symbol for an SCS. The lower gate is called the *cathode gate;* the upper gate is the *anode gate.* The SCS is a

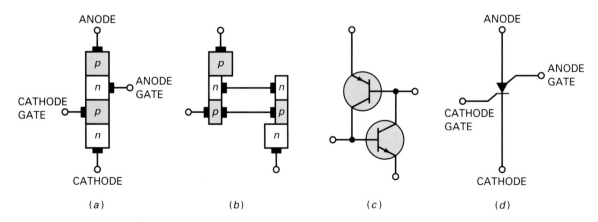

FIGURE 15-14 Silicon controlled switch.

low-power device compared with the SCR. It handles currents in milliamperes rather than amperes.

EXAMPLE 15-7

The circuit of Fig. 15-15a is in a dark room when the power is first turned on. What is the output voltage? If a bright light is turned on, what is the output voltage?

SOLUTION

The photo-SCR is in the nonconducting state. Therefore, the output voltage equals the supply voltage:

$$V_{\text{out}} = +25 \text{ V}$$

When the bright light comes on, the photo-SCR conducts and its output voltage is ideally

$$V_{\text{out}} = 0$$

To a second approximation,

$$V_{\text{out}} = 0.7 \text{ V}$$

EXAMPLE 15-8

The trigger voltage is 0.7 V in Fig. 15-15b. What is the gate current? If the trigger current is 5 mA, what is the approximate current through the 68 Ω?

SOLUTION

In the second approximation, we use 0.7 V for the gate voltage.

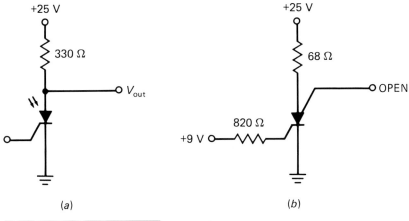

(a)   (b)

FIGURE 15-15   Example.

This means the voltage across the 820 Ω is

$$V = 9\,\text{V} - 0.7\,\text{V} = 8.3\,\text{V}$$

and the current through the 820 Ω is

$$I = \frac{8.3\,\text{V}}{820\,\Omega} = 10.1\,\text{mA}$$

This is the gate current.

Since the gate current is more than the trigger current, the SCS is conducting. Ideally, it has zero voltage across its output terminals. This means the current through the 68 Ω is

$$I = \frac{25\,\text{V}}{68\,\Omega} = 368\,\text{mA}$$

You can use the second approximation if you prefer, but it does little to improve the accuracy:

$$I = \frac{25\,\text{V} - 0.7\,\text{V}}{68\,\Omega} = 357\,\text{mA}$$

15-4   BIDIRECTIONAL THYRISTORS

Up until now, all devices have been unidirectional; current was in only one direction. This section discusses *bidirectional thyristors*, devices in which the current can flow in either direction.

Diac

The *diac* can have latch current in either direction. The equivalent circuit of a diac is a pair of four-layer diodes in parallel, as shown in Fig. 15-16*a*, ideally the same as the latches in Fig. 15-16*b*. The diac is nonconducting until the voltage across it tries to exceed the breakover voltage in either direction.

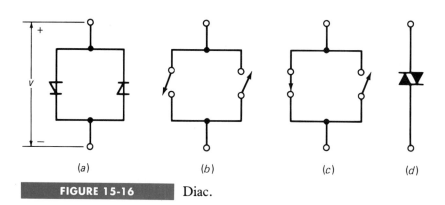

(a)   (b)   (c)   (d)

FIGURE 15-16   Diac.

For instance, if v has the polarity indicated in Fig. 15-16a, then the left diode conducts when v tries to exceed the breakover voltage. In this case, the left latch closes, as shown in Fig. 15-16c. However, if the polarity of v is opposite to that of Fig. 15-16a, the result would be that the right latch closes when v tries to exceed the breakover voltage.

Once the diac is conducting, the only way to open it is by low-current dropout. This means that you must reduce the current below the rated holding current of the device. Figure 15-16d shows the schematic symbol for a diac.

Triac

The *triac* acts like two SCRs in parallel (Fig. 15-17a), equivalent to the two latches of Fig. 15-17b. Because of this, the triac can control current in either direction. The breakover voltage is usually high, so that the normal way to turn on a triac is by applying a forward-bias trigger. Data sheets list the trigger voltage and trigger current needed to turn on a triac. If v has the polarity shown in Fig. 15-17a, we have to apply a positive trigger; this closes the left latch. When v has opposite polarity, a negative trigger is needed; it will close the right latch. Figure 15-17c is the schematic symbol for a triac.

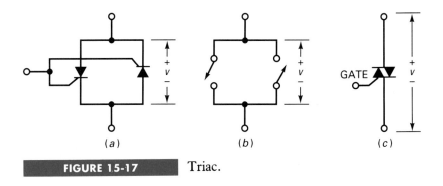

FIGURE 15-17   Triac.

Figure 15-18a shows a triac circuit that can be used to control the current through a heavy load. In this circuit, resistance R_1 and capacitance C shift the phase angle of the gate signal. Because of this phase shift, the gate voltage lags the line voltage by an angle between 0 and 90°. You can see these ideas in Fig. 15-18b and c. The line voltage has a phase angle of 0°, while the gate voltage lags the line voltage. When the gate voltage is large enough to supply the trigger current, the triac conducts. Once on, the triac continues to conduct until the line voltage returns to zero. The shaded portion of each half-cycle shows you when the triac is conducting. Because R_1 is variable, the phase angle of the gate voltage can be changed. This allows us to control the shaded portions of the line voltage. In other words, we can control the average load current. Control like this is useful in industrial heating, lighting, and other heavy-power applications.

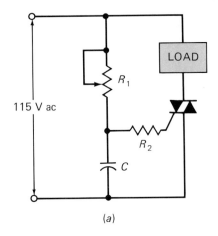

(a)

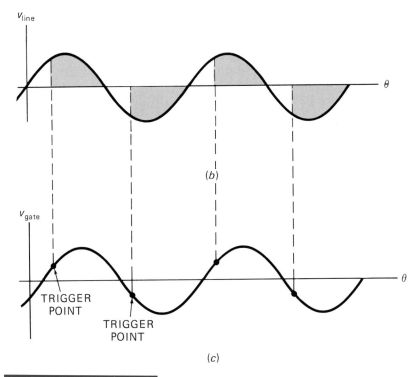

(b)

(c)

FIGURE 15-18   Triac circuit and waveforms.

EXAMPLE 15-9

In Fig. 15-19, the switch is closed. If the triac has fired, what is the approximate current through the 22 Ω?

SOLUTION

Ideally, the triac has 0 V across it when conducting. Therefore, the

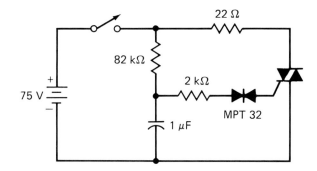

FIGURE 15-19   Example.

current through the 22 Ω is

$$I = \frac{75 \text{ V}}{22 \text{ }\Omega} = 3.41 \text{ A}$$

Even if the triac has 1 or 2 V across it, the current through the 22 Ω is still very close to 3.41 A because of the large supply voltage (75 V).

EXAMPLE 15-10

In Fig. 15-19, the switch is closed. The MPT32 is a diac with a breakover voltage of 32 V. If the triac has a trigger voltage of 1 V and a trigger current of 10 mA, what is the capacitor voltage that triggers the triac?

SOLUTION

As the capacitor charges, the voltage across the diac increases. When the diac voltage is slightly less than 32 V, the diac is on the verge of breakover. Since the triac has a trigger voltage of 1 V, the capacitor voltage is slightly less than 33 V. When the capacitor voltage becomes slightly greater than 33 V, the diac breaks over and triggers the triac. Expressed as an equation,

$$V_{\text{in}} = 32 \text{ V} + 1 \text{ V} = 33 \text{ V}$$

This is the minimum capacitor voltage that will trigger the triac.

15-5   THE UNIJUNCTION TRANSISTOR

The *unijunction transistor* (UJT) has two doped regions with three external leads (Fig. 15-20a). It has one emitter and two bases. The emitter is heavily doped, having many holes. The n region, however, is lightly doped. For this reason, the resistance between the bases is relatively high,

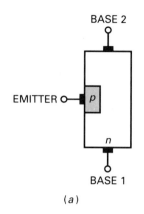

BASE 2

EMITTER o—

p

n

BASE 1

(a)

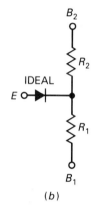

B_2

R_2

IDEAL

E o—

R_1

B_1

(b)

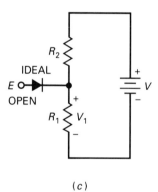

R_2

IDEAL

E o—

OPEN

R_1 V_1

V

(c)

FIGURE 15-20

UJT: (a) structure; (b) equivalent circuit; (c) standoff voltage.

typically 5 to 10 kΩ when the emitter is open. We call this the *interbase resistance*, symbolized by R_{BB}.

Intrinsic Standoff Ratio

Figure 15-20b shows the equivalent circuit of a UJT. The emitter diode drives the junction of two internal resistances R_1 and R_2. When the emitter diode is nonconducting, R_{BB} is the sum of R_1 and R_2. When a supply voltage is between the two bases, as shown in Fig. 15-20c, the voltage across R_1 is given by

$$V_1 = \frac{R_1}{R_1 + R_2}(V) = \frac{R_1}{R_{BB}}(V)$$

or

$$V_1 = \eta V \qquad (15\text{-}3)$$

where

$$\eta = \frac{R_1}{R_{BB}}$$

(The Greek letter η is pronounced "eta," e as the a in face and a as the a in about.)

The quantity η is called the *intrinsic standoff ratio*, which is nothing more than the voltage-divider factor. The typical range of η is from 0.5 to 0.8. For instance, a 2N2646 has an η of 0.65. If this UJT is used in Fig. 15-20c with a supply voltage of 10 V, then

$$V_1 = \eta V = 0.65(10 \text{ V}) = 6.5 \text{ V}$$

In Fig. 15-20c, V_1 is called the intrinsic standoff voltage because it keeps the emitter diode reverse-biased for all emitter voltages less than V_1. If V_1 equals 6.5 V, then ideally we have to apply slightly more than 6.5 V to the emitter to turn on the emitter diode.

How a UJT Works

In Fig. 15-21a, imagine that the emitter supply voltage is turned down to zero. Then the intrinsic standoff voltage reverse-biases the emitter

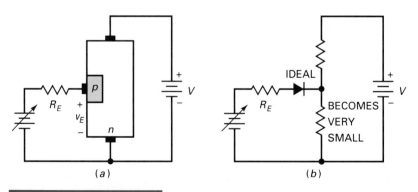

R_E v_E p n V

IDEAL R_E BECOMES VERY SMALL V

(a)   (b)

FIGURE 15-21   (a) UJT circuit; (b) R_1 approaches zero after emitter diode turns on.

diode. When we increase the emitter supply voltage, v_E increases until it is slightly greater than V_1. This turns on the emitter diode. Since the p region is heavily doped compared with the n region, holes are injected into the lower half of the UJT. The light doping of the n region gives these holes a long lifetime. These holes create a conducting path between the emitter and the lower base.

The flooding of the lower half of the UJT with holes drastically lowers resistance R_1 (Fig. 15-21b). Because R_1 is suddenly much lower in value, v_E suddenly drops to a low value, and the emitter current increases.

Latch-Equivalent Circuit

One way to remember how the UJT of Fig. 15-22a works is by relating it to the latch of Fig. 15-22b. With a positive voltage from B_2 to B_1, a standoff voltage V_1 appears across R_1. This keeps the emitter diode of Q_2 reverse-biased as long as the emitter input voltage is less than the standoff voltage. When the emitter input voltage is slightly greater than the standoff voltage, however, Q_2 turns on and positive feedback takes over. This drives both transistors into saturation, ideally shorting the emitter and the lower base.

Figure 15-22c is the schematic symbol for a UJT. The emitter arrow reminds us of the upper emitter in a latch. When the emitter voltage exceeds the standoff voltage, the latch between the emitter and the lower base closes. Ideally, you can visualize a short between E and B_1. To a second approximation, a low voltage called the *emitter saturation voltage* $V_{E(\text{sat})}$ appears between E and B_1.

The latch stays closed as long as the latch current (emitter current) is greater than the holding current. Data sheets specify a *valley current* I_V, which is equivalent to holding current. For instance, a 2N2646 has an I_V of 6 mA; to hold the latch closed, the emitter current must be greater than 6 mA.

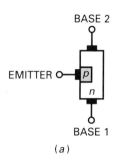

BASE 2

EMITTER

BASE 1

(a)

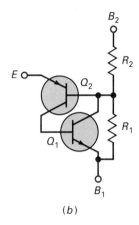

(b)

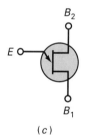

(c)

FIGURE 15-22

UJT: (*a*) Structure; (*b*) latch-equivalent circuit; (*c*) schematic symbol.

EXAMPLE 15-11

The 2N4871 of Fig. 15-23 has an η of 0.85. What is the ideal emitter current?

SOLUTION

The standoff voltage is

$$V_1 = 0.85(10 \text{ V}) = 8.5 \text{ V}$$

Ideally, v_E must be slightly greater than 8.5 V to turn on the emitter diode and close the latch. With the input switch closed, 20 V drives the 400 Ω. This is more than enough voltage to overcome the standoff voltage. Therefore, the latch is closed, and the emitter current equals

$$I_E = \frac{20 \text{ V}}{400 \text{ }\Omega} = 50 \text{ mA}$$

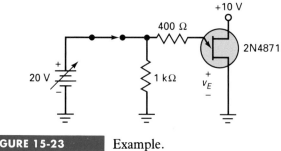

FIGURE 15-23   Example.

EXAMPLE 15-12

The valley current of a 2N4871 equals 7 mA, and the emitter voltage is 1 V at this point. At what value of the emitter supply voltage does the UJT in Fig. 15-23 open?

SOLUTION

As we reduce the emitter supply voltage, the emitter current decreases. At the point at which it equals 7 mA, v_E is 1 V and the latch is about to open. The emitter-supply voltage at this time is

$$V = 1 \text{ V} + (7 \text{ mA})(400 \text{ }\Omega) = 3.8 \text{ V}$$

When V is less than 3.8 V, the UJT opens. Then it is necessary to raise V above 8.5 V to close the UJT.

15-6 TROUBLESHOOTING

When you troubleshoot a circuit to find faulty resistors, diodes, transistors, etc., you are troubleshooting at the *component level*. The T-shooters of earlier chapters gave you troubleshooting practice at the component level. Troubleshooting at this level is an excellent foundation for troubleshooting at higher levels because it teaches you how to think logically, using Ohm's law as your guide. Component-level troubleshooting is equivalent to learning how to run before you can fly. Now we want to practice troubleshooting at the *system level*. This means thinking in terms of *functional blocks*, which are the smaller jobs or things being done by the different parts of the overall circuit.

To get the idea of this higher level of troubleshooting, look at the T-shooter shown in Fig. 15-43 (at the end of the problems). Here you see a block diagram of a power supply with an SCR crowbar. The power supply has been drawn in terms of its functional blocks. If you measure the voltages at the different points, you can often isolate the trouble to a particular block. Then you can continue troubleshooting at the component level, if necessary.

Often, a manufacturer's instruction manual includes block diagrams of the equipment where the function of each block is specified. For instance, a television receiver can be drawn in terms of its functional blocks. Once you know what the input and output signals of each block are supposed to be, you can troubleshoot the television receiver to isolate the defective block. After you isolate the defective block, you can either replace the entire block or continue troubleshooting at the component level.

The T-shooter of this chapter will give you practice in troubleshooting at the system level. This new type of troubleshooting is like flying instead of running. At the system level, you are looking down on the equipment from a higher viewpoint because you are thinking in terms of what chunks of the system do rather than how currents flow through components.

OPTIONAL TOPICS

The following material continues the earlier discussions at a more advanced and specialized level. All the topics are optional because they are not used in any of the basic discussions in later chapters. This section will be a useful reference when you are in industry because then you will probably want more advanced viewpoints.

15-7 MORE THYRISTOR APPLICATIONS

Earlier, we discussed the different kinds of thyristors including four-layer diodes, SCRs, diacs, triacs, etc. The thyristor is such an unusual control device that it can be used in all kinds of ways. Thyristors have become increasingly popular for controlling ac power to resistive and inductive loads, such as motors, solenoids, and heating elements. Compared with competing devices like relays, thyristors offer lower cost and better reliability. This section discusses some applications of thyristors to give you an idea of the variety of ways in which they can be used.

Overvoltage Detector

Figure 15-24 shows a circuit known as an *overvoltage detector*. Here is how it works: The four-layer diode has a breakover voltage of 10 V. As

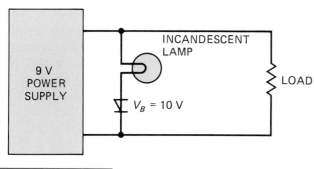

FIGURE 15-24 Overvoltage detector.

long as the power supply puts out 9 V, the four-layer diode is open and the lamp is dark. But if something goes wrong with the power supply and its voltage rises above 10 V, then the four-layer diode breaks over and the lamp comes on. Even if the supply should return to 9 V, the diode remains latched as a fixed indication of the overvoltage that has occurred. The only way to make the lamp go out is to turn off the supply.

Sawtooth Generator

Figure 15-25a shows a *sawtooth generator*. If the four-layer diode were not in the circuit, the capacitor would charge exponentially, and its voltage would follow the dashed curve of Fig. 15-25b. But the four-layer diode is in the circuit. Therefore, as soon as the capacitor voltage reaches 10 V, the diode breaks over and the latch closes. This discharges the capacitor, producing the *flyback* (sudden decrease) of capacitor voltage. At some point on the flyback, the current drops below the holding current, and the four-layer diode opens. The next cycle then begins.

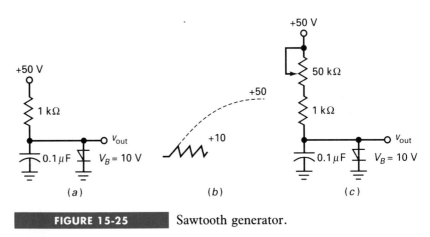

FIGURE 15-25 Sawtooth generator.

Figure 15-25a is an example of a *relaxation oscillator*, a circuit whose output depends on the charging and discharging of a capacitor (or inductor). If we increase the *RC* time constant, then the capacitor takes longer to charge to 10 V and the frequency of the sawtooth wave is lower. For instance, with the potentiometer of Fig. 15-25c we can get a 50:1 range in frequency.

SCR Crowbar

Because the knee of the zener diode is curved rather than sharp, the basic SCR crowbar discussed earlier has a *soft turn-on*. This crowbar circuit can be improved by adding some voltage gain as shown in Fig. 15-26. The transistor provides voltage gain, which produces a much sharper turn-on. When the voltage across R_4 exceeds approximately 0.7 V, the SCR turns on. An ordinary diode is included for temperature compensation of the transistor's base-emitter diode. The trigger adjust allows us to set the trip point of the circuit, typically around 10 to 15 percent above the normal voltage.

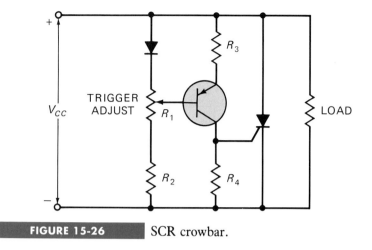

FIGURE 15-26   SCR crowbar.

Figure 15-27 shows an even better solution. The triangular box is an IC called a *comparator* (discussed in later chapters). This IC produces a very large voltage gain, typically 100,000 or more. The input to this IC is between the plus and minus input terminals. Because of its large voltage gain, the IC can detect the slightest overvoltage. The zener diode produces 10 V, which goes to the minus input of the comparator. The trigger adjust produces slightly less than 10 V for the plus input. As a result, the input voltage to the comparator is negative. The output of the comparator is also negative, which cannot trigger the SCR. If the supply voltage tries to rise above 20 V, the plus input of the comparator becomes greater than 10 V. Since the input voltage is positive, the output of the comparator becomes positive and drives the SCR into conduction. This rapidly shuts down the supply by crowbarring the load terminals.

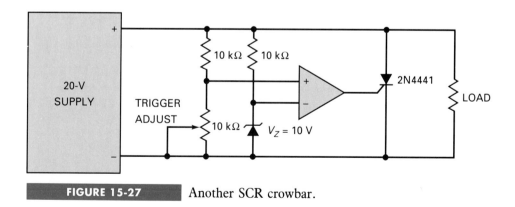

FIGURE 15-27   Another SCR crowbar.

The simplest solution when you need a crowbar is to use an IC crowbar as shown in Fig. 15-28. An IC crowbar is an integrated circuit with a zener diode for detection, transistors for voltage gain, and an SCR for crowbarring. The popular RCA SK9345 series is an example of what is commercially available. The SK9345 protects power supplies of +5 V,

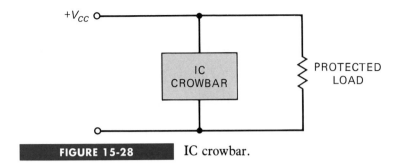

FIGURE 15-28   IC crowbar.

the SK9346 protects +12 V, and the SK9347 protects +15 V. For instance, if an SK9345 is used in Fig. 15-28, it will protect the load for a nominal supply voltage of +5 V. The data sheet of an SK9345 indicates that it fires at +6.6 V with a tolerance of ±0.2 V.

UJT Relaxation Oscillator

Figure 15-29*a* shows a UJT *relaxation oscillator*. The action is similar to that of the four-layer diode relaxation oscillator. The capacitor charges toward V_{CC}, but as soon as its voltage exceeds the standoff voltage, the UJT closes. This discharges the capacitor until low-current dropout occurs. As soon as the UJT opens, the next cycle begins. As a result, we get a sawtooth output.

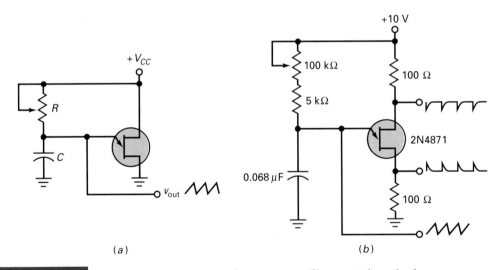

FIGURE 15-29   UJT circuits: (*a*) sawtooth generator; (*b*) sawtooth and trigger output.

If we add a small resistor to each base circuit, we can get three useful outputs: sawtooth waves, positive triggers, and negative triggers, as shown in Fig. 15-29*b*. The triggers appear during the flyback of the sawtooth because the UJT conducts heavily at this time. With the values of Fig. 15-29*b*, the frequency can be adjusted between 50 Hz and 1 kHz (approximately).

Automobile Ignition

Sharp trigger pulses out of a UJT relaxation oscillator can be used to trigger an SCR. For instance, Fig. 15-30 shows part of an automobile ignition system. With the distributor points open, the capacitor charges exponentially toward $+12$ V. As soon as the capacitor voltage exceeds the intrinsic standoff voltage, the UJT conducts heavily through the primary winding. The secondary voltage then triggers the SCR. When the SCR latches shut, the positive end of the output capacitor is suddenly grounded. As the output capacitor discharges through the ignition coil, a high-voltage pulse drives one of the spark plugs. When the points close, the circuit resets itself in preparation for the next cycle.

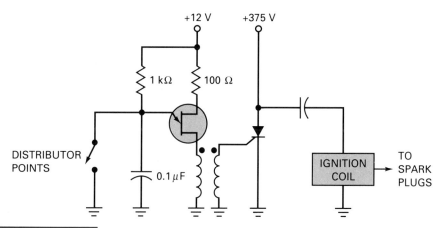

FIGURE 15-30 UJT triggers SCR to produce spark for automobile ignition.

Optocoupler Control

Figure 15-31 is an example of *optocoupler control*. When an input pulse turns on the LED (D_4), its light activates the photo-SCR (D_3). In turn, this produces a trigger voltage for the main SCR (D_2). In this way, we get isolated control of the positive half-cycles of line voltage. An ordinary diode D_1 is needed to protect the SCR from inductive kickback and transients that may occur during the reverse half-cycle.

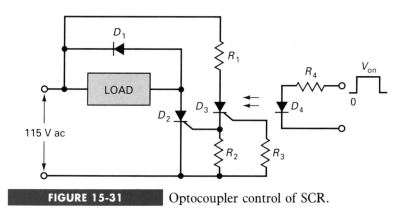

FIGURE 15-31 Optocoupler control of SCR.

Diac-Triggered SCR

In Fig. 15-32, the full-wave output from a bridge rectifier drives an SCR that is controlled by a diac and an *RC* charging circuit. By adjusting R_1, we can change the time constant and control the point at which the diac fires. Circuits like these can easily control several hundred watts of power to a lamp, heater, or other load. A four-layer diode could be used instead of a diac.

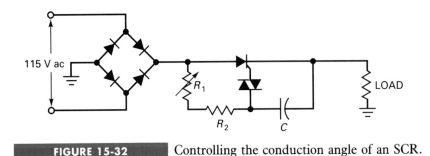

FIGURE 15-32   Controlling the conduction angle of an SCR.

UJT-Triggered SCR

Figure 15-33 shows another way to control an SCR, this time with a UJT relaxation oscillator. The load may be a motor, lamp, heater, or some other device. By varying R_1, we can change the *RC* time constant and alter the point at which the UJT fires. This allows us to control the conduction angle of the SCR, which means that we are controlling the load current. A circuit like this represents half-wave control because the SCR is off during the negative half-cycle.

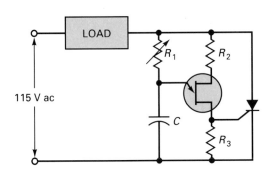

FIGURE 15-33   UJT relaxation oscillator controls conduction angle of SCR.

Full-Wave Control

The diac of Fig. 15-34 can trigger the triac on either half-cycle of line voltage. Variable resistance R_1 controls the *RC* time constant of the diac control circuit. Since this changes the point in the cycle at which the diac fires, we have control over the triac condition angle. In this way, we can change the large load current.

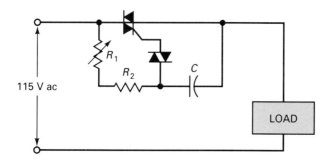

FIGURE 15-34   Controlling the conduction angle of a triad.

Microprocessor-Controlled SCR

In robotic systems, a microprocessor controls motors and other loads. Figure 15-35 is a simple example of how it is done. A rectangular pulse from a microprocessor drives an emitter follower, whose output controls the gate of an SCR. While the rectangular control pulse is high, the SCR latches during the positive half-cycles and shuts off during the negative half-cycles. The duration of the rectangular pulse from the microprocessor determines the number of positive half-cycles during which the load receives power.

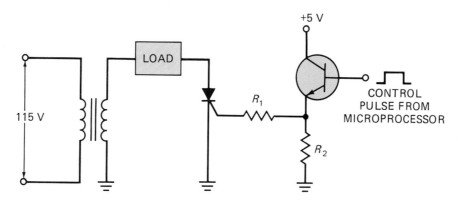

FIGURE 15-35   Microprocessor controls length of time. Power is applied to load.

STUDY AIDS

The following study aids will help to reinforce the ideas discussed in this chapter. For best results, use these study aids within 6 hours of reading the earlier material. Then review these study aids a week later and a month later to ensure that the concepts remain in your long-term memory.

SUMMARY

Sec. 15-1 The Four-Layer Diode

A thyristor is a semiconductor device that uses internal positive feedback to produce latching action. The main application is in controlling large amounts of load current. With a four-layer diode, the thyristor is closed by breakover and opened by low-current dropout.

Sec. 15-2 The Silicon Controlled Rectifier

This thyristor is more useful than a four-layer diode because it has an extra lead called the gate. By applying a trigger to the gate, we can turn on the SCR. One important application of SCRs is as crowbars in power supplies. Crowbarring is the standard way to protect delicate and expensive loads.

Sec. 15-3 Variations of the SCR

Other *pnpn* devices derived from the SCR are as follows. The photo-SCR has a gate that responds to the amount of light striking it. The GCS can be turned on by a positive trigger and turned off by a negative trigger. The SCS has two gates; a forward-bias trigger on either gate turns it on, and a reverse-bias trigger on either gate turns it off.

Sec. 15-4 Bidirectional Thyristors

The diac can latch current in either direction. The equivalent circuit for a diac is two four-layer diodes of opposite polarity in parallel. The triac acts like two parallel SCRs of opposite polarity. Because of this, the triac can control alternating current.

Sec. 15-5 The Unijunction Transistor

A UJT has two doped regions and three external leads, with one emitter and two bases. It is equivalent to a latch with a control input. The intrinsic standoff ratio indicates the firing point of the device.

VOCABULARY

In your own words, explain what each of the following terms means. Keep your answers short and to the point. If necessary, verify your answer by rereading the appropriate discussion or by looking at the end-of-book Glossary.

breakover	thyristor
crowbar	triac
holding current	trigger
latch	trigger current
positive feedback	trigger voltage
silicon controlled rectifier	unijunction transistor

STUDENT ASSIGNMENTS

QUESTIONS

Some questions have more than one right answer. Select the best answer, the one that is always true, or that most accurately describes the situation, etc.

1. A thyristor can be used as
   a. A resistor
   b. An amplifier
   c. A switch
   d. A power source

2. Positive feedback means the returning signal
   a. Opposes the original change
   b. Aids the original change
   c. Is equivalent to negative feedback
   d. Is amplified

3. A latch always uses
   a. Transistors
   b. Feedback
   c. Current
   d. Positive feedback

4. To turn on a four-layer diode, you need
   a. A positive trigger
   b. Low-current dropout
   c. Breakover
   d. Reverse-bias triggering

5. The minimum input current that can turn on a thyristor is called the
   a. Holding current
   b. Trigger current
   c. Breakover current
   d. Low-current dropout

6. The only way to stop a four-layer diode that is conducting is by
   a. A positive trigger
   b. Low-current dropout
   c. Breakover
   d. Reverse-bias triggering

7. The minimum anode current that keeps a thyristor turned on is called the
   a. Holding current
   b. Trigger current
   c. Breakover current
   d. Low-current dropout

8. A silicon controlled rectifier has
   a. Two external leads
   b. Three external leads
   c. Four external leads
   d. Three doped regions

9. An SCR is usually turned on by
   a. Breakover
   b. A gate trigger
   c. Breakdown
   d. Holding current

10. SCRs are
    a. Low-power devices
    b. Four-layer diodes
    c. High-current devices
    d. Bidirectional

11. The usual way to protect a load from excessive supply voltage is with a
    a. Crowbar
    b. Zener diode
    c. Four-layer diode
    d. Thyristor

12. An *RC* snubber protects an SCR against
    a. Supply overvoltages
    b. False triggering
    c. Breakover
    d. Crowbarring

13. When a crowbar is used with a power supply, the supply needs to have a fuse or
    a. Adequate trigger current
    b. Holding current
    c. Filtering
    d. Current limiting

14. The LASCR responds to
    a. Current   c. Humidity
    b. Voltage   d. Light

15. The diac is a
    a. Transistor
    b. Unidirectional device
    c. Three-layer device
    d. Bidirectional device

16. The triac is equivalent to
    a. A four-layer diode
    b. Two diacs in parallel
    c. A thyristor with a gate lead
    d. Two SCRs in parallel

17. The unijunction transistor acts as a
    a. Four-layer diode   c. Triac
    b. Diac   d. Latch

18. Any thyristor can be turned on with
    a. Breakover
    b. Forward-bias triggering
    c. Low-current dropout
    d. Reverse-bias triggering

19. A Shockley diode is the same as
    a. A four-layer diode   c. A diac
    b. An SCR   d. A triac

20. The trigger voltage of an SCR is closest to
    a. 0
    b. 0.7 V
    c. 2 V
    d. Breakover voltage

21. Any thyristor can be turned off with
    a. Breakover
    b. Forward-bias triggering
    c. Low-current dropout
    d. Reverse-bias triggering

22. Exceeding the critical rate of rise produces
    a. Excessive power dissipation
    b. False triggering
    c. Low-current dropout
    d. Reverse-bias triggering

23. A four-layer diode is sometimes called a
    a. Unijunction transistor
    b. Diac
    c. *pnpn* diode
    d. Switch

24. A latch is based on
    a. Negative feedback
    b. Positive feedback
    c. The four-layer diode
    d. SCR action

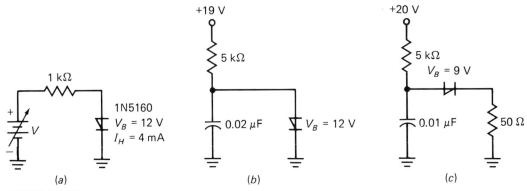

FIGURE 15-36

25. In Fig. 15-36*a* the battery voltage is 15 V, but the voltage across the four-layer diode is zero. The trouble is
    a. A shorted 1 kΩ
    b. An open 1N5160
    c. An open 1 kΩ
    d. An open ground on 1N5160

26. The input voltage of Fig. 15-38*a* changes from +12 to +20 V. If the load voltage is 20 V, the trouble is
    a. An open load resistor
    b. A shorted load resistor
    c. A shorted 2N4441
    d. An open 1N759

27. The input voltage of Fig. 15-38*b* changes from +15 to +22 V. If the load voltage is 22 V, the trouble is
    a. An open 1N965
    b. An open 2N4441
    c. 1N975 used instead of 1N965
    d. All the above

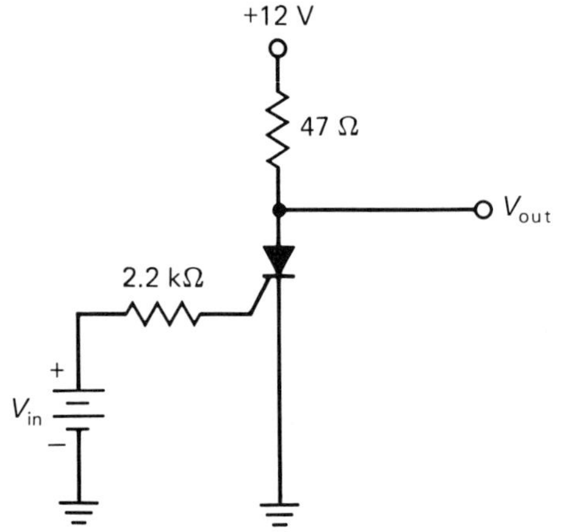

28. The crowbar of Fig. 15-38*b* is not working. The trouble is
    a. An open 56 Ω
    b. A ground open on 2N4441
    c. A shorted 2N4441
    d. An open load resistor

29. The circuit of Fig. 15-40*a* has a high output voltage even when it is exposed to sunlight. The trouble is
    a. A shorted thyristor
    b. An open 100 Ω
    c. A ground open on the thyristor
    d. No supply voltage

BASIC PROBLEMS

Sec. 15-1 The Four-Layer Diode

15-1. The 1N5160 of Fig. 15-36*a* is conducting. If we allow 0.7 V across the diode at the dropout point, what is the value of V at which low-current dropout occurs?

15-2. The capacitor of Fig. 15-36*b* charges from 0.7 to 12 V, causing the four-layer diode to break over. What is the current through the 5 kΩ just before the diode breaks over? The current through the 5 kΩ when it is conducting?

15-3. In Fig. 15-36*c*, the voltage across the capacitor is 9 V, and the voltage across the diode is 1 V. What is the current through the 50 Ω?

Sec. 15-2 The Silicon Controlled Rectifier

15-4. The SCR of Fig. 15-37 has $V_T = 0.7$ V, $I_T = 2$ mA, and $I_H = 2$ mA. What is

+12 V

47 Ω

V_{out}

2.2 kΩ

V_{in}

FIGURE 15-37

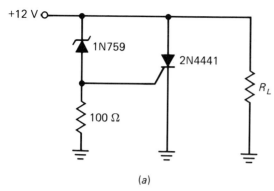

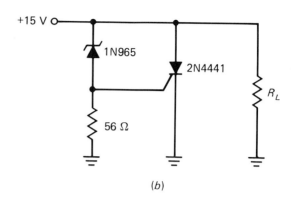

(a)   (b)

FIGURE 15-38

the output voltage when the SCR is off? What is input voltage that triggers the SCR? If V_{CC} is decreased until the SCR opens, what is the value of V_{CC}?

15-5. All resistances are doubled in Fig. 15-37. If the trigger current of the SCR is 1.5 mA, what is the input voltage that triggers the SCR?

15-6. The 1N759 of Fig. 15-38a has a breakdown voltage of 12 V. The 2N4441 has a trigger voltage of 0.75 V. Calculate the supply voltage that turns on the crowbar.

15-7. If the 1N759 of Fig. 15-38a has a tolerance of ±10 percent and the trigger voltage can be as high as 1.5 V, what is the maximum supply voltage where crowbarring takes place?

15-8. What is the typical supply voltage in Fig. 15-38b where crowbarring occurs?

15-9. If the 1N965 of Fig. 15-38b has a tolerance of ±10 percent and the trigger voltage can be as high as 1.5 V, what is the maximum supply voltage where the SCR turns on?

15-10. The 2N4216 of Fig. 15-39a has a trigger current of 0.1 mA. If we allow 0.8 V for the gate voltage, what is the value of V that turns on the SCR?

15-11. After the SCR fires in Fig. 15-39b, the voltage between the anode and ground is 1 V. What is the current through the 500 Ω?

15-12. In Fig. 15-39b, the voltage across the capacitor is 10 V, the voltage across the

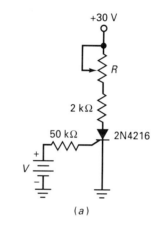

(a)

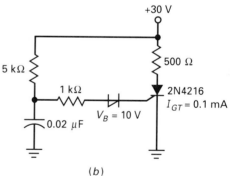

(b)

FIGURE 15-39

four-layer diode is 1 V, and the SCR gate voltage is 0.8 V. What is the current through the 1 kΩ?

Sec. 15-3 Variations of the SCR

15-13. The circuit of Fig. 15-40a is in a dark room. When a bright light is turned on, the thyristor fires. What is the approximate output voltage? If the bright light is turned off, what is the output voltage?

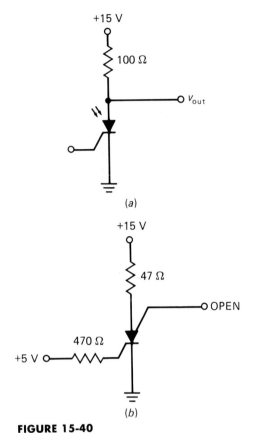

FIGURE 15-40

15-14. The trigger voltage is 0.7 V in Fig. 15-40*b*. What is the gate current? What is the approximate current through the 47 Ω?

Sec. 15-4 Bidirectional Thyristors

15-15. In Fig. 15-41*a*, the switch is closed. If the triac has fired, what is the approximate current through the 10 Ω?

15-16. After the switch closes in Fig. 15-41*a*, the capacitor charges toward 50 V. The MPT32 is a diac with a breakover voltage of 32 V. If the triac has a trigger voltage of 1 V, what is the minimum voltage across the capacitor that triggers the triac? When the MPT32 breaks over, its voltage suddenly drops from 32 to 1 V. What is the current through the 1 kΩ just after breakover occurs?

15-17. If the triac of Fig. 15-41*b* has 1 V across it when it is conducting, what is the maximum current through the 50 Ω?

15-18. The MPT32 of Fig. 15-41*b* is a diac with a breakover voltage of 32 V and an on voltage of 1 V. If the triac has a trigger voltage of 0.7 V, what is the current through the 2 kΩ just after the MPT32 breaks over?

Sec. 15-5 The Unijunction Transistor

15-19. The UJT of Fig. 15-42*a* has an η of 0.63. By allowing 0.7 V across the emitter diode, what value of *V* just turns on the UJT?

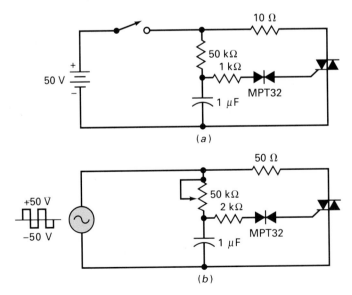

FIGURE 15-41

15-20. The UJT of Fig. 15-42a has a valley current of 5 mA and an emitter voltage of 0.8 V at this point. If the UJT is conducting, what value of the emitter supply voltage will open the UJT?

15-21. In Fig. 15-42b, the UJT has an η of 0.7. What is the voltage across the capacitor that turns on the UJT?

ADVANCED PROBLEMS

15-22. With a supply of 19 V, it takes the capacitor of Fig. 15-36b exactly one time constant to charge to 12 V, the breakover voltage of the diode. If we neglect the voltage across the diode when it is conducting, what is the frequency of the sawtooth output?

15-23. The current through the 50 Ω of Fig. 15-36c is maximum just after the diode latches. If we allow 1 V across the latched diode, what is the maximum current?

15-24. Will the output frequency in Fig. 15-36b increase, decrease, or stay the same for each of these troubles?
   a. Supply voltage is at +15 V.
   b. Resistor is 20 percent high.
   c. Capacitor is 0.01 μF.
   d. Breakover voltage is only 10 V.

15-25. Select a value of C in Fig. 15-36b that produces an output frequency of approximately 20 kHz.

15-26. The four-layer diode of Fig. 15-39b has a breakover voltage of 10 V. The SCR has a trigger current of 0.1 mA and a trigger voltage of 0.8 V. If the diode has a forward drop of approximately 0.7 V, what is the current through the diode just after it breaks over? The current through the 500 Ω after the SCR turns on?

15-27. The circuit of Fig. 15-42b generates a sawtooth wave. Calculate the maximum frequency of this signal for η = 0.63.

15-28. Figure 15-41a shows an alternative symbol for a diac. The MPT32 diac breaks over when the capacitor voltage reaches 32 V. It takes exactly one time constant for the capacitor to reach this voltage.

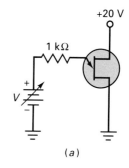

(a)

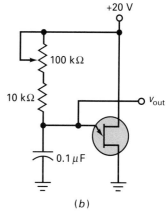

(b)

FIGURE 15-42

How long after the switch is closed does the triac turn on? What is the ideal value of the gate current when the diac breaks over? What is the load current after the triac has closed?

15-29. The frequency of the square wave in Fig. 15-41b is 10 kHz. It takes exactly one time constant for the capacitor to reach the breakover voltage of the diac. If the MPT32 breaks over at 32 V, what is the ideal value of the gate current at the instant the diac breaks over? What is the ideal load current?

15-30. The valley current of the UJT in Fig. 15-42a is 2 mA. If the UJT is latched, we have to reduce V to get low-current dropout. With 0.7 V across the emitter diode, what value of V just opens the UJT?

15-31. The intrinsic standoff ratio of the UJT in Fig. 15-42b is 0.63. Ignoring the drop across the emitter diode, what are the minimum and maximum output frequencies?

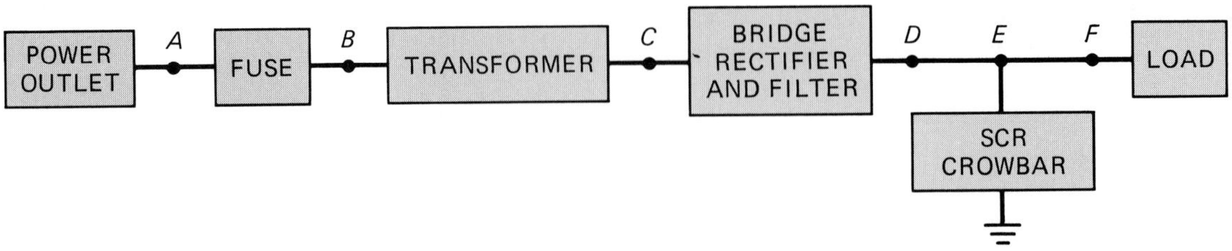

	OK	T1	T2	T3	T4	T5	T6
V_A	E7	B3	F1	B3	C2	G2	C2
V_B	G5	A1	A6	D4	D2	E1	A1
V_C	A4	B5	D2	F3	F1	D3	A4
V_D	B7	G6	B6	G4	A6	F7	B6
V_E	F5	D2	F3	B6	B2	C6	D2
V_F	E3	E5	B2	D5	F1	D7	E5
R_L	D6	F2	A7	C5	D5	B1	F2
SCR	A5	D1	C7	A3	G1	C3	A5

	T7	T8
V_A	G5	D4
V_B	C2	F1
V_C	E6	B6
V_D	B7	D2
V_E	G6	G7
V_F	F1	D5
R_L	D6	F2
SCR	E2	G1

FIGURE 15-43

T-Shooter™. (*Patent pending: Courtesy of Malvino Inc.*)

	1	2	3	4	5	6	7
A	115	18	Off	12.7	Off	0	100 Ω
B	100 Ω	0	115	On	12.7	0	18
C	18	115	Off	18	100 Ω	20.5	Off
D	Off	0	14.4	115	0	100 Ω	20.5
E	130	Off	18	100 Ω	0	12.7	115
F	0	100 Ω	0	0	18	115	20.5
G	Off	130	100 Ω	0	115	18	0

MEASUREMENTS

T-SHOOTER PROBLEMS

Use Fig. 15-43 for the remaining problems. This power supply has a bridge rectifier working into a capacitor-input filter. Therefore, the filtered dc voltage is approximately equal to the peak secondary voltage. All token values are in volts, unless otherwise indicated. Also, the measured voltages at points A, B, and C are given as rms values. The measured voltages at points D, E, and F are given as dc voltages. In this exercise, you are troubleshooting at the system level, meaning that you are to locate the most suspicious block for further testing. For instance, if the voltage is okay at point B but incorrect at point C, your answer should be "transformer."

15-32. Find Trouble 1.

15-33. Find Trouble 2.

15-34. Find Trouble 3.

15-35. Find Trouble 4.

15-36. Find Trouble 5.

15-37. Find Trouble 6.

15-38. Find Trouble 7.

15-39. Find Trouble 8.

16

FREQUENCY EFFECTS

In earlier chapters, we assumed all capacitors were open at low frequencies and shorted at high frequencies. Open capacitors at low frequencies gave us the dc equivalent circuit, and shorted capacitors at high frequencies gave us the ac equivalent circuit. This ac equivalent circuit contained only resistances. In other words, we restricted our analysis of amplifiers to the band of frequencies where only resistances appear in the ac equivalent circuit. Now we are ready to discuss the operation of amplifiers outside this normal range of frequencies.

16-1 FREQUENCY RESPONSE OF AN AMPLIFIER

Figure 16-1 shows the typical *frequency response* of an amplifier. This is a graph of the output voltage versus generator frequency, given a constant generator voltage. At low frequencies, the output voltage decreases because of the coupling and bypass capacitors. At high frequencies, the output voltage decreases because of transistor and stray-wiring capacitances. Transistor and stray-wiring capacitances are unwanted capacitances that provide shunt paths for the ac signal and prevent it from reaching the load resistor. This is why the output voltage decreases when the frequency is too high.

In the middle range of the frequencies, the amplifier is producing a maximum output voltage V_{max}. This band of frequencies represents the

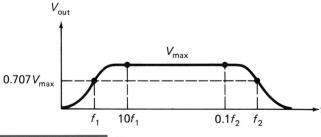

FIGURE 16-1   Frequency response.

frequencies where only resistances appear in the ac equivalent circuit of the amplifier. This middle range is where the amplifier is supposed to operate. The frequencies above and below this middle range are avoided in most applications.

The *critical frequencies* of an amplifier are the frequencies where the output voltage is 0.707 of V_{max}. An amplifier usually has two critical frequencies, f_1 and f_2. Coupling and bypass capacitors are responsible for the lower frequency f_1. Transistor and stray-wiring capacitances are responsible for the upper frequency f_2.

The middle range of frequencies is called the *midband*. This is the range of frequencies where the amplifier output is maximum. In Fig. 16-1, the midband is the band of frequencies between $10f_1$ and $0.1f_2$. In this range of frequencies, the amplifier is producing the maximum output voltage. The midband is where an amplifier is supposed to be operated. Normally, you don't operate an amplifier outside the midband because the output voltage is less than maximum.

What we really want to know about the frequency response is this: Given an amplifier, how can we find its two critical frequencies f_1 and f_2? These two frequencies are the key to the frequency response because we can easily calculate the midband after we have the two critical frequencies. The midband will simply be all frequencies between $10f_1$ and $0.1f_2$. So f_1 and f_2 are the key to frequency response. The sections that follow will show you how to find the f_1 and f_2 of an amplifier.

A final point. The critical frequencies are so important that they are also referred to by many other terms, depending on the application. Here are some of the alternative names for the critical frequencies: *cutoff frequencies, break frequencies, corner frequencies, half-power frequencies,* and *3-dB frequencies.*

EXAMPLE 16-1

An amplifier has these critical frequencies: $f_1 = 2$ Hz and $f_2 = 200$ kHz. What is the midband of the frequencies?

SOLUTION

If you recall our discussion of coupling and bypass capacitors, you will remember the high-frequency border, defined as 10 times the critical frequency. As derived earlier, this produces almost perfect coupling within 0.5 percent of maximum. For this reason, we use the same approximation to calculate the lower edge of the midband:

$$10f_1 = 10(2 \text{ Hz}) = 20 \text{ Hz}$$

A similar concept applies to the upper end of the midband, except that we multiply by 0.1 instead of by 10. This produces a frequency that is 10 times lower than f_2 and guarantees that the output voltage is within half a percent of its maximum value. Later discussions will clarify this point, but it should make some sense right now. The upper end of the midband is given by

$$0.1f_2 = 0.1(200 \text{ kHz}) = 20 \text{ kHz}$$

The midband of frequencies is from 20 Hz to 20 kHz. Incidentally, the frequencies between 20 Hz and 20 kHz are called the *audio* range of frequencies, the approximate range of human hearing. The amplifier of this example is called an audio amplifier because it amplifies audio frequencies. Frequencies above 20 kHz are called radio frequencies (abbreviated RF).

16-2 INPUT COUPLING CAPACITOR

The coupling circuit of Fig. 16-2a is one of the reasons for the decrease in the output voltage of an amplifier at low frequencies. As discussed in Chap. 9, the capacitive reactance is given by

$$X_C = \frac{1}{2\pi fC} \qquad (16\text{-}1)$$

At very low frequencies, X_C approaches infinity. At very high frequencies, X_C approaches zero. As we vary the generator frequency in Fig. 16-2a, the output voltage changes because of the coupling capacitor. Figure 16-2b shows the frequency response of the coupling circuit. At zero frequency, the output voltage is zero. As the frequency increases, the output voltage increases. When the frequency is high enough, the output voltage of the coupling circuit approaches its maximum value, as shown in Fig. 16-2b.

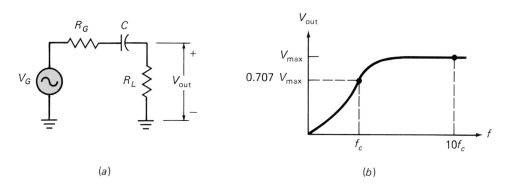

(a)   (b)

FIGURE 16-2   (a) Coupling circuits; (b) frequency response.

Critical Frequency

The current in Fig. 16-2a is given by

$$I = \frac{V_G}{\sqrt{R^2 + X_C^2}}$$

where $R = R_G + R_L$. The only frequency-sensitive term in the equation for current is X_C. Therefore, it determines the frequency response. At low frequencies, X_C approaches infinity and the current has a value of zero. At high frequencies, X_C approaches zero and the current has a maximum value of

$$I_{\max} = \frac{V_G}{R} \qquad (16\text{-}2)$$

This is the maximum value of the current because the capacitor appears like an ac short at high enough frequencies.

If you recall, in Chap. 9 we discussed the critical frequency. This is the generator frequency where $X_C = R$, or the frequency where the capacitive reactance equals the total resistance of the coupling circuit. For this condition, the current is

$$I = \frac{V_G}{\sqrt{R^2 + R^2}} = \frac{V_G}{\sqrt{2R^2}} = 0.707 I_{\max}$$

This says the current decreases to 70.7 percent of the maximum current. This idea is very important. To repeat, current equals $0.707 I_{\max}$ at the critical frequency.

When the frequency is 10 times as high as the critical frequency, $X_C = 0.1R$ and the current is

$$I = \frac{V_G}{\sqrt{R^2 + (0.1R)^2}} = \frac{V_G}{\sqrt{1.01R^2}} = 0.995 I_{\max}$$

This means the current is within 0.5 percent of the maximum current when the frequency is 10 times the critical frequency. Being within 0.5 percent of maximum allows us to approximate as follows. The current is approximately maximum at 10 times the critical frequency.

Here is what you need to remember to determine the effect of a coupling capacitor on the frequency response of an amplifier. As proved in Chap. 9, the critical frequency of the coupling circuit is given by

$$f_c = \frac{1}{2\pi RC} \qquad (16\text{-}3)$$

You can apply this basic equation to any coupling circuit, provided that you use the total resistance for R. At this frequency, the current in the coupling circuit is 70.7 percent of the maximum current. To operate in the normal range of the coupling circuit, you have to increase the generator frequency by a factor of 10. When the generator frequency is $10f_c$, the current in the coupling circuit is approximately maximum. In this case, almost all the ac signal is coupled to the load resistor.

Effect on Amplifier

Figure 16-3a shows the CE amplifier analyzed in earlier chapters. It has an input coupling capacitor and an output coupling capacitor. What we want to do now is to find the critical frequency produced by the input coupling capacitor. We postpone the effects of the output coupling capacitor and the emitter bypass capacitor until later. For now, let us

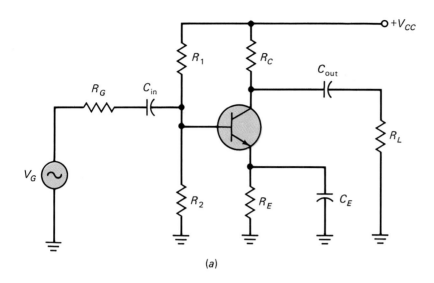

(a)

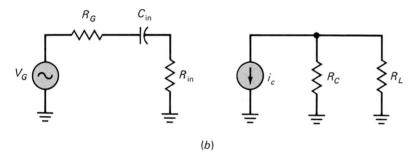

(b)

FIGURE 16-3 CE amplifier and equivalent circuit.

eliminate the output capacitor and emitter bypass capacitor from the analysis by assuming C_{out} and C_E are infinite capacitances. This allows us to draw the ac equivalent circuit as shown in Fig. 16-3b.

Now R_{in} is the input impedance of the stage in the midband of the amplifier. It equals

$$R_{\text{in}} = R_1 \parallel R_2 \parallel \beta r'_e$$

The input resistance R_{in} is identical to the z_{in} used in earlier chapters. The total resistance of the input coupling circuit is

$$R = R_G + R_{\text{in}} \tag{16-4}$$

Use this value of R to calculate the critical frequency.

A similar approach can be used with any amplifier (emitter follower, JFET amplifier, etc.). You can use

$$f_c = \frac{1}{2\pi RC} \tag{16-5}$$

to calculate the critical frequency. Just work out the value of R for the input coupling circuit of the particular amplifier, and substitute this value into Eq. (16-5) along with the value of the coupling capacitor. Here is the step-by-step process that you need to remember:

1. Work out the input resistance of the amplifier.

2. Get the total resistance of the input coupling circuit.

3. Calculate the critical frequency.

The following examples illustrate the process. Also, memorize Eq. (16-5). It is used throughout this chapter. It is one of the basic equations like Ohm's law that you will use a few thousand times.

EXAMPLE 16-2

If $\beta = 100$, what is the critical frequency of the input coupling circuit in Fig. 16-4a?

SOLUTION

First, get the input resistance of the amplifier. We already know the answer because we analyzed this CE amplifier many times in earlier chapters. Recall that $r'_e = 22.7\ \Omega$, $\beta r'_e = 2.27\ k\Omega$, $R_G = 600\ \Omega$, and $R_{in} = 1\ k\Omega$.

Second, get the total resistance of the input coupling circuit. It equals the sum of the generator resistance and the input resistance of the stage:

$$R = 600\ \Omega + 1\ k\Omega = 1.6\ k\Omega$$

Third, calculate the critical frequency, using the basic equation given earlier:

$$f_c = \frac{1}{2\pi(1.6\ k\Omega)(0.47\ \mu F)} = 212\ Hz$$

Since this is the critical frequency, good coupling starts at a frequency that is 10 times as high, approximately 2.12 kHz, as shown in Fig. 16-4b.

EXAMPLE 16-3

Calculate the critical frequency of the input coupling circuit in Fig. 16-5.

SOLUTION

This is an emitter follower, but the same basic strategy is used. You work out the input resistance of the stage. You add the generator resistance to get the total resistance. Then you calculate the critical frequency. Here are the detailed calculations.

The ac load resistance seen by the emitter is

$$r_e = 4.3\ k\Omega\ \|\ 620\ \Omega = 542\ \Omega$$

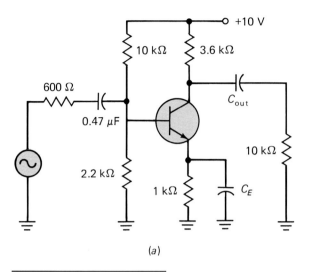

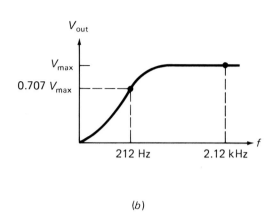

(a)

(b)

FIGURE 16-4 Example.

The dc base voltage is $+5$ V, so the dc emitter voltage is $+4.3$ V. With an R_E of 4.3 kΩ, the dc emitter current is 1 mA, which means r_e' is approximately 25 Ω. Therefore, the input impedance of the base is

$$z_{in(base)} = 100(542 \ \Omega + 25 \ \Omega) = 56.7 \ k\Omega$$

The input impedance of the stage is

$$z_{in} = 10 \ k\Omega \parallel 10 \ k\Omega \parallel 56.7 \ k\Omega = 4.59 \ k\Omega$$

This is identical to R_{in}. The total resistance and capacitance of the input coupling circuit are, respectively,

$$R = 3.6 \ k\Omega + 4.59 \ k\Omega = 8.19 \ k\Omega$$

and

$$C = 0.68 \ \mu F$$

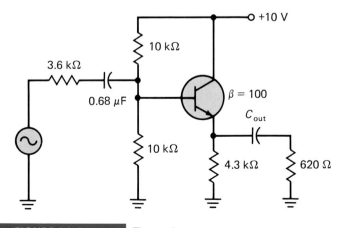

FIGURE 16-5 Example.

Now, calculate the critical frequency of the input coupling circuit:

$$f_c = \frac{1}{2\pi(8.19 \text{ k}\Omega)(0.68 \text{ }\mu\text{F})} = 28.6 \text{ Hz}$$

This means that the output voltage is 0.707 of the maximum output voltage when the generator frequency is 28.6 Hz. To get almost-perfect coupling, the generator frequency should be 10 times higher, or 286 Hz.

16-3 OUTPUT COUPLING CAPACITOR

The output coupling capacitor has an effect similar to that of the input coupling capacitor. But there is a snag. To get the critical frequency of the output coupling circuit, we need to use the *output impedance* of the amplifier instead of the input impedance. For the output impedance to make any sense, we need to use Thevenin's theorem.

Converting from Current Source to Voltage Source

There is an elegant way to analyze the output coupling circuit. Figure 16-6a shows the output side of a *CE* amplifier. In the midband of frequencies, the collector current source drives R_C in parallel with R_L. But below the midband the coupling capacitor has a large reactance. Therefore, we can no longer visualize R_C in parallel with R_L. Instead, we have to use another approach for frequencies below the midband.

The approach is this: Apply Thevenin's theorem to the circuit on the left of the coupling capacitor. In other words, break the connecting wire between the top of R_C and C_{out}. With the broken lead, all the current flows through R_C, giving a Thevenin voltage of

$$v_{\text{th}} = i_c R_C$$

Next, get the Thevenin resistance. Do this by reducing the source to zero, which is equivalent to opening the current source. With the current source open, all you see is R_C. This is the Thevenin resistance:

$$r_{\text{th}} = R_C \tag{16-6}$$

This Thevenin resistance is often referred to as the *output impedance* of the amplifier.

Figure 16-6b shows the equivalent circuit. Everything to the left of C_{out} has been thevenized. This equivalent circuit is analogous to the input coupling circuit, except that

$$R = R_C + R_L \tag{16-7}$$

and

$$C = C_{\text{out}}$$

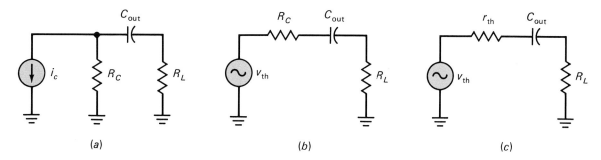

FIGURE 16-6   Output coupling capacitor and equivalent circuit.

Therefore, you still can use the basic equation to calculate the critical frequency:

$$f_c = \frac{1}{2\pi RC} \qquad (16\text{-}8)$$

Other Amplifiers

The idea of thevenizing the output side of an amplifier applies to any bipolar amplifier, JFET amplifier, etc. Any amplifier can be thevenized to get the equivalent circuit of Fig. 16-6c. Because of this, the total resistance in the output coupling circuit is

$$R = r_{th} + R_L$$

This means you can find the critical frequency of the output coupling circuit with the basic formula that applies to all coupling circuits:

$$f_c = \frac{1}{2\pi RC}$$

The only thing that varies among circuits is the values of R and C.

Here are the Thevenin resistances (output impedances) of all basic amplifiers discussed earlier:

CE Amplifier: $r_{th} = R_C$ (16-9)

Swamped CE: $r_{th} = R_C$ (16-10)

Emitter Follower: $r_{th} = R_E \parallel \left(r_e' + \dfrac{R_G \parallel R_1 \parallel R_2}{\beta} \right)$ (16-11)

CD Amplifier: $r_{th} = R_D$ (16-12)

Swamped CD: $r_{th} = R_D$ (16-13)

Source Follower: $r_{th} = R_S \parallel \dfrac{1}{g_m}$ (16-14)

The derivation of these output impedances is similar to that given for the *CE* amplifier.

The process for finding the critical frequency of the output coupling circuit is as follows:

1. Get the output impedance of the amplifier.

2. Find the total resistance of the output coupling circuit.

3. Calculate the critical frequency of the output coupling circuit.

The pattern is the same as before. You work out the total resistance of the coupling circuit and then calculate the critical frequency.

EXAMPLE 16-4

What is the critical frequency of the output coupling circuit in Fig. 16-7a? Ignore the input coupling capacitor and the emitter bypass capacitor.

SOLUTION

The output impedance of the CE amplifier equals the dc collector resistance R_C:

$$r_{th} = 3.6 \text{ k}\Omega$$

The total resistance of the output coupling circuit is

$$R = 3.6 \text{ k}\Omega + 10 \text{ k}\Omega = 13.6 \text{ k}\Omega$$

The output coupling capacitance is

$$C = 2.2 \text{ }\mu\text{F}$$

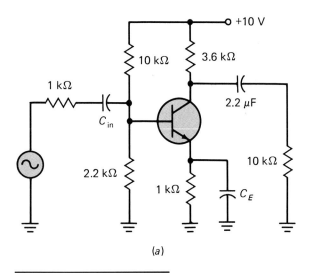

(a)

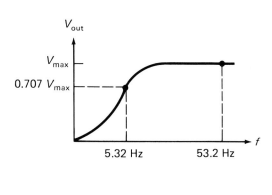

(b)

FIGURE 16-7 Example.

Now, calculate the critical frequency as follows:

$$f_c = \frac{1}{2\pi(13.6 \text{ k}\Omega)(2.2 \text{ }\mu\text{F})} = 5.32 \text{ Hz}$$

Good coupling starts at a frequency that is 10 times as high, approximately 53.2 Hz, as shown in Fig. 16-7b.

EXAMPLE 16-5

Calculate the critical frequency of the output coupling circuit in Fig. 16-8.

SOLUTION

This is an emitter follower, so we can use Eq. (16-11) to find the output impedance:

$$r_{th} = 4.3 \text{ k}\Omega \parallel \left(25 \text{ }\Omega + \frac{3.6 \text{ k}\Omega \parallel 10 \text{ k}\Omega \parallel 10 \text{ k}\Omega}{100} \right)$$

The equivalent resistance of $3.6 \text{ k}\Omega \parallel 10 \text{ k}\Omega \parallel 10 \text{ k}\Omega$ is $2.09 \text{ k}\Omega$. Therefore, the first step is simplifying the equation

$$r_{th} = 4.3 \text{ k}\Omega \parallel \left(25 \text{ }\Omega + \frac{2.09 \text{ k}\Omega}{100} \right) \quad \text{(step 1)}$$

Next, $2.09 \text{ k}\Omega$ divided by 100 gives $20.9 \text{ }\Omega$:

$$r_{th} = 4.3 \text{ k}\Omega \parallel (25 \text{ }\Omega + 20.9 \text{ }\Omega) \quad \text{(step 2)}$$

After adding 25 and $20.9 \text{ }\Omega$, you get

$$r_{th} = 4.3 \text{ k}\Omega \parallel 45.9 \text{ }\Omega \quad \text{(step 3)}$$

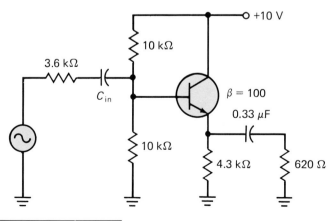

FIGURE 16-8   Example.

Because 4.3 kΩ is much larger than 45.9 Ω, the output impedance is only slightly less than 45.9 Ω:

$$r_{th} = 45.4 \ \Omega \qquad \text{(step 4)}$$

The total resistance and capacitance of the output coupling circuit are, respectively,

$$R = 45.4 \ \Omega + 620 \ \Omega = 665 \ \Omega \qquad \text{and} \qquad C = 0.33 \ \mu\text{F}$$

Now, calculate the critical frequency of the output coupling circuit:

$$f_c = \frac{1}{2\pi(665 \ \Omega)(0.33 \ \mu\text{F})} = 725 \ \text{Hz}$$

This means that the output voltage is 0.707 of the maximum output voltage when the generator frequency is 725 Hz. To get almost-perfect coupling, the generator frequency should be 10 times as high, or 7.25 kHz.

The most important calculation was the calculation of the output impedance. Each of the four steps has a physical meaning in the transistor circuit. If you understand the physical meaning of each step, you will be able to calculate the output impedance of an emitter follower from memory. Here is one way to remember the calculation: In Fig. 16-8, imagine that you are looking from the base back toward the generator. In the ac equivalent circuit, you see a Thevenin resistance of 3.6 kΩ in parallel with 10 kΩ in parallel with 10 kΩ. This is equivalent to 2.09 kΩ (step 1). As you move from the base to the emitter circuit, the Thevenin resistance is stepped down by the β factor, which decreases the Thevenin resistance to 20.9 Ω (step 2). Furthermore, you have to add r'_e as you move through the transistor to the emitter terminal. This adds 25 Ω to get 45.9 Ω (step 3). At this point you see 45.9 Ω in parallel with the emitter biasing resistor R_E. The final result is 45.4 Ω (step 4). This is the Thevenin resistance or output impedance that you would see if you were sitting on the left end of the output coupling capacitor looking back into the emitter follower.

A final point. In almost any emitter follower, R_E is large enough to ignore, and the output impedance is approximately equal to

$$r_{th} = r'_e + \frac{R_G \| R_1 \| R_2}{\beta} \qquad (16\text{-}15)$$

This says the output impedance equals r'_e plus the Thevenin resistance of the base divided by the ac current gain.

16-4 EMITTER BYPASS CAPACITOR

Figure 16-9*a* shows a *bypass circuit*, first discussed in Chap. 9. As we vary the generator frequency, the output voltage changes because of the bypass capacitor. Figure 16-9*c* shows the frequency response of the bypass circuit.

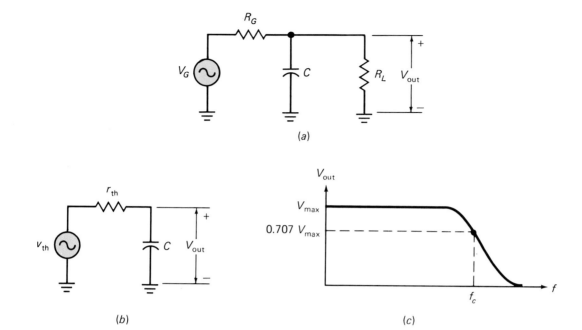

FIGURE 16-9 Bypass circuit.

At low frequencies, the output voltage is maximum. At high frequencies, the output voltage approaches zero. Again, the critical frequency is where the output voltage is 70.7 percent of the maximum value.

Critical Frequency

At low frequencies, the output voltage of Fig. 16-9a has a maximum value of

$$V_{max} = \frac{V_G}{R_G + R_L}(R_L) \qquad (16\text{-}16)$$

This is the maximum value of the output voltage because the capacitor appears open at low frequencies.

To get the critical frequency of the circuit, we thevenize the circuit facing the capacitor. We can do this as follows: Mentally disconnect the capacitor. The circuit that remains is a voltage divider with a voltage of

$$v_{th} = \frac{V_G}{R_G + R_L}(R_L)$$

Notice that this Thevenin voltage is the same as the maximum output voltage given by Eq. (16-16).

Next, get the Thevenin resistance facing the capacitor as follows. In Fig. 16-9a, disconnect the capacitor. Mentally reduce the voltage source to zero, which is equivalent to replacing it by a short. The Thevenin resistance seen by the capacitor is the parallel resistance of R_G and R_L:

$$r_{th} = R_G \parallel R_L \qquad (16\text{-}17)$$

Figure 16-9b shows the Thevenin equivalent circuit with the capacitor reconnected. In this equivalent circuit, the current is given by

$$I = \frac{v_{th}}{\sqrt{R^2 + X_C^2}}$$

where $R = r_{th}$. As before, the critical frequency is where $X_C = R$. This means we can use the same basic equation as before to calculate the critical frequency:

$$f_c = \frac{1}{2\pi RC}$$

All you have to remember is that the R of a bypass circuit equals the Thevenin resistance facing the capacitor.

Effect of Emitter Bypass Capacitor

Figure 16-10a shows a CE amplifier. It has an input coupling capacitor, an output coupling capacitor, and an emitter bypass capacitor. We already know how to find the critical frequency of the input coupling circuit and the output coupling circuit. Now we are ready to calculate the critical frequency produced by the emitter bypass capacitor. In the following discussion, assume the critical frequencies of the coupling circuits are much lower than the critical frequency of the emitter bypass circuit.

In the midband of the amplifier, the emitter is at ac ground, and the output voltage is maximum, as shown in Fig. 16-10b. As the frequency decreases to f_c, the output voltage decreases to 70.7 percent of the maximum output. What is happening is this. The emitter is no longer at

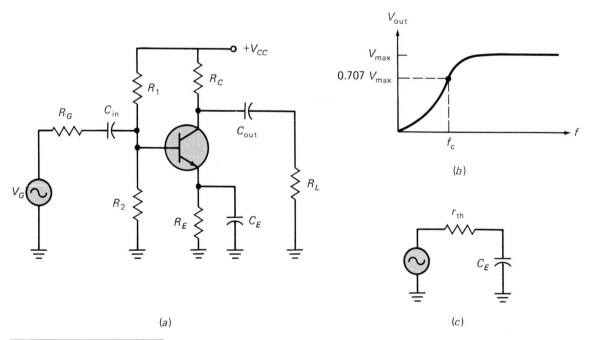

(a)

(b)

(c)

FIGURE 16-10 CE amplifier.

ac ground, so negative feedback is beginning to appear. As the frequency decreases even further, the negative feedback increases and this reduces the output voltage even more. For normal operation, the amplifier should be operated with a frequency that is at least 10 times the critical frequency. Then we can be sure that the emitter is at ac ground.

The critical frequency of the emitter bypass circuit is given by

$$f_c = \frac{1}{2\pi RC}$$

where R is the Thevenin resistance facing the emitter bypass capacitor (see Fig. 16-10c). This is identical to the output impedance of the emitter follower given by Eq. (16-15) and discussed in Example 16-5. Here is the step-by-step process that you need to remember:

1. Calculate the output impedance of the emitter.

2. Calculate the critical frequency.

EXAMPLE 16-6

If $\beta = 100$ in Fig. 16-11a, what is the critical frequency of the emitter bypass circuit? Ignore the effect of the input and output coupling capacitors throughout this example.

SOLUTION

We need to get the output impedance of the emitter. The Thevenin resistance of the base circuit (what you see looking from the base back toward the generator) is

$$R_G \| R_1 \| R_2 = 600\ \Omega \| 10\ k\Omega \| 2.2\ k\Omega = 450\ \Omega$$

When it is viewed from the emitter, this becomes

$$\frac{R_G \| R_1 \| R_2}{\beta} = \frac{450\ \Omega}{100} = 4.5\ \Omega$$

You have to add the r_e' of the circuit, which is 22.7 Ω:

$$r_e' + \frac{R_G \| R_1 \| R_2}{\beta} = 22.7\ \Omega + 4.5\ \Omega = 27.2\ \Omega$$

This is in parallel with 1 kΩ, which we can ignore because it is much larger. This means the output impedance of the emitter is

$$r_{\text{th}} = 27.2\ \Omega$$

The rest is easy. Once you have the Thevenin resistance facing the emitter bypass capacitor, you can use the basic equation for critical frequency:

$$f_c = \frac{1}{2\pi(27.2\ \Omega)(10\ \mu F)} = 585\ Hz$$

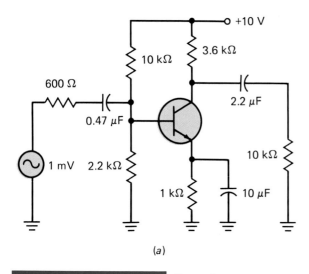

(a)

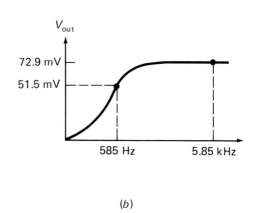

(b)

FIGURE 16-11   Example.

Figure 16-11*b* shows the frequency response. In the midband of the amplifier, the output voltage is 72.9 mV (calculated often in earlier chapters). The lowest frequency in the midband is 10 times the critical frequency, or 5.85 kHz. At this frequency, the output voltage is approximately maximum. When the generator frequency is less than 5.85 kHz, the output voltage will start to fall off. The decrease is definitely noticeable at the critical frequency, 585 Hz, where the output voltage is down to 0.707 of the maximum value.

EXAMPLE 16-7

If β = 100 in Fig. 16-11*a*, what are the three critical frequencies? What is the lowest frequency in the midband of the amplifier?

SOLUTION

We have already calculated all three critical frequencies. In Examples 6-2, 6-4, and 6-6, respectively, we found that

$$f_c = 212 \text{ Hz} \qquad \text{input coupling capacitor}$$

$$f_c = 5.32 \text{ Hz} \qquad \text{output coupling capacitor}$$

$$f_c = 585 \text{ Hz} \qquad \text{emitter bypass capacitor}$$

The weakest link in any chain is the most critical. When it comes to critical frequencies below the midband of the amplifier, the highest one is the most important. In this case, the most important critical frequency is 585 Hz. If you multiply this by 10, you get the lowest frequency in the midband of the amplifier.

EXAMPLE 16-8

Repeat Example 16-7 if C_{in} changes from 0.47 to 0.047 μF.

SOLUTION

If the input coupling capacitance decreases by a factor of 10, its critical frequency will increase by a factor of 10. In this case, the three critical frequencies are

$$f_c = 2.12 \text{ kHz} \quad \text{input coupling capacitor}$$

$$f_c = 5.32 \text{ Hz} \quad \text{output coupling capacitor}$$

$$f_c = 585 \text{ Hz} \quad \text{emitter bypass capacitor}$$

Now, the most important critical frequency is 2.12 kHz. If you multiply this by 10, you get 21.2 kHz, the lowest frequency in the midband of the amplifier.

In almost any design, one of the three critical frequencies will be higher than the others. Treat this highest critical frequency as the most important one because it determines the lowest frequency in the midband of the amplifier. If lightning strikes and the three critical frequencies are identical, you can still use the basic ×10 rule as an approximation. For instance, suppose an amplifier has these critical frequencies:

$$f_c = 2 \text{ Hz} \quad \text{input coupling capacitor}$$

$$f_c = 2 \text{ Hz} \quad \text{output coupling capacitor}$$

$$f_c = 2 \text{ Hz} \quad \text{emitter bypass capacitor}$$

The lowest frequency in the midband is approximately 20 Hz. At 20 Hz, the output voltage would be down slightly more than before because of the combined effect of all three critical frequencies (down about 2 percent instead of 0.5 percent).

16-5 COLLECTOR BYPASS CIRCUIT

Visualize a piece of wire as one plate of a capacitor and the chassis as the other plate. This unwanted capacitance is called the *stray wiring capacitance*. It is unwanted because it produces undesirable effects, to be discussed soon. The capacitance between a piece of wire and the chassis is really a capacitance because you can measure it with a capacitance meter. The longer the wire, the greater the stray wiring capacitance. Also, the closer the wire is to the chassis, the greater the capacitance. Stray wiring capacitance can seriously degrade the high-frequency response of an amplifier. This is why you have to keep leads as short as possible when building circuits that have to operate above 100 kHz. In the following discussion, we symbolize the stray wiring capacitance by C_{stray}.

The stray wiring capacitance is only the beginning of our problems at high frequency. An even more subtle capacitance is built into the transistor itself. The emitter diode has an r'_e, but it also has an internal capacitance symbolized by C'_e. Similarly, the collector diode has a capacitance symbolized by C'_c. Since these two internal capacitances are very small, they have little effect at frequencies below 100 kHz. But when you get above 100 kHz, you have to take the internal capacitances into account.

Figure 16-12a shows a CE amplifier with C'_c and C_{stray}. Dashed lines are used for these unwanted capacitances because they are invisible. These unwanted capacitances are typically in picofarads, so they have no effect at low frequencies. But when you get to high frequencies, the reactance of these capacitors becomes low enough to provide a shunt path to ground. Stated another way, these capacitances form unwanted bypass circuits that will ac ground the collector at high enough frequencies. When this happens, there is no output voltage and the amplifier is useless.

Figure 16-12b shows the equivalent circuit for the collector side of the transistor. This is what the circuit looks like above the midband. Since the output coupling capacitor is an ac short, R_C is in parallel with R_L. But the two capacitances C'_c and C_{stray} are also in parallel with R_L. When

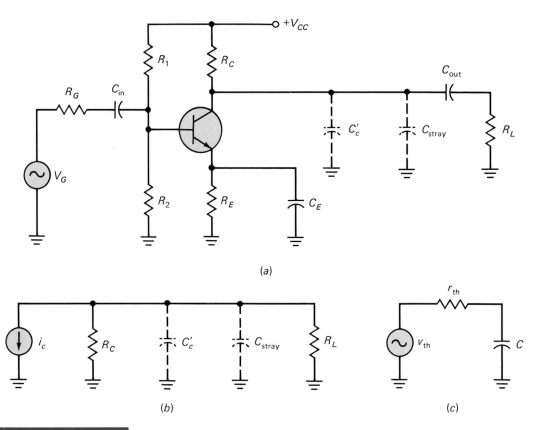

(a)

(b)

(c)

FIGURE 16-12   CE amplifier with stray capacitances.

the reactance of these capacitors is small, the ac collector current is shunted away from the load resistance. This means the output voltage decreases.

Now we are ready to find the critical frequency of the collector circuit. In Fig. 16-12b, the Thevenin resistance facing the two capacitances is

$$r_{\text{th}} = R_C \| R_L \qquad (16\text{-}18)$$

The total capacitance of the two capacitors in parallel is

$$C = C_c' + C_{\text{stray}} \qquad (16\text{-}19)$$

These are the quantities to use when you calculate the critical frequency of the collector bypass circuit. In other words, the final equivalent circuit for the collector bypass circuit is shown in Fig. 16-12c. At high frequencies, the output side of the transistor acts as an ordinary bypass circuit with an R and a C. Here is the process for calculating the critical frequency of the collector bypass circuit:

1. Work out the Thevenin resistance facing the shunt capacitances.
2. Add the collector capacitance and the stray-wiring capacitance.
3. Calculate the critical frequency.

EXAMPLE 16-9

The 2N3904 of Fig. 16-13 has $C_c' = 4$ pF. If the stray-wiring capacitance is 10 pF, what is the critical frequency of the collector bypass circuit?

SOLUTION

The calculations are similar to what we did before, except that the capacitances are a lot smaller. Because of this, the critical frequency

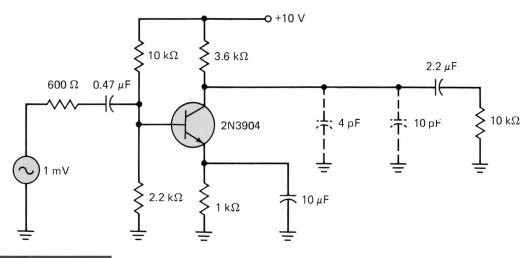

FIGURE 16-13   Example.

is going to be a lot higher. To begin, the Thevenin resistance facing the two capacitances is

$$r_{\text{th}} = 3.6 \text{ k}\Omega \parallel 10 \text{ k}\Omega = 2.65 \text{ k}\Omega$$

The total unwanted capacitance is

$$C = 4 \text{ pF} + 10 \text{ pF} = 14 \text{ pF}$$

The rest is easy. The critical frequency is

$$f_c = \frac{1}{2\pi(2.65 \text{ k}\Omega)(14 \text{ pF})} = 4.29 \text{ MHz}$$

16-6 MILLER'S THEOREM

The emitter diode has a capacitance of C_e'. Because of this capacitance, the base side of a transistor also has an unwanted bypass circuit that limits the high-frequency response of an amplifier. But before we can see what this bypass circuit looks like, we have to take a detour and talk about Miller's theorem.

Figure 16-14a shows an amplifier with a capacitor between the input and output terminals. The capacitor is sometimes called a *feedback capacitor* because the amplified output signal is fed back to the input. When A is large, this feedback can significantly change the operation of the amplifier. A circuit like this is difficult to analyze because the feedback capacitor affects the input and output circuits simultaneously.

Fortunately, there is a shortcut for determining the effect of the feedback capacitor. Known as *Miller's theorem*, the shortcut says that the

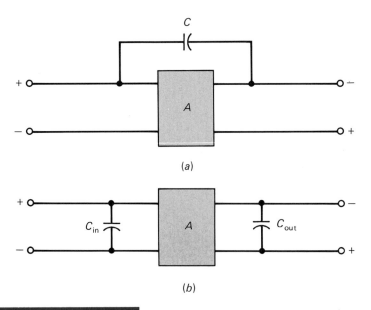

FIGURE 16-14   Miller's theorem: (a) before; (b) after.

original circuit can be replaced by the equivalent circuit of Fig. 16-14b. This equivalent circuit is much easier to work with because the feedback capacitor has been split into two new capacitances, C_{in} and C_{out}. With complex algebra, we can derive the values of the new capacitances. The input capacitance is

$$C_{in} = C(A + 1) \qquad (16\text{-}20)$$

and the output capacitance is

$$C_{out} = C\frac{A + 1}{A} \qquad (16\text{-}21)$$

The advantage of using Miller's theorem is that it splits the feedback capacitor into two equivalent capacitors, one for the input side and the other for the output side. This makes two simple problems out of one big one. Equations (16-20) and (16-21) are valid for any inverting amplifier such as a CE amplifier, swamped CE amplifier, etc. In these equations, A is the midband voltage gain. (The optional topics include the mathematical derivation of Miller's theorem.)

Memorize the basic idea of Miller's theorem: A feedback capacitor between the input and output of an inverting amplifier is equivalent to two new capacitances, one on the input side and the other on the output side. The input capacitance is larger than the feedback capacitance by a factor of $A + 1$. Let that sink in. It has led to a lot of inventions.

EXAMPLE 16-10

If $C = 4$ pF and $A = 117$ in Fig. 16-14a, what are the equivalent input and output capacitances of the amplifier?

SOLUTION

Use Eqs. (16-20) and (16-21) to get

$$C_{in} = (4\,\text{pF})(117 + 1) = 472\,\text{pF}$$

and

$$C_{out} = (4\,\text{pF})\left(\frac{117 + 1}{117}\right) = 4.05\,\text{pF}$$

Incidentally, the voltage gain of 117 is the gain of the CE amplifier discussed throughout earlier chapters. If a capacitance of 4 pF were connected between the collector and the base of this CE amplifier, the feedback capacitor would be equivalent to an input capacitance of 472 pF and an output capacitance of 4.05 pF.

Here is a practical point about the Miller theorem. Whenever the voltage gain is large, you can use these approximate equations for the two equivalent capacitances:

$$C_{in} = CA \qquad \text{and} \qquad C_{out} = C$$

16-7 HIGH-FREQUENCY BIPOLAR ANALYSIS

We have saved the best until last. What you are about to see is some very sophisticated high-frequency bipolar analysis. You may be surprised at how far you have come in a short time. The following discussion uses critical frequencies, Thevenin's theorem, Miller's theorem, etc. You may have to read this material a few times because it contains many subtle ideas. But it's worth the effort because it will pull everything together and give you more confidence about high-frequency analysis.

Unwanted Base Bypass Circuit

Figure 16-15a shows a signal generator V_G with an internal resistance R_G driving a CE amplifier. Figure 16-15b shows the ac equivalent circuit in the midband of the amplifier. Resistance r_g is the ac Thevenin resistance facing the base:

$$r_g = R_1 \parallel R_2 \parallel R_G$$

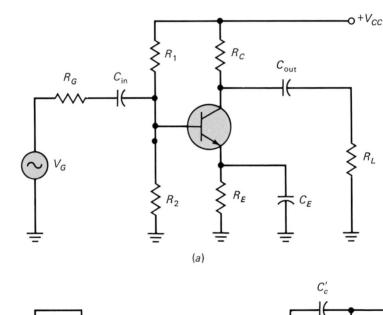

(a)

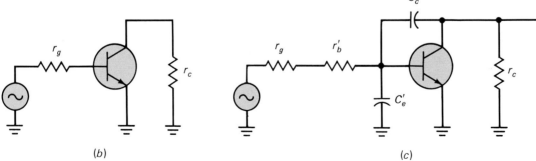

(b)　　　　　(c)

FIGURE 16-15　　(a) Bipolar amplifier; (b) midband equivalent circuit; (c) above the midband.

Resistance r_c is the ac resistance seen by the collector:

$$r_c = R_C \parallel R_L$$

In the midband of the amplifier, there are no capacitive effects. This is where an amplifier is supposed to operate to get the maximum output voltage. Below the midband, the coupling capacitors and emitter bypass capacitor decrease the output voltage. Above the midband, the transistor and stray-wiring capacitances decrease the output voltage.

Figure 16-15c shows the ac equivalent circuit above the midband of the amplifier, where C_e' is the capacitance of the emitter diode and C_c' is the capacitance of the collector diode. Notice that C_c' is a feedback capacitor because it is connected between the base and the collector. Also notice r_b'. This is the resistance of the base region. Earlier discussions ignored r_b' because it has little effect in the midband. But now it has to be included in the analysis because it has a large effect above the midband.

To find the critical frequencies for a bipolar amplifier, we have to identify the unwanted bypass circuits in the base and the collector. The first step is to get the two Miller capacitances. The midband voltage gain from the base to the collector is

$$A = \frac{r_c}{r_e'}$$

The input Miller capacitance therefore equals

$$C_{\text{in}} = C_c'\left(\frac{r_c}{r_e'} + 1\right)$$

and the output Miller capacitance is approximately

$$C_{\text{out}} = C_c'$$

The output Miller capacitance is approximately C_c' because the voltage gain A is usually large in a CE amplifier. Figure 16-16a shows the two Miller capacitances. The input Miller capacitance is in parallel with C_e', while the output Miller capacitance is in parallel with C_{stray}.

The total capacitance in the base circuit is

$$C = C_e' + C_c'\left(\frac{r_c}{r_e'} + 1\right) \tag{16-22}$$

as shown in Fig. 16-16b. To get the base circuit into the form of a bypass circuit, we have to thevenize the circuit driving the base capacitance to get Fig. 16-16c. The Thevenin resistance facing the base capacitance is

$$R = (r_g + r_b') \parallel \beta r_e' \tag{16-23}$$

In Fig. 16-16c, the base bypass circuit has a critical frequency of

$$f_c = \frac{1}{2\pi RC}$$

where R and C are given by Eqs. (16-23) and (16-22), respectively.

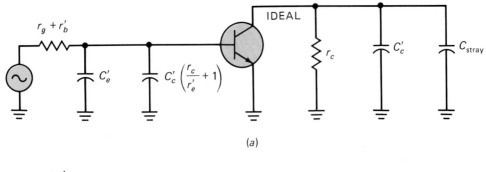

(a)

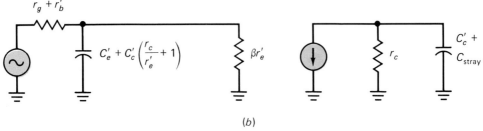

(b)

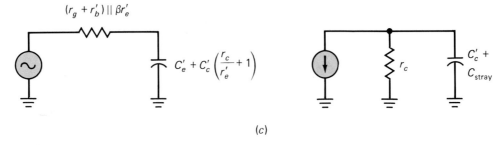

(c)

FIGURE 16-16   AC equivalent circuits of a bipolar amplifier.

Unwanted Collector Bypass Circuit

Look at the collector side of Fig. 16-16c. As discussed earlier, this unwanted collector bypass circuit has a Thevenin resistance of

$$R = r_c \tag{16-24}$$

and a capacitance of

$$C = C'_c + C_{\text{stray}} \tag{16-25}$$

You calculate the critical frequency of this bypass circuit with

$$f_c = \frac{1}{2\pi RC}$$

where R and C are given by Eqs. (16-24) and (16-25), respectively.

Dominant Critical Frequency

There are two critical frequencies above the midband: the critical frequency of the base bypass circuit and the critical frequency of the collector bypass circuit. When you analyze an amplifier, these two critical frequencies are almost always different. The one closer to the midband is more important. We call it the *dominant critical frequency*. As before, you have to change the frequency by a factor of 10 to get to the midband. This time, however, we divide by 10 rather than multiply.

For instance, assume f_2 is the dominant critical frequency. The midband is below f_2. The highest frequency in the midband is approximately $0.1f_2$. At $0.1f_2$ the output voltage is approximately equal to the maximum output voltage. As the frequency increases above $0.1f_2$, the output voltage gradually decreases until it equals $0.707V_{max}$ at f_2. The following examples will show you how to find the dominant critical frequency.

Estimates for C_e' and r_b'

There is no standard designation for C_c'. Data sheets may list it by any of the following equivalent symbols: C_c, C_{cb}, C_{ob}, and C_{obo}. For instance, the data sheet of a 2N3904 gives a maximum C_{ob} of 4 pF. This is the value of C_c' to use in high-frequency analysis.

Capacitance C_e' is not usually given on a data sheet because it is too difficult to measure directly. Instead, the manufacturer lists a value called the *current gain–bandwidth product*, designated f_T. This is the frequency at which the current gain of a transistor drops to unity. You can calculate C_e' by using

$$C_e' = \frac{1}{2\pi f_T r_e'} \tag{16-26}$$

Another quantity that is difficult to measure is r_b', the internal resistance of the base. This is why you may not find it on a data sheet. The value of this resistance changes with the collector current, ac current gain, etc. It can vary from a few ohms to several thousand ohms, depending on the particular transistor, the Q point, etc. Low values of r_b' allow a transistor to amplify higher frequencies. For transistors intended for high-frequency operation, data sheets often include a quantity called the *collector-base time constant*, symbolized $r_b'C_c'$. Given this quantity, you can calculate r_b' with

$$r_b' = \frac{r_b'C_c'}{C_c'} \tag{16-27}$$

This will give you an accurate value for r_b'.

Another way to estimate the value of r_b' is with this equation:

$$r_b' = h_{ie} - \beta r_e' \tag{16-28}$$

where h_{ie} is one of the h parameters usually listed on a data sheet as the small-signal input impedance. This gives a fairly accurate approximation, provided you use the correct values of h_{ie} and β. If you can find typical values on a data sheet, Eq. (16-28) usually gives reasonably accurate results.

If you cannot use Eq. (16-27) or (16-28) because data sheets don't include the necessary data, you can get a rough estimate of r_b' with this equation:

$$r_b' = 0.2\beta r_e' \tag{16-29}$$

This says you take 20 percent of $\beta r_e'$ to find a rough value of r_b'. This approximation works quite well for some transistors, but it is way off for others. Remember, electronics is not an exact science, so we often have to settle for estimates.

EXAMPLE 16-11

The data sheet of a 2N5208 lists $f_T = 1200$ MHz, $r_b'C_c' = 10$ ps for $I_E = 2$ mA, and $C_c' = 1$ pF. What is the value of C_e' for $I_E = 2$ mA? What is r_b' for $I_E = 2$ mA?

SOLUTION

When $I_E = 2$ mA, $r_e' = 12.5\ \Omega$. With Eq. (16-26),

$$C_e' = \frac{1}{2\pi(1200\ \text{MHz})(12.5\ \Omega)} = 10.6\ \text{pF}$$

With Eq. (16-27),

$$r_b' = \frac{10\ \text{ps}}{1\ \text{pF}} = 10\ \Omega$$

This is a relatively low value of base resistance because the 2N5208 is designed for high-frequency operation.

EXAMPLE 16-12

The data sheet of a 2N3904 lists $f_T = 300$ MHz. What is the value of C_e' for $I_E = 5$ mA? What is r_b' at 5 mA?

SOLUTION

When $I_E = 5$ mA, $r_e' = 5\ \Omega$. With Eq. (16-26),

$$C_e' = \frac{1}{2\pi(300\ \text{MHz})(5\ \Omega)} = 106\ \text{pF}$$

The 2N3904 is not considered a high-frequency transistor, which is why the data sheet does not include the value of $r_b'C_c'$. But you will find graphs of current gain and input impedance (Figs. 11 and 13 in the Appendix). These graphs show typical values. By reading the approximate values of β and h_{ie} at 5 mA, we get 150 and 870 Ω. With Eq. (16-28), we can get an estimate of r_b':

$$r_b' = 870\ \Omega - (150)(5\ \Omega) = 120\ \Omega$$

EXAMPLE 16-13

Find the dominant critical frequency of Fig. 16-17 above the midband.

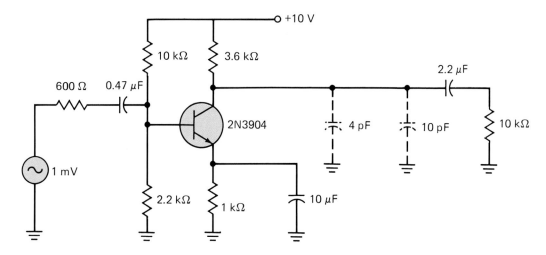

FIGURE 16-17 Example of high-frequency analysis.

SOLUTION

With Eq. (16-26),

$$C'_e = \frac{1}{2\pi(300 \text{ MHz})(22.7 \text{ }\Omega)} = 23.4 \text{ pF}$$

In this equation, we are using $f_T = 300$ MHz and $r'_e = 22.7 \text{ }\Omega$ (calculated earlier).

With Eq. (16-28),

$$r'_b = 3.5 \text{ k}\Omega - (125)(22.7 \text{ }\Omega) = 663 \text{ }\Omega$$

The values used in this equation come from the graphs of current gain and input impedance on the data sheet of a 2N3904 (see Appendix): At $I_C = 1.1$ mA, the typical values are $h_{ie} = 3.5$ kΩ and $\beta = 125$.

In Fig. 16-16c, the quantity r_g is the parallel of R_G, R_1, and R_2. It equals

$$r_g = 600 \text{ }\Omega \parallel 10 \text{ k}\Omega \parallel 2.2 \text{ k}\Omega = 450 \text{ }\Omega$$

As seen in Fig. 16-16c, you have to add r'_b to this to get

$$r_g + r'_b = 450 \text{ }\Omega + 663 \text{ }\Omega = 1.11 \text{ k}\Omega$$

Then you have to work out the Thevenin resistance of the base bypass circuit:

$$R = 1.11 \text{ k}\Omega \parallel 125(22.7 \text{ }\Omega) = 799 \text{ }\Omega$$

The last calculation can also be done by substituting directly in Eq. (16-23).

Next, recall that we analyzed this amplifier several times in earlier chapters and have already worked out its voltage gain in the midband.

It is
$$A = 117$$

This is identical to r_c/r'_e. Since $r_c/r'_e = 117$, the total capacitance in the base bypass circuit of Fig. 16-16c is

$$C = 23.4 \text{ pF} + (4 \text{ pF})(117 + 1) = 495 \text{ pF}$$

This gives us a critical frequency of

$$f_c = \frac{1}{2\pi(799 \ \Omega)(495 \text{ pF})} = 402 \text{ kHz}$$

One down, and one to go. The critical frequency of the collector bypass circuit is much easier to find. In Fig. 16-17, the Thevenin resistance is

$$R = 3.6 \text{ k}\Omega \parallel 10 \text{ k}\Omega = 2.65 \text{ k}\Omega$$

and the total capacitance is

$$C = 4 \text{ pF} + 10 \text{ pF} = 14 \text{ pF}$$

The critical frequency is

$$f_c = \frac{1}{2\pi(2.65 \text{ k}\Omega)(14 \text{ pF})} = 4.29 \text{ MHz}$$

The dominant critical frequency is 402 kHz. At this frequency, the output voltage is 0.707 of the maximum value. When we analyzed this amplifier in earlier examples, we found the output voltage was 72.9 mV. Therefore, the output voltage is

$$V_{\text{out}} = 0.707(72.9 \text{ mV}) = 51.5 \text{ mV}$$

when the generator frequency is 402 kHz. If the generator frequency is reduced to 40.2 kHz, the output voltage will increase to approximately 72.9 mV.

16-8 TOTAL FREQUENCY RESPONSE

Figure 16-18 shows the frequency response of the amplifier just analyzed. Here you see the output voltage versus the generator frequency, given a constant generator voltage of 1 mV. We found 585 Hz in Example 16-7 and 402 kHz in Example 16-13. A rough sketch like this showing the maximum output voltage, the two dominant critical frequencies, and the 0.707 voltage is adequate for a basic idea of what the amplifier is doing.

The decrease in output voltage is gradual as the frequency moves outside the midband. For this reason, the amplifier can still provide usable voltage gain outside the midband. If you need to calculate the output voltage outside the midband, here is a useful formula:

$$V_{\text{out}} = \frac{V_{\text{max}}}{\sqrt{1 + (f_1/f)^2} \ \sqrt{1 + (f/f_2)^2}} \qquad (16\text{-}30)$$

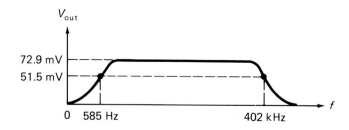

FIGURE 16-18   Total frequency response.

This comes from advanced mathematics. At first, it may appear over-whelming, but it is simple if you break it down into the three separate ranges: the midband, below the midband, and above the midband.

In the midband, f is at least 10 times greater than f_1 and at least 10 times smaller than f_2. As a result, both radicals disappear because f_1/f and f/f_2 are each approximately zero. This leaves an output voltage in the midband of

$$V_{out} = V_{max}$$

Given any amplifier, you can analyze it as we did in earlier chapters to get the output voltage in the midband. This gives you the value of V_{max}.

Below the midband, f is at least 10 times smaller than f_2. As a result, the second radical of Eq. (16-30) disappears because f/f_2 is approximately zero. This leaves an output voltage below the midband of

$$V_{out} = \frac{V_{max}}{\sqrt{1 + (f_1/f)^2}} \tag{16-31}$$

Above the midband, f is at least 10 times greater than f_1. As a result, the first radical of Eq. (16-30) disappears because f_1/f is approximately zero. This leaves an output voltage above the midband of

$$V_{out} = \frac{V_{max}}{\sqrt{1 + (f/f_2)^2}} \tag{16-32}$$

We can use Eqs. (16-31) and (16-32) to calculate the output voltage outside the midband. These equations assume one dominant critical frequency below the midband and one dominant critical frequency above the midband.

EXAMPLE 16-14

What is the output voltage of Fig. 16-17 when $f = 1$ kHz and when $f = 100$ kHz?

SOLUTION

In Fig. 16-18, the critical frequencies are $f_1 = 585$ Hz and $f_2 = 402$ kHz.

When $f = 1$ kHz, the frequency ratio is

$$\frac{f_1}{f} = \frac{585 \text{ Hz}}{1 \text{ kHz}} = 0.585$$

With Eq. (16-31), the output voltage at 1 kHz is

$$V_{out} = \frac{72.9 \text{ mV}}{\sqrt{1 + 0.585^2}} = 62.9 \text{ mV}$$

When $f = 100$ kHz, the frequency ratio is

$$\frac{f}{f_2} = \frac{100 \text{ kHz}}{402 \text{ kHz}} = 0.249$$

With Eq. (16-32), the output voltage at 100 kHz is

$$V_{out} = \frac{72.9 \text{ mV}}{\sqrt{1 + 0.249^2}} = 70.7 \text{ mV}$$

16-9 DECIBELS

We are about to discuss an important subject called *decibels*. But before we do, it will help if we review something you learned in basic mathematics. The something is called *logarithms*.

Review of Logarithms

Suppose we are given this equation:

$$y = 10^x \qquad (16\text{-}33)$$

Each x value produces a different y value. For instance, here are some basic calculations:

When $x = 1$ $\quad y = 10^1 = 10$
When $x = 2$ $\quad y = 10^2 = 100$
When $x = 3$ $\quad y = 10^3 = 1000$

Notice that y increases by a factor of 10 for each time x increases by 1.
 We can also calculate y values, given negative values of x, as follows:

When $x = -1$ $\quad y = 10^{-1} = 0.1$
When $x = -2$ $\quad y = 10^{-2} = 0.01$
When $x = -3$ $\quad y = 10^{-3} = 0.001$

This time, y decreases by a factor of 10 each time x decreases by 1.
 If we solve Eq. (16-33) for x in terms of y, we get

$$x = \log_{10} y$$

This says x is the logarithm (or exponent) of 10 that gives y. Usually, the 10 is omitted, and the equation is written as

$$x = \log y \qquad (16\text{-}34)$$

An equation like this forces you to think in reverse. The relation between x and y is still the same as before because Eq. (16-34) is nothing more than a disguised form of Eq. (16-33). If you ever have trouble understanding what Eq. (16-34) means, return to Eq. (16-33).

If you have a scientific calculator or a table of logarithms, you can find the x value for any y value. For instance, here are some basic calculations for $x = \log y$:

When $y = 10$ $x = \log 10 = 1$
When $y = 100$ $x = \log 100 = 2$
When $y = 1000$ $x = \log 1000 = 3$

You can see that x increases by 1 each time y increases by a factor of 10.

We can also calculate x values, given decimal values of y, as follows:

When $y = 0.1$ $x = \log 0.1 = -1$
When $y = 0.01$ $x = \log 0.01 = -2$
When $y = 0.001$ $x = \log 0.001 = -3$

Notice that x decreases by 1 each time y decreases by a factor of 10.

Decibel Power Gain

The *ordinary power gain* G of an amplifier is defined as the ratio of the output power to the input power:

$$G = \frac{P_2}{P_1}$$

The *decibel power gain* is defined as

$$G' = 10 \log G \qquad (16\text{-}35)$$

There are several reasons why the decibel power gain has advantages over ordinary power gain. We discuss these later. For now, let us look at some examples of how to calculate the decibel power gain.

Incidentally, G is the ratio of output power to input power. Therefore, G has no units or dimensions. When you take the logarithm of G, you wind up with a quantity that still has no units or dimensions. But to make sure that G' is never confused with G, we attach the unit *decibel* (abbreviated dB) to all answers for G.

EXAMPLE 16-15

Calculate the decibel power gain for power gains of 1, 10, 100, and 1000.

SOLUTION

When the ordinary power gain is 1, then

$$G' = 10 \log 1 = 0 \text{ dB}$$

When the ordinary power gain is 10, the decibel power gain is

$$G' = 10 \log 10 = 10 \text{ dB}$$

When the ordinary power gain is 100,

$$G' = 10 \log 100 = 20 \text{ dB}$$

When the ordinary power gain is 1000, then

$$G' = 10 \log 1000 = 30 \text{ dB}$$

Notice that each time the ordinary power gain increases by a factor of 10, the decibel power gain increases by 10 dB.

EXAMPLE 16-16

Calculate the decibel power gain for power gains of 1, 0.1, 0.01, and 0.001.

SOLUTION

When the ordinary power gain is 1, the decibel power gain is

$$G' = 10 \log 1 = 0 \text{ dB}$$

When the ordinary power gain is 0.1,

$$G' = 10 \log 0.1 = -10 \text{ dB}$$

When the ordinary power gain is 0.01,

$$G' = 10 \log 0.01 = -20 \text{ dB}$$

When the ordinary power gain is 0.001,

$$G' = 10 \log 0.001 = -30 \text{ dB}$$

In this case, each time the ordinary power gain decreases by a factor of 10, the decibel power gain decreases by 10 dB.

EXAMPLE 16-17

Calculate the decibel power gain for the following power gains: 100, 200, 400, and 800.

SOLUTION

When the ordinary power gain is 100, the decibel power gain is

$$G' = 10 \log 100 = 20 \text{ dB}$$

When the ordinary power gain is 200,

$$G' = 10 \log 200 = 23 \text{ dB}$$

When the ordinary power gain is 400,

$$G' = 10 \log 400 = 26 \text{ dB}$$

When the ordinary power gain is 800,

$$G' = 10 \log 800 = 29 \text{ dB}$$

Can you see a pattern in these numbers? Of course you can. Each time the ordinary power gain increases by a factor of 2, the decibel power gain increases by 3 dB. This property also applies to decreases. For example, if the ordinary power gain decreases from 500 to 250, the decibel power gain decreases from

$$G' = 10 \log 500 = 27 \text{ dB}$$

to

$$G' = 10 \log 250 = 24 \text{ dB}$$

which is a change of -3 dB.

Let us summarize the important properties of the decibel power gain that have been demonstrated:

1. Each time the ordinary power gain increases (decreases) by a factor of 10, the decibel power gain increases (decreases) by 10 dB.

2. Each time the ordinary power gain increases (decreases) by a factor of 2, the decibel power gain increases (decreases) by 3 dB.

16-10 DECIBEL·VOLTAGE GAIN

Voltage measurements are more common than power measurements. For this reason, decibels are even more useful with voltage gain. The *decibel voltage gain* is defined as

$$A' = 20 \log A \qquad (16\text{-}36)$$

The reason for using 20 instead of 10 occurs because power is proportional to the square of voltage. For more details, see the optional topics. Otherwise, accept Eq. (16-36) as the basic definition of the decibel voltage gain. Remember, definitions don't have to be proved. They are made up, rather than being derived from other equations. Think of Eq. (16-36) as a starting point, something that you memorize just as you would memorize the meaning of a foreign word.

As a process, the defining equation says

1. Take the logarithm of the ordinary voltage gain.

2. Multiply by 20 to get the decibel voltage gain.

Basic Rules for Voltage Gain

In Eq. (16-36), multiplying by 20 instead of by 10 means we can modify the basic rules summarized in Example 16-17 to get these useful rules for decibel voltage gain:

1. Each time the ordinary voltage gain increases (decreases) by a factor of 10, the decibel voltage gain increases (decreases) by 20 dB.

2. Each time the ordinary voltage gain increases (decreases) by a factor of 2, the decibel voltage gain increases (decreases) by 6 dB.

For instance, suppose the voltage gain decreases from 1000 to 100. This is a factor of 10 in the ordinary voltage gain. It means the decibel voltage gain decreases by 20 dB. Here is the proof:

When $A = 1000$ $\qquad A' = 20 \log 1000 = 60$ dB
When $A = 100$ $\qquad A' = 20 \log 100 = 40$ dB

As you can see, the ordinary voltage gain has decreased by a factor of 10, and the decibel voltage gain has decreased by 20 dB.

Cascaded Stages

As proved earlier, the total voltage gain of a two-stage amplifier is the product of the individual voltage gains:

$$A = A_1 A_2 \qquad\qquad (16\text{-}37)$$

Something unusual happens when we use the decibel voltage gain instead of the ordinary voltage gain. The mathematical phenomenon that you are

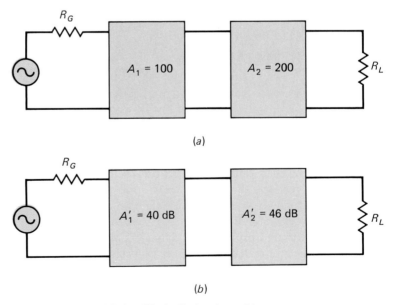

(a)

(b)

FIGURE 16-19   (a) Ordinary gains multiply; (b) decibel gains add.

about to see is one of the reasons that decibels are widely used in electronics. In Eq. (16-37), take the logarithm of both sides to get

$$\log A = \log A_1A_2 = \log A_1 + \log A_2$$

Multiply both sides by 20 to get

$$20 \log A = 20 \log A_1 + 20 \log A_2$$

But this equation can be rewritten in terms of the decibel voltage gains:

$$A' = A'_1 + A'_2 \qquad (16\text{-}38)$$

Look deeply into this equation. It says that the total decibel voltage gain of two cascaded stages equals the sum of the individual decibel voltage gains. The same idea applies to any number of stages. This additive property of decibel gain is one reason for its popularity.

EXAMPLE 16-18

Use Fig. 16-19a and b to demonstrate the validity of Eq. (16-38).

SOLUTION

If you did not understand the formal proof of Eq. (16-38), the following numerical proof may help. In Fig. 16-19a, the total voltage gain is

$$A = (100)(200) = 20,000$$

which is equivalent to a decibel voltage gain of

$$A' = 20 \log 20,000 = 86 \text{ dB}$$

Next, calculate the decibel voltage gain of each stage as follows:

$$A'_1 = 20 \log 100 = 40 \text{ dB} \qquad \text{and} \qquad A'_2 = 20 \log 200 = 46 \text{ dB}$$

Notice that 40 dB plus 46 dB does add up to 86 dB. In other words, we can calculate the total decibel voltage gain either with

$$A' = 20 \log 20,000 = 86 \text{ dB}$$

or with

$$A' = 40 \text{ dB} + 46 \text{ dB} = 86 \text{ dB}$$

This proves that Eq. (16-38) is valid.

16-11 VOLTAGE GAIN OUTSIDE THE MIDBAND

When we divide both sides of Eq. (16-30) by V_G, we get

$$A = \frac{A_{\text{mid}}}{\sqrt{1 + (f_1/f)^2} \sqrt{1 + (f/f_2)^2}} \qquad (16\text{-}39)$$

where A is the voltage gain from the generator to the output. Remember that this equation assumes two critical frequencies, one below the midband at f_1 and the other above the midband at f_2. Because the equation is complicated, we can break it down into three simpler equations as follows. In the midband, f is much larger than f_1 and much smaller than f_2. This means f_1/f and f/f_2 are approximately zero, and the equation simplifies to

$$A = A_{\text{mid}}$$

for any frequency in the midband. This is any f between $10f_1$ and $0.1f_2$. In this range, the voltage gain is approximately equal to the midband voltage gain.

Below the midband, f is much smaller than f_2, which means f/f_2 is approximately zero. In this case, Eq. (16-39) simplifies to

$$A = \frac{A_{\text{mid}}}{\sqrt{1 + (f_1/f)^2}} \tag{16-40}$$

With this equation, we can calculate the voltage gain for frequencies below the midband.

Above the midband, f is much larger than f_1, which means f_1/f is approximately zero. In this case, Eq. (16-39) simplifies to

$$A = \frac{A_{\text{mid}}}{\sqrt{1 + (f/f_2)^2}} \tag{16-41}$$

With this equation, we can calculate the voltage gain for frequencies above the midband.

Octaves and Decades

The middle C on a piano has a frequency of 256 Hz. The next higher C is an *octave* higher, and it has a frequency of 512 Hz. The next higher C has a frequency of 1024 Hz, and so on. In music, the word *octave* refers to a doubling of the frequency. Every time you go up one octave, you have doubled the frequency.

In electronics, an octave has a similar meaning for ratios like f_1/f and f/f_2. For instance, if $f_1 = 100$ Hz and $f = 50$ Hz, the f_1/f ratio is

$$\frac{f_1}{f} = \frac{100 \text{ Hz}}{50 \text{ Hz}} = 2$$

We can describe this by saying f is one octave below f_1. As another example, suppose $f = 400$ kHz and $f_2 = 200$ kHz. Then

$$\frac{f}{f_2} = \frac{400 \text{ kHz}}{200 \text{ kHz}} = 2$$

This means f is one octave above f_2.

A *decade* has a similar meaning for ratios like f_1/f and f/f_2, except that a factor of 10 is used instead of a factor of 2. For instance, if $f_1 = 500$ Hz and $f = 50$ Hz, the f_1/f ratio is

$$\frac{f_1}{f} = \frac{500 \text{ Hz}}{50 \text{ Hz}} = 10$$

We can describe this by saying f is one decade below f_1. As another example, suppose $f = 2$ MHz and $f_2 = 200$ kHz. Then

$$\frac{f}{f_2} = \frac{2\,\text{MHz}}{200\,\text{kHz}} = 10$$

This means f is one decade above f_2.

Linear and Logarithmic Scales

Ordinary graph paper has a *linear scale* on both axes. This means the spaces between the numbers are the same for all numbers, as shown in Fig. 16-20a. With a linear scale, you start at 0 and proceed in uniform steps toward higher numbers. All the graphs discussed up to now have used linear scales. For instance, the vertical and horizontal axes of Fig. 16-18 are understood to be linear.

Sometimes we prefer to use a *logarithmic scale* because it compresses distances and allows us to see over many decades. Figure 16-20b shows a logarithmic scale. Notice the numbering begins with 1. The space between 1 and 2 is much larger than the space between 9 and 10. By compressing the scale logarithmically as shown here, we can take advantage of certain properties of logarithms and decibels to be discussed.

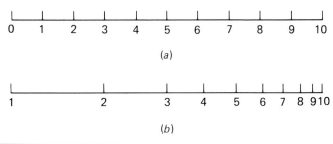

(a)

(b)

FIGURE 16-20   (*a*) Linear scale; (*b*) logarithmic scale.

Besides ordinary graph paper, there is semilogarithmic paper. This type of graph paper has a linear scale on the vertical axis and a logarithmic scale on the horizontal axis. People use semilogarithmic paper when they want to graph a quantity like voltage gain over many decades of frequency.

Graph of Decibel Voltage Gain

Figure 16-21a shows the frequency response of a typical amplifier. The graph is similar to Fig. 16-18, but this time we are looking at the decibel voltage gain versus frequency as it would appear on semilogarithmic paper. A graph like this is called a *Bode plot*. The vertical axis uses a linear scale, and the horizontal axis uses a logarithmic scale. As shown, the decibel voltage gain is maximum in the midband of frequencies. At each critical frequency, the decibel voltage gain is down 3 dB from the maximum value. Below the midband, the decibel voltage gain decreases 20 dB per decade. Above the midband, the decibel voltage gain decreases 20 dB per decade.

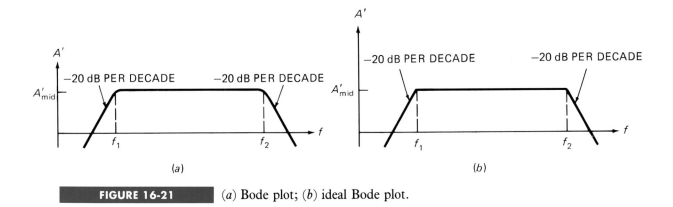

FIGURE 16-21 (*a*) Bode plot; (*b*) ideal Bode plot.

Figure 16-21*b* shows the same graph in ideal form. Many people prefer using this ideal Bode plot because it is easier to draw and gives approximately the same information. Anyone looking at this ideal graph knows the decibel voltage gain is 3 dB lower at the critical frequencies. So the graph really contains all the original information when this correction of 3 dB is mentally included.

In Fig. 16-21*b*, the decibel voltage gain decreases at a rate of 20 dB per decade at very low and at very high frequencies. An alternative way to describe the decreased voltage gain is to say 6 dB per octave. The two statements are equivalent.

EXAMPLE 16-19

An amplifier has $f_1 = 585$ Hz and $f_2 = 402$ kHz. The voltage gain in the midband is 72.9. What is the voltage gain at 253 Hz? What is the voltage gain at 349 kHz?

SOLUTION

This is a good time to use Eqs. (16-40) and (16-41) since nothing else will work. We can calculate the voltage gain at 253 Hz as follows: The frequency ratio f_1/f is equal to

$$\frac{f_1}{f} = \frac{585\,\text{Hz}}{253\,\text{Hz}} = 2.31$$

Substitute this into Eq. (16-40) to get

$$A = \frac{72.9}{\sqrt{1 + (2.31)^2}} = 29$$

Next, get the voltage gain at 349 kHz as follows: The frequency ratio f/f_2 is equal to

$$\frac{f}{f_2} = \frac{349\,\text{kHz}}{402\,\text{kHz}} = 0.868$$

Substitute this into Eq. (16-40) to get

$$A = \frac{72.9}{\sqrt{1 + 0.868^2}} = 55.1$$

EXAMPLE 16-20

Earlier, we analyzed the amplifier of Fig. 16-17 and found that it had the frequency response shown in Fig. 16-18. Show the Bode plot of the frequency response

SOLUTION

The first thing to do is to calculate the midband voltage gain from the generator to the output. In Fig. 16-17, $V_G = 1$ mV and $V_{max} = 72.9$ mV. The midband voltage gain is

$$A_{mid} = \frac{72.9\,\text{mV}}{1\,\text{mV}} = 72.9$$

Next, work out the decibel voltage gain:

$$A'_{mid} = 20 \log 72.9 = 37.3\,\text{dB}$$

The two dominant critical frequencies are $f_1 = 585$ Hz and $f_2 = 402$ kHz.

Figure 16-22 shows the graph of decibel voltage gain versus frequency. The midband voltage gain is 37.3 dB with critical frequencies of 585 Hz and 402 kHz. The graph is ideal because we have to include the correction factor of -3 dB at the critical

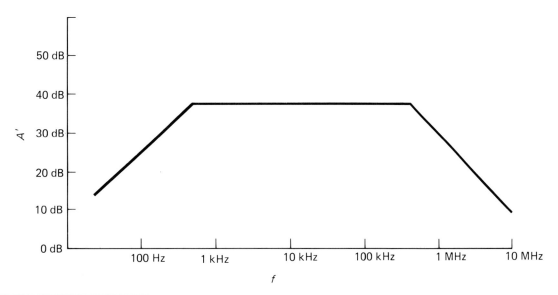

FIGURE 16-22 Example of ideal Bode plot.

frequencies. In reality, the voltage gain is down 3 dB at the critical frequencies, which means the voltage gain is actually

$$A' = 37.3 \text{ dB} - 3 \text{ dB} = 34.3 \text{ dB}$$

Beyond the critical frequencies, the voltage gain decreases 20 dB per decade. This means the decibel voltage gain is

$$A' = 37.3 \text{ dB} - 20 \text{ dB} = 17.3 \text{ dB}$$

when $f = 58.5$ Hz or when $f = 4.02$ MHz. The gain keeps dropping at the same rate, which means it ideally equals

$$A' = 17.3 \text{ dB} - 20 \text{ dB} = -2.7 \text{ dB}$$

when $f = 5.85$ Hz or when $f = 40.2$ MHz. In other words, the ideal frequency responses looks like this:

$$A' = \begin{cases} 37.3 \text{ dB} & \text{for 585 Hz to 402 kHz} \\ 17.3 \text{ dB} & \text{for 58.5 Hz and 4.02 MHz} \\ -2.7 \text{ dB} & \text{for 5.85 Hz and 40.2 MHz} \\ -22.7 \text{ dB} & \text{for 0.585 Hz and 402 MHz} \end{cases}$$

Ideal Bode plots like Fig. 16-22 give us a quick visual summary of what an amplifier can do. We immediately can see what the midband voltage gain is, where the dominant critical frequencies are, how fast the gain drops off, etc. Graphs like these are very popular in industry. Incidentally, many technicians and engineers use the term *corner frequency* instead of *critical frequency*. This is because the ideal Bode plot has a sharp corner at each critical frequency. Another term often used instead of *critical frequency* is *break frequency*. This is because the graph breaks at each critical frequency where it then decreases at a rate of 20 dB per decade.

☐ OPTIONAL TOPICS

The following material continues the earlier discussions at a more advanced and specialized level. All the topics are optional because they are not used in any of the basic discussions in later chapters. This section will be a useful reference when you are in industry because then you will probably want more advanced viewpoints.

☐ 16-12 PROOF OF MILLER'S THEOREM

In Fig. 16-23*a*, the feedback capacitor has an alternating current given by

$$I_C = \frac{V_{\text{in}} - (-V_{\text{out}})}{-jX_C}$$

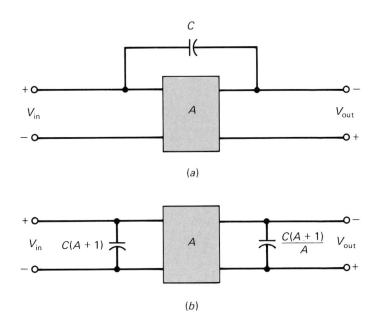

(a)

(b)

Proof of Miller's theorem.

Since $V_{out} = AV_{in}$, we can rewrite the equation as

$$I_C = \frac{V_{in}(1 + A)}{-jX_C}$$

or

$$\frac{V_{in}}{I_C} = \frac{-jX_C}{A + 1} = \frac{-j1}{2\pi fC(A + 1)}$$

The ratio V_{in}/I_C is the impedance of the capacitor seen from the input side of the amplifier. The denominator contains $2\pi f$ times an equivalent capacitance of $C(A + 1)$. We define this equivalent capacitance as the *input Miller capacitance*:

$$C_{in(Miller)} = C(A + 1)$$

This input Miller capacitance appears in parallel with the input terminals of the amplifier (see Fig. 16-23*b*).

The output capacitance can be derived as follows. The current through the capacitor is

$$I_C = \frac{V_{out} - (-V_{in})}{-jX_C} = \frac{(1 + 1/A)V_{out}}{-jX_C}$$

or

$$\frac{V_{out}}{I_C} = \frac{-jX_C}{(A + 1)/A} = \frac{-j1}{2\pi fC(A + 1)/A}$$

The ratio V_{out}/I_C is the impedance of the capacitor seen from the output terminals. The denominator contains $2\pi f$ times an equivalent capacitance of $C(A + 1)/A$. We define this equivalent capacitance as the *output Miller capacitance*:

$$C_{\text{out(Miller)}} = \frac{C(A + 1)}{A}$$

as shown in Fig. 16-23b. When A is large, this capacitance is approximately equal to C, the feedback capacitance.

16-13 HIGH-FREQUENCY FET ANALYSIS

Because of the similarity between bipolar transistors and FETs, the high-frequency analysis is similar but not identical. The gate and drain will each have an unwanted bypass circuit. The following discussion will show you how to find the equivalent resistance and capacitance of each bypass circuit. After calculating the R and C for each bypass circuit, you can calculate the critical frequency with the basic equation used before.

Midband

Figure 16-24a shows a signal generator V_G with an internal resistance R_G driving a FET amplifier with voltage-divider bias. At high frequencies, coupling and bypass capacitors act as ac shorts. For this reason, the ac equivalent circuit appears as shown in Fig. 16-24b. Resistance r_d is the ac resistance seen by the drain, the parallel combination of R_D and R_L:

$$r_d = R_D \parallel R_L$$

Resistance r_g is the ac Thevenin resistance facing the gate terminal of the FET. This resistance includes the biasing resistors in parallel with the generator resistance. For instance, with voltage-divider bias,

$$r_g = R_1 \parallel R_2 \parallel R_G$$

In the midband of the amplifier, the voltage gain is

$$A = g_m r_d$$

Above the midband, internal FET capacitances and stray wiring capacitances form unwanted bypass networks that cause the voltage gain to decrease. Here C_{gs} is the internal capacitance between the gate and the source. C_{gd} is the capacitance between the gate and the drain, and C_{ds} is the capacitance between the drain and the source. Figure 16-24c shows these capacitances in the ac equivalent circuit. A stray capacitance C_{stray} also appears across the drain-ground terminals, as shown in Fig. 16-24c.

Gate Bypass Network

In Fig. 16-24c, C_{gd} is a feedback capacitor. With Miller's theorem,

$$C_{\text{in(Miller)}} = C_{gd}(A + 1)$$

or

$$C_{\text{in(Miller)}} = C_{gd}(g_m r_d + 1)$$

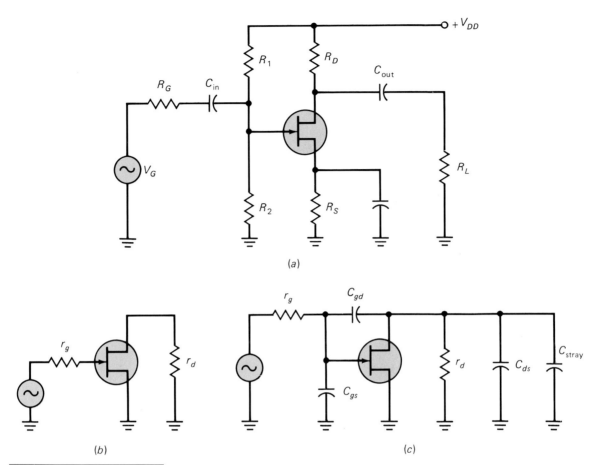

(a)

(b)   (c)

FIGURE 16-24   High-frequency FET analysis.

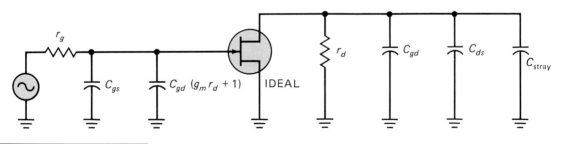

FIGURE 16-25   AC equivalent circuit.

Figure 16-25 shows this input Miller capacitance.

In most cases, A is large enough to approximate the output Miller capacitance as equal to the feedback capacitance:

$$C_{\text{out(Miller)}} = C_{gd}$$

As shown in Fig. 16-25, the output Miller capacitance is in parallel with C_{ds} and C_{stray}.

The FET amplifier of Fig. 16-25 has two unwanted bypass circuits, one on the gate side and the other on the drain side. The total capacitance in the gate circuit is

$$C = C_{gs} + C_{gd}(g_m r_d + 1) \tag{16-42}$$

The resistance facing this capacitance is

$$R = R_1 \parallel R_2 \parallel R_G \tag{16-43}$$

Therefore, the critical frequency of the gate bypass circuit is

$$f_c = \frac{1}{2\pi RC}$$

where R and C are given by Eqs. (16-43) and (16-42), respectively.

Drain Bypass Network

The drain acts as a current source driving ac resistance r_d in parallel with capacitances C_{gd}, C_{ds}, and C_{stray}. The total capacitance in the drain circuit is

$$C = C_{gd} + C_{ds} + C_{\text{stray}} \tag{16-44}$$

The total resistance facing this capacitance is r_d, which we can write as

$$R = R_D \parallel R_L \tag{16-45}$$

So the critical frequency of the drain bypass circuit is

$$f_c = \frac{1}{2\pi RC}$$

where R and C are given by Eqs. (16-45) and (16-44), respectively.

Capacitances on a Data Sheet

The FET capacitances (C_{gd}, C_{gs}, and C_{ds}) are difficult to measure. This is why you will not find them on the typical FET data sheet. Instead you will find these quantities: C_{iss}, C_{oss}, and C_{rss}. These are called *short-circuit capacitances*, and they have the advantage of being easy to measure. These capacitances are defined as follows:

$$C_{iss} = C_{gs} + C_{gd}$$

$$C_{oss} = C_{ds} + C_{gd}$$

$$C_{rss} = C_{gd}$$

Solving these equations simultaneously gives these conversion formulas:

$$C_{gd} = C_{rss} \tag{16-46}$$

$$C_{gs} = C_{iss} - C_{rss} \tag{16-47}$$

$$C_{ds} = C_{oss} - C_{rss} \tag{16-48}$$

With these formulas, we can calculate the capacitances needed to analyze the lag networks of a FET amplifier.

16-14 dBm

Decibels are sometimes used to indicate the power level with respect to 1 mW. In this case, the label *dBm* is used instead of *dB*; the m at the end of dBm reminds you of the milliwatt reference. The dBm formula is

$$P' = 10 \log \frac{P}{1 \text{ mW}} \tag{16-49}$$

where P' = power in dBm and P = power in watts. For instance, if the power is 2 W, then

$$P' = 10 \log \frac{2 \text{ W}}{1 \text{ mW}} = 10 \log 2000 = 33 \text{ dBm}$$

Here are some more equivalences:

$$P' = \begin{cases} 0 \text{ dBm} & \text{when } P = 1 \text{ mW} \\ 10 \text{ dBm} & \text{when } P = 10 \text{ mW} \\ 20 \text{ dBm} & \text{when } P = 100 \text{ mW} \\ 30 \text{ dBm} & \text{when } P = 1 \text{ W} \end{cases}$$

You can have negative dBm as follows:

$$P' = \begin{cases} -10 \text{ dBm} & \text{when } P = 100 \text{ } \mu\text{W} \\ -20 \text{ dBm} & \text{when } P = 10 \text{ } \mu\text{W} \\ -30 \text{ dBm} & \text{when } P = 1 \text{ } \mu\text{W} \end{cases}$$

Sometimes, you will see power meters calibrated in dBm. For instance, if you read 23 dBm, the power will be 200 mW.

16-15 POWER AND VOLTAGE GAINS

The input power to the amplifier is

$$P_1 = \frac{V_1^2}{R_1}$$

and the output power is

$$P_2 = \frac{V_2^2}{R_2}$$

The power gain from input to output equals

$$G = \frac{P_2}{P_1} = \frac{V_2^2/R_2}{V_1^2/R_1}$$

Since the ratio V_2/V_1 is the voltage gain A, we can rewrite the power gain as

$$G = A^2 \left(\frac{R_1}{R_2} \right) \tag{16-50}$$

Impedance-Matched Case

In many systems (for example, microwave and telephone), the input and load impedances are matched, meaning $R_1 = R_2$. For this condition, Eq. (16-50) simplifies to

$$G = A^2 \qquad (16\text{-}51)$$

This says that the power gain equals the square of the voltage gain. For instance, if $A = 100$, then $G = 10,000$.

Taking the logarithm of both sides of Eq. (16-51) gives

$$\log G = \log A^2 = 2 \log A$$

Multiplying both sides by 10 yields

$$10 \log G = 10(2 \log A) = 20 \log A$$

which can be written as

$$G' = A' \qquad (16\text{-}52)$$

This says that the decibel power gain equals the decibel voltage gain, a relation that is true for all impedance-matched systems.

Impedances Not Matched

In most amplifiers, the impedances are not matched because the load impedance does not equal the input impedance. In this case, G' does not equal A', and we must calculate each separately. In other words, we have to use

$$G' = 10 \log G$$

for the decibel power gain and

$$A' = 20 \log A$$

for the decibel voltage gain.

Both types of decibel gain are widely used. The decibel power gain predominates in communications, microwaves, and other systems in which power is important. The decibel voltage gain is preferred in areas of electronics in which measuring voltage is more convenient.

◻ 16-16 BODE PLOTS

In the basic discussion of amplifier response, we introduced the Bode plot of voltage gain versus frequency. What follows is a more mathematical treatment of the subject, including Bode plots of the phase angle. When a capacitor is present, there may be a difference between the phase of the input and output signals. In fact, a coupling circuit is often referred to as a *lead network* because the output voltage leads the input voltage. A bypass circuit is often referred to as a *lag network* because the output voltage lags the input voltage. Also, the critical frequency may be referred to by any of these terms: *cutoff frequency, break frequency, corner frequency, half-power frequency,* or *3-dB frequency.*

In complex numbers, the voltage gain of a lag network is

$$\frac{V_{out}}{V_{in}} = \frac{-jX_C}{R - jX_C}$$

This can be converted to a magnitude of

$$\frac{V_{out}}{V_{in}} = \frac{X_C}{\sqrt{R^2 + X_C^2}} \qquad (16\text{-}53)$$

and a phase angle of

$$\phi = -\arctan\frac{R}{X_C} \qquad (16\text{-}54)$$

Decibel Voltage Gain

Equation (16-53) can be rewritten as

$$A = \frac{V_{out}}{V_{in}} = \frac{X_C}{\sqrt{R^2 + X_C^2}} = \frac{1}{\sqrt{1 + (R/X_C)^2}}$$

Since

$$\frac{R}{X_C} = 2\pi fRC = \frac{f}{f_c}$$

the voltage gain can be written as

$$A = \frac{1}{\sqrt{1 + (f/f_c)^2}}$$

The decibel voltage gain is

$$A' = 20\log\frac{1}{\sqrt{1 + (f/f_c)^2}} \qquad (16\text{-}55)$$

Using Eq. (16-55), we can calculate the decibel voltage gain of a lag network.

For instance, when $f/f_c = 0.1$, Eq. (16-55) gives

$$A' = 20\log\frac{1}{\sqrt{1 + 0.1^2}} = -0.0432 \text{ dB} \approx 0 \text{ dB}$$

When $f/f_c = 1$, the decibel voltage gain is

$$A' = 20\log\frac{1}{\sqrt{1 + 1^2}} = -3.01 \text{ dB} \approx -3 \text{ dB}$$

When $f/f_c = 10$,

$$A' = 20\log\frac{1}{\sqrt{1 + 10^2}} = -20 \text{ dB}$$

When $f/f_c = 100$,

$$A' = 20\log\frac{1}{\sqrt{1 + 100^2}} = -40 \text{ dB}$$

When $f/f_c = 1000$,

$$A' = 20 \log \frac{1}{\sqrt{1 + 1000^2}} = -60 \, \text{dB}$$

The decibel voltage gains we have just calculated tell a story. Here is what we have found:

1. When the input frequency to a lag (bypass) network is a decade (a factor of 10) below the cutoff frequency, the decibel voltage gain is approximately zero.

2. When the input frequency equals the cutoff frequency, the decibel voltage gain equals $-3 \, \text{dB}$ (the half-power point).

3. When the input frequency is a decade above the cutoff frequency, the decibel voltage gain is $-20 \, \text{dB}$.

4. Beyond this point, each time the frequency increases by a decade, the decibel voltage gain decreases by 20 dB.

Bode Plot of the Voltage Gain

Whenever we can reduce a circuit to a lag network, we can calculate the cutoff frequency. Then we know that the decibel voltage gain is down 20 dB when the input frequency is one decade above the cutoff frequency, down 40 dB when the input frequency is two decades above the cutoff frequency, down 60 dB when the input frequency is three decades above the cutoff frequency, and so forth.

Figure 16-26a shows the decibel voltage gain of a lag network. This is called a *Bode plot* of the voltage gain. To see the graph over several decades of frequency, we have compressed the horizontal scale by showing equal distances between each decade in frequency. (Semilogarithmic graph paper does the same thing.) The advantage of marking off the horizontal scale in decades rather than units is this: Well above the cutoff frequency, the decibel voltage gain decreases 20 dB for each decade increase in frequency; therefore, the graph of voltage gain will be a straight line with a slope of $-20 \, \text{dB}$ per decade.

Figure 16-26b shows the ideal Bode plot of the voltage gain. To a first approximation, we neglect the $-3 \, \text{dB}$ at the cutoff frequency and draw a straight line with slope of $-20 \, \text{dB}$ per decade. Ideal Bode plots are used for preliminary analysis because they are easy to draw.

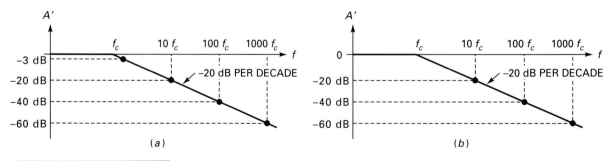

FIGURE 16-26   Bode plot of lag network.

6 dB per Octave

Above the cutoff frequency, the decibel voltage gain of a lag network decreases 6 dB per octave. This is easy to prove. When $f/f_c = 10$, $A' = -20$ dB. When $f/f_c = 20$ (an octave change),

$$A' = 20 \log \frac{1}{\sqrt{1 + 20^2}} = -26 \text{ dB}$$

As you can see, the decibel voltage gain has decreased 6 dB.

In other words, you can describe the frequency response of a lag network above the cutoff frequency in either of two ways. You can say that the decibel voltage gain decreases at a rate of 20 dB per decade, or you can say that it decreases at a rate of 6 dB per octave. Both roll-off rates are used in industry.

Phase Angle

Equation (16-54) can be written as

$$\phi = -\arctan \frac{f}{f_c} \qquad (16\text{-}56)$$

For instance, when $f/f_c = 0.1$, Eq. (16-56) gives

$$\phi = -\arctan 0.1 = -6°$$

When $f/f_c = 1$,

$$\phi = -\arctan 1 = -45°$$

When $f/f_c = 10$,

$$\phi = -\arctan 10 = -84°$$

Figure 16-27a shows how the phase angle of a lag network varies with the frequency. At very low frequencies the phase angle is zero. When $f = 0.1f_c$, the phase angle is $-6°$. When the input frequency equals the cutoff frequency, the phase angle equals $-45°$. For an input frequency that is 10 times the critical frequency, the phase angle is $-84°$. Further increases in frequency produce little change because the limiting value is $-90°$. As you can see, the phase angle of a lag network is between 0 and $-90°$. This means that the output voltage lags the input voltage.

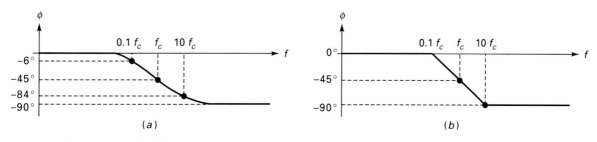

FIGURE 16-27   Bode plot of phase angle for lag network; (a) exact; (b) ideal.

Bode Plot of the Phase Angle

A graph like Fig. 16-27a is a Bode plot of the phase angle. Knowing that the phase angle is $-6°$ at $0.1f_c$ and $-84°$ at $10f_c$ is of little value except to indicate how close the phase angle is to its limiting value. The ideal Bode plot of Fig. 16-27b is much more useful. This is the one to remember because it emphasizes these ideas:

1. When $f = 0.1f_c$, the phase angle is approximately zero.

2. When $f = f_c$, the phase angle is $-45°$.

3. When $f = 10f_c$, the phase angle is approximately $-90°$.

Another way to summarize the Bode plot of the phase angle is this: At the cutoff frequency, the phase angle equals $-45°$. A decade below the cutoff frequency, the phase angle is approximately $0°$; a decade above the cutoff frequency, the phase angle is approximately $-90°$.

Lead Network

The analysis of a lead network is similar to that for a lag network. The decibel voltage gain is

$$A' = 20 \log \frac{1}{\sqrt{1 + (f_c/f)^2}} \tag{16-57}$$

The phase angle is

$$\phi = \arctan \frac{f_c}{f} \tag{16-58}$$

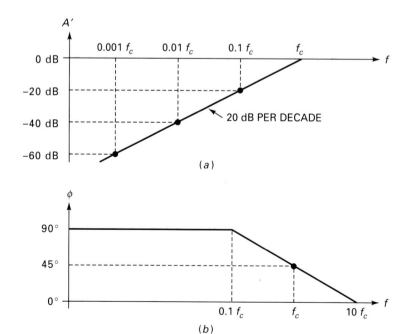

(a)

(b)

FIGURE 16-28   Bode plots: (a) gain; (b) angle.

If you compare these equations with those of a lag network, you will see an inverse symmetry. This is why the ideal Bode plots of decibel voltage gain and phase angle appear as shown in Fig. 16-28. Below the cutoff frequency, the decibel voltage gain rolls off at a rate of 20 dB per decade, equivalent to 6 dB per octave. The phase angle is between 0 and 90°, which means that the output leads the input.

16-17 MORE ABOUT AMPLIFIER RESPONSE

In the midband of an amplifier, the voltage gain is maximum. Below the midband, coupling and bypass capacitors cause the voltage gain to decrease. Above the midband, internal and stray capacitances cause the voltage gain to decrease.

AC Amplifier

Figure 16-29 shows the frequency response of an ac amplifier. The lower critical frequency is f_1, and the upper critical frequency is f_2. The *bandwidth B* of an amplifier is defined as

$$B = f_2 - f_1 \qquad (16\text{-}59)$$

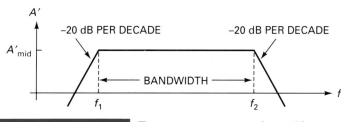

FIGURE 16-29   Frequency response of amplifier.

For instance, if an ac amplifier has a lower critical frequency of 1 MHz and an upper critical frequency of 3 MHz, then its bandwidth is

$$B = 3\text{ MHz} - 1\text{ MHz} = 2\text{ MHz}$$

A tuned amplifier (one with resonant circuits) is often called a *narrowband amplifier* because f_1 and f_2 are close in value. For example, if $f_1 = 450$ kHz and $f_2 = 460$ kHz, then

$$B = 460\text{ kHz} - 450\text{ kHz} = 10\text{ kHz}$$

This is a narrowband amplifier because f_2 is only slightly larger than f_1.

On the other hand, an untuned amplifier is often called a *wideband amplifier* because f_2 is much greater than f_1. For example, if $f_1 = 200$ Hz and $f_2 = 10$ MHz, then

$$B = 10\text{ MHz} - 200\text{ Hz} = 10\text{ MHz}$$

This is a wideband amplifier because f_2 is much greater than f_1.

DC Amplifier

As mentioned earlier, we can design an amplifier with no coupling or bypass capacitors. An amplifier like this is called a *dc amplifier* (or *dc amp*). Figure 16-30*a* shows the ideal Bode plot of the decibel voltage gain for a dc amp. As you can see, the dc amp has a decibel voltage gain of A_{mid} up to the critical frequency f_2. Beyond this, the decibel voltage gain drops 20 dB per decade.

The Bode plot of Fig. 16-30*a* assumes one dominant bypass circuit. This is why the slope is -20 dB per decade. If another bypass circuit had a critical frequency near the first, the Bode plot would break again, and the gain would roll off at a rate of 40 dB per decade.

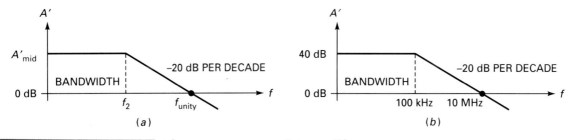

FIGURE 16-30   Frequency response of dc amplifier.

To avoid oscillations (discussed later), the voltage gain of a typical dc amplifier is designed to roll off at a rate of 20 dB per decade until the graph crosses the horizontal axis at f_{unity} (Fig. 16-30*a*). The subscript *unity* means the ordinary voltage gain equals 1 at this frequency, which is equivalent to a decibel voltage gain of 0 dB. For instance, Fig. 16-30*b* shows the Bode plot of a dc amp whose decibel voltage gain drops 20 dB per decade until it crosses the horizontal axis at 10 MHz. Beyond this frequency, other unwanted bypass circuits will reach their critical frequencies, but we are not interested in them.

Since there is no lower critical frequency in a dc amplifier, the bandwidth is

$$B = f_2 \qquad (16\text{-}60)$$

In Fig. 16-30*b*, the bandwidth equals 100 kHz.

Bandwidth of Cascaded Stages

The overall bandwidth of cascaded stages is less than the bandwidth of any stage. For instance, suppose we have two dc-coupled stages, each with a bandwidth of 10 kHz. Then at 10 kHz the response of each stage is down 3 dB, and the overall response is down 6 dB. This implies that the overall bandwidth is less than 10 kHz.

With advanced mathematics, we can prove that the overall bandwidth of *n* identical cascaded stages is given by

$$B_n = B\sqrt{2^{1/n} - 1} \qquad (16\text{-}61)$$

where B_n = overall bandwidth and B = bandwidth of one stage. For instance, if there are two identical stages, each with a bandwidth of 10 kHz, the overall bandwidth is

$$B_2 = 10 \, \text{kHz} \, \sqrt{2^{1/2} - 1} = 6.44 \, \text{kHz}$$

Three cascaded stages, each with a bandwidth of 10 kHz, produce an overall bandwidth of

$$B_3 = 10 \, \text{kHz} \, \sqrt{2^{1/3} - 1} = 5.1 \, \text{kHz}$$

Try to remember this idea because you often have to cascade stages: If the stages are not identical, you will still get bandwidth shrinkage because of the cumulative effects of the stages. In other words, the overall bandwidth is always smaller than the smallest bandwidth of any stage.

16-18 RISETIME-BANDWIDTH RELATION

Sine-wave testing of an amplifier means that we drive it with a sinusoidal input and measure the sinusoidal output voltage. To find the upper critical frequency, we have to vary the input frequency until the voltage gain drops 3 dB from the midband value. Sine-wave testing is a common approach. But there is a faster and simpler way to test an amplifier by using a square-wave input instead of a sine-wave input.

Risetime

Given a bypass circuit like Fig. 16-31a, basic circuit theory tells us what happens after the switch is closed. If the capacitor is initially uncharged,

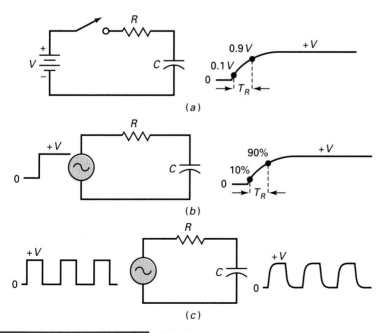

(a)

(b)

(c)

FIGURE 16-31   Risetime.

the voltage will rise exponentially toward the supply voltage V. The risetime T_R is the amount of time it takes the capacitor voltage to go from $0.1V$ (the 10 percent point) to $0.9V$ (the 90 percent point). If it takes 10 μs for the exponential waveform to go from the 10 percent point to the 90 percent point, the waveform has a risetime of 10 μs.

Instead of using a switch to apply the sudden step in voltage, we can use a square-wave generator. For instance, Fig. 16-31b shows the leading edge of a square wave driving the same RC circuit as before. The risetime is still the time it takes for the voltage to go from the 10 percent point to the 90 percent point.

Figure 16-31c shows how several cycles will look. The input voltage changes suddenly from one voltage level to another. The output voltage takes longer to make its transitions. It cannot suddenly step because the capacitor has to charge and discharge through the resistance.

Relation between T_R and RC

Basic courses prove that

$$v_C = V(1 - e^{-t/RC})$$

where v_C = capacitor voltage
$\quad\quad V$ = total change in input voltage
$\quad\quad t$ = time after input transition
$\quad RC$ = time constant of bypass circuit

At the 10 percent point, $v_{out} = 0.1V$. At the 90 percent point, $v_{out} = 0.9V$. Substituting these voltages and solving for the time difference give the risetime:

$$T_R = 2.2RC \quad\quad\quad (16\text{-}62)$$

where T_R = risetime of the capacitor voltage and RC = time constant of the bypass circuit. For instance, if R equals 10 kΩ and C is 50 pF, then

$$RC = (10 \text{ k}\Omega)(50 \text{ pF}) = 0.5 \text{ μs}$$

and the risetime of the output waveform equals

$$T_R = 2.2RC = 2.2(0.5 \text{ μs}) = 1.1 \text{ μs}$$

An Important Relation

As mentioned earlier, a dc amplifier typically has one dominant bypass circuit that rolls off the voltage gain at a rate of 20 dB per decade until f_{unity} is reached. The critical frequency of this bypass circuit is given by

$$f_2 = \frac{1}{2\pi RC}$$

By rearranging,

$$RC = \frac{1}{2\pi f_2}$$

Substituting this into Eq. (16-62) and rearranging gives

$$f_2 = \frac{0.35}{T_R} \qquad (16\text{-}63)$$

where f_2 = upper critical frequency of the amplifier and T_R = risetime of the amplifier output voltage.

This is an important result because it relates sinusoidal and square-wave operation of an amplifier. It means that we can test an amplifier with either a sine wave or a square wave. For instance, Fig. 16-32*a* shows the leading edge of a square wave driving a dc amplifier. If we measure a risetime of 1 μs, we can calculate a sinusoidal critical frequency of

$$f_2 = \frac{0.35}{1\ \mu s} = 350\ \text{kHz}$$

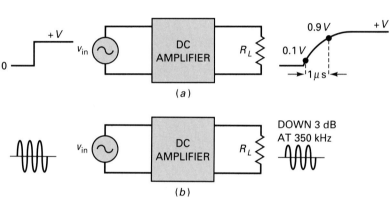

FIGURE 16-32 Relation between risetime and critical frequency.

Alternatively, we can measure the critical frequency directly by driving the dc amplifier with a sine wave, as shown in Fig. 16-32*b*. If we vary the frequency until the output is down 3 dB, we will measure a critical frequency of 350 kHz.

16-19 STRAY EFFECTS

Stray wiring capacitance can seriously degrade the high-frequency response of an amplifier. This is why you have to keep leads as short as possible when building circuits to operate above 100 kHz. In addition to stray wiring capacitance, other unwanted effects can also degrade the high-frequency response.

Equivalent Circuits

Every resistor has a small amount of inductance and capacitance. At lower frequencies, the unwanted L and C have negligible effects. But as the frequency increases, the resistor no longer acts as pure resistance. Figure 16-33*a* shows the equivalent circuit of a resistor with its inductance and

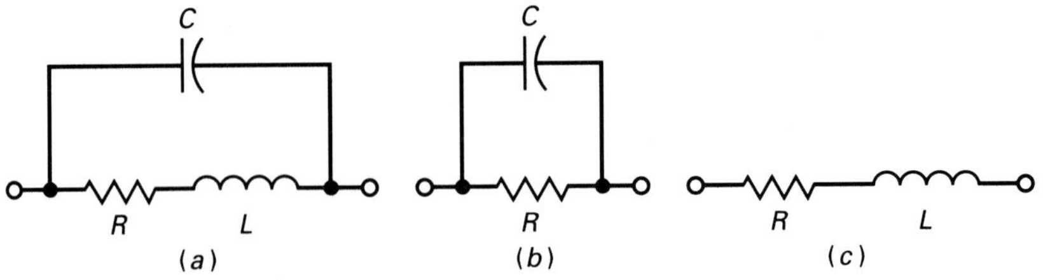

FIGURE 16-33   (*a*) Resistor equivalent circuit; (*b*) resistor with stray capacitances; (*c*) resistor with lead inductance.

capacitance. At lower frequencies, the inductive reactance approaches zero, and the capacitive reactance approaches infinity. In other words, the inductor appears shorted, and the capacitor appears open. In this case, the resistor acts as a pure resistance.

We refer to the inductance as the *lead inductance* because it is produced by the leads going into the resistor. And we refer to the capacitance as the *stray capacitance* because it is the stray wiring capacitance between the ends of the resistor. For frequencies less that 100 MHz, either the inductive or the capacitive effect dominates. This means that a resistor has an equivalent circuit like Fig. 16-33*b* or *c*.

Stray Capacitance

The stray capacitance of a typical resistor (⅛ to 2 W) is in the vicinity of 1 pF, with the exact value determined by the length of the leads, the size of the resistor body, and other factors. In most applications, you can neglect stray capacitance when

$$\frac{X_C}{R} > 10$$

For instance, if a 10-kΩ resistor has 1 pF of stray capacitance, then X_C at 1 MHz equals

$$X_C = \frac{1}{2\pi(1\text{ MHz})(1\text{ pF})} = 159\text{ k}\Omega$$

The ratio of reactance to resistance is

$$\frac{X_C}{R} = \frac{159\text{ k}\Omega}{10\text{ k}\Omega} = 15.9$$

This is greater than 10; therefore, we can neglect the stray capacitance of a 10-kΩ resistor operating at 1 MHz.

Lead Inductance

The lead inductance of a typical resistor is approximately 0.02 μH/in. You can neglect the lead inductance when

$$\frac{R}{X_L} > 10$$

For example, suppose the leads of a 1-kΩ resistor are cut to ½ in on each end. Then the total lead length is 1 in, equivalent to approximately 0.02 μH. At 300 MHz, the reactance is

$$X_L = 2\pi(300 \text{ MHz})(0.02 \text{ μH}) = 37.7 \text{ Ω}$$

and the ratio of resistance to reactance is

$$\frac{R}{X_L} = \frac{1000}{37.7} = 26.5$$

Therefore, even at 300 MHz we can neglect the lead inductance of a 1-kΩ resistor.

A Useful Graph

By setting $X_C/R = 10$ and $R/X_L = 10$, we can plot frequency versus resistance, as shown in Fig. 16-34. This graph gives the dividing lines between the resistive, inductive, and capacitive approximations, for a stray capacitance of 1 pF and a lead inductance of 0.02 μH.

Here is how to use the graph. For any point below the two lines, you can idealize the resistor; that is, you can neglect its capacitance and inductance. But if a point falls above either line, you may need to include the lead inductance or stray capacitance. For instance, a 10-kΩ resistor operating at 1 MHz can be idealized because it falls in the ideal region of Fig. 16-34. But if this 10-kΩ resistor is operating at 5 MHz, you must

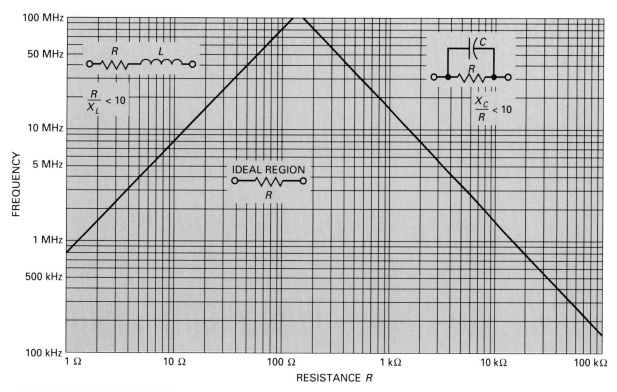

FIGURE 16-34   Approximation guide for resistors.

include the stray capacitance in precise calculations. Similarly, a 10-Ω resistor acts ideally up to 16 MHz, but beyond this frequency, the lead inductance becomes important.

Don't lean too heavily on Fig. 16-34; it is only a guide to help you determine when to include the stray capacitance or lead inductance in your calculations. If you are working at high frequencies where precise analysis is required, you may have to measure the stray capacitance and the lead inductance with a high-frequency *RLC* bridge or *Q* meter.

16-20 SEVERAL CRITICAL FREQUENCIES

Earlier discussions ignored the critical frequencies that were farther from the midband than the two dominant critical frequencies. If you recall, the one-stage amplifier that we analyzed had three critical frequencies below the midband and two critical frequencies above the midband. What happens if we include the other critical frequencies in our analysis?

Figure 16-35 shows the Bode plot for an amplifier with critical frequencies of 5.32 Hz, 212 Hz, 585 Hz, 402 kHz, and 4.29 MHz. These are all the critical frequencies of the amplifier we analyzed earlier (Fig. 16-17). Below the midband, the voltage gain breaks first at 585 Hz. It decreases at a 20 dB per decade until it reaches 212 Hz, where it breaks again and decreases 40 dB per decade. It continues decreasing until it reaches 5.32 Hz, where it breaks for a third time and then decreases 60 dB per decade.

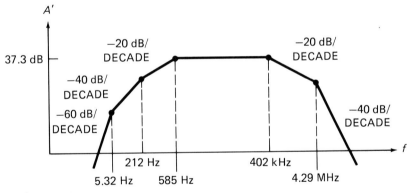

FIGURE 16-35 Bode plot of amplifier with several critical frequencies.

On the high side, the voltage gain breaks at 402 kHz and decreases 20 dB per decade until it hits 4.29 MHz. Then it breaks again and decreases at a rate of 40 dB per decade. Now you can see how useful the Bode plot is. It is an excellent way to summarize all the critical frequencies of an amplifier and how they affect the amplifier gain.

The following study aids will help to reinforce the ideas discussed in this chapter. For best results, use these study aids within 6 hours of reading the earlier material. Then review these study aids a week later and a month later to ensure that the concepts remain in your long-term memory.

SUMMARY

Sec. 16-1 Frequency Response of an Amplifier

An amplifier produces maximum voltage gain in the midband of frequencies. Below the midband, the voltage gain decreases because of coupling and emitter bypass capacitors. Above the midband, the voltage gain decreases because of transistor and stray wiring capacitances.

Sec. 16-2 Input Coupling Capacitor

At low frequencies, the reactance of the input coupling capacitor approaches infinity, and the signal is poorly coupled. This means less input voltage appears across the input of the amplifier. The total resistance of the input circuit and the coupling capacitor determines the critical frequency of the input coupling circuit.

Sec. 16-3 Output Coupling Capacitor

At low frequencies, the reactance of the output coupling capacitor approaches infinity, and the signal is poorly coupled. This means less output voltage appears across the load resistor. The total resistance of the output circuit and the coupling capacitor determines the critical frequency of the output coupling circuit.

Sec. 16-4 Emitter Bypass Capacitor

At low frequencies, the reactance of the emitter bypass capacitor approaches infinity, and the amplifier changes from a CE amplifier to a swamped amplifier. This results in less voltage gain below the critical frequency. The total resistance facing the emitter bypass capacitor and the bypass capacitance determines the critical frequency of the emitter bypass circuit.

Sec. 16-5 Collector Bypass Circuit

Every piece of connecting wire in a circuit has some capacitance to ground. This unwanted capacitance is called the stray wiring capacitance. The transistor also has a collector capacitance built into the collector diode. The total capacitance in the collector circuit is the sum of C_c' and C_{stray}. This capacitance forms an unwanted bypass circuit with the ac collector resistance.

Sec. 16-6 Miller's Theorem

When a capacitor is connected between the input and the output of an amplifier, it becomes a feedback capacitor. Miller's theorem tells us this feedback capacitor affects both the input and the output of an amplifier because the feedback capacitor is equivalent to two new capacitances. The equivalent input capacitance is approximately A times larger than the feedback capacitance. The equivalent output capacitance is approximately equal to the feedback capacitance.

Sec. 16-7 High-Frequency Bipolar Analysis

Because of the transistor and stray wiring capacitances, a bipolar stage has two unwanted bypass circuits, one in the base and the other in the collector. The critical frequencies of each are found with the basic equation where R is the Thevenin resistance facing a total capacitance of C. Almost always, one of two critical frequencies emerges as the dominant critical frequency. At this dominant critical frequency, the output voltage has decreased to 0.707 of its maximum value.

Sec. 16-8 Total Frequency Response

An amplifier usually has two dominant critical frequencies, one below the midband and the other above it. At each of these frequencies, the output voltage equals 0.707 of its maximum value. The midband is the range of frequencies between $10f_1$ and $0.1f_2$. Between the midband and the dominant frequencies, the voltage gain gradually decreases from its maximum value to $0.707V_{max}$.

Sec. 16-9 Decibels

The ordinary power gain is defined as the output power divided by the input power. The decibel power gain is defined as 10 times the logarithm of the ordinary power gain. Each time the ordinary power gain increases (decreases) by a factor of 10, the decibel power gain increases (decreases) by 10 dB. Each time the ordinary power gain increases (decreases) by a factor of 2, the decibel power gain increases (decreases) by 3 dB.

Sec. 16-10 Decibel Voltage Gain

The ordinary voltage gain is defined as the output voltage divided by the input voltage. The decibel voltage gain is defined as 20 times the logarithm of the ordinary power gain. Each time the ordinary voltage gain increases (decreases) by a factor of 10, the decibel voltage gain increases (decreases) by 20 dB. Each time the ordinary voltage gain increases (decreases) by a factor of 2, the decibel voltage gain increases (decreases) by 6 dB. The total decibel voltage gain of cascaded stages equals the sum of the individual decibel voltage gains.

Sec. 16-11 Voltage Gain outside the Midband

Bode plots show the decibel voltage gain versus frequency, plotted on semilogarithmic paper. The vertical axis uses a linear scale, and the horizontal axis uses a logarithmic scale. Because of the properties of logarithms, decibels, and semilogarithmic paper, the voltage gain outside the midband can be drawn as straight lines. Assuming two dominant critical frequencies, the voltage gain decreases 20 dB per decade beyond the break frequencies.

VOCABULARY

In your own words, explain what each of the following terms means. Keep your answers short and to the point. If necessary, verify your answer by rereading the appropriate discussion or by looking at the end-of-book Glossary.

coupling circuit	Miller's theorem
critical frequency	octave
decade	output impedance
decibel voltage gain	stray wiring capacitance
frequency response	Thevenin's theorem
midband	unwanted bypass circuit

IMPORTANT EQUATIONS

Here are some important equations. Say each of the following equations in symbols, then say each in words. Try to explain what the equation means and how it is used. Then read the description that follows.

Eq. 16-3 Critical Frequency

$$f_c = \frac{1}{2\pi RC}$$

This is as basic as Ohm's law because it applies to all kinds of low- and high-frequency circuits. After you find the values of R and C for a particular circuit, you multiply by 2π and take the reciprocal. The result is the critical frequency, which is the frequency where R equals X_C.

Eq. 16-9 Output Impedance of CE Amplifier

$$r_{th} = R_C$$

This is the Thevenin or output impedance of a CE amplifier.

Eq. 16-11 Output Impedance of Emitter Follower

$$r_{th} = R_E \| \left(r_e' + \frac{R_G \| R_1 \| R_2}{\beta} \right)$$

When you are looking from the load back to an emitter follower, you see the dc emitter resistor R_E in parallel with the impedance looking back into the emitter. This impedance consists of r_e' plus the base impedance divided by the ac current gain.

Eq. 16-14 Output Impedance of Source Follower

$$r_{th} = R_S \| \frac{1}{g_m}$$

When looking from the load back to a source follower, you see the dc source resistor R_S in parallel with the impedance looking back into the source. This impedance equals the reciprocal of g_m.

Eq. 16-20 Input Miller Capacitance

$$C_{in} = C(A + 1)$$

This is very important. It says the input Miller capacitance equals the feedback capacitance times the voltage gain plus 1. This effect is one of the major limitations on the high-frequency response of an amplifier. This effect is also the basis of many inventions because it allows us to create much higher capacitances than are possible by any other means.

Eq. 16-22 Total Capacitance of Base Circuit

$$C = C_e' + C_c'\left(\frac{r_c}{r_e'} + 1\right)$$

Because of the Miller effect, the unwanted base bypass circuit has a much larger capacitance than without the Miller effect. Use this total capacitance in calculating the critical frequency of the base bypass circuit.

Eq. 16-23 Total Capacitance of Collector Circuit

$$C = C_c' + C_{\text{stray}}$$

Use this total capacitance to calculate the critical frequency of the unwanted collector bypass circuit.

Eq. 16-36 Decibel Voltage Gain

$$A' = 20 \log A$$

This is the defining formula for the decibel voltage gain. You cannot derive it from other equations because it is a starting point. It says you take the logarithm of ordinary voltage gain and multiply by 20 to get the decibel voltage gain.

Eq. 16-38 Cascaded Decibel Voltage Gain

$$A' = A_1' + A_2'$$

The total decibel voltage gain of a two-stage amplifier equals the sum of the individual decibel voltage gains.

IMPORTANT PROCESSES

Each of the following is important for analyzing frequency effects. Try to remember each step's basic ideas in order to solve problems more easily.

Critical Frequency of Input Coupling Circuit

1. Work out the input resistance of the amplifier.
2. Add the generator resistance to get R.
3. Calculate the critical frequency with $1/(2\pi RC)$.

Critical Frequency of Output Coupling Circuit

1. Find the output impedance of the amplifier.
2. Add the load resistance to get R.
3. Calculate the critical frequency with $1/(2\pi RC)$.

Critical Frequency of Emitter Bypass Circuit

1. Work out the Thevenin resistance R facing the emitter bypass capacitor.
2. Calculate the critical frequency with $1/(2\pi RC)$.

Critical Frequency of Collector Bypass Circuit

1. Work out the Thevenin resistance facing the shunt capacitances.
2. Add the collector capacitance and the stray wiring capacitance.
3. Calculate the critical frequency with $1/(2\pi RC)$.

Critical Frequency of Base Bypass Circuit

1. Work out the Thevenin resistance facing the shunt capacitances.
2. Add the emitter capacitance and the Miller input capacitance.
3. Calculate the critical frequency with $1/(2\pi RC)$.

Decibel Voltage Gain

1. Take the logarithm of the ordinary voltage gain.
2. Multiply by 20 for decibel voltage gain.

STUDENT ASSIGNMENTS

QUESTIONS

Some questions have more than one right answer. Select the best answer, the one that is always true, or that most accurately describes the situation, etc.

1. Frequency response is a graph of voltage gain versus
   a. Frequency
   b. Midband
   c. Input voltage
   d. Voltage gain

2. The audio range of frequencies is from
   a. 20 Hz to 2 kHz
   b. 2 Hz to 20 kHz
   c. 20 Hz to 20 kHz
   d. 200 Hz to 200 kHz

3. At low frequencies, the input coupling capacitor produces a decrease in
   a. Input resistance
   b. Input voltage
   c. Midband voltage gain
   d. Generator voltage

4. The higher the input coupling capacitance, the
   a. Higher the output voltage
   b. Lower the critical frequency
   c. Higher the input resistance
   d. Lower the input voltage

5. If the generator resistance doubles, the critical frequency of the input coupling circuit
   a. Drops in half
   b. Doubles
   c. Stays the same
   d. Decreases

6. At low frequencies, the output coupling capacitor produces a decrease in the
   a. Output resistance
   b. Output voltage
   c. Midband voltage gain
   d. Generator voltage

7. The higher the output coupling capacitance, the
   a. Higher the output voltage
   b. Lower the critical frequency
   c. Higher the output resistance
   d. Lower the output voltage

8. If the collector resistance doubles, the critical frequency of the output coupling circuit
   a. Drops in half
   b. Doubles
   c. Stays the same
   d. Decreases

9. At low frequencies, the emitter bypass capacitor
   a. Reduces the input voltage
   b. Increases the voltage gain
   c. Increases the critical frequency
   d. Is no longer an ac short

10. In the midband of the amplifier, the emitter bypass capacitor
    a. Reduces the input voltage
    b. Decreases the voltage gain
    c. Increases the critical frequency
    d. Is an ac short

11. The larger the stray wiring capacitance, the
    a. Lower the critical frequency of the collector circuit
    b. Higher the critical frequency of the base circuit
    c. Smaller the input resistance
    d. Larger the output impedance

12. When an inverting amplifier has a feedback capacitance of C, the input Miller capacitance is approximately equal to
    a. C c. $1/(2\pi RC)$
    b. AC d. $C/(A + 1)$

13. At high frequencies, a bipolar stage has two unwanted
    a. Lead networks
    b. Bypass circuits
    c. Coupling circuits
    d. Capacitances

14. The emitter capacitance C_e' is
    a. Given on data sheets
    b. Inversely proportional to f_T
    c. Measured in ohms
    d. Solely responsible for the critical frequency

15. The base-spreading resistance r_b' is
    a. Fixed in value
    b. Inversely proportional to f_T
    c. Typically large
    d. β-dependent

16. If the power gain doubles, the decibel power gain increases by
    a. A factor of 2   c. 6 dB
    b. 3 dB   d. 10 dB

17. If the voltage gain doubles, the decibel voltage gain increases by
    a. A factor of 2   c. 6 dB
    b. 3 dB   d. 10 dB

18. If the voltage gain is 10, the decibel voltage gain is
    a. 6 dB   c. 40 dB
    b. 20 dB   d. 60 dB

19. If the voltage gain is 100, the decibel voltage gain is
    a. 6 dB   c. 40 dB
    b. 20 dB   d. 60 dB

20. If the voltage gain is 2000, the decibel voltage gain is
    a. 40 dB   c. 66 dB
    b. 46 dB   d. 86 dB

21. Two stages have decibel voltage gains of 20 and 40 dB. The total ordinary voltage gain is
    a. 1   c. 100
    b. 10   d. 1000

22. Two stages have voltage gains of 100 and 200. The total decibel voltage gain is
    a. 46 dB   c. 86 dB
    b. 66 dB   d. 106 dB

23. One frequency is 8 times another frequency. How many octaves apart are the two frequencies?
    a. 1   c. 3
    b. 2   d. 4

24. Frequency $f = 1$ MHz, and frequency $f_2 = 10$ Hz. The ratio f/f_2 represents how many decades?
    a. 2   c. 4
    b. 3   d. 5

25. Semilogarithmic paper means
    a. One axis is linear, and the other is logarithmic
    b. One axis is linear, and the other is semilogarithmic
    c. Both axes are semilogarithmic
    d. Neither axis is linear

26. The corner frequency is the same as
    a. Cutoff frequency
    b. Critical frequency
    c. Break frequency
    d. All the above

27. If you want to improve the high-frequency response of an amplifier, which of these would you try?
    a. Decrease the coupling capacitances.
    b. Increase the emitter bypass capacitance.
    c. Shorten leads as much as possible.
    d. Increase the generator resistance.

28. If you decrease the collector resistance, the critical frequency of the collector bypass circuit
    a. Decreases
    b. Increases
    c. Stays the same
    d. Cannot be determined

29. The voltage gain of an amplifier decreases 20 dB per decade above 30 kHz. If the midband voltage gain is 86 dB, what is the decibel voltage gain at 3 MHz?
    a. 26 dB   c. 66 dB
    b. 46 dB   d. 86 dB

30. The voltage gain of an amplifier decreases 20 dB per decade above 20 kHz. If the midband voltage gain is 86 dB, what is the ordinary voltage gain at 20 MHz?
    a. 20   c. 2000
    b. 200   d. 20,000

BASIC PROBLEMS

Sec. 16-1 Frequency Response of an Amplifier

16-1. An amplifier has these two critical frequencies: $f_1 = 250$ Hz and $f_2 = 5$ MHz. What is the midband of the amplifier?

16-2. The two critical frequencies of an amplifier are $f_1 = 127$ Hz and $f_2 = 2.45$ MHz. What is the midband of the amplifier? If the output voltage is 224 mV in the midband, what is the output voltage at each critical frequency?

Sec. 16-2 Input Coupling Capacitor

16-3. If $\beta = 100$ in Fig. 16-36, what is the critical frequency of the input coupling circuit?

16-4. What is the critical frequency of the input coupling circuit in Fig. 16-36 if $\beta = 300$? If $\beta = 50$?

16-5. If $\beta = 100$ in Fig. 16-37, what is the critical frequency of the input coupling circuit?

16-6. What is the critical frequency of the input coupling circuit in Fig. 16-37 if $\beta = 300$? If $\beta = 50$?

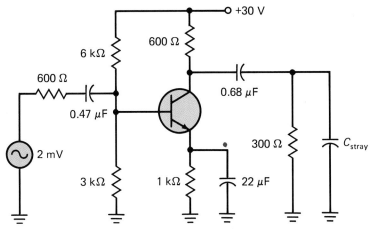

FIGURE 16-36

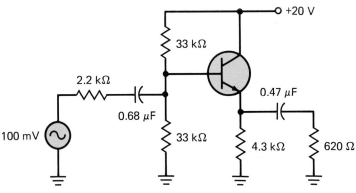

FIGURE 16-37

Sec. 16-3 Output Coupling Capacitor

16-7. What is the critical frequency of the output coupling circuit in Fig. 16-36?

16-8. Calculate the critical frequency of the output coupling circuit for $\beta = 100$ in Fig. 16-37.

16-9. Repeat Prob. 16-8 for $\beta = 50$ and $\beta = 300$.

Sec. 16-4 Emitter Bypass Capacitor

16-10. If $\beta = 100$ in Fig. 16-36, what is the critical frequency of the emitter bypass circuit?

16-11. Repeat Prob. 16-10 for $\beta = 50$ and $\beta = 300$.

16-12. If $\beta = 150$ in Fig. 16-36, what are the critical frequencies? What is the lowest frequency in the midband of the amplifier?

16-13. In Fig. 16-37, all capacitors are increased by a factor of 10. What is the lowest frequency in the midband of the amplifier for $\beta = 200$?

Sec. 16-5 Collector Bypass Circuit

16-14. The transistor of Fig. 16-36 has $C_c' = 6$ pF. What is the critical frequency of the collector bypass circuit if $C_{stray} = 15$ pF?

16-15. If $C_c' = 3$ pF and $C_{stray} = 7$ pF in Fig. 16-36, what is the critical frequency of the collector bypass circuit?

16-16. In Fig. 16-36, $C_c' = 4$ pF and $C_{stray} = 10$ pF. What is the critical frequency of the collector bypass circuit? What happens to the critical frequency if all resistances are increased by a factor of 10?

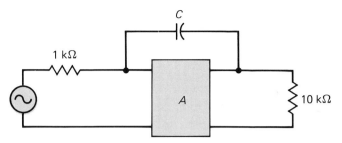

FIGURE 16-38

If all resistances are decreased by a factor of 10?

Sec. 16-6 Miller's Theorem

16-17. In Fig. 16-38, $A = 200$ and $C = 5$ pF. What are the input and output Miller capacitances of the amplifier?

16-18. If $A = 10,000$ and $C = 100$ pF in Fig. 16-38, what are the input and output Miller capacitances of the amplifier?

16-19. The input Miller capacitance of Fig. 16-38 creates a bypass circuit on the input side. If $A = 300$ and $C = 10$ pF, what is the critical frequency of this bypass circuit?

Sec. 16-7 High-Frequency Bipolar Analysis

16-20. The data sheet of a transistor lists $f_T = 500$ MHz, $r_b'C_c' = 25$ ps for $I_E = 1$ mA, and $C_c' = 1$ pF. What is the value of C_e' for $I_E = 1$ mA? What is r_b' for $I_E = 1$ mA?

16-21. A transistor has $f_T = 250$ MHz. What is the value of C_e' for $I_E = 10$ mA?

16-22. In Fig. 16-36, $f_T = 500$ MHz, $r_b' = 100\ \Omega$, $\beta = 200$, $C_c' = 5$ pF, and $C_{\text{stray}} = 15$ pF. What are the two critical frequencies above the midband?

16-23. In Fig. 16-36, $f_T = 750$ MHz, $r_b'C_c' = 240$ ps, $\beta = 100$, $C_c' = 4$ pF, and $C_{\text{stray}} = 10$ pF. What is the dominant critical frequency above the midband?

Sec. 16-8 Total Frequency Response

16-24. An audio amplifier has $f_1 = 12$ Hz and $f_2 = 15$ kHz. Assuming one dominant critical frequency above and below the midband, what is the voltage gain when $f = 20$ Hz and when $f = 20$ kHz?

16-25. An amplifier has an upper critical frequency of $f_2 = 100$ kHz. Assume no other bypass circuits. If the amplifier has a voltage gain of 100 in the midband, what is the voltage gain at 200 kHz, 400 kHz, 1 MHz, and 10 MHz?

Sec. 16-9 Decibels

16-26. What is the decibel power gain of an amplifier if it has an ordinary power gain of 438?

16-27. An amplifier has an input power of 2 mW and an output power of 345 mW. What is the its decibel power gain?

16-28. An amplifier has a decibel power gain of 32 dB. What is its ordinary power gain?

Sec. 16-10 Decibel Voltage Gain

16-29. What is the decibel voltage gain of an amplifier if it has an ordinary voltage gain of 100,000?

16-30. A two-stage amplifier has these stage gains: $A_1 = 25.8$ and $A_2 = 117$. What is the decibel voltage gain of each stage? The total decibel voltage gain?

16-31. A three-stage amplifier has these stage gains: 20, 32, and 46 dB. What are the total decibel voltage gain and the total ordinary voltage gain?

Sec. 16-11 Voltage Gain Outside the Midband

16-32. An amplifier has two dominant critical frequencies: $f_1 = 20$ Hz and $f_2 = 20$ kHz. Describe the ideal decibel graph of the frequency response.

16-33. The two dominant critical frequencies of an amplifier are $f_1 = 50$ Hz and $f_2 = 100$ kHz. Sketch the ideal decibel graph of the frequency response.

16-34. An amplifier has dominant critical frequencies of $f_1 = 125$ Hz and $f_2 = 450$ kHz. Graph the ideal decibel graph of the frequency response, using semilogarithmic paper.

TROUBLESHOOTING PROBLEMS

16-35. Somebody uses a coupling capacitor of 0.1 μF instead of 10 μF when building an amplifier. What effect will this have on the frequency response of the amplifier?

16-36. An oscilloscope has an input capacitance of 10 pF. What effect does this have on the frequency response of an amplifier if you use the oscilloscope to look at the output signal?

16-37. Somebody builds an amplifier with long leads. What effect does this have on the frequency response of the amplifier?

ADVANCED PROBLEMS

16-38. What is the critical frequency of the output coupling circuit in Fig. 16-39?

16-39. The FET of Fig. 16-39 has these capacitances: $C_{gs} = 6$ pF, $C_{gd} = 4$ pF, and $C_{ds} = 1$ pF. Calculate the critical frequencies of the gate and drain bypass circuits.

16-40. What are the critical frequencies of the coupling circuits in Fig. 16-40?

16-41. How much power does 54 dBm represent?

16-42. In Fig. 16-41a, what is the decibel voltage gain in the midband? What is the bandwidth? What is the decibel voltage gain when $f = 10$ kHz?

16-43. In Fig. 16-41b, the voltage gain breaks a second time and rolls off a -40 dB per decade above 10 kHz. What is the decibel voltage gain at 100 kHz? At 1 MHz?

16-44. The amplifier of Fig. 16-42a has a midband voltage gain of 100. If V_{in} equals 20 mV, what is the output voltage at the 90 percent point? What is the upper cutoff frequency of the amplifier?

16-45. The negative input step voltage produces a positive-going output in the ac equivalent circuit of Fig. 16-42b. What risetime does the output waveform have?

16-46. A dc amplifier has a decibel voltage gain of 60 dB and a cutoff frequency of 10 kHz. If a square wave drives the amplifier, what is the risetime of the output?

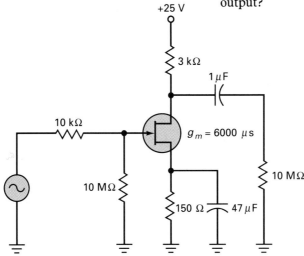

FIGURE 16-39

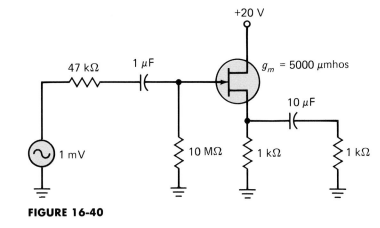

FIGURE 16-40

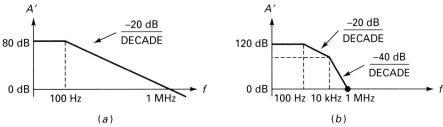

(a)   *(b)*

FIGURE 16-41

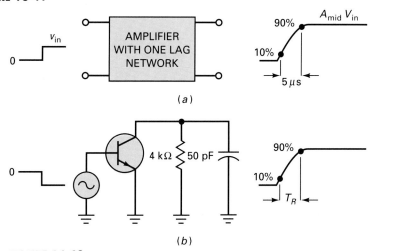

(a)

(b)

FIGURE 16-42

16-47. You have data sheets for two different dc amplifiers. The first shows a cutoff frequency of 1 MHz. The second has a risetime of 1 μs. Which amplifier has the greater bandwidth?

16-48. For an amplifier with three critical frequencies below the midband and two above it, the voltage gain is given by

where f_1, f_2, and f_3 are the low critical frequencies and f_4 and f_5 are the high critical frequencies. An amplifier has critical frequencies of 5.32 Hz, 212 Hz, 585 Hz, 402 kHz, and 4.29 MHz. What is the voltage gain 10 Hz, 300 Hz, 600 Hz, 300 kHz, 1 MHz, and 5 MHz?

$$A = \frac{A_{\text{mid}}}{\sqrt{1 + (f_1/f)^2}\,\sqrt{1 + (f_2/f)^2}\,\sqrt{1 + (f_3/f)^2}\,\sqrt{1 + (f/f_4)^2}\,\sqrt{1 + (f/f_5)^2}}$$

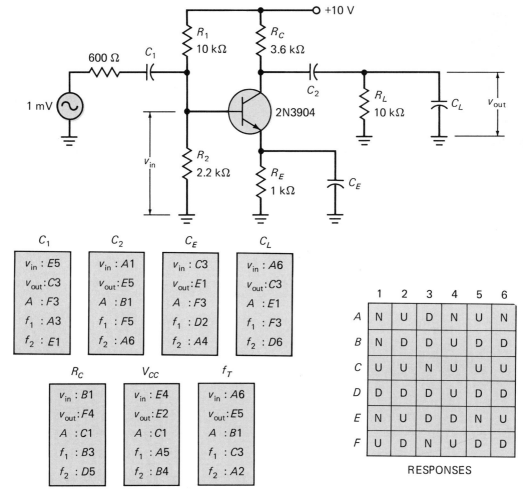

FIGURE 16-43 Software Engine™. (*Patent pending: Courtesy of Malvino Inc.*)

SOFTWARE ENGINE PROBLEMS

Use Fig. 16-43 for the remaining problems. Assume increases of approximately 10 percent in the independent variable. A response should be an N (no change) if the change in a dependent variable is so small that you would have difficulty measuring it. For this exercise, v_{in}, v_{out}, and A are measured in the midband of the amplifier. Also, a change in C_1, C_2, or C_E will affect the lower critical frequency of the amplifier. Similarly, a change in C_L will affect the upper critical frequency f_2 of the amplifier.

16-49. Try to predict the response of each dependent variable in the box labeled C_1. Check your answers. Then answer the following question as simply and directly as possible: What effect does an increase in C_1 have on the dependent variables of the circuit?

16-50. Predict the response of each dependent variable in the box labeled C_2. Check your answers. Then summarize your findings in one or two sentences.

For each of the following problems predict the response of each dependent variable in the labeled boxes. Check your answers. List the dependent variables that decrease. Explain why they decrease.

16-51. Box labeled C_E

16-52. Box labeled C_L

16-53. Box labeled R_C

For each of the following problems predict the response of each dependent variable in the labeled boxes. List the dependent variables that increase. Explain why they show an increase.

16-54. Box labeled V_{CC} **16-55.** Box labeled f_T

17

OP-AMP THEORY

Up to now, we have been discussing *discrete* circuits. The word *discrete* means separate or distinct. It refers to using separate resistors and transistors in building a circuit. A discrete circuit is one where all the components have been soldered together or otherwise mechanically connected.

The invention of the *integrated circuit* (IC) during the 1960s was a major breakthrough because it did away with the need to mechanically connect discrete components. To begin with, an IC is a device that includes its own resistors and transistors. These internal components are not discrete; they are *integrated*. This means they are produced and connected during the manufacturing process itself. The final product, such as a multistage amplifier or a crowbar circuit, can carry out a complete function. Because the integrated components are microscopically small, a manufacturer can place hundreds of these components in the space occupied by a single discrete transistor.

One of the first ICs to be manufactured was the *operational amplifier* (op amp). The typical op amp is a high-gain dc amplifier usable from 0 to over 1 MHz. An IC op amp is like a magical black box with external pins or connecting points. By connecting these pins to supply voltages, signal generators, and load resistances, you can quickly and easily build a superb amplifier. The trick, however, is knowing which pins to connect to what. It also helps to know a little bit about what is inside the black box, because then you will be in a better position to troubleshoot, analyze, or design circuits with ICs.

17-1 INTEGRATED CIRCUITS

At one time, op amps were built as discrete circuits. The term *operational amplifier* refers to an amplifier that carries out a mathematical operation. Historically, the first op amps were used in analog computers, where they carried out mathematical operations like integration and differentiation.

Now, most op amps are produced as integrated circuits. Before we discuss op-amp circuits and other topics, let's take a brief look at how some bipolar integrated circuits are made. The process described here is only one of the many ways in which ICs are made. All you need is the general idea of how an IC is produced. This background knowledge will make it easier to understand more advanced ideas about op amps.

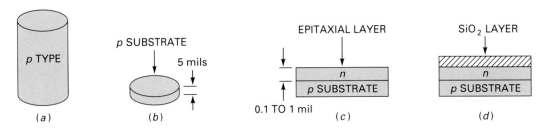

(a) P-crystal; (b) wafer; (c) expitaxial layer; (d) insulating layer.

FIGURE 17-1

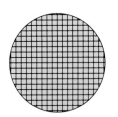

FIGURE 17-2

Cutting the wafer into chips.

Basic Idea

First, the manufacturer produces a p crystal several inches long (Fig. 17-1a). This is sliced into many thin wafers as in Fig. 17-1b. One side of the wafer is lapped and polished to get rid of surface imperfections. This wafer is called the p *substrate;* it will be used as a chassis for the integrated components. Next, the wafers are put into a furnace. A gas mixture of silicon atoms and pentavalent atoms passes over the wafers. This forms a thin layer of n-type semiconductor on the heated surface of the substrate (see Fig. 17-1c). We call this thin layer an *epitaxial layer*. As shown in Fig. 17-1c, the epitaxial layer is about 0.1 to 1 mil thick.

To prevent contamination of the epitaxial layer, pure oxygen is blown over the surface. The oxygen atoms combine with the silicon atoms to form a layer of silicon dioxide (SiO_2) on the surface, as shown in Fig. 17-1d. This glasslike layer of SiO_2 seals off the surface and prevents further chemical reactions. Sealing off the surface like this is known as *passivation*. The wafer is then cut into the rectangular areas shown in Fig. 17-2. Each of these areas will be a separate *chip* after the wafer is cut. But before the wafer is cut, the manufacturer produces hundreds of circuits on the wafer, one on each area of Fig. 17-2. This simultaneous mass production is the reason for the low cost of integrated circuits.

Here is how an integrated transistor is formed. Part of the SiO_2 is etched off, exposing the epitaxial layer (see Fig. 17-3a). The wafer is then put into a furnace and trivalent atoms are diffused into the epitaxial layer. The concentration of trivalent atoms is enough to change the exposed epitaxial layer from n material to p material. Therefore, we get an island of n material under the SiO_2 layer (Fig. 17-3b). Oxygen is again blown over the surface to form the complete SiO_2 layer shown in Fig. 17-3c. A hole is now etched in the center of the SiO_2 layer. This exposes the n epitaxial layer (Fig. 17-3d). The hole in the SiO_2 layer is called a window. We are now looking down at what will be the collector of the transistor. To get the base, we pass trivalent atoms through this window; these impurities diffuse into the epitaxial layer and form an island of p-type material (Fig. 17-3e). Then, the SiO_2 layer is re-formed by passing oxygen over the wafer (Fig. 17-3f). To form the emitter, we etch a window in the SiO_2 layer and expose the p island (Fig. 17-3g). By diffusing pentavalent atoms into the p island, we can form the small n island shown in Fig. 17-3h. We then passivate the structure by blowing oxygen over the wafer (Fig. 17-3i). By etching windows in the SiO_2 layer, we can deposit metal to make electrical contact with the emitter, base, and collector. This gives us the integrated transistor of Fig. 17-4a.

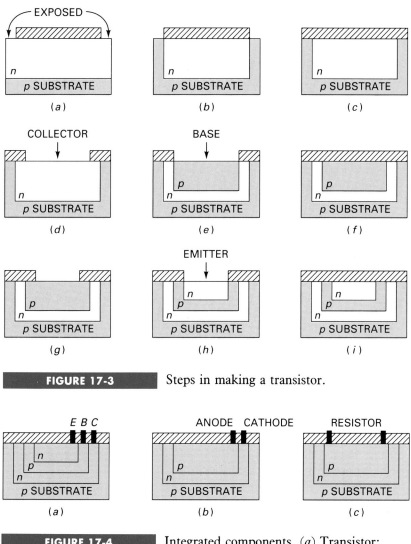

FIGURE 17-3   Steps in making a transistor.

FIGURE 17-4   Integrated components. (*a*) Transistor; (*b*) diode; (*c*) resistor.

To get a diode, we follow the same steps up to the point at which the *p* island has been formed and sealed off (Fig. 17-3*f*). Then, we etch windows to expose the *p* and *n* islands. By depositing metal through these windows, we make electrical contact with the cathode and anode of the integrated diode (Fig. 17-4*b*). By etching two windows above the *p* island of Fig. 17-3*f*, we can make metallic contact with this *p* island; this gives us an integrated resistor (Fig. 17-4*c*).

Transistors, diodes, and resistors are easy to fabricate on a chip. For this reason, almost all integrated circuits use these components. Inductors and large capacitors are not practical to integrate on the surface of a chip.

A Simple Example

To give you an idea of how a circuit is produced, look at the simple three-component circuit of Fig. 17-5*a*. To fabricate this circuit, we would simultaneously produce hundreds of circuits like this on a wafer. Each

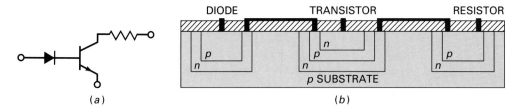

DIODE   TRANSISTOR   RESISTOR

(a)   (b)

FIGURE 17-5 Simple integrated circuit.

chip area would resemble Fig. 17-5b. The diode and resistor would be formed at the point mentioned earlier. At a later step, the emitter of the transistor would be formed. Then we would etch windows and deposit metal to connect the diode, transistor, and resistor as shown in Fig. 17-5b.

Regardless of how complicated a circuit may be, producing it is mainly a process of etching windows, forming p and n islands, and connecting the integrated components. The p substrate isolates the integrated components from each other. In Fig. 17-5b, there are depletion layers between the p substrate and the three n islands that touch it. Because the depletion layers have essentially no current carriers, the integrated components are insulated from one another. This kind of insulation is known as *depletion-layer* isolation.

Types of ICs

The integrated circuits we have described are called *monolithic* ICs. The word *monolithic* is from Greek and means "one stone." The word is appropriate because the components are part of one chip. Monolithic ICs are the most common type of integrated circuit. Since their invention, manufacturers have been producing monolithic ICs to carry out all kinds of functions. Commercially available types can be used as amplifiers, voltage regulators, crowbars, AM receivers, TV circuits, and computer circuits. But the monolithic IC has power limitations. Since most monolithic ICs are about the size of a discrete small-signal transistor, they typically have a maximum power rating of less than 1 W. This limits their use to low-power applications.

When higher power is needed, you can use *thin-film* and *thick-film* ICs. These devices are larger than monolithic ICs but smaller than discrete circuits. With a thin- or thick-film IC, the passive components like resistors and capacitors are integrated, but the transistors and diodes are connected as discrete components to form a complete circuit. Therefore, commercially available thin- and thick-film circuits are combinations of integrated and discrete components.

Another popular IC used in high-power applications is the *hybrid* IC. Hybrid ICs combine two or more monolithic ICs in one package or they combine monolithic ICs with thin- or thick-film circuits. Hybrid ICs are widely used for high-power audio-amplifier applications from 5 W to more than 50 W.

Levels of Integration

Figure 17-5b is an example of *small-scale integration* (SSI); only a few components have been integrated to form a complete circuit. As a guide,

SSI refers to ICs with less than 12 integrated components. Most SSI chips use integrated resistors, diodes, and bipolar transistors.

Medium-scale integration (MSI) refers to ICs that have from 12 to 100 integrated components per chip. Either bipolar transistors or MOS transistors (enhancement-mode MOSFETs) can be used as the integrated transistors of an IC. Again, most of the MSI chips use bipolar components.

Large-scale integration (LSI) refers to ICs with more than a hundred components. Since it takes fewer steps to make an integrated MOS transistor, a manufacturer can produce more of them on a chip than bipolar transistors. For this reason, most LSI chips are MOS types. Today's personal computers use LSI chips with thousands of MOS transistors.

17-2 THE DIFFERENTIAL AMPLIFIER

Transistors, diodes, and resistors are the only practical components that can be produced on a chip. Capacitors have been fabricated on a chip, but they are usually less than 50 pF. Therefore, IC designers don't use coupling and bypass capacitors the way a discrete-circuit designer does. Instead, the stages of a monolithic IC are direct coupled. This means the output of one stage is connected directly to the input of the next stage without using a capacitor. Because of this, dc as well as ac is coupled between the stages. Furthermore, the emitter is no longer bypassed as in the ordinary CE amplifier.

One of the best direct-coupled stages available to the IC designer is the *differential amplifier* (diff amp). This amplifier is widely used as the input stage of an op amp as shown in Fig. 17-6. After the signal is amplified by the diff amp, it goes to the intermediate stages where it receives more voltage gain. The final stage of the op amp is typically a class B push-pull emitter follower. This produces power gain as well as a low output impedance. By studying a diff amp and its characteristics, we will be studying the input characteristics of a typical op amp. Similarly, by reviewing the class B push-pull amplifier, we will learn about the output characteristics of a typical op amp.

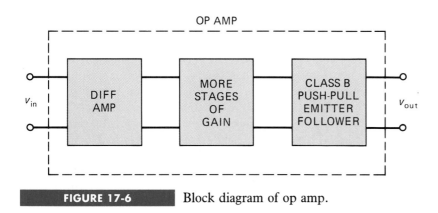

FIGURE 17-6   Block diagram of op amp.

General Form

Figure 17-7*a* shows the original form of a diff amp as it first appeared historically. It has two inputs: V_1 and V_2. Because there are no coupling

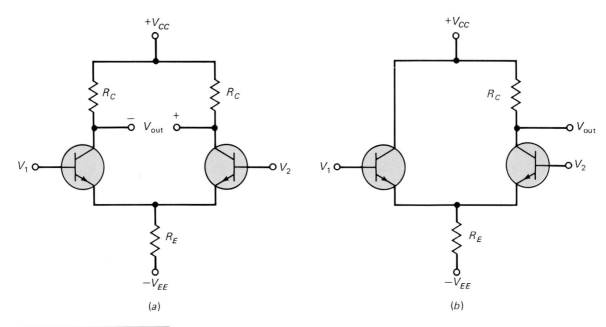

FIGURE 17-7 Differential amps. (a) Double-ended input, double-ended output (b) double-ended input, single-ended output.

or bypass capacitors, the input signals can have frequencies all the way down to zero, equivalent to dc. The output voltage V_{out} is the voltage between the collectors. Ideally, the circuit is symmetrical with identical transistors and collector resistors. As a result, the output voltage is zero when the two input voltages are equal. When V_1 is greater than V_2, an output voltage with the polarity shown appears. When V_1 is less than V_2, the output voltage has the opposite polarity.

Circuit Used in IC Op Amps

Figure 17-7b shows the modified form of a diff amp that is used in IC op amps. This is the circuit we want to study because it will tell us a great deal about the input characteristics of the typical IC op amp. Here is the basic idea of what happens in this diff amp. When V_1 increases, the emitter current of the left transistor increases. This raises the voltage at the top of R_E, equivalent to decreasing the V_{BE} of the right transistor. Less V_{BE} on the right transistor means less collector current in this transistor, which increases the output voltage. You have just seen an increase in V_1 produce an increase in output voltage. This is why the V_1 input voltage is called the *noninverting* input. The output voltage is in phase with V_1.

On the other hand, suppose input V_2 increases. This will increase the collector current in the right transistor, which means the output voltage will decrease. This is why the V_2 input voltage is called the *inverting* input. The output voltage is 180° out of phase with V_1.

Tail Current

Figure 17-8 shows a diff amp with base resistors. If you look at this circuit with a little imagination, you can visualize the emitter resistor as

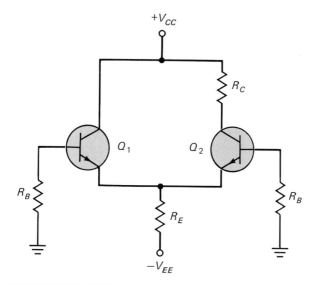

FIGURE 17-8   Diff amp with base resistors.

a tail. The current through the emitter resistor is called the *tail current*. When the transistors are identical, the tail current splits equally between Q_1 and Q_2. Given the tail current, we can divide it by 2 to get the current through each transistor.

How can we find the tail current? The key is to find the voltage across R_E. That's what we need first. After we have this voltage, we can use Ohm's law to calculate the tail current. Does the bias of Fig. 17-8 remind you of anything? It should. It is nothing more than two-supply emitter bias. Since we already know something about this type of bias, let us see how an experienced troubleshooter would find the voltage across R_E:

> This circuit uses two-supply emitter bias. That means each base is approximately at ground potential because $I_B R_B$ is very small. Moving from either base to the emitter, the potential decreases about 0.7 V. This means the top of R_E has a node voltage of -0.7 V. The voltage across the resistor is therefore $V_{EE} - 0.7$ V. Since V_{EE} is typically 15 V, I can ignore the 0.7 V without losing too much accuracy. This means that I can approximate the voltage across R_E as V_{EE}.

The rest is easy. With Ohm's law, the ideal equation for the tail current is

$$I_T = \frac{V_{EE}}{R_E} \qquad (17\text{-}1)$$

This is all you need for preliminary analysis. The answer you get is adequate for most diff amps because the answer is usually within 5 percent of the exact answer. For this reason, most troubleshooters get the tail current of Fig. 17-8 like this:

> If the circuit is typical, I can visualize a ground at the top of R_E. Then it's obvious that all of the emitter supply voltage is across R_E. To get the tail current, all I have to do is divide V_{EE} by R_E.

If you want a second approximation of the tail current, you can include the V_{BE} drop in your calculations as follows:

$$I_T = \frac{V_{EE} - V_{BE}}{R_E} \qquad (17\text{-}2)$$

For silicon transistors, V_{BE} is approximately 0.7 V. Typically, the emitter supply voltage is 15 V, so including 0.7 V in your calculations reduces the ideal current by about 5 percent. In Fig. 17-8, it helps to visualize -0.7 V at the top of the R_E. Then, the voltage across R_E is the algebraic difference of the top and bottom:

$$-0.7 \text{ V} - (-V_{EE}) = -0.7 \text{ V} + V_{EE} = V_{EE} - 0.7 \text{ V}$$

In other words, the numerator of Eq. (17-2) is always smaller than the magnitude of V_{EE}. Rather than get mixed up in minus signs, it's best to work with magnitudes of voltage. For instance, if the emitter supply voltage is -15 V, take magnitude to get 15 V. Then subtract 0.7 from this magnitude as follows to get the voltage across R_E:

$$15 \text{ V} - 0.7 \text{ V} = 14.3 \text{ V}$$

If you want an almost exact answer, you can include the effect of the base resistance in your calculations by using this equation:

$$I_T = \frac{V_{EE} - V_{BE}}{R_E + R_B/2\beta_{dc}} \qquad (17\text{-}3)$$

Here you see an additional correction term, $R_B/(2\beta_{dc})$. You can derive this equation by writing a loop equation around the emitter and base paths. It's fairly easy to do. Try it if you want to see where Eq. (17-3) comes from.

When you have the tail current, you can divide by 2 to get the emitter current in each transistor. If you divide by β_{dc}, you get the base current in each transistor. When the two transistors are identical, the base currents are equal. If the transistors have different β_{dc}, the two base currents will be different. As before, the collector current in each transistor is approximately equal to the emitter current.

EXAMPLE 17-1

In Fig. 17-9, the transistors are identical with $\beta_{dc} = 100$. What is the output voltage?

SOLUTION

Let us do an ideal analysis, then correct the answer to improve the accuracy. Ideally, the top of the emitter resistor is at ground potential, which means there is 15 V across the emitter resistor. The ideal tail current is

$$I_T = \frac{15 \text{ V}}{15 \text{ k}\Omega} = 1 \text{ mA}$$

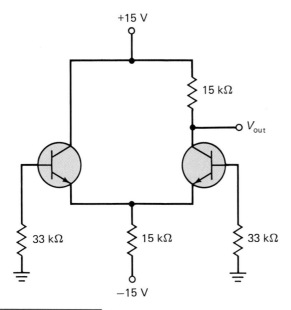

+15 V

15 kΩ

V_{out}

33 kΩ   15 kΩ   33 kΩ

−15 V

FIGURE 17-9   Example.

Each transistor gets half this tail current, which means the emitter current is

$$I_E = 0.5(1 \text{ mA}) = 0.5 \text{ mA}$$

The collector current in the right transistor is approximately 0.5 mA, so the output voltage is

$$V_{out} = 15 \text{ V} - (0.5 \text{ mA})(15 \text{ k}\Omega) = 7.5 \text{ V}$$

To a second approximation, the tail current is

$$I_T = \frac{15 \text{ V} - 0.7 \text{ V}}{15 \text{ k}\Omega} = 0.953 \text{ mA}$$

The emitter current in each transistor is

$$I_E = 0.5(0.953 \text{ mA}) = 0.477 \text{ mA}$$

and the output voltage is

$$V_{out} = 15 \text{ V} - (0.477 \text{ mA})(15 \text{ k}\Omega) = 7.85 \text{ V}$$

For an almost exact answer, use Eq. (17-3) to calculate the tail current:

$$I_T = \frac{15 \text{ V} - 0.7 \text{ V}}{15 \text{ k}\Omega + (33 \text{ k}\Omega)/200} = 0.943 \text{ mA}$$

The emitter current in each transistor is

$$I_E = 0.5(0.943 \text{ mA}) = 0.472 \text{ mA}$$

and the output voltage is

$$V_{\text{out}} = 15\ \text{V} - (0.472\ \text{mA})(15\ \text{k}\Omega) = 7.92\ \text{V}$$

You can see the answers improving with each refinement, but the improvement is not that great. That's typical of a diff amp because the emitter supply voltage is usually 15 V, which is approximately 20 times larger than 0.7 V. Because of this 20:1 ratio, ignoring 0.7 V produces an error of less than 5 percent. Furthermore, the designer usually selects the base resistors to get only a small voltage across them, equivalent to a small $R_B/(2\beta_{\text{dc}})$ correction term. Suggestion: use the ideal analysis unless you are designing a diff amp or have a good reason for wanting more accuracy.

EXAMPLE 17-2

If $\beta_{\text{dc}} = 100$ for both transistors in Fig. 17-9, what is the base current? The base voltage?

SOLUTION

The ideal tail current is 1 mA, found in the preceding example. Half of this is 0.5 mA. The base current in either transistor is

$$I_B = \frac{0.5\ \text{mA}}{100} = 5\ \mu\text{A}$$

The voltage at either base is

$$V_B = -(5\ \mu\text{A})(33\ \text{k}\Omega) = -0.165\ \text{V}$$

EXAMPLE 17-3

In Fig. 17-9, the transistors are identical except for β_{dc}. The left transistor has $\beta_{\text{dc}} = 90$ and the right transistor has $\beta_{\text{dc}} = 110$. What are the base currents? The base voltages?

SOLUTION

The ideal tail current is 1 mA. Half of this is 0.5 mA. The base current in the left transistor is

$$I_{B1} = \frac{0.5\ \text{mA}}{90} = 5.56\ \mu\text{A}$$

The base current in the right transistor is

$$I_{B2} = \frac{0.5\ \text{mA}}{110} = 4.55\ \mu\text{A}$$

The voltage on the left base is

$$V_{B1} = -(5.56\ \mu\text{A})(33\ \text{k}\Omega) = -0.183\ \text{V}$$

and the voltage on the right base is

$$V_{B2} = -(4.55 \text{ μA})(33 \text{ kΩ}) = -0.15 \text{ V}$$

The point of the example is to show you that the two base currents and voltages are different when the transistors are not identical. Besides differing in β_{dc}, the transistors can differ in their V_{BE} values and bulk resistances. Therefore, whenever the transistors are not identical (which is almost always the case), there will be a slight difference in the two base currents and the resulting base voltages.

17-3 TWO INPUT CHARACTERISTICS

Because IC op amps usually have a diff amp as the first stage, we are now in a position to describe two important input characteristics known as the input offset current and the input bias current.

Input Offset Current

Base currents I_{B1} and I_{B2} flow through the base resistors of a diff amp. The *input offset current* is defined as the difference between the base currents. In symbols,

$$I_{\text{in(off)}} = I_{B1} - I_{B2} \qquad (17\text{-}4)$$

This difference in base currents indicates how closely matched the transistors are. If the transistors are identical, the input offset current is zero because both base currents will be equal. But almost always, the two transistors are different and the two base currents are not equal.

As an example, suppose $I_{B1} = 85 \text{ μA}$ and $I_{B2} = 75 \text{ μA}$. Then,

$$I_{\text{in(off)}} = 85 \text{ μA} - 75 \text{ μA} = 10 \text{ μA}$$

The Q_1 transistor has 10 μA more base current than the Q_2 transistor. This can cause a problem if the base resistances are very large. The larger the base resistances, the greater the base voltages. Why this can be a problem will be discussed later. All you need now is this idea: the smaller the input offset current, the better. In a perfect world, the input offset current is zero.

Input Bias Current

The *input bias current* is defined as the average of the two base currents:

$$I_{\text{in(bias)}} = \frac{I_{B1} + I_{B2}}{2} \qquad (17\text{-}5)$$

For instance, if $I_{B1} = 85 \text{ μA}$ and $I_{B2} = 75 \text{ μA}$, the input bias current is

$$I_{\text{in(bias)}} = \frac{85 \text{ μA} + 75 \text{ μA}}{2} = 80 \text{ μA}$$

Because the input bias current is the average of the two base currents, it always has a value half way between the two base currents. This makes it easy to mentally estimate the input bias current. For instance, given base currents of 6.8 μA and 7.4 μA, you go half way between the two values to get 7.1 μA. This is the input bias current.

Base Currents

Data sheets for op amps always include the values of input bias current and input offset current, but they never include the values of base currents. This is why we have to calculate the base currents using these two equations:

$$I_{B1} = I_{\text{in(bias)}} + \frac{I_{\text{in(off)}}}{2} \qquad (17\text{-}6)$$

$$I_{B2} = I_{\text{in(bias)}} - \frac{I_{\text{in(off)}}}{2} \qquad (17\text{-}7)$$

Where do they come from? In Eqs. (17-4) and (17-5), I_{B1} and I_{B2} are the unknowns. The quantities $I_{\text{in(bias)}}$ and $I_{\text{in(offset)}}$ are known quantities given on a data sheet. Equations (17-4) and (17-5) can be solved simultaneously to get Eqs. (17-6) and (17-7). Try it if you want to see how the equations are derived.

A final point, Equations (17-6) and (17-7) assume I_{B1} is greater than I_{B2}. When the opposite is true, the two values are reversed. See Example 17-7 for more details.

EXAMPLE 17-4

Assume the transistors are identical in Fig. 17-10 and calculate the output voltage.

SOLUTION

The ideal tail current is

$$I_T = \frac{15\,\text{V}}{1\,\text{M}\Omega} = 15\,\mu\text{A}$$

Each transistor gets half this tail current, which means the emitter current is

$$I_E = 0.5(15\,\mu\text{A}) = 7.5\,\mu\text{A}$$

The collector current in the right transistor is approximately equal to 7.5 μA, so the output voltage is

$$V_{\text{out}} = 15\,\text{V} - (7.5\,\mu\text{A})(1\,\text{M}\Omega) = 7.5\,\text{V}$$

There's nothing here that you did not see in an earlier example, except for the much smaller currents. Numbers like these are common with

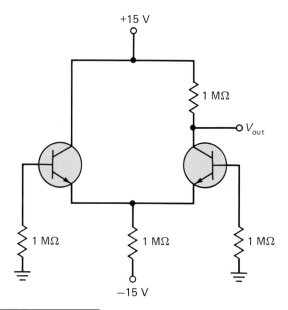

+15 V

1 MΩ

V_{out}

1 MΩ

1 MΩ

1 MΩ

−15 V

FIGURE 17-10   Example.

integrated circuits because the power dissipation of the entire IC is typically 500 mW. Since many integrated transistors are placed in the same area that a discrete transistor occupies, the individual transistors have to use much smaller currents.

EXAMPLE 17-5

In Fig. 17-10, the left transistor has β_{dc} = 90 and the right transistor has β_{dc} = 110. What are the two base currents? The two base voltages?

SOLUTION

The ideal tail current is 15 μA. Half of this is 7.5 μA. The base current in the left transistor is

$$I_{B1} = \frac{7.5 \, \mu A}{90} = 83.3 \, nA$$

The base current in the right transistor is

$$I_{B2} = \frac{7.5 \, \mu A}{110} = 68.2 \, nA$$

The voltage on the left base is

$$V_{B1} = -(83.3 \, nA)(1 \, M\Omega) = -0.0833 \, V$$

and the voltage on the right base is

$$V_{B2} = -(68.2 \, nA)(1 \, M\Omega) = -0.0682 \, V$$

The point of the example is this. Notice how small the base currents are. This is why we can use a resistance of 1 MΩ on each base and still have very small base voltages. A designer tries to keep the base voltage less than 0.1 V if possible. This makes the ideal tail current an accurate approximation for the exact tail current.

EXAMPLE 17-6

In the previous example, what is the input offset current? The input bias current?

SOLUTION

The input offset current is the difference between the two base currents:

$$I_{in(offset)} = 83.3 \text{ nA} - 68.2 \text{ nA} = 15.1 \text{ nA}$$

The input bias current is the average of the two base currents:

$$I_{in(bias)} = \frac{83.3 \text{ nA} + 68.2 \text{ nA}}{2} = 75.8 \text{ nA}$$

EXAMPLE 17-7

The data sheet of an IC op amp gives these values: $I_{in(offset)} = 20 \text{ nA}$ and $I_{in(bias)} = 80 \text{ nA}$. What are the two base currents?

SOLUTION

Data sheets give only absolute values for $I_{in(offset)}$. An $I_{in(offset)}$ of 20 nA means one base current is 20 nA greater than the other. We cannot tell which base current is larger for a particular device. It can go either way in mass production. With this in mind, here is how to find the two base currents. Assume I_{B1} is larger than I_{B2}. Then Eqs. (17-6) and (17-7) give

$$I_{B1} = 80 \text{ nA} + \frac{20 \text{ nA}}{2} = 90 \text{ nA}$$

$$I_{B2} = 80 \text{ nA} - \frac{20 \text{ nA}}{2} = 70 \text{ nA}$$

If I_{B2} is greater than I_{B1}, the values are reversed to get $I_{B1} = 70 \text{ nA}$ and $I_{B2} = 90 \text{ nA}$. If we were mass-producing circuits using this op amp, either base current could be in range of 70 nA to 90 nA with every possible combination occurring sooner or later.

17-4 AC ANALYSIS OF A DIFF AMP

The diff amp has a noninverting input and an inverting input. The viewpoint of two separate inputs is useful in some situations but it is a handicap in others. Figure 17-11 shows another way to visualize the input

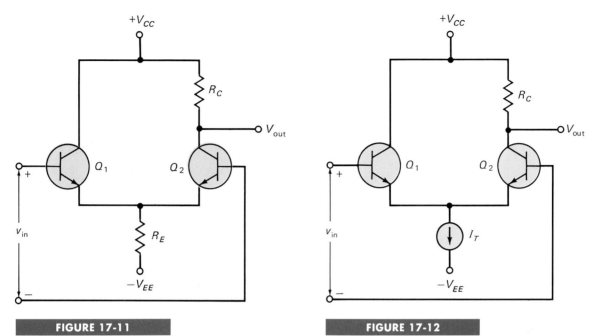

FIGURE 17-11

Analysis of diff amp.

FIGURE 17-12

Diff amp with current source producing tail current.

and output voltages of a diff amp. Forget about V_1 and V_2. Think only about the voltage between the two input terminals. This voltage, labeled v_{in}, is what the diff amp actually responds to. This is the voltage that the diff amp amplifies to produce an output signal. What we want to derive in this section are the equations for voltage gain and input impedance.

Tail Is a Current Source

As discussed in Chap. 8, the whole point of emitter bias is to produce a rock-solid emitter current. In Fig. 17-11, the ideal emitter current is

$$I_E = \frac{V_{EE}}{R_E}$$

Once the values of V_{EE} and R_E are set, the emitter current is constant. Ideally, it remains constant even though temperature changes and transistors are replaced. To simplify ac analysis, therefore, we can replace the tail by a current source as shown in Fig. 17-12. This will simplify the ac analysis of the circuit with almost no loss in accuracy.

In Fig. 17-12 the current source is producing a constant current of I_T. If the transistors are identical, this tail current splits equally between the transistors. The output voltage then equals the collector supply voltage minus the voltage across the collector resistor. If v_{in} is zero, this output voltage will remain constant. But when v_{in} is greater than zero, the output voltage will increase. When v_{in} is less than zero, the output voltage will decrease.

Here is why the output voltage changes when v_{in} is not zero. The tail current is the sum of the two emitter currents:

$$I_T = I_{E1} + I_{E2}$$

Assume v_{in} is greater than zero. Then the emitter current in Q_1 increases. Since the tail current is constant, the emitter current in Q_2 has to decrease because

$$I_{E2} = I_T - I_{E1}$$

Less emitter current through Q_2 means less collector current and more output voltage. On the other hand, when v_{in} is less than zero or negative, the emitter current in Q_1 decreases, which forces the emitter current in Q_2 to increase. This means the collector current is larger, which produces a lower output voltage.

AC Equivalent Circuit

In a diff amp, a signal is any change from a quiescent value. This definition is used because a diff amp can amplify dc as well as ac signals. In fact, a dc signal can be treated like an ac signal of zero frequency. Because of this, we can draw the ac equivalent circuit of a diff amp using the rules established in earlier chapters. The rules were to short all capacitors and reduce all dc sources to zero. Since a diff amp has no capacitors, all that's left are the dc sources. Reducing a voltage source to zero is equivalent to replacing it by a short. Reducing a current source to zero is equivalent to opening it. In Fig. 17-12, this means ground the V_{CC} point, ground the V_{EE} point, and open the current source.

Figure 17-13 shows the ac equivalent circuit of a diff amp. This is how a diff amp looks to an ac signal. Store a photograph of this picture in your mind because it is a valuable summary of how the input side of many op amps appears to ac signals. With this equivalent circuit, we can easily derive the voltage gain and input impedance of the diff amp as follows. Because the two r'_es are in series, the same ac emitter current exists in both transistors. This ac emitter current is given by

$$i_e = \frac{v_{in}}{2r'_e}$$

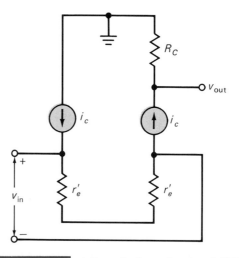

FIGURE 17-13   AC equivalent circuit of diff amp.

This is easy to remember because it is almost identical to a CE amplifier where $i_e = v_{in}/r'_e$. The only difference is the factor of 2 because a diff amp uses two transistors.

The ac collector current is approximately equal to the ac emitter current. Therefore, the ac output voltage is

$$v_{out} = i_c R_C = \frac{v_{in}}{2r'_e} R_C$$

or

$$\frac{v_{out}}{v_{in}} = \frac{R_C}{2r'_e}$$

Since voltage gain is usually symbolized by A, we can write

$$A = \frac{R_C}{2r'_e} \tag{17-8}$$

You will have no trouble remembering this. It is identical to the voltage gain of a CE amplifier, except for the factor of 2.

Input Impedance

Look at Fig. 17-13. Someone who never analyzed a CE amp might think the input impedance is just $2r'_e$. But you know better. You know that the ac current gain of a transistor increases the impedance of the emitter circuit by a factor of β. This is why the input impedance of the diff amp is given by

$$z_{in} = 2\beta r'_e \tag{17-9}$$

This is identical to the $z_{in(base)}$ of a CE amp, except for the factor of 2.

If you are not convinced by the logic used in deriving Eq. (17-9), then here comes the ultimate weapon: mathematical derivation. The ac emitter current is given by

$$i_e = \frac{v_{in}}{2r'_e} \approx \beta i_b$$

We can rewrite the foregoing equation like this:

$$\frac{v_{in}}{i_b} = 2\beta r'_e$$

What do you think v_{in} divided by i_b is? It is the left side of Eq. (17-9). In other words, i_b is the ac input current to the diff amp. Therefore, v_{in}/i_b is the input impedance.

Notation

Notice that the ac voltages and currents are symbolized by lowercase letters. In the ac equivalent circuit of Fig. 17-13, we are using v_{in} and v_{out}. In the complete circuits discussed earlier, we used V_1, V_2, and V_{out}. These voltages are capitalized because they are total voltages. For instance, the output voltage of Fig. 17-10 is ideally 7.5 V. This is the total output voltage without an input signal. When there is an input signal, the total output voltage changes. The ac signal is defined as the change in the total output voltage:

$$v_{\text{out}} = \Delta V_{\text{out}} \tag{17-10}$$

where Δ is our shorthand notation for "the change in." For instance, if an input signal causes V_{out} to change from 7.5 V to 8 V, the ac output voltage is

$$v_{\text{out}} = 8 \text{ V} - 7.5 \text{ V} = 0.5 \text{ V}$$

The ac input voltage, v_{in}, is the difference between the two total base voltages:

$$v_{\text{in}} = V_1 - V_2 \tag{17-11}$$

Because of this, V_1 and V_2 may be greater than zero, but v_{in} may still be zero. For instance, if $V_1 = 0.2$ V and $V_2 = 0.2$ V,

$$v_{\text{in}} = 0.2 \text{ V} - 0.2 \text{ V} = 0$$

It's the difference that makes the difference. The diff amp amplifies the difference between the noninverting and inverting inputs.

EXAMPLE 17-8

What is V_{out} when $v_{\text{in}} = 0$ in Fig. 17-14? When $v_{\text{in}} = 1$ mV? When $v_{\text{in}} = -1$ mV?

SOLUTION

From earlier examples, we already know this amplifier has an ideal tail current of 15 μA. Assuming identical transistors, each has a dc emitter current of 7.5 μA. The ac resistance of each emitter diode is

$$r'_e = \frac{25 \text{ mV}}{7.5 \text{ μA}} = 3.33 \text{ kΩ}$$

The voltage gain is

$$A = \frac{1 \text{ MΩ}}{2(3.33 \text{ kΩ})} = 150$$

When v_{in} is zero, the ac output voltage is zero. Therefore, the total voltage V_{out} is at the quiescent value found earlier:

$$V_{\text{out}} = 7.5 \text{ V}$$

When $v_{\text{in}} = 1$ mV, the ac output voltage is

$$v_{\text{out}} = 150(1 \text{ mV}) = 0.15 \text{ V}$$

Therefore, V_{out} increases by 0.15 V to

$$V_{\text{out}} = 7.5 \text{ V} + 0.15 \text{ V} = 7.65 \text{ V}$$

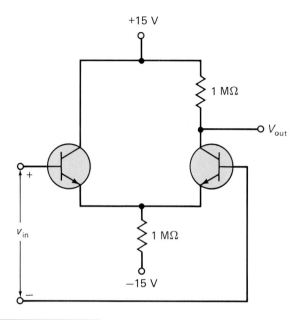

+15 V

1 MΩ

V_{out}

v_{in}

1 MΩ

−15 V

FIGURE 17-14   Example.

When $v_{in} = -1$ mV, the ac output voltage is

$$v_{out} = 150(-1 \text{ mV}) = -0.15 \text{ V}$$

and the total output voltage is

$$V_{out} = 7.5 \text{ V} - 0.15 \text{ V} = 7.35 \text{ V}$$

EXAMPLE 17-9

If $\beta = 300$, what is the input impedance in Fig. 17-14?

SOLUTION

In the previous example, we found that $r_e' = 3.33$ kΩ. The input impedance is

$$z_{in} = 2(300)(3.33 \text{ k}\Omega) = 2 \text{ M}\Omega$$

As you will see later, the 741 is a widely used op amp whose input diff amp has a tail current of 15 μA, an r_e' of 3.33 kΩ, and an input impedance of 2 MΩ.

17-5   OUTPUT OFFSET VOLTAGE

With integrated circuits, it is possible to get almost perfect matches between the two transistors of a diff amp. But almost is not good enough because even the smallest difference between the two transistors is amplified to produce an *output offset voltage*. This output offset voltage is a false output signal. This section discusses the sources of output offset voltage and how to minimize them.

Ideal Output Voltage

Figure 17-15a shows a diff amp with both bases grounded. Grounding both bases is never done under normal conditions because it would prevent input signals from having any effect on the diff amp. We have grounded the bases temporarily for this discussion. With the bases grounded, the tail current will split equally between the two transistors. This assumes the transistors are identical in every respect. With half the tail current flowing through the right transistor, the output voltage is

$$V_{\text{out}} = V_{CC} - \frac{I_T}{2} R_C \tag{17-12}$$

The value of V_{out} is ideal because it is based on two identical transistors. In a typical design, V_{out} equals half of V_{CC}.

As discussed in the preceding section, the input signal may be dc or ac. The diff amp amplifies the input signal to get a change in the output voltage. This change in the output voltage is the ac output voltage. In symbols,

$$v_{\text{out}} = \Delta V_{\text{out}}$$

This says the ac output voltage is any difference or change from the ideal output voltage. The ac output voltage is related to the ac input voltage through this expression:

$$v_{\text{out}} = A v_{\text{in}}$$

This equation makes one thing clear. When the diff amp is perfect, the only way to get an ac output signal is to apply a desired input signal.

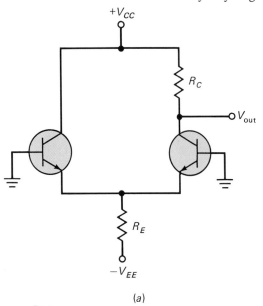

(a)

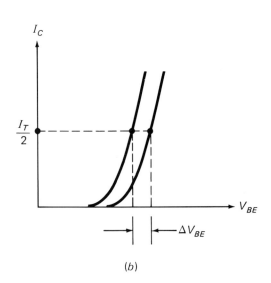

(b)

FIGURE 17-15 V_{BE} values may be different.

Different V_{BE} Values

What happens when the diff amp is not perfect, when the two transistors are not identical? You get an *output offset voltage*, an unwanted deviation from the ideal output voltage. In Fig. 17-15a, both bases are grounded to eliminate the effect of base current. This eliminates the problem of different β_{dc} values, but it does not eliminate the problem of different I_C versus V_{BE} curves. In Fig. 17-15a, both transistors have the same V_{BE} value because the emitter diodes are in parallel. If the I_C versus V_{BE} curves are different, the two collector currents must be different. Different collector currents mean that the current in the right transistor is no longer half of the tail current. Therefore, the output voltage will be different from the ideal output voltage given by Eq. (17-12).

With integrated circuits, the difference between the I_C versus V_{BE} curves will be small. For instance, suppose the two curves have the same current as shown in Fig. 17-15b. Because the curves are different, there is a difference between the two V_{BE} values. This difference acts like a small ac input signal of

$$v_{in} = \Delta V_{BE}$$

where Δ stands for "the difference in." As a result, the diff amp amplifies the unwanted difference in the V_{BE} values to get an output voltage of

$$v_{out} = A(\Delta V_{BE})$$

One way to eliminate the output offset voltage is by applying a small input voltage equal to the difference in V_{BE} values. For instance, assume the V_{BE} values differ by 2 mV. This means we can apply 2 mV to either transistor to eliminate the output offset voltage. Figure 17-16a shows a +2 mV being applied to the left transistor. If this doesn't eliminate the output offset voltage, we have to reverse the input voltage as shown in

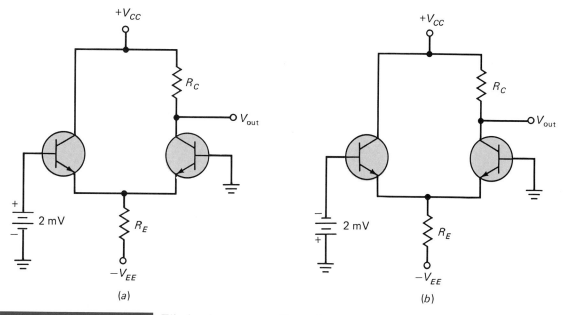

FIGURE 17-16   Eliminating output offset voltage.

Fig. 17-16b. In general, given a difference of 2 mV between the two V_{BE} values of a diff amp, either Fig. 17-16a or b will eliminate the output offset voltage.

Effect of Base Current

Different V_{BE} values are one possible source of output offset voltage. Now, we want to discuss other sources of output offset voltage. To simplify the discussion, let us temporarily assume the two V_{BE} values are identical. This eliminates the differences in V_{BE} as a problem.

What's left? The differences in the base currents. Some diff amps are operated with a base resistor on one side and a ground on the other as shown in Fig. 17-17a. This produces an output offset voltage even when there are no differences in the V_{BE} values. Why? Because the base current through R_B produces a voltage at the noninverting input given by

$$v_{\text{in}} = I_{B1}R_B$$

This voltage has the same effect as a genuine input signal. The diff amp amplifies this false input to get an ac output voltage of

$$v_{\text{out}} = A(I_{B1}R_B)$$

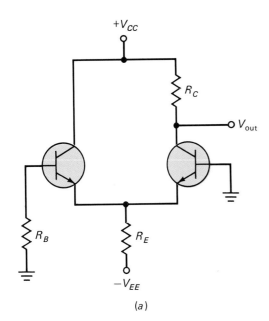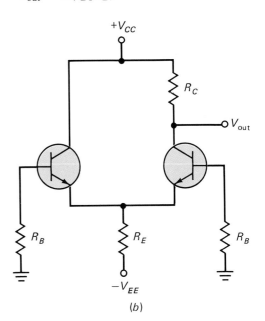

(a)   (b)

FIGURE 17-17   Effect of base current.

One way to reduce the output offset voltage is by using an equal base resistance on the other side of the diff amp (see Fig. 17-17b). In this case, the unwanted input voltage decreases to

$$v_{\text{in}} = I_{B1}R_B - I_{B2}R_B = R_B(I_{B1} - I_{B2})$$

or

$$v_{\text{in}} = I_{\text{in(offset)}}R_B \qquad (17\text{-}13)$$

The most general case uses two different base resistances of R_{B1} and R_{B2}. In this case, the unwanted input voltage is

$$v_{\text{in}} = I_{B1}R_{B1} - I_{B2}R_{B2} \qquad (17\text{-}14)$$

In some designs, R_{B1} or R_{B2} can be adjusted to eliminate this source of unwanted input voltage.

Combined Effects

As you have seen, there are two fundamental sources of output offset voltage. First, there are the differences in the V_{BE} values. This occurs whether the bases are grounded or not. Second, there are the differences in base voltages produced by the base currents flowing through base resistors. These differences add to the V_{BE} differences to get a total unwanted input of

$$v_{\text{in}} = \Delta V_{BE} + I_{B1}R_{B1} - I_{B2}R_{B2} \qquad (17\text{-}15)$$

This assumes positive differences in V_{BE}. If the differences are negative, then

$$v_{\text{in}} = -\Delta V_{BE} + I_{B1}R_{B1} - I_{B2}R_{B2} \qquad (17\text{-}16)$$

You can use these two formulas to calculate the total input signal produced by unwanted differences in the transistors. Then you can multiply by A to get the total output offset voltage.

Nulling the Output Offset Voltage

We can null or eliminate the output offset voltage by applying a small dc input voltage given by

$$v_{\text{in}} = \frac{v_{\text{out}}}{A} \qquad (17\text{-}17)$$

In this equation, v_{out} is the total output offset voltage produced by all causes. If we divide this by the voltage gain of the diff amp, we get the required input voltage needed to eliminate the output offset voltage. Some diff amps and op amps are designed with a potentiometer that allows you to adjust the input voltage that eliminates the output offset voltage.

EXAMPLE 17-10

In Fig. 17-18a, what is the output offset voltage if $I_{\text{in(bias)}} = 80$ nA and $I_{\text{in(offset)}} = 20$ nA?

SOLUTION

Assume that the I_C versus V_{BE} curves are identical. This does not mean that the two β_{dc} values are the same. Differences in V_{BE} values and β_{dc} values are two independent problems. In this example, we are assuming that only the β_{dc} differences exist.

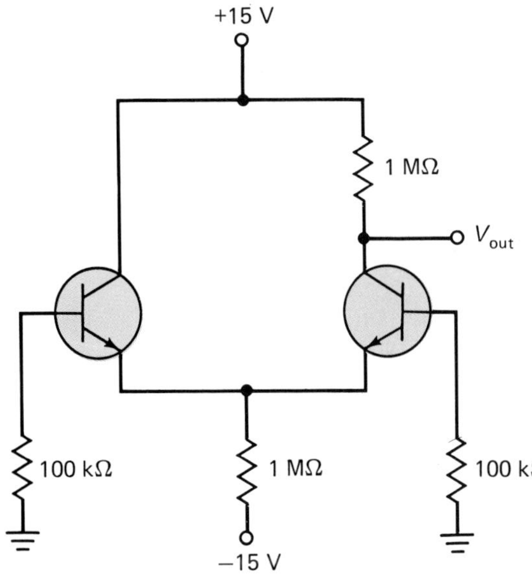

FIGURE 17-18   Example.

When the two base resistors are equal, we can use Eq. (17-13) to calculate the input voltage:

$$v_{\text{in}} = (20 \text{ nA})(100 \text{ k}\Omega) = 2 \text{ mV}$$

Since the voltage gain is 150 (found in Example 17-8), the output offset voltage is

$$v_{\text{out}} = 150(2 \text{ mV}) = 0.3 \text{ V}$$

The ideal output voltage is 7.5 V (found in Example 17-4). This means the total output voltage is 7.5 V $\pm$ 0.3 V, depending on which of the two base currents is larger.

EXAMPLE 17-11

In Fig. 17-18, the two base resistors have a tolerance of ± 10 percent. What is the output offset voltage if $I_{\text{in(bias)}} = 80$ nA and $I_{\text{in(offset)}} = 20$ nA?

SOLUTION

In Example 17-7 we worked out the base currents and found these two worst-case possibilities:

$$I_{B1} = 70 \text{ nA and } I_{B2} = 90 \text{ nA}$$

or

$$I_{B1} = 90 \text{ nA and } I_{B2} = 70 \text{ nA}$$

The two worst-case resistor possibilities are

$$R_{B1} = 90 \text{ k}\Omega \text{ and } R_{B2} = 110 \text{ k}\Omega$$

or

$$R_{B1} = 110 \text{ k}\Omega \text{ and } R_{B2} = 90 \text{ k}\Omega$$

The input voltage is given by Eq. (17-14):

$$v_{\text{in}} = I_{B1}R_{B1} - I_{B2}R_{B2}$$

In mass production, you must be prepared for the worst of all possible worlds because sooner or later it will turn up. In this case, the largest positive input voltage is

$$v_{\text{in}} = (90 \text{ nA})(110 \text{ k}\Omega) - (70 \text{ nA})(90 \text{ k}\Omega) = 3.6 \text{ mV}$$

At the other extreme, the largest negative input voltage is

$$v_{\text{in}} = (70 \text{ nA})(90 \text{ k}\Omega) - (90 \text{ nA})(110 \text{ k}\Omega) = -3.6 \text{ mV}$$

The output offset voltage is

$$v_{\text{out}} = 150(\pm 3.6 \text{ mV}) = \pm 0.54 \text{ V}$$

EXAMPLE 17-12

Repeat the preceding example. This time include $\Delta V_{BE} = \pm 2 \text{ mV}$.

SOLUTION

All we have to do is add or subtract ΔV_{BE} in the worst possible way. Since the maximum positive input voltage was 6.3 mV, the total unwanted input voltage may be as large as

$$v_{\text{in}} = 3.6 \text{ mV} + 2 \text{ mV} = 5.6 \text{ mV}$$

By a similar argument, the maximum negative value is -5.6 mV. The total output offset voltage is

$$v_{\text{out}} = 150(\pm 5.6 \text{ mV}) = \pm 0.84 \text{ V}$$

To null this unwanted output voltage, we can apply a separate input voltage. Since the total unwanted input voltage is ± 5.6 mV, we need a separate dc input voltage of the same magnitude but of the opposite phase. In mass production, this would require an adjustment that produces an input voltage in the range of -5.6 mV to $+5.6$ mV.

17-6   COMMON-MODE GAIN

Assume that the transistors of Fig. 17-19a are identical, so that there is no output offset voltage. Notice that the same input voltage, $v_{\text{in(CM)}}$, is being applied to each base. This voltage is called a *common-mode signal*.

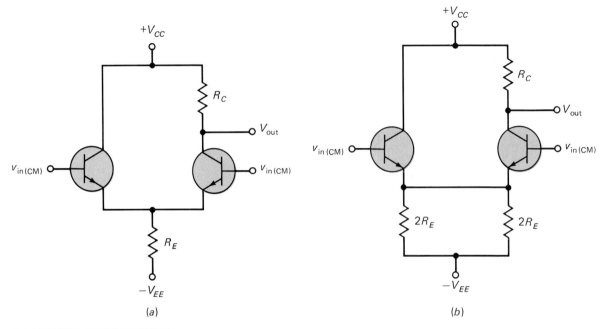

(a)   (b)

FIGURE 17-19   Common-mode gain.

Ideally, there is no ac output voltage with a common-mode input signal because the voltage between the bases is zero. To a second approximation, however, there will be a small ac output voltage. To understand why, we have to switch our viewpoint, so that we can see things in a different way. What you are about to see is a good example of how the "many right answers" approach can produce an elegant solution to a difficult problem.

In Fig. 17-19a, equal voltages are applied to the noninverting and inverting inputs. Nobody in his right mind would deliberately use a diff amp this way because the output voltage is ideally zero. Then why are we bothering to discuss this possibility? Because most static, interference, and other kinds of undesirable pickup are common-mode signals. What happens is this. The connecting wires on the input bases act like small antennas. If the diff amp is operating in an environment with a lot of electromagnetic interferences, each base acts like a small antenna and picks up an unwanted signal voltage. As you will see, one of the reasons the diff amp is so popular is because it discriminates against these common-mode signals. In other words, a diff amp refuses to amplify common-mode signals. Because of this, you don't get a lot of unwanted interference at the output.

Nothing prevents us from redrawing the circuit as shown in Fig. 17-19b. In this equivalent circuit, the two parallel resistances of $2R_E$ produce an equivalent resistance of R_E. Therefore, this equivalent circuit will not affect the output voltage. In Fig. 17-19b, an equal voltage $v_{in(CM)}$ drives both inputs simultaneously. Assuming identical transistors, the equal base voltages produce equal emitter currents. These equal emitter currents flow through the emitter resistors and produce the same voltage across the emitter resistors.

Since the two emitter currents are equal, there is no current through the wire between the emitters. Therefore, we can remove the connecting wire as shown in Fig. 17-20a without disturbing any of the currents or voltages. As far as a common-mode input signal is concerned, the circuit acts the same with or without the connecting wire.

To get the ac equivalent circuit, we can reduce both supply voltages to zero, equivalent to grounding each supply point. Replacing the transistors by their ac equivalent circuits gives us the ac equivalent circuit of Fig. 17-20b. Here is what it means. When a common-mode signal drives a diff amp, a large unbypassed emitter resistance appears in the ac equivalent circuit. The negative feedback in the emitter becomes very large, the same as a *swamped* CE amplifier. Therefore, the voltage gain for a common-mode signal is

$$\frac{v_{\text{out}}}{v_{\text{in(CM)}}} = \frac{R_C}{r'_e + 2R_E}$$

Since R_E is always much greater than r'_e, we can approximate the common-mode voltage gain as

$$A_{\text{CM}} = \frac{R_C}{2R_E} \tag{17-18}$$

This says that the common-mode voltage gain is very small.

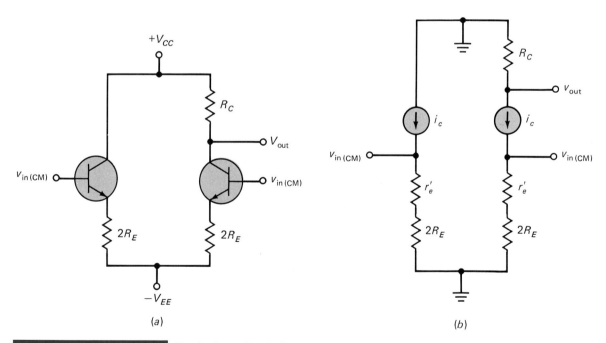

Equivalent circuit for common-mode gain.

To distinguish the ordinary voltage gain discussed earlier from the common-mode voltage gain, we will sometimes refer to the ordinary voltage gain as the *differential* voltage gain where the word *differential* reminds us that we are talking about amplifying the difference between the two base voltages. In other words, A stands for differential voltage gain, and A_{CM} stands for common-mode voltage gain.

Why Flexibility Is Important

You have to admit the foregoing derivation is brilliant. Without this clever approach, we would have had to endure a nightmare of mathematics to get the same result because the standard engineering approach is to write several loop equations and solve them simultaneously to get Eq. (17-18). But you don't have to do things the hard way if you can find an easy way. We avoided a very difficult mathematical derivation by changing the original circuit into a circuit discussed in earlier chapters. Splitting the emitter resistor into the two separate resistors was the key. The ability to do things like this once in a while leads to outstanding success. The way you will learn to do things like this is by becoming flexible, using different equivalent circuits, and trying to solve problems in original ways.

Any fool can learn enough mathematics to do things the hard way. The wise person looks for alternatives, for new ways to solve old problems. What we did here echos what was said earlier: there are many right answers and many ways to find them. That's what anyone at the top of his or her profession will tell you. Electronics is as much an art as it is a science. Because of this, you have a lot of freedom to do things your way. As long as you don't contradict fundamental laws, any solution is acceptable. For major problems, the best strategy is to find as many right answers as you can, then select the best answer for the particular situation.

Common-Mode Rejection Ratio

Data sheets list the *common-mode rejection ratio* (CMRR). It is defined as the ratio of differential voltage gain to common-mode voltage gain. In symbols,

$$CMRR = \frac{A}{A_{CM}} \tag{17-19}$$

If a diff amp were perfect, CMRR would be infinite because A_{CM} would be zero. Data sheets often specify CMRR in decibels, using the following formula for the decibel conversion:

$$CMRR' = 20 \log CMRR \tag{17-20}$$

Data sheets usually list the values of A and CMRR′, but not the value of A_{CM}. If you ever need the value of A_{CM}, you can rearrange Eq. (17-19) as follows:

$$A_{CM} = \frac{A}{CMRR} \tag{17-21}$$

Given the value CMRR′ on a data sheet, we can calculate the value of CMRR. Then we can divide A by CMRR to get A_{CM}.

EXAMPLE 17-13

In Fig. 17-21, what is the common-mode voltage gain? The output voltage?

SOLUTION

Because 1 mV is being applied to both bases, we have a common-mode input signal. We can find the common-mode voltage gain as follows. Mentally split the tail resistor into separate emitter resistors of 2 MΩ each. Then, you can see that the right transistor has a swamping resistance of 2 MΩ. As a swamped amplifier, it has a voltage gain of

$$A_{CM} = \frac{1 \text{ M}\Omega}{2 \text{ M}\Omega} = 0.5$$

You can get the same result by direct substitution into Eq. (17-18), but you will get better results if you don't rely on formulas. Visualizing the 1 MΩ as two separate resistors of 2 MΩ is a better approach than plugging numbers into a formula.

The output voltage equals

$$v_{out} = 0.5(1 \text{ mV}) = 0.5 \text{ mV}$$

The diff amp refuses to amplify the common-mode signal. Here, we get *attenuation* rather than voltage gain. (Attenuation means weakening the signal, equivalent to a voltage gain of less than one.)

EXAMPLE 17-14

In Fig. 17-22, the base leads are picking up an unwanted common-mode signal of 1 mV. A desired input signal v_{in} is also driving the diff amp. What is the output voltage in Fig. 17-22 if the desired input voltage is also 1 mV?

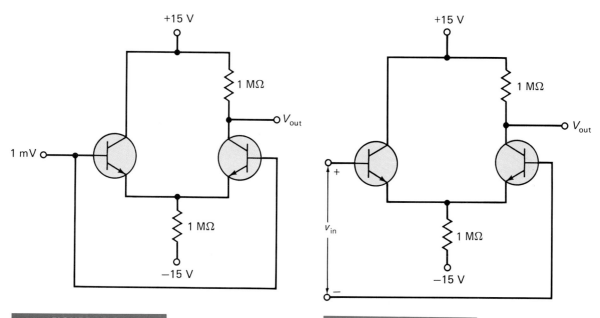

FIGURE 17-21
Example.

FIGURE 17-22
Example.

SOLUTION

The first thing to realize is this. We have two equal signals, a desired signal, v_{in}, and an undesired signal, $v_{in(CM)}$. Each of these signals has a value of 1 mV. The undesired signal, $v_{in(CM)}$, is being electromagnetically induced in the base leads. Because the base leads are in series with v_{in}, the common-mode signal is *added* to the desired signal. The total input signal to the diff amp is

$$v_{in(total)} = v_{in} + v_{in(CM)}$$

Each component of this input signal is treated differently by the diff amp. The desired signal is amplified by a factor of A, but the undesired signal is amplified by a factor of A_{CM}. This is why the ac output voltage is

$$v_{out} = Av_{in} + A_{CM}v_{in(CM)}$$

In earlier examples, we found that $A = 150$ and $A_{CM} = 0.5$. Therefore, the first term in the output expression is much larger than the second term. Here are the calculations:

$$Av_{in} = 150(1 \text{ mV}) = 150 \text{ mV}$$

and

$$A_{CM}v_{in(CM)} = 0.5(1 \text{ mV}) = 0.5 \text{ mV}$$

The desired output is 150 mV versus an undesired output of 0.5 mV, despite the fact that both signals were equal on the input side of the diff amp. That's why the diff amp is very popular with designers. They know that it discriminates between the desired input signal and a common-mode signal that is picked up through stray electromagnetic radiation. This is a distinct advantage over the ordinary CE amplifier, which amplifies a stray pickup signal the same as a desired signal.

EXAMPLE 17-15

What is the common-mode rejection ratio in Fig. 17-22? Express this in decibels.

SOLUTION

From earlier examples, we already know that $A = 150$ and $A_{CM} = 0.5$. Therefore, the common-mode rejection ratio is

$$CMRR = \frac{150}{0.5} = 300$$

and the decibel equivalent is

$$CMRR' = 20 \log 300 = 49.5 \text{ dB}$$

EXAMPLE 17-16

The data sheet of an op amp gives these typical values: $A = 200,000$ and $CMRR' = 90$ dB. What is the common-mode voltage gain?

SOLUTION

First, get the value of CMRR. What does 90 dB stand for? We know that each 20 dB represents a factor of 10. Therefore, 80 dB stands for 10,000. Add another 6 dB to get 86 dB and 20,000. Add another 6 dB to get 92 dB and 40,000. Therefore, CMRR is between 20,000 and 40,000. Let's estimate 30,000.

Here is how to get the exact answer. Since

$$90 \text{ dB} = 20 \log CMRR$$

we can solve in terms of the antilog as follows:

$$CMRR = \text{antilog} \frac{90 \text{ dB}}{20} = 31,623$$

If you do this on a calculator, enter 90 and divide by 20. Then take the inverse logarithm to get the answer.

The common-mode voltage gain is

$$A_{CM} = \frac{200,000}{31,623} = 6.32$$

This gives you a clue as to how an op amp discriminates against common-mode signals. The desired signal is amplified by a factor of 200,000, while a common-mode signal is amplified by only 6.32. If a desired input signal and a common-mode input signal are each 1 μV at the input to the op amp, they will have values of 200 mV and 6.32 μV at the output of the op amp. That is real discrimination.

OPTIONAL TOPICS

The following material continues the earlier discussions at a more advanced and specialized level. All the topics are optional because they are not used in any of the basic discussions in later chapters. This section will be a useful reference when you are in industry because then you will probably want more advanced viewpoints.

17-7 THE CURRENT MIRROR

There is a way to improve the diff amp, a way to increase its voltage gain and common-mode rejection ratio. What you are about to see is extremely clever, another example of what happens when someone looks for more than the right answer, when someone practices the art of electronics.

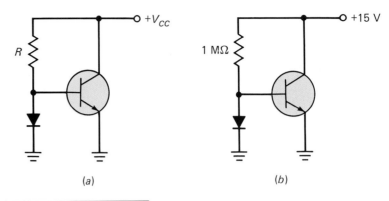

FIGURE 17-23 The current mirror.

In Fig. 17-23a the current through the resistor is given by

$$I_R = \frac{V_{CC} - V_{BE}}{R} \qquad (17\text{-}22)$$

Assume the external diode and the emitter diode have identical current-versus-voltage curves. Since the external diode and emitter diode are in parallel, they have equal voltages. And since their curves are identical, they have equal currents. This means the collector current of the transistor is equal to the current through the diode.

The current through the diode is the difference of the resistor current and the base current:

$$I_D = I_R - I_B$$

Since $I_C = I_D$ for matched diodes, we can write

$$I_C = I_R - I_B$$

Because the base current is much smaller than the collector current, we can approximate as follows:

$$I_C = I_R \qquad (17\text{-}23)$$

This approximation is accurate to within 1 percent when β_{dc} is greater than 100.

Equation (17-23) is deceptively simple but enormously important. It is a gold mine for a designer. Why? Because it says the collector current equals the resistor current. A circuit like the one in Fig. 17-23a is called a *current mirror* because the collector current is a reflection of the current through the resistor, almost like a mirror image. This means we can build a current source with a value of I_R.

There is a catch. It is almost impossible with discrete circuits to find a diode that exactly matches the emitter diode. But this is not the case with integrated circuits. With integrated circuits, it is possible to closely match the current-voltage curves of the ordinary diode and the emitter diode. This is why the current mirror is rarely used in discrete circuits, but is heavily used in integrated circuits. Whenever you see a current mirror used in an integrated circuit, remember this key idea: the collector current is equal to whatever the current is through the resistor.

A final point. The external diode is often called a *compensating diode* because it automatically compensates for changes in temperature. When we say the diode curves are matched, we mean at all temperatures as well as voltages. When the temperature increases, the voltage across the emitter diode decreases approximately 2 mV per degree. Since the voltage across the compensating diode also decreases by 2 mV per degree, the collector current shows little change with temperature increase.

EXAMPLE 17-17

What is the collector current in Fig. 17-23b?

SOLUTION

Since V_{CC} is much greater than V_{BE}, we can approximate the voltage across the 1 MΩ as 15 V. Then the ideal current through the resistor is

$$I_R = \frac{15\ V}{1\ M\Omega} = 15\ \mu A$$

This is the approximate value of current through the diode. It is also the value of collector current.

When you see a circuit like this one combined with a diff amp, remember this mental trick: In Fig. 17-23b, visualize 15 μA through the resistor. A current mirror reflects this 15 μA into the collector, so that the collector current is 15 μA.

Incidentally, a more accurate answer includes the V_{BE} drop as follows:

$$I_R = \frac{15\ V - 0.7\ V}{1\ M\Omega} = 14.3\ \mu A$$

This is only 5 percent more accurate than the ideal answer. Unless you are designing a circuit, use the ideal current. It prevents getting bogged down in more accurate analysis than is necessary.

17-8 DIFF AMP WITH CURRENT MIRRORS

With an emitter-biased diff amp, the differential voltage gain is

$$A = \frac{R_C}{2r_e'}$$

and the common-mode voltage gain is

$$A_{CM} = \frac{R_C}{2R_E}$$

Dividing A by A_{CM} gives the common-mode rejection ratio:

$$\text{CMRR} = \frac{R_C/2r'_e}{R_C/2R_E} = \frac{R_E}{r'_e}$$

From this, it is clear that the higher we can make R_E, the better the CMRR.

Current-Sourcing the Tail Current

One way to get a much higher equivalent R_E is to use a current mirror to produce the tail current, as shown in Fig. 17-24. This is typical for the first stage of an integrated op amp. Here you see a current mirror driving the emitters of a diff amp. The current through Q_3 is given by

$$I_R = \frac{V_{CC} + V_{EE} - V_{BE}}{R} \qquad (17\text{-}24)$$

This is the value of tail current produced by Q_4. Since Q_4 acts like a current source, it ideally appears to have an output impedance of infinity. In reality, this means that the equivalent R_E of the diff amp is in tens of megohms and the CMRR is dramatically improved.

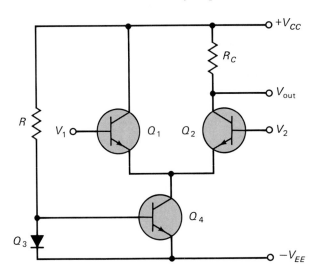

FIGURE 17-24 Current mirror sources tail current.

Active Load

Earlier, we derived a differential voltage gain of

$$A = \frac{R_C}{2r'_e}$$

The larger we can make R_C, the greater the differential voltage gain. But we have to be careful. An R_C that is too large will saturate the right transistor. As a rule, the designer selects an R_C to get a quiescent voltage that is about half of V_{CC}. For example, if the collector supply voltage is

+15 V, then R_C is selected to get a V_C of +7.5 V. This limits the size of R_C for a given collector current, equivalent to limiting the voltage gain.

One way around the problem is to use an active load. Figure 17-25 shows a current mirror used as a load resistor. Since Q_5 acts like a compensating diode, it has a very low impedance and the load on Q_1 still appears almost like an ac short. On the other hand, Q_6 acts like a *pnp* current source. Therefore, Q_2 sees an equivalent R_C that is ideally infinite. In reality, the equivalent R_C is in tens of megohms. As a result, the differential voltage gain is much higher with an active load than with an ordinary resistor. Active loading like this is typical of most op amps.

EXAMPLE 17-18

In Fig. 17-25, the supply voltages are ±15 V. If R = 2 MΩ, what is the Q_2 collector current?

SOLUTION

Assume V_1 and V_2 are zero. Then, the voltage at the top of the resistor is +15 V. If we ignore the 0.7 V across the compensating diode, the voltage at the bottom of the resistor is −15 V. This means a total of 30 V is across the resistor. With Ohm's law, the current through the resistor is

$$I_R = \frac{30 \text{ V}}{2 \text{ M}\Omega} = 15 \text{ }\mu\text{A}$$

This is the value of the tail current. In other words, Q_4 acts like a current source that produces a tail current of 15 μA.

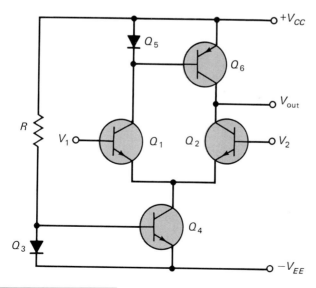

FIGURE 17-25 Current mirror is active load.

Because Q_1 and Q_2 are identical, the tail current splits equally with 7.5 μA at the Q_1 and Q_2 collectors. The 7.5 μA from Q_1 flows through the compensating diode, Q_5. The voltage across Q_5 forces Q_6 to produce a collector current of 7.5 μA. But 7.5 μA is exactly what the collector of Q_2 is producing. At this point, you can see why Q_6 has to be a *pnp* transistor. Conventional current flows out of the Q_6 collector down to the Q_2 collector. If you prefer electron flow, the electron flow out of the Q_2 transistor flows into the Q_6 collector.

STUDY AIDS

The following study aids will help to reinforce the ideas discussed in this chapter. For best results, use these study aids within 6 hours of reading the earlier material. Then review these study aids a week later and a month later to ensure that the concepts remain in your long-term memory.

SUMMARY

Sec. 17-1 Integrated Circuits
Monolithic ICs are complete circuit functions on a single chip such as amplifiers, voltage regulators and computer circuits. Monolithic ICs usually have power dissipations under a watt. For higher power applications, thin-film, thick-film, and hybrid ICs may be used. SSI refers to fewer than 10 integrated components, MSI to between 10 and 100 integrated components, and LSI to more than 100 integrated components.

Sec. 17-2 The Differential Amplifier
The diff amp is the typical input stage of an op amp. It has no coupling or bypass capacitors, which means it is direct coupled. Because of this, it can amplify frequencies all the way down to dc, which is equivalent to an ac signal of zero frequency. The tail current in a diff amp splits equally between the transistors when they are identical.

Sec. 17-3 Two Input Characteristics
When the two transistors of a diff amp are not identical, the two base currents are different. The input offset current is defined as the difference between the two base currents. The input bias current is defined as the average of the two base currents. Data sheets list $I_{in(offset)}$ and $I_{in(bias)}$.

Sec. 17-4 AC Analysis of a Diff Amp
Because the tail current is ideally constant, an increase in the emitter current of one transistor produces a decrease in the emitter current of the other transistor. The ac input voltage between the two bases appears across the $2r'_e$. Because of this, the voltage gain of a diff amp is $R_C/2r'_e$, and the input impedance is $2\beta r'_e$.

Sec. 17-5 Output Offset Voltage
The output offset voltage is any deviation or change from the ideal output voltage. The two sources of output offset voltage are the differences in V_{BE} values and the differences in β_{dc} values. Each of these independent causes produces the equivalent of an unwanted input voltage. The diff amp amplifies these unwanted input signals to get the output offset voltage. One way to null the output offset voltage is by applying an input voltage of the same magnitude as the unwanted input voltages but of opposite phase.

Sec. 17-6 Common-Mode Gain
Unwanted pickup can produce an equal voltage at each base. The diff amp discriminates against this unwanted voltage because it has low voltage gain for a common-mode signal. The common-mode rejection ratio is the differential voltage gain divided by the common-mode voltage gain. The higher CMRR is, the better.

VOCABULARY

In your own words, explain what each of the following terms means. Keep your answers short and to the point. If necessary, verify your answer by rereading the appropriate discussion or by looking at the end-of-book Glossary.

common-mode signal	LSI
diff amp	MSI
discrete circuit	monolithic IC
input bias current	noninverting input
input offset current	op amp
integrated circuit	output offset voltage
inverting input	SSI

IMPORTANT EQUATIONS

Formulas are dangerous things. Improperly used, they make bridges fall down and circuits go up in smoke. Repeat each formula in symbols, then in words. Try to explain what the equation means and how it is used. Then read the comments that follow the equation. Your chances of remembering are much better if you concentrate on meaning rather than formulas.

Eq. 17-1 Tail Current

$$I_T = \frac{V_{EE}}{R_E}$$

This is nothing more than Ohm's law applied to the tail resistor of a diff amp. It is an ideal approximation because it assumes all of the V_{EE} supply voltage is across the tail resistor. You can subtract 0.7 V from V_{EE} if you want to improve the answer slightly.

Eq. 17-4 Input Offset Current

$$I_{in(off)} = I_{B1} - I_{B2}$$

This is the definition of input offset current. It says the input offset current equals the difference between the two base currents. This difference is an indication of how different the two β_{dc} values are. When a diff amp is perfect, the input offset current is zero.

Eq. 17-5 Input Bias Current

$$I_{in(bias)} = \frac{I_{B1} + I_{B2}}{2}$$

This is the definition of input bias current. It says the input bias current equals the average of the two base currents.

Eq. 17-8 Voltage Gain of Diff Amp

$$A = \frac{R_C}{2r'_e}$$

The voltage gain of a diff amp equals the ac collector resistance divided by two times the ac resistance of the emitter diode. This voltage gain is half that of a similar CE amplifier. The factor of 2 occurs because the ac input voltage is across two r'_es in series.

Eq. 17-9 Input Impedance of Diff Amp

$$z_{in} = 2\beta r'_e$$

The input impedance of a diff amp equals two times the ac current gain times the ac resistance of the emitter diode. Since the ac input voltage is across $2r'_e$, the input impedance is increased by β when viewed from the base terminals.

Eqs. 17-15 and 17-16 Total Unwanted Input Voltages

$$v_{in} = \pm\Delta V_{BE} + I_{B1}R_{B1} - I_{B2}R_{B2}$$

There is a false input voltage produced by the differences in the transistors. It has two components. First, there are the differences in the two V_{BE} values. Second, there are the differences in the two base voltages. The unwanted input voltage is amplified to produce the output offset voltage.

Eq. 17-19 Common-Mode Rejection Ratio

$$CMRR = \frac{A}{A_{CM}}$$

This is a large number because it equals the differential voltage gain divided by the common-mode voltage gain. The value indicates how effectively a diff amp or op amp discriminates against a common-mode signal.

IMPORTANT PROCESSES

Each of the following processes is important for analyzing diff amps. If you can remember the basic ideas in each step, you will be able to solve problems more easily.

Ideal Tail Current

1. Visualize a ground at the top of R_E.
2. Visualize all of V_{EE} across R_E.
3. Use Ohm's law to get the tail current.

Ideal Collector Current

1. Calculate the tail current.
2. Assume identical transistors.
3. Divide the tail current by 2.

Base Voltages

1. Find each emitter current.
2. Divide by each β_{dc}.
3. Multiply each base current by the base resistance.

Differential Voltage Gain

1. Visualize input voltage across two $r'_e s$ in series.
2. Realize effective voltage is half of CE amp.
3. Use $R_C/2r'_e$ to get voltage gain.

Common-Mode Voltage Gain

1. Split R_E into two resistors of $2R_E$ each.
2. Recall swamped amplifier.
3. Use $R_C/2R_E$ to get voltage gain.

STUDENT ASSIGNMENTS

QUESTIONS

Some questions have more than one right answer. Select the best answer, the one that is always true, or that most accurately describes the situation.

1. Monolithic ICs are
   a. Forms of discrete circuits
   b. On a single chip
   c. Combinations of thin-film and thick-film circuits
   d. Also called hybrid ICs

2. The op amp can amplify
   a. Ac signals only
   b. Dc signals only
   c. Both ac and dc signals
   d. Neither ac nor dc signals

3. Components are soldered together in
   a. Discrete circuits
   b. Integrated circuits
   c. SSI
   d. Monolithic ICs

4. The tail current of a diff amp is
   a. Half of either collector current
   b. Equal to either collector current
   c. Two times either collector current
   d. Equal to the difference in base currents

5. The node voltage at the top of the tail resistor is closest to
   a. The collector supply voltage
   b. Zero
   c. Emitter supply voltage
   d. Tail current times base resistance

6. The input offset current equals the
   a. Difference between two base currents
   b. Average of two base currents
   c. Collector current divided by current gain
   d. Difference between two base-emitter voltages

7. The tail current equals the
   a. Difference between two emitter currents
   b. Sum of two emitter currents
   c. Collector current divided by current gain
   d. Collector voltage divided by collector resistance

8. The differential voltage gain of a diff amp is equal to R_C divided by
   a. r'_e
   b. $r'_e/2$
   c. $2r'_e$
   d. R_E

9. The input impedance of a diff amp equals r_e' times
   a. β
   b. R_C
   c. R_E
   d. 2β

10. A dc signal has a frequency of
    a. 0
    b. 60 Hz
    c. 0 to over 1 MHz
    d. 1 MHz

11. When the two input terminals of a diff amp are grounded,
    a. The base currents are equal
    b. The collector currents are equal
    c. An output offset voltage may exist
    d. The ac output voltage is zero

12. One source of output offset voltage is
    a. Input bias current
    b. Difference in emitter diode curves
    c. Tail current
    d. Common-mode voltage gain

13. A common-mode signal is applied to
    a. The noninverting input
    b. The inverting input
    c. Both inputs
    d. Top of the tail resistor

14. The common-mode voltage gain is
    a. Smaller than differential voltage gain
    b. Equal to differential voltage gain
    c. Greater than differential gain
    d. None of the above

15. The input stage of an op amp is usually a
    a. Diff amp
    b. Class B push-pull amplifier
    c. CE amplifier
    d. Swamped amplifier

16. The tail of a diff amp acts like a
    a. Battery
    b. Current source
    c. Transistor
    d. Diode

17. The common-mode voltage gain of a diff amp is equal to R_C divided by
    a. r_e'
    b. $r_e'/2$
    c. $2r_e'$
    d. $2R_E$

18. When the two bases are grounded in a diff amp, the voltage across each emitter diode is
    a. Zero
    b. 0.7 V
    c. The same
    d. High

BASIC PROBLEMS

Sec. 17-2 The Differential Amplifier

17-1. In Fig. 17-26, the transistors are identical with β_{dc} = 200. What is the output voltage?

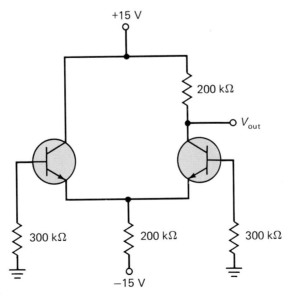

FIGURE 17-26

17-2. What are base voltages in Fig. 17-26 if each transistor has $\beta_{dc} = 300$?

17-3. In Fig. 17-26, the left transistor has $\beta_{dc} = 225$ and the right transistor has $\beta_{dc} = 275$. What are the base voltages?

Sec. 17-3 Two Input Characteristics

17-4. The base currents are 70 and 50 nA. What is the input offset current? The input bias current?

17-5. A data sheet gives an input bias current of 20 nA and an input offset current of 3 nA. What are the base currents?

17-6. In Fig. 17-26, the left transistor has $\beta_{dc} = 180$ and the right transistor has $\beta_{dc} = 220$. What is the input offset current? The input bias current?

17-7. A diff amp uses JFETs instead of bipolar transistors. The input bias current is 20 pA and the input offset current is 3 pA. What are the two gate currents?

Sec. 17-4 AC Analysis of a Diff Amp

17-8. In Fig. 17-27, what is V_{out} when $v_{in} = 0$? When $v_{in} = 2$ mV? When $v_{in} = -2$ mV?

17-9. If $\beta = 200$, what is the input impedance in Fig. 17-27?

17-10. The resistors of Fig. 17-27 have a tolerance of ± 10 percent. What is the minimum voltage gain? The maximum voltage gain?

17-11. If the supply voltages of Fig. 17-27 are increased to ± 20 V, what is V_{out} when $v_{in} = 0$? When $v_{in} = 2$ mV? When $v_{in} = -2$ mV?

Sec. 17-5 Output Offset Voltage

17-12. In Fig. 17-26, the transistors have identical V_{BE} curves but their β_{dc} values are different. The left transistor has $\beta_{dc} = 175$ and the right transistor has $\beta_{dc} = 225$. What is the output offset voltage?

17-13. The diff amp of Fig. 17-26 has an input bias current of 375 nA and an input offset current of 50 nA. If the V_{BE} curves are identical, what is the output offset voltage?

17-14. Repeat the preceding problem, except for this: the base resistors have a tolerance of ± 10 percent.

17-15. The diff amp of Fig. 17-26 has an input bias current of 200 nA and an input offset current of 20 nA. If $\Delta V_{BE} = \pm 3$ mV, what is the output offset voltage?

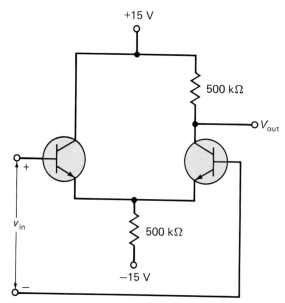

FIGURE 17-27

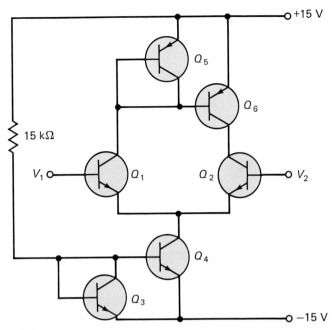

FIGURE 17-28

Sec. 17-6 Common-Mode Gain

17-16. What is the common-mode voltage gain of Fig. 17-27? If a common-mode voltage of 20 μV exists on both bases, what is the common-mode output voltage?

17-17. In Fig. 17-27, v_{in} = 2 mV and $v_{in(CM)}$ = 5 mV. What is the ac output voltage?

17-18. What is the common-mode rejection ratio of Fig. 17-27? Express the answer in decibels.

17-19. The data sheet of an op amp gives A = 100,000 and CMRR = 80 dB. What is the common-mode voltage gain?

TROUBLESHOOTING PROBLEMS

17-20. Somebody builds the diff amp of Fig. 17-26 without base resistors so that the two bases are floating. What does the output voltage equal? Based on your preceding answer, what does any diff amp need to work properly?

17-21. In Fig. 17-26, 20 kΩ is mistakenly used for the upper 200 kΩ. What does the output voltage equal?

17-22. In Fig. 17-26, V_{out} is almost zero. The input bias current is 80 nA. Which of the following is the trouble?
a. Upper 200 kΩ shorted
b. Lower 200 kΩ open
c. Left 300 kΩ open
d. Both inputs shorted together

ADVANCED PROBLEMS

17-23. In Fig. 17-28, transistors Q_3 and Q_5 are connected to act like compensating diodes for Q_4 and Q_6. What is the tail current? The current through the active load?

17-24. The 15 kΩ of Fig. 17-28 is changed to get a tail current of 15 μA. What is the new value of resistance?

17-25. At room temperature, the output voltage of Fig. 17-26 has a value of 7.5 V. As the temperature increases, the V_{BE} of each emitter diode decreases. If the left V_{BE} decreases 2 mV per degree and the right V_{BE} decreases 2.1 mV per degree, what is the output voltage at 75°C?

17-26. The dc resistance of each signal source in Fig. 17-29a is zero. What is the r'_e of

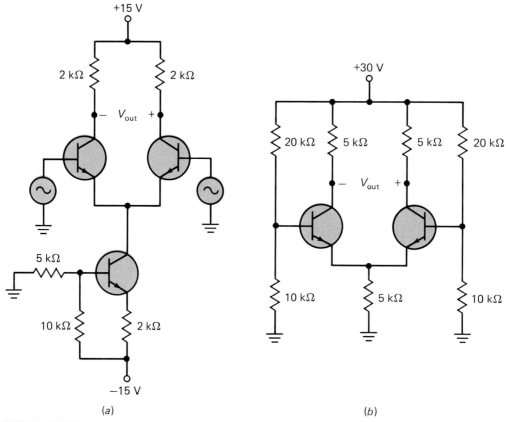

FIGURE 17-29

each transistor? If the ac output voltage is between the collectors, what is the differential voltage gain?

17-27. If the transistors are identical in Fig. 17-29b, what is the tail current? The voltage between the left collector and ground? Between the right collector and ground?

SOFTWARE ENGINE PROBLEMS

Use Fig. 17-30 for the remaining problems. Assume increases of approximately 10 percent in the independent variable. A response should be an N (no change) if the change in a dependent variable is so small that you would have difficulty measuring it. In this *pnp* version of the diff amp, the base currents produce positive dc voltages at the bases, $+V_{B1}$ and $+V_{B2}$. The difference between these two base voltages is the input voltage V_{in}. In the last three boxes (R_E, R_C, and $\pm V_{CC}$), assume that the transistors are identical.

17-28. Try to predict the response of each dependent variable in the box labeled I_{B1}. Check your answers. Then, answer the following question as simply and directly as possible. What effect does an increase in I_{B1} have on the dependent variables of the circuit?

17-29. Predict the response of each dependent variable in the box labeled I_{B2}. Check your answers. Then summarize your findings in one or two sentences.

17-30. Predict the response of each dependent variable in the box labeled R_E. Check your answers. List the dependent variables that decrease. Explain why these variables decrease.

17-31. Predict the response of each dependent variable in the box labeled R_C. List the dependent variables that decrease. Explain why these variables decrease.

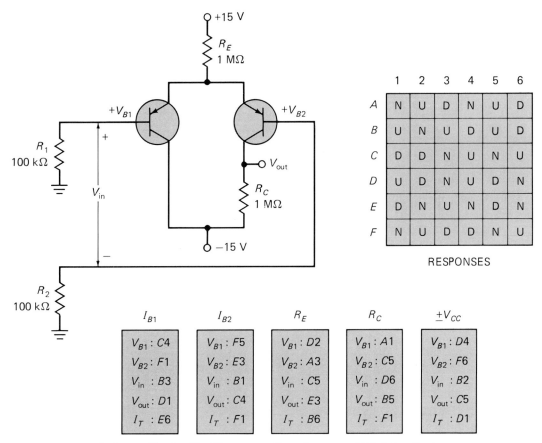

	1	2	3	4	5	6
A	N	U	D	N	U	D
B	U	N	U	D	U	D
C	D	D	N	U	N	U
D	U	D	N	U	D	N
E	D	N	U	N	D	N
F	N	U	D	D	N	U

RESPONSES

I_{B1}

| V_{B1} : C4 |
| V_{B2} : F1 |
| V_{in} : B3 |
| V_{out} : D1 |
| I_T : E6 |

I_{B2}

| V_{B1} : F5 |
| V_{B2} : E3 |
| V_{in} : B1 |
| V_{out} : C4 |
| I_T : F1 |

R_E

| V_{B1} : D2 |
| V_{B2} : A3 |
| V_{in} : C5 |
| V_{out} : E3 |
| I_T : B6 |

R_C

| V_{B1} : A1 |
| V_{B2} : C5 |
| V_{in} : D6 |
| V_{out} : B5 |
| I_T : F1 |

$\pm V_{CC}$

| V_{B1} : D4 |
| V_{B2} : F6 |
| V_{in} : B2 |
| V_{out} : C5 |
| I_T : D1 |

FIGURE 17-30 Software Engine™. (*Patent pending: Courtesy of Malvino Inc.*)

17-32. Predict the response of each dependent variable in the box labeled $\pm V_{CC}$. List the dependent variables that increase. Explain why these variables show an increase.

18

MORE OP-AMP THEORY

This chapter continues our discussion of IC op amps. About a third of all linear ICs are operational amplifiers (op amps). By connecting external resistors to an IC op amp, you can adjust the voltage gain and bandwidth to your requirements. There are over 2000 types of commercially available op amps. Most are low-power devices because their power dissipations are less than a watt. Whenever you need voltage gain for a low-power application, check the available op amps. You can almost always find an op amp to do the job.

18-1 SMALL-SIGNAL FREQUENCY RESPONSE

Figure 18-1 shows the typical stages of an IC op amp. The input stage is a diff amp, followed by more stages of gain and a class-B push-pull emitter follower. Because the diff amp is the first stage, it becomes extremely important in determining the input characteristics of the op amp. As you saw in the preceding chapter, any differences between the two input transistors are amplified to produce an output offset voltage. As you are about to see, the diff amp also controls the frequency response.

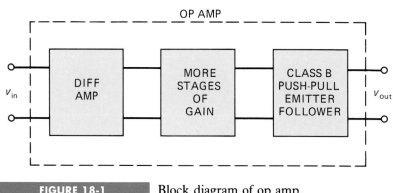

FIGURE 18-1   Block diagram of op amp.

Figure 18-2 shows a diff amp. Since it has no coupling or bypass capacitors, there are no critical frequencies below the midband. But there are critical frequencies above the midband produced by transistor and stray capacitances. In an IC op amp, these critical frequencies are well into the megahertz region. But there is another critical frequency that is much lower. It is produced by the capacitor C_C connected across the output of Fig. 18-2.

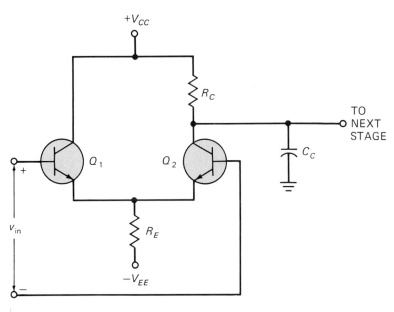

| **FIGURE 18-2** | Effect of compensating capacitor.

Capacitor C_C is called a *frequency compensating capacitor*. Why is it used? Rather than let the critical frequency of an op amp be determined by transistor and stray capacitances, a designer inserts an extra capacitor, C_C, into the circuit. C_C is selected to produce a critical frequency that is much lower than the critical frequencies of the unwanted base and collector bypass circuits. This has two advantages. First, it allows the designer to control the frequency response of the op amp. Second, it prevents *oscillations* (discussed in a later chapter). Under certain conditions, a high-gain amplifier can produce an unwanted high-frequency signal called an oscillation. For reasons given later, the compensating capacitor prevents unwanted oscillations.

In Fig. 18-2, the collector resistor and the compensating capacitor form a bypass circuit because a Thevenin resistance of R_C faces a capacitor of C_C. This bypass circuit has a critical frequency of

$$f_c = \frac{1}{2\pi RC} \tag{18-1}$$

where $R = R_C$ and $C = C_C$. The midband of the diff amp is between 0 and $0.1f_c$. In this range of frequencies, the diff amp has maximum voltage gain. Above the midband, the voltage gain decreases until it equals

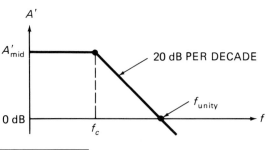

FIGURE 18-3 Frequency response.

0.707 of maximum at the critical frequency. Beyond the critical frequency, the voltage gain decreases at a rate of 20 dB per decade as shown in Fig. 18-3.

Not all data sheets list the critical frequency of an op amp. But many will list a frequency symbolized by f_{unity}. This is called the *unity-gain frequency* because at this frequency, the voltage gain equals 1 (equivalent to 0 dB). When you have the midband voltage gain and the unity-gain frequency, you can calculate the critical frequency with

$$f_c = \frac{f_{\text{unity}}}{A_{\text{mid}}} \qquad (18\text{-}2)$$

By cross multiplying the sides of the equation, we get this alternative form:

$$f_{\text{unity}} = A_{\text{mid}} f_c \qquad (18\text{-}3)$$

Both of these equations are very useful with op amps.

Where does Eq. (18-2) come from? It is derived as follows. The voltage gain above the midband with one dominant critical frequency is given by

$$A = \frac{A_{\text{mid}}}{\sqrt{1 + (f/f_c)^2}}$$

When $f = f_{\text{unity}}$, $A = 1$. Substituting these values gives

$$1 = \frac{A_{\text{mid}}}{\sqrt{1 + (f_{\text{unity}}/f_c)^2}}$$

In a typical op amp, f_{unity} is much greater than f_c. Therefore, the equation simplifies to

$$1 = \frac{A_{\text{mid}}}{f_{\text{unity}}/f_c}$$

which can be rearranged into Eq. (18-2):

$$f_c = \frac{f_{\text{unity}}}{A_{\text{mid}}}$$

This equation is used so often with op amps that you should memorize it.

EXAMPLE 18-1

What is the critical frequency in Fig. 18-4? The unity-gain frequency?

SOLUTION

The Thevenin resistance is 1 MΩ and the capacitance is 3000 pF. We can ignore the transistor and stray capacitances because they are much smaller. The critical frequency is

$$f_c = \frac{1}{2\pi(1 \text{ M}\Omega)(3000 \text{ pF})} = 53.1 \text{ Hz}$$

The unity-gain frequency is the product of the midband voltage gain and the critical frequency:

$$f_{unity} = 150(53.1 \text{ Hz}) = 7.97 \text{ kHz}$$

EXAMPLE 18-2

The data sheet of an op amp gives $A_{mid} = 100,000$ and $f_{unity} = 1$ MHz. What is the critical frequency of the op amp?

SOLUTION

The critical frequency equals the unity-gain frequency divided by the midband voltage gain:

$$f_c = \frac{1 \text{ MHz}}{100,000} = 10 \text{ Hz}$$

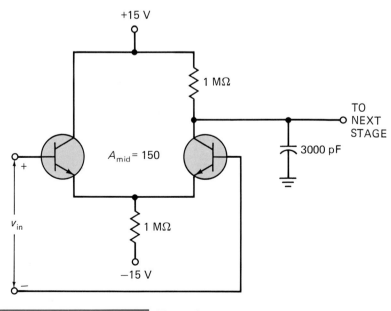

FIGURE 18-4   Example.

18-2 LARGE-SIGNAL FREQUENCY RESPONSE

In Chap. 16, we analyzed the frequency response of amplifiers. Although it was never stated, we were assuming small-signal operation. When a signal is small, all devices are linear and the critical frequencies are calculated as discussed earlier. But when the signal is large enough, a new effect occurs because of the nonlinearity. We are now ready to discuss this new effect.

Basic Idea

Suppose the input voltage of Fig. 18-4 steps from 0 to 1 V. Since the differential voltage gain is 150, the ac output voltage is ideally

$$v_{\text{out}} = 150(1 \text{ V}) = 150 \text{ V}$$

This result is impossible because the diff amp saturates long before the ac output voltage reaches 150 V. In other words, the voltage gain of 150 is valid only as long as the transistors are not saturated or cut off.

When the input voltage steps from 0 to 1 V in Fig. 18-4, it drives the left transistor into saturation and the right transistor into cutoff. The compensating capacitor then begins charging. As it does, its voltage increases exponentially from $+7.5$ V to $+15$ V. In other words, the capacitor prevents the output voltage from instantaneously stepping between 7.5 and 15 V. Instead of the ideal response shown in Fig. 18-5, we get the exponential response that increases slowly between the initial and final voltage.

The input signal does not have to be 1 V. Any input larger than 7.5 V/150, or 50 mV, will drive the diff amp into saturation. Then it will respond with the same exponential wave shown in Fig. 18-5. The time constant for this charging waveform is the product of R and C. In the case of a diff amp, R equals R_C and C equals C_C.

Slew rate refers to the initial slope of the exponential charging shown in Fig. 18-5. As you know, charging of a capacitor is rapid at first, but then it slows down as the capacitor voltage builds up. The slew rate or initial slope represents the fastest response an amplifier is capable of. Because the amplifier's fastest response is limited by its slew rate, an amplifier will also distort sine waves when their amplitude and frequency are too large. This gives you the basic idea behind slew rate. What follows is a mathematical discussion to nail down the specific details of this new concept.

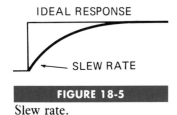

IDEAL RESPONSE

← SLEW RATE

FIGURE 18-5

Slew rate.

Rate of Change

Of all the specifications affecting the ac operation of an op amp, slew rate is one of the most important because it limits the size of the output voltage at higher frequencies. To enhance our understanding of slew rate, we have to discuss some basic circuit theory. The charging current in a capacitor is given by

$$i = C \frac{dv}{dt}$$

where dv/dt is the symbol for the rate of voltage change across the capacitor. We can rearrange this basic equation to get

$$\frac{dv}{dt} = \frac{i}{C}$$

This says that the rate of voltage change equals the charging current divided by the capacitance. The greater the charging current, the faster the capacitor charges. If for any reason the charging current is limited to a maximum value, the rate of voltage change is also limited to a maximum value.

Figure 18-6a brings out the idea of current limiting and its effect on output voltage. A current of I_{max} charges the capacitor. Because this current is constant, the capacitor voltage increases linearly, as shown in Fig. 18-6b. The rate of voltage change with respect to time is

$$\frac{dv_{\text{out}}}{dt} = \frac{I_{\text{max}}}{C_C} \qquad (18\text{-}4)$$

For instance, if $I_{\text{max}} = 60\ \mu\text{A}$ and $C_C = 30\ \text{pF}$ (see Fig. 18-6c), the maximum rate of voltage change is

$$\frac{dv_{\text{out}}}{dt} = \frac{60\ \mu\text{A}}{30\ \text{pF}} = 2\ \text{V}/\mu\text{s}$$

This answer equals 2,000,000 volts per second. Data sheets use the equivalent value of 2V/μs because it is more convenient when using an oscilloscope to measure slew rate. This says that the output voltage across the capacitor changes at a maximum rate of 2 V/μs (Fig. 18-6d). The voltage cannot change faster than this unless we can increase I_{max} or decrease C_C.

Slew rate S_R is defined as the maximum rate of output voltage change. Because of this, we can rewrite Eq. (18-4) as

$$S_R = \frac{I_{\text{max}}}{C_C} \qquad (18\text{-}5)$$

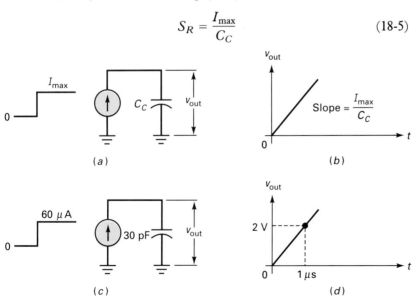

FIGURE 18-6 Slew rate equals maximum charging current divided by capacitance.

In Fig. 18-6a, slew rate limits the rate at which the output voltage can change. If $I_{max} = 60\ \mu A$ and $C_C = 30$ pF, the circuit can slew no faster than 2 V/μs.

Slew-Rate Distortion

We can also get slew-rate limiting with a sinusoidal signal. Fig. 18-7a shows a large sinusoidal output with a peak voltage of 7.5 V. A diff amp can produce this large output signal as long as the initial slope of the sine wave is less than or equal to S_R. For instance, if the sine wave has an initial slope of 1 V/μs and the slew rate is 2 V/μs, the diff amp can respond fast enough to produce the required output slope of 1 V/μs.

But when the initial slope of the sine wave is greater than S_R, we get the slew-rate distortion shown in Fig. 18-7b. For instance, if the initial slope of the sine wave is 4 V/μs and the slew rate is only 2 V/μs, then the amplifier cannot respond fast enough to produce the required output slope of 4 V/μs. In this case, the slew-rate distortion makes the output signal look more triangular than sinusoidal. If there were no slew-rate problem, the output would follow the dashed waveform. When there is a slew-rate problem, the actual output gets distorted into a triangular wave as shown here. As the frequency increases further, the signal swing becomes smaller and the shape becomes even more triangular.

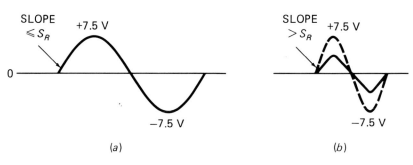

(a)   (b)

FIGURE 18-7   (a) Initial slope of sine wave cannot exceed slew rate; (b) slew rate distortion.

With calculus, it is possible to derive this equation for the initial slope of a sine wave:

$$S_S = 2\pi f V_p \qquad (18\text{-}6)$$

where S_S is the initial slope of the sine wave, f is its frequency, and V_p is its peak value. To avoid slew-rate distortion, S_S has to be less than S_R.

EXAMPLE 18-3

A compensating capacitor of 1000 pF has a maximum charging current of 1 mA. What is the slew rate?

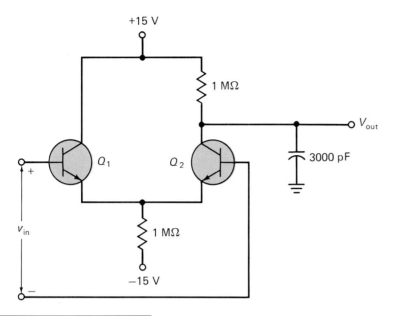

FIGURE 18-8   Example.

SOLUTION

Divide the maximum current by the capacitance for the slew rate:

$$S_R = \frac{1\,\text{mA}}{1000\,\text{pF}} = 1\,\text{V}/\mu\text{s}$$

EXAMPLE 18-4

What is the slew rate in Fig. 18-8?

SOLUTION

We have to find the maximum current before we can calculate the slew rate. With no input signal, the output voltage is 7.5 V. When a large input step drives the diff amp, Q_1 saturates and Q_2 cuts off. Visualize Q_2 open in Fig. 18-8. Then you can see that the 3000 pF will charge toward $+15$ V. Since the initial voltage is 7.5, the maximum current is

$$I_{\text{max}} = \frac{15\,\text{V} - 7.5\,\text{V}}{1\,\text{M}\Omega} = 7.5\,\mu\text{A}$$

The slew rate is

$$S_R = \frac{7.5\,\mu\text{A}}{3000\,\text{pF}} = 2.5\,\text{V/ms}$$

EXAMPLE 18-5

An op amp has a slew rate of 0.5 V/μs. If an input signal has a frequency of 8 kHz and a peak value of 5 V, will slew-rate distortion occur?

SOLUTION

With Eq. (18-6), we can calculate the initial slope of the sine wave:

$$S_S = 2\pi(8 \text{ kHz})(5 \text{ V}) = 0.251 \text{ V/μs}$$

Since the slew rate is 0.5/μs, there is no slew-rate distortion.

EXAMPLE 18-6

An op amp has a slew rate of 2 V/μs. If an input signal has a frequency of 100 kHz and a peak value of 10 V, will slew-rate distortion occur?

SOLUTION

With Eq. (18-6), the initial slope of the sine wave is

$$S_S = 2\pi(100 \text{ kHz})(10 \text{ V}) = 6.28 \text{ V/μs}$$

Since the slew rate is 2 V/μs, there is slew-rate distortion.

There are two ways to eliminate the slew-rate distortion. First, we can reduce the frequency from 100 kHz to a lower frequency. Second, we can reduce the peak value from 10 V to a lower value. In either case, we have to reduce S_S to less than 2 V/μs. Since S_S is now 6.28 V/μs, we need to reduce either f or V_p by a factor of slightly more than 3 to eliminate the slew-rate distortion.

18-3 POWER BANDWIDTH

Slew-rate distortion of a sine wave occurs when the initial slope of the sine wave is greater than the slew rate. Because a distorted output signal is not usable in most applications, slew-rate distortion is one of the major limitations on the large-signal performance of any op amp. The critical condition that separates normal operation from distorted operation is this:

$$S_S = S_R$$

This says that the initial slope of the sine wave equals the slew rate. You cannot force an op amp beyond this point without getting slew-rate distortion.

A Useful Equation

With Eq. (18-6), we can write the foregoing equation as

$$2\pi f V_p = S_R$$

Solve this equation for f to get

$$f = \frac{S_R}{2\pi V_p}$$

In this equation, f is the maximum frequency without slew-rate distortion. As a reminder, we will rewrite the equation like this:

$$f_{max} = \frac{S_R}{2\pi V_p} \qquad (18\text{-}7)$$

Above this frequency, you will begin to notice slew-rate distortion on an oscilloscope.

Frequency f_{max} is sometimes called the *power bandwidth* (also the large-signal bandwidth) of an op amp. If we try to amplify higher frequencies of the same peak value, the output voltage decreases and distortion appears. Data sheets normally include the power bandwidth of an op amp because it immediately tells you what the op amp can do with signals of high frequency and large peak value.

Tradeoff

Figure 18-9 is a graph of Eq. (18-7) for three different op amps. Suppose we are using the middle device. It has a slew rate of 5 V/μs. To get an undistorted output peak voltage of 10 V, the frequency can be no higher than 80 kHz. One way to increase the f_{max} is to accept less output voltage.

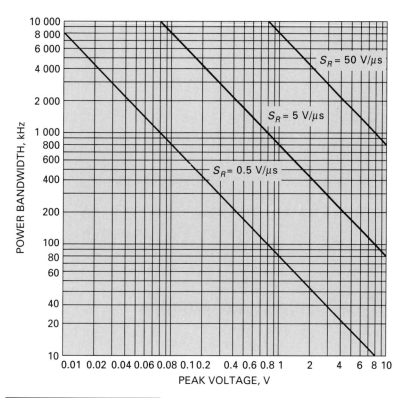

FIGURE 18-9   Graphs to trade off peak amplitude for power bandwidth.

By trading off peak value for frequency, we can improve the power bandwidth. For instance, if our application can accept a peak output voltage of 1 V, then the f_{max} of the middle device increases to 800 kHz.

The top device has a slew rate of 50 V/μs. Because of this, it has a higher f_{max} than other devices for the same peak output voltage. For instance, if we want a peak output voltage of 10 V, the power bandwidth becomes 800 kHz. If a peak output voltage of 1 V is acceptable, the power bandwidth increases to 8 MHz.

EXAMPLE 18-7

An op amp has $V_p = 10$ V and $S_R = 0.5$ V/μs. What is the power bandwidth?

SOLUTION

The maximum frequency for the undistorted large-signal operation is

$$f_{max} = \frac{0.5 \text{ V/μs}}{2\pi(10 \text{ V})} = 7.96 \text{ kHz}$$

EXAMPLE 18-8

What is the power bandwidth in the preceding example if $V_p = 1$ V?

SOLUTION

The maximum frequency for the undistorted large-signal operation is

$$f_{max} = \frac{0.5 \text{ V/μs}}{2\pi(1 \text{ V})} = 79.6 \text{ kHz}$$

This illustrates the idea of trading off some of the peak value for more frequency. By reducing the peak output voltage from 10 to 1 V, we increase the power bandwidth from 7.96 to 79.6 kHz. The bottom curve of Fig. 18-9 gives you approximately the same information.

18-4 THE OPERATIONAL AMPLIFIER

In 1965, Fairchild Semiconductor introduced the μA709, the first widely used monolithic op amp. Although successful, this first-generation op amp had many disadvantages. This led to an improved op amp known as the μA741. Because it is inexpensive and easy to use, the μA741 has been an enormous success. Other 741 designs have appeared from various manufacturers. For instance, Motorola produces the MC1741, National Semiconductor the LM741, and Texas Instruments the SN72741. All these monolithic op amps are equivalent to the μA741 because they have the same specifications on their data sheets. For convenience, most people drop the prefixes and refer to this widely used op amp simply as the 741.

The 741 has become an industry standard. As a rule, you try to use it first in your designs. In those cases where you cannot meet a design specification with a 741, you upgrade to a better op amp. Because of its great importance, we will use the 741 as a basic device in our discussions. Once you understand the 741, you can branch out to other op amps.

Incidentally, the 741 has different versions numbered 741, 741A, 741C, 741E, 741N, and so on. These differ in their voltage gain, temperature range, noise level, and other characteristics. The 741C (the C stands for commercial grade) is the least expensive and most widely used. It has an input impedance of 2 MΩ, a voltage gain of 100,000, and an output impedance of 75 Ω.

Schematic Diagram of the 741

Figure 18-10 is a simplified schematic diagram of the 741. This circuit is equivalent to the 741 and many later-generation op amps. You do not

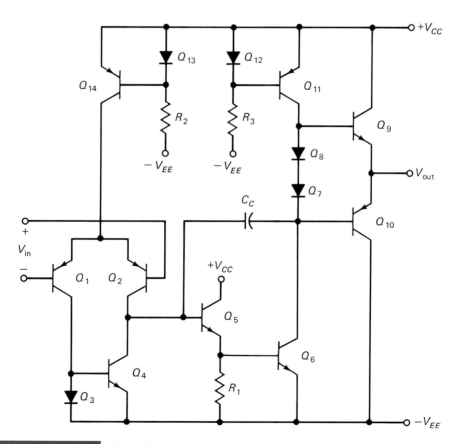

FIGURE 18-10 Simplified schematic diagram for 741 and similar op amps.

need to understand every detail about the circuit design, but you should have a general idea of how the op amp works. With that in mind, here is the basic idea behind a 741.

The input stage is a diff amp using *pnp* transistors (Q_1 and Q_2). As you know, the tail resistor is supposed to act like a current source. In the 741, Q_{14} is a current source that replaces the tail resistor. R_2 and Q_{13} control the bias on Q_{14}, which produces the tail current of the diff amp. Instead of using an ordinary resistor as the collector resistor of the diff amp, the 741 uses an active load resistor. This active load, Q_4, acts like a current source with an extremely high impedance. Because of this, the voltage gain of the diff amp is much higher than before.

The amplified signal from the diff amp drives the base of Q_5, which is an emitter follower. This stage steps up the impedance level to avoid loading down the diff amp. The signal out of Q_5 goes to Q_6. Diodes Q_7 and Q_8 are part of the biasing for the final stage. Q_{11} is an active load resistor for Q_6. Therefore, Q_6 and Q_{11} are like a CE stage with a very high voltage gain.

The amplified signal out of the CE stage goes to the final stage, which is a class B push-pull emitter follower (Q_9 and Q_{10}). Because of the split supply (equal positive and negative voltages), the quiescent output is ideally 0 V when the input voltage is zero. Any deviation from 0 V is called the output offset voltage. When there is an input voltage, V_{in}, with the polarity shown, the output voltage, V_{out}, is positive. If V_{in} has the opposite polarity from that shown in Fig. 18-10, V_{out} is negative. Ideally, V_{out} can be as positive as $+V_{CC}$ and as negative as $-V_{EE}$ before clipping occurs. To second approximation, the output swing is within 1 to 2 V of each supply voltage because of voltage drops inside the 741.

In Fig. 18-10, we have two examples of *active loading* (using transistors instead of resistors for loads). First, there is an active load, Q_4, on the diff amp. Second, there is an active load, Q_{11}, on the CE driver stage. Because current sources have high impedances, active loads produce much higher voltage gain than is possible with resistors. These active loads produce a typical voltage gain of 100,000 for the 741C.

Active loading is very popular in integrated circuits because it is easier and less expensive to fabricate transistors on a chip than it is to fabricate resistors. MOS digital integrated circuits use active loading almost exclusively. In MOS ICs, one MOSFET is the active load for another.

In Fig. 18-10, C_C is a compensating capacitor. Because of the Miller effect, this small capacitor (typically 30 pF) is multiplied by the voltage gain of Q_5 and Q_6 to get a much larger equivalent capacitance of

$$C_{in(Miller)} = (A + 1)C_C$$

where A is the voltage gain of the Q_5 and Q_6 stages. The resistance facing this Miller capacitance is the output impedance of the diff amp. Therefore, we have a bypass circuit as previously described. This bypass circuit produces the dominant critical frequency of the op amp. In other words, the voltage gain of the op amp is 0.707 of the midband voltage gain at the critical frequency of this bypass circuit. The voltage gain decreases approximately 20 dB per decade for input frequencies above the critical frequency.

DC Return Paths for Input Bases

In Fig. 18-10, notice that the input bases are floating. The op amp cannot possibly work unless each input has a base resistor R_B or an equivalent dc return path to ground. These return paths can be provided by Thevenin resistances of the circuits driving the op amp. If the driving circuits are capacitively coupled, you have to insert separate base return resistors. The key thing to remember is there must be a dc path from each input base to ground. If a base does not have a dc path to ground, its transistor goes into cutoff.

Assuming there is a dc path to ground for each base, we still have an offset problem to worry about. Because the input transistors are not quite identical, an unwanted offset voltage will exist at the output of the op amp. As discussed earlier, one way to eliminate the output offset voltage is by using a small input voltage of the correct magnitude and phase.

Input Impedance

Recall that the input impedance of a diff amp is

$$r_{in} = 2\beta r_e'$$

With a small tail current in the input diff amp, a bipolar op amp can have a fairly high input impedance. For instance, the input diff amp of a 741 has a tail current of approximately 15 μA. Since each emitter gets half of this,

$$r_e' = \frac{25\,\text{mV}}{7.5\,\mu\text{A}} = 3.33\,\text{k}\Omega$$

In a 741, each input transistor has a typical β of 300, which gives an input impedance of

$$r_{in} = 2(300)(3.3\,\text{k}\Omega) = 2\,\text{M}\Omega$$

This is the value of input impedance listed on the data sheet of a 741.

If extremely high input impedance is required, a designer may use a *BIFET op amp*. This is an op amp that combines FETs and bipolar transistors. For instance, the LF13741 is a modified 741 where JFET source followers have been added to the front end of the circuit. The output of the JFET source followers drives an ordinary 741 op amp. This combination produces input advantages of JFET source followers with the other characteristics of a 741. Because of this, LF13741 is a "drop-in replacement" for the standard 741.

Schematic Symbol

Figure 18-11a shows the schematic symbol of an op amp. A is the voltage gain. The noninverting input is v_1, and the inverting input is v_2. The differential input is

$$v_{in} = v_1 - v_2$$

NONINVERTING

INVERTING

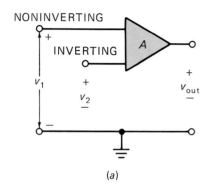

(a)

NONINVERTING

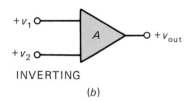

INVERTING

(b)

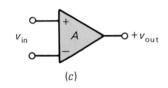

(c)

FIGURE 18-11

Symbols for op amp.

Notice voltages v_1, v_2, and v_{out} are node voltages. This means they are always measured with respect to ground. The differential input v_{in} is the difference of two node voltages, v_1 and v_2.

Most of the time, we don't bother to draw the ground line as shown in Fig. 18-11b. The main thing to remember here is that the voltages are with respect to ground, even though no ground is shown.

Figure 18-11c shows another symbol for an op amp. The noninverting input has a plus sign because no phase inversion occurs with this input. On the other hand, the inverting input has a minus sign, a reminder of the phase inversion that takes place with this input. The differential input voltage v_{in} appears between the noninverting and inverting inputs. The output voltage is given by

$$v_{out} = Av_{in} \qquad (18\text{-}8)$$

Given the input voltage, you can multiply by the voltage gain to get the output voltage.

By rearranging Eq. (18-8), we get

$$v_{in} = \frac{v_{out}}{A} \qquad (18\text{-}9)$$

This is useful because sometimes you can easily measure the output voltage but not the input voltage. In this case, you are measuring v_{out} to calculate v_{in}.

When you look at Fig. 18-11c, the most important things to remember are the input impedance, voltage gain, and output impedance. In other words, you want to use your imagination and visualize Fig. 18-12 as a rock-bottom summary for an op amp. In Fig. 18-12, z_{in} is the input impedance of the op amp, approximately 2 MΩ for a 741. As long as the op amp is operating in its linear region (unsaturated output transistors), the output of an op amp is equivalent to a Thevenin circuit, as shown in Fig. 18-12. The Thevenin output voltage is

$$v_{th} = Av_{in}$$

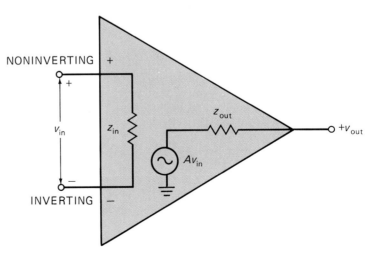

FIGURE 18-12   Input impedance and Thevenin output circuit.

For a 741C, A is 100,000 and z_{out} is 75 Ω. Because z_{out} is usually much smaller than the load resistance connected to the output of an op amp, v_{out} is approximately equal to v_{th}.

EXAMPLE 18-9

A 741C has an input voltage of 1 μV. What is the output voltage?

SOLUTION

Multiply the input voltage by the voltage gain. Since a 741C has a voltage gain of 100,000, the output voltage is

$$v_{out} = 100,000(1 \text{ μV}) = 0.1 \text{ V}$$

This answer assumes that no load resistance is connected to the op amp.

If there is a load resistance, some of the Thevenin output voltage in Fig. 18-12 will be dropped across the output impedance of the op amp. You can ignore this internal drop when the load resistance is at least 100 times greater than the output impedance. With a 741C, the output impedance is 75 Ω, so a load resistance greater than 7.5 kΩ will produce negligible loading effect.

EXAMPLE 18-10

A 741C has an output voltage of 5 V. What is the input voltage if the voltage gain is 100,000?

SOLUTION

Divide the output voltage by the voltage gain:

$$v_{in} = \frac{5 \text{ V}}{100,000} = 50 \text{ μV}$$

18-5 OP-AMP CHARACTERISTICS

Because an op amp is a dc amplifier, you have to consider both dc and ac characteristics when troubleshooting, analyzing, and designing op-amp circuits. In this section, we take a closer look at the offset problem, as well as discuss other characteristics that affect op-amp performance.

Input Offset Voltage

When the inputs of an op amp are grounded, there is almost always an output offset voltage, as shown in Fig. 18-13a, because the input transistors have different V_{BE} values. The input offset voltage is caused by the differences in V_{BE} curves. For instance, the data sheet of a typical 741C

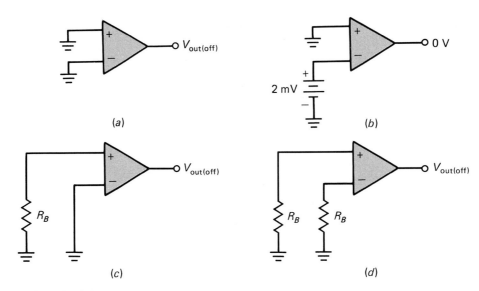

FIGURE 18-13 (a) Output offset voltage; (b) nulling the output offset voltage; (c) return resistor may produce output offset voltage; (d) equal return resistors reduce output offset voltage.

lists an input offset voltage of ± 2 mV. This difference of 2 mV is an unwanted input signal that is amplified to produce an output offset voltage. To eliminate the output offset voltage, we can apply a voltage of 2 mV to the inverting input, as shown in Fig. 18-13b. Then the output offset voltage decreases to zero. Since the offset can have either polarity, it might be necessary to reverse the polarity of the 2 mV.

Input Offset Current

Suppose we get lucky and happen to use an op amp whose input transistors have equal V_{BE} curves. Then the input offset voltage is zero. But a problem can still arise because of the bias currents. If either input to the op amp has a lot of resistance in its return path, an output offset voltage can exist. For instance, Fig. 18-13c shows a resistance R_B between the noninverting input and ground. Since there is a base current I_{B1} through R_B, a voltage appears at the noninverting input, given by

$$v_1 = I_{B1}R_B$$

Since the inverting input is grounded, $v_2 = 0$. Therefore,

$$v_{\text{in}} = I_{B1}R_B \tag{18-10}$$

This unwanted input voltage is amplified to produce an output offset voltage. If R_B is small enough, the resulting output offset voltage may be small enough to ignore.

One way to reduce the output offset voltage is to add an equal resistance to the other input as shown in Fig. 18-13. In this case,

$$v_{\text{in}} = I_{B1}R_B - I_{B2}R_B$$

or

$$v_{\text{in}} = I_{\text{in(offset)}}R_B \tag{18-11}$$

Because $I_{in(offset)}$ is much smaller than I_{B1}, the unwanted input voltage is much smaller.

In the op-amp circuits to be discussed later, the base resistors may be unequal, as shown in Fig. 18-14. Because the input offset voltage can have either polarity and because either input current may be larger than the other, the differential input voltage is given by

$$v_{in} = \pm\Delta V_{BE} + I_{B1}R_{B1} - I_{B2}R_{B2} \qquad (18\text{-}12)$$

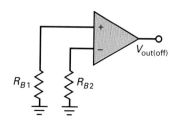

Unequal base resistors.

This equation for the total unwanted input voltage includes the effects of different V_{BE} curves and different β_{dc} values. Data sheets call the first term, ΔV_{BE}, the input offset voltage. The next two terms, $I_{B1}R_{B1}$ and $I_{B2}R_{B2}$, are calculated as discussed in the preceding chapter. Given $I_{in(bias)}$ and $I_{in(offset)}$, we can calculate I_{B1} and I_{B2}. When the two base resistances are equal as shown in Fig. 18-13d, Eq. (18-12) simplifies to

$$v_{in} = \pm\Delta V_{BE} \pm I_{in(offset)}R_B \qquad (18\text{-}13)$$

Three More Characteristics

The *common-mode rejection ratio* was defined earlier. For a 741C, CMRR′ = 90 dB at lower frequencies. Given equal signals, one a desired signal and the other a common-mode signal, the desired signal will be 90 dB larger at the output than the common-mode signal. In ordinary numbers, this means that the desired signal will be approximately 30,000 times larger than the common-mode signal. At higher frequencies, reactive effects degrade CMRR′, as shown in Fig. 18-15a. Notice that CMRR′ is approximately 75 dB at 1 kHz, 56 dB at 10 kHz, and so on.

The *MPP value* of an amplifier is the maximum peak-to-peak unclipped output that the amplifier can produce. Since the quiescent output of an op amp is ideally zero, the ac output voltage can swing positively or negatively. For load resistances that are much larger than z_{out}, the output voltage can swing almost to the supply voltages. For instance, if $V_{CC} = +15$ V and $V_{EE} = -15$ V, the MPP value with a load resistance of 10 kΩ is ideally 30 V.

In reality, the output cannot swing all the way to the value of the supply voltages because there are some small voltage drops in the final stages of the op amp. Furthermore, when the load resistance is not large compared to z_{out}, some of the amplified voltage is dropped across z_{out}, which means the final output voltage is smaller. Figure 18-15b shows the MPP versus load resistance for a 741C. Notice that the MPP is approximately 27 V for an R_L of 10 kΩ, 25 V for 1 kΩ, and 7 V for 100 Ω.

In some applications, an op amp may drive a load resistance of approximately zero. In this case, you need to know the value of *short-circuit output current*. The data sheet of a 741C lists a short-circuit output current of 25 mA. If you are using small load resistors (less than 75 Ω), don't expect to get a large output voltage because the voltage cannot be greater than the 25 mA times the load resistance.

Frequency Response

Figure 18-15c shows the small-signal frequency response of a 741C. In the midband, the voltage gain is 100,000. The 741C has a critical frequency

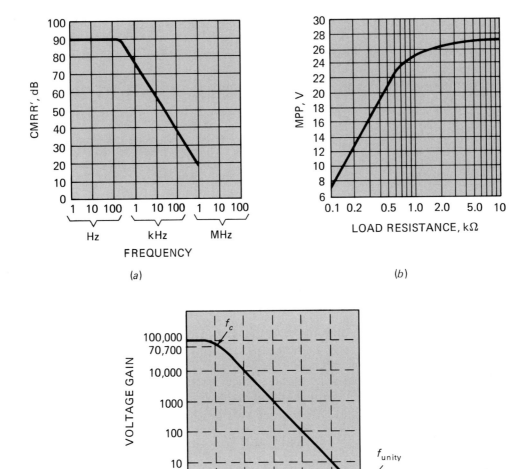

FIGURE 18-15 (*a*) Common-mode rejection ratio; (*b*) maximum peak-to-peak output; (*c*) frequency response.

f_c of 10 Hz. As indicated, the voltage gain is 70,700 (down 3 dB) at 10 Hz. Above the critical frequency, the voltage gain decreases at a rate of 20 dB per decade.

The unity-gain frequency is the frequency where the voltage gain equals 1. In Fig. 18-15*c*, f_{unity} is 1 MHz. Data sheets usually specify the value of f_{unity} because it represents the upper limit on the useful gain of an op amp. For instance, the data sheet of a 741C lists an f_{unity} of 1 MHz. This means that the 741C can amplify signals up to 1 MHz. Beyond 1 MHz, the voltage gain is less than 1 and the 741C is useless. If a designer needs a higher f_{unity}, other op amps are available. For instance, the LM318 has an f_{unity} of 15 MHz, which means it can produce usable voltage gain all the way to 15 MHz.

Slew-Rate Distortion

Because of the compensating capacitor in a 741, the output of the diff amp can change no faster than the slew rate given by

$$S_R = \frac{I_T}{C_C}$$

In a 741C, $I_T = 15$ μA and $C_C = 30$ pF. Therefore, the slew rate of a 741C is

$$S_R = \frac{15 \text{ μA}}{30 \text{ pF}} = 0.5 \text{ V/μs}$$

This is the large-signal limit of a 741C. Its output voltage can change no faster than 0.5 V/μs.

As you know, slew rate limits the high-frequency large-signal response of an op amp. If the initial slope of the amplified sine wave is greater than the slew rate of the op amp, the output is smaller than it should be and it looks triangular instead of sinusoidal. Earlier, we derived this equation for the power bandwidth:

$$f_{\max} = \frac{S_R}{2\pi V_p} \qquad (18\text{-}14)$$

This says the highest frequency without slew-rate distortion is given by the slew rate divided by 2π times the peak voltage. A useful alternative is

$$V_p = \frac{S_R}{2\pi f_{\max}} \qquad (18\text{-}15)$$

Incidentally, Fig. 18-9 on page 687 shows three different graphs of Eq. (18-14) for slew rates of 0.5 V/μs, 5 V/μs, and 50 V/μs. Since the slew rate of a 741C is 0.5 V/μs, the bottom curve of Fig. 18-9 applies to a 741C. This curve will tell you at a glance what the power bandwidth is for any output peak value. In other words, you don't have to use Eq. (18-14) if the slew rate is 0.5 V/μs (bottom curve), 5 V/μs (middle curve), or 50 V/μs (top curve). For other values of slew rate, you can always fall back on Eq. (18-14) to calculate the power bandwidth.

EXAMPLE 18-11

Figure 18-16 shows a 741C with its pin numbers. Pin 3 is the non-inverting input, pin 2 is the inverting input, pins 7 and 4 are for the supply voltages, and pin 6 is the output. A 741C has these worst-case values given on the data sheet: $\Delta V_{BE} = 2$ mV,

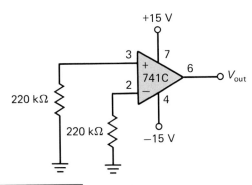

FIGURE 18-16 Example.

$I_{in(bias)} = 80$ nA, and $I_{in(offset)} = 20$ nA. What is the total unwanted input voltage in the worst case? The output offset voltage?

SOLUTION

There are two distinct components in the unwanted input voltage. First, there is the effect of different V_{BE} curves. Second, there is the difference in β_{dc} values, which translates into a difference in the two base voltages, pins 3 and 2. With Eq. (18-13),

$$v_{in} = \pm 2 \text{ mV} \pm (20 \text{ nA})(220 \text{ k}\Omega) = \pm 6.4 \text{ mV}$$

This means that the unwanted input voltage can be anywhere in the range of -6.4 mV to $+6.4$ mV. In the worst case, its magnitude can be as large as 6.4 mV.

When a 741C is operating linearly, it has a voltage gain of 100,000. Assume that the voltage gain is 100,000 and calculate the output offset voltage as follows:

$$v_{out} = 100,000(\pm 6.4 \text{ mV}) = \pm 640 \text{ V}$$

This answer is impossible, a good example of *reductio ad absurdum* (Chap. 7). Because we got an absurd answer, we have to reexamine our assumptions. We assumed that the voltage gain was 100,000. This is true only when the op amp is operating in the linear region. Since the answer is impossible, the op amp must be saturated.

Since a 741C has an MPP value of 27 V (see Fig. 18-15b), it can swing to $+13.5$ V on the positive side and to -13.5 V on the negative side. When the input voltage is $+6.4$ mV, the output of the op amp goes to 13.5 V. When the input voltage is -6.4 mV, the output goes to -13.5 V.

EXAMPLE 18-12

Use the data of the preceding example. What is the unwanted input voltage that just produces saturation of the op amp?

SOLUTION

On the positive side, the op amp can swing to $+13.5$ V before saturation occurs. The op amp has a voltage gain of 100,000 up to this level. Therefore, the input voltage that just produces saturation is

$$v_{\text{in}} = \frac{13.5 \text{ V}}{100,000} = 0.135 \text{ mV}$$

This is much smaller than the worst case of 6.4 mV. This means that if you build the circuit of Fig. 18-16, it will be saturated at least 95 times out of 100. Therefore, a circuit like Fig. 18-16 is useless in its current design. We have to do something to eliminate the severity of the output offset problem. The next chapter will tell you what the "something" is.

EXAMPLE 18-13

A 741C has a slew rate of 0.5 V/μs. What is the power bandwidth if the output voltage has a peak value of 10 V?

SOLUTION

With Eq. (18-14), we can calculate the maximum frequency without slew-rate distortion:

$$f_{\text{max}} = \frac{0.5 \text{ V/μs}}{2\pi(10 \text{ V})} = 7.96 \text{ kHz}$$

At this frequency, the op amp can produce an undistorted sinusoidal output signal with a peak value of 10 V. If you increase the frequency above 7.96 kHz, the output signal starts to shrink and becomes triangular instead of sinusoidal.

EXAMPLE 18-14

What is the largest undistorted output signal from a 741C if the frequency is 50 kHz?

SOLUTION

With Eq. (18-15),

$$V_p = \frac{0.5 \text{ V/μs}}{2\pi(50 \text{ kHz})} = 1.59 \text{ V}$$

This means the op amp is producing an undistorted output signal with a peak value of 1.59 V and a frequency of 50 kHz. If you increase the input signal in an attempt to get more output voltage, the output signal will be distorted and will look triangular instead of sinusoidal.

18-6 POPULAR OP AMPS

Table 18-1 lists some popular op amps. The LF351 through LF13741 and the TL071 through TL074 are BIFET op amps. Notice how low the input bias and offset currents are for these devices. The LM10C through NE531 are bipolar op amps. As indicated, the LM741C has a typical input offset voltage of 2 mV, an input bias current of 80 nA, an input offset current of 20 nA, and so on. If the dc return resistances on the noninverting and inverting inputs must be high, a 741C may produce too much output offset voltage. In a case like this, you can upgrade to an op amp such as the LF13741, a general-purpose BIFET op amp that is a replacement for the bipolar 741.

The table also includes maximum output current, unity-gain frequency, and slew rate. Sometimes a 741C may not slew fast enough to produce

Table 18-1 Typical Parameters of Popular Op Amps

Number	$V_{in(off)}$, mV	$I_{in(bias)}$, nA	$I_{in(off)}$, nA	$I_{out(max)}$, mA	f_{unity}, MHz	Slew rate, $V/\mu s$
LF351	5	0.05	0.025	20	4	13
LF353	5	0.05	0.025	20	4	13
LF355	3	0.03	0.003	20	2.5	5
LF356	3	0.03	0.003	20	5	12
LF13741	5	0.05	0.01	25	1	0.5
LM10C	0.5	12	0.4	20	0.1	0.12
LM11C	0.1	0.025	0.0005	2	0.5	0.3
LM301C	2	70	3	10	1	0.5
LM307	2	70	3	10	1	0.5
LM308	2	1.5	0.2	5	0.3	0.15
LM318	4	150	30	21	15	70
LM324	2	45	5	20	1	0.5
LM348	1	30	4	25	1	0.5
LM358	2	45	5	40	1	0.5
LM709	2	300	100	42	*	0.25
LM739	1	300	50	1.5	6	1
LM741C	2	80	20	25	1	0.5
LM747C	2	80	20	25	1	0.5
LM748	2	80	20	27	*	*
LM1458	1	200	80	20	1	0.5
LM4250	3–5	*	*	*	*	*
LM13080	3	*	*	250	1	*
NE 531	2	400	50	20	1	35
TL071	3	0.03	0.005	10	3	13
TL072	3	0.03	0.005	10	3	13
TL074	5	0.05	0.025	17	4	13

* Externally controlled by resistors or capacitors.

an adequate power bandwidth for your application. In this case, you can upgrade to a fast slew-rate device like the TL071, an inexpensive BIFET op amp. The ultimate weapon for slew-rate problems is the LM318; it has a slew rate of 70 V/μs.

All data are typical. For worst-case values and other specifications, you will have to refer to manufacturers' data sheets. For the devices listed in Table 18-1, CMRR is from 80 to 100 dB, and voltage gain from 100,000 to 300,000. Some of the devices are rather unusual. The LM4250, for instance, has a string of asterisks. An asterisk means that the quantity can be varied by the user. In other words, the LM4250 is programmable by a single external resistor that allows you to vary input bias and offset currents, slew rate, unity-gain frequency, and so on.

OPTIONAL TOPICS

The following material continues the earlier discussions at a more advanced and specialized level. All the topics are optional because they are not used in any of the basic discussions in later chapters. This section will be a useful reference when you are in industry because then you will probably want more advanced viewpoints.

18-7 OTHER LINEAR ICs

Although the op amp is the most important linear IC, you will encounter other linear ICs in various applications. This section briefly examines some of these ICs. Our survey covers only the main types.

Audio Amplifiers

Preamplifiers (preamps) are audio amplifiers with less than 50 mW of output power. Preamps are optimized for low noise because they are used at the front end of audio systems, where they amplify weak signals from photograph cartridges, magnetic tape heads, microphones, and so on.

An example of an IC preamp is the LM381, a low-noise dual preamplifier. Each amplifier is completely independent of the other. The LM381 has a voltage gain of 112 dB and a 10-V power bandwidth of 75 kHz. It operates from a positive supply of 9 to 40 V. Its input impedance is 100 kΩ, and its output impedance is 150 Ω. The LM381's input stage is a diff amp, which allows differential or single-ended input.

Medium-level audio amplifiers have output powers from 50 to 500 mW. These are useful near the output end of small audio systems like transistor radios or signal generators. An example is the MHC4000P, which has an output power of 250 mW.

Audio power amplifiers deliver more than 500 mW of output power. They are used in phonograph amplifiers, intercoms, AM-FM radios, and other applications. The LM380 is an example. It has a voltage gain of 34 dB, a bandwidth of 100 kHz, and an output power of 2 W. As another example, the LM2002 has a voltage gain of 40 dB, a bandwidth of 100 kHz, and an output power of 8 W.

Figure 18-17 shows a simplified schematic diagram of the LM380. The input diff amp uses *pnp* inputs. The signal can be directly coupled, which is an advantage with transducers. The diff amp drives a current-mirror load (Q_5 and Q_6). The output of the current mirror goes to an emitter follower (Q_7) and CE driver (Q_8). The output stage is a class B push-pull emitter follower (Q_{13} and Q_{14}).

There is an internal compensating capacitor of 10 pF that rolls off the decibel voltage gain at a rate of 20 dB per decade. This capacitor produces a slew rate of approximately 5 V/μs.

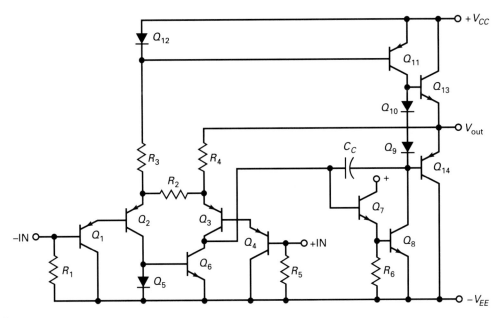

FIGURE 18-17   Simplified schematic diagram for LM380 and similar audio ICs.

Video Amplifiers

A video or wideband amplifier has a flat response (constant decibel voltage gain) over a very broad range of frequencies. Typical bandwidths are well into the megahertz region. Video amps are not necessarily dc amps, but they often do have a response that extends down to zero frequency. They are used in applications in which the range of input frequencies is very large. For instance, many oscilloscopes handle frequencies from 0 to over 10 MHz; instruments like these use video amps to increase the signal strength before applying it to the cathode-ray tube. As another example, a television receiver uses a video amp to handle frequencies from near zero to about 4 MHz.

IC video amps have voltage gains and bandwidths that you can adjust by connecting different external resistors. For instance, the μA702 has a decibel voltage gain of 40 dB and a critical frequency of 5 MHz; by changing external components, you can get useful gain to 30 MHz. The MC1553 has a decibel voltage gain of 52 dB and a bandwidth of 20 MHz;

these are adjustable by changing external components. The LM733 has a very wide bandwidth; it can be set up to give 20 dB gain and a bandwidth of 120 MHz.

RF and IF Amplifiers

A radio-frequency (RF) amplifier is usually the first stage in an AM, FM, or TV receiver. Intermediate-frequency (IF) amplifiers typically are the middle stages. ICs like the LM703 include RF and IF amplifiers on the same chip. The amplifiers are tuned (resonant) so that they amplify only a narrow band of frequencies. This allows the receiver to tune into a desired signal from a particular radio or television station. As mentioned earlier, it is impractical to integrate inductors and large capacitors on a chip. For this reason, you have to connect external Ls and Cs to the chip to get tuned amplifiers.

Voltage Regulators

Chapter 4 discussed rectifiers and power supplies. After filtering, we have a dc voltage with ripple. This dc voltage is proportional to the line voltage; that is, it will change 10 percent if the line voltage changes 10 percent. In most applications, a 10 percent change in dc voltage is too much, and so voltage regulation is necessary. Typical of the new IC voltage regulators is the LM340 series. Chips of this type can hold the output dc voltage to within 0.01 percent for normal changes in line voltage and load resistance. Other features include positive or negative output, adjustable output voltage, and short-circuit protection.

STUDY AIDS

The following study aids will help to reinforce the ideas discussed in this chapter. For best results, use these study aids within 6 hours of reading the earlier material. Then review these study aids a week later and a month later to ensure that the concepts remain in your long-term memory.

SUMMARY

Sec. 18-1 Small-Signal Frequency Response
The compensating capacitor produces a dominant critical frequency that is much lower than the critical frequencies produced by internal transistor and stray capacitances. Because of this, the voltage gain of a diff amp decreases approximately 20 dB per decade above the dominant critical frequency. Besides the critical frequency, the unity-gain frequency is important. At this frequency, the voltage gain equals one.

Sec. 18-2 Large-Signal Frequency Response
Because of the compensating capacitor, the output voltage cannot suddenly change from one voltage level to another. The maximum rate of change in voltage occurs at the beginning of the capacitor charge and is called the slew rate. When the initial slope of a sine wave is less than the slew rate, no distortion occurs. But when the initial slope is greater than the slew rate, the output signal is distorted and looks triangular instead of sinusoidal.

Sec. 18-3 Power Bandwidth
The initial slope of a sine wave is determined by two things: frequency and peak value. Because of this, maximum undistorted frequency out of an op amp will depend on the slew rate and the peak value. The power bandwidth is the highest frequency that an op amp can amplify without distortion for a specified peak voltage. By trading off peak value, you can get more power bandwidth.

Sec. 18-4 The Operational Amplifier

The 741C is a second-generation op amp that has become an industry standard. It has a voltage gain of 100,000, an input impedance of 2 MΩ, and an output impedance of 75 Ω. The device uses a diff amp for the input stage, followed by additional stages of voltage gain. The output stage is a class B push-pull emitter follower. Besides a compensating capacitor to produce the dominant critical frequency, the 741C uses transistors as active load resistors. If a higher input impedance is required, the BIFET op amp like the LF13741 is available. It uses JFET source followers on the front end to increase the impedance level.

Sec. 18-5 Op-Amp Characteristics

In the worst case, the 741C has an input offset voltage of ±2 mV. Also, the two base currents usually flow through resistors, which produces another unwanted input voltage. The 741C has an MPP value or output voltage swing of approximately 27 V when the supply voltages are ±15 V. The CMRR' is 90 dB, the short-circuit current is 25 mA, and it has a slew rate of 0.5 V/μs. Finally, it has a critical frequency of 10 Hz and a unity-gain frequency of 1 MHz.

VOCABULARY

In your own words, explain what each of the following terms means. Keep your answers short and to the point. If necessary, verify your answer by rereading the appropriate discussion or by looking at the end-of-book Glossary.

active loading	op amp
BIFET op amp	oscillations
compensating capacitor	power bandwidth
dc return	short-circuit output current
initial slope of sine wave	slew rate
input offset voltage	unity-gain frequency
MPP value	

IMPORTANT EQUATIONS

Here are some important equations. Say each of the following equations in symbols, then say each of them in words. Try to explain what the equation

means and how it is used. Then, read the description that follows.

Eq. 18-2 Critical Frequency

$$f_c = \frac{f_{\text{unity}}}{A_{\text{mid}}}$$

This is quite important and should be memorized. We will use it a lot in future chapters. It says the critical frequency equals unity-gain frequency divided by the midband voltage gain.

Eq. 18-3 Unity-Gain Frequency

$$f_{\text{unity}} = A_{\text{mid}}f_c$$

A rewrite of Eq. (18-2), it says that the unity-gain frequency of an op amp equals the midband voltage gain times the critical frequency.

Eq. 18-6 Initial Slope of Sine Wave

$$S_S = 2\pi f V_p$$

If you can remember this equation, it will tell you if slew-rate distortion will occur. The initial slope of a sine wave equals 2π times the frequency times the peak value. This initial slope must be less than the slew rate of the op amp to avoid distortion.

Eq. 18-12 Unwanted Input Voltage

$$v_{\text{in}} = \pm \Delta V_{BE} + I_{B1}R_{B1} - I_{B2}R_{B2}$$

To eliminate the output offset voltage, a designer has to use a compensating input voltage equal in magnitude and opposite in phase to the input voltage given by this equation. The first term in this equation is the input offset voltage given on a data sheet. The next two terms are voltages across base resistors. You take the worst combination to get the total unwanted input voltage.

Eq. 18-14 Power Bandwidth

$$f_{\text{max}} = \frac{S_R}{2\pi V_p}$$

An equation derived from equating the initial slope of a sine wave to the slew rate. It says that a signal with a peak value of V_p can have a frequency no higher than f_{max} if slew-rate distortion is to be avoided. The equation shows you that f_{max} is inversely proportional to the peak value. If you want a higher power bandwidth, therefore, you must accept a smaller peak value.

STUDENT ASSIGNMENTS

QUESTIONS

Some questions have more than one right answer. Select the best answer, the one that is always true, or that most accurately describes the situation.

1. What controls the dominant critical frequency of an op amp?
   a. Stray-wiring capacitance
   b. Base-emitter capacitance
   c. Collector-base capacitance
   d. Compensating capacitance

2. A compensating capacitor prevents
   a. Voltage gain
   b. Oscillations
   c. Input offset current
   d. Power bandwidth

3. At f_{unity}, the voltage gain is
   a. One
   b. R_C/r_e'
   c. A_{mid}
   d. Zero

4. The critical frequency of an op amp equals the unity-gain frequency divided by
   a. f_c
   b. A_{mid}
   c. 1
   d. A_{CM}

5. If the critical frequency is 15 Hz and the midband voltage gain is 1,000,000, the unity-gain frequency is
   a. 25 Hz
   b. 1 MHz
   c. 1.5 MHz
   d. 15 MHz

6. If the unity-gain frequency is 5 MHz and the midband voltage gain is 200,000, the critical frequency is
   a. 25 Hz
   b. 1 MHz
   c. 1.5 MHz
   d. 15 MHz

7. Slew rate equals current divided by
   a. Voltage
   b. Resistance
   c. Capacitance
   d. Power

8. The initial slope of a sine wave is directly proportional to
   a. Slew rate
   b. Frequency
   c. Voltage gain
   d. Capacitance

9. When the initial slope of a sine wave is greater than the slew rate,
   a. Distortion occurs
   b. Linear operation occurs
   c. Voltage gain is maximum
   d. The op amp works best

10. The power bandwidth increases when
    a. Frequency decreases
    b. Peak value decreases
    c. Initial slope decreases
    d. Voltage gain increases

11. A 741 uses
    a. Discrete resistors
    b. Passive loading
    c. Active loading
    d. A small coupling capacitor

12. A 741 cannot work without
    a. Discrete resistors
    b. Passive loading
    c. Dc returns on the two bases
    d. A small coupling capacitor

13. The input impedance of a BIFET op amp is
    a. Low
    b. Medium
    c. High
    d. Extremely high

14. An LF13741 is a
    a. Diff amp
    b. Source follower
    c. Bipolar op amp
    d. BIFET op amp

15. If the two supply voltages are ±15 V, the MPP value of an op amp is ideally
    a. 0
    b. +15 V
    c. −15 V
    d. 30 V

16. The dominant critical frequency of a 741 is controlled by
   a. A coupling capacitor
   b. The output short circuit current
   c. The power bandwidth
   d. A compensating capacitor

17. The 741C has a unity-gain frequency of
   a. 10 Hz
   b. 20 kHz
   c. 1 MHz
   d. 15 MHz

18. The unity-gain frequency equals the product of critical frequency and
   a. Compensating capacitance
   b. Tail current
   c. Midband voltage gain
   d. Load resistance

19. If f_{unity} = 10 MHz and A_{mid} = 1,000,000, the critical frequency of the op amp is
   a. 10 Hz
   b. 20 Hz
   c. 50 Hz
   d. 100 Hz

20. The initial slope of a sine wave increases when
   a. Frequency decreases
   b. Peak value increases
   c. π increases
   d. Slew rate decreases

21. If the frequency is greater than the power bandwidth,
   a. Slew-rate distortion occurs
   b. A normal output signal occurs
   c. Output offset voltage increases
   d. Distortion may occur

22. An op amp has an open base resistor. The output voltage will be
   a. Zero
   b. Slightly different from zero
   c. Maximum positive or negative
   d. An amplified sine wave

23. An op amp has a voltage gain of 500,000. If the output voltage is 1 V, the input voltage is
   a. 2 μV
   b. 5 mV
   c. 10 mV
   d. 1 V

24. A 741C has supply voltages of ±15 V. If the load resistance is large, the MPP value is
   a. 0
   b. +15 V
   c. 27 V
   d. 30 V

25. Above the critical frequency, the voltage gain of a 741C decreases approximately
   a. 10 dB per decade
   b. 20 dB per octave
   c. 10 dB per octave
   d. 20 dB per decade

26. The voltage gain of an op amp is one at the
   a. Critical frequency
   b. Unity-gain frequency
   c. Generator frequency
   d. Power bandwidth

27. When slew-rate distortion of a sine wave occurs, the output
   a. Is larger
   b. Appears triangular
   c. Is normal
   d. Has no offset

28. A 741C has
   a. A voltage gain of 100,000
   b. An input impedance of 2 MΩ
   c. An output impedance of 75 Ω
   d. All of the above

BASIC PROBLEMS

Sec. 18-1 Small-Signal Frequency Response

18-1. What is the critical frequency in Fig. 18-18? The unity-gain frequency?

18-2. The data sheet of an op amp gives A_{mid} = 400,000 and f_{unity} = 2 MHz. What is the critical frequency of the op amp?

18-3. An op amp has a critical frequency of 20 Hz and a midband voltage gain of 500,000. What is its unity-gain frequency?

18-4. The tail and collector resistances of Fig. 18-18 are changed to 47 kΩ. What is the critical frequency? The unity-gain frequency?

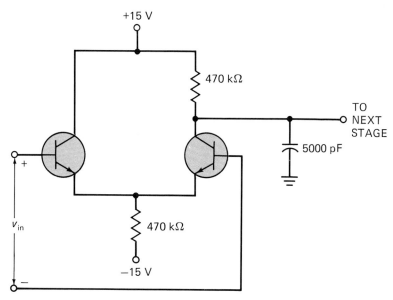

+15 V

470 kΩ

TO
NEXT
STAGE

5000 pF

v_{in}

470 kΩ

−15 V

FIGURE 18-18

18-5. A 741C has a midband voltage gain of 100,000 and a unity-gain frequency of 1 MHz. What is its critical frequency?

18-6. What is the critical frequency of an LM308 if the midband voltage gain is 300,000 and the unity-gain frequency is 0.3 MHz?

18-7. An LM318 has a midband voltage gain of 200,000 and a unity-gain frequency of 15 MHz. What is its critical frequency?

Sec. 18-2 Large-Signal Frequency Response

18-8. A compensating capacitor of 470 pF has a maximum charging current of 500 μA. What is the slew rate?

18-9. To get a slew rate of 1 V/μs with a maximum charging current of 20 μA, what size should the compensating capacitor be?

18-10. What is the slew rate in Fig. 18-18?

18-11. An op amp has a slew rate of 2.5 V/μs. If an input signal has a frequency of 20 kHz and a peak value of 5 V, will slew-rate distortion occur?

18-12. What is the initial slope of a sine wave with a frequency of 20 kHz and a peak value of 1 V? What happens to the initial slope if the peak value increases to 10 V?

18-13. What is the initial slope of a sine wave with a frequency of 15 kHz and a peak value of 2 V? What happens to the initial slope if the frequency increases to 30 kHz?

18-14. A 741C has a slew rate of 0.5 V/μs. Will slew-rate distortion occur if the frequency is 100 kHz and the peak value is 1 V?

18-15. An LM308 has a slew rate of 0.15 V/μs. Will slew-rate distortion occur if the sinusoidal signal has a frequency of 10 kHz and 2 V?

18-16. An LM318 has a slew rate of 70 V/μs. Will slew-rate distortion occur if the sinusoidal signal has a frequency of 500 kHz and 2 V?

Sec. 18-3 Power Bandwidth

18-17. An op amp has $V_p = 5$ V and $S_R = 3$ V/μs. What is the power bandwidth?

18-18. What is the power bandwidth if $V_p = 10$ V and $S_R = 1$ V/μs? What happens to the power bandwidth if V_p decreases to 1 V? If V_p decreases to 0.1 V?

18-19. What is the power bandwidth if $V_p = 5$ V and $S_R = 1$ V/μs? What happens to

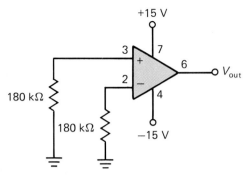

FIGURE 18-19

the power bandwidth if S_R increases to 8 V/µs?

18-20. An LF351 has a slew rate of 13 V/µs. If the peak-to-peak output voltage is 20 V, what is the highest frequency this op amp can amplify without distortion?

18-21. An LM318 has S_R = 70 V/µs. What is the power bandwidth for an output peak voltage of 10 V?

18-22. A 741C has a slew rate of 0.5 V/µs. What is the power bandwidth for each of these peak values: 0.1 V, 1 V, and 10 V?

18-23. An LM308 has a slew rate of 0.15 V/µs. What is the power bandwidth for these peak values: 0.1 V, 1 V, and 10 V?

18-24. An LM318 has a slew rate of 70 V/µs. What is the power bandwidth for these peak values: 0.1 V, 1 V, and 10 V?

Sec. 18-4 The Operational Amplifier

18-25. An op amp has a voltage gain of 500,000. If the input voltage is 12 µV, what is the output voltage?

18-26. What is the input voltage of an op amp with an output voltage of 10 V and a voltage gain of 200,000?

18-27. A 741C has a voltage gain of 100,000. What is the input voltage for these output voltages: 0.1 V, 1 V, and 10 V?

18-28. An LM318 has a voltage gain of 200,000. What is the input voltage for these output voltages: 0.1 V, 1 V, and 10 V?

Sec. 18-5 Op-Amp Characteristics

18-29. The op amp of Fig. 18-19 has ΔV_{BE} = ±4 mV, $I_{in(bias)}$ = 150 nA, and $I_{in(offset)}$ = 30 nA. What is the total unwanted input voltage in the worst case? The output offset voltage if A = 100,000?

18-30. If the op amp of Fig. 18-19 has a voltage gain of 20,000, what is the unwanted input voltage that just produces saturation?

18-31. An LF353 has a slew rate of 13 V/µs. What is the power bandwidth for these peak output voltages: 0.5 V and 5 V?

18-32. What is the largest undistorted output signal from an LF353 if the frequency is 360 kHz and the slew rate is 13 V/µs?

Sec. 18-6 Popular Op Amps

18-33. What is the unity-gain frequency for each of these op amps: LF356, LM301C, LM1458, and TL074?

18-34. If you want an op amp with minimum input offset voltage, which op amp would you select from Table 18-1?

18-35. If you want an op amp with minimum input offset current, which op amp would you select from Table 18-1?

ADVANCED PROBLEMS

18-36. The amplifier of Fig. 18-20a has a z_{in} of 2 MΩ, a z_{out} of 75 Ω, and an A of 100,000. What is the output voltage across the 100 Ω?

18-37. In Fig. 18-20a, A' equals 92 dB and z_{out} is 75 Ω. How much output voltage is there?

18-38. If A is 100,000 and z_{out} is 75 Ω in Fig. 18-20b, what is the output voltage?

18-39. In Fig. 18-20c, the input base currents are $I_{B1} = 90$ nA and $I_{B2} = 70$ nA. What does the input bias current equal? The input offset current? How much voltage is there at the inverting input? If $A = 100,000$, what does the output offset voltage equal?

18-40. In Fig. 18-20c, $v_{in(off)} = 0$, $I_{in(bias)} = 80$ nA, and $I_{in(off)} = 20$ nA. If $A = 100,000$, what is the maximum output offset voltage?

18-41. $I_{in(off)} = 20$ nA in Fig. 18-20d. If $\Delta V_{BE} = 0$, what is the differential input voltage? If $A = 100,000$, what does the output offset voltage equal?

18-42. Refer to Fig. 18-15 to answer these questions:
a. What is the CMRR' of a 741 C at 100 kHz?
b. What is the MPP value when the load resistance is 500 Ω?

c. What is the voltage gain of a 741 C at 1 kHz?

SOFTWARE ENGINE PROBLEMS

Use Fig. 18-21 for the remaining problems. Assume increases of approximately 10 percent in the independent variable. A response should be an N (no change) if the change in a dependent variable is so small that you would have difficulty measuring it. A circuit like this is impractical for mass production because the large voltage gain of the 741C will amplify the input offset voltage and drive the output into positive or negative saturation. In this exercise, however, assume that we have hand-selected a 741C to get a zero output offset voltage.

18-43. Try to predict the response of each dependent variable in the box labeled I_{B1}. Check your answers. Then, answer the following question as simply and directly as possible. What effect does an increase in I_{B1} have on the dependent variables of the circuit?

18-44. Predict the response of each dependent variable in the box labeled I_{B2}. Check your answers. Then summarize your findings in one or two sentences.

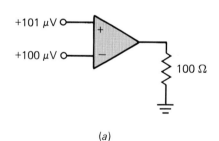

(a)

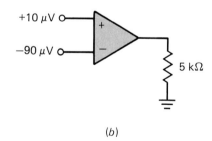

(b)

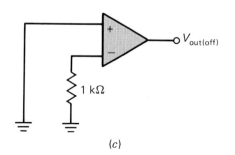

(c)

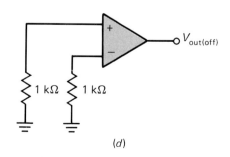

(d)

FIGURE 18-20

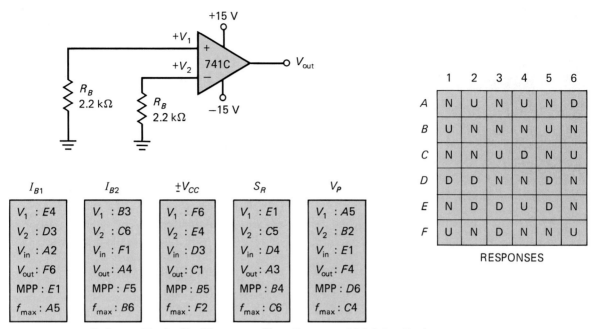

FIGURE 18-21 Software Engine™. (*Patent pending: Courtesy of Malvino Inc.*)

18-45. Predict the response of each dependent variable in the box labeled $\pm V_{CC}$. Check your answers. List the dependent variables that increase. Explain why these variables increase.

18-46. Predict the response of each dependent variable in the box labeled S_R. List the dependent variables that increase. Explain why these variables increase.

18-47. Predict the response of each dependent variable in the box labeled V_p. List the dependent variables that decrease. Explain why they show a decrease.

OP-AMP NEGATIVE FEEDBACK

Sometimes you have a great idea that others don't understand and some even ridicule. It happened to the Wright brothers, it happened to Marconi, and it happened to H. S. Black. When he tried to patent a *negative-feedback* amplifier in 1928, the U. S. Patent Office classified his idea as "another one of those perpetual-motion follies." As it turned out, negative feedback has become one of the most valuable ideas ever discovered in electronics.

In a negative-feedback amplifier, the output is sampled and part of it is returned to the input. Negative feedback means that the returning signal has a phase that opposes the input signal. The negative-feedback signal produces remarkable changes in circuit performance. The advantages of negative feedback are stable gain, less distortion, and higher frequency response.

19-1 NONINVERTING VOLTAGE FEEDBACK

There are four types of negative feedback. The most basic type is *noninverting voltage feedback*. With this type of feedback, the input signal drives the noninverting input of an amplifier. A fraction of the output voltage is then fed back to the inverting input. An amplifier with noninverting voltage feedback tends to act like an ideal voltage amplifier, one with infinite input impedance, zero output impedance, and constant voltage gain.

Error Voltage

Figure 19-1 shows an op amp connected to external resistors. Notice that the output voltage is being *sampled* by a voltage divider. Therefore, a voltage is being fed back to the inverting input of the op amp. This feedback voltage is proportional to the output voltage because it equals

$$v_2 = \frac{v_{\text{out}}}{R_1 + R_2} R_2 \qquad (19\text{-}1)$$

The values of R_1 and R_2 determine the size of the feedback voltage, which must have a value between 0 and v_{out}.

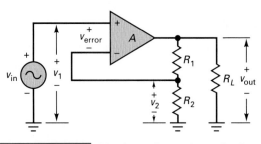

FIGURE 19-1   Noninverting voltage feedback.

In this type of negative-feedback amplifier, the input voltage is applied to the noninverting input of the op amp, so that

$$v_1 = v_{in}$$

To avoid confusion, the differential input to the op amp is renamed the *error voltage*. In symbols,

$$v_{error} = v_1 - v_2 \qquad (19\text{-}2)$$

The op amp amplifies this error voltage as previously described to get an output voltage of

$$v_{out} = Av_{error} \qquad (19\text{-}3)$$

Typically, A is very large and v_{error} is very small. For normal operation, v_{out} must be less than half the MPP value of the op amp. Stated another way, v_{out} must be smaller than either supply voltage (less a volt or two).

Stable Voltage Gain

In Fig. 19-1, there are two voltage gains. First, there is the voltage gain given by v_{out}/v_{in}, the output voltage divided by the generator voltage. Second, there is the voltage gain given by v_{out}/v_{error}, the output voltage divided by the differential input voltage. Usually, we are interested in the first voltage gain, v_{out}/v_{in}, because it tells us how much amplification the total input signal receives. To keep the two gains distinct, we will call v_{out}/v_{in} the voltage gain, and v_{out}/v_{error} the differential voltage gain.

The voltage gain of a negative-feedback amplifier is approximately constant, even though the differential voltage gain may change. This is one of the major advantages of a negative-feedback amplifier. Why is the voltage gain constant? Suppose the differential voltage gain, A, increases for some reason (temperature change, op-amp replacement, etc.). Then the output voltage will try to increase. This means that more voltage is fed back to the inverting input, causing the error voltage to decrease. This decrease in error voltage almost completely offsets the attempted increase in output voltage.

A similar argument applies to a decrease in differential voltage gain. If A decreases, the output voltage tries to decrease. In turn, the feedback voltage decreases causing v_{error} to increase. The increase in error voltage almost completely eliminates the attempted decrease in A.

Mathematical Analysis

When the input voltage to a negative-feedback amplifier is constant, a fraction of any change in output voltage is fed back to the input, producing an error voltage that opposes the attempted change in output voltage. Let's pin this idea down mathematically to see how effective the negative feedback really is.

Most op amps have a very large voltage gain A, a very high input impedance z_{in}, a very low output impedance z_{out}. For instance, the 741C has typical values of $A = 100,000$, $z_{in} = 2$ MΩ, and $z_{out} = 75$ Ω. In Fig. 19-1, the voltage divider returns a fraction of output voltage to the inverting input. If the voltage divider appears stiff to the inverting input, the feedback voltage is

$$v_2 = \frac{v_{out}}{R_1 + R_2} R_2$$

or as it is often written,

$$v_2 = \frac{R_2}{R_1 + R_2} v_{out} \tag{19-4}$$

This is usually written as

$$v_2 = Bv_{out}$$

where B is the fraction of output voltage fed back to the input. In symbols,

$$B = \frac{R_2}{R_1 + R_2} \tag{19-5}$$

The error voltage to the amplifier is

$$v_{error} = v_1 - v_2$$

Since $v_1 = v_{in}$ and $v_2 = Bv_{out}$, we can write

$$v_{error} = v_{in} - Bv_{out}$$

This error voltage is amplified to get an output voltage of approximately

$$v_{out} = A(v_{in} - Bv_{out})$$

By rearranging,

$$\frac{v_{out}}{v_{in}} = \frac{A}{1 + AB} \tag{19-6}$$

This famous formula shows you exactly what effect negative feedback has on the amplifier. Here you can see that the voltage gain with negative feedback is less than the differential voltage gain of the op amp. The fraction B is the key to how much effect the negative feedback has. When B is very small, the negative feedback is small and the voltage gain approaches A. But when B is large, the negative feedback is large and the voltage gain is much smaller than A.

Ideal Voltage Gain

The product AB is called the *loop gain* because it represents the voltage gain going all the way around the circuit, from input to output and back to input. A is the differential voltage gain of the op amp. B is the voltage gain of the voltage divider. The product AB is the combined voltage gain of the op amp and the voltage divider.

For noninverting voltage feedback to be effective, the designer must deliberately make the loop gain AB much greater than 1. When this condition is satisfied, Eq. (19-6) reduces to

$$\frac{v_{\text{out}}}{v_{\text{in}}} = \frac{1}{B} \tag{19-7}$$

This equation is deceptively simple, but enormously important. The equation says that the voltage gain equals the reciprocal of B, the feedback fraction. Since the feedback fraction depends only on resistances R_1 and R_2, the voltage gain no longer depends on the exact value of A. Since A does not appear in this equation, it can change with temperature or op amp replacement without affecting the voltage gain. Sounds incredible, doesn't it?

Remember the crucial requirement for Eq. (19-7) to be valid: *AB must be much greater than 1*. When a designer satisfies this condition, he or she knows that the voltage gain of the negative-feedback amplifier depends only on the values of the feedback resistors. If precision resistors with a tolerance of ± 1 percent are used, the voltage gain will be predictable to within ± 2 percent.

Ideal Approximation

Here is a simple way to remember Eq. (19-7). If it could be built, an ideal op amp would have infinite differential voltage gain, infinite input impedance, and zero output impedance. The IC op amps discussed in the preceding chapters approach ideal op amps because they do have extremely high differential voltage gain, high input impedance, and low output impedance. To get an ideal approximation of what is going on, assume the differential voltage gain is infinite. The error voltage is

$$v_{\text{error}} = \frac{v_{\text{out}}}{\infty} = 0$$

where ∞ represents a differential voltage gain approaching infinity. But an error voltage of zero means that

$$v_1 - v_2 = 0$$

or

$$v_1 = v_2 \tag{19-8}$$

This says that ideally the noninverting input voltage equals the inverting input voltage. Since $v_1 = v_{\text{in}}$ and $v_2 = Bv_{\text{out}}$, we can write

$$v_{\text{in}} = Bv_{\text{out}}$$

or

$$\frac{v_{\text{out}}}{v_{\text{in}}} = \frac{1}{B}$$

This is the same result as before, but derived with a lot less work.

The main idea to remember is this. Ideally, the noninverting and inverting input voltages are equal. Remember this idea. It will save you a lot of time when troubleshooting and analyzing op amp circuits. For this ideal approximation to be valid, the op amp must operate in its linear region. If the op amp is saturated on either the positive or negative side, the two inputs will no longer be equal. With this one caution in mind, you will see how this ideal approximation will simplify the later discussions of different kinds of op-amp circuits.

EXAMPLE 19-1

If the 741C of Fig. 19-2 has a differential voltage gain of 100,000, what is the voltage gain?

SOLUTION

The voltage divider has a feedback fraction of

$$B = \frac{2\text{ k}\Omega}{100\text{ k}\Omega} = 0.02$$

The loop gain is

$$AB = (100{,}000)(0.02) = 2000$$

This is much greater than 1, so we can use $1/B$ as the approximate voltage gain:

$$\frac{v_{\text{out}}}{v_{\text{in}}} = \frac{1}{0.02} = 50$$

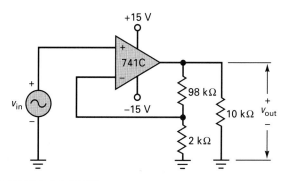

FIGURE 19-2   Example.

What is the exact answer? With Eq. (19-6),

$$\frac{v_{\text{out}}}{v_{\text{in}}} = \frac{100{,}000}{1 + 100{,}000(0.02)} = 49.975$$

Look how close this is to 50. If we rounded this off to three places as we usually do, we would get 50. The point is that $1/B$ is a very accurate approximation for the voltage gain of an op amp that uses noninverting voltage feedback.

EXAMPLE 19-2

If $v_{\text{in}} = 1$ mV, what do the output and error voltages equal in the preceding example?

SOLUTION

If $v_{\text{in}} = 1$ mV, the output voltage is

$$v_{\text{out}} = 50(1 \text{ mV}) = 50 \text{ mV}$$

The error voltage is

$$v_{\text{error}} = \frac{50 \text{ mV}}{100{,}000} = 0.5 \text{ μV}$$

Notice how small the error voltage is. This is typical of op-amp feedback amplifiers because the differential voltage gain is quite high. Ideally, v_{error} is zero because the differential voltage gain approaches infinity.

EXAMPLE 19-3

Suppose the 741C of Fig. 19-2 is replaced by another 741C that has a voltage gain of only 20,000 (worst-case value on the data sheet). What is the new voltage gain?

SOLUTION

The feedback fraction is unchanged. It still equals 0.02 as before. But the loop gain decreases to

$$AB = (20{,}000)(0.02) = 400$$

This loop gain is still much greater than 1. Therefore, the ideal voltage gain $1/B$ still gives an answer of 50. Ideally, a drop in differential voltage gain from 100,000 to 20,000 has no effect on the overall voltage gain.

If you don't believe it, then use Eq. (19-6) which gives exact answers:

$$\frac{v_{\text{out}}}{v_{\text{in}}} = \frac{20{,}000}{1 + 20{,}000(0.02)} = 49.875$$

The closed-loop gain is still extremely close to 50, despite the huge drop in differential voltage gain.

Now, you can see why negative feedback is useful, why the U. S. Patent Office was crazy for refusing H. S. Black's patent application. Without negative feedback, the overall voltage gain drops from 100,000 to 20,000, a decrease of 80 percent. With negative feedback, the voltage gain changes from 49.975 to 49.875, a decrease of only 0.2 percent. We have a lot less voltage gain, but what we do have is ultrastable.

In most applications, a designer will gladly trade off excess voltage gain to improve the stability of the voltage gain. Mass production requires circuits whose performance is predictable and repeatable. Negative feedback is perfect for mass production because you can predict with great accuracy what the voltage gain will be, no matter how many thousands of op amps are used, who manufactured them, or what the temperature is. The voltage gain becomes as stable as the resistors used in the voltage divider. This is why negative feedback is widely used with op amps.

EXAMPLE 19-4

What is the error voltage in the preceding example?

SOLUTION

Since the voltage gain is 50 to a close approximation, the output voltage is

$$v_{\text{out}} = 50(1 \text{ mV}) = 50 \text{ mV}$$

The error voltage changes to

$$v_{\text{error}} = \frac{50 \text{ mV}}{20{,}000} = 2.5 \text{ }\mu\text{V}$$

This is the error voltage when $A = 20{,}000$.

As shown in Example 19-2, the error voltage is 0.5 μV when A is 100,000. Therefore, the error voltage increases from 0.5 μV to 2.5 μV when A decreases from 100,000 to 20,000. Do you understand what this means? Think about it for awhile before you read what comes next.

When we replace op amps, the differential gain decreases from 100,000 to 20,000, a factor of 5. At the same time, the error voltage increases from 0.5 μV to 2.5 μV, also a factor of 5. Since the changes are in opposite directions, the output voltage remains at 50 mV. This echoes the earlier explanation. Negative feedback in an op amp automatically changes the error voltage as needed to hold the output voltage constant.

19-2 OPEN-LOOP AND CLOSED-LOOP VOLTAGE GAINS

Data sheets define the *open-loop voltage gain* A_{OL} as the ratio v_{out}/v_{in} with the feedback path opened, as shown in Fig. 19-3. When calculating A_{OL}, the impedances on each terminal must not be disturbed. This is why the inverting input terminal is returned to ground through an equivalent resistance of

$$R_B = R_1 \| R_2$$

and why the output terminal is loaded by an equivalent resistance of

$$R_L' = (R_1 + R_2) \| R_L$$

Usually, R_L' is much greater than the output impedance of the amplifier, so that the open-loop voltage gain A_{OL} is approximately equal to the differential voltage gain A. With a 741C, this means that the open-loop voltage gain is typically 100,000.

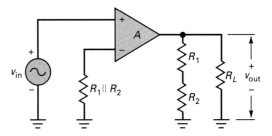

Open-loop connection.

The closed-loop voltage gain is the voltage gain when the feedback path is closed. Equation (19-6) is sometimes written as

$$A_{CL} = \frac{A_{OL}}{1 + A_{OL}B} \tag{19-9}$$

In most feedback amplifiers, the loop gain $A_{OL}B$ is much greater than 1, and Eq. (19-9) simplifies to

$$A_{CL} = \frac{1}{B} \tag{19-10}$$

Since $B = R_2/(R_1 + R_2)$, an alternative form is

$$A_{CL} = \frac{R_1 + R_2}{R_2}$$

which is often written as

$$A_{CL} = \frac{R_1}{R_2} + 1 \tag{19-11}$$

The closed-loop voltage gain is the same as v_{out}/v_{in} discussed in the preceding section. Likewise, the open-loop voltage gain is the same as the differential voltage gain A. The only reason for introducing the terms *closed-loop voltage gain* and *open-loop voltage gain* is because these terms are normally listed on a data sheet. From now on, we will use these terms, plus their symbols A_{CL} and A_{OL}.

After you have a value for the closed-loop voltage gain, you can calculate the output voltage with this formula:

$$v_{out} = A_{CL}v_{in}$$

But this is not the only way to find the output voltage. There is another way based on common sense and Ohm's law that you can read about in Example 19-5. Remember, the more ways you can solve a problem, the better. If you have several methods at your disposal, you will be in a better position to analyze new circuits that you've never seen before.

EXAMPLE 19-5

What is the output voltage in Fig. 19-4?

SOLUTION

One way to get a right answer is to work out the feedback fraction and take the reciprocal. The calculations look like this:

$$B = \frac{1\ k\Omega}{220\ k\Omega + 1\ k\Omega} = 0.00452$$

and

$$A_{CL} = \frac{1}{0.00452} = 221$$

Then the output voltage is

$$v_{out} = 221(1\ mV)$$

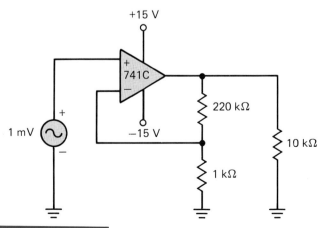

FIGURE 19-4   Example.

Another way is to use Eq. (19-11). Take the ratio of the two resistances and add 1 to get

$$A_{CL} = \frac{220 \text{ k}\Omega}{1 \text{ k}\Omega} + 1 = 221$$

The output voltage is

$$v_{\text{out}} = 221 \ (1 \text{ mV}) = 221 \text{ mV}$$

When the resistance ratio R_1/R_2 is large, you can estimate the closed-loop voltage gain by R_1/R_2. Looking at Fig. 19-4, a troubleshooter would estimate the closed-loop voltage gain as 220 and the output voltage as 220 mV.

Here is a third way to get the same answer. It uses common sense and Ohm's law. In Fig. 19-4, the 1 mV is applied between the noninverting input and ground. Since the error voltage approaches zero, the voltage between the inverting input and ground is approximately equal to 1 mV. This means 1 mV is across the 1 kΩ. Therefore, we can calculate the current through the voltage divider with Ohm's law:

$$i = \frac{1 \text{ mV}}{1 \text{ k}\Omega} = 1 \text{ } \mu\text{A}$$

This same current flows through the 220 kΩ and produces a voltage of

$$v = (1 \text{ } \mu\text{A})(220 \text{ k}\Omega) = 220 \text{ mV}$$

The output voltage equals the voltage across the voltage divider, so

$$v_{\text{out}} = 220 \text{ mV} + 1 \text{ mV} = 221 \text{ mV}$$

19-3 INPUT AND OUTPUT IMPEDANCES

Figure 19-5 shows an amplifier with noninverting voltage feedback. The op amp has an open-loop input impedance of approximately z_{in}. The overall amplifier has a closed-looped input impedance of $z_{\text{in}(CL)}$. The closed-loop impedance $z_{\text{in}(CL)}$ is larger than the open-loop impedance z_{in}.

How much larger is the closed-loop input impedance? To find out, we have to derive an expression for $v_{\text{in}}/i_{\text{in}}$. In Fig. 19-5,

$$v_{\text{in}} = v_{\text{error}} + Bv_{\text{out}}$$

or

$$v_{\text{in}} = v_{\text{error}} + ABv_{\text{error}} = (1 + AB)v_{\text{error}}$$

Because $v_{\text{error}} = i_{\text{in}}z_{\text{in}}$,

$$v_{\text{in}} = (1 + AB)i_{\text{in}}z_{\text{in}}$$

or

$$\frac{v_{\text{in}}}{i_{\text{in}}} = (1 + AB)z_{\text{in}}$$

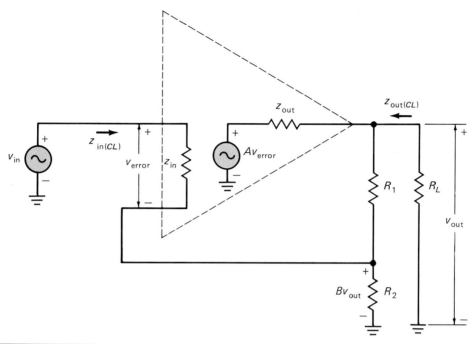

FIGURE 19-5   Input and output impedances.

The ratio of v_{in}/i_{in} is the input impedance seen by the source. Therefore, we can write

$$z_{in(CL)} = (1 + AB)z_{in} \qquad (19\text{-}12)$$

This is the closed-loop input impedance of the amplifier. In most feedback amplifiers, AB is much larger than 1, which means that the $z_{in(CL)}$ is much larger than z_{in}. The noninverting voltage feedback with op amps produces input impedances that approach infinity.

The op amp of Fig. 19-5 has an open-loop output impedance of z_{out}. The overall amplifier, however, has a closed-loop impedance $z_{out(CL)}$. The closed-loop output impedance is lower than the open-loop output impedance. Why? In Fig. 19-5, the op-amp output is equivalent to a Thevenin voltage of Av_{error} and an output impedance of z_{out}. If R_L decreases, more output current flows, producing a larger internal voltage drop across z_{out}. This implies that v_{out} will try to decrease. Since less voltage is fed back to the input, v_{error} increases. This produces a larger Thevenin output voltage, which almost completely offsets the additional voltage drop across z_{out}. The effect is equivalent to decreasing the output impedance of the feedback amplifier.

The following equation can be derived for the closed-loop output impedance:

$$z_{out(CL)} = \frac{z_{out}}{1 + AB} \qquad (19\text{-}13)$$

When the loop gain is much greater than unity, $z_{out(CL)}$ is much smaller than z_{out}. In fact, noninverting voltage feedback with op amps results in output impedances that approach zero, the ideal case for a voltage amplifier.

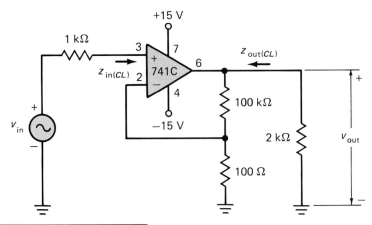

FIGURE 19-6   Example.

EXAMPLE 19-6

Figure 19-6 shows a 741C with pin numbers. If $A = 100,000$, $z_{in} = 2\ M\Omega$, and $z_{out} = 75\ \Omega$, what are the closed-loop input and output impedances?

SOLUTION

The feedback fraction is

$$B = \frac{100}{100,100} = 0.000999$$

The loop gain is

$$AB = (100,000)(0.000999) = 99.9$$

The quantity $1 + AB$ has a value of

$$1 + AB = 1 + 99.9 = 101$$

The quantity $1 + AB$ appears in Eqs. (19-12) and (19-13). With Eq. (19-12), the closed-loop input impedance is

$$z_{in(CL)} = (101)(2\ M\Omega) = 202\ M\Omega$$

With Eq. (19-13), the closed-loop output impedance is

$$z_{out(CL)} = \frac{75\ \Omega}{101} = 0.743\ \Omega$$

As you can see, closed-loop input impedance approaches infinity and the closed-loop output impedance approaches zero.

19-4 OTHER BENEFITS OF NEGATIVE FEEDBACK

Negative feedback has a way of improving almost everything. You have already seen it stabilize the voltage gain, increase the input impedance, and decrease the output impedance. This section will show you how negative feedback reduces distortion and output offset voltage.

Distortion

The final stage of an op amp has *nonlinear distortion* when the signal swings over most of the ac load line. Large swings in current cause the r_e' of a transistor to change during the cycle. This means the open-loop voltage gain varies throughout the cycle because r_e' is changing. It is this changing voltage gain that is a source of the nonlinear distortion.

Noninverting voltage feedback reduces nonlinear distortion because the feedback stabilizes the closed-loop voltage gain, making it almost independent of the changes in open-loop voltage gain. As long as the loop gain is much greater than 1, the output voltage equals $1/B$ times the input voltage. This implies that the output will be a faithful reproduction of the input. And this is exactly what happens when we use noninverting voltage feedback.

Figure 19-7 shows why distortion is reduced. Under large-signal conditions, the op amp produces a distortion voltage, designated v_{dist}. We can visualize v_{dist} as a new voltage source in series with the original source Av_{error}. Without negative feedback, all the distortion voltage v_{dist} would appear at the output. But with negative feedback, a fraction of the distortion voltage is fed back to the inverting input. This is amplified and arrives at the output with inverted phase, almost completely canceling the original distortion produced by the output stage.

How much improvement is there? In Fig. 19-7, the output voltage is

$$v_{out} = Av_{error} + v_{dist}$$

Since $v_{error} = v_{in} - Bv_{out}$, we can write

$$v_{out} = A(v_{in} - Bv_{out}) + v_{dist}$$

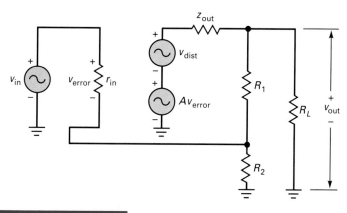

| FIGURE 19-7 | Nonlinear distortion is reduced.

Solving for v_{out} gives

$$v_{\text{out}} = \frac{A}{1 + AB} v_{\text{in}} + \frac{v_{\text{dist}}}{1 + AB}$$

The first term is what we want because it represents the amplified input voltage. The second term is the closed-loop distortion that appears at the final output. This second term can be written as

$$v_{\text{dist}(CL)} = \frac{v_{\text{dist}}}{1 + AB} \tag{19-14}$$

When the loop gain is much greater than 1, the closed-loop distortion is much smaller than the open-loop distortion.

Reduced Output Offset Voltage

In Chap. 17, we saw that an output offset voltage can exist even though the input voltage is zero. Figure 19-8 shows a feedback amplifier with an output offset voltage in series with the original source Av_{error}. This new voltage source represents the output offset voltage without feedback. The actual output offset voltage with negative feedback is much smaller.

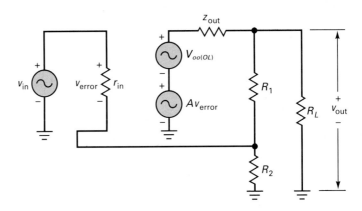

FIGURE 19-8   Output offset voltage is reduced.

Why? The reasoning is similar to that given for distortion. Some of the output offset voltage is fed back to the inverting input. After amplification, an out-of-phase voltage arrives at the output, canceling most of the original output offset voltage. It can be proved that

$$V_{oo(CL)} = \frac{V_{oo(OL)}}{1 + AB} \tag{19-15}$$

where $V_{oo(CL)}$ and $V_{oo(OL)}$ are the closed-loop and open-loop output offset voltages. When the loop gain is much greater than 1, the closed-loop output offset voltage is much smaller than the open-loop output offset voltage.

Desensitivity

The closed-loop voltage gain with noninverting voltage feedback is

$$A_{CL} = \frac{A}{1 + AB} \tag{19-16}$$

The quantity $1 + AB$ has appeared in many equations. It deserves a name. Some people call it the *desensitivity* of a feedback amplifier because it indicates how much the voltage gain is reduced by the negative feedback. For instance, if $A = 100,000$ and $B = 0.02$, then

$$1 + AB = 1 + 100,000(0.02) = 2001$$

The desensitivity is 2001, meaning the voltage gain is reduced by a factor of 2001:

$$A_{CL} = \frac{100,000}{2001} = 50$$

We can rearrange Eq. (19-16) to get

$$1 + AB = \frac{A}{A_{CL}} \tag{19-17}$$

This says that desensitivity equals the ratio of open-loop voltage gain to closed-loop voltage gain. For instance, if $A = 100,000$ and $A_{CL} = 250$, the desensitivity is

$$1 + AB = \frac{100,000}{250} = 400$$

Equation (19-17) is convenient to use when you know the values of A and A_{CL}, but not B.

The quantity $1 + AB$ is also known as the *sacrifice factor*. It represents how much voltage gain you have sacrificed to improve other qualities of the amplifier. For instance, if the sacrifice factor is 1000, it means you will improve the input impedance, output impedance, distortion, and output offset voltage by a factor of 1000.

Table 19-1 summarizes the effects of noninverting voltage feedback. As you can see, the desensitivity (sacrifice factor) appears in most of the formulas. This is why it is important to remember how to calculate desensitivity. You can calculate its value either with $1 + AB$ or A/A_{CL}.

Table 19-1 Noninverting Voltage Feedback

Quantity	Symbol	Effect	Formula
Voltage gain	A_{CL}	Decreases	$1/B$
Input impedance	$z_{in(CL)}$	Increases	$(1 + AB)z_{in}$
Output impedance	$z_{out(CL)}$	Decreases	$z_{out}/(1 + AB)$
Distortion	$v_{dist(CL)}$	Decreases	$v_{dist}/(1 + AB)$
Output offset	$V_{oo(CL)}$	Decreases	$V_{oo(OL)}/(1 + AB)$

Study this table carefully because it is very important. Noninverting voltage feedback is one of the most important types of negative feedback. With this type of negative feedback, the input impedance increases, the output impedance decreases, the distortion decreases, and the output offset voltage decreases.

EXAMPLE 19-7

The 741C of Fig. 19-9 has $I_{in(bias)} = 80$ nA, $I_{in(off)} = 20$ nA, and $v_{in(off)} = 2$ mV. What is the output offset voltage?

SOLUTION

As shown in Example 17-7, the two base currents may be anywhere in the following ranges:

$$70 \text{ nA} < I_{B1} < 90 \text{ nA}$$

$$70 \text{ nA} < I_{B2} < 90 \text{ nA}$$

Base current I_{B1} flows through a resistance of 1 kΩ, and base current I_{B2} flows through a resistance of approximately 100 Ω. The worst combination is 90 nA through the 1 kΩ and 70 nA through the 100 Ω, because this produces maximum output offset voltage. With Eq. (17-15), the maximum input offset voltage is

$$v_1 - v_2 = 2 \text{ mV} + (90 \text{ nA})(1 \text{ k}\Omega) - (70 \text{ nA})(100 \text{ }\Omega) = 2.08 \text{ mV}$$

Example 19-6 calculated a desensitivity of 101 for this circuit. With Eq. (19-15), the closed-loop output offset voltage is

$$V_{oo(CL)} = \frac{100,000(2.08 \text{ mV})}{101} = 2.06 \text{ V}$$

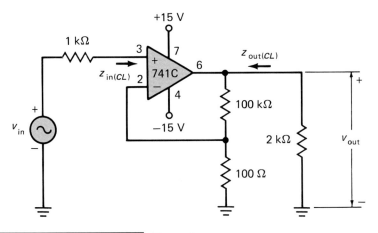

FIGURE 19-9   Example.

With supply voltages of ± 15 V, the output voltage swing of a 741C is ± 13.5 V. With an output offset of approximately 2 V, the available positive swing is from $+2$ V to $+13.5$ V. This may be acceptable in some applications.

If necessary, a designer can reduce the closed-loop output offset voltage in three ways. First, he can reduce the closed-loop voltage gain to 100 (done by changing feedback resistors). Then the desensitivity increases to

$$1 + AB = \frac{A}{A_{CL}} = \frac{100,000}{100} = 1000$$

and the closed-loop output offset voltage drops to approximately

$$V_{oo(CL)} = \frac{100,000(2.08 \text{ mV})}{1000} = 0.208 \text{ V}$$

The second option is to upgrade to a better op amp, like the LM11C (see Table 18-1). It has an input offset voltage of 0.1 mV, an input bias current of 25 pA, and an input offset current of 0.5 pA. Because the LM11C has much smaller input offsets, the output offset voltage would be lower.

The third alternative is described on the data sheet of a 741C. It involves connecting a 10-kΩ potentiometer between pins 1 and 5 with the wiper tied to the negative supply voltage, as shown in Fig. 19-10. By adjusting this potentiometer, we can null or zero the output offset voltage.

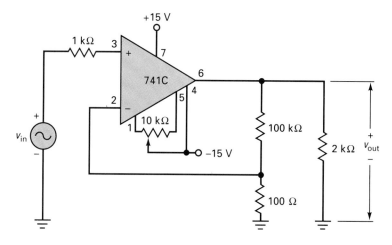

FIGURE 19-10   Nulling the output offset voltage.

Without negative feedback, the typical op amp saturates immediately because the input offset voltage multiplied by the open-loop gain drives the output stage into saturation. This is important to remember. Monolithic op amps are intended to be used with some form of feedback. Running wide open, they have too much voltage gain to be of any use because they hang up in the saturation region.

A final point. The basic equation

$$V_{oo(CL)} = \frac{V_{oo(OL)}}{1 + AB}$$

can be rewritten as

$$V_{oo(CL)} = A_{CL}V_{in(off)}$$

where $V_{in(off)}$ includes all input offset voltages, including those caused by input bias and offset currents.

EXAMPLE 19-8

Figure 19-11 shows a circuit called a *voltage follower*. What is its closed-loop voltage gain? Its closed-loop input and output impedances? The output offset voltage? Use typical 741C parameters: $A_{OL} = 100,000$, $z_{in} = 2$ MΩ, $z_{out} = 75$ Ω, $\Delta V_{BE} = 2$ mV, $I_{in(bias)} = 80$ nA, and $I_{in(off)} = 20$ nA.

SOLUTION

All of the output voltage is fed back to the inverting input because R_1 is zero and R_2 is infinite. Therefore, the feedback fraction is

$$B = 1$$

and

$$A_{CL} = 1$$

This is massive negative feedback, the most you can have. In this case, the closed-loop voltage gain equals 1 to a very close approximation.

The closed-loop output impedance of a voltage follower is

$$z_{in(CL)} = (1 + AB)z_{in} = (1 + A)z_{in} = 100,001 \, (2 \text{ M}\Omega) = 2(10^{11})\Omega$$

The closed-loop output impedance of a voltage follower is

$$z_{out(CL)} = \frac{z_{out}}{1 + AB} = \frac{z_{out}}{1 + A} = \frac{75 \, \Omega}{100,001} = 0.00075 \, \Omega$$

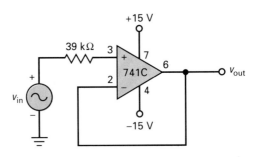

FIGURE 19-11   Voltage follower.

As you can see, $z_{in(CL)}$ approaches infinity and $z_{out/(CL)}$ approaches zero. A voltage follower is a great buffer amplifier because of its high input impedance, low output impedance, and unity voltage gain.

Since the closed-loop voltage gain is 1, the desensitivity equals

$$\frac{A}{A_{CL}} = 100,000$$

In the worst case, the maximum input offset voltage is

$$V_{in(off)} = 2 \text{ mV} + (90 \text{ nA})(39 \text{ k}\Omega) = 5.51 \text{ mV}$$

So, the output offset voltage is

$$V_{oo(CL)} = \frac{100,000(5.51 \text{ mV})}{100,000} = 5.51 \text{ mV}$$

In other words, the voltage follower is almost immune to offset problems. Because it has unity voltage gain, the output offset voltage equals the input offset voltage.

19-5 INVERTING VOLTAGE FEEDBACK

Figure 19-12a shows an amplifier with the noninverting input grounded. The input signal drives the inverting input. The output voltage drives a feedback resistor, which is also connected to the inverting input. This produces another type of negative feedback called *inverting voltage feedback*.

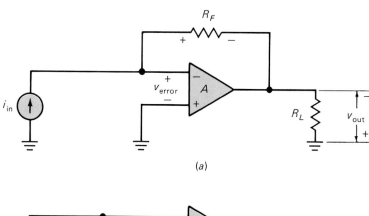

(a)

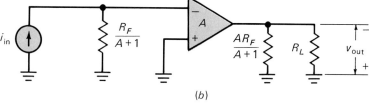

(b)

FIGURE 19-12   Inverting voltage feedback.

An amplifier with inverting voltage feedback acts differently from an amplifier with noninverting voltage feedback. Instead of acting like an ideal voltage amplifier, an amplifier with inverting voltage feedback acts like an ideal *current-to-voltage converter*, a device with zero input impedance, zero output impedance, and a constant *transresistance*.

Mathematical Analysis

The easiest way to analyze inverting voltage feedback is by driving the input with a current source i_{in} and calculating the output voltage v_{out} that results. With the input signal driving the inverting input, the polarity of output voltage is reversed, as shown in Fig. 19-12a. This output voltage is given by

$$v_{out} = Av_{error}$$

To avoid excessive output offset voltage, the feedback resistor R_F is usually less than 100 kΩ. Since the input resistance of the typical op amp is in megohms, almost all of the input current flows through R_F, which is the path of least resistance. Because of this, we can sum the voltages around the circuit as follows:

$$v_{out} - i_{in}R_F + v_{error} = 0$$

or

$$v_{out} - i_{in}R_F + \frac{v_{out}}{A} = 0$$

Now we can rearrange this as

$$\frac{v_{out}}{i_{in}} = \frac{AR_F}{A + 1} \tag{19-18}$$

Since the differential gain is much greater than 1, the foregoing simplifies to

$$\frac{v_{out}}{i_{in}} = R_F \tag{19-19}$$

or

$$v_{out} = i_{in}R_F \tag{19-20}$$

In Eq. (19-19), the ratio of v_{out} to i_{in} is sometimes referred to as the *transresistance* because it equals the output voltage divided by the input current. A feedback amplifier with inverting voltage feedback is sometimes called a *transresistance amplifier*. A circuit like this is also called a *current-to-voltage converter* because an input current controls the output voltage.

Equation (19-20) says the output voltage equals the input current times the feedback resistance. Given a fixed input current, we can control the size of the output voltage by changing the value of feedback resistance. Among other things, a circuit like this can be used as an electronic ammeter. In general, whenever you have a current that you want converted to a voltage, this is the circuit to use.

Input Impedance

Chapter 16 discussed Miller's theorem. The idea was to split the feedback capacitor of an inverting amplifier into an input Miller capacitance and an output Miller capacitance. The same idea applies to a feedback resistor. Miller's theorem says you can split the feedback resistance of an inverting amplifier into an input Miller resistance and an output Miller resistance. The input Miller resistance is

$$R_{\text{in(Miller)}} = \frac{R_F}{A + 1} \qquad (19\text{-}21)$$

and the output Miller resistance is

$$R_{\text{out(Miller)}} = \frac{AR_F}{A + 1} \qquad (19\text{-}22)$$

Whenever you see a feedback resistor between the input and output of an inverting amplifier (Fig. 19-12a), you can split it into an input Miller resistance and an output Miller resistance (Fig. 19-12b). Because A is extremely high, the input Miller resistance approaches zero and the output Miller resistance approaches R_F.

What is the closed-loop input impedance of this type of amplifier? Figure 19-12b is shouting the answer: it is the input Miller resistance in parallel with the input impedance of the op amp. Since the z_{in} of the op amp is typically 2 MΩ or more, it is completely out of the picture, leaving only the input Miller resistance. This means the closed-loop input impedance is:

$$z_{\text{in(CL)}} = \frac{R_F}{A + 1} \qquad (19\text{-}23)$$

Since $A + 1$ has a very large value, $z_{\text{in(CL)}}$ has a very small value, ideally zero. In other words, when a troubleshooter looks at a circuit like the one in Fig. 19-12a, he doesn't hesitate to say that the input impedance is zero. In most current-to-voltage converters, R_F is less than 100 kΩ and A is more than 100,000. Therefore, $z_{\text{in(CL)}}$ is less than 1 Ω.

Other Benefits

Inverting voltage feedback differs from noninverting voltage feedback primarily because of the way it affects the input characteristics of the amplifier. The input impedance is ideally zero rather than infinity. But on the output side, inverting voltage feedback has the same effects as noninverting voltage feedback. That is, it decreases the output impedance, distortion, and output offset voltage. Table 19-2 summarizes this new type of negative feedback.

Virtual Ground

There is a widely-used shortcut for analyzing any current-to-voltage converter. Assume the op amp is ideal. Then, the voltage gain and input impedance of the op amp are both infinite. This allows us to conclude

Table 19-2	Inverting Voltage Feedback		
Quantity	Symbol	Effect	Formula
Transresistance	v_{out}/i_{in}	Stabilizes	R_F
Input impedance	$z_{in(CL)}$	Decreases	$R_F/(1 + A)$
Output impedance	$z_{out(CL)}$	Decreases	$z_{out}/(1 + A)$
Distortion	$v_{dist(CL)}$	Decreases	$v_{dist}/(1 + A)$
Output offset	$V_{oo(CL)}$	Decreases	$V_{oo(OL)}/(1 + A)$

the following about any current-to-voltage converter with an ideal op amp:

1. The error voltage is zero

2. The current into the op amp is zero.

These two key ideas can be captured beautifully by a concept called *virtual ground*. This new type of ground is defined as a node that has zero voltage with respect to ground, but is not mechanically grounded.

Recall the two types of grounds discussed earlier. First, there was the *mechanical ground*. This is what you get when you connect a wire between a node and ground. A mechanical ground shorts the node to ground at all frequencies. Second, there was the *ac ground*. Because it is produced by a bypass capacitor, this type of ground exists only at high frequencies. Both the mechanical ground and the ac ground provide a path for current when they are grounding the node.

Virtual ground is different. This third type of ground does not provide a path for current between the node and ground. And yet, the node always has zero voltage with respect to ground. It's like being a ground for voltage, but not for current. It's like half of a ground. To remind us of this half-ground quality, we can visualize a dashed line between the inverting input and ground as shown in Fig. 19-13. The ground symbol means the node has a potential of 0 V, but no current can flow to ground because of the broken line.

Here is the shortcut for analyzing an amplifier with inverting voltage feedback. Because of the virtual ground, all the input current i_{in} has to pass through R_F. Why? Because the ideal op amp has infinite input

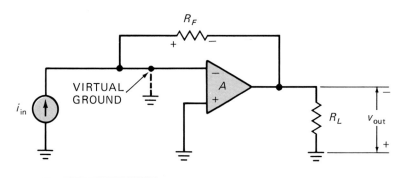

FIGURE 19-13   Virtual ground.

impedance. The only path left for the input current is through the feedback resistor. This crucial idea is worth repeating: *The only path for the input current is through the feedback resistor.* Once this idea is clear, the rest is easy. Because of the virtual ground, R_F is in parallel with R_L. Therefore, the output voltage equals the voltage across R_F. With Ohm's law, the voltage across R_F is

$$v_{\text{out}} = i_{\text{in}}R_F \qquad (19\text{-}24)$$

The equation says that the output voltage equals the input current times the feedback resistance. This equation is identical to Eq. (19-20), which we derived earlier.

Here are the key steps that we used to derive Eq. (19-24):

1. Visualize a virtual ground on the inverting input.

2. Visualize all of the input current passing through the feedback resistor.

3. Realize that the output voltage is across the feedback resistor as well as the load resistor.

4. Use Ohm's law to calculate the output voltage across the feedback resistor.

You should read these steps until you are convinced you understand them fully. Understanding this process will help you to understand a large number of op-amp circuits used in industry.

EXAMPLE 19-9

What does the output voltage equal in Fig. 19-14?

SOLUTION

Visualize a virtual ground on the inverting input of the op amp. Next, visualize all of the input current passing through the 100 kΩ.

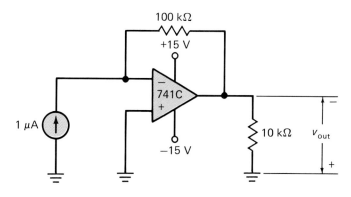

FIGURE 19-14   Example.

Now, use Ohm's law to calculate the voltage across the feedback resistor:

$$v_{out} = (1 \text{ μA})(100 \text{ kΩ}) = 0.1 \text{ V}$$

This is the voltage across the feedback resistor. It is also the voltage across the load resistor because the virtual ground places the feedback resistor in parallel with the load resistor.

You see how simple it is when you know what you're doing. You don't have to memorize a lot of formulas because you know exactly what you are doing and why you are doing it. You should avoid plugging numbers into formulas if at all possible. It is much better to rely on Ohm's law and a solid understanding of a circuit than it is to rely on finding the right formula. Of course, there are many times when we have no choice but to use a formula. The main thing to remember is that formulas should be a last resort, what you use when you can see no other way to solve the problem.

EXAMPLE 19-10

Figure 19-15 shows how to build an electronic ammeter, a circuit that can measure current and display its value on a voltmeter. What does the output voltage equal when the input current is 50 μA? When the input current is 10 μA?

SOLUTION

Again, visualize all of the input current passing through the feedback resistor. When the input current is 50 μA, the output voltage is

$$v_{out} = (50 \text{ μA})(100 \text{ kΩ}) = 5 \text{ V}$$

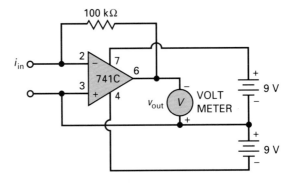

FIGURE 19-15   Electronic ammeter.

When the input current is 10 μA, the output voltage is

$$v_{out} = (10 \text{ μA})(100 \text{ kΩ}) = 1 \text{ V}$$

You can see that the circuit is converting an input current into an output voltage. This output voltage is directly proportional to the input current. The feedback resistor sets up the proportion. By changing this resistor, we can change the proportion.

EXAMPLE 19-11

In Fig. 19-15, what is the input impedance of the electronic ammeter?

SOLUTION

Ideally, an ammeter has an input resistance of zero. This way, you can insert an ammeter in series with any branch without changing the current in the branch. Here is how to find the input impedance of the electronic ammeter. Visualize the feedback resistor split into its two Miller components. The output Miller resistance appears in parallel with the voltmeter, which has no meaningful effect. But the input Miller resistance appears across the input terminals and has an impedance of

$$z_{in(CL)} = \frac{100 \text{ kΩ}}{100,000 + 1} = 1 \text{ Ω}$$

We use the typical voltage gain of a 741C in this calculation. As you see, the input impedance is only 1 Ω. This means the electronic ammeter adds only 1 Ω of resistance to the branch it is in. This will have negligible effect in most circuits.

EXAMPLE 19-12

In Fig. 19-16, what is the output voltage?

SOLUTION

The first stage is a current-to-voltage converter. All of the 1 μA of input current passes through the feedback resistor to produce an output voltage of

$$v_{out} = (1 \text{ μA})(10 \text{ kΩ}) = 10 \text{ mV}$$

The second stage is a voltage amplifier with a voltage gain of

$$A_{CL} = \frac{99 \text{ kΩ}}{1 \text{ kΩ}} + 1 = 100$$

Therefore, the final output voltage is

$$v_{out} = 100(10 \text{ mV}) = 1 \text{ V}$$

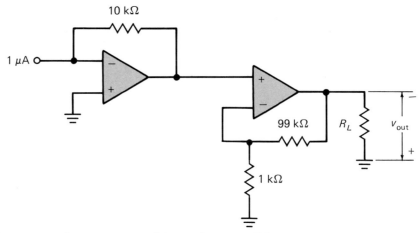

FIGURE 19-16 Current-to-voltage converter drives voltage amplifier.

19-6 BANDWIDTH

In Fig. 19-17a, the *bandwidth B* of an amplifier is defined as the difference between its lower and upper critical frequencies. As an equation,

$$B = f_2 - f_1 \qquad (19\text{-}25)$$

Because an op amp is direct coupled, it has no lower critical frequency. This is why its frequency response looks like Fig. 19-17b. This is a graph of the open-loop voltage gain versus frequency. For an op amp, the open-loop bandwidth is

$$B = f_2 \qquad (19\text{-}26)$$

For instance, the 741C has an f_2 of 10 Hz. This means its open-loop bandwidth is 10 Hz.

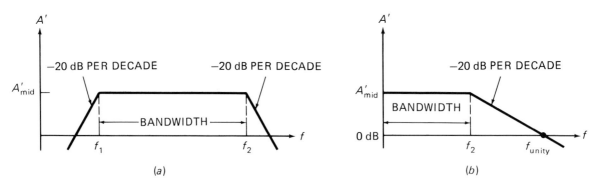

FIGURE 19-17 Frequency response. (*a*) ac amplifier. (*b*) dc amplifier.

Closed-Loop Bandwidth

It is possible to derive this equation:

$$f_{2(CL)} = (1 + AB)f_2 \qquad (19\text{-}27)$$

where A is the differential voltage gain of op amp in the midband. When the loop gain AB is much greater than 1, the closed-loop cutoff frequency is much greater than the open-loop cutoff frequency. Again, the desensitivity $1 + AB$ appears in an equation. If you recall, $1 + AB$ keeps showing up in almost all of the formulas for negative feedback. This time, we see that the bandwidth is increased by the desensitivity or sacrifice factor of the amplifier.

Figure 19-18 illustrates the effect of negative feedback on bandwidth. When there is no negative feedback, the voltage gain has a maximum value of 100,000 and a bandwidth of 10 Hz. At 10 Hz, the voltage gain is 70,700. Beyond this frequency, the voltage gain decreases at a rate of 20 dB per decade.

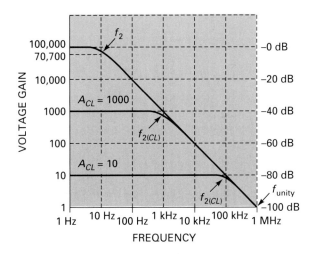

FIGURE 19-18   Open-loop and closed-loop responses.

When there is negative feedback, however, there is less voltage gain. For instance, a closed-loop voltage gain of 1000 means that we have decreased the voltage gain by a factor of 100. Therefore, Eq. (19-27) means that the critical frequency increases to

$$f_{2(CL)} = 100(10 \text{ Hz}) = 1 \text{ kHz}$$

This is the critical frequency or bandwidth of the amplifier when the closed-loop voltage gain is 1000. You can see this in Fig. 19-18.

When the negative feedback is really heavy, the voltage gain is small but the bandwidth is large. For instance, when the closed-loop voltage gain is 10, the sacrifice factor is 10,000. In this case, the bandwidth increases to

$$f_{2(CL)} = 10,000(10 \text{ Hz}) = 100 \text{ kHz}$$

Gain-Bandwidth Product

When you look at Fig. 19-18, you would swear there is something more to be said about the voltage gain and the bandwidth. If you feel like this, you are quite right. Figure 19-18 is often summarized by saying that the *gain-bandwidth product is a constant.* Let us find out what this means.

As derived earlier, the closed-loop voltage gain is given by

$$A_{CL} = \frac{A}{1 + AB}$$

We can rearrange this to get

$$1 + AB = \frac{A}{A_{CL}}$$

When we substitute this into Eq. (19-27), we get

$$f_{2(CL)} = \frac{A}{A_{CL}} f_2 \tag{19-28}$$

or

$$A_{CL} f_{2(CL)} = A f_2 \tag{19-29}$$

The right-hand side of this equation is called the *open-loop gain-bandwidth product* because it is the product of open-loop gain and bandwidth. For a 741C, the typical voltage gain is 100,000 and the critical frequency is 10 Hz. Therefore, a 741C has an open-loop gain-bandwidth product of

$$A f_2 = 100,000(10 \text{ Hz}) = 1 \text{ MHz}$$

The left-hand side of Eq. (19-29) is called the *closed-loop gain-bandwidth product* because it is a product of closed-loop gain and bandwidth. Equation (19-29) tells us that the gain-bandwidth product is the same whether we calculate it with open-loop quantities or closed-loop quantities. Given a typical 741C, the product of A_{CL} and $f_{2(CL)}$ equals 1 MHz. Equation (19-29) means that the gain-bandwidth product is a constant. Therefore, even though A_{CL} and $f_{2(CL)}$ change when we change external resistors, the product of these two quantities remains constant and equal to 1 MHz.

Figure 19-18 echoes the idea of a constant gain-bandwidth product. When the gain is 100,000, the gain-bandwidth product is

$$A_{CL} f_{2(CL)} = 100,000(10 \text{ Hz}) = 1 \text{ MHz}$$

When the gain is 1000, the gain-bandwidth product is

$$A_{CL} f_{2(CL)} = 1000(1 \text{ kHz}) = 1 \text{ MHz}$$

When the gain is 10, the gain-bandwidth product is

$$A_{CL} f_{2(CL)} = 10(100 \text{ kHz}) = 1 \text{ MHz}$$

Unity-Gain Frequency

If $A_{CL} = 1$, $f_{2(CL)} = f_{\text{unity}}$, and Eq. (19-29) reduces to

$$f_{\text{unity}} = A f_2 \tag{19-30}$$

This says that the unity-gain frequency equals the gain-bandwidth product. Data sheets usually list the value of f_{unity} because it equals the gain-bandwidth product. The higher the f_{unity}, the larger the gain-bandwidth product of the op amp. For instance, the 741C has an f_{unity} of 1 MHz. The LM318 has an f_{unity} of 15 MHz. Although it costs more, the LM318 may be a better choice if you need more gain-bandwidth product. An f_{unity} of 15 MHz means the LM318 has 15 times more voltage gain than a 741C with the same bandwidth.

The gain-bandwidth product gives us a fast way of comparing amplifiers. The greater the gain-bandwidth product, the higher we can go in frequency and still have usable gain. With Eq. (19-29) and (19-30), we can derive this useful relation:

$$f_{2(CL)} = \frac{f_{unity}}{A_{CL}} \tag{19-31}$$

This equation is valid when the voltage gain rolls off at a rate of 20 dB per decade. The 741C and most other op amps have a compensating capacitor to satisfy this condition.

Slew Rate and Power Bandwidth

Negative feedback has no effect on slew rate or power bandwidth. Until the output voltage has changed, there is no feedback signal and no benefits of negative feedback. In other words, slew rate and power bandwidth are the same with or without negative feedback. As before, you have to calculate the power bandwidth using this equation:

$$f_{max} = \frac{S_R}{2\pi V_p}$$

When f_{max} is greater than $f_{2(CL)}$, there is no slew-rate distortion. When f_{max} is less than $f_{2(CL)}$, you will see this type of distortion appear on the output signal. One way to eliminate it is to reduce the peak output voltage as needed to make f_{max} greater than $f_{2(CL)}$. By substituting $f_{2(CL)}$ for f_{max} and rearranging the equation we get

$$V_{p(max)} = \frac{S_R}{2\pi f_{2(CL)}} \tag{19-32}$$

As long as the output peak voltage is less than $V_{p(max)}$, the amplifier has no slew-rate distortion at any frequency between 0 and $f_{2(CL)}$.

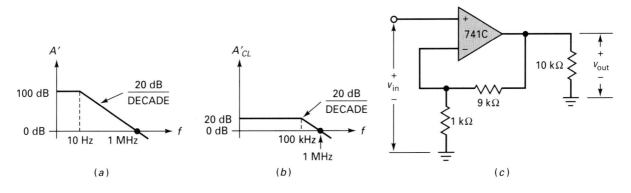

FIGURE 19-19   Gain-bandwidth product is constant.

EXAMPLE 19-13

Figure 19-19*a* shows the frequency response for a 741C. What is the bandwidth in Fig. 19-19*c*?

SOLUTION

In Fig. 19-19*a*, the midband decibel voltage gain of the 741C is 100 dB, equivalent to an open-loop voltage gain of

$$A = 100,000$$

Since the critical frequency is 10 Hz, the gain-bandwidth product is

$$Af_2 = 100,000(10 \text{ Hz}) = 1 \text{ MHz}$$

Another way to get the gain-bandwidth product is to read the value of f_{unity} in Fig. 19-19*a*:

$$f_{\text{unity}} = 1 \text{ MHz}$$

Either way, the 741C has a gain-bandwidth product of 1 MHz.

In Fig. 19-19*c*, the closed-loop voltage gain equals

$$A_{CL} = \frac{9 \text{ k}\Omega}{1 \text{ k}\Omega} + 1 = 10$$

The gain-bandwidth product is

$$A_{CL} f_{2(CL)} = 1 \text{ MHz}$$

Since $A_{CL} = 10$, we can divide 1 MHz by 10 to get the closed-loop bandwidth:

$$f_{2(CL)} = \frac{1 \text{ MHz}}{10} = 100 \text{ kHz}$$

Figure 19-19*b* shows the decibel voltage gain of the negative-feedback amplifier. In the midband of the feedback amplifier, the decibel gain is 20 db. The closed-loop bandwidth is 100 kHz, the roll-off rate is 20 db per decade, and the unity-gain frequency is 1 MHz.

EXAMPLE 19-14

In Fig. 19-19*c*, what is the maximum undistorted output peak voltage at $f_{2(CL)}$? Use $S_R = 0.5 \text{ V/}\mu\text{s}$.

SOLUTION

In the preceding example, we calculated $A_{CL} = 10$ and $f_{2(CL)} = 100 \text{ kHz}$. With Eq. (19-32), we can calculate the maximum output without slew-rate distortion:

$$V_{p(\text{max})} = \frac{0.5 \text{ V/}\mu\text{s}}{2\pi(100 \text{ kHz})} = 0.796 \text{ V}$$

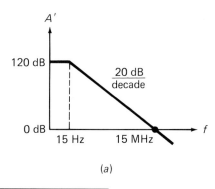

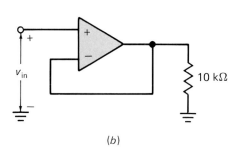

(a)   (b)

FIGURE 19-20   Example.

As long as you keep the output peak voltage less than 0.796 V, you will have a closed-loop voltage gain of 10, a closed-loop bandwidth of 100 kHz, and no slew-rate distortion at any frequency between 0 and 100 kHz.

EXAMPLE 19-15

Suppose an op amp with the graph of Fig. 19-20a is used in the voltage follower of Fig. 19-20b. What is the closed-loop bandwidth?

SOLUTION

The gain-bandwidth product is

$$A_{CL}f_{2(CL)} = 15 \text{ MHz}$$

Since a voltage follower has a closed-loop voltage gain of 1, the closed-loop bandwidth is

$$f_{2(CL)} = \frac{15 \text{ MHz}}{1} = 15 \text{ MHz}$$

Because its voltage gain is only 1, the voltage follower has maximum bandwidth.

OPTIONAL TOPICS

The following material continues the earlier discussions at a more advanced and specialized level. All the topics are optional because they are not used in any of the basic discussions in later chapters. This section will be a useful reference when you are in industry because then you will probably want more advanced viewpoints.

19-7 EXACT VALUE OF FEEDBACK FRACTION

When we derived the feedback fraction of an amplifier with noninverting voltage feedback, we got

$$B = \frac{R_2}{R_1 + R_2}$$

This assumes that the z_{in} of the op amp is much greater than R_2. The exact expression includes the z_{in} of the op amp in parallel with R_2:

$$B = \frac{R_2 \| z_{in}}{R_1 + R_2 \| z_{in}} \tag{19-33}$$

19-8 MORE ON INVERTING VOLTAGE FEEDBACK

We can redraw Fig. 19-12a as shown in Fig. 19-21. Now it resembles a noninverting voltage-feedback amplifier, except that a current source is driving the inverting terminal. Since the current source ideally has infinite impedance, the feedback fraction B is approximately equal to unity, and the circuit has maximum negative feedback. With derivations identical to those given earlier, we can get formulas for closed-loop output impedance, nonlinear distortion, and output offset voltage.

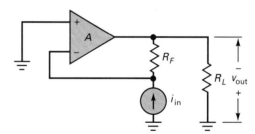

FIGURE 19-21 Inverting voltage feedback redrawn.

In most circuits the noninverting input is grounded for dc as well as ac signals. Therefore, the virtual ground has a dc potential of zero with respect to ground. But don't get the idea that a virtual ground is always at dc ground potential. There are some circuits in which the noninverting input is biased at a positive or negative dc level. A bypass capacitor is then used to ac ground the noninverting input. In this case, the virtual ground has a dc potential. But it is still a virtual ground to ac signals, meaning that it has zero ac voltage and draws no ac current.

19-9 NONINVERTING CURRENT FEEDBACK

With *noninverting current feedback*, an input voltage drives the noninverting input of an amplifier, and the output current is sampled to get a feedback voltage. An amplifier with noninverting current feedback tends to act like an ideal voltage-to-current converter, one with infinite input impedance, infinite output impedance, and a stable transconductance.

AC Equivalent Circuit

Figure 19-22 shows the ac equivalent circuit for a feedback amplifier with noninverting current feedback. The load resistor and feedback resistor are in series. Because of this, the load current passes through the feedback resistor. The feedback voltage is proportional to load current because

$$v_2 = i_{\text{out}} R_F$$

Whenever the feedback voltage is proportional to output current, the circuit has current feedback.

Current feedback stabilizes the output current. This means that a constant input voltage produces an almost constant output current, despite changes in open-loop gain and load resistance. For instance, suppose the open-loop voltage gain decreases. Then the output current tries to decrease. This results in less feedback voltage and more error voltage. In turn, this means more output voltage and current. The increased error voltage almost completely offsets the decrease in open-loop voltage gain, so that the output current remains almost constant. A similar argument applies to an increase in open-loop gain. Attempted increases in output current are almost eliminated by the negative feedback.

Mathematical Analysis

The feedback amplifier of Fig. 19-22 has a closed-loop voltage gain of

$$A_{CL} = \frac{A}{1 + AB}$$

where

$$B = \frac{R_F}{R_L + R_F}$$

The output curent is

$$i_{\text{out}} = \frac{v_{\text{out}}}{R_L + R_F} = \frac{A_{CL} v_{\text{in}}}{R_L + R_F}$$

which can be rearranged as

$$\frac{i_{\text{out}}}{v_{\text{in}}} = \frac{A_{CL}}{R_L + R_F} \tag{19-34}$$

When the loop gain is high, A_{CL} approximately equals $1/B$ and Eq. (19-34) gives

$$\frac{i_{\text{out}}}{v_{\text{in}}} = \frac{(R_L + R_F)/R_F}{R_L + R_F}$$

or

$$\frac{i_{\text{out}}}{v_{\text{in}}} = \frac{1}{R_F} \tag{19-35}$$

This says that the ratio of output current to input voltage equals the reciprocal of R_F. Since R_F is an external resistor, i_{out}/v_{in} has a constant value independent of the open-loop voltage gain and load resistance.

Transconductance

A feedback amplifier with noninverting current feedback is often called a *transconductance amplifier*, and Eq. (19-35) is written as

$$g_m = \frac{1}{R_F} \tag{19-36}$$

The circuit of Fig. 19-22 is also called a *voltage-to-current converter* because an input voltage controls an output current. If we rearrange Eq. (19-35), the output current is

$$i_{out} = \frac{v_{in}}{R_F} \tag{19-37}$$

Other Benefits

Compare the circuit of Fig. 19-22 with that of Fig. 19-1. The only difference is the location of the load resistor. In Fig. 19-22, the load resistor is floating to allow output current to pass through the feedback resistor. In Fig. 19-1, the load resistor is connected between the output and ground. Because the circuits are similar except for the location of the load resistor, the negative feedback again reduces distortion and output offset voltage. Also, the input impedance approaches infinity.

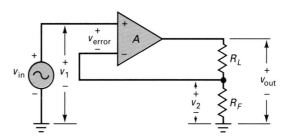

FIGURE 19-22   Noninverting current feedback.

The one quantity that is different with noninverting current feedback is the closed-loop output impedance. Since the load is no longer grounded but is part of the feedback circuit, it sees a different Thevenin output impedance from before. It can be shown that

$$z_{out(CL)} = (1 + A)R_F \tag{19-38}$$

Since A is very large, $z_{out(CL)}$ approaches infinity. From now on, remember that voltage feedback produces a low output impedance, while current feedback produces a high output impedance. Table 19-3 summarizes noninverting current feedback. As indicated, transconductance is stabilized, input impedance increases, output impedance increases, and so on.

Table 19-3 Noninverting Current Feedback

Quantity	Symbol	Effect	Formula
Transconductance	i_{out}/v_{in}	Decreases	$1/R_F$
Input impedance	$z_{in(CL)}$	Increases	$(1 + AB)z_{in}$
Output impedance	$z_{out(CL)}$	Increases	$(1 + A)R_F$
Distortion	$v_{dist(CL)}$	Decreases	$v_{dist}/(1 + AB)$
Output offset	$V_{oo(CL)}$	Decreases	$V_{oo(OL)}/(1 + AB)$

Simplified Viewpoint

There is a simple way of looking at the circuit of Fig. 19-22. Since the inverting input voltage is within microvolts of the noninverting input voltage,

$$v_2 = v_{in}$$

Therefore,

$$i_{out} = \frac{v_2}{R_F} = \frac{v_{in}}{R_F}$$

This is the same result we got earlier with formal mathematics.

19-10 INVERTING CURRENT FEEDBACK

In Fig. 19-23, the input signal drives the inverting input, and the output current is sampled. This produces *inverting current feedback*. An amplifier with inverting current feedback tends to act like an ideal current amplifier, one that has zero input impedance, infinite output impedance, and a constant current gain.

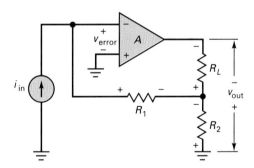

FIGURE 19-23 Inverting current feedback.

With the input signal driving the inverting input, the polarity of output voltage is reversed, as shown in Fig. 19-23. It can be shown that the current gain is:

$$\frac{i_{out}}{i_{in}} = \frac{R_1}{R_2} + 1 \tag{19-39}$$

Equation (19-39) says that the current gain ideally depends on the ratio of two resistances, both of which can be of precision tolerance. An amplifier with inverting current feedback is called a current amplifier because the current gain has been stabilized.

Inverting current feedback decreases the input impedance, increases the output impedance, decreases distortion, and decreases output offset voltage. With this type of negative feedback,

$$B = \frac{R_2}{R_1 + R_2} \tag{19-40}$$

Table 19-4 summarizes the properties of this type of negative feedback.

Table 19-4 **Inverting Current Feedback**

Quantity	Symbol	Effect	Formula
Current gain	i_{out}/i_{in}	Stabilizes	$1/B$
Input impedance	$z_{in(CL)}$	Decreases	$R_1/(1 + AB)$
Output impedance	$z_{out(CL)}$	Increases	$(1 + A)R_2$
Distortion	$v_{dist(CL)}$	Decreases	$v_{dist}/(1 + AB)$
Output offset	$V_{oo(CL)}$	Decreases	$V_{oo(OL)}/(1 + AB)$

19-11 MORE ON CLOSED-LOOP BANDWIDTH

Suppose an amplifier has a lower and upper critical frequency. Negative feedback improves each of these. As usual, the desensitivity is the key factor in the improvement. With complex numbers, it can be shown that the two critical frequencies are

$$f_{1(CL)} = \frac{f_1}{1 + AB} \tag{19-41}$$

and

$$f_{2(CL)} = (1 + AB)f_2 \tag{19-42}$$

As you see, the lower critical frequency is decreased and the upper critical frequency is increased. These formulas apply to noninverting voltage feedback.

Negative feedback increases the bandwidth of all feedback amplifiers. By derivations similar to those for noninverting voltage feedback, we can derive the formulas given in Table 19-5. With inverting feedback, the feedback fraction $B = 1$ because we assume the current source has an infinite impedance. This is the reason the desensitivity is $1 + A$ for the inverting types of negative feedback. The last two entries mean that the output voltage is down 3 dB at $(1 + A)f_2$. In other words, if you drive an inverting-feedback amplifier with a current source, the output voltage breaks at approximately f_{unity}.

Table 19-5 Bandwidth with Negative Feedback		
Type	$f_{1(CL)}$	$f_{2(CL)}$
Noninverting voltage feedback	$f_1/(1 + AB)$	$(1 + AB)f_2$
Noninverting current feedback	$f_1/(1 + AB)$	$(1 + AB)f_2$
Inverting voltage feedback	$f_1/(1 + A)$	$(1 + A)f_2$
Inverting current feedback	$f_1/(1 + A)$	$(1 + A)f_2$

STUDY AIDS

The following study aids will help to reinforce the ideas discussed in this chapter. For best results, use these study aids within 6 hours of reading the earlier material. Then review these study aids a week later and a month later to ensure that the concepts remain in your long-term memory.

SUMMARY

Sec. 19-1 Noninverting Voltage Feedback

There are four types of negative feedback. Noninverting voltage feedback is the most basic type. The input signal drives the noninverting input of an op amp. The amplified output voltage is sampled and part of it is returned to the inverting input of the op amp. The differential input to the op amp is therefore the difference between the input voltage and a fraction of the output voltage. An amplifier with this type of feedback approximates an ideal voltage amplifier, one with infinite input impedance, zero output impedance, and constant voltage gain.

Sec. 19-2 Open-Loop and Closed-Loop Voltage Gains

The open-loop voltage gain is the voltage gain when the feedback path is open. This results in maximum voltage gain. The closed-loop voltage gain is the voltage gain when there is negative feedback. This voltage gain is usually much smaller than the open-loop voltage gain.

Sec. 19-3 Input and Output Impedances

Noninverting voltage feedback increases the input impedance and decreases the output impedance.

The improvement factor is $1 + AB$. This means you multiply z_{in} by $1 + AB$ to get the closed-loop input impedance. You divide z_{out} by $1 + AB$ to get the closed-loop output impedance.

Sec. 19-4 Other Benefits of Negative Feedback

Negative feedback of any type will decrease the distortion and output offset voltages. Again, the improvement factor is $1 + AB$. You divide the open-loop distortion and the open-loop output offset voltage by $1 + AB$ to get the closed-loop quantities. Since $1 + AB$ is usually large, the closed-loop distortion and closed-loop output offset voltage are much smaller than without the negative feedback.

Sec. 19-5 Inverting Voltage Feedback

This is another important type of negative feedback used with op amps. The input signal is applied to the inverting input, while the noninverting input is grounded. With this type of feedback, the amplifier acts like an ideal current-to-voltage converter, one with zero input impedance, zero output impedance, and constant transresistance. The inverting input is at virtual ground which is like a half ground; a short to ground for voltage but an open for current.

Sec. 19-6 Bandwidth

This is defined as the difference between the dominant lower and upper critical frequencies. Since an op amp has no lower critical frequency, the bandwidth equals the upper critical frequency. The gain-bandwidth product is a constant, meaning the product of voltage gain and bandwidth is fixed for a given op amp. This implies that less closed-loop voltage gain results in more closed-loop bandwidth.

VOCABULARY

In your own words, explain what each of the following terms means. Keep your answers short and to the point. If necessary, verify your answer by rereading the appropriate discussion or by looking at the end-of-book Glossary.

bandwidth	open-loop quantity
closed-loop quantity	unity-gain frequency
error voltage	virtual ground
gain-bandwidth product	voltage feedback
loop gain	voltage follower
negative feedback	

IMPORTANT EQUATIONS

Here are some important equations. Say each of the following equations in symbols, then say each in words. Try to explain what the equation means and how it is used. Then read the description that follows.

Eq. 19-5 Feedback Fraction

$$B = \frac{R_2}{R_1 + R_2}$$

This gives the fraction of output voltage that is returned to the inverting input. It is sometimes called the voltage gain of the voltage divider. B is always a number between 0 and 1.

Eq. 19-6 Exact Voltage Gain with Negative Feedback

$$\frac{v_{out}}{v_{in}} = \frac{A}{1 + AB}$$

As equations go, this is a classic. It is famous in the history of negative feedback. The denominator is especially important because it keeps reappearing in the formulas for most closed-loop quantities. It says that the voltage gain with negative feedback equals the differential voltage gain of the op amp divided by the desensitivity or sacrifice factor.

Eq. 19-7 Voltage Gain with Negative Feedback

$$\frac{v_{out}}{v_{in}} = \frac{1}{B}$$

Here is the magic of negative feedback. When the sacrifice factor is large, the voltage gain of the amplifier no longer depends on the differential voltage gain of the op amp. It depends only on the feedback fraction. Since B is determined by fixed resistances, the voltage gain is very stable. When you first encounter this phenomenon, it does seem like magic. You take an op amp with very large voltage gain, connect some external resistors, and then you've got an amplifier whose gain no longer depends on the gain of the op amp. If that's not magic, what is?

Eq. 19-17 Desensitivity

$$1 + AB = \frac{A}{A_{CL}}$$

The factor $1 + AB$ appears in many closed-loop equations. It is important to realize that this factor equals differential voltage gain divided by the closed-loop voltage gain. Why? Because in some problems it is difficult to calculate the value of $1 + AB$ but easy to calculate A/A_{CL}.

Eq. 19-24 Current-to-Voltage Converter

$$v_{out} = i_{in}R_F$$

Extremely important. It says the output voltage equals the input current times the feedback resistance. In the next chapter you will see several circuits that use this type of feedback. It's easy to remember this important equation if you remember the concept of virtual ground. Because of the virtual ground, the input current has nowhere to go except through the feedback resistor. Ohm's law gives the voltage across the feedback resistor, which is in parallel with the load resistor.

Eq. 19-25 Bandwidth

$$B = f_2 - f_1$$

A fundamental definition that should be memorized because it is used in many different areas of electronics. It says the bandwidth is the difference of the dominant critical frequencies of the amplifier.

Eq. 19-29 Gain-Bandwidth Product Is a Constant

$$A_{CL}f_{2(CL)} = Af_2$$

A major concept. It says that the closed-loop gain-bandwidth product equals the open-loop gain-bandwidth product. Since Af_2 is constant for a given op amp, it means $A_{CL}f_{2(CL)}$ is also constant. In turn, this means a decrease in A_{CL} produces an increase in $f_{2(CL)}$.

Eq. 19-31 Closed-Loop Bandwidth

$$f_{2(CL)} = \frac{f_{unity}}{A_{CL}}$$

A bread-and-butter formula. You can get the value of f_{unity} from a data sheet. When you divide this value by A_{CL}, you have the bandwidth of the amplifier.

IMPORTANT PROCESSES

Each of the following processes is important for analyzing circuits with negative feedback. If you can remember the basic ideas in each step, you will be able to solve problems more easily.

Exact Noninverting Voltage Gain

1. Work out feedback fraction.
2. Multiply by voltage gain of op amp.
3. Add 1 to get desensitivity.
4. Divide voltage gain of op amp by desensitivity.

Approximate Noninverting Voltage Gain

1. Work out feedback fraction.
2. Take reciprocal to get voltage gain.

Output Voltage with Noninverting Voltage Feedback

1. Multiply v_{in} by A_{CL} to get v_{out}

   or

2. Visualize v_{in} across lower feedback resistor.
3. Calculate current through this resistor.
4. Calculate voltage across upper resistor.
5. Add two voltages to get v_{out}.

Output Voltage with Inverting Voltage Feedback

1. Visualize a virtual ground on inverting input.
2. Visualize input current passing through feedback resistor.
3. Realize feedback resistor is in parallel with load resistor.
4. Use Ohm's law to calculate voltage across feedback resistor.

Gain-Bandwidth Product

1. Multiply A by f_2

   or

2. Read f_{unity} on data sheet

   or

3. Multiply A_{CL} by $f_{2(CL)}$

Closed-loop Bandwidth

1. Find gain-bandwidth product.
2. Find closed-loop voltage gain.
3. Divide gain-bandwidth product by A_{CL}.

STUDENT ASSIGNMENTS

QUESTIONS

Some questions have more than one right answer. Select the best answer, the one that is always true, or that most accurately describes the situation.

1. With negative feedback, the returning signal
   a. Aids the input signal
   b. Opposes the input signal
   c. Is proportional to output current
   d. Is proportional to differential voltage gain

2. How many types of negative feedback are there?
   a. One
   b. Two
   c. Three
   d. Four

3. With noninverting voltage feedback, the circuit approximates an ideal
   a. Voltage amplifier
   b. Current-to-voltage converter
   c. Voltage-to-current converter
   d. Current amplifier

4. The error voltage of an ideal op amp is
   a. Zero
   b. Very small
   c. Very large
   d. Equal to the input voltage

5. The error voltage of a real op amp is
   a. Zero
   b. Very small
   c. Very large
   d. Equal to the input voltage

6. The feedback fraction B
   a. Is always less than 1
   b. Is usually greater than 1
   c. May equal 1
   d. May not equal 1

7. The loop gain AB
   a. Is usually much smaller than 1
   b. Is usually much greater than 1
   c. May not equal 1
   d. Is between 0 and 1

8. With noninverting voltage feedback, any decrease in differential voltage gain produces an increase in
   a. Output voltage
   b. Error voltage
   c. Feedback voltage
   d. Input voltage

9. Negative feedback is
   a. Another of those perpetual-motion follies
   b. A mediocre phenomenon
   c. Useless in mass production
   d. A fantastic invention that borders on magic

10. The open-loop voltage gain equals the
    a. Gain with negative feedback
    b. Differential voltage gain of the op amp
    c. Gain when B is 1
    d. Gain at f_{unity}

11. The closed-loop input impedance with non-inverting voltage feedback is
    a. Usually larger than the open-loop input impedance
    b. Equal to the open-loop input impedance
    c. Sometimes less than the open-loop input impedance
    d. Ideally zero

12. The closed-loop input impedance with inverting voltage feedback is
    a. Usually larger than the open-loop input impedance
    b. Equal to the open-loop input impedance
    c. Sometimes less than the open-loop impedance
    d. Ideally zero

13. With inverting voltage feedback, the circuit approximates an ideal
    a. Voltage amplifier
    b. Current-to-voltage converter
    c. Voltage-to-current converter
    d. Current amplifier

14. Negative feedback reduces the
    a. Feedback fraction
    b. Distortion
    c. Input offset voltage
    d. Error voltage

15. The desensitivity equals
    a. AB c. A_{CL}/A
    b. $1 + AB$ d. f_{unity}

16. The sacrifice factor equals
    a. AB c. A/A_{CL}
    b. $1 - AB$ d. f_{unity}

17. A voltage follower has a voltage gain of
    a. Less than 1
    b. 1
    c. More than 1
    d. A

18. When an op amp is not saturated, the voltages at the noninverting and inverting inputs are
    a. Almost equal
    b. Distinctly different
    c. Equal to the output voltage
    d. Equal to ± 15 V

19. The transresistance of an amplifier is the ratio of its
    a. Output current to input voltage
    b. Input voltage to output current
    c. Output voltage to input voltage
    d. Output voltage to input current

20. Current cannot flow to ground through
    a. A mechanical ground
    b. An ac ground
    c. A virtual ground
    d. An ordinary ground

21. A node that has 0 V with respect to ground without being mechanically grounded is called
    a. A mechanical ground
    b. An ac ground
    c. A virtual ground
    d. An ordinary ground

22. In a current-to-voltage converter, the input current flows
    a. Through the input impedance of the op amp
    b. Through the feedback resistor
    c. To ground
    d. Through the load resistor

23. The feedback resistance of a transresistance amplifier is
    a. Approximately zero
    b. Very large
    c. In parallel with the load resistor
    d. Equal to the input impedance of the op amp

24. The input impedance of a current-to-voltage converter is
    a. Small
    b. Large
    c. Ideally zero
    d. Ideally infinite

25. The open-loop bandwidth equals
    a. f_{unity}
    b. f_2
    c. f_{unity}/A_{CL}
    d. f_1

26. The closed-loop bandwidth equals
    a. f_{unity}
    b. f_2
    c. f_{unity}/A_{CL}
    d. f_1

27. For a given op amp, which of these is constant?
    a. Error voltage
    b. Feedback voltage
    c. A_{CL}
    d. $A_{CL}f_{2(CL)}$

28. Negative feedback does not improve
    a. Stability of voltage gain
    b. Distortion
    c. Output offset voltage
    d. Power bandwidth

29. An op amp with inverting voltage feedback is saturated. A possible trouble is
    a. No supply voltages
    b. Open feedback resistor
    c. No input voltage
    d. Open load resistor

30. An op amp with noninverting voltage feedback has no output voltage. A possible trouble is
    a. Shorted load resistor
    b. Open feedback resistor
    c. Excessive input voltage
    d. Open load resistor

31. An op amp with noninverting voltage feedback is saturated. A possible trouble is
    a. Shorted load resistor
    b. Upper feedback resistor open
    c. No input voltage
    d. Open load resistor

32. An op amp with noninverting voltage feedback is saturated. A possible trouble is
    a. Shorted load resistor
    b. Lower feedback resistor shorted
    c. No input voltage
    d. Open load resistor

33. An op amp with inverting voltage feedback has no output voltage. A possible trouble is
    a. No positive supply voltage
    b. Open feedback resistor
    c. No feedback voltage
    d. Shorted load resistor

34. An op amp with inverting voltage feedback has no output voltage. A possible trouble is
   a. No negative supply voltage
   b. Shorted feedback resistor
   c. No feedback voltage
   d. Open load resistor

BASIC PROBLEMS

Sec. 19-1 Noninverting Voltage Feedback

19-1. If the op amp of Fig. 19-24a has a differential voltage gain of 200,000, what is the voltage gain?

19-2. In Fig. 19-24a, what do the output and error voltages equal if the op amp has a voltage gain of 300,000?

19-3. Suppose the op amp voltage gain of Fig. 19-24a changes from 200,000 to 50,000. What is the original error voltage? The new error voltage?

19-4. If the op amp of Fig. 19-24b has a differential voltage gain of 80,000, what is the voltage gain?

19-5. In Fig. 19-24b, what do the output and error voltages equal if the op amp has a voltage gain of 50,000?

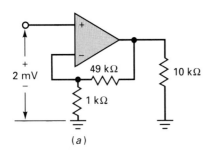

(a)

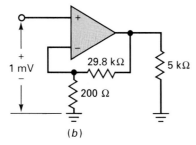

(b)

FIGURE 19-24

19-6. Suppose the op amp voltage gain of Fig. 19-24b changes from 150,000 to 50,000. What is the original error voltage? The new error voltage?

19-7. In Fig. 19-25, what is the voltage gain of the amplifier for each position of the switch?

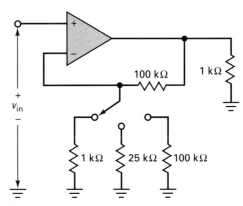

FIGURE 19-25

Sec. 19-2 Open-Loop and Closed-Loop Voltage Gains

19-8. What is the output voltage in Fig. 19-24a?

19-9. What is the output voltage in Fig. 19-24b?

19-10. What is the voltage across the 1 kΩ of Fig. 19-24a? The current through the 49 kΩ?

19-11. What is the voltage across the 200 Ω of Fig. 19-24b? The current through the 29.8 kΩ of Fig. 19-24b?

19-12. In Fig. 19-25, what is the output voltage for each position of the switch if the input voltage is 1 mV?

Sec. 19-3 Input and Output Impedances

19-13. In Fig. 19-24a, $A = 200,000$, $z_{in} = 5$ MΩ, and $z_{out} = 50$ Ω. What are the closed-loop input and output impedances?

19-14. In Fig. 19-24b, $A = 350,000$, $z_{in} = 4$ MΩ, and $z_{out} = 65$ Ω. What are the closed-loop input and output impedances?

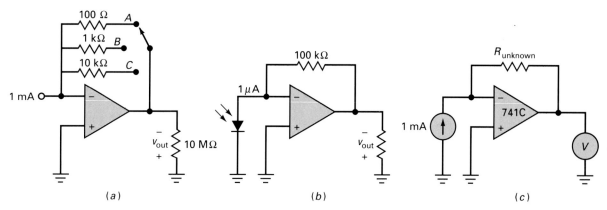

FIGURE 19-26

19-15. A 741C with $A = 100{,}000$, $z_{in} = 2\ M\Omega$, and $z_{out} = 75\ \Omega$ is used in Fig. 19-25. What are the closed-loop input and output impedances for each switch position?

Sec. 19-4 Other Benefits of Negative Feedback

19-16. The op amp of Fig. 19-24a has $A = 200{,}000$, $I_{in(bias)} = 50\ nA$, $I_{in(off)} = 10\ nA$, and $v_{in(off)} = 0.5\ mV$. What is the output offset voltage?

19-17. The op amp of Fig. 19-24b has $A = 300{,}000$, $I_{in(bias)} = 100\ nA$, $I_{in(off)} = 25\ nA$, and $\Delta V_{BE} = 3\ mV$. What is the output offset voltage?

19-18. A 741C with $A = 100{,}000$ and $I_{in(bias)} = 80\ nA$, $I_{in(off)} = 20\ nA$, and $v_{in(off)} = 2\ mV$ is used in Fig. 19-25. What is the output offset voltage for each position of the switch?

Sec. 19-5 Inverting Voltage Feedback

19-19. A current-to-voltage converter has an input current of 1 mA and a feedback resistance of 2 kΩ. What does the output voltage equal?

19-20. What does the output voltage equal in Fig. 19-26a for each position of the switch?

19-21. The photodiode of Fig. 19-26b produces a current of 1 μA. What is the output voltage?

19-22. The unknown resistor of Fig. 19-26c has a value of 3.3 kΩ. What is the output voltage?

19-23. If the output voltage is 2 V in Fig. 19-26c, what is the value of the unknown resistance?

Sec. 19-6 Bandwidth

19-24. An op amp has $A = 100{,}000$ and $f_2 = 10\ Hz$. If the closed-loop voltage gain is 150, what is the closed-loop bandwidth?

19-25. An op amp has $f_{unity} = 2\ MHz$. What is the closed-loop bandwidth if $A_{CL} = 400$?

19-26. Given Fig. 19-27a, what is the open-loop voltage gain? The value of f_{unity}? The gain-bandwidth product?

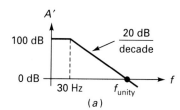

(a)

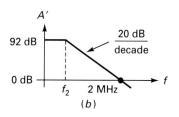

(b)

FIGURE 19-27

19-27. An op amp has the graph of Fig. 19-27a. What is the closed-loop bandwidth if $A_{CL} = 200$?

19-28. Given Fig. 19-27b, what is the value of f_2? The gain-bandwidth product?

19-29. The op amp of Fig. 19-24a has the graph of Fig. 19-27a. What is the maximum undistorted output peak voltage at $f_{2(CL)}$ if the slew rate is 1 V/μs?

19-30. A voltage follower has the graph of Fig. 19-27a. What is the closed-loop bandwidth?

TROUBLESHOOTING PROBLEMS

19-31. In Fig. 19-26c, the output voltmeter reads five times too high for all values of R_{unknown}. Which of the following is a possible trouble:
  a. Inverting input grounded by solder bridge
  b. 741C has an open-loop gain of 500,000 instead of 100,000
  c. Current source produces 5 mA instead of 1 mA
  d. Noninverting input is open

19-32. In Fig. 19-11, a large output offset voltage exists. Which of the following is a possible trouble:
  a. 741C defective
  b. 39 kΩ shorted
  c. Supplies are 10 V instead of 15 V.
  d. Noninverting input grounded by solder bridge

19-33. In Fig. 19-26a, there is no output voltage. Which of these is a possible trouble:
  a. Inverting input grounded by solder bridge
  b. 741C has an open-loop gain of 500,000 instead of 100,000
  c. Current source produces 5 mA instead of 1 mA
  d. Noninverting input is open

19-34. The 100 kΩ of Fig. 19-26b is open. Which of these is the most likely output voltage?
  a. Zero
  b. Maximum positive or negative
  c. Slightly different from zero
  d. +15 V

19-35. The amplifier of Fig. 19-25 has no output in the middle switch position. Name some of the possible troubles.

ADVANCED PROBLEMS

19-36. The inverting input of Fig. 19-28a is a virtual ground. What is the output voltage?

19-37. What is the output voltage in Fig. 19-28b?

19-38. Derive a formula for the voltage gain $v_{\text{out}}/v_{\text{in}}$ of Fig. 19-28a.

19-39. The feedback resistor of Fig. 19-29 has a resistance that is controlled by sound waves. If the feedback resistance varies sinusoidally between 9 kΩ and 11 kΩ, what is the output voltage?

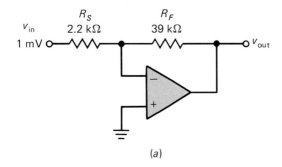

(a)

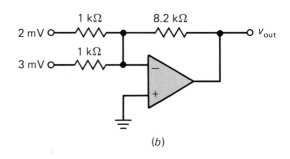

(b)

FIGURE 19-28

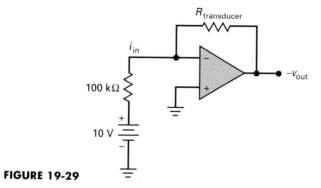

FIGURE 19-29

19-40. Temperature controls the feedback resistance of Fig. 19-29. If the feedback resistance varies from 1 kΩ to 10 kΩ, what is the range of output voltage?

19-41. Figure 19-30 shows a sensitive dc voltmeter that uses a BIFET op amp. Assume the output voltage has been nulled with the zero adjustment. What is the input voltage that produces full-scale deflection for each switch position?

T-SHOOTER PROBLEMS

Use Fig. 19-31 for the remaining problems. Any resistor R_2 through R_4 may be open or shorted.

Also, connecting wires AB, CD, or FG may be open.

19-42. Find Trouble 1.

19-43. Find Trouble 2.

19-44. Find Trouble 3.

19-45. Find Trouble 4.

19-46. Find Trouble 5.

19-47. Find Trouble 6.

19-48. Find Trouble 7.

19-49. Find Trouble 8.

19-50. Find Trouble 9.

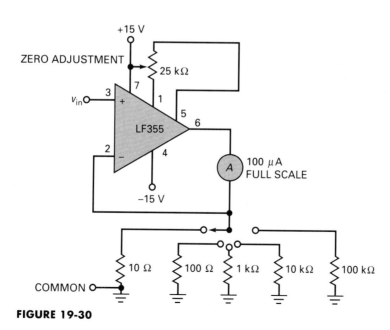

FIGURE 19-30

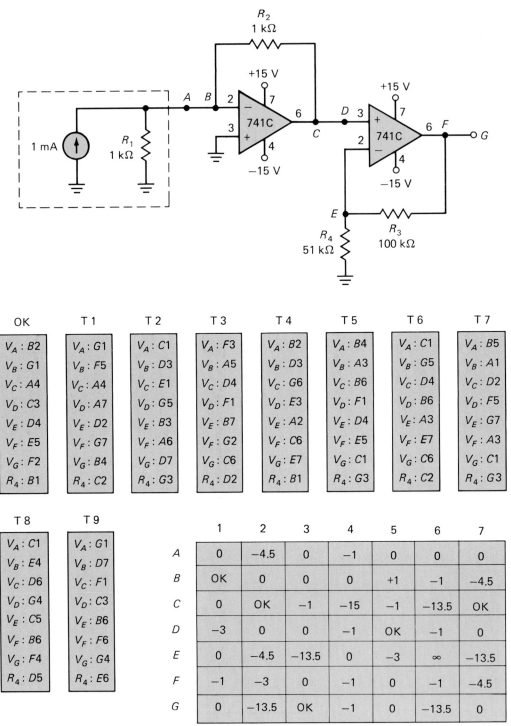

FIGURE 19-31 T-Shooter™. (*Patent pending: Courtesy of Malvino Inc.*)

20 LINEAR OP-AMP CIRCUITS

Previous chapters discussed the theory of op amps and negative feedback. Now, you are ready to see how op amps are used in practical applications. Don't be intimidated by the variety of circuits you see in this chapter. You don't have to understand every detail of every circuit. All that's necessary is that you get the basic idea of how op amps can be used.

This chapter is about *linear* op-amp circuits, the kind of circuits where the op amp never saturates under normal operating conditions. The amplified output from a linear op-amp circuit preserves the shape of the input signal. For instance, if the input signal is sinusoidal, the output signal is also sinusoidal. The chapter covers voltage amplifiers, current sources, differential amplifiers, and a variety of other op-amp circuits.

20-1 NONINVERTING VOLTAGE AMPLIFIERS

A noninverting voltage-feedback amplifier is approximately an ideal voltage amplifier because of its high input impedance, low output impedance, and stable voltage gain. Let's take a look at some circuits that utilize noninverting voltage feedback.

Basic Circuit

Figure 20-1 shows the basic circuit for a noninverting voltage-feedback amplifier. As you know, it has a closed-loop voltage gain of

$$A_{CL} = \frac{R_1}{R_2} + 1 \qquad (20\text{-}1)$$

and a bandwidth of

$$f_{2(CL)} = \frac{f_{unity}}{A_{CL}} \qquad (20\text{-}2)$$

Since the gain-bandwidth product is a constant, you have to decrease the closed-loop voltage gain if you want more bandwidth.

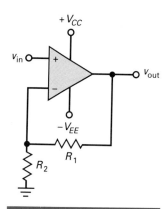

FIGURE 20-1
Amplifier with noninverting voltage feedback.

There is a basic idea about any op-amp circuit that is worth memorizing. As long as the operation is linear (op amp not saturated), the error voltage approaches zero. This means that the inverting input voltage is within microvolts of the noninverting input voltage. Because of this, a trouble-shooter treats the two voltages as equal. This idea applies not only to troubleshooting, but to all preliminary analysis and design of op-amp circuits. It is a crucial idea, one that will let you sail through the analysis of new circuits with minimum difficulty. So, memorize this approximation: *If the op amp is not saturated, its two input voltages are equal.*

AC Amplifier

In some applications, you don't need a response that extends down to zero frequency because only ac signals drive the input. In this case, you can insert a bypass capacitor on R_2, as shown in Fig. 20-2. This has the advantage of minimizing the output offset voltage. Here's why. In the midband of the amplifier, the bypass capacitor appears shorted and places an ac ground at the bottom of R_2. Then the feedback fraction is

$$B = \frac{R_2}{R_1 + R_2}$$

In this case, the circuit amplifies the input voltage as previously described. In the midband of the amplifier, therefore, the circuit acts the same as before.

The reason for using the bypass capacitor is this. It almost eliminates the output offset voltage. Why? When the frequency is zero, the bypass capacitor is open and the feedback fraction B increases

$$B = \frac{\infty}{R_1 + \infty} = 1$$

A pure mathematician would shudder at the sight of this equation, but it is valid if we define ∞ as an extremely large value. And that's actually what we have with a capacitor at zero frequency. With B equal to 1, the desensitivity becomes $1 + A$, the maximum value it can have. This reduces the output offset voltage to

$$V_{oo(CL)} = \frac{V_{oo(OL)}}{1 + A} \tag{20-3}$$

This is the minimum possible value of output offset voltage. With a 741C, it is only a couple of millivolts.

We can also use input and output coupling capacitors in Fig. 20-2. The advantage of this is dc isolation of the amplifier from the source and the load. If the source or the load has a dc voltage with respect to ground, the coupling capacitor will prevent this dc voltage from interfering with the operation of the amplifier.

If you ever have to analyze or design an ac amplifier of this type, you can calculate the critical frequencies as follows. The basic equation for critical frequency is

$$f = \frac{1}{2\pi RC} \tag{20-4}$$

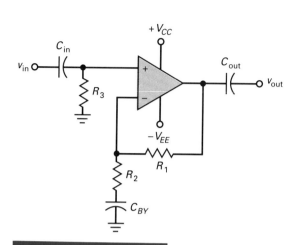

FIGURE 20-2

Ac-coupled voltage amplifier.

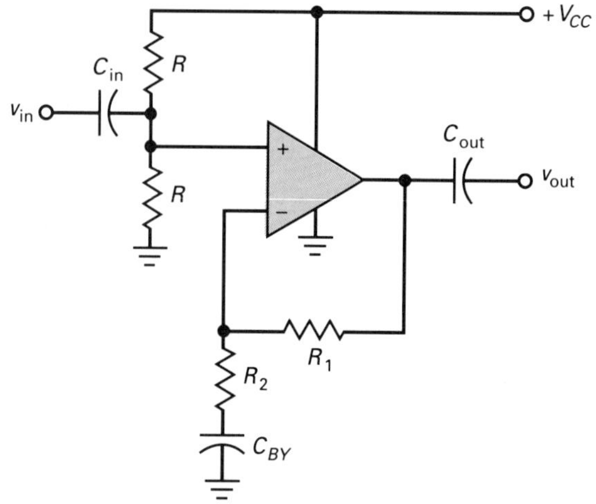

FIGURE 20-3

Single supply ac-coupled voltage amplifier.

Now you have to figure out the values of R and C for each coupling and by-pass circuit. For the input coupling capacitor, assume a generator with a resistance of R_G faces the input coupling capacitor. Then $R = R_G + R_3$ and $C = C_{in}$. For the output coupling capacitor, assume the closed-loop output impedance is approximately zero and assume a load resistance of R_L. Then $R = R_L$ and $C = C_{out}$. For the bypass capacitor, the inverting input is a virtual ground, which gives $R = R_2$ and $C = C_{BY}$.

Single-Supply Operation

Most op-amp circuits use dual or split supplies, such as $V_{CC} = +15$ V and $V_{EE} = -15$ V. But sometimes you will see an op-amp circuit running off a single supply, as shown in Fig. 20-3. Notice that the V_{EE} input is grounded. To get maximum output swing, you need to bias the noninverting input at half the supply voltage, which is conveniently done with an equal-resistor voltage divider. This produces a dc input of $+0.5V_{CC}$ at the noninverting input. Because the negative feedback forces v_{error} to be approximately zero, the inverting input is automatically pulled up to a quiescent value of $+0.5V_{CC}$. The ac operation is the same as with Fig. 20-2, except that the output swing is limited to a few volts less than V_{CC}. For $V_{CC} = +15$ V, this means a maximum peak-to-peak unclipped output of approximately 12 to 13 V.

Audio Amplifier

Figure 20-4 shows another single-supply design. The collector of the bipolar stage typically has a quiescent voltage of approximately half of V_{CC}. Therefore, we can direct-couple to the noninverting input. This neatly eliminates the coupling capacitor and the voltage divider shown earlier, while providing additional voltage gain.

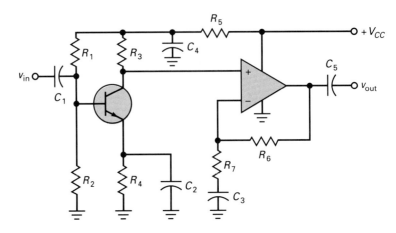

FIGURE 20-4 Bipolar CE stage is direct-coupled to op-amp stage.

Most of the components for the bipolar stage are familiar from earlier discussions. For instance, R_1 and R_2 provide voltage-divider bias, with C_2 bypassing the emitter to ground for maximum voltage gain. The only new components are R_5 and C_4. This bypass circuit has a low critical frequency to prevent oscillations caused by unwanted feedback between stages (discussed in Chap. 22).

An audio amplifier covers frequencies from 20 Hz to 20 kHz. If a 741C is used for the op amp, a closed-loop voltage gain of 50 produces an upper cutoff frequency of 20 kHz. With a supply of $+15$ V, the bipolar stage will have a large voltage gain. Assuming a bipolar gain of 200, the audio amplifier has a total voltage gain of 10,000, equivalent to 80 dB.

JFET-Switched Voltage Gain

Some applications require a change in closed-loop voltage gain. Figure 20-5 shows a JFET-controlled amplifier. The control voltage for the JFET switch comes from another circuit that produces a two-level output, either 0 V or a voltage that is equal to $V_{GS(off)}$. When the control voltage equals $V_{GS(off)}$, the JFET switch is open and the closed-loop voltage gain is

$$A_{CL} = \frac{R_1}{R_2} + 1$$

When the control voltage is zero, the JFET switch is closed and R_3 is placed in parallel with R_2. In this case, the closed-loop voltage gain decreases to

$$A_{CL} = \frac{R_1}{R_2 \| R_3} + 1 \tag{20-5}$$

A typical JFET for an application like this is the 2N4860, which has a maximum $r_{ds(on)}$ of 40 Ω. In most designs, R_3 is made much larger than $r_{ds(on)}$ to prevent $r_{ds(on)}$ from affecting the closed-loop voltage gain. Often, you will see several resistors and JFET switches in parallel with R_2 to provide a selection of closed-loop voltage gains.

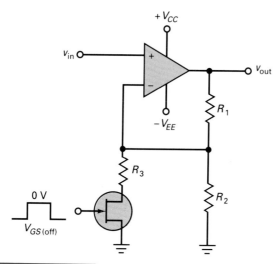

FIGURE 20-5   JFET switch controls voltage gain of op-amp circuit.

EXAMPLE 20-1

What is the output voltage in the midband of Fig. 20-6? What is the upper closed-loop critical frequency? The three lower critical frequencies?

SOLUTION

Since $R_1 = 100 \text{ k}\Omega$ and $R_2 = 1 \text{ k}\Omega$, the closed-loop voltage gain is

$$A_{CL} = \frac{100 \text{ k}\Omega}{1 \text{ k}\Omega} + 1 = 101$$

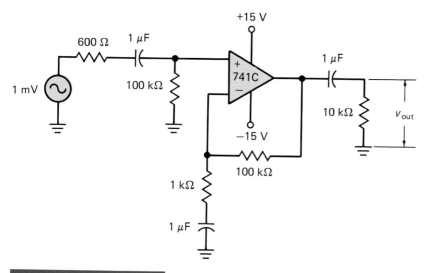

FIGURE 20-6   Example.

Since $A_{CL} = 101$, the closed-loop upper critical frequency is

$$f_{2(CL)} = \frac{1\text{ MHz}}{101} = 9.9\text{ kHz}$$

The input coupling capacitor sees a total series resistance of approximately 100 kΩ. Therefore, it has a critical frequency of

$$f_c = \frac{1}{2\pi(100\text{ k}\Omega)(1\ \mu\text{F})} = 1.59\text{ Hz}$$

The output coupling capacitor sees a total series resistance of 10 kΩ because the closed-loop output impedance approaches zero. Therefore, it has a critical frequency of

$$f_c = \frac{1}{2\pi(10\text{ k}\Omega)(1\ \mu\text{F})} = 15.9\text{ Hz}$$

The bypass capacitor sees a resistance of 1 kΩ because the inverting input is a virtual ground. Therefore, it has a critical frequency of

$$f_c = \frac{1}{2\pi(1\text{ k}\Omega)(1\ \mu\text{F})} = 159\text{ Hz}$$

EXAMPLE 20-2

The JFET of Fig. 20-7 is off when $V_{GS} = -5$ V. It has a drain resistance of 40 Ω when turned on. What is the minimum output voltage of the circuit? The maximum?

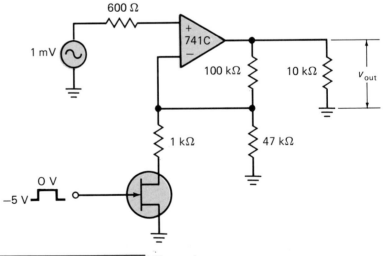

FIGURE 20-7 Example.

SOLUTION

When the JFET has a gate voltage of -5 V, the JFET is off and the closed-loop voltage gain is

$$A_{CL} = \frac{100 \text{ k}\Omega}{47 \text{ k}\Omega} + 1 = 3.13$$

and the output voltage is

$$v_{\text{out}} = 3.13 \,(1 \text{ mV}) = 3.13 \text{ mV}$$

When the JFET has a gate voltage of 0 V, the JFET turns on and has a drain resistance of 40 Ω. This is much smaller than 1 kΩ, so we will ignore it in this analysis. In effect, the bottom of the 1 kΩ is grounded. The equivalent resistance of 1 kΩ in parallel with 47 kΩ is approximately 1 kΩ. Therefore, the closed-loop voltage gain increases to

$$A_{CL} = \frac{100 \text{ k}\Omega}{1 \text{ k}\Omega} + 1 = 101$$

and the output voltage is

$$v_{\text{out}} = 101(1 \text{ mV}) = 101 \text{ mV}$$

20-2 THE INVERTING VOLTAGE AMPLIFIER

The noninverting voltage amplifier produces an output voltage that is in phase with the input voltage, which is desirable in many applications. But there are many other applications where we prefer an inverted output signal. Figure 20-8 shows an inverting voltage amplifier. The circuit is a combination of voltage source and a current-to-voltage converter. Because of the series resistor R_S, the feedback fraction changes and some feedback properties change.

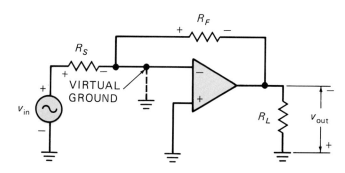

FIGURE 20-8 Inverting voltage amplifier using inverting voltage feedback.

Simplified Analysis

The inverting input of a current-to-voltage converter is a virtual ground. Since the virtual ground has a node voltage of 0 V, all of the input voltage appears across the series resistor. With Ohm's law, the input current is

$$i_{\text{in}} = \frac{v_{\text{in}}}{R_S} \tag{20-6}$$

Because the virtual ground can *sink* (accept) no current, all the input current passes through R_F, producing an output voltage of

$$v_{\text{out}} = i_{\text{in}}R_F$$

which can be rewritten as

$$v_{\text{out}} = \frac{v_{\text{in}}R_F}{R_S} \tag{20-7}$$

By rearranging, we get

$$\frac{v_{\text{out}}}{v_{\text{in}}} = \frac{R_F}{R_S}$$

or simply

$$A_{CL} = \frac{R_F}{R_S} \tag{20-8}$$

This says that the closed-loop voltage gain equals the ratio of the feedback resistor to the series resistor.

You should memorize Eq. (20-8) because you will use it a lot. Also, you should remember the basic process of calculating the output voltage directly from the circuit quantities. The process for analyzing any circuit in the form of Fig. 20-8 is this:

1. Visualize a virtual ground on the inverting input.

2. Visualize all of the input voltage across the series resistor.

3. Use Ohm's law to get the input current through the series resistor.

4. Visualize all of the input current passing through the feedback resistor.

5. Realize that the output voltage is across the feedback resistor as well as the load resistor.

6. Use Ohm's law to calculate the output voltage across the feedback resistor.

Each step in this process should be absolutely clear if you want to solve op-amp circuits quickly and easily.

Impedances

Because of the virtual ground, the right end of R_S appears grounded. This is why the source sees a closed-loop input impedance of

$$z_{\text{in}(CL)} = R_S \tag{20-9}$$

This is an advantage in some applications. The equation says that the input impedance of an inverting voltage amplifier equals R_S. This allows a designer to select a desired value of R_S and thereby set up a specific value of input impedance. One of the reasons the inverting voltage amplifier is very popular is that it allows a designer to control input impedance as well as voltage gain and bandwidth.

We can redraw an inverting voltage amplifier as shown in Fig. 20-9. In this form, it is easier to see the value of the feedback fraction:

$$B = \frac{R_S}{R_F + R_S} \tag{20-10}$$

With this value, we can find the closed-loop output impedance:

$$z_{out(CL)} = \frac{z_{out}}{1 + AB} \tag{20-11}$$

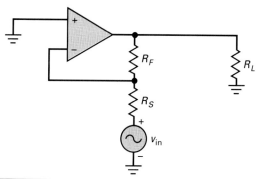

FIGURE 20-9   Inverting voltage amplifier redrawn.

Closed-Loop Gain-Bandwidth Product

Because of the negative feedback, the closed-loop bandwidth of an inverting amplifier is given by

$$f_{2(CL)} = (1 + AB)f_2 \tag{20-12}$$

As usual, the desensitivity factor is the key to the improvement. The larger it is, the greater the closed-loop bandwidth. When the loop gain AB is much greater than 1, Eq. (20-12) is approximated by

$$f_{2(CL)} = ABf_2$$

Since $Af_2 = f_{unity}$, the equation can be rewritten as

$$f_{2(CL)} = Bf_{unity} \tag{20-13}$$

This is useful because it relates the closed-loop bandwidth to the feedback fraction and the unity-gain frequency.

To get the gain-bandwidth product, multiply both sides by A_{CL} to get

$$A_{CL}f_{2(CL)} = A_{CL}Bf_{unity}$$

Since $A_{CL} = R_F/R_S$ and $B = R_S/(R_F + R_S)$, the equation becomes

$$A_{CL}f_{2(CL)} = \frac{R_F}{R_F + R_S} f_{unity}$$

which may be rewritten as

$$A_{CL}f_{2(CL)} = \frac{A_{CL}}{A_{CL} + 1} f_{unity} \qquad (20\text{-}14)$$

When a circuit has A_{CL} much greater than 1, the equation simplifies to

$$A_{CL}f_{2(CL)} = f_{unity} \qquad (20\text{-}15)$$

Equation (20-15) says that the gain product is a constant, the same result that we got with noninverting voltage feedback. But remember the restriction that exists with an inverting voltage amplifier. It is constant only when A_{CL} is much greater than 1. Most designs will satisfy this condition. If you are ever in doubt, fall back on Eq. (20-14) to get a more accurate value for the closed-loop gain-bandwidth product.

Offset Caused by Input Bias Current

In Fig. 20-10a, there is an input offset voltage caused by the base current flowing through the Thevenin resistance facing the inverting input:

$$v_2 = I_{B2}(R_S \parallel R_F)$$

In some designs, a resistor is added between the noninverting input and ground, as shown in Fig. 20-10b. This cancels out much of the unwanted offset because the differential input now is

$$v_1 - v_2 = I_{B1}(R_S \parallel R_F) - I_{B2}(R_S \parallel R_F)$$

or

$$v_1 - v_2 = I_{in(off)}(R_S \parallel R_F)$$

Since $I_{in(off)}$ is usually much smaller than $I_{in(bias)}$, the offset is minimized. The added resistor has no effect on the closed-loop voltage gain because there is no ac voltage across it.

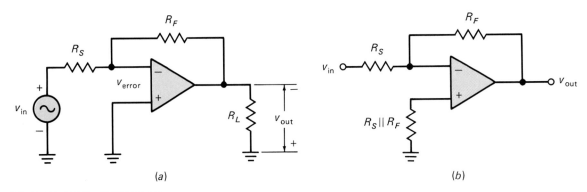

(a)   (b)

FIGURE 20-10   Inverting voltage amplifier.

Properties of an Inverting Voltage Amplifier

Table 20-1 summarizes the properties of an inverting voltage amplifier. The negative feedback stabilizes the voltage gain and input impedance.

Table 20-1 Inverting Voltage Amplifier

Quantity	Effect	Formula
v_{out}/v_{in}	Stabilizes	R_F/R_S
Input impedance	Stabilizes	R_S
Output impedance	Decreases	$z_{out}/(1 + AB)$
Distortion	Decreases	$v_{dist}(1 + AB)$
Output offset	Decreases	$V_{oo(OL)}/(1 + AB)$
Bandwidth	Increases	Bf_{unity}
Gain-bandwidth	May decrease	$A_{CL}f_{unity}/(A_{CL} + 1)$

Because of the voltage feedback, the amplifier has a very low output impedance. As usual, negative feedback decreases distortion and output offset voltage. Notice that gain-bandwidth product may decrease if A_{CL} is not large compared to 1.

EXAMPLE 20-3

Calculate the output voltage, the closed-loop input impedance, and the closed-loop bandwidth in Fig. 20-11. The 741C has an f_{unity} of 1 MHz.

SOLUTION

The closed-loop voltage gain equals the ratio of the feedback to series resistance:

$$A_{CL} = \frac{100 \text{ k}\Omega}{1 \text{ k}\Omega} = 100$$

So, the output voltage is

$$v_{out} = 100(5 \text{ mV}) = 500 \text{ mV}$$

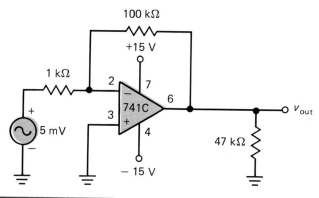

FIGURE 20-11   Example.

The closed-loop input impedance seen by the ac source is

$$z_{\text{in}(CL)} = 1 \text{ k}\Omega$$

Since the closed-loop voltage gain is much greater than 1, we can use Eq. (20-15):

$$A_{CL}f_{2(CL)} = 1 \text{ MHz}$$

With $A_{CL} = 100$, the closed-loop bandwidth is

$$f_{2(CL)} = \frac{1 \text{ MHz}}{100} = 10 \text{ kHz}$$

This is a close approximation to the closed-loop bandwidth.

EXAMPLE 20-4

Solve the problem of the preceding example in a different way.

SOLUTION

Here is another way to find the output voltage. Because of the virtual ground, all of the input voltage (5 mV) is across the series resistor (1 kΩ). This gives an input current of

$$i_{\text{in}} = \frac{5 \text{ mV}}{1 \text{ k}\Omega} = 5 \text{ μA}$$

Since a virtual ground can sink no current, all of the input current must flow through the feedback resistor. This means the feedback resistor has a voltage of

$$v_{\text{out}} = (5 \text{ μA})(100 \text{ k}\Omega) = 500 \text{ mV}$$

Since the right end of the 1 kΩ has zero voltage because of the inverting ground, the input impedance seen by the source is

$$z_{\text{in}(CL)} = 1 \text{ k}\Omega$$

The feedback fraction is

$$B = \frac{1 \text{ k}\Omega}{101 \text{ k}\Omega} = 0.0099$$

Therefore, the closed-loop bandwidth is

$$f_{2(CL)} = 0.0099(1 \text{ MHz}) = 9.9 \text{ kHz}$$

Notice that this bandwidth is slightly less than the bandwidth calculated in the preceding example because we used the more accurate Bf_{unity} to calculate it.

20-3 OP-AMP INVERTING CIRCUITS

The inverting voltage amplifier has a stable voltage gain and input impedance. These properties allow designers to come up with a variety of inverting op-amp circuits for different applications. In analyzing these circuits, remember that the inverting input is within microvolts of the noninverting input because the open-loop voltage gain is extremely high.

Switchable Inverter

Figure 20-12a shows an op amp that can function as either an inverter or noninverter. With the switch in the lower position, the noninverting input is grounded. Since the feedback and series resistances are equal, we have an inverting voltage amplifier with a closed-loop voltage gain of 1.

When the switch is moved to the upper position, the input signal drives the noninverting input. Since the inverting input is within microvolts of the noninverting input, there is approximately zero current through the series resistance. But if there is zero current through the series resistor, it can be opened either physically or mentally. In other words, the circuit

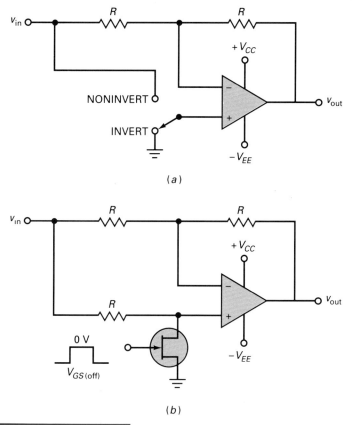

FIGURE 20-12 (a) Switchable inverter/noninverter; (b) JFET switch for inverter/noninverter.

will work the same whether the series resistance is present or not. Visualize the series resistor removed from the circuit. Then we have a voltage follower, which means we have a noninverting voltage amplifier with a closed-loop voltage gain of 1.

JFET-Controlled Switchable Inverter

Figure 20-12b is a modification of Fig. 20-12a. Recall that the drain curves of a JFET extend on both sides of the origin. For this reason, no dc supply voltage is needed. The ac signal voltage on the drain is sufficient. The JFET has either a very low or a very high resistance, depending on the gate voltage.

When the gate voltage is at 0 V, the JFET has a low resistance and the noninverting input is approximately grounded. In this case, the circuit acts like an inverting voltage amplifier with a closed-loop voltage gain of 1. On the other hand, when the gate voltage is at $V_{GS(off)}$, the JFET switch is open and all of the input signal arrives at the noninverting input. The circuit now acts like a nonverting voltage amplifier with a closed-loop voltage gain of 1. For proper operation, R should be at least 100 times greater than $r_{ds(on)}$ of the JFET.

Adjustable Bandwidth

Sometimes we would like to change the closed-loop bandwidth of an inverting voltage amplifier without changing the closed-loop voltage gain. Sound impossible? Not at all. Figure 20-13a shows an adjustable resistor, R, connected between the inverting input and ground. When R is varied, the bandwidth will change but the voltage gain will remain constant. Take a few minutes and try to figure out why this is true before you read what comes next.

Figure 20-13b shows an equivalent circuit with the input side thevenized. The effective source resistance driving the inverting input is now R_S in parallel with R. For this reason, the feedback fraction is

$$B = \frac{R_S \parallel R}{R_S \parallel R + R_F} \qquad (20\text{-}16)$$

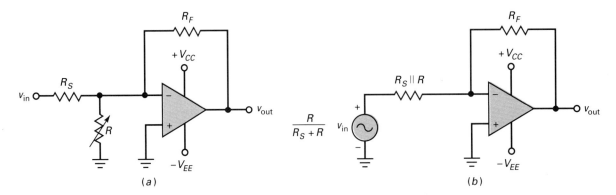

FIGURE 20-13 Circuit with constant voltage gain but adjustable bandwith.

Just by looking at this equation, you can see that B will vary when R varies. This means the closed-loop bandwidth also will vary because

$$f_{2(CL)} = Bf_{unity}$$

On the other hand, the output voltage is

$$v_{out} = \frac{R_F}{R_S \| R} \frac{R}{R_S + R} v_{in}$$

which reduces to

$$v_{out} = \frac{R_F}{R_S} v_{in} \qquad (20\text{-}17)$$

This equation no longer contains R because it has cancelled out. Since R_F and R_S are constants, the closed-loop voltage gain remains constant even though R varies.

Single-Supply Operation

Figure 20-14 shows a single-supply inverting voltage amplifier that can be used only with ac signals. The V_{EE} supply is grounded and half the V_{CC} supply is applied to the noninverting input. Because the two inputs differ by only a few microvolts, the inverting input has a quiescent voltage of approximately $+V_{CC}/2$. Since the input coupling capacitor is open at zero frequency, the circuit appears to be a voltage follower at zero frequency and the closed-loop voltage gain is 1, which means minimum output offset voltage.

For ac signals, the circuit is an inverting amplifier with a critical frequency of

$$f_{in} = \frac{1}{2\pi RC} \qquad (20\text{-}18)$$

where $R = R_S$ and $C = C_{in}$. If you visualize a virtual ground on the inverting input, you will see why these values are correct for the coupling

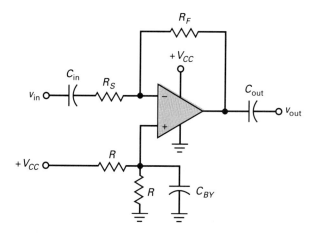

FIGURE 20-14   Single-supply inverting amplifier.

circuit. Ten times above this frequency the closed-loop voltage gain is within a half percent of the midband value, R_F/R_S.

A bypass capacitor is used on the noninverting input, as shown in Fig. 20-14. This reduces the power-supply ripple and noise appearing at the noninverting input. To be effective, the cutoff frequency of this bypass circuit should be much lower than the ripple frequency out of the power supply. You calculate the critical frequency of this bypass circuit with $R = R/2$ and $C = C_{BY}$. Where does $R = R/2$ come from? It is the Thevenin resistance facing C_{BY}.

Adjustable Inverter

When the adjustable resistor of Fig. 20-15a is reduced to zero, the noninverting input is grounded and the circuit becomes an inverting amplifier with a maximum voltage gain of R_F/R_S. When the adjustable resistor is increased to R_F, equal voltages drive the noninverting and inverting inputs. Because of the common-mode rejection, the output voltage is zero. Therefore, the circuit of Fig. 20-15a has an adjustable voltage gain from approximately zero to R_F/R_S.

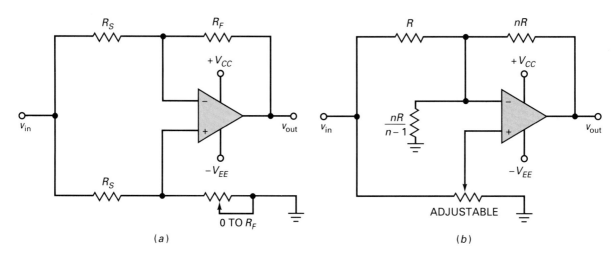

FIGURE 20-15 (a) Inverter with adjustable gain; (b) circuit with gain adjustable from $+1$ to -1.

Adjustable Inverter/Noninverter

Figure 20-15b shows a circuit that allows us to adjust the voltage gain between $-n$ and $+n$. When the adjustable resistor is at zero, the noninverting input is grounded and the circuit becomes an inverting amplifier with a closed-loop voltage gain of

$$A_{CL} = \frac{nR}{R} = n \qquad (20\text{-}19)$$

When the adjustable resistor is at the other extreme, the input voltage is applied directly to the noninverting input. Because the two inputs are within microvolts of each other, the inverting input voltage is approximately the same as the noninverting input. This means that there is zero voltage across R, equivalent to zero current. Again, we can physically or mentally open resistor R. In this case, the feedback fraction is

$$B = \frac{nR/(n-1)}{nR/(n-1) + nR} = \frac{1}{n}$$

Therefore, the closed-loop gain is

$$A_{CL} = \frac{1}{1/n} = n \qquad\qquad (20\text{-}20)$$

EXAMPLE 20-5

What is the output voltage and bandwidth in Fig. 20-16?

SOLUTION

This is the circuit with the constant voltage gain and adjustable bandwidth. The voltage gain equals the feedback resistance divided by the series resistance:

$$A_{CL} = \frac{100\text{ k}\Omega}{2\text{ k}\Omega} = 50$$

The output voltage is

$$v_{\text{out}} = 50\ (4\text{ mV}) = 200\text{ mV}$$

With Eq. (20-16), the feedback fraction is given by

$$B = \frac{R_S \| R}{R_S \| R + R_F}$$

The potentiometer is connected as a variable resistance with a maximum value of 5 kΩ. When variable resistance is zero, the feedback fraction is zero. When the variable resistance is 5 kΩ, the feedback fraction is

$$B = \frac{2\text{ k}\Omega \| 5\text{ k}\Omega}{2\text{ k}\Omega \| 5\text{ k}\Omega + 100\text{ k}\Omega} = \frac{1.43\text{ k}\Omega}{1.43\text{ k}\Omega + 100\text{ k}\Omega} = 0.0141$$

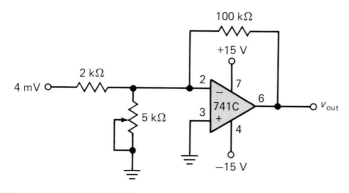

FIGURE 20-16   Example.

The closed-loop voltage gain is given by

$$f_{2(CL)} = Bf_{unity}$$

When B is zero, the closed-loop bandwidth is zero. When $B = 0.0141$, the closed-loop bandwidth is

$$f_{2(CL)} = 0.0141(1 \text{ MHz}) = 14.1 \text{ kHz}$$

In summary, when the potentiometer is varied, the voltage gain remains constant with a value of 50 but the bandwidth varies from 0 to 14.1 kHz.

EXAMPLE 20-6

What is the output voltage in the midband of Fig. 20-17? Find the critical frequencies of the input coupling capacitor, output coupling capacitor, and noninverting bypass capacitor.

SOLUTION

In the midband, all capacitors are ac shorts and the circuit becomes an ordinary inverting amplifier with a voltage gain of

$$A_{CL} = \frac{100 \text{ k}\Omega}{2 \text{ k}\Omega} = 50$$

The input coupling circuit has $R = 2 \text{ k}\Omega$ and $C = 10 \text{ μF}$. Its critical frequency is

$$f_c = \frac{1}{2\pi(2 \text{ k}\Omega)(10 \text{ μF})} = 7.96 \text{ Hz}$$

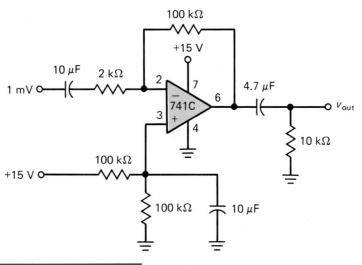

FIGURE 20-17   Example.

The output coupling circuit has $R = 10 \text{ k}\Omega$ and $C = 4.7 \text{ }\mu\text{F}$. Its critical frequency is

$$f_c = \frac{1}{2\pi(10 \text{ k}\Omega)(4.7 \text{ }\mu\text{F})} = 3.39 \text{ Hz}$$

The noninverting bypass circuit has $R = 50 \text{ k}\Omega$ and $C = 10 \text{ }\mu\text{F}$. Its critical frequency is

$$f_c = \frac{1}{2\pi(50 \text{ k}\Omega)(10 \text{ }\mu\text{F})} = 0.318 \text{ Hz}$$

Since the closed gain is 50, the upper critical frequency is

$$f_{2(CL)} = \frac{1 \text{ MHz}}{50} = 20 \text{ kHz}$$

The dominant lower critical frequency is

$$f_{1(CL)} = 7.96 \text{ Hz}$$

Therefore, the amplifier response is down approximately 3 dB at 7.96 Hz and at 20 kHz.

Also, the bypass capacitor has a critical frequency of 0.318 Hz, which is well below 120 Hz, the typical ripple frequency out of a full-wave bridge rectifier. Since 0.318 Hz is almost three decades lower, the ripple on the supply line will be decreased by almost 60 dB before it reaches the noninverting input.

20-4 THE SUMMING AMPLIFIER

Another major advantage of the inverting voltage amplifier is its ability to amplify more than one signal at a time. To understand why, look at Fig. 20-18a. Because of the virtual ground, both input resistors are effectively grounded on the right side. The input current through R_1 is

$$i_1 = \frac{v_1}{R_1}$$

and the input current through R_2 is

$$i_2 = \frac{v_2}{R_2}$$

The total input current is

$$i = i_1 + i_2$$

This current flows through the feedback resistor. Therefore, the output voltage is

$$v_{\text{out}} = (i_1 + i_2)R_3$$

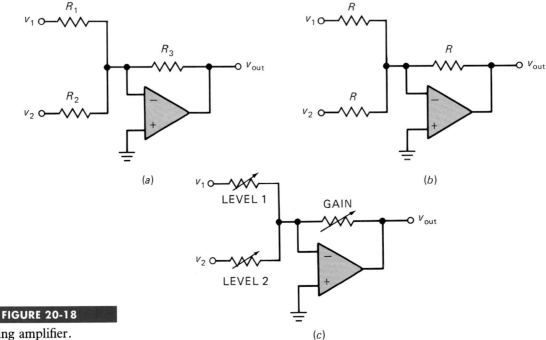

FIGURE 20-18

Summing amplifier.

or

$$v_{\text{out}} = \frac{R_3}{R_1} v_1 + \frac{R_3}{R_2} v_2 \qquad (20\text{-}21)$$

This means that we can have a different voltage gain for each input signal; the output is the sum of the amplified inputs. The same idea applies to any number of inputs because we can add another resistor for each new input signal.

Often, we need a circuit that adds two or more input signals. In this case, we can use a *summer,* an inverting amplifier with several inputs, each with unity voltage gain. Figure 20-18b shows a summer with two inputs. Because all resistors are equal, each input has unity voltage gain and the output is given by

$$v_{\text{out}} = v_1 + v_2$$

Figure 20-18c shows a *mixer,* a convenient way to mix two audio signals from different sources. The adjustable resistors allow us to set the level of each input, and the gain control allows us to adjust the output volume. By decreasing LEVEL 1, we can make the v_1 signal louder at the output. By decreasing LEVEL 2, we make the v_2 signal louder. By increasing GAIN, we can make both signals louder.

EXAMPLE 20-7

What is the output voltage in Fig. 20-19?

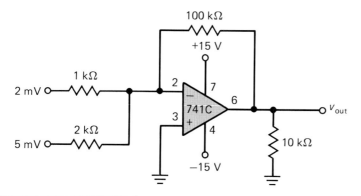

FIGURE 20-19   Example.

SOLUTION

The voltage gain for the upper input signal is

$$A_1 = \frac{100 \text{ k}\Omega}{1 \text{ k}\Omega} = 100$$

and the voltage gain for the lower input signal is

$$A_2 = \frac{100 \text{ k}\Omega}{2 \text{ k}\Omega} = 50$$

The output is the combined sum of the two amplified input signals:

$$v_{\text{out}} = 100(2 \text{ mV}) + 50(5 \text{ mV}) = 450 \text{ mV}$$

20-5 CURRENT BOOSTERS FOR VOLTAGE AMPLIFIERS

The maximum output current of a typical op amp is limited. For instance, the 741C has a maximum output current of 25 mA. If the load requires more than this, you can add a *current booster* to output. Figure 20-20 shows one way to increase the maximum load current. The output of an op amp now drives an emitter follower. Since the circuit is a noninverting voltage-feedback amplifier, the closed-loop voltage gain is

$$A_{CL} = \frac{R_1}{R_2} + 1 \qquad (20\text{-}22)$$

and the output impedance is

$$z_{\text{out}(CL)} = \frac{z_{\text{out}}}{1 + AB} \qquad (20\text{-}23)$$

where z_{out} is the open-loop output impedance looking back into the emitter. Unlike previous voltage amplifiers, the op amp no longer has to

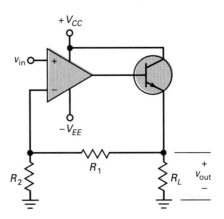

FIGURE 20-20   Bipolar transistor is current booster for op amp.

supply the load current. Instead, it only has to supply base current to the transistor. Because of the transistor current gain, the maximum load current is increased by a factor of β_{dc}. If $\beta_{dc} = 100$ and $I_{max} = 25$ mA, the new maximum load current is 2.5 A.

The major disadvantage of Fig. 20-20 is its unidirectional load current. One way to get bidirectional load current is with a class B push-pull emitter follower, as shown in Fig. 20-21. In this case, the closed-loop voltage gain is

$$A_{CL} = \frac{R_F}{R_S} \qquad (20\text{-}24)$$

The output impedance is still $z_{out}/(1 + AB)$ because of the voltage feedback.

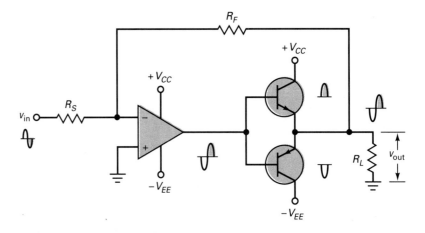

FIGURE 20-21   Class B current booster for op amp.

The negative feedback automatically adjusts the V_{BE} values to whatever they need to be. When the input voltage goes positive, the lower transistor is conducting and the load voltage is negative. On the other hand, when the input goes negative, the upper transistor conducts and the output voltage is positive.

EXAMPLE 20-8

In Fig. 20-20, $\beta_{dc} = 50$ and an LF351 is used for the op amp. What is the maximum load current?

SOLUTION

In Table 18-1, you will find a maximum output current of 20 mA for an LF351. Since the current gain is 50, the maximum load current is

$$I_{max} = 50(20 \text{ mA}) = 1 \text{ A}$$

20-6 VOLTAGE-CONTROLLED CURRENT SOURCES

Figure 20-22a shows a circuit that produces a controlled load current. The load may be a resistor, a relay, or a motor. Because the inverting input is within microvolts of the noninverting input, voltage v_{in} appears across R. Ohm's law then gives us the load current:

$$i_{out} = \frac{v_{in}}{R} \tag{20-25}$$

The value of load resistance does not appear in this equation. Therefore, the output current is independent of the load resistance. Stated another way, the load appears to be driven by a very stiff current source.

Grounded Load

If a *floating load* is all right, a circuit like Fig. 20-22a works quite well. But if the load needs to be grounded on one end (the usual case), we can modify the basic circuit, as shown in Fig. 20-22b. Since the collector and emitter currents of the transistor are almost equal, the current through R is approximately equal to the load current. Because the inverting voltage is within microvolts of the noninverting voltage, the inverting input voltage approximately equals v_{in}. This means that the voltage across R equals V_{CC} minus v_{in}. Therefore, the current through R equals

$$i_{out} = \frac{V_{CC} - v_{in}}{R} \tag{20-26}$$

There is a limit to this output current. The base current in the transistor equals i_{out}/β_{dc}. Since the op amp has to supply this base current, i_{out}/β_{dc} must be less than the $I_{out(max)}$ of the op amp, typically 10 to 25 mA.

There is also a limit on the output voltage in Fig. 20-22b. When the load resistance increases, the load voltage increases and eventually the transistor goes into saturation. Since the emitter is at v_{in} with respect to

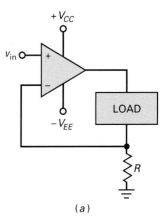

(a)

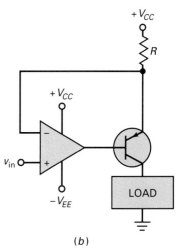

(b)

FIGURE 20-22
Voltage-controlled current sources. (a) Floating load; (b) load grounded on one end.

ground, the maximum load voltage is slightly less than v_{in} when the transistor is saturated. Therefore, the second thing to check with this kind of circuit is to make certain that the output current multiplied by the largest load resistance does not exceed v_{in}.

Grounded Voltage-to-Current Converter

In Eq. (20-26), the load current decreases when the input voltage increases. Figure 20-23 shows a circuit in which the load current is directly proportional to the input voltage. Because the inverting input is within microvolts of the noninverting input, the inverting input of the first op amp has a voltage of v_{in}. The current through the first transistor is

$$i = \frac{v_{in}}{R} \qquad (20\text{-}27)$$

This current produces a collector voltage of

$$V_C = V_{CC} - v_{in} \qquad (20\text{-}28)$$

Since this voltage drives the noninverting input of the second op amp, the inverting voltage is approximately $V_{CC} - v_{in}$. This means that the voltage across the final R is

$$V_{CC} - (V_{CC} - v_{in}) = v_{in}$$

and the output current is

$$i_{out} = \frac{v_{in}}{R} \qquad (20\text{-}29)$$

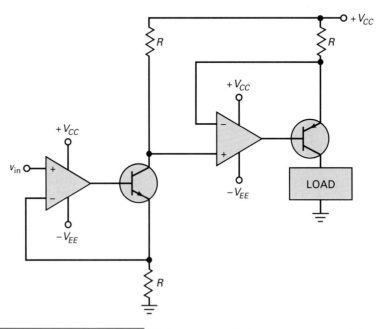

FIGURE 20-23   Voltage-controlled current source with output current proportional to input voltage.

As before, this output current must satisfy the condition: i_{out}/β_{dc} must be less than the $I_{out(max)}$ of the op amp. Furthermore, the load voltage cannot exceed $V_{CC} - v_{in}$ because of transistor saturation. Therefore, $i_{out}R_L$ must be less than $V_{CC} - v_{in}$.

The circuit of Fig. 20-23 is a *voltage-to-current* converter that drives a grounded load. A grounded load is used a lot more than a floating load, so try to remember this circuit. It is very useful and practical whenever you want to convert an input voltage to an output current.

EXAMPLE 20-9

What is the load current in Fig. 20-24a?

SOLUTION

Visualize the input voltage of 2 V appearing across the 1 kΩ. Then the current through the 1 kΩ is

$$i_{out} = \frac{2\ V}{1\ k\Omega} = 2\ mA$$

Because no current can flow into or out of the inverting input, all of the 2 mA must flow through the load resistance of 100 Ω.

You can see that the load resistance has no effect on the output current. If the load resistance is changed to 200 Ω, there is still

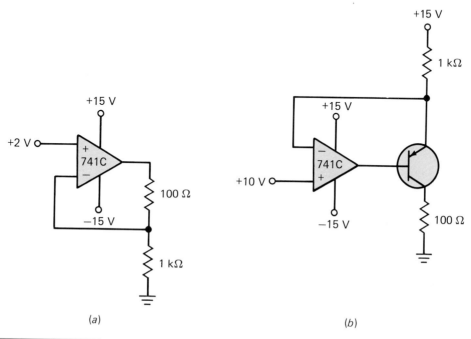

(a)   (b)

FIGURE 20-24   Example.

2 V across the 1 kΩ. Therefore, the output current still equals 2 V divided by 1 kΩ. This means the output current is independent of the load resistance. Put it this way. If the load resistor had a mind, it would think it was being driven by a stiff current source because the load current is constant.

EXAMPLE 20-10

Calculate the output current in Fig. 20-24*b*. Also, work out the value of load voltage.

SOLUTION

The inverting input voltage approximately equals the noninverting input voltage of +10 V. This is why +10 V appears at the bottom of the 1 kΩ. Since 5 V is across the 1 kΩ, the output current is approximately

$$i_{\text{out}} = \frac{5\,\text{V}}{1\,\text{k}\Omega} = 5\,\text{mA}$$

When this current flows through the load resistor, the load voltage is

$$v_{\text{out}} = (5\,\text{mA})(100\,\Omega) = 0.5\,\text{V}$$

EXAMPLE 20-11

What is the maximum load resistance that we can use in Fig. 20-24*b* without saturating the transistor?

SOLUTION

As it now stands, the transistor has +10 V on the emitter and +0.5 V on the collector with a load resistance of 100 Ω. This means that V_{CE} is 9.5 V, which is nowhere near saturation. When the load resistance is increased, the collector voltage will increase and this will decrease V_{CE}. When V_{CE} is approximately zero, the transistor is saturated. So, the question becomes this: What is the load resistance that produces a load voltage of 10 V? When we answer this question, we will have the maximum load resistance.

Since the output current is constant and has a value of 5 mA, the maximum load resistance is

$$R_L = \frac{10\,\text{V}}{5\,\text{mA}} = 2\,\text{k}\Omega$$

20-7 DIFFERENTIAL AND INSTRUMENTATION AMPLIFIERS

Figure 20-25a shows an op amp connected as a differential amplifier. It amplifies v_{in}; the difference between v_1 and v_2. The output voltage is given by

$$v_{out} = \frac{R_1}{R_2} v_{in} \qquad (20\text{-}30)$$

We can derive this equation as follows. When v_2 is zero, the circuit becomes an inverting amplifier with an output voltage of

$$v_{out(1)} = \frac{R_1}{R_2} v_1 \qquad (20\text{-}31)$$

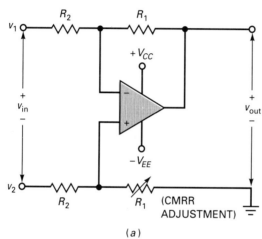

(a)

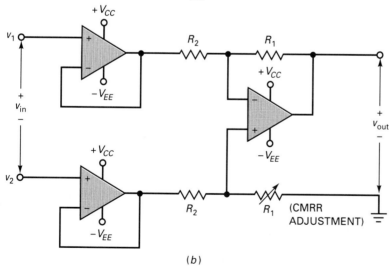

(b)

FIGURE 20-25 (a) Differential amplifier; (b) instrumentation amplifier.

When v_1 is zero, the circuit becomes a noninverting amplifier with a voltage gain of

$$A_{CL} = \frac{R_1}{R_2} + 1 = \frac{R_1 + R_2}{R_2}$$

Because of the lower voltage divider, voltage at the noninverting input is

$$v_2' = \frac{R_1}{R_1 + R_2} v_2$$

Therefore, the output voltage is

$$v_{\text{out}(2)} = A_{CL} v_2' = \frac{R_1}{R_2} v_2 \tag{20-32}$$

When both inputs are present, the output voltage is the difference between $v_{\text{out}(1)}$ and $v_{\text{out}(2)}$:

$$v_{\text{out}} = v_{\text{out}(1)} - v_{\text{out}(2)}$$

or

$$v_{\text{out}} = \frac{R_1}{R_2} (v_1 - v_2) = \frac{R_1}{R_2} v_{\text{in}}$$

which proves Eq. (20-30).

Incidentally, the adjustable resistor of Fig. 20-25a allows us to null or zero common-mode output signals. In this way, we can get a maximum common-mode rejection ratio. A circuit like this amplifies the input voltage v_{in} to get an output voltage of v_{out}. The voltage gain is R_1/R_2. By using precision resistors, we can build a differential amplifier with a stable voltage gain.

Figure 20-25b is an example of an *instrumentation* amplifier; a differential amplifier optimized for high input impedance and high CMRR. An instrumentation amplifier is typically used in applications in which a small differential voltage and a large common-mode voltage are the inputs. In this example of an instrumentation amplifier, voltage followers on each input produce a very high input impedance. Again, a CMRR adjustment is included to balance out the common-mode signals.

Manufacturers can put voltage followers and differential amplifiers on a single chip to get an IC instrumentation amplifier. Examples include the LH0036, LF352, and AD521. The LF352 is an example of a BIFET device, with JFETs used for the input voltage followers and bipolar transistors for the diff amp. This results in an input impedance of approximately $2(10^{12})\ \Omega$ and input bias currents of only 3 pA. The JFETs have extremely low noise, an essential characteristic of a good instrumentation amplifier. The LF352 has other outstanding features, like a CMRR of at least 110 dB, a supply current of only 1 mA, and a single external resistor to control gain.

20-8 ACTIVE FILTERS

At lower frequencies, inductors are bulky and expensive. By using op amps, it is possible to build *active RC* filters. There are many possible filter designs. They are known as Butterworth, Chebyshev, Bessel, and others. You can find entire books on the subject of filter design. In this section, we will discuss only the most popular active filters, known as the Butterworth or maximally flat filters. In the following discussion, we will use the term *cutoff* frequency instead of critical frequency because cutoff frequency is commonly used in filter discussions.

Low-Pass Filter

In Fig. 20-26, a bypass circuit is on the input side of a noninverting voltage amplifier. In the midband of the amplifier, the closed-loop voltage gain is

$$A_{CL} = \frac{R_1}{R_2} + 1$$

This is the gain from the noninverting input to the output. If the cutoff frequency, f_c, of the bypass circuit is much lower than $f_{2(CL)}$, the overall voltage gain v_{out}/v_{in} will be down 3 dB at

$$f_c = \frac{1}{2\pi RC} \tag{20-33}$$

This is the cutoff frequency of the bypass circuit.

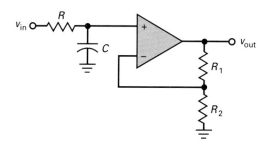

FIGURE 20-26   One-pole low-pass filter.

Above the cutoff frequency, the voltage gain decreases at a rate of 20 dB per decade, equivalent to 6 dB per octave. The mathematical expression for voltage gain is

$$\frac{v_{out}}{v_{in}} = \frac{A_{CL}}{1 + jf/f_c} \tag{20-34}$$

The active filter of Fig. 20-26 passes all frequencies from zero to the cutoff frequency. A filter like this is called a *low-pass* filter. You can recognize a low-pass filter because it usually has one or more bypass circuits. Another way to recognize a low-pass filter is the presence of $1 + jf/f_c$ factors in the denominator of the *transfer function* (the formula for voltage gain).

You often see the word *pole* used with active filters. To understand the complete meaning of a pole, you have to use an advanced topic called the *complex-number plane*. We do not have time for this. So, we will use a simpler definition of a pole. For preliminary discussion, a pole is a bypass circuit that appears in an active filter. For instance, the circuit of Fig. 20-26 has one pole because it has one bypass circuit. Mathematically, each pole in an active filter produces one j factor in the transfer function. Since Fig. 20-26 has only one bypass circuit, Eq. (20-34) has only one j factor.

Two-Pole Low-Pass Filter

Figure 20-27 is a two-pole low-pass filter because it has two bypass circuits. The feedback capacitor is part of a bypass circuit and the other capacitor is part of another bypass circuit. The two poles of this circuit modify the cutoff frequency and the response of the active filter. A mathematical analysis reveals that a closed-loop voltage gain A_{CL} of 1.586 is a critical value. When the gain is 1.586, you get the flattest possible response in the midband. This response is called the *Butterworth* or maximally-flat response and is the one that is most popular.

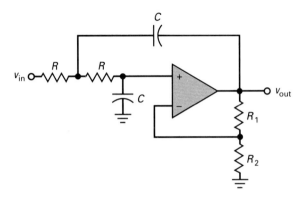

FIGURE 20-27   Two-pole low-pass filter.

Since the closed-loop voltage gain must be 1.586 for a Butterworth response,

$$1.586 = \frac{R_1}{R_2} + 1$$

or

$$R_1 = 0.586R_2$$

If $R_1 = 1 \text{ k}\Omega$, then $R_2 = 0.586 \text{ k}\Omega$. By using the nearest standard value, 560 Ω, we get approximately a maximally flat response.

When $A_{CL} = 1.586$, the cutoff frequency is

$$f_c = \frac{1}{2\pi RC} \tag{20-35}$$

A two-pole Butterworth filter like the one in Fig. 20-27 has the advantage of using equal-value components. At the cutoff frequency, the overall voltage gain is down 3 dB. Above the cutoff frequency, the voltage gain decreases 40 dB per decade, equivalent to 12 dB per octave. This rolloff is twice as fast as before. The reason is that we have a two-pole filter, with each bypass circuit producing a decrease of 20 dB per decade.

In general, a three-pole filter produces 60 dB per decade, a four-pole produces 80 dB per decade, and so on. The simplest way to build a three-pole low-pass filter is by cascading a one-pole filter with a two-pole. To maintain a maximally-flat response, the voltage gain of each section has to be exactly correct. Butterworth tables are available for designing filters with any number of poles. If you are interested in reading more about active filters, see the "Optional Topics" section on the next page.

High-Pass Filters

You can change a low-pass Butterworth filter into a high-pass Butterworth filter by using coupling circuits instead of bypass networks. In this case, a pole is any coupling circuit that appears in the circuit. Figure 20-28a shows a one-pole high-pass filter. Instead of a bypass circuit, we use a coupling circuit with a resistance of R and a capacitance of C. A circuit like this passes the high frequencies but blocks the low frequencies. The cutoff frequency is still given by $1/(2\pi RC)$. Below this frequency, the output voltage decreases 20 dB per decade.

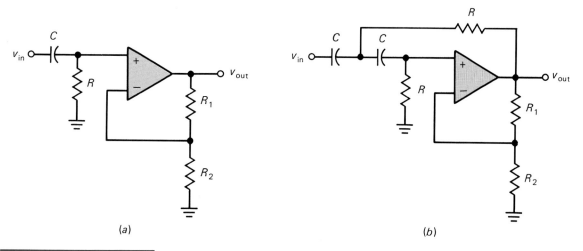

(a)   (b)

FIGURE 20-28   (a) One-pole high-pass filter; (b) two-pole high-pass filter.

Similarly, Figure 20-28b shows a two-pole high-pass filter. With a filter like this, the voltage gain is down 3 dB at the cutoff frequency. Below the cutoff frequency, the voltage gain rolls off at a rate of 40 dB per decade. Again, the midband voltage gain has to be set to 1.586 to get a maximally-flat response in the passband.

EXAMPLE 20-12

Figure 20-29 shows a two-pole filter. What is the output voltage in the midband? The cutoff frequency?

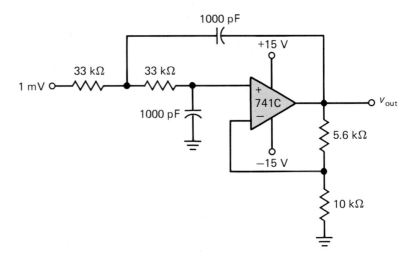

Example.

SOLUTION

The voltage gain in the midband is

$$A_{CL} = \frac{5.6\,\text{k}\Omega}{10\,\text{k}\Omega} + 1 = 1.56$$

The output voltage is

$$v_{\text{out}} = 1.56(1\text{ mV}) = 1.56\text{ mV}$$

Notice that the voltage gain of 1.56 is very close to the theoretical requirement of 1.586, the value needed for a Butterworth response.
The cutoff frequency of the filter is

$$f = \frac{1}{2\pi(33\text{ k}\Omega)(1000\text{ pF})} = 4.82\text{ kHz}$$

At this frequency the output voltage is down 3 dB, which means

$$v_{\text{out}} = 0.707(1.56\text{ mV}) = 1.1\text{ mV}$$

Above 4.82 kHz, the output voltage decreases approximately 40 dB per decade.

OPTIONAL TOPICS

The following material continues the earlier discussions at a more advanced and specialized level. All the topics are optional because they are not used in any of the basic discussions in later chapters. This section will be a useful reference when you are in industry because then you will probably want more advanced viewpoints.

20-9 AUTOMATIC GAIN CONTROL

AGC stands for *automatic gain control*. In many applications like radio and television, we want the voltage gain to change automatically when the input signal changes. Specifically, when the input signal increases, we want the voltage gain to decrease. In this way, the output voltage will be approximately constant. The reason for wanting AGC in a radio or television is to keep the sound signal from changing abruptly when we tune in different stations. We don't want our ears blasted by excessive volume just because the input signal is a lot stronger from one station than from another.

Audio AGC

Figure 20-30 shows an audio AGC circuit. Q_1 is a JFET used as a voltage-variable resistance. For small-signal operation with drain voltages near zero, the JFET operates in the ohmic region and has a resistance of $r_{ds(on)}$ to ac signals. The $r_{ds(on)}$ of a JFET can be controlled by the gate voltage. The more negative V_{GS} is, the larger $r_{ds(on)}$ becomes. With a JFET like the 2N4861, $r_{ds(on)}$ can vary from 100 Ω to more than 10 MΩ. The combination of R_3 and Q_1 acts like a voltage divider whose output varies between $0.001v_{in}$ and v_{in}. Therefore, the noninverting input voltage is between $0.001v_{in}$ and v_{in}, a 60-dB range. The amplified output voltage is $R_1/R_2 + 1$ times this input voltage.

In Fig. 20-30, the output voltage is coupled to the base of Q_2. For peak-to-peak outputs less than 1.4 V, Q_2 is cut off because there is no

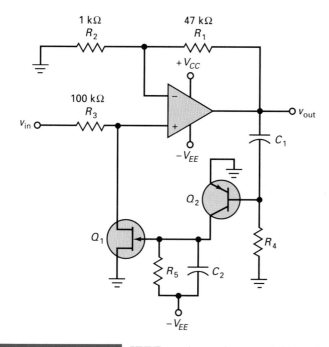

FIGURE 20-30 JFET used as voltage-variable resistance in AGC circuit.

bias on it. In this case, capacitor C_2 is uncharged and the gate of Q_1 is at $-V_{EE}$, enough to cut off the JFET. This means almost all of the input voltage reaches the noninverting input. In other words, an output voltage of less than 1.4 V peak-to-peak means that the circuit acts like an ordinary noninverting voltage amplifier with a maximum input signal.

When the output has peak-to-peak voltage greater than 1.4 V, Q_2 conducts during part of the negative half cycle. This charges capacitor C_2 and increases the gate voltage above the quiescent level of $-V_{EE}$. When this happens, $r_{ds(on)}$ decreases. As it does, the output of the R_3 and Q_1 voltage divider decreases, meaning there is less input voltage at the noninverting input. Stated another way, the overall voltage gain of the circuit decreases when the peak-to-peak output voltage gets above 1.4 V.

The whole purpose of the AGC circuit is to change the voltage gain as needed to keep the output voltage approximately constant. This way, the voltage gain decreases if the input voltage increases, and vice versa. One reason for using AGC is to prevent sudden increases in signal level from overdriving a loudspeaker. For instance, if you're listening to a radio, you don't want an unexpected increase in signal to bombard your hearing. In summary, even though the input voltage of Fig. 20-30 varies over a 60-dB range, the peak-to-peak output is only slightly more than 1.4 V.

Low-Level Video AGC

Figure 20-31a shows a standard technique for *video* AGC that has been used for frequencies up to 10 MHz. In this circuit the JFET acts like a voltage-variable resistance. When the AGC voltage is zero, the JFET is cut off by the negative bias, and its $r_{ds(on)}$ is maximum. As the AGC voltage increases, the $r_{ds(on)}$ of the JFET decreases. The signal driving the inverting voltage amplifier is

$$v_A = \frac{R_2 + r_{ds(on)}}{R_1 + R_2 + r_{ds(on)}} v_{in}$$

The output voltage from the inverting amplifier is

$$v_{out} = \frac{R_F}{R_S} v_A$$

In this circuit the JFET acts like a voltage-variable resistance controlled by $+V_{AGC}$. The more positive the AGC voltage, the smaller the value of $r_{ds(on)}$ and the lower the voltage to the inverting amplifier. This means that the AGC voltage controls the overall voltage gain of the circuit.

With a wideband op amp, the circuit works well for input signals up to approximately 100 mV. Beyond this level, the JFET resistance becomes a function of the signal level in addition to the AGC voltage. This is undesirable because only the AGC voltage should control the overall voltage gain.

High-Level Video AGC

For high-level video signals, we can replace the JFET by a LED-photoresistor combination like Fig. 20-31b. The resistance, R_P, of the

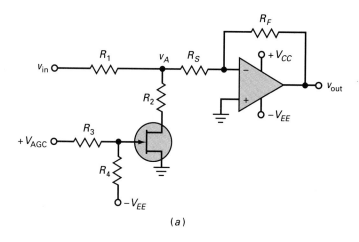

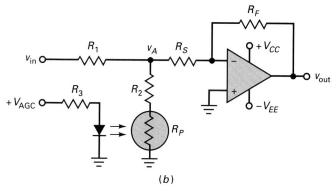

FIGURE 20-31   (a) Low-level video AGC;
(b) high-level video AGC.

photoresistor decreases as the amount of light increases. Therefore, the
larger the AGC voltage, the lower the value of R_P. As before, the input
voltage divider controls the amount of voltage driving the inverting voltage
amplifier. This voltage is given by

$$v_A = \frac{R_2 + R_P}{R_1 + R_2 + R_P} v_{in}$$

The circuit can handle high-level input voltages up to 10 V because the
photocell resistance is unaffected by larger voltages and is only a function
of V_{AGC}. Also note that there is almost total isolation between the AGC
voltage and the input voltage v_{in}.

20-10 HOWLAND CURRENT SOURCE

The current source of Fig. 20-23 produces a unidirectional load current.
Figure 20-32 shows a *Howland* current source. It can produce a bidirectional load current. The circuit is difficult to analyze mathematically. By

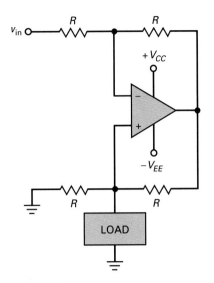

FIGURE 20-32 Howland current source can supply bidirectional current.

writing four loop equations and rearranging as needed, it is possible to prove that

$$i_{out} = \frac{v_{in}}{R} \qquad (20\text{-}36)$$

The maximum load current is approximately V_{CC}/R. One way to see this is to short the load. Then, the noninverting input is grounded, and we have a standard inverting voltage amplifier with a closed-loop voltage gain of 1. This means that the output voltage equals v_{in}, where v_{in} can be positive or negative. Since the maximum output voltage of an op amp is within 1 or 2 V of V_{CC}, the maximum current through the lower right R is roughly V_{CC}/R. A circuit like this typically has a load resistance that is much smaller than R.

⬚ 20-11 MORE ACTIVE FILTERS

This section continues the discussion of active filters, particularly Butterworth filters. As you recall, these are filters that have a maximally flat response to the cutoff frequency. Beyond the cutoff frequency, the output voltage decreases at a rate of $20n$ dB per decade, where n is the number of poles.

Three-Pole Low-Pass Filter

The simplest way to build a three-pole low-pass filter is by cascading a one-pole filter (first section) with a two-pole (second section), as shown in Fig. 20-33. The voltage gain of the first section is optional; you can set it to what you want. The voltage gain of the second section, however, affects the flatness of the overall response. If we keep the closed-loop gain

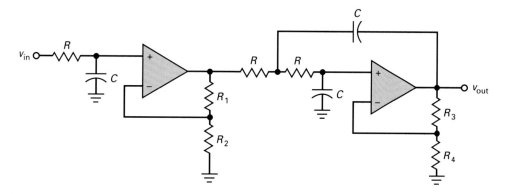

FIGURE 20-33   Three-pole low-pass filter.

at 1.586, then the overall gain will be down 6 dB (3 dB for each section) at the cutoff frequency. By increasing the voltage gain of the second section slightly, we can offset this cumulative loss of voltage gain. Using an advanced mathematical derivation, we can prove that an A_{CL} of 2 is the critical value needed for a maximally flat response. In this case,

$$R_3 = R_4$$

When $A_{CL} = 2$, the cutoff frequency is

$$f_c = \frac{1}{2\pi RC} \qquad (20\text{-}37)$$

where R and C are the resistance and capacitance of each section. At the cutoff frequency, the overall voltage gain is down 3 dB. Above the cutoff frequency, the voltage gain decreases at a rate of 60 dB per decade, equivalent to 18 dB per octave.

More Poles

Figure 20-34 shows a four-pole low-pass filter, a cascade of a two-pole section and a two-pole section. If we try to use an A_{CL} of 1.586 for both

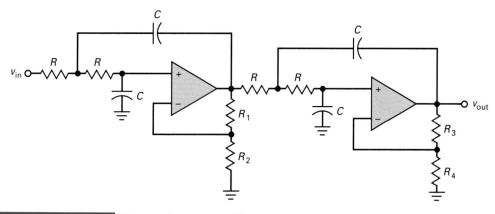

FIGURE 20-34   Four-pole low-pass filter.

sections, the voltage gain will be down 6 dB at the cutoff frequency. By using different gains for each section, we can strike a compromise that produces a maximally flat response. An advanced derivation shows that we need to use $A_{CL} = 1.152$ for the first section and $A_{CL} = 2.235$ for the second section. In all our Butterworth designs, the cutoff frequency is given by $1/(2\pi RC)$.

Butterworth Table

Table 20-2 lists the voltage gains you need to build low-pass Butterworth filters. As indicated, a one-pole filter has an optional A_{CL}. A two-pole filter needs an A_{CL} of 1.586, as previously discussed. A three-pole filter requires two sections, the first a one-pole with an optional A_{CL}, and the second a two-pole with an A_{CL} of 2.

Table 20-2 Gains for Butterworth Filters

Poles	Rolloff (decade)	1st section (1 or 2 poles)	2d section (2 poles)	3d section (2 poles)
1	20 dB	optional		
2	40 dB	1.586		
3	60 dB	optional	2	
4	80 dB	1.152	2.235	
5	100 dB	optional	1.382	2.382
6	120 dB	1.068	1.586	2.482

A four-pole filter has two sections. The first section is a two-pole with a gain of 1.152, and the second is a two-pole with a gain of 2.235. A five-pole filter has three sections, as shown in Fig. 20-35a. As indicated in Table 20-2, the first section is a one-pole with an optional A_{CL}, the second section is a two-pole with a gain of 1.382, and the third section is a two-pole with a gain of 2.382. A six-pole filter is a cascade of three two-pole sections, as shown in Fig. 20-35b. With Table 20-2, the first section needs an A_{CL} of 1.068, the second a gain of 1.586, and the third a gain of 2.482. This Butterworth design produces a roll-off rate of 120 dB per decade.

In all filters, the same resistance and capacitance values are used in the bypass networks, a definite convenience in selection of components and ease of construction. Furthermore, the 3-dB cutoff frequency is always the same, given by

$$f_c = \frac{1}{2\pi RC}$$

High-Pass Filters

To change a low-pass Butterworth filter into a high-pass Butterworth filter you use coupling circuits instead of bypass circuits. The cutoff frequency is still given by $1/2\pi RC$ and the voltage gains are the same as those listed in Table 20-2. For instance, Fig. 20-36 shows a four-pole high-pass filter.

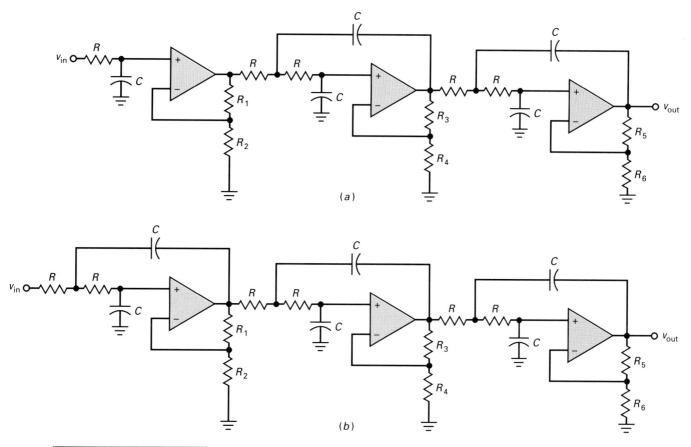

Low-pass filters. (*a*) Five-pole; (*b*) six-pole.

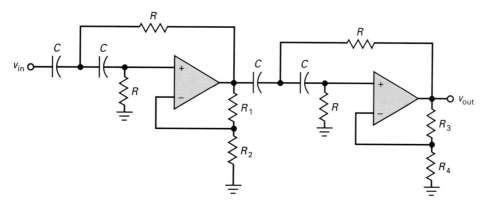

Four-pole high-pass filter.

Instead of bypass circuits, we use coupling circuits with resistances of R and capacitances of C. From Table 20-2, the first section needs an A_{CL} of 1.152 and the second section an A_{CL} of 2.235. With a filter like this, the overall voltage gain is down 3 dB at the cutoff frequency. Below the cutoff frequency, the voltage gain rolls off at a rate of 80 dB per decade.

The following study aids will help to reinforce the ideas discussed in this chapter. For best results, use these study aids within 6 hours of reading the earlier material. Then review these study aids a week later and a month later to ensure that the concepts remain in your long-term memory.

SUMMARY

Sec. 20-1 Noninverting Voltage Amplifiers

A noninverting voltage-feedback amplifier is approximately an ideal voltage amplifier because of its high input impedance, low output impedance, and stable voltage gain. If you don't need a frequency response all the way down to zero frequency, you can use coupling and bypass capacitors.

Sec. 20-2 The Inverting Voltage Amplifier

An inverting voltage amplifier is a combination of a series resistor and a current-to-voltage converter. Because of the series resistor R_S, the feedback fraction changes and some feedback properties change. In this circuit, all of the input current passes through R_S and R_F. This is why the closed-loop voltage gain equals R_F/R_S.

Sec. 20-3 Op-Amp Inverting Circuits

The advantages of an inverting voltage amplifier are its stable voltage gain and controllable input impedance. This basic circuit allows us to come up with a variety of useful designs such as a switchable inverter, adjustable-bandwidth amplifier, single-supply inverter, adjustable inverter, adjustable inverter/noninverter.

Sec. 20-4 The Summing Amplifier

Another major advantage of the inverting voltage amplifier is its ability to amplify more than one signal at a time. A summer is an inverting amplifier with several inputs, each with unity voltage gain. A mixer is a convenient way to mix two audio signals from different sources. The adjustable resistors allow us to set the level of each input, and the gain control allows us to adjust the output volume.

Sec. 20-5 Current Boosters for Voltage Amplifiers

The maximum output current of a typical op amp is limited. If the load requires more current than an op amp can deliver, you can add a power transistor to the output of the op amp. The current gain of this transistor increases the maximum load current that the circuit can produce.

Sec. 20-6 Voltage-Controlled Current Sources

Op amps allow us to build practical current sources. Most of the designs are basically voltage-to-current converters. An input voltage controls the output current. Because of negative feedback, the ratio of i_{out} to v_{in} is a constant. When the circuit is operating normally, the load resistance has no effect on the output current. This is equivalent to saying that the load resistance is driven by a stiff current source.

Sec. 20-7 Differential and Instrumentation Amplifiers

With external resistors, op amps can be used as differential amplifiers with a stable voltage gain. An instrumentation amplifier is a differential amplifier optimized for high input impedance and high CMRR. They are typically used in applications where a small differential voltage and a large common-mode voltage are the inputs.

Sec. 20-8 Active Filters

Op amps can be used to build filters. The low-pass filter will pass all frequencies from zero to the cutoff frequency. A high-pass filter passes all frequencies from the cutoff frequency to infinity. The number of poles in an active filter equals the number of coupling or bypass circuits. Each pole produces a rolloff of 20 dB per decade.

VOCABULARY

In your own words, explain what each of the following terms means. Keep your answers short and to the point. If necessary, verify your answer by rereading the appropriate discussion or by looking at the end-of-book Glossary.

active filter

audio amplifier

Butterworth filter

current booster

cutoff frequency

floating load

instrumentation
amplifier

inverting voltage
amplifier

linear op-amp
circuit

mixer

sink

summer

transfer function

voltage-to-current
converter

IMPORTANT EQUATIONS

Here are some important equations. Say each of the following equations in symbols, then say each in words. Try to explain what the equation means and how it is used. Then read the description that follows.

Eq. 20-6 Input Current

$$i_{in} = \frac{v_{in}}{R_S}$$

This is the input current to an inverting voltage amplifier. You can remember it by visualizing a virtual ground on the inverting input. Then, the series resistor appears grounded on one end. Since v_{in} is applied to the other end, the input current equals the input voltage divided by the series resistance.

Eq. 20-8 Inverting Voltage Gain

$$A_{CL} = \frac{R_F}{R_S}$$

This is the closed-loop voltage gain of an inverting voltage amplifier. It's easy to remember because the input current flows through both resistors. Because of the virtual ground, the voltage gain is the ratio of resistances.

Eq. 20-12 Inverting Amplifier Bandwidth

$$f_{2(CL)} = (1 + AB)f_2$$

This equation is identical for all types of negative feedback. The only difference from circuit to circuit is the value of B. If you had to remember one equation for bandwidth, this would be the one to remember because of its universal application.

Eq. 20-13 Inverting Amplifier Bandwidth

$$f_{2(CL)} = Bf_{unity}$$

When AB is much greater than 1, this equation can be used to find the closed-loop bandwidth of an inverting voltage amplifier. The equation says the closed-loop bandwidth equals the feedback fraction times the f_{unity} of the op amp.

Eq. 20-14 Inverting Gain-Bandwidth Product

$$A_{CL} f_{2(CL)} = \frac{A_{CL}}{A_{CL} + 1} f_{unity}$$

This equation gives you the closed-loop gain-bandwidth product of an inverting voltage amplifier. As you can see, it is not a constant because it depends on the value of A_{CL}. The equation shows you that the gain-bandwidth product shrinks as the closed-loop voltage gain becomes smaller.

Eq. 20-15 Inverting Gain-Bandwidth Product

$$A_{CL} f_{2(CL)} = f_{unity}$$

This approximation of the preceding equation is valid when A_{CL} is much greater than 1. For instance, if A_{CL} is more than 10, you can use this approximation with an error of less than 10 percent. The equation says that the closed-loop gain-bandwidth product equals the f_{unity} of the op amp. This is identical to the equation for a noninverting voltage amplifier.

IMPORTANT PROCESSES

Each of the following processes is important for analyzing op-amp circuits with negative feedback.

If you can remember the basic ideas in each step, you will be able to solve problems more easily.

Output Voltage of an Inverting Amplifier

1. Visualize a virtual ground on the inverting input.
2. Use Ohm's law to get the input current through the series resistor.
3. Visualize all of the input current passing through the feedback resistor.
4. Use Ohm's law to calculate the output voltage across the feedback resistor.

Output Voltage of an Inverting Amplifier (Alternative)

1. Locate the feedback resistor and the series resistor.

2. Divide the feedback resistance by the series resistance.
3. Multiply the input voltage by A_{CL}.

Finding Either Input Voltage

1. Remember that v_{error} is almost zero in a linear op-amp circuit.
2. Realize that v_1 and v_2 are ideally equal.
3. Given the value of either, you automatically have the other.

Bandwidth of an Inverting Amplifier (Approximate)

1. Get f_{unity} from data sheet.
2. Realize this is the approximate gain-bandwidth product.
3. Divide f_{unity} by A_{CL}.

STUDENT ASSIGNMENTS

QUESTIONS

The following questions may have more than one right answer. Select the best answer. This is the one that is always true, covers more situations, or fits the context. In some questions, the abbreviation IVA is used for an inverting voltage amplifier similar to Fig. 20-8.

1. In a linear op-amp circuit, the
   a. Signals are always sine waves
   b. Op amp never saturates
   c. Input impedance is ideally infinite
   d. Gain-bandwidth product is constant

2. In the ac amplifier using an op amp with coupling and bypass capacitors, the output offset voltage is
   a. Zero
   b. Minimum
   c. Maximum
   d. Unchanged

3. To use an op amp, you need at least
   a. One supply voltage
   b. Two supply voltages
   c. One coupling capacitor
   d. One bypass capacitor

4. An audio amplifier ideally amplifies signals with frequencies from
   a. 2 Hz to 2 kHz
   b. 20 Hz to 20 kHz
   c. 200 Hz to 200 kHz
   d. 20 Hz to 200 kHz

5. The virtual ground of an op-amp circuit
   a. Cannot have zero voltage
   b. Cannot sink current
   c. Is always positive
   d. Is the same as ac ground

6. The input current to an inverting voltage amplifier (IVA) flows
   a. Through the series resistor
   b. Into the noninverting input
   c. Into ground
   d. Through the load resistor

7. The input current to an IVA equals the input voltage divided by the
   a. Series resistance
   b. Input impedance of the op amp
   c. Feedback resistance
   d. Load resistor

8. The output voltage of an IVA equals the input current times the
   a. Series resistance
   b. Input impedance of the op amp
   c. Feedback resistance
   d. Load resistor

9. A summer is an op-amp circuit that
   a. Adds the input voltages to get the output voltage
   b. Mixes two audio signals
   c. Works best in the summertime
   d. Uses transistors

10. The closed-loop voltage gain of an IVA equals the
    a. Ratio of load resistance to feedback resistance
    b. Feedback resistance divided by the load resistance
    c. Feedback resistance divided by the series resistance
    d. Series resistance divided by the feedback resistance

11. The closed-loop output impedance of an IVA is
    a. Ideally infinite
    b. Ideally zero
    c. Equal to the z_{out} of the op amp
    d. Equal to the series resistance

12. The closed-loop input impedance of an IVA is
    a. Ideally infinite
    b. Ideally zero
    c. Equal to the z_{out} of the op amp
    d. Equal to the series resistance

13. In a controlled current source with op amps, the circuit acts like a
    a. Voltage amplifier
    b. Current-to-voltage converter
    c. Voltage-to-current converter
    d. Current amplifier

14. The closed-loop bandwidth of an IVA is equal to f_{unity} times
    a. $1 + AB$
    b. A
    c. B
    d. A_{CL}

15. The gain-bandwidth product of an IVA is
    a. Constant
    b. Constant when AB is much greater than 1
    c. Constant when A_{CL} is much greater than 1
    d. Equal to f_2

16. An instrumentation amplifier has a high
    a. Output impedance
    b. Power gain
    c. CMRR
    d. Supply voltage

17. A resistance of $R_S \parallel R_F$ may be used as a dc return on the noninverting input of an IVA to reduce the
    a. Bandwidth
    b. Output voltage
    c. Closed-loop voltage gain
    d. Output offset voltage

18. The closed-loop distortion of an IVA is
    a. Ideally infinite
    b. More than the open-loop distortion
    c. Less than the open-loop distortion
    d. Stable

19. A low-pass filter will
    a. Pass high frequencies
    b. Block low frequencies
    c. Pass dc
    d. Block dc

20. A high-pass filter will
    a. Stop high frequencies
    b. Pass low frequencies
    c. Block dc
    d. Pass dc

21. Above the cutoff frequency of a two-pole low-pass filter, the output voltage decreases at a rate of
    a. 20 dB per decade
    b. 40 dB per decade
    c. 60 dB per decade
    d. 80 dB per decade

22. Below the cutoff frequency of a four-pole high-pass filter, the output voltage decreases at a rate of
    a. 20 dB per decade
    b. 40 dB per decade
    c. 60 dB per decade
    d. 80 dB per decade

23. A two-pole low-pass filter has a cutoff frequency of 20 kHz. At 2 MHz, the output voltage is down
   a. 20 dB
   b. 40 dB
   c. 60 dB
   d. 80 dB

24. A current booster on the output of an op amp will increase the maximum allowable load current by
   a. A_{CL}
   b. β_{dc}
   c. f_{unity}
   d. R_F/R_S

25. When an op amp is not saturated, the two input voltages are
   a. Ideally equal
   b. Exactly equal
   c. Different
   d. Quite a bit different

26. When an op amp is saturated, the two input voltages are usually
   a. Ideally equal
   b. Exactly equal
   c. Different
   d. Quite a bit different

27. The current flowing into the inverting input of an op amp is
   a. Ideally zero
   b. Equal to the input current
   c. Equal to the load current
   d. Large

BASIC PROBLEMS

Sec. 20-1 Noninverting Voltage Amplifiers

20-1. What is the output voltage in the midband of Fig. 20-37? What is the upper closed-loop critical frequency? The three lower critical frequencies?

20-2. The JFET of Fig. 20-38 has $V_{GS(off)} = -4$ V and a drain resistance of 30 Ω when turned on. What is the minimum output voltage of the circuit? The maximum?

20-3. The resistors of Fig. 20-37 have a tolerance of ± 5 percent. What is the output voltage in the midband of Fig. 20-37?

20-4. What is the closed-loop bandwidth in Fig. 20-38?

Sec. 20-2 The Inverting Voltage Amplifier

20-5. Calculate the output voltage, the closed-loop input impedance, and the closed-loop bandwidth in Fig. 20-39. The 741C has an f_{unity} of 1 MHz.

20-6. The resistors of Fig. 20-39 have a tolerance of ± 5 percent. What is the output voltage in the midband of Fig. 20-39?

20-7. To get an output voltage of 0.5 V in Fig. 20-39, what changes can you make to one resistor?

20-8. What is the output offset voltage in Fig. 20-39? The 741C has input bias and offset currents of 80 nA and 20 nA.

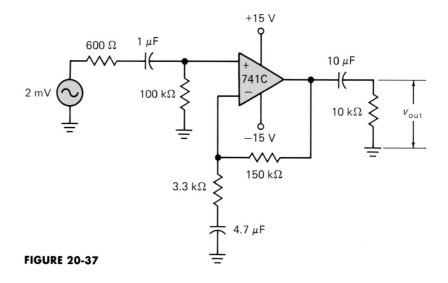

FIGURE 20-37

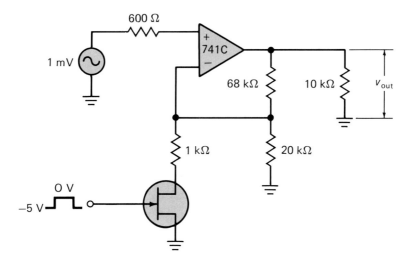

FIGURE 20-38

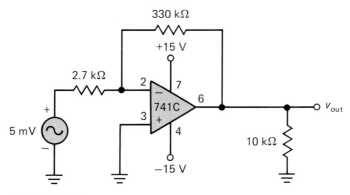

FIGURE 20-39

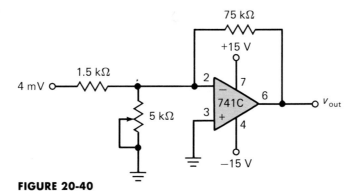

FIGURE 20-40

Sec. 20-3 Op-Amp Inverting Circuits

20-9. What is the output voltage and bandwidth in Fig. 20-40?

20-10. What is the output voltage in the midband of Fig. 20-41? Find the critical frequencies of the input coupling capac-

itor, output coupling capacitor, and non-inverting bypass capacitor.

20-11. If the pot is changed to 10 kΩ in Fig. 20-40, what happens to the output voltage and bandwidth?

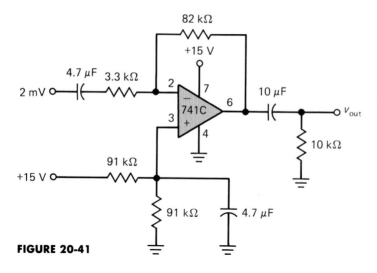

FIGURE 20-41

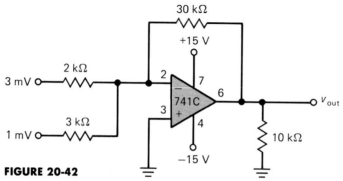

FIGURE 20-42

20-12. All resistances are doubled in Fig. 20-41. Find the critical frequencies of the input coupling capacitor, output coupling capacitor, and noninverting bypass capacitor.

Sec. 20-4 The Summing Amplifier

20-13. What is the output voltage in Fig. 20-42?

20-14. The two inputs of Fig. 20-42 are changed from 3 mV and 1 mV to 2 mV and 4 mV. What is the output voltage?

20-15. What is the output voltage in Fig. 20-42 if the 30 kΩ is doubled?

Sec. 20-5 Current Boosters for Voltage Amplifiers

20-16. An op amp has a maximum output current of 30 mA. If a current booster with

a current gain of 80 is used, what is the new value of maximum output current?

20-17. In Fig. 20-43, $\beta_{dc} = 100$ for the transistor. If $v_{in} = 100$ mV, what is the output voltage? The load current?

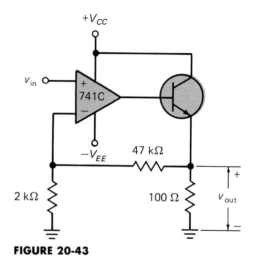

FIGURE 20-43

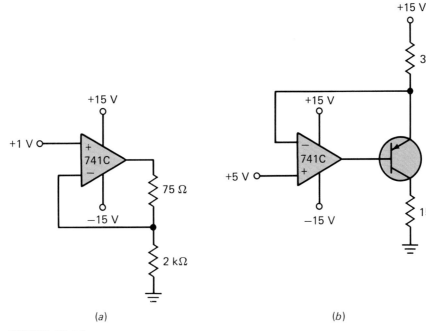

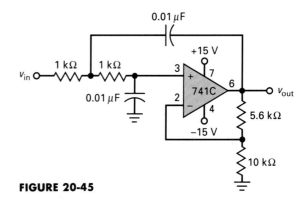

(a)   (b)

FIGURE 20-44

20-18. If the load resistance of Fig. 20-43 is reduced to zero, what is the base current? The load current?

Sec. 20-6 Voltage-Controlled Current Sources

20-19. What is the load current in Fig. 20-44*a*?

20-20. Calculate the output current in Fig. 20-44*b*. Also, work out the value of load voltage.

20-21. What is the maximum load resistance that we can use in Fig. 20-44*b* without saturating the transistor?

20-22. What can you do in Fig. 20-44*a* to produce a load current of 1 mA?

20-23. What can you do in Fig. 20-44*b* to produce a load current of 5 mA?

Sec. 20-8 Active Filters

20-24. Figure 20-45 shows a two-pole filter. What is the output voltage in the midband? The cutoff frequency?

20-25. All resistances are doubled in Fig. 20-45. What is the output voltage in the midband? The cutoff frequency?

FIGURE 20-45

TROUBLESHOOTING PROBLEMS

20-26. If you measure a voltage gain of about 46 dB for Fig. 20-46, which of the following is a possible cause?
   a. No supply voltage
   b. Emitter bypass capacitor open
   c. Feedback bypass capacitor open
   d. Transistor open

20-27. What will typically happen to the output voltage of Fig. 20-47 if the 10 kΩ opens?

20-28. For a v_{in} of 5 V in Fig. 20-47, there is no current through the load. Name at least three possible causes.

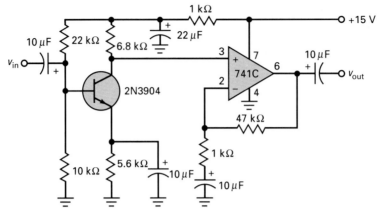

FIGURE 20-46

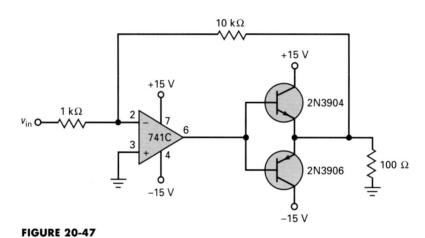

FIGURE 20-47

ADVANCED PROBLEMS

20-29. In Fig. 20-15b, $R = 1$ kΩ, $nR = 10$ kΩ, and the other resistance is 1.1 kΩ. The potentiometer has a maximum value of 1 kΩ. What is the minimum voltage gain? The maximum voltage gain?

20-30. Design a two-pole low-pass Butterworth filter with a cutoff frequency of 20 kHz.

20-31. What is the voltage gain in Fig. 20-46?

20-32. Assume $R_S = 0$, $R_L = 10$ kΩ, and

$\beta = 150$. Calculate the input, output, and bypass critical frequencies in Fig. 20-46 (use an f_{unity} of 1 MHz for the 741C).

20-33. The transistors of Fig. 20-47 have $\beta_{dc} = 75$. If v_{in} is 0.5 V, what is the base current to the conducting transistor?

20-34. In Fig. 20-47, v_{in} is 1 V. Calculate the approximate value of the load current.

20-35. What is the smallest load resistance that can be used in Fig. 20-47 if the input voltage is 1 V?

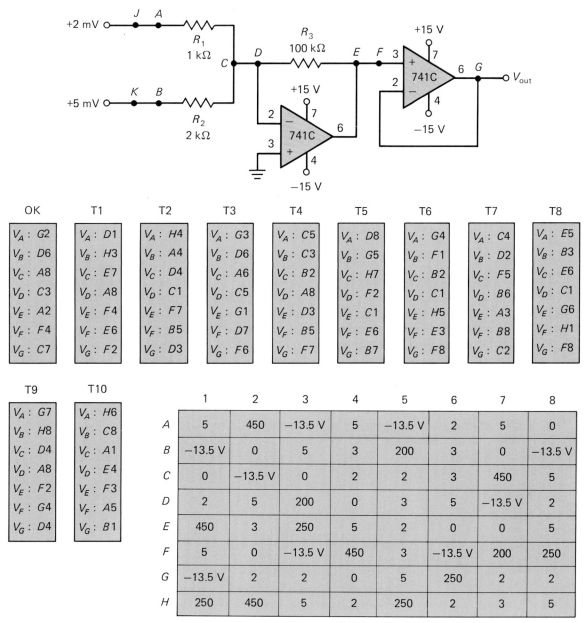

FIGURE 20-48 T-Shooter™. (*Patent pending: Courtesy of Malvino Inc.*)

T-SHOOTER PROBLEMS

Use Fig. 20-48 for the remaining problems. Any resistor may be open or shorted. Also, connecting wires CD, EF, JA, or KB may be open.

20-36. Find Trouble 1.   **20-37.** Find Trouble 2.

20-38. Find Troubles 3 and 4.

20-39. Find Troubles 5 and 6.

20-40. Find Troubles 7 and 8.

20-41. Find Troubles 9 and 10.

21

NONLINEAR OP-AMP CIRCUITS

Monolithic op amps are inexpensive, versatile, and reliable. For this reason, they can be used not only for voltage amplifiers, current sources, and active filters, but also for active diode circuits, comparators, and waveshaping. This chapter is about nonlinear op-amp circuits, the kind where the output signal shape is different from the input. Nonlinear op-amp circuits are slightly more complicated than linear op-amp circuits. Why? Because a linear op-amp circuit never saturates the op amp under normal operating conditions. The nonlinear op-amp circuit, however, may saturate the op amp or open the feedback loop during part of the cycle. Because of this, you have to analyze two different modes or regions of operation to see what happens during an entire cycle of the output signal.

This chapter will discuss a wide variety of circuits that will show you the many different tricks used in op-amp circuits. Again, you don't have to understand every detail of every circuit. All that's necessary is to get the basic idea of how op amps are used in nonlinear applications. This will be valuable background for troubleshooting, analysis, and design.

21-1 ACTIVE DIODE CIRCUITS

Op amps can enhance the performance of diode circuits. For one thing, an op amp with negative feedback reduces the effect of the knee voltage, allowing us to rectify, peak-detect, clip, and clamp low-level signals (those with amplitudes less than the knee voltage). And because of their buffering action, op amps can eliminate the effects of the source and load on diode circuits.

Half-Wave Rectifier

Figure 21-1 is an *active half-wave rectifier*. When the input signal to the noninverting input goes positive, the output goes positive and turns on the diode. The circuit then acts as a voltage follower, and the positive half-cycle appears across the load resistor. However, when the input goes negative, the op-amp output goes negative and turns off the diode. Since the diode is open, no voltage appears across the load resistor. This is why the final output is almost a perfect half-wave signal.

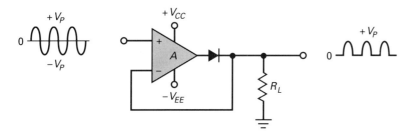

FIGURE 21-1   Active half-wave rectifier.

There are two distinct modes or regions of operation. First, when the input voltage is positive, the diode is conducting and the operation is linear. In this case, the output voltage is fed back to the input, and we have negative feedback as before. Second, when the input voltage is negative, the diode is nonconducting and the feedback path is open. In this case, there is no negative feedback, and the op-amp output is isolated from the load resistor. In Fig. 21-1, therefore, the positive half-cycle of input voltage makes the circuit act as a voltage follower driving the load, while the negative half-cycle isolates the op amp from the load. This is why the output voltage is a half-wave-rectified sine wave.

The high gain of the op amp almost eliminates the effect of the knee voltage. For instance, if the knee voltage is 0.7 V and A is 100,000, the input voltage that just turns on the diode is

$$v_{in} = \frac{0.7\,V}{100,000} = 7\,\mu V$$

When the input is greater than 7 μV, the output voltage is greater than 0.7 V and the diode turns on. Then the circuit acts as a voltage follower. The effect is equivalent to reducing the knee voltage by a factor of $1 + AB$, which is approximately equal to A. In symbols,

$$V'_K = \frac{V_K}{A} \tag{21-1}$$

where V'_K = closed-loop knee voltage and V_K = knee voltage. Because V'_K is so small, the active half-wave rectifier may be used with low-level signals in the millivolt region. Think about that for a bit.

Up to now, we have been using passive half-wave rectifiers. These are the ordinary kind of rectifiers where we need an input voltage that is much greater than the knee voltage of a silicon diode. A passive rectifier is fine for power supplies because there a large input voltage is coming from the secondary winding of a transformer. But when we have low-level or small signals whose peak values are less than 0.7 V, we need to use another approach. One right answer is to amplify the low-level signal before rectification. Another one is to use an active half-wave rectifier.

Active Peak Detector

To peak-detect small signals, we can use an *active peak detector* like Fig. 21-2a. Again, the closed-loop knee voltage is in the microvolt region,

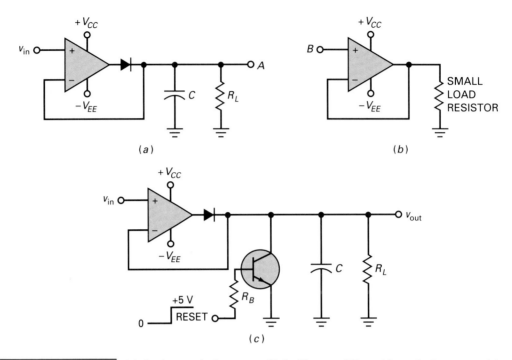

FIGURE 21-2 (*a*) Active peak detector; (*b*) buffer amplifier; (*c*) peak detector with reset.

which means that we can peak-detect low-level signals. When the diode is on, the heavy noninverting voltage feedback produces a Thevenin output impedance that approaches zero. This means that the charging time constant is very low, so the capacitor can quickly charge to the positive peak value. However, when the diode is off, the capacitor has to discharge through R_L. Because the discharging time constant $R_L C$ can be made much longer than the period of the input signal, we can get almost perfect peak detection of low-level signals.

There are two distinct regions of operation. First, when the input voltage is positive, the diode is conducting and the operation is linear. In this case, the capacitor charges to the peak of the input voltage. Second, when the input voltage is negative, the diode is nonconducting and the feedback path is open. In this case, the capacitor discharges through the load resistor. As long as the discharging time constant is much greater than the period of the input signal, the output voltage will be approximately equal to the peak value of the input voltage.

If the peak-detected signal has to drive a small load, we can avoid loading effects by using an op-amp buffer. For instance, if we connect point A of Fig. 21-2*a* to point B of Fig. 21-2*b*, then the voltage follower isolates the small load resistor from the peak detector. This prevents the small load resistor from discharging the capacitor too quickly.

At a minimum, the $R_L C$ time constant should be at least 10 times longer than the period T of the lowest input frequency. In symbols,

$$R_L C > 10T \qquad (21\text{-}2)$$

If this condition is satisfied, the output voltage will be within 5 percent of the peak input. For instance, if the lowest frequency is 1 kHz, the period is 1 ms. In this case, the R_LC time constant should be at least 10 ms if you want an error of less than 5 percent.

Often, a *reset* is included with an active peak detector, as shown in Fig. 21-2c. When the reset input is low, the transistor switch is off. This allows the circuit to work as previously described. When the reset input is high, the transistor switch is closed; this rapidly discharges the capacitor. The reason you may need a reset is because the long discharge time constant means that the capacitor will hold its charge for a long time, even though the input signal is removed. By using a high reset input, we can quickly discharge the capacitor in preparation for another input signal with a different peak value.

Active Positive Limiter

Figure 21-3a is an *active positive limiter*. This is a circuit that will clip off part of the signal. With the wiper all the way to the left, v_{ref} is zero and the noninverting input is grounded. When v_{in} goes slightly positive, the error voltage drives the op-amp output negative and turns on the diode. This produces heavy negative feedback because the feedback resistance

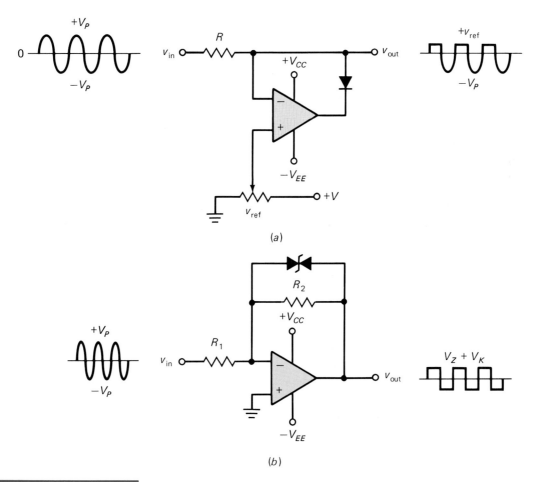

(a)

(b)

FIGURE 21-3   (a) Active positive limiter; (b) zener diodes produce rectangular wave.

is zero. Since R_F is zero, the output node is at virtual ground for all positive values of v_{in}.

When v_{in} goes negative, the op-amp output is positive, which turns off the diode and opens the loop. As this happens, the virtual ground is lost, and the final output v_{out} is free to follow the negative half-cycle of input voltage. This is why the negative half-cycle appears at the output as shown.

Here are the two regions of operation. First, when the input voltage is positive, the diode is conducting and the output node is at virtual ground, which means zero output voltage. Second, when the input voltage is negative, the diode is nonconducting and the feedback path is open. In this case, the output voltage is free to follow the input voltage.

To change the clipping level, all we do is adjust v_{ref} as needed. In this case, clipping occurs at v_{ref}, as shown in Fig. 21-3a. As usual, the knee voltage is reduced to V_K' at the input, which means that the circuit is suitable for low-level inputs.

Figure 21-3b shows an active circuit that clips on both half-cycles. Notice the back-to-back zener diodes in the feedback loop. Below the zener voltage, the circuit has a closed-loop gain of R_2/R_1. When the output tries to exceed the zener voltage plus one forward diode drop, the zener diode breaks down and the output voltage is $V_Z + V_K$ away from virtual ground. This is why the output is clipped as shown.

Active Positive Clamper

Figure 21-4 is an *active positive clamper*. This is a circuit that adds a dc component to the input signal. As a consequence, the output has the same size and shape as the input signal, except for the dc shift.

Here is how the circuit works. The first negative-input half-cycle is coupled through the uncharged capacitor and produces a positive op-amp output which turns on the diode. Because of the virtual ground, the capacitor charges to the peak value of the negative-input half-cycle with the polarity shown in Fig. 21-4. Just beyond the negative input peak, the diode turns off, the loop opens, and the virtual ground is lost. In this case, the output voltage is the sum of the input voltage and the capacitor voltage:

$$v_{\text{out}} = v_{\text{in}} + V_P \tag{21-3}$$

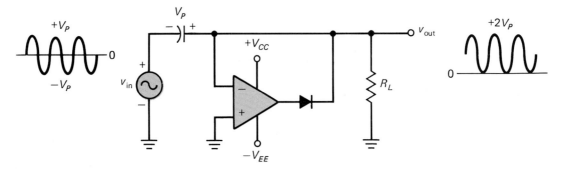

FIGURE 21-4   Active positive clamper.

If you don't believe it, look at Fig. 21-4 and put your finger on the lower end of the input generator. Move your finger through the generator and through the charged capacitor. Then your finger is on the output node. This means the output voltage equals the input voltage plus the capacitor voltage.

Since V_P is being added to a sinusoidal input voltage, the final output waveform is shifted positively through V_P, as shown in Fig. 21-4. A waveform like this is called a *positively clamped waveform*. Notice that the clamped waveform swings from 0 to $2V_P$, which means it has a peak-to-peak value of $2V_P$, the same as the input. Again, the negative feedback reduces the knee voltage by a factor of approximately A, which means we can build excellent clampers for low-level inputs.

Figure 21-4 shows the op-amp output. During most of the cycle, the op amp operates in negative saturation. Right at the negative input peak, however, the op amp produces a sharp positive-going pulse that replaces any charge lost by the clamping capacitor between negative input peaks.

EXAMPLE 21-1

A sinusoidal voltage of 100 mV rms is the input in Fig. 21-5. If the frequency is 50 kHz, what is the output?

SOLUTION

The circuit is an active peak detector when the frequency is high enough. What is a high enough frequency? It is the one whose period is at least 10 times the RC time constant. The period of the input signal is

$$T = \frac{1}{50 \text{ kHz}} = 20 \text{ μs}$$

The time constant of the capacitor input filter is

$$R_L C = (10 \text{ kΩ})(1 \text{ μF}) = 10 \text{ ms}$$

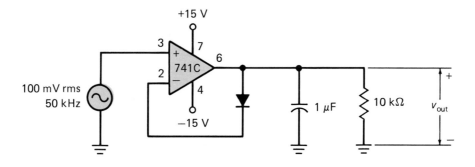

FIGURE 21-5 Example.

The ratio of the time constant to the period is

$$\frac{R_LC}{T} = \frac{10\text{ ms}}{20\text{ }\mu\text{s}} = 500$$

Because the time constant is 500 times greater than the period of the input signal, the capacitor has almost no time to discharge. Therefore, the output voltage equals the peak input voltage to a close approximation.

The input voltage is 100 mV rms, which means it has a peak value of

$$V_P = 1.414(100\text{ mV}) = 141\text{ mV}$$

Therefore,

$$v_{\text{out}} = 141\text{ mV}$$

EXAMPLE 21-2

What is the lowest recommended input frequency in Fig. 21-5?

SOLUTION

The basic rule is to keep the time constant at least 10 times longer than the period of the input signal. In Example 21-1 we calculated a time constant of 10 ms. One-tenth of this is

$$T = 1\text{ ms}$$

Therefore, the minimum frequency is

$$f = \frac{1}{1\text{ ms}} = 1\text{ kHz}$$

21-2 COMPARATORS

Often we want to compare one voltage with another to see which is larger. In this situation, a *comparator* may be the perfect solution. This circuit has two input voltages (noninverting and inverting) and one output voltage. When the noninverting voltage is larger than the inverting voltage, the comparator produces a high output voltage. When the noninverting input is less than the inverting input, the comparator produces a low output voltage. The question "Is the noninverting input greater than the inverting input?" has the answer yes when the output voltage is high and the answer no when the output voltage is low.

Basic Circuit

The simplest way to build a comparator is to connect an op amp without feedback resistors, as shown in Fig. 21-6a. When the inverting input is

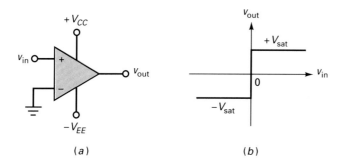

FIGURE 21-6 (*a*) Op amp used as comparator; (*b*) transfer characteristic.

grounded, the slightest input voltage (in fractions of a millivolt) is enough to saturate the op amp. If A is the differential voltage gain of the op amp, then the minimum input voltage that produces saturation is

$$v_{\text{in}} = \frac{V_{\text{sat}}}{A} \qquad (21\text{-}4)$$

For instance, if the supplies are ± 15 V, the output swing is from approximately -13.5 to $+13.5$ V. With a 741C, the open-loop voltage gain is typically 100,000. Therefore, the input voltage needed to produce positive saturation is

$$v_{\text{in}} = \frac{13.5 \text{ V}}{100,000} = 135 \text{ } \mu\text{V}$$

This is so small that the graph of Fig. 21-6*b* has an almost a vertical transition or change at $v_{\text{in}} = 0$. As you see, the slightest input voltage saturates the op amp. For instance, it takes only $+135$ μV of input voltage to positively saturate a 741C and only -135 μV to negatively saturate it.

Moving the Trip Point

The *trip point* (also called the *threshold*, the *reference*, etc.) of a comparator is the input voltage where the output switches states (low to high, or vice versa). In Fig. 21-6*a*, the trip point is zero, because this is the value of input voltage where the output switches states. When v_{in} is greater than the trip point, the output is high. When v_{in} is less than the trip point, the output is low. A circuit like Fig. 21-6*a* is often called a *zero-crossing detector*.

In Fig. 21-7*a*, a reference voltage is applied to the inverting input:

$$v_{\text{ref}} = \frac{R_2}{R_1 + R_2} V_{CC} \qquad (21\text{-}5)$$

When v_{in} is greater than v_{ref}, the error voltage is positive and the output voltage is high. When v_{in} is less than v_{ref}, the error voltage is negative and the output voltage is low.

Incidentally, a bypass capacitor is typically used on the inverting input, as shown in Fig. 21-7*a*. This reduces the amount of power-supply ripple and noise appearing at the inverting input. For it to be effective, the

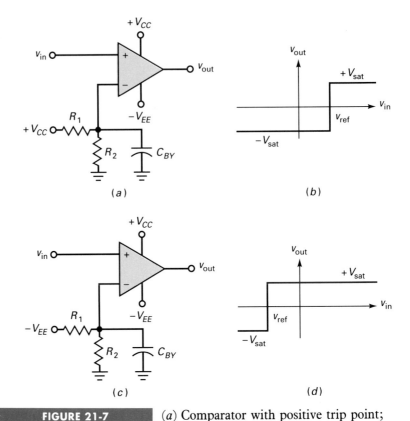

FIGURE 21-7   (*a*) Comparator with positive trip point; (*b*) transfer characteristic; (*c*) comparator with negative trip point; (*d*) transfer characteristic.

critical frequency of this bypass circuit should be much lower than the ripple frequency of the power supply.

Figure 21-7*b* shows the transfer characteristic (a graph of output versus input). The trip point is now equal to v_{ref}. When v_{in} is larger than v_{ref}, the output of the comparator goes into positive saturation. When v_{in} is less than v_{ref}, the output goes into negative saturation. A comparator like this is sometimes called a *limit detector* because a positive output indicates that the input voltage exceeds a specific limit. With different values of R_1 and R_2, we can set the positive trip point anywhere between 0 and V_{CC}.

If a negative trip point is preferred, connect $-V_{EE}$ to the voltage divider, as shown in Fig. 21-7*c*. Now a negative reference voltage is applied to the inverting input. When v_{in} is more positive than v_{ref}, the error voltage is positive and the output is high, as shown in Fig. 21-7*d*. When v_{in} is more negative than v_{ref}, the output is low.

Single-Supply Comparator

As you know, a typical op amp like the 741C can run on a single positive supply by grounding the $-V_{EE}$ pin, as shown in Fig. 21-8*a*. Now the output voltage has only one polarity, either a low or a high positive voltage. For instance, with V_{CC} equal to $+15$ V, the output swing is from approximately 1.5 V (low state) to around 13.5 V (high state).

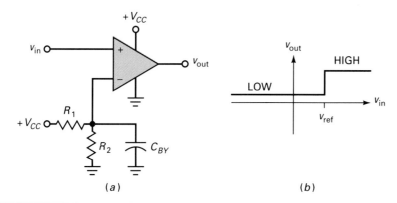

FIGURE 21-8 (*a*) Single-supply comparator; (*b*) transfer characteristic.

The reference voltage applied to the inverting input is positive and equal to

$$v_{\text{ref}} = \frac{R_2}{R_1 + R_2} V_{CC} \qquad (21\text{-}6)$$

When v_{in} is greater than v_{ref}, the output is high, as shown in Fig. 21-8*b*. When v_{in} is less than v_{ref}, the output is low. In either case, the output has a positive polarity. For most digital applications, this kind of positive output is preferred.

IC Comparators

An op amp like a 741C can be used as a comparator, but it has speed limitations. As you know, the slew rate limits the rate of output voltage change. With a 741C, the output can change no faster than 0.5 V/μs. Because of this, a 741C takes more than 50 μs to switch between a low output of -13.5 V and a high output of $+13.5$ V. One right answer to this slew-rate problem is to use a faster slew-rate op amp like the LM318. Since it has a slew rate of 70 V/μs, it can switch from -13.5 to $+13.5$ V in approximately 0.3 μs.

Another right answer is to eliminate the compensating capacitor found in a typical op amp. For linear op-amp circuits, this capacitor is essential because it rolls off the open-loop voltage gain at a rate of 20 dB per decade and prevents oscillations. A comparator is always used as a nonlinear circuit, so there is no need to include a compensating capacitor. A manufacturer can redesign the typical op amp by deleting the compensating capacitor. When an IC has been optimized for use as a comparator, the device is listed in a separate section of the manufacturer's catalog. This is why you will find a section on op amps and another section on comparators in a typical manufacturer's data book.

Figure 21-9*a* is a simplified schematic diagram for an IC comparator. The input stage is a diff amp (Q_1 and Q_2). A current source Q_6 supplies the tail current. The diff amp drives an active load Q_4. The output stage is a single transistor Q_5 with an open collector. The manufacturer often leaves this collector open, as shown. This allows the user to control the output swing of the comparator.

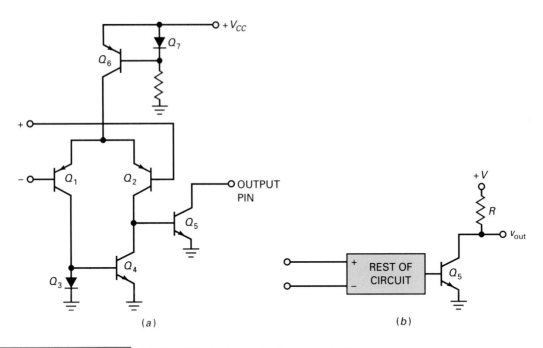

FIGURE 21-9   (*a*) Simplified schematic diagram of IC comparator.

For the circuit to work, the user has to connect the open collector of Q_5 to an external resistor and supply voltage, as shown in Fig. 21-9*b*, The resistor is called a *pullup resistor* because it pulls the output voltage up to the supply voltage when Q_5 is cut off. When Q_5 is saturated, the output voltage is low. Basically, the output stage is a transistor switch. This is why the comparator produces a *two-state output*, either a low or a high voltage.

With no compensating capacitor in the circuit, the output in Fig. 21-9*a* can slew very rapidly because only small stray capacitances remain in the circuit. One limitation on the switching speed is the amount of capacitance across Q_5. This output capacitance is the sum of the collector capacitance and the stray wiring capacitance. The output time constant is the product of the pullup resistance and the output capacitance. For this reason, the smaller the pullup resistance in Fig. 21-9*b*, the faster the output voltage can change. Typically, R is from a couple of hundred to a couple of thousand ohms.

Examples of IC comparators are the LM311, LM339, and NE529. These all have an open-collector output stage, which means you have to connect the output pin to a pullup resistor and a positive supply voltage. Because of their high slew rates, these IC comparators can switch output states in a microsecond or less. The LM339 is a *quad comparator*, or four comparators in a single IC package. Because it is inexpensive and easy to use, the LM339 has become a popular comparator for general-purpose applications.

Figure 21-10*a* shows how an LM339 can be connected to interface with TTL devices (ICs used in digital computers and other circuits). Notice that the open-collector output is connected to a supply of +5 V through a pullup resistor of 1 kΩ. Because of this, the output can be

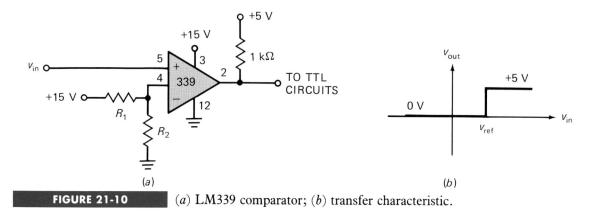

FIGURE 21-10 (a) LM339 comparator; (b) transfer characteristic.

either 0 or +5 V, as shown in Fig. 21-10b. This output drive is ideal for TTL devices because they are designed to work with supplies of +5 V.

One final point. Most IC comparators have an output stage that requires you to connect a separate pullup resistor and a supply voltage. But some comparators have an output stage that can swing between the positive and negative supply voltages, similar to an op amp.

EXAMPLE 21-3

If R_1 = 10 kΩ and R_2 = 2 kΩ, what is the reference voltage in Fig. 21-10a?

SOLUTION

With Eq. (21-6),

$$v_{ref} = \frac{2\,k\Omega}{10\,k\Omega + 2\,k\Omega}\,15\,V = 2.5\,V$$

An alternative solution is to use Ohm's law. The total resistance of the voltage divider of Fig. 21-10 is the sum of R_1 and R_2, which is 12 kΩ. The current through the voltage divider is

$$I = \frac{15\,V}{12\,k\Omega} = 1.25\,mA$$

This current flows through R_2 to produce a voltage of

$$v_{ref} = (1.25\,mA)(2\,k\Omega) = 2.5\,V$$

When the input voltage is less than 2.5 V, the output voltage is 0 V. When the input voltage is greater than 2.5 V, the output voltage is +5 V. This kind of circuit action is useful when you want use the output voltage as a *flag*, a signal that tells you when an event has taken place. In this example, the event is an input voltage exceeding 2.5 V. The input voltage could be coming from a photodiode, a thermistor, etc. Then the comparator would indicate when the light, temperature, etc., exceeded a certain limit.

EXAMPLE 21-4

A bypass capacitor is used on the inverting input of Fig. 21-10a similar to Fig. 21-8a. If $R_1 = 10$ kΩ, $R_2 = 2$ kΩ, and $C = 68$ μF, what is the critical frequency of the bypass circuit? The supply voltage is 15 V with a peak-to-peak ripple of 0.5 V. Estimate the ripple at the inverting input for troubleshooting purposes.

SOLUTION

Use the standard formula with $R = 10$ kΩ ∥ 2 kΩ and $C = 68$ μF. By now, you should know that the resistance to use is the Thevenin resistance facing the capacitor:

$$R = 10 \text{ k}\Omega \parallel 2 \text{ k}\Omega = 1.67 \text{ k}\Omega$$

The critical frequency is

$$f_c = \frac{1}{2\pi(1.67 \text{ k}\Omega)(68 \text{ μF})} = 1.4 \text{ Hz}$$

This is approximately two decades below 120 Hz, the ripple frequency of the power supply. Therefore, the bypass circuit will produce approximately 40 dB of attenuation for the ripple at the inverting input. If the bypass capacitor is open, the ripple at the inverting input is reduced as it passes through the voltage divider to

$$V_R = \frac{2 \text{ k}\Omega}{12 \text{ k}\Omega}(0.5 \text{ V}) = 0.083 \text{ V} = 83 \text{ mV}$$

When the bypass capacitor is connected, the foregoing ripple is further reduced by approximately 40 dB (a factor of 100). This gives

$$V_R = \frac{83 \text{ mV}}{100} = 0.83 \text{ mV} = 830 \text{ μV}$$

A troubleshooter would probably estimate 1 mV of peak-to-peak ripple at the inverting input. If the ripple measures approximately 100 mV, she or he knows the bypass capacitor is open.

21-3 THE SCHMITT TRIGGER

If the input to a comparator contains noise, the output may be erratic when v_{in} is near a trip point. For instance, with a zero-crossing detector, the output is high when v_{in} is positive and low when v_{in} is negative. If the input contains a noise voltage with a peak of 1 mV or more, the comparator will detect the zero crossings produced by the noise. Something similar happens when the input is near the trip points of a limit detector. The noise causes the output to jump back and forth between its low and high states. We can avoid this unwanted noise triggering by using a

Schmitt trigger, which is an op amp or a comparator used with positive feedback.

Basic Circuit

Figure 21-11*a* shows a Schmitt trigger. Notice that the input voltage is applied to the inverting input. The circuit uses *positive* voltage feedback instead of negative feedback. This means the feedback voltage is aiding the input voltage rather than opposing it. For instance, assume the inverting input voltage is slightly positive. This will produce a negative output voltage. The voltage divider feeds back a negative voltage to the noninverting input, which results in a larger negative output voltage. This feeds back more negative voltage until the comparator is driven into negative saturation. If the input voltage were slightly negative instead of positive, the comparator would be driven into positive saturation.

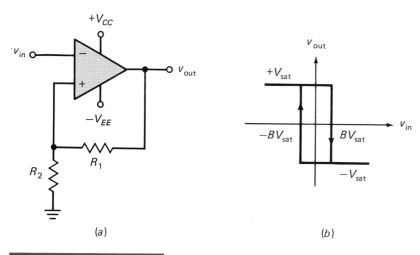

FIGURE 21-11   Schmitt trigger.

When the comparator is positively saturated, a positive voltage is fed back, to the noninverting input. This positive input holds the output in the high state. Similarly, when the output voltage is negatively saturated, a negative voltage is fed back to the noninverting input, holding the output in the low state. In either case, the positive feedback reinforces the existing output state.

The feedback fraction is

$$B = \frac{R_2}{R_1 + R_2} \tag{21-7}$$

When the output is positively saturated, the reference voltage applied to the noninverting input is

$$v_{\text{ref}} = + BV_{\text{sat}} \tag{21-8}$$

When the output is negatively saturated, the reference voltage is

$$v_{\text{ref}} = - BV_{\text{sat}} \tag{21-9}$$

The output voltage will remain in a given state until the input voltage exceeds the reference voltage for that state. For instance, if the output is positively saturated, the reference voltage is $+BV_{sat}$. The input voltage v_{in} must be increased to slightly more than $+BV_{sat}$ to switch the output voltage from positive to negative, as shown in Fig. 21-11b. Once the output is in the negative state, it will remain there indefinitely until the input voltage becomes more negative than $-BV_{sat}$. Then the output switches from negative to positive (Fig. 21-11b).

Hysteresis

The unusual graph of Fig. 21-11b has a useful property called *hysteresis*. To understand this concept, put your finger on the upper end of the graph where it says $+V_{sat}$; this is the value of the output voltage. Move your finger along the horizontal line. Along this horizontal line, the input voltage is changing but the output voltage is still equal to $+V_{sat}$. When you reach the upper right corner, v_{in} equals $+BV_{sat}$. When v_{in} increases to slightly more than $+BV_{sat}$, the output voltage switches from $+V_{sat}$ to $-V_{sat}$. If you move your finger along the vertical line with the down arrow, you will simulate the switching of the output voltage from high to low. When your finger is on the lower horizontal line, the output voltage is negatively saturated and equal to $-V_{sat}$. The output voltage stays at this level anywhere on the lower horizontal line. Move your finger until it reaches the lower left corner. At this point, v_{in} equals $-BV_{sat}$. When v_{in} becomes slightly more negative than $-BV_{sat}$, the output voltage switches from $-V_{sat}$ to $+V_{sat}$. If you move your finger along the vertical line with the up arrow, you will simulate the switching of output voltage from low to high.

In Fig. 21-11b, the *trip points* are defined as the two input voltages where the output voltage changes states. The upper trip point (abbreviated UTP) has a value

$$UTP = BV_{sat}$$

and the lower trip point (LTP) has a value

$$LTP = -BV_{sat}$$

The difference between the trip points is the value of hysteresis H:

$$H = BV_{sat} - (-BV_{sat})$$

or

$$H = 2BV_{sat} \qquad (21\text{-}10)$$

Positive feedback causes the hysteresis of Fig. 21-11b. If there were no positive feedback, B would equal zero and the hysteresis would disappear, because the trip points would both equal zero.

Hysteresis is desirable in a Schmitt trigger because it prevents noise from causing false triggering. Visualize a Schmitt trigger with no hysteresis, equivalent to Fig. 21-11b with $B = 0$. Then any noise voltage at the input of the Schmitt trigger will cause the output voltage to randomly switch from the low to the high state, and vice versa. Next, visualize a Schmitt trigger with hysteresis as shown in Fig. 21-11b. If the peak-to-peak noise voltage is less than the hysteresis, the noise cannot produce false triggering. For instance, if UTP = +1 V and LTP = −1 V, then H = 2 V. In this case, the Schmitt trigger is immune to false triggering as long as the peak-to-peak noise voltage is less than 2 V.

Noninverting Circuit

Figure 21-12a shows a *noninverting Schmitt trigger*. Its graph of output voltage versus input voltage again has a hysteresis loop, as shown in Fig. 21-12b. Again, you can get a better feel for how the circuit works by moving your finger along the horizontal and vertical parts of the graph. When you move your finger along a horizontal part, it is equivalent to the input voltage changing but the output voltage remaining constant. When you move your finger along a vertical part in the direction of the arrows, it is equivalent to the output voltage changing states. Low-to-high switching occurs at the lower right corner of the hysteresis loop, and high-to-low switching occurs at the upper left corner.

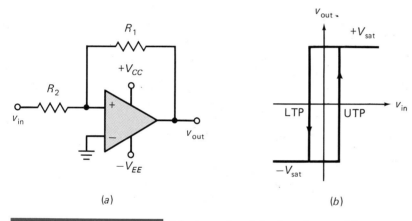

(a)   (b)

| **FIGURE 21-12** | Noninverting Schmitt trigger with zero |

center voltage.

Here is how it works. Assume that the output is positively saturated in Fig. 21-12a. Then the feedback voltage to the noninverting input is positive, which reinforces the positive saturation. Similarly, when the output is negatively saturated, the feedback voltage to the noninverting input is negative, which reinforces the negative saturation.

Assume the output is negatively saturated. The feedback voltage will hold the output in negative saturation until the input voltage becomes positive enough to make the error voltage positive. When this happens, the output goes into positive saturation. Once the output is in positive saturation, it stays there until the input voltage becomes negative enough to make the error voltage negative. When it does, the output can change back to the negative state.

Here is how to derive the upper trip point. Assume the output is negatively saturated. Then

$$v_{\text{out}} = -V_{\text{sat}}$$

The output will change states from low to high when v_{error} passes through 0. When v_{error} is exactly 0,

$$\text{UTP} = i_{\text{in}} R_2 \qquad (21\text{-}11)$$

All the input current passes through R_1, so we can write

$$i_{\text{in}} = \frac{V_{\text{sat}}}{R_1}$$

Substituting this expression into Eq. (21-11) gives

$$\text{UTP} = \frac{R_2}{R_1} V_{\text{sat}} \qquad (21\text{-}12)$$

The lower trip point has the same value, with opposite sign.

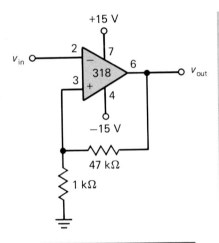

FIGURE 21-13

Example.

EXAMPLE 21-5

If $V_{\text{sat}} = 13.5$ V, what are the trip points and hysteresis in Fig. 21-13?

SOLUTION

The feedback fraction is

$$B = \frac{1 \text{ k}\Omega}{48 \text{ k}\Omega} = 0.0208$$

The upper trip point is

$$\text{UTP} = 0.0208(13.5 \text{ V}) = 0.281 \text{ V}$$

The lower trip has the same value with opposite sign:

$$\text{LTP} = -0.281 \text{ V}$$

The hysteresis is the difference of the two trip points:

$$H = 0.281 \text{ V} - (-0.281 \text{ V}) = 0.562 \text{ V}$$

This means the Schmitt trigger of Fig. 21-13 can withstand a peak-to-peak noise voltage of up to 0.562 V without false triggering. It also means that a desired input voltage will have to be more positive than 0.281 V to switch the output from high to low. Also it will have to be more negative than -0.281 V to switch the output from low to high.

EXAMPLE 21-6

Figure 21-14 shows a noninverting Schmitt trigger. If $V_{\text{sat}} = 13.5$ V, what are the trip points and hysteresis?

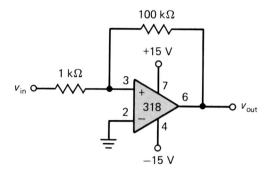

FIGURE 21-14   Noninverting Schmitt trigger with zero center voltage.

SOLUTION

With Eq. (21-12),

$$\text{UTP} = \frac{1\text{ k}\Omega}{100\text{ k}\Omega}\,13.5\text{ V} = 0.135\text{ V}$$

The hysteresis is double the foregoing amount:

$$H = 2(0.135\text{ V}) = 0.27\text{ V}$$

21-4 THE INTEGRATOR

An *integrator* is a circuit that performs a mathematical operation called integration. The most popular application of an integrator is to produce a *ramp* of output voltage, which is a linearly increasing or decreasing voltage.

Basic Circuit

Figure 21-15*a* is an op-amp integrator. As you can see, the feedback component is a capacitor instead of a resistor. The usual input is a rectangular pulse like Fig. 21-15*b*. When the pulse is low, $v_{\text{in}} = 0$. When the pulse is high, $v_{\text{in}} = V_{\text{in}}$. Visualize this pulse applied to the left end of R. Because of the virtual ground, a high input voltage produces an input current of

$$I_{\text{in}} = \frac{V_{\text{in}}}{R}$$

All this input current goes to the capacitor. As a result, the capacitor will charge and its voltage will increase with the polarity shown. The virtual ground implies that the output voltage equals the voltage across the

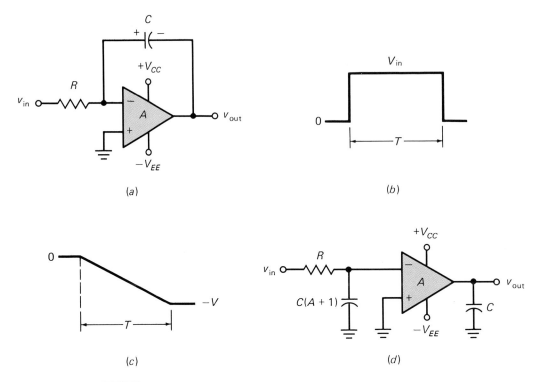

FIGURE 21-15 (*a*) Integrator; (*b*) rectangular input; (*c*) typical output ramp; (*d*) equivalent circuit with Miller capacitances.

capacitor. For a positive input voltage, the output voltage will be negative and increasing.

Here is how we can prove the output voltage is a ramp. The basic capacitor law says that

$$C = \frac{Q}{V}$$

or

$$V = \frac{Q}{C} \qquad (21\text{-}13)$$

Since a constant current is flowing to the capacitor, the charge Q increases linearly with respect to time. This means that the capacitor voltage increases linearly, equivalent to a negative ramp of output voltage, as shown in Fig. 21-15*c*. At the end of the pulse period, the input voltage returns to zero, and the charging current stops. Because the capacitor retains its charge, the output voltage remains constant at a negative voltage of $-V$.

To get a formula for the output voltage, divide both sides of Eq. (21-13) by T:

$$\frac{V}{T} = \frac{Q/T}{C}$$

Since the charging current is constant, we can write

$$\frac{V}{T} = \frac{I}{C}$$

or

$$V = \frac{IT}{C} \tag{21-14}$$

This is the voltage across the capacitor at the end of the pulse.

A final point. Because of the Miller effect, we can split the feedback capacitor into two equivalent capacitances, as shown in Fig. 21-15d. The closed-loop time constant RC' for the input bypass circuit is

$$RC' = RC(A + 1) \tag{21-15}$$

For the integrator to work properly, this time constant should be much greater than the width of the input pulse (at least 10 times greater). As a formula,

$$RC' > 10T \tag{21-16}$$

In the typical op-amp integrator, the closed-loop time constant is extremely long, so you rarely have any problem satisfying the 10:1 condition.

Eliminating Output Offset

The circuit of Fig. 21-15a needs a slight modification to make it practical. Because a capacitor is open to dc signals, there is no negative feedback at zero frequency. Without negative feedback, the circuit treats any input offset voltage as a valid input voltage. The result is that the capacitor charges and the output goes into positive or negative saturation, where it stays indefinitely.

One way to reduce the effect of input offset voltage is to decrease the voltage gain at zero frequency by inserting a resistor in parallel with the capacitor, as shown in Fig. 21-16a. This resistor should be at least 10 times larger than the input resistor. If the added resistance equals $10R$, the closed-loop voltage gain is 10 and the output offset voltage is satisfactorily reduced to an acceptable level. When a valid input voltage is present, the additional resistor has almost no effect on the charging of a capacitor, so that the output voltage is still almost a perfect ramp.

Another way to suppress the effect of input offset voltage is to use a JFET switch, as shown in Fig. 21-16b. The reset voltage on the gate of the JFET is either 0 V or $-V_{CC}$, which is enough to cut off the JFET. Therefore, we can set the JFET to a low resistance when the integrator is idle and to a high resistance when the integrator is active.

The JFET discharges the capacitor in preparation for the next input pulse. Just before the beginning of the next input pulse, the reset voltage is made equal to 0 V. This discharges the capacitor. At the instant the next pulse begins, the reset voltage becomes $-V_{CC}$, which cuts off the JFET. The integrator then does its thing and produces an output voltage ramp.

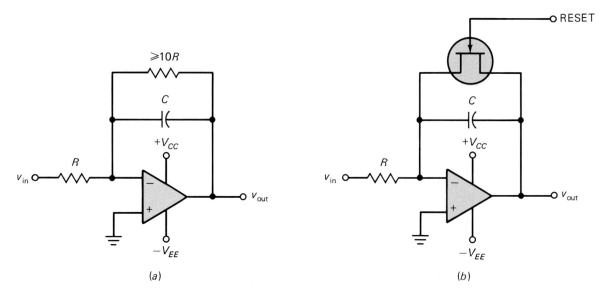

FIGURE 21-16 (a) Resistor across capacitor reduces output offset voltage; (b) JFET used to reset integrator.

EXAMPLE 21-7

In Fig. 21-17, what is the capacitor voltage at the end of the pulse? If the 741C has a differential voltage gain of 100,000, what is the closed-loop time constant?

SOLUTION

Because of the virtual ground on the inverting input, the input current is

$$I = \frac{8\,\text{V}}{2\,\text{k}\Omega} = 4\,\text{mA}$$

With Eq. (12-14), the output voltage at the end of the pulse is

$$V = \frac{(4\,\text{mA})(1\,\text{ms})}{1\,\mu\text{F}} = 4\,\text{V}$$

Visualize an output voltage that starts at 0 V and decreases linearly to -4 V. This is what comes out of the integrator. The waveform is like Fig. 21-15c.

Because of the Miller effect, the closed-loop time constant is

$$RC' = (2\,\text{k}\Omega)(1\,\mu\text{F})(100,000) = 200\,\text{s}$$

This is Eq. (21-15) with $A + 1$ equal to approximately 100,000 for a 741C. Since the closed-loop time constant (200 s) is much greater than the pulse width (1 ms), only the very earliest part of the

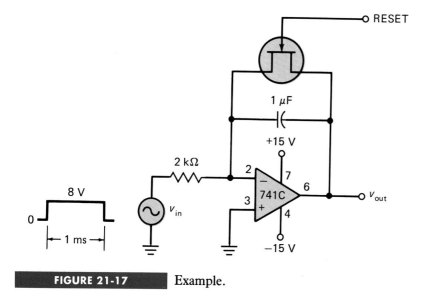

FIGURE 21-17 Example.

exponential charge is involved. Because of this, the output voltage is almost a perfect ramp. This is how the linear sweep voltages of an oscilloscope are produced.

21-5 WAVEFORM CONVERSION

With op amps we can convert sine waves to rectangular waves, rectangular waves to triangular waves, and so on. This section is about some basic circuits that convert an input waveform to an output waveform of a different shape.

Sine to Rectangular

Figure 21-18a shows a Schmitt trigger, and Fig. 21-18b is the graph of output voltage versus input voltage. When the input signal is *periodic* (repeating cycles), the Schmitt trigger produces a rectangular output, as shown. This assumes that the input signal is large enough to pass through both trip points of Fig. 21-18c. When the input voltage exceeds the UTP on the upward swing of the positive half-cycle, the output voltage switches to $-V_{sat}$. One half-cycle later, the input voltage becomes more negative than LTP, and the output switches back to $+V_{sat}$.

A Schmitt trigger always produces a rectangular output, regardless of the shape of the input signal. In other words, the input voltage does not have to be sinusoidal, as shown in Fig 21-18a. As long as the waveform is periodic and has an amplitude large enough to pass through the trip points, we get a rectangular output from the Schmitt trigger. This rectangular wave has the same frequency as the input signal (evident in Fig. 21-18c).

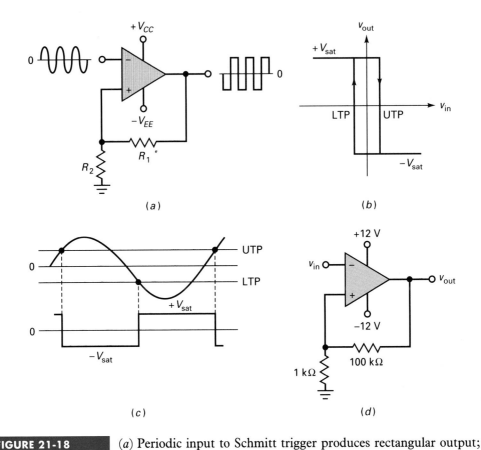

FIGURE 21-18 (*a*) Periodic input to Schmitt trigger produces rectangular output; (*b*) inverting transfer characteristic; (*c*) output transitions occur when input hits trip points; (*d*) example.

As an example, Fig. 21-18*d* shows a Schmitt trigger with trip points of approximately UTP = +0.1 V and LTP = −0.1 V. If the input voltage is repetitive and has a peak-to-peak value greater than 0.2 V, the output voltage is a rectangular wave with a peak-to-peak value of approximately 20 V.

Rectangular to Triangular

In Fig. 21-19*a*, a rectangular wave is the input to an integrator. Since the input voltage has a dc or average value of zero, the dc or average value of the output is also zero. As shown in Fig. 21-19*b*, the ramp is decreasing during the positive half-cycle of input voltage and increasing during the negative half-cycle. Therefore, the output is a triangular wave with the same frequency as the input. By analyzing the voltage change in a ramp, we can prove that the output voltage is given by

$$v_{\text{out}} = \frac{v_{\text{in}}}{4fRC} \tag{21-17}$$

where v_{in} and v_{out} are expressed in peak-to-peak values.

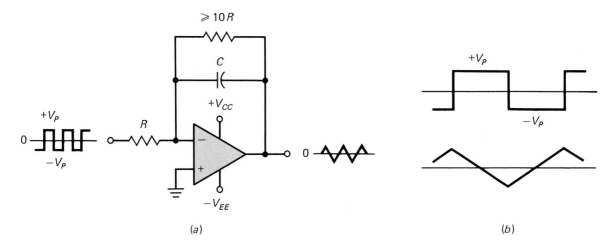

(a)   (b)

FIGURE 21-19   (a) Rectangular input to integrator produces rectangular output; (b) input and output waveforms; (c) example.

Triangle to Pulse

Figure 21-20a shows a circuit that converts a triangular input to a rectangular output. By varying R_2, we can change the width of the output pulses, equivalent to varying the *duty cycle*. In Fig. 21-20b, W represents the width of the pulse and T is the period. The duty cycle D is defined as the width of the pulse divided by the period:

$$D = \frac{W}{T} (100\%) \tag{21-18}$$

For instance, if the output has $W = 1$ ms and $T = 4$ ms, the duty cycle is

$$D = \frac{1 \text{ ms}}{4 \text{ ms}} (100\%) = 25\%$$

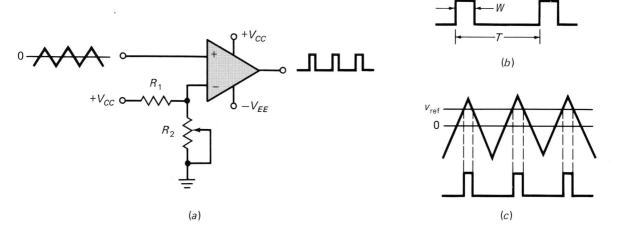

(a)   (c)

FIGURE 21-20   (a) Output has adjustable duty cycle; (b) waveform of triangular input and pulse output.

In some applications, we want to produce a variable duty-cycle pulse. The circuit of Fig. 21-20a is ideal for this job. Basically, it is a limit detector that switches output states when the input voltage crosses a certain level. The comparator has an adjustable reference voltage on the inverting input. This allows us to move the trip point from zero to a positive level. When the triangular input voltage exceeds the reference voltage, the output is high, as shown in Fig. 21-20c. Since v_{ref} is adjustable, we can vary the width of the output pulse, which is equivalent to changing the duty cycle. With a circuit like this, we can vary the duty cycle from approximately 0 to 50 percent.

EXAMPLE 21-8

A rectangular input drives the integrator of Fig. 21-21. If the frequency is 1 kHz and the peak-to-peak input voltage is 10 V, what is the output voltage?

SOLUTION

With Eq. (21-17), the output is a triangular wave with a peak-to-peak voltage of

$$v_{\text{out}} = \frac{10 \text{ V}}{4(1 \text{ kHz})(1 \text{ k}\Omega)(10 \text{ }\mu\text{F})} = 0.25 \text{ V}$$

EXAMPLE 21-9

A triangular input drives the circuit of Fig. 21-22a. If the frequency is 1 kHz, what is the frequency of the output wave? What is the duty cycle when the wiper is at the middle of its range?

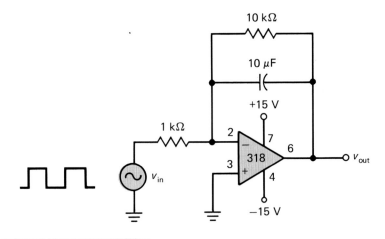

FIGURE 21-21   Example of integrator.

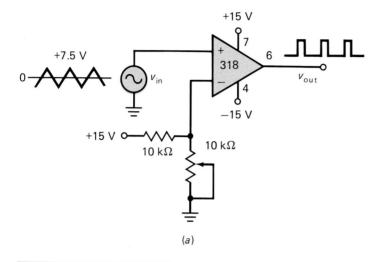

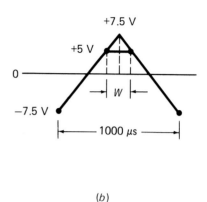

(a)

(b)

FIGURE 21-22   Example.

SOLUTION

Each output pulse occurs during the positive half cycle of input voltage. Therefore, the output frequency must be 1 kHz, the same as the input frequency.

On a schematic diagram, the value listed next to a potentiometer is usually the maximum resistance. Therefore, when the wiper is in the middle of its range, the resistance is 5 kΩ. This means the reference voltage is

$$v_{ref} = \frac{15 \text{ V}}{15 \text{ k}\Omega} 5 \text{ k}\Omega = 5 \text{ V}$$

The period of the input and output voltage signal is

$$T = \frac{1}{1 \text{ kHz}} = 1 \text{ ms} = 1000 \text{ μs}$$

Figure 21-22b shows this value. It takes 500 μs for the input voltage to increase from −7.5 to +7.5 V because this is half of the cycle. The trip point of the comparator is at +5 V. This means the output pulse has a width W, as shown in Fig. 21-22b.

Because of the geometry of Fig. 21-22b, we can set up a proportion between voltage and time as follows:

$$\frac{W/2}{500 \text{ μs}} = \frac{7.5 \text{ V} - 5 \text{ V}}{15 \text{ V}}$$

Solve for W to get

$$W = 167 \text{ μs}$$

The duty cycle is

$$D = \frac{167\ \mu s}{1000\ \mu s}(100\%) = 16.7\%$$

In Fig. 21-22a, moving the wiper down will increase the reference voltage and decrease the output duty cycle. Moving the wiper up will decrease the reference voltage and increase the output duty cycle. For the values given in Fig. 21-22a, the duty cycle can vary from 0 to 50 percent.

21-6 WAVEFORM GENERATION

With positive feedback, it is also possible to build *oscillators*—circuits that generate or create an output signal with no external input signal. This section takes a brief look at some op-amp circuits that can generate nonsinusoidal signals. Chapter 20 discusses sinusoidal oscillators and more advanced nonsinusoidal oscillators.

Relaxation Oscillator

In Fig. 21-23a, there is no input signal. Nevertheless, the circuit generates an output rectangular wave. How is that possible? Assume that the output is in positive saturation. The capacitor will charge exponentially toward $+V_{sat}$. It never reaches $+V_{sat}$ because its voltage hits the UTP, as shown in Fig. 21-23b. When this happens, the output voltage switches to $-V_{sat}$ and the capacitor reverses its charging direction. The capacitor voltage then decreases as shown. When the capacitor voltage hits the LTP, the output switches back to $+V_{sat}$. The cycle then repeats. Because of the continuous charging and discharging of the capacitor, the output is a rectangular wave with a duty cycle of 50 percent.

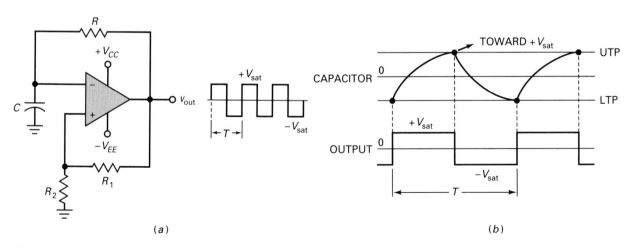

FIGURE 21-23　(*a*) Relaxation oscillator; (*b*) capacitor and output waveforms.

By analyzing the exponential charge and discharge of the capacitor, we can derive this formula for the period of the rectangular output:

$$T = 2RC \ln \frac{1 + B}{1 - B} \qquad (21\text{-}19)$$

This equation uses the natural logarithm, which is a logarithm to base e. A scientific calculator or table of natural logarithms needs to be used with this equation.

Figure 21-23a is called a *relaxation oscillator*, defined as a circuit that generates an output signal whose frequency depends on the charging of a capacitor. If we increase the RC time constant, it takes longer for the capacitor voltage to reach the trip points. Therefore, the frequency is lower. By making R adjustable, we can easily get a 50:1 tuning range.

This circuit really makes you appreciate electronics. The only inputs are the dc supply voltages. The circuit is *creating* an ac output signal without an ac input signal. It's almost like magic. The circuit creates something out of nothing. Or does it? From the standpoint of energy, there is a price to be paid: Part of the dc power from the supply is converted to the ac energy for the output signal. This aside, the circuit is fascinating. It really does create an ac signal. Chapter 22 will continue the discussion of *oscillators*—circuits that create their own ac output signals without ac input signals.

Generating Triangular Waves

When we cascade a relaxation oscillator and an integrator, we get a circuit that produces a triangular output as shown in Fig. 21-24. The rectangular wave out of the relaxation oscillator drives the integrator, which produces a triangular output waveform. The rectangular wave swings between $+V_{\text{sat}}$ and $-V_{\text{sat}}$. You can calculate its period with Eq. (21-19). The triangular wave has the same period and frequency. You can calculate its peak-to-peak value with Eq. (21-17).

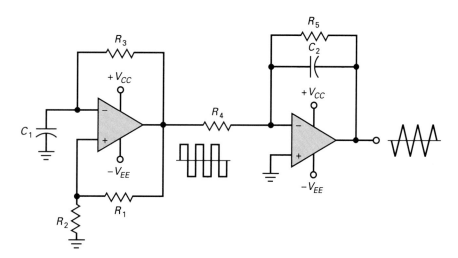

FIGURE 21-24   Relaxation oscillator drives integrator to produce triangular output.

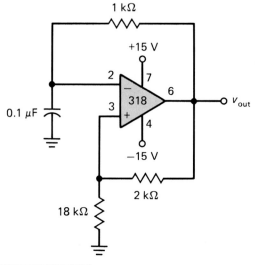

1 kΩ

+15 V

2

7

6

3

318

+

4

0.1 μF

−15 V

V_{out}

2 kΩ

18 kΩ

FIGURE 21-25 Example.

EXAMPLE 21-10

What is the frequency of the output signal in Fig. 21-25?

SOLUTION

In Fig. 21-25, locate the feedback voltage divider. Notice that it contains an R_2 of 18 kΩ and an R_1 of 2 kΩ. With these values the feedback fraction is

$$B = \frac{18\text{ k}\Omega}{20\text{ k}\Omega} = 0.9$$

The RC time constant is

$$RC = (1\text{ k}\Omega)(0.1\ \mu\text{F}) = 100\ \mu\text{s}$$

With Eq. (21-19), the period of the output signal is

$$T = 2(100\ \mu\text{s})\ln\frac{1.9}{0.1} = 589\ \mu\text{s}$$

and the frequency is

$$f = \frac{1}{589\ \mu\text{s}} = 1.7\text{ kHz}$$

The ac output voltage has a frequency of 1.7 kHz and a peak-to-peak value of $2V_{sat}$, approximately 27 V for the circuit of Fig. 21-25.

OPTIONAL TOPICS

The following material continues the earlier discussions at a more advanced and specialized level. All the topics are optional because they are not used in any of the basic discussions in later chapters. This section will be a useful reference when you are in industry because then you will probably want more advanced viewpoints.

21-7 A/D CONVERTER

Figure 21-26 is part of an *analog-to-digital converter* (an A/D converter) used in digital voltmeters and many other applications. The input voltage being measured or converted is applied to the noninverting input. A staircase voltage drives the inverting input. As the inverting voltage increases, the error voltage becomes less positive. Somewhere along the staircase, the inverting input becomes more positive than the noninverting input. When this happens, the output of the comparator switches to the low state. The time it takes the staircase voltage to exceed v_{in} is the key to how the circuit works. The larger v_{in} is, the more time it takes for the staircase voltage to exceed v_{in}. In other words, the time is directly proportional to v_{in}. With other circuits not shown, we can measure this time and display the voltage with a seven-segment indicator.

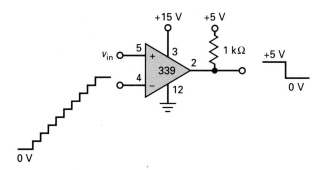

FIGURE 21-26 Comparator as part of analog-to-digital converter.

21-8 WINDOW COMPARATOR

An ordinary comparator indicates when the input voltage exceeds a certain limit or threshold. A *window comparator* (also called a *double-ended limit detector*) detects when the input voltage is between two limits. In this section we discuss two examples of window comparators.

Op-Amp Example

Figure 21-27a is an example of a window comparator that uses an op amp. The noninverting input is referenced by a Thevenin voltage of $+V_{CC}/3$ and the inverting input by a Thevenin voltage of $+V_{CC}/4$. Since

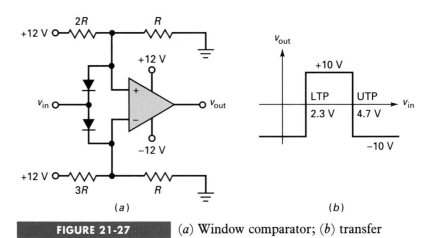

FIGURE 21-27 (*a*) Window comparator; (*b*) transfer characteristic.

V_{CC} is 12 V, the Thevenin references are $+4$ V for the noninverting input and $+3$ V for the inverting input.

When the input voltage is zero the upper diode is on and the lower diode is off. Since the noninverting input is clamped one diode drop above the input voltage, the noninverting input is $+0.7$ V. The inverting input, however, is at $+3$ V. Therefore, the error voltage is negative, and the comparator output is low.

As the input voltage increases, the noninverting input also increases, remaining 0.7 V higher than v_{in}. When v_{in} reaches $+2.3$ V, the noninverting input is clamped at $+3$ V. Since the inverting input is still at $+3$ V, the error voltage is now zero. If the input voltage v_{in} rises above $+2.3$ V, the output of the comparator goes high. An input of $+2.3$ V is a critical value because the output of the comparator is on the verge of switching from low to high. This input voltage is called the *lower trip point* (LTP). When v_{in} is greater than the LTP, the output voltage switches into the high state as shown in Fig. 21-27*b*.

As the input voltage increases, the comparator output stays high until v_{in} equals $+$ 4.7 V. At this value of input voltage, the lower diode is on and the inverting input is at $+4$ V; therefore, the error voltage is again at zero. Once again, the comparator is on the verge of switching its output. When v_{in} is greater than $+4.7$ V, the error voltage goes negative, driving the output into the low state. An input of $+4.7$ V is the UTP because at slightly above this level, the output switches back to the low state.

The transfer characteristic of Fig. 21-27*b* is called a *window* because the output is high only when the input is between the LTP and the UTP. With a V_{CC} of $+12$ V, the window comparator of Fig 21-27*a* has an LTP of $+2.3$ V and a UTP of $+4.7$ V. By changing the voltage dividers, we can change the width of the window. The window comparator is a useful circuit whenever we are trying to check if the input is between two limits.

Using the LM339

Figure 21-28 shows another design for a window comparator. It uses an LM339, which is a dual comparator. With a positive supply of $+12$ V, the reference voltages are $+4$ V for the upper comparator and $+3$ V for

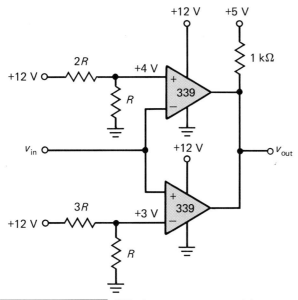

FIGURE 21-28   Window comparator with two 339s.

the lower comparator. When v_{in} is between $+3$ and $+4$ V, both comparators have a positive error voltage and their output transistors are open. Because of this, the final output is high. When v_{in} is less than $+3$ V or more than $+4$ V, one of the comparators will have a saturated transistor and the other will have a cutoff transistor. The saturated transistor pulls the output voltage down to a low level. The transfer characteristic has an LTP of $+3$ V and a UTP of $+4$ V.

21-9 ADDITIONAL SCHMITT TRIGGER

The Schmitt trigger can be improved in several ways. To begin, the stray capacitance will slow down the switching speed of the circuit because voltages cannot change until capacitances have been charged. This section shows you the classical solution to this problem. We will discuss moving the trip points.

Speed-up Capacitor

Besides suppressing the effects of noise, positive feedback speeds up the switching of output states. When the output voltage begins to change, this change is fed back to the noninverting input and amplified, forcing the output to change faster. Sometimes a capacitor C_1 is connected in parallel with R_1, as shown in Fig. 21-29a. Known as a *speed-up capacitor*, it helps to cancel the bypass circuit formed by the stray capacitance C_2 across R_2. This stray capacitance has to be charged before the noninverting input voltage can change. The speed-up capacitor supplies this charge.

To neutralize the stray capacitance, the capacitive voltage divider formed by C_2 and C_1 must have the same impedance ratio as the resistive voltage divider:

$$\frac{X_{C2}}{X_{C1}} = \frac{R_2}{R_1}$$

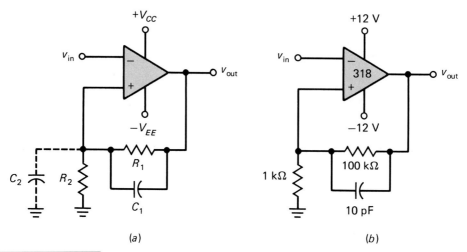

FIGURE 21-29 (*a*) Speed-up capacitor compensates from stray capacitance; (*b*) example.

Since, $X_C = 1/(2\pi f C)$, the equation may be simplified and rearranged to

$$C_1 = \frac{R_2}{R_1} C_2 \tag{21-20}$$

The value of C_1 given by this equation is the minimum value that neutralizes the bypass effects of stray capacitance C_2. As long as C_1 is equal to or greater than the value given by Eq. (21-20), the output will switch states at maximum speed. Since you often have to estimate the value of the stray capacitance, it's best to make C_1 at least 2 times larger than the value given by Eq. (21-20). In typical circuits, C_1 is from 10 to 100 pF.

As an example, Fig. 21-29*b* shows a 318 comparator connected as a zero-crossing detector with hysteresis. Because the supply voltages are + 12 V, voltage V_{sat} is approximately 10 V. With B approximately equal to 0.01, the UTP is + 0.1 V and the LTP is − 0.1 V. A speedup capacitor of 10 pF is used to neutralize the stray capacitance across R_2.

An alternative form of Eq. (21-20) is

$$R_1 C_1 \geq R_2 C_2 \tag{21-21}$$

This says that the time constant of the speedup section must be equal to or greater than the time constant of the stray-capacitance section.

Moving the Trip Points

Figure 21-30 shows how to move the trip points of an inverting Schmitt trigger. An additional resistor R_3 is connected between the noninverting input and $+V_{CC}$. This sets up the center of the hysteresis loop:

$$v_{\text{cen}} = \frac{R_2}{R_2 + R_3} V_{CC} \tag{21-22}$$

The positive feedback spreads the trip point on each side of the center voltage. To understand why this happens, apply Thevenin's theorem to

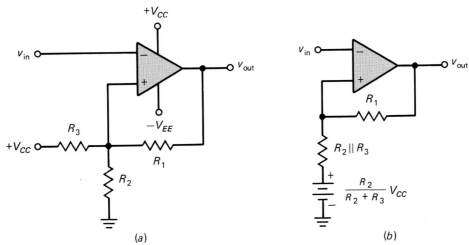

FIGURE 21-30 (*a*) Schmitt trigger with positive reference voltage; (*b*) equivalent circuit.

get Fig. 21-30*b*. The feedback fraction is

$$B = \frac{R_2 \| R_3}{R_1 + R_2 \| R_3} \qquad (21\text{-}23)$$

When the output is positively saturated, the noninverting reference voltage is

$$\text{UTP} = v_{\text{cen}} + BV_{\text{sat}} \qquad (21\text{-}24)$$

When the output is negatively saturated, the noninverting voltage is

$$\text{LTP} = v_{\text{cen}} - BV_{\text{sat}} \qquad (21\text{-}25)$$

If you want to move the trip points of a noninverting Schmitt trigger, then apply a reference voltage to the inverting input, as shown in Fig. 21-31. The reference voltage equals

$$v_{\text{ref}} = \frac{R_4}{R_3 + R_4} V_{CC} \qquad (21\text{-}26)$$

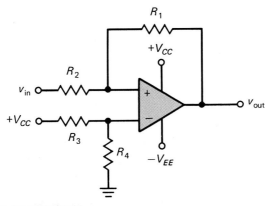

FIGURE 21-31 Noninverting Schmitt trigger with positive center voltage.

By a derivation similar to that given earlier, we can prove that the center voltage is

$$v_{\text{cen}} = 1 + \frac{R_2}{R_1} v_{\text{ref}} \qquad (21\text{-}27)$$

The trips points are spread on both sides of this center:

$$\text{UTP} = v_{\text{cen}} + \frac{R_2}{R_1} V_{\text{sat}} \qquad (21\text{-}28)$$

and

$$\text{LTP} = v_{\text{cen}} - \frac{R_2}{R_1} V_{\text{sat}} \qquad (21\text{-}29)$$

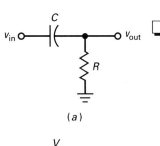

(a)

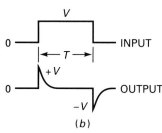

(b)

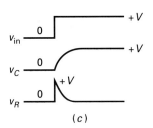

(c)

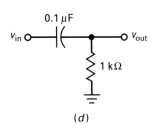

(d)

FIGURE 21-32

(a) RC differentiator; (b) rectangular input pulse produces narrow output spikes; (c) voltage waveforms; (d) example.

21-10 THE DIFFERENTIATOR

A *differentiator* is a circuit that performs a calculus operation called differentiation. It produces an output voltage proportional to the instantaneous rate of change of the input voltage. Common applications of a differentiator are to detect the leading and trailing edges of a rectangular pulse or to produce a rectangular output from a ramp input.

RC Differentiator

A coupling circuit like Fig. 21-32a can be used to differentiate the input signal. Instead of a sinusoidal signal, the typical input is a rectangular pulse, as shown in Fig. 21-32b. The output of the circuit is a series of positive and negative spikes. The positive spike occurs at the same instant as the leading edge of the input; the negative spike occurs at the same instant as the trailing edge. Spikes like these are useful signals because they can tell other circuits when the rectangular input signal starts and ends.

To understand how the *RC* differentiator works, look at Fig. 21-32c. When the input voltage changes from 0 to V, the capacitor begins to charge exponentially, as shown. After 5 time constants, the capacitor voltage is within 1 percent of the final voltage V. To satisfy Kirchhoff's voltage law, the voltage across the resistor of Fig. 21-32a is

$$v_R = v_{\text{in}} - v_C$$

Since v_C is initially zero, the output voltage suddenly jumps from 0 to V, then decays exponentially, as shown in Fig. 21-32c. By a similar argument, the trailing edge of a rectangular pulse produces a negative spike. Incidentally, notice that each spike in Fig. 21-32b has a peak value of approximately V, the size of the voltage step.

If an *RC* differentiator is to produce narrow spikes, the time constant should be at least 10 times smaller than the pulse width T:

$$RC < 10T$$

For instance, if the pulse width is 1 ms, the *RC* time constant should be less than 0.1 ms. Figure 21-32d shows an *RC* differentiator with a time constant of 0.1 ms. If you drive this circuit with any rectangular pulse whose T is greater than 1 ms, the output is a series of sharp positive and negative voltage spikes.

Op-Amp Differentiator

Figure 21-33a shows an op-amp differentiator. Notice the similarity to the op-amp integrator. The difference is that the resistor and capacitor are interchanged. Because of the virtual ground, the capacitor current passes through the feedback resistor, producing a voltage. An input often used with op-amp differentiators is a ramp like the top waveform of Fig. 21-33b. Because of the virtual ground, all the input voltage appears across the capacitor. The ramp of voltage implies that the capacitor current is constant. Why? Because the capacitor current is given by this fundamental relation:

$$i = C \frac{dv}{dt}$$

The quantity dv/dt has a value that is equal to the slope of the input voltage. If the input voltage is a ramp, the slope is a constant. This makes the current a constant, as shown in Fig. 21-33b (second waveform). Since all this constant current flows through the feedback resistor, we get an inverted pulse at the output, as shown in Fig. 21-33b (third waveform).

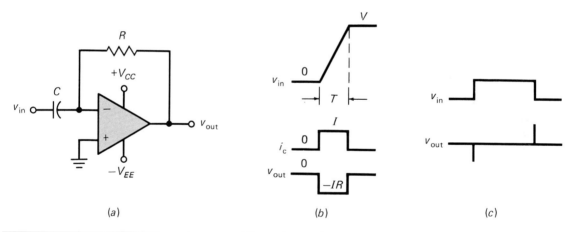

(a) (b) (c)

FIGURE 21-33 (a) Op-amp differentiator; (b) ramp input produces rectangular output; (c) rectangular input produces spiked output.

Here is how to derive the current. At the end of the ramp, the capacitor voltage is

$$V = \frac{Q}{C}$$

Dividing both sides by the ramp time gives

$$\frac{V}{T} = \frac{Q/T}{C}$$

or

$$\frac{V}{T} = \frac{I}{C}$$

Solving for current, we get

$$I = \frac{CV}{T} \tag{21-30}$$

This current is the key. Once you have calculated it, you can get the output voltage with

$$v_{\text{out}} = IR \tag{21-31}$$

On an oscilloscope, the leading edge of a rectangular pulse may look perfectly vertical. But if you shorten the time base enough, you will see that the leading edge is usually a rising exponential wave. As an approximation, we can treat this rising exponential as a positive ramp. One common application of the op-amp differentiator is to produce very narrow spikes, as shown in Fig. 21-33c. The leading edge of the pulse is approximately a positive ramp, so that the output will be a negative-going spike of very short duration. Similarly, the trailing edge of the input pulse is approximately a negative ramp, so that the output is a very narrow positive spike. The advantage of an op-amp differentiator over a simple RC differentiator is that the spikes are coming from a low-impedance source, which makes driving typical load resistances easier.

Practical Op-Amp Differentiator

The op-amp differentiator of Fig. 21-33a has a tendency to oscillate. To avoid this, a practical op-amp differentiator usually includes some resistance in series with the capacitor, as shown in Fig. 21-34. A typical value for this added resistance is between $0.01R$ and $0.1R$. With this resistor, the closed-loop voltage gain is between 10 and 100. The effect is to limit the closed-loop voltage gain at higher frequencies, where the oscillation problem arises.

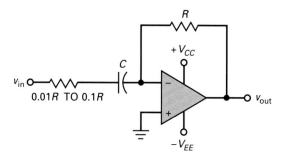

FIGURE 21-34 Resistance in series with capacitor prevents high-frequency oscillations.

Incidentally, the source driving the op-amp differentiator has an output impedance. If this is a resistance between $0.01R$ and $0.1R$, you don't have to include an extra resistor because the source impedance supplies it.

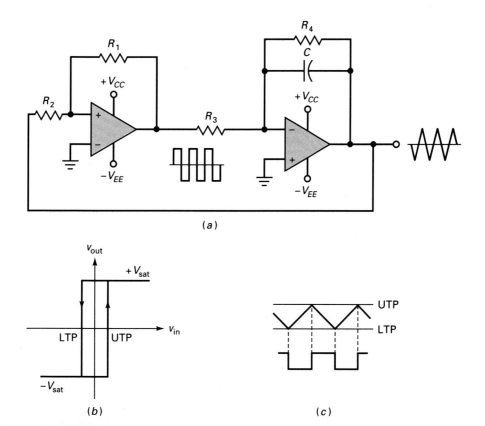

FIGURE 21-35 Feedback circuit with Schmitt trigger and integrator produces triangular output.

21-11 ANOTHER TRIANGULAR GENERATOR

In Fig. 21-35a, the output of a noninverting Schmitt trigger is a rectangular wave that drives an integrator. The output of the integrator is a triangular wave. This triangular wave is fed back and used to drive the Schmitt trigger. So we have a very interesting circuit. The first stage drives the second, and the second drives the first.

Figure 21-35b is the transfer characteristic of the Schmitt trigger. When the output is low, the input must increase to the UTP to switch the output to high. Likewise, when the output is high, the input must decrease to the LTP to switch the output to low.

The triangular wave from the integrator is perfect for driving the Schmitt trigger. When the Schmitt trigger output is low in Fig. 21-35a, the integrator produces a positive ramp. This positive ramp increases until it reaches the UTP, as shown in Fig. 21-35c. At this point, the output of the Schmitt trigger switches to the high state and forces the triangular wave to reverse direction. The negative ramp now decreases until it reaches the LTP, where another Schmitt output change takes place.

There is one missing concept in the previous explanation: How does the circuit get started in the first place? When you first power up, the

Schmitt-trigger output must be either low or high. If it is low, the integrator produces a rising ramp; if it is high, the integrator produces a falling ramp. Either way, the triangular waveform has started, and the positive feedback will keep it going.

STUDY AIDS

The following study aids will help to reinforce the ideas discussed in this chapter. For best results, use these study aids within 6 hours of reading the earlier material. Then review these study aids a week later and a month later to ensure that the concepts remain in your long-term memory.

SUMMARY

Sec. 21-1 Active Diode Circuits
The active diode circuits include the half-wave rectifier, peak detector, positive limiter, and positive clamper. In all the circuits, an op amp is used to reduce the knee voltage by a factor of A. The result is an active circuit that can process low-level signals because the closed-loop knee voltage is only V_K/A.

Sec. 21-2 Comparators
A comparator is a circuit with a low or a high output. It indicates when the input voltage is greater than the trip point. A comparator can be made from op amps, but the usual comparator is an IC that has been optimized for this application. An IC comparator is similar to an IC op amp, except that the compensating capacitor is omitted and the output stage may require an external pullup resistor and supply voltage.

Sec. 21-3 The Schmitt Trigger
A Schmitt trigger is a comparator with hysteresis. Instead of a single trip point, it has two trip points. The hysteresis prevents false triggering by unwanted noise voltages at the input.

Sec. 21-4 The Integrator
An integrator produces a ramp of output voltage when the input is a rectangular pulse. The key to its operation is the closed-loop time constant, which is much larger than the open-loop time constant.

Because of the Miller effect, the charging of the capacitor appears to use only the earliest part of the exponential wave. The result is an output that is almost a perfect ramp.

Sec. 21-5 Waveform Conversion
Op amps can be used to convert waveforms to different shapes. A Schmitt trigger changes an input of any shape to a rectangular wave. An integrator changes a rectangular wave to a triangular wave. A comparator with an adjustable trip point changes a triangular wave to a series of pulses whose duty cycle depends on the value of the trip point.

Sec. 21-6 Waveform Generation
A relaxation oscillator generates a rectangular output voltage. This is a comparator that uses positive feedback. The frequency is controlled by the charging and discharging of a capacitor. By cascading a relaxation oscillator and an integrator, we can generate triangular waves.

VOCABULARY

In your own words, explain what each of the following terms means. Keep your answers short and to the point. If necessary, verify your answer by rereading the appropriate discussion or by looking at the end-of-book Glossary.

comparator	positive clamper
duty cycle	positive feedback
flag	positive limiter
hysteresis	pullup resistor
integrator	relaxation oscillator
peak detector	Schmitt trigger
periodic	trip point

IMPORTANT EQUATIONS

Here are some important equations. Say each of the following equations in symbols, then say each in words. Try to explain what the equation means and how it is used. Then read the description that follows.

Eq. 21-1 Closed-Loop Knee Voltage

$$V'_K = \frac{V_K}{A}$$

This applies to all active diode circuits such as the half-wave rectifier, peak detector, limiter, and clamper. It tells you why an active circuit can process a low-level signal, one whose peak voltage is much smaller than 0.7 V.

Eq. 21-2 Peak Detection

$$R_L C > 10T$$

For a peak detector to work properly, you have to satisfy this condition. It says that you must make the discharging time constant at least 10 times greater than the period of the input signal. When this condition is satisfied, the capacitor can discharge only slightly. Thus its voltage remains near the peak of the input voltage.

Eq. 21-14 Ramp Voltage

$$V = \frac{IT}{C}$$

This gives you the output voltage of an op-amp integrator. It says the voltage equals the current into the capacitor multiplied by the time and then divided by the capacitance.

Eq. 21-18 Duty Cycle

$$D = \frac{W}{T}(100\%)$$

This is the defining formula for the duty cycle. The value of D can vary from 0 to 100 percent.

STUDENT ASSIGNMENTS

QUESTIONS

The following questions may have more than one right answer. Select the best answer. This is the one that is always true, or covers more situations, or fits the context, etc.

1. In a nonlinear op-amp circuit, the
   a. Op amp never saturates
   b. Feedback loop is never opened
   c. Output shape is the same as the input shape
   d. Op amp may saturate

2. An active half-wave rectifier has a knee voltage of
   a. V_K c. More than 0.7 V
   b. 0.7 V   d. Much less than 0.7 V

3. In an active peak detector, the discharging time constant is
   a. Much longer than the period
   b. Much shorter than the period
   c. Equal to the period
   d. The same as the charging time constant

4. If the reference voltage is zero, the output of an active positive limiter is
   a. Positive
   b. Negative
   c. Either positive or negative
   d. A ramp

5. The output of an active positive clamper is
   a. Positive
   b. Negative
   c. Either positive or negative
   d. A ramp

6. The positive clamper adds
   a. A positive dc voltage to the input
   b. A negative dc voltage to the input
   c. An ac signal to the output
   d. A trip point to the input

7. If you want a circuit that detects when the input is greater than a particular value, you can use a
   a. Comparator c. Limiter
   b. Clamper   d. Relaxation oscillator

8. A flag is
   a. A voltage that is low or high
   b. A sine wave
   c. The output of a relaxation oscillator
   d. Something hanging from a pole

9. The voltage out of a Schmitt trigger is
   a. A low voltage
   b. A high voltage
   c. Either a low or a high voltage
   d. A sine wave

10. Hysteresis prevents false triggering associated with
    a. A sinusoidal input
    b. Unwanted noise voltages
    c. Stray capacitances
    d. Trip points

11. If the input is a rectangular pulse, the output of an integrator is a
    a. Sine wave
    b. Square wave
    c. Ramp
    d. Rectangular pulse

12. When a large sine wave drives a Schmitt trigger, the output is a
    a. Rectangular wave
    b. Triangular wave
    c. Rectified sine wave
    d. Series of ramps

13. When the width of the pulse decreases and the period stays the same, the duty cycle
    a. Decreases
    b. Stays the same
    c. Increases
    d. Is zero

14. The output of a relaxation oscillator is a
    a. Sine wave
    b. Square wave
    c. Ramp
    d. Spike

15. The op amp in an active half-wave rectifier has a voltage gain of 200,000. The closed-loop knee voltage is
    a. 1 μV
    b. 3.5 μV
    c. 7 μV
    d. 14 μV

16. The input to a peak detector is a triangular wave with a peak-to-peak value of 8 V and an average value of 0. The output is
    a. 0   c. 8 V
    b. 4 V   d. 16 V

17. The input to a positive limiter is a triangular wave with a peak-to-peak value of 8 V and an average value of 0. If the reference level is 2 V, the output has a peak-to-peak value of
    a. 0   c. 6 V
    b. 2 V   d. 8 V

18. The discharging time constant of a peak detector is 10 ms. The lowest frequency you should use is
    a. 10 Hz   c. 1 kHz
    b. 100 Hz   d. 10 kHz

19. A comparator with a trip point of zero is sometimes called a
    a. Threshold detector
    b. Zero-crossing detector
    c. Positive limit detector
    d. Half-wave detector

20. To work properly, many IC comparators need an external
    a. Compensating capacitor
    b. Pullup resistor
    c. Bypass circuit
    d. Output stage

21. A Schmitt trigger uses
    a. Positive feedback
    b. Negative feedback
    c. Compensating capacitors
    d. Pullup resistors

22. A Schmitt trigger
    a. Is a zero-crossing detector
    b. Has two trip points
    c. Produces triangular output waves
    d. Is designed to trigger on noise voltages

23. A relaxation oscillator depends on the charging of a capacitor through a
    a. Resistor   c. Capacitor
    b. Inductor   d. Noninverting input

24. A ramp of voltage
    a. Always increases
    b. Is a rectangular pulse
    c. Increases or decreases at a linear rate
    d. Is produced by hysteresis

25. The op-amp integrator uses
    a. Inductors
    b. The Miller effect
    c. Sinusoidal inputs
    d. Hysteresis

26. The trip point of a comparator is the input voltage that causes the
    a. Circuit to oscillate
    b. Peak detection of the input signal
    c. Output to switch states
    d. Clamping to occur

27. In an op-amp integrator, the current through the input resistor flows into the
    a. Inverting input
    b. Noninverting input
    c. Bypass capacitor
    d. Feedback capacitor

BASIC PROBLEMS

Sec. 21-1 Active Diode Circuits
21-1. What is the output voltage in Fig. 21-36?

21-2. What is the lowest recommended frequency in Fig. 21-36?

21-3. Suppose the diode of Fig. 21-36 is reversed. What is the output voltage?

21-4. The input voltage of Fig. 21-36 is changed from 75 mV rms to 150 mV peak to peak. What is the output voltage?

21-5. A positive clamper like Fig. 21-4, on page 810, has $R_L = 10$ kΩ and $C = 4.7$ μF. What is the lowest recommended frequency for this clamper?

Sec. 21-2 Comparators
21-6. What is the reference voltage in Fig. 21-37?

21-7. A bypass capacitor of 47 μF is connected across the 3.3 kΩ of Fig. 21-37. What is the critical frequency of the bypass circuit? If the supply ripple is 1 V rms, what is the approximate ripple at the inverting input?

21-8. Suggest one or more changes in Fig. 21-37 to get a reference voltage of 1 V.

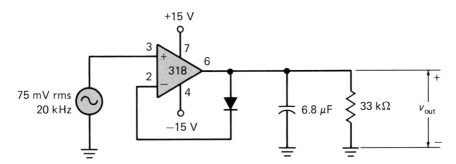

FIGURE 21-36 Example.

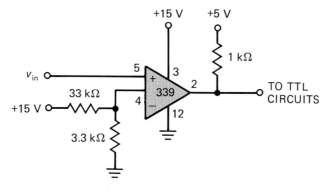

FIGURE 21-37 LM339 comparator.

Sec. 21-3 The Schmitt Trigger

21-9. If $V_{sat} = 13.5$ V in Fig. 21-38, what are the trip points and hysteresis?

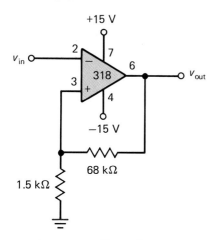

FIGURE 21-38 Example.

21-10. What are the trip points and hysteresis if $V_{sat} = 13.5$ V in Fig. 21-39?

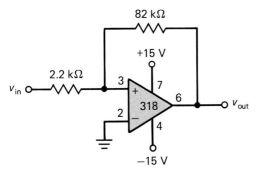

FIGURE 21-39 Noninverting Schmitt trigger with zero center voltage.

21-11. The resistors of Fig. 21-38 have a tolerance of ± 5 percent. What is the minimum hysteresis?

21-12. The noise voltage at the input of Fig. 21-39 may be as large as 1 V peak-to-peak. Suggest one or more changes that make the circuit immune to noise voltage.

Sec. 21-4 The Integrator

21-13. What is the capacitor charging current in Fig. 21-40 when the input pulse is high?

21-14. In Fig. 21-40, the output voltage is reset just before the pulse begins. What is the output voltage at the end of the pulse?

21-15. The input voltage is changed from 5 to 1 V in Fig. 21-40. The capacitance of Fig. 21-40 is changed to each of these values: 0.1, 1, 10, and 100 μF. What is the output voltage at the end of the pulse for each capacitance?

21-16. We want to produce ramp output voltages in Fig. 21-40 that swing from 0 to $+10$ V with times of 0.1, 1, and 10 ms. What changes can you make in the circuit to accomplish this? (Many right answers are possible.)

Sec. 21-5 Waveform Conversion

21-17. What is the output voltage in Fig. 21-41? If the capacitance is 0.68 μF, what is the output voltage?

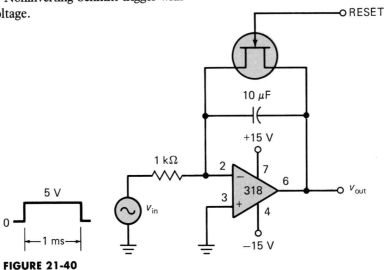

FIGURE 21-40

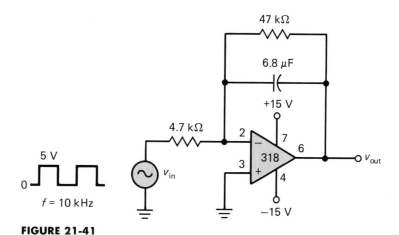

FIGURE 21-41

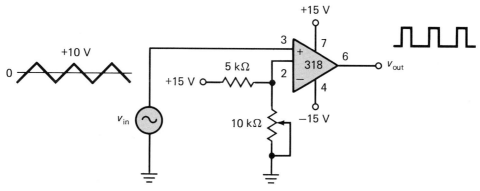

FIGURE 21-42

21-18. In Fig. 21-41, what happens to the output voltage if the frequency changes to 5 kHz? To 20 kHz?

21-19. What is the duty cycle in Fig. 21-42 when the wiper is at the top? At the bottom?

21-20. What is the duty cycle in Fig. 21-42 when the wiper is one-quarter of the way from the top?

Sec. 21-6 Waveform Generation

21-21. What is the frequency of the output signal in Fig. 21-43?

21-22. If all resistors are doubled in Fig. 21-43, what happens to the frequency?

21-23. We want the output frequency of Fig. 21-43 to be 20 kHz. Suggest some changes that will accomplish this.

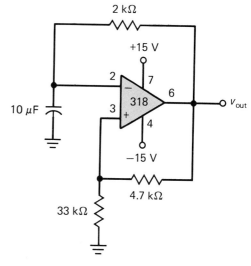

FIGURE 21-43

ADVANCED PROBLEMS

21-24. In Fig. 21-26, a staircase voltage drives the inverting input, and a constant voltage v_{in} drives the noninverting input. The staircase has 256 steps between 0 and +5 V. Each step has a duration of 1 μs. If v_{in} is +2.75 V, how long after the staircase starts does the output change states?

21-25. Company XYZ is mass-producing relaxation oscillators. The output voltage is supposed to be at least 10 V peak to peak. Suggest some ways to check the output of each unit to see if it is at least 10 V peak to peak. (There are many right answers here. See how many you can think of. You can use any device or circuit in this and earlier chapters.)

21-26. How can you build a circuit that turns on the lights when it gets dark and turns them off when it gets light? (Use this and earlier chapters to find as many right answers as you can think of.)

21-27. You have some electronics equipment that malfunctions when the line voltage is too low. Suggest one or more ways to set off an audible alarm when the line voltage is less than 105 V rms.

21-28. Radar waves travel at 186,000 mi/s. A transmitter on earth sends a radar wave to the moon, and an echo of this radar wave returns to earth. In Fig. 21-40, 1 kΩ is changed to 1 MΩ. The input rectangular pulse starts at the instant the radar wave is sent to the moon, and the pulse ends at the instant the radar wave arrives back at earth. If the output ramp has decreased from 0 to a final voltage of −1.23 V, how far away is the moon?

T-SHOOTER PROBLEMS

Use Fig. 21-44 for the remaining problems. In this T-shooter you are troubleshooting at the system level, meaning that you are to locate the most suspicious block for further testing. For instance, if the waveform is okay at point B but incorrect at point C, your answer should be "peak detector." As an example of using this T-shooter, look at the box labeled "OK." To see the voltage waveform at point E, read the token next to V_E, which is $G7$. In the big box labeled "Waveforms," translate token $G7$ to a value of L. Then look for waveform L, and you will see the normal output of the integrator.

Here is what the system does. The relaxation oscillator produces a rectangular output that swings from +5 to −5 V. The positive clamper shifts this waveform upward, so that it swings from +10 to 0 V. The peak detector produces a dc output of +10 V, which drives the comparator into positive saturation. The comparator output then drives the integrator to produce a negative ramp. Before you start troubleshooting, look at all the normal waveforms given in the OK box.

21-29. Find Trouble 1.

21-30. Find Trouble 2.

21-31. Find Troubles 3 and 4.

21-32. Find Troubles 5 and 6.

21-33. Find Troubles 7 and 8.

21-34. Find Troubles 9 and 10.

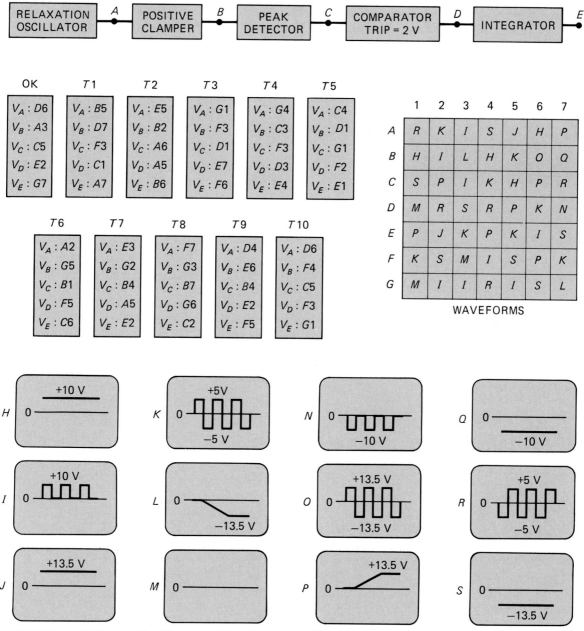

| | RELAXATION OSCILLATOR | A | POSITIVE CLAMPER | B | PEAK DETECTOR | C | COMPARATOR TRIP = 2 V | D | INTEGRATOR | E |

OK
V_A : D6
V_B : A3
V_C : C5
V_D : E2
V_E : G7

T1
V_A : B5
V_B : D7
V_C : F3
V_D : C1
V_E : A7

T2
V_A : E5
V_B : B2
V_C : A6
V_D : A5
V_E : B6

T3
V_A : G1
V_B : F3
V_C : D1
V_D : E7
V_E : F6

T4
V_A : G4
V_B : C3
V_C : F3
V_D : D3
V_E : E4

T5
V_A : C4
V_B : D1
V_C : G1
V_D : F2
V_E : E1

T6
V_A : A2
V_B : G5
V_C : B1
V_D : F5
V_E : C6

T7
V_A : E3
V_B : G2
V_C : B4
V_D : A5
V_E : E2

T8
V_A : F7
V_B : G3
V_C : B7
V_D : G6
V_E : C2

T9
V_A : D4
V_B : E6
V_C : B4
V_D : E2
V_E : F5

T10
V_A : D6
V_B : F4
V_C : C5
V_D : F3
V_E : G1

	1	2	3	4	5	6	7
A	R	K	I	S	J	H	P
B	H	I	L	H	K	O	Q
C	S	P	I	K	H	P	R
D	M	R	S	R	P	K	N
E	P	J	K	P	K	I	S
F	K	S	M	I	S	P	K
G	M	I	I	R	I	S	L

WAVEFORMS

FIGURE 21-44 T-Shooter™. (*Patent pending: Courtesy of Malvino Inc.*)

OSCILLATORS

Chapter 21 introduced the relaxation oscillator, a circuit that generates a rectangular output even though it has no input signal. As you recall, the positive feedback alternately drives the output into positive and negative saturation. The principle behind a relaxation oscillator is to let the charging and discharging of a capacitor determine the frequency of the rectangular output. This chapter continues the discussion of oscillators. In particular, you will be reading about oscillators with sinusoidal outputs. These include *RC* oscillators, *LC* oscillators, and quartz-crystal oscillators.

22-1 THEORY OF SINUSOIDAL OSCILLATION

To build a sinusoidal oscillator, we need to use an amplifier with positive feedback. The idea is to use the feedback signal in place of the usual input signal to the amplifier. If the feedback signal is large enough and has the correct phase, there will be an output signal even though there is no external input signal. In other words, an oscillator is an amplifier that is modified by positive feedback to supply its own input signal. This may sound like perpetual motion, and in a way, it is—but not as far as energy is concerned. The oscillator does not create energy. It only changes dc energy from the power supply to ac energy for the output signal. Even that is quite an accomplishment.

Loop Gain and Phase

Figure 22-1a shows a voltage source v_{in} driving the input terminals of an amplifier. The amplified output voltage is

$$v_{out} = Av_{in}$$

This voltage drives a feedback circuit that is usually a *resonant circuit*. Because of this, we get maximum feedback at one frequency. The feedback voltage returning to point x is given by

$$v_f = ABv_{in}$$

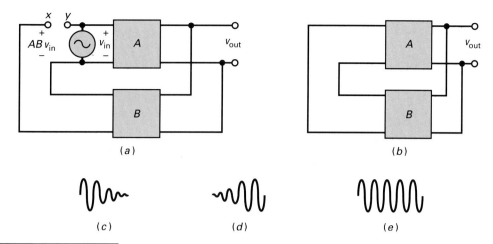

FIGURE 22-1 (*a*) Positive feedback returns a voltage to point *x*; (*b*) connecting points *x* and *y*; (*c*) oscillations die out; (*d*) oscillations increase; (*e*) oscillations are fixed in amplitude.

If the phase shift through the amplifier and feedback circuit is 0°, then ABv_{in} is in phase with the signal v_{in} that drives the input terminals of the amplifier.

Suppose we connect point *x* to point *y* and simultaneously remove voltage source v_{in}. Then the feedback voltage ABv_{in} drives the input terminals of the amplifier, as shown in Fig. 22-1*b*. What happens to the output voltage? If AB is less than 1, then ABv_{in} is less than v_{in} and the output signal will die out, as shown in Fig. 22-1*c*. However, if AB is greater than 1, then ABv_{in} is greater than v_{in} and the output voltage builds up (Fig. 22-1*d*). If AB equals 1, then ABv_{in} equals v_{in} and the output voltage is a steady sine wave like Fig. 22-1*e*. In this case, the circuit supplies its own input signal and produces a sine-wave output.

In an oscillator the value of loop gain AB is greater than 1 when the power is first turned on. A small starting voltage is applied to the input terminals, and the output voltage builds up, as shown in Fig. 22-1*d*. After the output voltage reaches a desired level, the value of AB automatically decreases to 1, and the peak-to-peak output becomes constant (Fig. 22-1*e*).

The Starting Voltage

Where does the *starting voltage* for an oscillator come from? Every resistor contains some free electrons. Because of the ambient temperature, these free electrons move randomly in different directions and generate a noise voltage across the resistor. The motion is so random that it contains frequencies to over 1000 GHz. You can think of each resistor as a small ac voltage source producing all frequencies.

In Fig. 22-1*b*, here is what happens. When you first turn on the power, the only signals in the system are the noise voltages generated by the resistors. These noise voltages are amplified and appear at the output terminals. The amplified noise drives the resonant feedback circuit. By deliberate design, we can make the phase shift around the loop equal 0°

at the resonant frequency. Above and below the resonant frequency, the phase shift is different from 0°. In this way, we get oscillations at only one frequency, the resonant frequency of the feedback circuit.

In other words, the amplified noise is filtered so that there is only one sinusoidal component with exactly the right phase for positive feedback. When loop gain AB is greater than 1, the oscillations build up at this frequency (Fig. 22-1d). After a suitable level is reached, AB decreases to 1, and we get a constant-amplitude output signal (Fig. 22-1e).

AB Decreases to Unity

There are two ways in which AB can decrease to 1: either A can decrease, or B can decrease. In some oscillators, the signal is allowed to build up until clipping occurs because of saturation and cutoff; this is equivalent to reducing voltage gain A. In other oscillators, the signal builds up and causes B to decrease before clipping occurs. In either case, the product AB decreases until it equals 1.

Here are the key ideas behind any feedback oscillator.

1. Initially, loop gain AB must be greater than 1 at the frequency at which the loop phase shift is 0°.

2. After the desired output level is reached, AB must decrease to 1 through reduction of either A or B.

22-2 THE WIEN-BRIDGE OSCILLATOR

The *Wien-bridge oscillator* is the standard oscillator circuit for low to moderate frequencies, in the range of 5 Hz to about 1 MHz. It is almost always used in commercial audio generators and is usually preferred for other low-frequency applications.

Lag Circuit

In complex numbers, the voltage gain of the bypass circuit of Fig. 22-2a is

$$\frac{\mathbf{V}_{\text{out}}}{\mathbf{V}_{\text{in}}} = \frac{-jX_C}{R - jX_C}$$

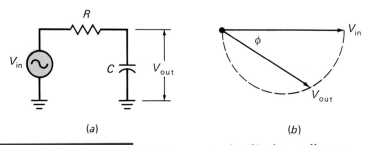

(a)	(b)

FIGURE 22-2   (a) Bypass circuit; (b) phasor diagram.

This can be converted to a magnitude of

$$\frac{V_{\text{out}}}{V_{\text{in}}} = \frac{X_C}{\sqrt{R^2 + X_C^2}}$$

and a phase angle of

$$\phi = -\arctan\frac{R}{X_C}$$

where ϕ is the phase angle between the output and the input.

Notice the minus sign in this equation for phase angle. It means that the output voltage lags the input voltage, as shown in Fig. 22-2b. Because of this, a bypass circuit is also called a *lag circuit*. In Fig. 22-2b, the half circle shows the possible positions of the output phasor voltage. This implies that the output phasor can lag the input phasor by an angle between $0°$ and $-90°$.

Lead Circuit

Figure 22-3a shows a coupling circuit. In complex numbers, the voltage gain in this circuit is

$$\frac{\mathbf{V}_{\text{out}}}{\mathbf{V}_{\text{in}}} = \frac{R}{R - jX_C}$$

This can be converted to a magnitude of

$$\frac{V_{\text{out}}}{V_{\text{in}}} = \frac{R}{\sqrt{R^2 + X_C^2}}$$

and a phase angle of

$$\phi = \arctan\frac{X_C}{R}$$

where ϕ is the phase angle between the output and the input.

Notice that the phase angle is positive. It means that the output voltage leads the input voltage as shown in Fig. 22-3b. Because of this, a coupling circuit is also called a *lead circuit*. In Fig. 22-3b, the half circle shows the possible positions of the output phasor voltage. This implies that the output phasor can lead the input phasor by an angle between $0°$ and $+90°$.

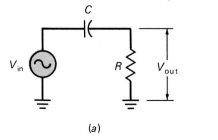

(a)

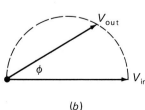

(b)

FIGURE 22-3   (*a*) Coupling circuit; (*b*) phasor diagram.

Coupling and bypass circuits are examples of phase-shifting circuits. These circuits shift the phase of the output signal either positive (leading) or negative (lagging) with respect to the input signal. Sinusoidal oscillators always use some kind of phase-shifting circuits to produce oscillation at one frequency. (If you want more mathematical analysis of the phase angle of lag and lead circuits, see Sec. 16-16, "Optional Topics.")

Lead-Lag Circuit

The Wien-bridge oscillator uses a feedback circuit called a *lead-lag circuit* (Fig. 22-4a). At very low frequencies, the series capacitor looks open to the input signal, and there is no output signal. At very high frequencies, the shunt capacitor looks shorted, and there is no output. In between these extremes, the output voltage from the lead-lag circuit reaches a maximum value (see Fig. 22-4b). The frequency where the output is maximum is called the *resonant frequency* f_r. At this frequency, the feedback fraction reaches a maximum value of ⅓.

Figure 22-4c shows the phase angle of the output voltage with respect to the input voltage. At very low frequencies, the phase angle is positive (leading). But at very high frequencies, the phase angle is negative (lagging). In between, there is a resonant frequency f_r where the phase shift is 0°. Figure 22-4d shows the phasor diagram of the input and output voltages. The tip of the phasor can lie anywhere on the dashed circle. Because of this, the phase angle may vary from $+90°$ to $-90°$.

The lead-lag circuit of Fig. 22-4a acts as a resonant circuit. At the resonant frequency f_r, the feedback fraction reaches a maximum value of ⅓, and the phase angle equals 0°. Above and below the resonant

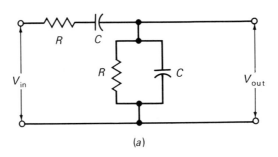

(a)

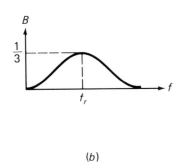

(b)

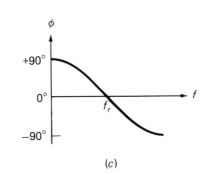

(c)

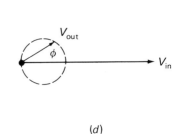

(d)

FIGURE 22-4 (a) Lead-lag network; (b) voltage gain; (c) phase shift; (d) phasor diagram.

frequency, the feedback fraction is less than $\frac{1}{3}$, and the phase angle no longer equals $0°$.

Formula for Resonant Frequency

In Fig. 22-4a, the output of the lead-lag circuit is

$$V_{out} = \frac{R \| (-jX_C)}{R - jX_C + R \| (-jX_C)} V_{in}$$

By expanding and simplifying, the foregoing equation leads to these two formulas:

$$B = \frac{1}{\sqrt{9 + (X_C/R - R/X_C)^2}} \qquad (22\text{-}1)$$

and

$$\phi = \arctan \frac{X_C/R - R/X_C}{3} \qquad (22\text{-}2)$$

Graphing these equations produces Fig. 22-4b and c.

Equation (22-1) has a maximum when $X_C = R$. For this condition, $B = \frac{1}{3}$ and $\phi = 0°$. This represents the resonant frequency of the lead-lag circuit. Since $X_C = R$, we can write

$$\frac{1}{2\pi f_r C} = R$$

or

$$f_r = \frac{1}{2\pi RC} \qquad (22\text{-}3)$$

How It Works

Figure 22-5a shows a Wien-bridge oscillator. It uses positive and negative feedback because there are two paths for feedback. There is a path for positive feedback from the output through the lead-lag circuit to the noninverting input. There is also a path for negative feedback from the output through the voltage divider to the inverting input.

Initially, there is more positive feedback than negative feedback. This helps the oscillations to build up when the power is turned on. After the output signal reaches the desired level, the negative feedback reduces the loop gain to 1. How does this happen? At power-up, the tungsten lamp has a low resistance, and the negative feedback is small. For this reason, the loop gain is greater than 1, and the oscillations can build up at the resonant frequency. As the oscillations build up, the tungsten lamp heats slightly and its resistance increases. In most circuits, the current through the lamp is not enough to make the lamp glow, but it is enough to increase the resistance. At some high output level, the tungsten lamp has a resistance of exactly R'. At this point, the closed-loop voltage gain from the noninverting input to the output decreases to

$$A_{CL} = \frac{R_1}{R_2} + 1 = \frac{2R'}{R'} + 1 = 3$$

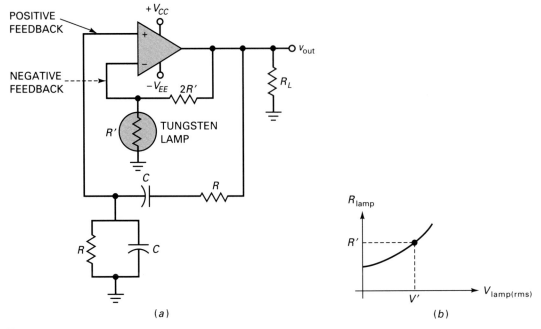

POSITIVE FEEDBACK

NEGATIVE FEEDBACK

$+V_{CC}$

$-V_{EE}$ 2R'

v_{out}

R_L

R' TUNGSTEN LAMP

C

R

R C

(a)

R_{lamp}

R'

V'

$V_{lamp(rms)}$

(b)

FIGURE 22-5 (a) Wien-bridge oscillator; (b) resistance of tungsten lamp increases with voltage.

Since the lead-lag circuit has a B of $\frac{1}{3}$, the loop gain is shown as follows:

$$A_{CL}B = 3(\frac{1}{3}) = 1$$

The closed-loop voltage gain from the noninverting input to the output is greater than 3 when the power is first turned on. Because of this, $A_{CL}B$ is greater than 1 initially. As the oscillations build up, the peak-to-peak output becomes large enough to increase the resistance of the tungsten lamp. When its resistance equals R', the loop gain $A_{CL}B$ is exactly equal to 1. At this point, the oscillations become stable, and the output voltage has a constant peak-to-peak value.

Initial Conditions

At power-up, the output voltage is zero, and the resistance of the tungsten lamp is less than R' as shown in Fig. 22-5b. When the output voltage increases, the resistance of the lamp increases as shown in the graph. At some voltage V' the tungsten lamp has a resistance of R'. This means that A_{CL} has a value of 3 and the loop gain becomes 1. When this happens, the output amplitude levels off and becomes constant.

Amplifier Phase Shift

The phase shift around the loop has to be 0°; otherwise, the circuit will not oscillate. In a Wien-bridge oscillator, the phase shift of the lead-lag circuit equals 0° when the oscillations have a frequency of

$$f_r = \frac{1}{2\pi RC}$$

Because of this, we can adjust the frequency by varying the value of R or C. This assumes that the phase shift of the amplifier is negligibly small, ideally 0°. Stated another way, the amplifier must have a closed-loop critical frequency well above the resonant frequency f_r. Then the amplifier introduces no additional phase shift. If the amplifier did introduce phase shift, the neat formula $f_r = 1/(2\pi RC)$ would be an approximation.

Why Is It Called a Wien-Bridge Oscillator?

Figure 22-6 shows another way to draw the Wien-bridge oscillator. The lead-lag circuit is the left side of a bridge, and the voltage divider is the right side. This ac bridge, called a *Wien bridge*, is used in other applications besides oscillators. The error voltage is the output of the bridge. When the bridge approaches balance, the error voltage approaches zero.

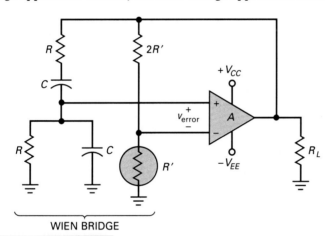

WIEN BRIDGE

FIGURE 22-6   Wien-bridge oscillator.

The Wien bridge is sometimes referred to as a *notch filter*—a circuit with zero output at one particular frequency. For a Wien bridge, the notch frequency equals

$$f_r = \frac{1}{2\pi RC}$$

Because the required error voltage for the op amp is so small, the Wien bridge is almost perfectly balanced, and the oscillation frequency equals f_r to a close approximation.

EXAMPLE 22-1

Calculate the minimum and maximum frequencies in the Wien-bridge oscillator of Fig. 22-7a.

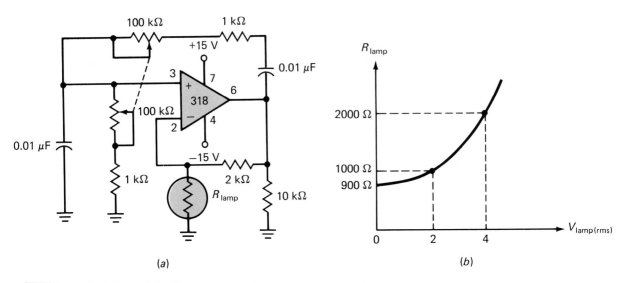

(a)   (b)

FIGURE 22-7   Example of Wien-bridge oscillator.

SOLUTION

The two rheostats are ganged, meaning they each change together and have the same values for any setting of the wipers. Since each has a maximum resistance of 100 kΩ, R varies from 1 to 101 kΩ. The minimum frequency of oscillation is

$$f_{min} = \frac{1}{2\pi(101\ \text{k}\Omega)(0.01\ \mu\text{F})} = 158\ \text{Hz}$$

and the maximum frequency is

$$f_{max} = \frac{1}{2\pi(1\ \text{k}\Omega)(0.01\ \mu\text{F})} = 15.9\ \text{kHz}$$

EXAMPLE 22-2

Figure 22-7b shows the lamp resistance of Fig. 22-7a. Calculate the output voltage.

SOLUTION

In Fig. 22-7a, the peak-to-peak output voltage becomes constant when the lamp resistance equals 1 kΩ. In Fig. 22-7b, this means that the lamp voltage is 2 V rms. The lamp current is

$$I = \frac{2\ \text{V}}{1\ \text{k}\Omega} = 2\ \text{mA}$$

This current also flows through the 2 kΩ, which means that the output voltage is

$$V_{out} = (2\ \text{mA})(1\ \text{k}\Omega + 2\ \text{k}\Omega) = 6\ \text{V rms}$$

which is equivalent to a peak-to-peak voltage of

$$v_{out} = 2(1.414)(6 \text{ V}) = 17 \text{ V}$$

22-3 OTHER *RC* OSCILLATORS

Although the Wien-bridge oscillator is the industry standard for frequencies up to 1 MHz, you occasionally see different *RC* oscillators. This section discusses two other basic designs, called the *twin-T oscillator* and the *phase-shift oscillator*.

Twin-T Oscillator

Figure 22-8a is a *twin-T filter*. A mathematical analysis of this circuit shows that it acts as a lead-lag circuit with a phase angle, as shown in Fig. 22-8b. Again, there is a frequency f_r at which the phase shift equals 0°. In Fig. 22-8c, the voltage gain equals 1 at low and high frequencies. In between, there is a frequency f_r at which the voltage gain drops to 0. The twin-T filter is another example of a notch filter because it can notch out, or block, those frequencies near f_r. The resonant frequency of a twin-T filter is given by a familiar equation:

$$f_r = \frac{1}{2\pi RC} \tag{22-4}$$

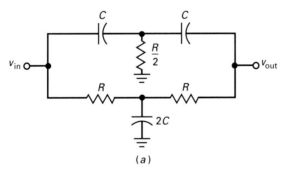

(a)

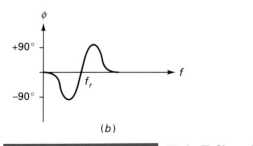

(b)

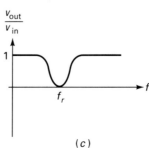

(c)

FIGURE 22-8 Twin-T filter: (*a*) circuit; (*b*) phase shift; (*c*) voltage gain.

Figure 22-9 shows a twin-T oscillator. The positive feedback is through the voltage divider to the noninverting input. The negative feedback is through the twin-T filter. When power is first turned on, the lamp resistance R_1 is low, and the positive feedback is maximum. As the oscillations build up, the lamp resistance increases and the positive feedback decreases. As the feedback decreases, the oscillations level off and become constant. In this way, the lamp stabilizes the level of the output voltage.

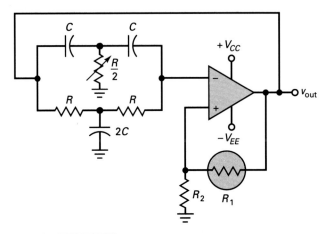

FIGURE 22-9　Twin-T oscillator.

In the twin-T filter, resistance $R/2$ is adjusted. This is necessary because the circuit oscillates at a frequency slightly different from the ideal resonant frequency of Eq. (22-4). To ensure that the oscillation frequency is close to the notch frequency, the voltage divider should have R_1 much larger than R_2. As a guide, R_1/R_2 is in the range of 10 to 1000. This forces the oscillator to operate at a frequency near the notch frequency.

Although it is occasionally used, the twin-T oscillator is not a popular circuit because it works well only at one frequency; that is, it cannot be easily adjusted over a large frequency range as the Wien-bridge oscillator can.

Phase-Shift Oscillator

Figure 22-10*a* is a *phase-shift oscillator* with three lead circuits in the feedback path. As you recall, a lead circuit produces a phase shift between 0° and 90°, depending on the frequency. Therefore, at some frequency, the total phase shift of the three lead circuits equals 180° (approximately 60° each). The amplifier has an additional 180° of phase shift because the signal drives the inverting input. As a result, the phase shift around the loop will be 360°, equivalent to 0°. If *AB* is greater than 1 at this particular frequency, oscillations can start.

Figure 22-10*b* shows an alternative design. It uses three lag circuits. The operation is similar. The amplifier produces 180° of phase shift, and the lag circuits contribute another 180° at some higher frequency. If *AB* is greater than 1 at this frequency, oscillations can start.

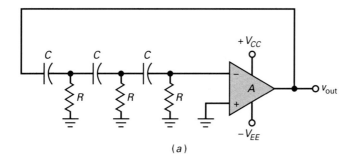

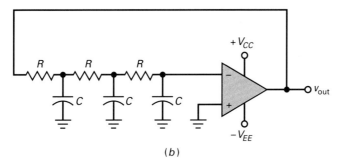

FIGURE 22-10   Phase-shift oscillators: (*a*) Lead networks; (*b*) lag networks.

The phase-shift oscillator is not a popular circuit. Again, the main problem with the circuit is that it cannot be easily adjusted over a large frequency range. The reason for introducing it is that you may accidentally build a phase-shift oscillator when you are trying to build an amplifier. This is discussed later in the chapter.

22-4 THE COLPITTS OSCILLATOR

Although it is superb at low frequencies, the Wien-bridge oscillator is not suited to high frequencies (well above 1 MHz). The main problem is the phase shift through the amplifier. This phase shift adds to the phase shift of the lead-lag circuit and causes resonance to occur far away from the ideal resonant frequency. One alternative is an *LC* oscillator, a circuit that can be used for frequencies between 1 and 500 MHz. This frequency range is beyond the f_{unity} of most op amps. This is why a bipolar transistor or a FET is typically used for the amplifier.

With an amplifier and *LC* tank circuit, we can feed back a signal with the right amplitude and phase to sustain oscillations. The analysis and the design of high-frequency oscillators border on black magic. Why? Because at higher frequencies, stray capacitances and lead inductances become very important in determining the oscillation frequency, feedback fraction, output power, and other ac quantities. For this reason, the analysis turns into a nightmare. This is why most people use ballpark approximations for an initial design and adjust the built-up oscillator as needed to get the desired performance. In this section, we examine the *Colpitts oscillator*, one of the most widely used *LC* oscillators.

CE Connection

Figure 22-11a shows a Colpitts oscillator. The voltage-divider bias sets up a quiescent operating point. The circuit then has a low-frequency voltage gain of r_c/r'_e, where r_c is the ac resistance seen by the collector. Because of the base and collector lag circuits, the high-frequency voltage gain is less than r_c/r'_e.

Figure 22-11b is a simplified ac equivalent circuit. The circulating or loop current in the tank flows through C_1 in series with C_2. Notice that v_{out} equals the ac voltage across C_1. Also, the feedback voltage v_f appears across C_2. This feedback voltage drives the base and sustains the oscillations developed across the tank circuit, provided there is enough voltage gain at the oscillation frequency. Since the emitter is at ac ground, the circuit is a CE connection.

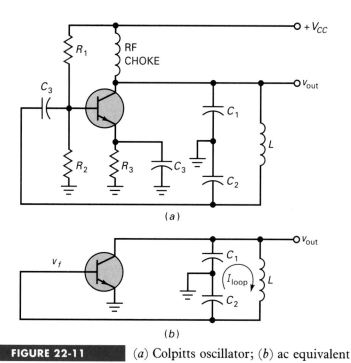

FIGURE 22-11   (a) Colpitts oscillator; (b) ac equivalent circuit.

You will encounter many variations of the Colpitts oscillator. One way to recognize it is by the capacitive voltage divider formed by C_1 and C_2. This capacitive voltage divider produces the feedback voltage necessary for oscillations. In other kinds of oscillators, the feedback voltage is produced by transformers, inductive voltage dividers, and so on.

Resonant Frequency

Most LC oscillators use tank circuits with a Q greater than 10. Because of this, we can calculate the approximate *resonant frequency* as

$$f_r = \frac{1}{2\pi\sqrt{LC}} \tag{22-5}$$

This is accurate to better than 1 percent when Q is greater than 10.

The capacitance to use in Eq. (22-5) is the equivalent capacitance that the circulating current passes through. In the Colpitts tank circuit of Fig. 22-11b, the circulating current flows through C_1 in series with C_2. Therefore, the equivalent capacitance is

$$C = \frac{C_1 C_2}{C_1 + C_2} \qquad (22\text{-}6)$$

For instance, if C_1 and C_2 are 100 pF each, you would use 50 pF in Eq. (22-5).

Starting Condition

The required starting condition for any oscillator is

$$AB > 1$$

at the resonant frequency of the tank circuit. This is equivalent to

$$A > \frac{1}{B}$$

The voltage gain A in this expression is the voltage gain at the oscillation frequency. In Fig. 22-11b, the output voltage appears across C_1 and the feedback voltage across C_2. Since the circulating current is the same for both capacitors,

$$B = \frac{v_f}{v_{\text{out}}} = \frac{X_{C_2}}{X_{C_1}} = \frac{1/(2\pi f C_2)}{1/(2\pi f C_1)}$$

or

$$B = \frac{C_1}{C_2}$$

Therefore, the starting condition is

$$A > \frac{C_2}{C_1} \qquad (22\text{-}7)$$

Remember that this is a crude approximation because it ignores the impedance looking into the base. An exact analysis would take the base impedance into account because it is in parallel with C_2. This complicates the analysis considerably because then you have to use complex numbers.

What does A equal? This depends on the upper critical frequencies of the amplifier. As you recall, there are base and collector bypass circuits in a bipolar amplifier. If the critical frequencies of these bypass circuits are greater than the oscillation frequency, then A is approximately equal to r_c/r'_e. If the critical frequencies are lower than the oscillation frequency, the voltage gain is less than r_c/r'_e and there is additional phase shift through the amplifier. This complicates the whole situation. In fact, the oscillator may not even work. If you ever have a high-frequency oscillator that won't work, you will know where to look for the trouble: It won't start because the loop gain is less than 1 when the phase shift is 0°.

Output Voltage

With *light feedback* (small *B*), the value *A* is only slightly larger than 1/*B*, and the operation is approximately class A. When you first turn on the power, the oscillations build up, and the signal swings over more and more of the ac load line. With this increased signal swing, the operation changes from a small-signal to a large-signal one. As this happens, the voltage gain decreases slightly. With light feedback, the value of *AB* can decrease to 1 without excessive clipping.

With *heavy feedback* (large *B*), the large feedback signal drives the base of Fig. 22-11*a* into saturation and cutoff. This charges capacitor C_3, producing negative dc clamping at the base. The negative clamping automatically adjusts the value of *AB* to 1. If the feedback is too heavy, you may lose some of the output voltage because of stray power losses.

When you build an oscillator, you can adjust the amount of feedback to maximize the output voltage. The trick is to use enough feedback to start under all conditions (different transistors, temperature, voltage, etc.), but not so much that you lose more output than necessary. Designing reliable high-frequency oscillators is a big challenge because the usual approximations don't work. As a last resort, some designers use computers to model high-frequency oscillators. This approach is too advanced and mathematical to discuss here.

Coupling to a Load

The exact frequency of oscillation depends on the *Q* of the circuit and is given by

$$f_r = \frac{1}{2\pi\sqrt{LC}}\sqrt{\frac{Q^2}{1 + Q^2}} \qquad (22\text{-}8)$$

In most cases, *Q* is greater than 10, and this exact equation simplifies to the ideal value given by Eq. (22-5). If *Q* is less than 10, the frequency is pulled lower than the ideal value. Furthermore, a low *Q* may prevent the oscillator from starting by lowering the high-frequency gain below 1/*B*.

Figure 22-12*a* shows one way to couple to the load resistance. If the load resistance is large, it will not load down the resonant circuit too much and the *Q* will be greater than 10. But if the load resistance is small, *Q* drops under 10 and the oscillations may not start. One solution to a small load resistance is to use a small capacitance C_4, one whose X_C is large compared with the load resistance. This prevents excessive loading of the tank circuit.

Figure 22-12*b* shows *link coupling*, another way of coupling the signal to a small load resistance. Link coupling means using only a few turns on the secondary winding of an RF transformer. This light coupling ensures that the load resistance will not lower the *Q* of the tank circuit to such a point that the oscillator will not start.

Whether capacitive or link coupling is used, the loading effect should be kept as small as possible. In this way, the high *Q* of the tank ensures an undistorted sinusoidal output with a reliable start for the oscillations.

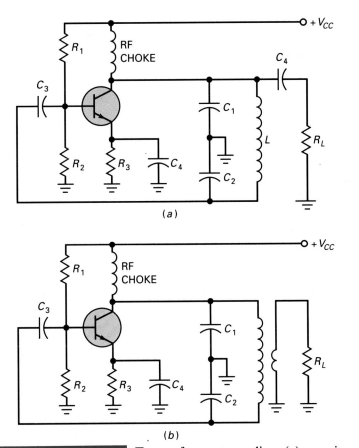

FIGURE 22-12 Types of output coupling: (*a*) capacitor; (*b*) link.

CB Connection

When the feedback signal in an oscillator drives the base, a large Miller capacitance appears across the input. This produces a relatively low critical frequency, which means the voltage gain may be too low at the resonant frequency we want. To get a higher critical frequency, the feedback signal can be applied to the emitter, as shown in Fig. 22-13. Capacitor C_3 ac grounds the base, and so the transistor acts as a common-base (*CB*) amplifier. A circuit like this can oscillate at higher frequencies because its high-frequency gain is larger than that of a comparable *CE* oscillator. With link coupling on the output, the tank is lightly loaded, and the resonant frequency is still given by Eq. (22-5).

The feedback fraction is slightly different. The output voltage appears across C_1 and C_2 in series, while the feedback voltage appears across C_2. Ideally, the feedback fraction is

$$B = \frac{v_f}{v_{\text{out}}} = \frac{X_{C_2}}{X_{C_1} + X_{C_2}}$$

After expanding and simplifying, this becomes

$$B = \frac{C_1}{C_1 + C_2}$$

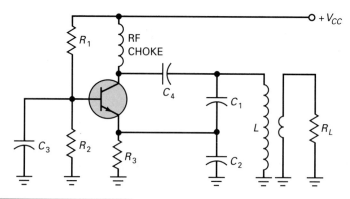

FIGURE 22-13 *CB* oscillator can oscillate at higher frequencies than *CE* oscillator.

For the oscillations to start, A must be greater than $1/B$. As an approximation, this means that

$$A > \frac{C_1 + C_2}{C_1} \tag{22-9}$$

This is a rough approximation because it ignores the input impedance of the emitter, which is in parallel with C_2. An exact analysis would use complex numbers to include the emitter impedance.

FET Colpitts Oscillator

Figure 22-14 is an example of a FET Colpitts oscillator in which the feedback signal is applied to the gate. Since the gate has a high input resistance, the loading effect on the tank circuit is much less than with a bipolar transistor. In other words, the approximation

$$B = \frac{C_1}{C_2}$$

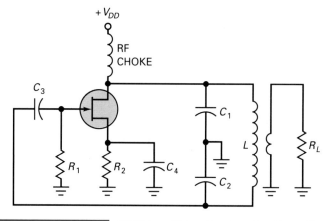

FIGURE 22-14 JFET oscillator has less loading effect on tank circuit.

is more accurate with a FET because the impedance looking into the gate is higher. The starting condition for this FET oscillator is

$$A > \frac{C_2}{C_1} \tag{22-10}$$

In a FET oscillator, the low-frequency voltage gain is $g_m r_d$. Above the critical frequency of the FET amplifier, the voltage gain decreases. In Eq. (22-10), A is the gain at the oscillation frequency. As a rule, try to keep the oscillation frequency lower than the critical frequency of the FET amplifier. Otherwise, the additional phase shift through the amplifier may prevent the oscillator from starting.

EXAMPLE 22-3

What is the frequency of oscillation in Fig. 22-15? What is the feedback fraction? How much voltage gain does the circuit need to start oscillating?

SOLUTION

The equivalent capacitance of the tank circuit equals the product over the sum of tank capacitances:

$$C = \frac{(0.001 \ \mu F)(0.01 \ \mu F)}{0.001 \ \mu F + 0.01 \ \mu F} = 909 \ \text{pF}$$

The inductance is 15 μH; therefore, the frequency of oscillation is

$$f_r = \frac{1}{2\pi \sqrt{(15 \ \mu H)(909 \ pF)}} = 1.36 \ \text{MHz}$$

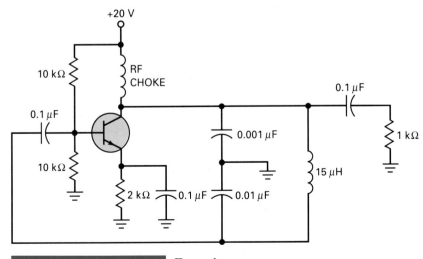

FIGURE 22-15   Example.

The feedback fraction is

$$B = \frac{0.001 \ \mu F}{0.01 \ \mu F} = 0.1$$

For the oscillator to start, the voltage gain must be

$$A > \frac{1}{0.1} = 10$$

at 1.36 MHz.

22-5 OTHER *LC* OSCILLATORS

The Colpitts is the most widely used *LC* oscillator. The capacitive voltage divider in the resonant circuit is a convenient way to develop the feedback voltage. But other kinds of oscillators are also used. In this section, we briefly discuss the Armstrong, Hartley, Clapp, and crystal oscillators.

Armstrong Oscillator

Figure 22-16*a* is an example of an *Armstrong oscillator*. In this circuit, the collector drives an *LC* resonant tank. The feedback signal is taken from a small secondary winding and fed back to the base. There is a phase shift of 180° in the transformer, which means that the phase shift around the loop is zero. Stated another way, the feedback is positive. By ignoring the loading effect of the base, the feedback fraction is

$$B = \frac{M}{L} \tag{22-11}$$

where M is the mutual inductance and L is the primary inductance. For the Armstrong oscillator to start, the voltage gain must be greater than $1/B$.

An Armstrong oscillator uses transformer coupling for the feedback signal. This is how you can recognize variations of this basic circuit. The secondary winding is sometimes called a *tickler coil* because it feeds back the signal that sustains the oscillations. The resonant frequency is given by Eq. (22-5), using the L and C shown in Fig. 22-16*a*. As a rule, you don't see the Armstrong oscillator used much because most designers avoid transformers whenever possible. Transformers are not as easy to work with as capacitors.

Hartley Oscillator

Figure 22-16*b* is an example of a *Hartley oscillator*. When the *LC* tank is resonant, the circulating current flows through L_1 in series with L_2. So the equivalent L to use in Eq. (22-5) is

$$L = L_1 + L_2 \tag{22-12}$$

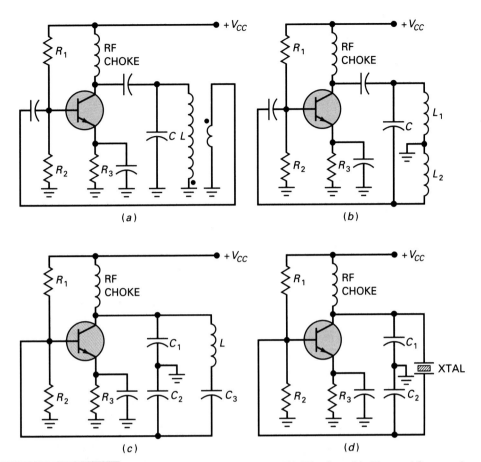

FIGURE 22-16   Oscillators: (*a*) Armstrong; (*b*) Hartley; (*c*) Clapp; (*d*) crystal.

In a Hartley oscillator, the feedback voltage is developed by the inductive voltage divider, L_1 and L_2. Since the output voltage appears across L_1 and the feedback voltage across L_2, the feedback fraction is

$$B = \frac{v_f}{v_{\text{out}}} = \frac{X_{L_2}}{X_{L_1}}$$

or

$$B = \frac{L_2}{L_1} \tag{22-13}$$

As usual, this ignores the loading effects of the base. For oscillations to start, the voltage gain must be greater than $1/B$.

Often a Hartley oscillator uses a single tapped inductor instead of two separate inductors. The action is basically the same either way. Another variation sends the feedback signal to the emitter instead of to the base. Also, you may see a FET used instead of a bipolar transistor. The output signal can be either capacitively or link coupled.

Clapp Oscillator

The *Clapp oscillator* of Fig. 22-16c is a refinement of the Colpitts oscillator. The capacitive voltage divider produces the feedback signal as before. An additional capacitor C_3 is in series with the inductor. Since the circulating tank current flows through C_1, C_2, and C_3 in series, the equivalent capacitance used to calculate the resonant frequency is

$$C = \frac{1}{1/C_1 + 1/C_2 + 1/C_3} \tag{22-14}$$

In a Clapp oscillator, C_3 is much smaller than C_1 and C_2. As a result, C is approximately equal to C_3, and the resonant frequency is given by

$$f_r = \frac{1}{2\pi\sqrt{LC_3}} \tag{22-15}$$

Why is this important? Because C_1 and C_2 are shunted by transistor and stray capacitances. These extra capacitances alter the values of C_1 and C_2 slightly. In a Colpitts oscillator, the resonant frequency therefore depends on the transistor and stray capacitances. But in a Clapp oscillator, the transistor and stray capacitances have no effect on C_3, so the oscillation frequency is more stable and accurate. This is why you occasionally see the Clapp oscillator used instead of a Colpitts oscillator.

Crystal Oscillator

When accuracy and stability of the oscillation frequency are important, a *quartz-crystal oscillator* is used. In Fig. 22-16d, the feedback signal comes from a capacitive tap. As will be discussed in the next section, the crystal (abbreviated XTAL) acts as a large inductor in series with a small capacitor (similar to the Clapp). Because of this, the resonant frequency is almost totally unaffected by transistor and stray capacitances.

EXAMPLE 22-4

If 50 pF is added in series with the 15-μH inductor of Fig. 22-15, on page 869, the circuit becomes a Clapp oscillator. What is the frequency of oscillation?

SOLUTION

With Eq. (22-14),

$$C = \frac{1}{1/0.001\,\mu\text{F} + 1/0.01\,\mu\text{F} + 1/50\,\text{pF}} = 50\,\text{pF}$$

The approximate oscillation frequency is

$$f_r = \frac{1}{2\pi\sqrt{(15\,\mu\text{H})(50\,\text{pF})}} = 5.81\,\text{MHz}$$

22-6 QUARTZ CRYSTALS

Some crystals found in nature exhibit the *piezoelectric effect*. When you apply an ac voltage across them, they vibrate at the frequency of the applied voltage. Conversely, if you mechanically force them to vibrate, they generate an ac voltage of the same frequency. The main substances that produce this piezoelectric effect are quartz, Rochelle salts, and tourmaline.

Rochelle salts have the greatest piezoelectric activity. For a given ac voltage, they vibrate more than quartz or tourmaline. Mechanically, they are the weakest because they break easily. Rochelle salts have been used to make microphones, phonograph pickups, headsets, and loudspeakers. Tourmaline shows the least piezoelectric activity but is the strongest of the three. It is also the most expensive. It is occasionally used at very high frequencies. Quartz is a compromise between the piezoelectric activity of Rochelle salts and the strength of tourmaline. Because it is inexpensive and readily available in nature, quartz is widely used for RF oscillators and filters.

Crystal Cuts

The natural shape of a quartz crystal is a hexagonal prism with pyramids at the ends (see Fig. 22-17a). To get a usable crystal out of this, we have to slice a rectangular slab out of the natural crystal. Figure 22-17b shows this slab with thickness t. The number of slabs we can get from a natural crystal depends on the size of the slabs and the angle of the cut.

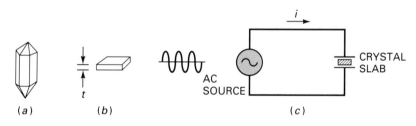

(a)   (b)   (c)

FIGURE 22-17   (a) Natural quartz crystal; (b) slab; (c) input current reaches maximum at crystal resonance.

There are a number of different ways to cut the natural crystal. These cuts have names like X cut, XY cut, and AT cut. For our purposes, all we need to know is that different cuts have different piezoelectric properties. Manufacturers' catalogs are usually the best source of information on different cuts and their properties.

For use in electronic circuits, the slab must be mounted between two metal plates, as shown in Fig. 22-17c. In this circuit the amount of crystal vibration depends on the frequency of the applied voltage. By changing the frequency, we can find resonant frequencies at which the crystal vibrations reach a maximum. Since the energy for the vibrations must be supplied by the ac source, the ac current is maximum at each resonant frequency.

Fundamental Frequency and Overtones

Most of the time, the crystal is cut and mounted to vibrate best at one of its resonant frequencies, usually the *fundamental frequency*, or lowest frequency. Higher resonant frequencies, called *overtones*, are almost exact multiples of the fundamental frequency. As an example, a crystal with a fundamental frequency of 1 MHz has a first overtone of approximately 2 MHz, a second overtone of approximately 3 MHz, and so on.

The formula for the fundamental frequency of a crystal is

$$f = \frac{K}{t} \tag{22-16}$$

where K is a constant that depends on the cut and other factors and t is the thickness of the crystal. Clearly the fundamental frequency is inversely proportional to the thickness. For this reason, there is a practical limit to how high in frequency we can go. The thinner the crystal, the more fragile it becomes and the more likely it is to break because of vibrations.

Quartz crystals work well up to 10 MHz on the fundamental frequency. To reach higher frequencies, we can use a crystal that vibrates on overtones. In this way, we can reach frequencies up to 100 MHz. Occasionally, the more expensive but stronger tourmaline is used at higher frequencies.

AC Equivalent Circuit

What does the crystal look like as far as the ac source is concerned? When the crystal of Fig. 22-18a is not vibrating, it is equivalent to a capacitance C_m because it has two metal plates separated by a dielectric. And C_m is known as the *mounting capacitance*.

When the crystal is vibrating, the equivalent circuit becomes more interesting. A vibrating crystal acts as a tuned circuit. Figure 22-18b shows the ac equivalent circuit of a crystal vibrating at or near its fundamental frequency. Typical values are L in henrys, C_s in fractions of a picofarad, R in hundreds of ohms, and C_m in picofarads. As an example, here are the values for one available crystal: $L = 3$ H, $C_s = 0.05$ pF, $R = 2$ kΩ, and $C_m = 10$ pF. Among other things, the cut, thickness, and mounting of the slab affect these values.

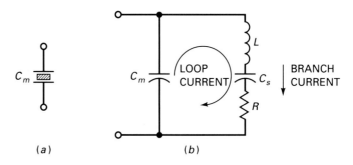

(a) (b)

FIGURE 22-18 (a) Mounting capacitance; (b) ac equivalent circuit of vibrating crystal.

Crystals have an incredibly high Q. For the values of $L = 3$ H, $C_s = 0.05$ pF, $R = 2$ kΩ, and $C_m = 10$ pF, we can calculate a Q of over 3000. The Q's of crystals can easily be over 10,000. The extremely high Q of a crystal leads to oscillators with very stable frequency values. You can see this when you look at the exact equation given earlier:

$$f_r = \frac{1}{2\pi\sqrt{LC}} \sqrt{\frac{Q^2}{1 + Q^2}}$$

When Q approaches infinity, as it does in a crystal, the resonant frequency approaches the ideal value determined by L and C. These L and C values are precisely determined in a crystal. By comparison, the ordinary LC tank has L and C with large tolerances, which is why its frequency is not as precisely controlled as in a crysal oscillator.

Series and Parallel Resonance

Besides the Q, L, C_s, R, and C_m of the crystal, there are two other characteristics to know about. The *series resonant frequency f_s* of a crystal is the resonant frequency of the LCR branch in Fig 22-18*b*. At this frequency, the branch current reaches a maximum value because L resonates with C_s. The formula for this resonant frequency is

$$f_s = \frac{1}{2\pi\sqrt{LC_s}} \tag{22-17}$$

The *parallel resonant frequency f_p* of the crystal is the frequency at which the circulating or loop current of Fig. 22-18*b* reaches a maximum value. Since this loop current must flow through the series combination of C_s and C_m, the equivalent C_{loop} is

$$C_{\text{loop}} = \frac{C_m C_s}{C_m + C_s} \tag{22-18}$$

and the parallel resonant frequency is

$$f_p = \frac{1}{2\pi\sqrt{LC_{\text{loop}}}} \tag{22-19}$$

Two capacitances in series always produce a capacitance smaller than either. Therefore, C_{loop} is less than C_s, and f_p is greater than f_s.

In any crystal, C_s is much smaller than C_m. For instance, with the values given earlier, C_s was 0.05 pF and C_m was 10 pF. Because of this, Eq. (22-18) gives a value of C_{loop} just slightly less than C_s. This means that f_p is only slightly more than f_s. When you use a crystal in an ac equivalent circuit like Fig. 22-19, the additional circuit capacitances appear in shunt with C_m. Because of this, the oscillation frequency will lie between f_s and f_p. This is the advantage of knowing the values of f_s and f_p; they set lower and upper limits on the frequency of the crystal oscillator. (See Example 22-5.)

Crystal Stability

The frequency of an oscillator tends to change slightly with time. This *drift* is produced by temperature, aging, and other causes. In a crystal

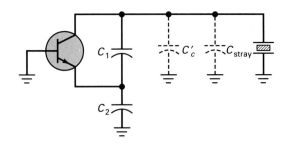

FIGURE 22-19 Stray circuit capacitances are in parallel with mounting capacitance.

oscillator, the frequency drift with time is very small, typically less than 1 part in 10^6 (0.0001 percent) per day. Stability like this is important in electronic wristwatches because they use quartz-crystal oscillators as the basic timing device.

By using crystal oscillators in temperature-controlled ovens, crystal oscillators have been built that have frequency drift of less than 1 part in 10^{10} per day. Stability like this is needed in frequency and time standards.

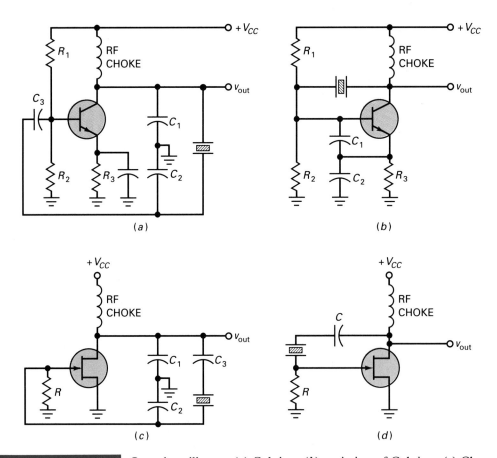

FIGURE 22-20 Crystal oscillators: (*a*) Colpitts; (*b*) variation of Colpitts; (*c*) Clapp; (*d*) Pierce.

To give you an idea of how precise 1 part in 10^{10} is, a clock with this drift will take 300 years to gain or lose 1 s.

Crystal Oscillators

Briefly, here are some of the different designs for a crystal oscillator. Figure 22-20a is a Colpitts crystal oscillator. The capacitive voltage divider produces the feedback voltage for the base of the transistor. The crystal acts as an inductor that resonates with C_1 and C_2. The oscillation frequency is between the series and parallel resonant values. Figure 22-20b is a variation of the Colpitts crystal oscillator. The feedback signal is applied to the emitter instead of the base. This variation allows the circuit to work at higher resonant frequencies. Figure 22-20c is a FET Clapp oscillator. The intention is to improve the frequency stability by reducing the effect of stray capacitances. Figure 22-20d is a circuit called a *Pierce crystal oscillator*. Its main advantage is simplicity.

EXAMPLE 22-5

A crystal has these values: $L = 3$ H, $C_s = 0.05$ pF, $R = 2$ kΩ, and $C_m = 10$ pF. Calculate the f_s and f_p of the crystal.

SOLUTION

With Eq. (22-17),

$$f_s = \frac{1}{2\pi\sqrt{(3\text{ H})(0.05\text{ pF})}} = 411 \text{ kHz}$$

With Eq. (22-18),

$$C_{\text{loop}} = \frac{(10\text{ pF})(0.05\text{ pF})}{10\text{ pF} + 0.05\text{ pF}} = 0.0498 \text{ pF}$$

With Eq. (22-19),

$$f_p = \frac{1}{2\pi\sqrt{(3\text{ H})(0.0498\text{ pF})}} = 412 \text{ kHz}$$

If this crystal is used in any oscillator, the frequency of oscillation is guaranteed to be between 411 and 412 kHz.

22-7 UNWANTED OSCILLATIONS

Oscillators can be frustrating. Sometimes when you are trying to build an oscillator, you wind up with an amplifier because the loop gain is less than 1 at the resonant frequency. After some tinkering, you may finally get the circuit to oscillate. On the other hand, when you are trying to build an amplifier, you often get a circuit that oscillates. Both of these unwanted results are common experiences for practicing technicians and engineers.

Low-Frequency Oscillations

Look at Fig. 22-21a. It is just another three-stage amplifier with a total gain of $A_1A_2A_3$. Furthermore, there is no feedback path from output to input. Therefore, the circuit cannot oscillate. Right? Wrong! There are invisible feedback paths that are producing positive feedback. When the feedback signal has the right amplitude and phase, any amplifier will produce unwanted oscillations.

Motorboating is a putt-putt sound from a loudspeaker connected to an amplifier like Fig. 22-21a. This sound represents very low-frequency oscillations, like a few hertz. The feedback path exists because of the power supply. Ideally, the power supply looks like a perfect ac short to ground. But to a second approximation, the power supply is an ideal

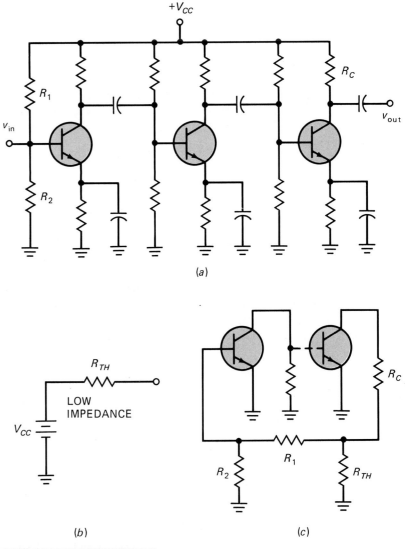

(a)

(b)   (c)

FIGURE 22-21   (a) Cascaded stages; (b) power supply has low impedance; (c) current feedback through power supply impedance.

voltage source in series with a Thevenin resistance, as shown in Fig. 22-21b. This equivalent resistance may be extremely small, but it is not zero. Because of the Thevenin resistance, the supply line is not quite an ac ground. In other words, part of the amplified output voltage in the final stage appears across R_{TH} (see Fig. 22-21c). This small ac feedback voltage drives the voltage divider consisting of R_1 and R_2. In turn, the ac voltage across R_2 appears on the base of the first transistor.

The frequency of the oscillation is determined by the lead circuits in the amplifier and the reactance of the power supply. At some frequency below the midband of the amplifier, the phase shift produced by the lead circuits and the supply reactance is exactly 0°. If the loop gain AB is greater than 1 at this frequency, motorboating occurs. Motorboating is a very distinct putt-putt sound. If you have ever heard motorboating, you will never forget it. Even if you have not heard it, you will probably recognize when you do hear it for the first time.

What is the cure for motorboating? Some people try adding a bypass capacitor to the power supply line. The idea is to reduce the Thevenin impedance to a lower value. But this usually doesn't work. The best solution to motorboating is use a power supply with an extremely small Thevenin resistance. Then the feedback voltage is too small to permit oscillations. Specifically, if you get motorboating, use a regulated power supply. This kind of supply has a Thevenin resistance under 0.1 Ω, sometimes as low as 0.0005 Ω. Chapter 23 will tell you more about regulated power supplies.

Unwanted Feedback

You can get unwanted oscillations above the midband of the amplifier. At higher frequencies, the stray capacitance between the last stage and the first stage may couple large enough feedback voltages to produce oscillations. Figure 22-22a illustrates this idea. The output collector lead acts like one plate of a capacitor, and the input base lead acts like the other plate. Although this feedback capacitance is very small, it can easily feed back enough signal to produce oscillations at high enough frequencies.

This is not a fairy tale. After all, a four-stage amplifier has a huge voltage gain. If each stage has a voltage gain of 100, the total voltage gain is

$$A = 100^4 = 10^8 = 100,000,000$$

If the capacitance between the output collector lead and the input base lead is 0.01 pF, the reactance at 15.9 MHz is

$$X_C = \frac{1}{2\pi(15.9\text{ MHz})(0.01\text{ pF})} = 1\text{ M}\Omega$$

If the input impedance of the first stage is 1 kΩ, the feedback fraction is

$$B = \frac{1\text{ k}\Omega}{1\text{ M}\Omega} = 0.001$$

Therefore, the loop gain at 15.9 MHz is

$$AB = 100,000,000(0.001) = 100,000$$

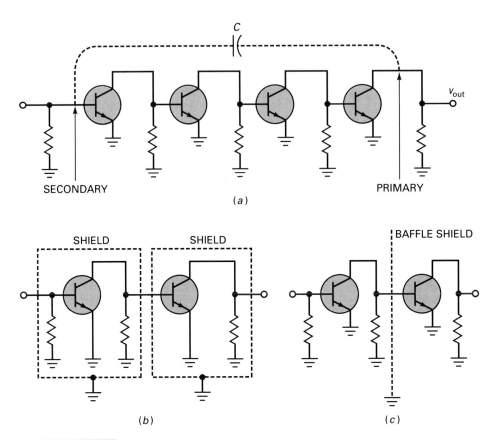

FIGURE 22-22 (*a*) Capacitive and magnetic coupling from output to input; (*b*) shields enclose each stage and prevent oscillations; (*c*) baffle shield prevents capacitive coupling.

We can also show that *AB* is greater than 1 for many other frequencies and feedback capacitances.

The point is this: At some high frequency, the phase shift may be 0°, and the loop gain may be greater than 1. As a result, the four-stage amplifier will break into oscillations. The exact frequency where this happens depends on the distance between the stages, length of the leads, etc. But one thing is sure: The amplifier will automatically find one or more frequencies where it will oscillate. If you don't believe it, try building a four-stage amplifier, and look at the output with an oscilloscope.

Magnetic coupling between the last stage and the first stage is also possible. The output wire labeled "Primary" in Fig. 22-22*a* can act as a primary winding of a transformer. The input wire labeled "Secondary" can act as a secondary winding. As a result, an ac signal in the primary can induce an ac signal in the secondary. If the feedback signal is strong enough and the phase shift is correct, we get oscillations because of magnetic feedback.

What is the cure for unwanted capacitive and magnetic feedback? One approach is to increase the *distance* between stages. This decreases both types of coupling. If this is not practical, you can enclose each stage in a *shield* or metallic container (see Fig. 22-22*b*). Shielding like this is common in many high-frequency applications because it blocks high-frequency electric and magnetic fields. If only capacitive coupling is a problem, a *baffle shield* (a metallic plate) between stages may eliminate high-frequency oscillations (Fig. 22-22c).

Ground Loops

Another subtle cause of high-frequency oscillations is a *ground loop*, an unwanted difference of potential between two ground points. In Fig. 22-22a, all ac grounds are ideally at the same potential. But in reality, the chassis or whatever serves as ground has a very small impedance between ground points. Therefore, if alternating currents from the last stage happen to flow through part of the chassis being used by an earlier stage, we can get enough unwanted positive feedback to cause oscillations.

The solution to a ground-loop problem is proper layout of the stages to prevent ac ground currents of later stages from flowing through ground paths of earlier stages. One way to accomplish this is to use a *single* ground point, as shown in Fig. 22-23. When this approach is used, there is no difference of potential between two ground points because there is only one ground point.

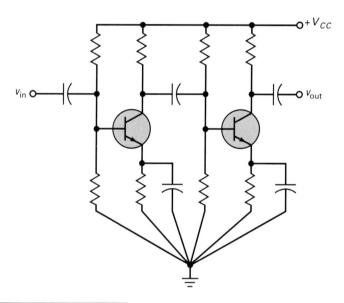

FIGURE 22-23   Single ground point prevents ground loops.

Supply Bypassing

Watch out for lead inductance between the power supply and the circuit (see Fig. 22-24). A long lead may have enough inductance to result in positive feedback at high frequencies. Even if the power supply is regulated and has a very small Thevenin resistance, the inductive reactance of the long lead produces an ac voltage that is fed back to the earlier stages. The solution is to add a large *bypass* capacitor across the circuit, as shown in Fig. 22-24. Bypassing the supplies is almost always essential with ICs. Depending on the particular IC, bypass capacitors from about 0.1 to over

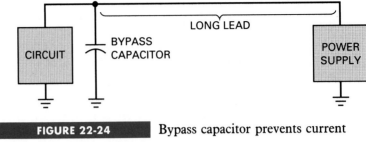

FIGURE 22-24   Bypass capacitor prevents current feedback caused by long lead.

1 μF may be needed to prevent oscillations. The bypass capacitors should be located as close as possible to the IC.

Parasitic Oscillations

The small transistor capacitances and lead inductances distributed throughout a circuit may form unwanted Colpitts or Hartley oscillators. The resulting oscillations are called *parasitic oscillations*. Usually, these oscillations occur at a very high frequency and have a weak amplitude because the feedback is very small. Parasitic oscillations tend to make circuits act erratically. Oscillators produce more than one frequency, op amps have too much offset, power supplies have unexplained ripple, amplifiers produce distorted signals, and video displays contain snow. An old troubleshooting trick is to touch low-voltage parts of the circuit that are suspected of having parasitic oscillations. If the trouble clears, you almost certainly have parasitic oscillations.

What is the cure for parasitic oscillations? You can reduce the positive feedback by adding small resistors to the base leads of transistors. Values like 10 Ω may do the job, but you have to experiment here to see whether it will work. Another approach that has been used is to slip a *ferrite* bead over each base lead. This absorbs enough ac energy in the parasitic oscillation and usually stops these unwanted oscillations. In either case, the feedback fraction is reduced, or the phase shift is changed enough to kill the parasitic oscillations.

Negative-Feedback Amplifiers

Figure 22-25*a* shows a three-stage negative-feedback amplifier. In the midband of the internal amplifier, the phase shift is 180° because there are an odd number of inverting stages. Therefore, the phase shift around the entire loop is 180°, and the feedback is negative. But outside the midband the internal bypass circuits of the amplifier produce additional phase shifts. Because of this, at some high frequency the phase shift of the amplifier is an additional 180°, which means the phase shift around the entire loop is 360° or 0°. In this case, the feedback becomes positive. In other words, Fig. 22-25*a* acts as a phase-shift oscillator discussed earlier. At some higher frequency, the three bypass circuits of the amplifier will produce a phase shift of 180° (60° each if the circuits are identical).

The only way to prevent oscillations outside the midband of a negative-feedback amplifier is to make sure that the loop gain *AB* is less than 1

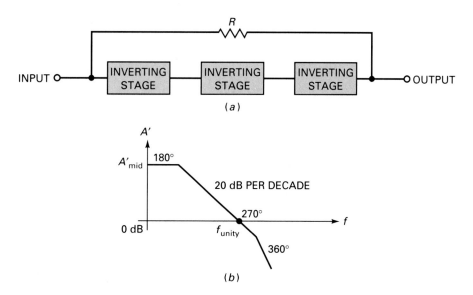

FIGURE 22-25 (a) Feedback around three inverting stages; (b) one dominant lag network prevents oscillations.

when the phase shift reaches 0°. The safest and most widely used method is this: Make one of the bypass circuits dominant enough to produce a 20-dB roll-off until the loop gain crosses the 0-dB axis. A 20-dB roll-off at the horizontal crossing implies that only one bypass circuit is working beyond its critical frequency; all others are still operating below the critical frequency. This means that the loop phase shift is around 270° at f_{unity}, which makes it impossible to have oscillations.

Figure 22-25b illustrates the idea of one dominant bypass circuit. In the midband the gain is high, and the phase shift is 180°. One of the stages has a dominant bypass circuit, so that the gain breaks at a low frequency and rolls off at 20 dB per decade. A decade above this frequency, the phase shift is 270°. It stays at approximately 270° until the gain crosses the horizontal axis at f_{unity}. Beyond this point, oscillations are impossible because the loop gain is less than 1.

With monolithic op amps, the dominant bypass circuit is usually integrated on the chip and automatically provides the 20-dB roll-off until f_{unity} is reached. For instance, the 741C uses a compensating capacitor of 30 pF. This makes the voltage gain break at 10 Hz and roll off at 20 dB per decade until an f_{unity} of 1 MHz is reached.

❑ **OPTIONAL TOPICS**

The following material continues the earlier discussions at a more advanced and specialized level. All the topics are optional because they are not used in any of the basic discussions in later chapters. This section will be a useful reference when you are in industry because then you will probably want more advanced viewpoints.

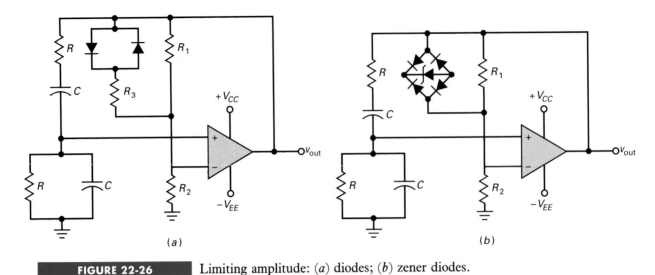

FIGURE 22-26 Limiting amplitude: (*a*) diodes; (*b*) zener diodes.

22-8 OTHER WAYS TO REDUCE *AB* TO UNITY

A low-power incandescent lamp is the standard method of reducing AB to unity in Wien-bridge oscillators. There are alternatives to an incandescent lamp. Figure 22-26a shows a Wien-bridge oscillator that relies on diodes to limit the amplitude of the output signal. On power-up, the diodes are off, and the feedback fraction is less than $\frac{1}{3}$ because the ratio R_1/R_2 is more than 2. This allows the output signal to build up.

After the desired output level is reached, the diodes conduct on alternate half-cycles. This places R_3 in parallel with R_1 and increases the feedback fraction to $\frac{1}{3}$. The output voltage then stabilizes. Sometimes LEDs are used instead of ordinary diodes. This is a clever trick because the LEDs are lit only when the circuit is oscillating.

In Fig. 22-26b, a zener diode is the limiting element. At power-up, the bridge diodes are off, and the feedback fraction is less than $\frac{1}{3}$ because R_1/R_2 is more than 2. As the output builds up, the bridge diodes are forward-biased, but nothing happens below zener breakdown. At some high output level, the zener diode breaks down, and the output level stabilizes.

Figure 22-27 shows another approach. This time, a JFET acting as a voltage-variable resistance limits the output amplitude. At power-up, the JFET has a minimum resistance because its gate voltage is zero. By design, the feedback fraction is less than $\frac{1}{3}$, and so the oscillations can start. When the output level exceeds the zener voltage plus one diode drop, we get negative peak detection, and the gate voltage goes negative. When this happens, the $r_{ds(\text{on})}$ of the JFET increases, which increases the feedback fraction until it equals $\frac{1}{3}$. The output then stabilizes.

Figure 22-28 shows an alternative method of limiting the output level. In this circuit, a JFET is used as a voltage-variable resistance. The gate of the JFET is connected to the output of a negative peak detector. At some output level, the negative voltage out of the peak detector increases

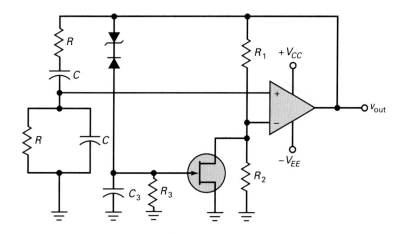

FIGURE 22-27 JFET used as voltage-variable resistance to limit output amplitude.

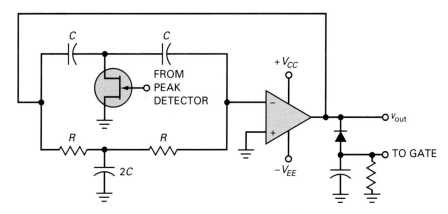

FIGURE 22-28 JFET used as voltage-variable resistance to limit output amplitude.

the $r_{ds(\text{on})}$ to approximately $R/2$. At this point, the twin-T filter is resonant, and the oscillator output stabilizes.

22-9 THE 555 TIMER

The *555 timer* combines a relaxation oscillator, two comparators, an *RS* flip-flop, and a discharge transistor. This versatile IC has many applications. Designers are constantly finding new uses for this amazing IC.

RS Flip-Flop

Figure 22-29*a* shows a pair of cross-coupled transistors. Each collector drives the opposite base through a resistance R_B. In a circuit like this, one transistor is saturated, and the other is cut off. For instance, if the right transistor is saturated, its collector voltage is approximately zero. This means no base drive for the left transistor, and so it goes into cutoff

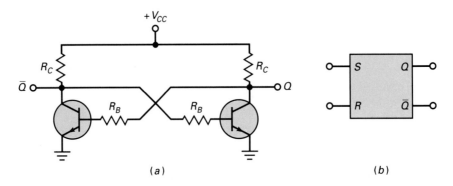

FIGURE 22-29 (a) Part of an *RS* flip-flop; (b) symbol for *RS* flip-flop.

and its collector voltage approaches $+V_{CC}$. This high voltage produces enough base current to keep the right transistor in saturation.

But if the right transistor is cut off, its collector voltage drives the left transistor into saturation. The low collector voltage out of this left transistor then keeps the right transistor in cutoff.

Depending on which transistor is saturated, the *Q* output is either low or high. By adding more components to the circuit, we get an *RS flip-flop*, a circuit that can set the *Q* output to high or reset it to low. Incidentally, a complementary (opposite) output *Q* is available from the collector of the other transistor.

Figure 22-29b shows the schematic symbol for an *RS* flip-flop of any design. Whenever you see this symbol, remember the action: The circuit latches in either of two states. A high *S* input sets *Q* to high; a high *R* input resets *Q* to low. Output *Q* remains in a given state until it is triggered into the opposite state.

Basic Timing Concept

Figure 22-30a illustrates some basic ideas that we will need in our later discussion of the 555 timer. Assume that output *Q* is high. This saturates the transistor and clamps the capacitor voltage at ground. In other words, the capacitor is shorted and cannot charge.

The noninverting input voltage of the comparator is called the *threshold voltage*, and the inverting input voltage is referred to as the *control voltage*. With the *RS* flip-flop set, the saturated transistor holds the threshold voltage at zero. The control voltage, however, is fixed at $+10$ V because of the voltage divider.

Suppose we apply a high voltage to the *R* input. This resets the *RS* flip-flop. Output *Q* goes low and cuts off the transistor. Capacitor *C* is now free to charge. As the capacitor charges, the threshold voltage increases. Eventually, the threshold voltage becomes slightly greater than the control voltage ($+10$ V). The output of the comparator then goes high, forcing the *RS* flip-flop to set. The high *Q* output saturates the transistor, and this quickly discharges the capacitor. Notice the two waveforms in Fig. 22-30b. An exponential rise is across the capacitor, and a positive-going pulse appears at the $\overline{Q}$ output.

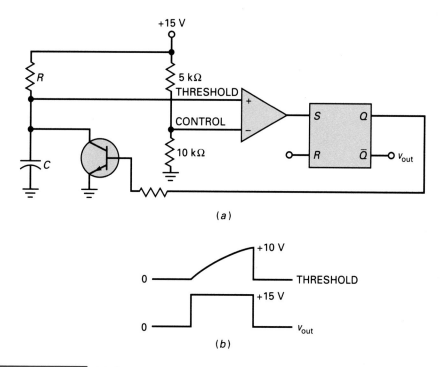

FIGURE 22-30 (*a*) Basic timing circuit; (*b*) capacitor voltage is exponential, and output is rectangular.

555 Block Diagram

Figure 22-31 is a simplified block diagram of the NE555 timer, an eight-pin IC timer introduced by Signetics Corporation. Notice that the upper comparator has a threshold input (pin 6) and a control input (pin 5). In

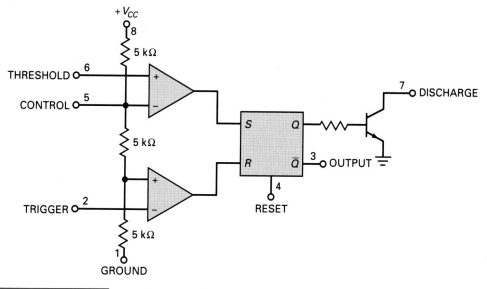

FIGURE 22-31 Simplified schematic diagram of 555 timer.

most applications, the control input is not used, so that the control voltage equals $+2V_{CC}/3$. As before, whenever the threshold voltage exceeds the control voltage, the high output from the comparator will set the flip-flop.

The collector of the discharge transistor goes to pin 7. When this pin is connected to an external timing capacitor, a high Q output from the flip-flop will saturate the transistor and discharge the capacitor. When Q is low, the transistor opens and the capacitor can charge as previously described.

The complementary signal out of the flip-flop goes to pin 3, the output. When the external reset (pin 4) is grounded, it inhibits the device (prevents it from working). This on/off feature is sometimes useful. In most applications, however, the external reset is not used, and pin 4 is tied directly to the supply voltage.

Notice the lower comparator. Its inverting input is called the *trigger* (pin 2). Because of the voltage divider, the noninverting input has a fixed voltage of $+V_{CC}/3$. When the trigger input voltage is slightly less than $+V_{CC}/3$, the op-amp output goes high and resets the slip-flop.

Finally, pin 1 is the chip ground, while pin 8 is the supply pin. The 555 timer will work with any supply voltage between 4.5 and 16 V.

Monostable Operation

Figure 22-32a shows the 555 timer connected for *monostable operation* (also called one-shot operation). The circuit works as follows: When the trigger input is slightly less than $+V_{CC}/3$, the lower comparator has a high output and resets the flip-flop. This cuts off the transistor, allowing the capacitor to charge. When the capacitor voltage is slightly greater than $+2V_{CC}/3$, the upper comparator has a high output, which sets the flip-flop. As soon as Q goes high, it turns on the transistor; this quickly discharges the capacitor.

Figure 22-32b shows typical waveforms. The trigger input is a narrow pulse with a quiescent value of $+V_{CC}$. The pulse must drop below $+V_{CC}/3$ to reset the flip-flop and allow the capacitor to charge. When the threshold voltage slightly exceeds $+2V_{CC}/3$, the flip-flop sets; this saturates the transistor and discharges the capacitor. As a result, we get one rectangular output pulse.

The capacitor C has to charge through resistance R. The larger the RC time constant, the longer it takes for the capacitor voltage to reach $+2V_{CC}/3$. In other words, the RC time constant controls the width of the output pulse. It is possible to derive this formula for the pulse width:

$$W = 1.1RC \qquad (22\text{-}20)$$

For instance, if $R = 22\ \text{k}\Omega$ and $C = 0.068\ \mu\text{F}$, then

$$W = 1.1(22\ \text{k}\Omega)(0.068\ \mu\text{F}) = 1.65\ \text{ms}$$

Normally, a schematic diagram does not show the comparators, flip-flop, and other components inside the 555 timer. Rather, you will see a schematic diagram like Fig. 22-33 for the monostable 555 circuit. Only the pins and external components are shown. Incidentally, notice that pin 5 (control) is bypassed to ground through a small capacitor, typically

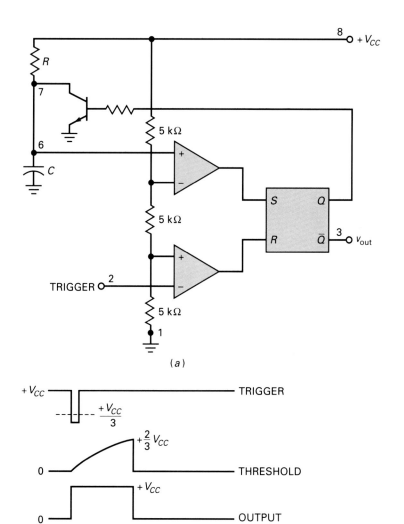

(a)

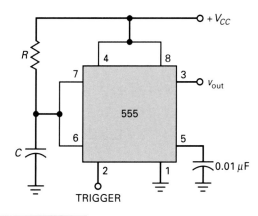

(b)

FIGURE 22-32 (a) 555 timer connected as monostable multivibrator; (b) waveforms.

FIGURE 22-33 Monostable timer circuit.

0.01 μF. This provides noise filtering for the control voltage. Recall that grounding pin 4 inhibits the 555 timer. To avoid accidental reset, pin 4 is usually tied to the supply voltage, as shown in Fig. 22-33.

In summary, the monostable 555 timer produces a single pulse whose width is determined by the external R and C used in Fig. 22-33. The pulse begins with the leading edge of the negative trigger input. One-shot operation like this has a number of applications, as you will see in later studies.

Astable Operation

Figure 22-34a shows the 555 timer connected for *astable* (free-running) operation. When Q is low, the transistor is cut off and the capacitor is charging through a total resistance of $R_A + R_B$. Because of this, the

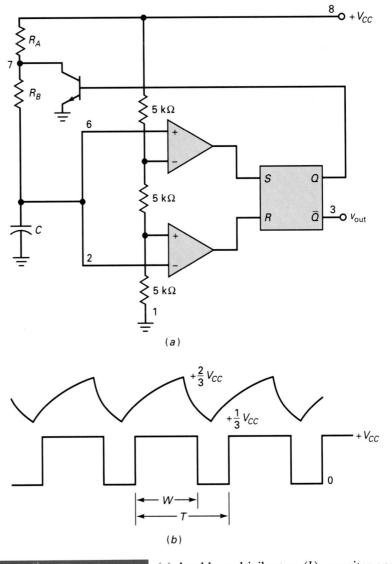

(a)

(b)

FIGURE 22-34   (a) Astable multivibrator; (b) capacitor and output waveforms.

charging time constant is $(R_A + R_B)C$. As the capacitor charges, the threshold voltage (pin 6) increases. Eventually, the threshold voltage exceeds $+2V_{CC}/3$; then the upper comparator has a high output, and this sets the flip-flop. With Q high, the transistor saturates and grounds pin 7. Now the capacitor discharges through R_B. Therefore, the discharging time constant is $R_B C$. When the capacitor voltage drops slightly below $+V_{CC}/3$, the lower comparator has a high output, and this resets the flip-flop.

Figure 22-34b illustrates the waveforms. As you can see, the timing capacitor has an exponentially rising and falling voltage. The output is a rectangular wave. Since the charging time constant is longer than the discharging time constant, the output is not symmetric; the high output state lasts longer than the low output state. To specify how unsymmetric the output is, we will use the *duty cycle*, defined as

$$D = \frac{W}{T}(100\%) \tag{22-21}$$

As an example, if $W = 2$ ms and $T = 2.5$ ms, then

$$D = \frac{2 \text{ ms}}{2.5 \text{ ms}}(100\%) = 80\%$$

Depending on resistances R_A and R_B, the duty cycle is between 50 and 100 percent.

A mathematical solution to the charging and discharging equations gives the following formulas. The output frequency is

$$f = \frac{1.44}{(R_A + 2R_B)C} \tag{22-22}$$

and the duty cycle is

$$D = \frac{R_A + R_B}{R_A + 2R_B}(100\%) \tag{22-23}$$

If R_A is much smaller than R_B, the duty cycle approaches 50 percent.

Figure 22-35 shows the astable 555 timer as it usually appears on a schematic diagram. Again notice how pin 4 (reset) is tied to the supply

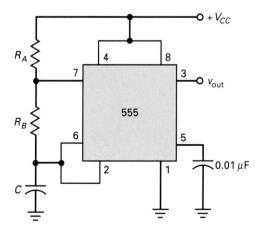

FIGURE 22-35   Astable timer circuit.

voltage and how pin 5 (control) is bypassed to ground through a 0.01-µF capacitor. An astable 555 timer is often called a *free-running multivibrator* because it produces a continuous train of rectangular pulses.

Voltage-Controlled Oscillator

Figure 22-36*a* shows a *voltage-controlled oscillator* (VCO), one application for a 555 timer. The circuit is sometimes called a *voltage-to-frequency converter* because an input voltage can change the output frequency. Here is how the circuit works. Recall that pin 5 (control) connects to the inverting input of the upper comparator. Normally, the control voltage is $+2V_{CC}/3$ because of the internal voltage divider. In Fig. 22-36*a*, however, the voltage from an external potentiometer overrides the internal voltage. In other words, by adjusting the potentiometer, we can change the control voltage.

Figure 22-36*b* illustrates the voltage across the timing capacitor. Notice that it varies between $+V_{con}/2$ and $+V_{con}$. If we increase V_{con}, it takes the capacitor longer to charge and discharge; therefore, the frequency decreases. As a result, we can change the frequency of the circuit by varying the control voltage. Incidentally, the control voltage may come from a potentiometer, or it may be the output of a transistor circuit, op amp, or some other device.

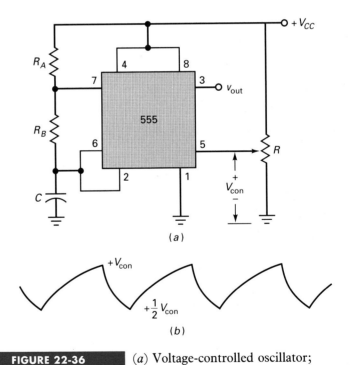

(a)

(b)

FIGURE 22-36 (*a*) Voltage-controlled oscillator; (*b*) capacitor timing waveforms.

Ramp Generator

Here is another application for the 555 timer. Charging a capacitor through a resistor produces an exponential waveform. If we use a constant current

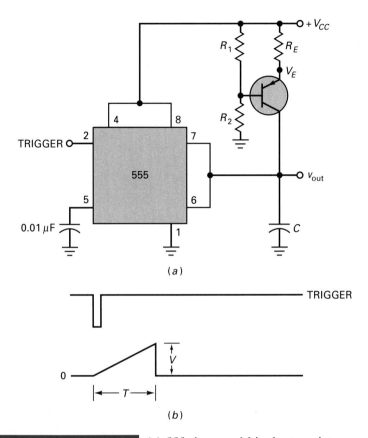

FIGURE 22-37 (*a*) 555 timer and bipolar transistor produce ramp output; (*b*) trigger and ramp waveforms.

source to charge a capacitor, however, we get a *ramp*. This is the idea behind the circuit of Fig. 22-37a. Here we have replaced the resistor of previous circuits with a *pnp* current source that produces a constant charging current of

$$I_C = \frac{V_{CC} - V_E}{R_E} \qquad (22\text{-}24)$$

where

$$V_E = \frac{R_2}{R_1 + R_2} V_{CC} + V_{BE} \qquad (22\text{-}25)$$

For instance, if $V_{CC} = 15$ V, $R_E = 20$ kΩ, $R_1 = 5$ kΩ, $R_2 = 10$ kΩ, and $V_{BE} = 0.7$ V, then

$$V_E = 10 \text{ V} + 0.7 \text{ V} = 10.7 \text{ V}$$

and

$$I_C = \frac{15 \text{ V} - 10.7 \text{ V}}{20 \text{ k}\Omega} = 0.215 \text{ mA}$$

When a trigger starts the monostable 555 timer of Fig. 22-37a, the *pnp* current source forces a constant charging current into the capacitor.

Therefore, the voltage across the capacitor is a ramp, as shown in Fig. 22-37b. As derived earlier, the slope S of the ramp is given by

$$S = \frac{I}{C} \qquad (22\text{-}26)$$

If the charging current is 0.215 mA and the capacitance is 0.022 μF, the ramp will have a slope of

$$S = \frac{0.215 \text{ mA}}{0.022 \text{ μF}} = 9.77 \text{ V/ms}$$

STUDY AIDS

The following study aids will help to reinforce the ideas discussed in this chapter. For best results, use these study aids within 6 hours of reading the earlier material. Then review these study aids a week later and a month later to ensure that the concepts remain in your long-term memory.

SUMMARY

Sec. 22-1 Theory of Sinusoidal Oscillation
To build a sinusoidal oscillator, we need to use an amplifier with positive feedback. For the oscillator to start, the loop gain must be greater than 1 when the phase shift around the loop is 0°. As the output voltage increases in peak-to-peak value, the feedback fraction automatically decreases until the loop gain is 1. At this point, the peak-to-peak output becomes constant.

Sec. 22-2 The Wien-Bridge Oscillator
This is the standard oscillator for low to moderate frequencies in the range of 5 Hz to 1 MHz. It is almost always used in commercial audio generators and is usually preferred for other low-frequency applications. The Wien-bridge oscillator produces an almost perfect output sine wave. As with any oscillator, it starts because the loop gain is greater than 1 at the resonant frequency. A tungsten lamp or other nonlinear resistance is used to decrease the loop gain to 1.

Sec. 22-3 Other RC Oscillators
The twin-T oscillator uses an amplifier and RC circuits to produce the required loop gain and phase shift at the resonant frequency. It works well at one frequency but is not suitable for an adjustable-frequency oscillator. The phase-shift oscillator also uses an amplifier and RC circuits to produce oscillations. Like the twin-T oscillator, it works well at one frequency but not over a frequency range. The phase-shift oscillator is like an unwelcome relative—it shows up at the worst times. Specifically, a phase-shift oscillator lurks inside every amplifier you build because of the stray lead and lag circuits in each stage.

Sec. 22-4 The Colpitts Oscillator
Above 1 MHz RC oscillators usually don't work well. This is why LC oscillators are preferred for frequencies between 1 and 500 MHz. This frequency range is beyond the f_{unity} of most op amps, which is why a bipolar transistor or FET is commonly used for the amplifying device. An LC oscillator uses an LC tank circuit to determine the resonant frequency. The Colpitts oscillator is one of the most widely used LC oscillators. You can recognize the circuit by the capacitive voltage divider used to produce the feedback signal.

Sec. 22-5 Other LC Oscillators
The Armstrong oscillator uses a transformer to produce the feedback signal. The Hartley oscillator uses an inductive voltage divider to produce the feedback signal. The Clapp oscillator has a small series capacitor in the inductive branch of the resonant tank circuit. This reduces the effect of stray capacitances elsewhere in the circuit.

Sec. 22-6 Quartz Crystals
Some crystals in nature exhibit the piezoelectric effect. Because of this effect, a vibrating crystal

acts as an *LC* resonant circuit with an extremely high *Q*. Quartz is the most important crystal with the piezoelectric effect. It is used in crystal oscillators where a precise and reliable frequency is needed. One common application is the electronic wristwatch.

Sec. 22-7 Unwanted Oscillations

Every time you build an amplifier, you are building a potential oscillator. As you saw with the phase-shift oscillator, all it takes to get oscillations is an amplifier and three lead or lag circuits. Low-frequency oscillations, sometimes called motorboating, occur because a feedback voltage is developed across the Thevenin impedance of the power supply. High-frequency oscillations may be caused by capacitive or magnetic feedback, ground loops, unbypassed supply voltages, and stray capacitances with lead inductances. Amplifiers with negative feedback typically use a bypass circuit with a very low critical frequency. This decreases the loop gain to less than 1 at the frequency where the phase shift around the loop is 0°.

VOCABULARY

In your own words, explain what each of the following terms means. Keep your answers short and to the point. If necessary, verify your answer by rereading the appropriate discussion or by looking at the end-of-book Glossary.

Colpitts oscillator	motorboating
compensating capacitor	notch filter
ground loop	parasitic oscillations
lag circuit	phase shift
lead circuit	resonant frequency
lead-lag circuit	Wien-bridge oscillator
loop gain	

IMPORTANT EQUATIONS

Here are some important equations. Say each of the following equations in symbols, then say each

in words. Try to explain what the equation means and how it is used. Then read the description that follows.

Eq. 22-3 Resonant Frequency of *RC* Oscillators

$$f_r = \frac{1}{2\pi RC}$$

This is a familiar equation. It is the critical or resonant frequency of a lead-lag circuit. This ideal frequency is the frequency out of a Wien-bridge oscillator. The formula is exact when the amplifier is operating in its midband.

Eq. 22-5 Resonant Frequency of *LC* Oscillators

$$f_r = \frac{1}{2\pi\sqrt{LC}}$$

This equation gives you the resonant frequency of an *LC* tank circuit. When the tank circuit is part of an *LC* oscillator, this equation is approximately equal to the frequency of oscillation.

Eq. 22-7 Colpitts Starting Condition

$$A > \frac{C_2}{C_1}$$

The Colpitts tank has an inductor in parallel with two series capacitors. The voltage across C_2 is fed back to the input of the transistor or FET. This equation is an approximation for the minimum voltage gain needed to start a Colpitts oscillator.

Eq. 22-8 Effect of *Q* on Resonant Frequency

$$f_r = \frac{1}{2\pi\sqrt{LC}}\sqrt{\frac{Q^2}{1 + Q^2}}$$

This tells you that the frequency of oscillation is less than the ideal resonant frequency given by Eq. 22-5. For a resonant tank circuit with a *Q* greater than 10, the frequency of oscillation will be within 1 percent of the ideal resonant frequency.

QUESTIONS

The following questions may have more than one right answer. Select the best answer. This is the one that is always true, or covers more situations, or fits the context, etc.

1. An oscillator always needs an amplifier with
   a. Positive feedback
   b. Negative feedback
   c. Both types of feedback
   d. An *LC* tank circuit

2. The voltage that starts an oscillator is caused by
   a. Ripple from the power supply
   b. Noise voltage in resistors
   c. The input signal from a generator
   d. Positive feedback

3. The Wien-bridge oscillator is useful
   a. At low frequencies
   b. At high frequencies
   c. With *LC* tank circuits
   d. At small input signals

4. A lag circuit has a phase angle that is
   a. Between $0°$ and $+90°$
   b. Greater than $90°$
   c. Between $0°$ and $-90°$
   d. The same as the input voltage

5. A coupling circuit is a
   a. Lag circuit
   b. Lead circuit
   c. Lead-lag circuit
   d. Resonant circuit

6. A lead circuit has a phase angle that is
   a. Between $0°$ and $+90°$
   b. Greater than $90°$
   c. Between $0°$ and $-90°$
   d. The same as the input voltage

7. A Wien-bridge oscillator uses
   a. Positive feedback
   b. Negative feedback
   c. Both types of feedback
   d. An *LC* tank circuit

8. Initially, the loop gain of Wien-bridge oscillator is
   a. Less than 1
   b. Equal to 1
   c. Greater than 1
   d. Small

9. A Wien bridge is sometimes called a
   a. Notch filter
   b. Twin-T oscillator
   c. Phase shifter
   d. Wheatstone bridge

10. To vary the frequency of a Wien bridge, you can vary
    a. One resistor
    b. Two resistors
    c. Three resistors
    d. One capacitor

11. The phase-shift oscillator usually has
    a. Two lead or lag circuits
    b. Three lead or lag circuits
    c. A lead-lag circuit
    d. A twin-T filter

12. For oscillations to start in a circuit, the loop gain must be greater than 1 when the phase shift around the loop is
    a. $0°$
    b. $360°$
    c. $720°$
    d. All the above

13. The most widely used *LC* oscillator is the
    a. Armstrong
    b. Clapp
    c. Colpitts
    d. Hartley

14. Heavy feedback in an *LC* oscillator
    a. Prevents the circuit from starting
    b. Causes saturation and cutoff
    c. Produces maximum output voltage
    d. Means *B* is small

15. When *Q* decreases in a Colpitts oscillator, the frequency of oscillation
    a. Decreases
    b. Remains the same
    c. Increases
    d. Becomes erratic

16. Link coupling refers to
    a. Capacitive coupling
    b. Transformer coupling
    c. Resistive coupling
    d. Power coupling

17. The Hartley oscillator uses
    a. Negative feedback
    b. Two inductors
    c. A tungsten lamp
    d. A tickler coil

18. To vary the frequency of an *LC* oscillator, you can vary
    a. One resistor
    b. Two resistors
    c. Three resistors
    d. One capacitor

19. Of the following, the one with the most stable frequency is the
    a. Armstrong
    b. Clapp
    c. Colpitts
    d. Hartley

20. The material with the piezoelectric effect is
    a. Quartz   c. Tourmaline
    b. Rochelle salts d. All the above

21. Crystals have a very
    a. Low *Q*   c. Small inductance
    b. High *Q*   d. Large resistance

22. The series and parallel resonant frequencies of a crystal are
    a. Very close together
    b. Very far apart
    c. Equal
    d. Low frequencies

23. The kind of oscillator found in an electronic wristwatch is the
    a. Armstrong
    b. Clapp
    c. Colpitts
    d. Quartz crystal

24. An amplifier
    a. Uses negative feedback
    b. May oscillate
    c. Always has a voltage gain greater than 1
    d. Uses positive feedback

25. Motorboating is an unwanted oscillation of
    a. Low frequency
    b. High frequency
    c. Very high frequency
    d. Magnetic origins

26. The capacitance between the last stage and the first stage of an amplifier may produce oscillations of
    a. Low frequency
    b. High frequency
    c. Transformer coupling
    d. Magnetic origin

27. High-frequency oscillations are sometimes prevented by using a bypass capacitor on the
    a. Base of the first stage
    b. Collector of the last stage
    c. Supply line
    d. Emitter of all stages

28. If you have too many grounds in an amplifier, you may get oscillations because of
    a. Magnetic feedback
    b. Capacitive feedback
    c. A ground loop
    d. Supply impedance

29. Parasitic oscillations are unwanted oscillations of
    a. Low frequency
    b. High frequency
    c. Very high frequency
    d. Magnetic origin

30. An amplifier with a negative-feedback path at midband frequencies
    a. Cannot oscillate because of the negative feedback
    b. Can easily oscillate at higher frequencies
    c. Will have magnetic oscillations
    d. Will have a phase shift of 270° if uncompensated

BASIC PROBLEMS

Sec. 22-2 The Wien-Bridge Oscillator
22-1. The Wien-bridge oscillator of Fig. 22-38*a* uses a lamp with the characteristics of Fig. 22-38*b*. How much output voltage is there?

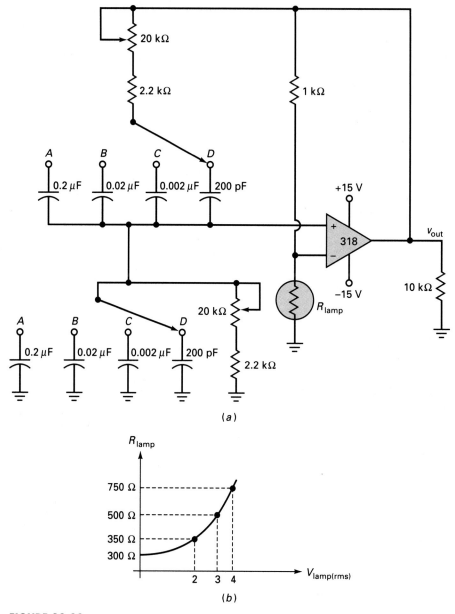

(a)

(b)

FIGURE 22-38

22-2. Position D in Fig. 22-38a is the highest frequency range of the oscillator. We can vary the frequency by ganged rheostats. What are the minimum and maximum frequencies of oscillation on this range?

22-3. Calculate the minimum and maximum frequency of oscillation for each position of the ganged switch of Fig. 22-38a.

22-4. To change the output voltage of Fig. 22-38a to a value of 6 V rms, what change can you make?

22-5. In Fig. 22-38a, the critical frequency of the amplifier with negative feedback is at least one decade above the highest frequency of oscillation. What is the critical frequency?

Sec. 22-3 Other RC Oscillators

22-6. The twin-T oscillator of Fig. 22-9, on page 862, has $R = 100\,\text{k}\Omega$ and $C = 0.01\,\mu\text{F}$. What is the frequency of oscillation?

22-7. If the values in Prob. 22-6 are doubled,

what happens to the frequency of oscillation?

Sec. 22-4 The Colpitts Oscillator

22-8. What is the approximate value of the dc
emitter current in Fig. 22-39? What is
the dc voltage from the collector to the
emitter?

22-9. What is the approximate frequency of
oscillation in Fig. 22-39? The value of
B? For the oscillator to start, what is the
minimum value of A?

22-10. If the oscillator of Fig. 22-39 is rede-
signed to get a CB amplifier similar to
Fig. 22-13, what is the feedback fraction?

22-11. If the value of L is doubled in Fig. 22-
39, what is the frequency of oscillation?

22-12. What can you do to the inductance of
Fig. 22-39 to double the frequency of
oscillation?

Sec. 22-5 Other *LC* Oscillators

22-13. If 47 pF is connected in series with the
5 µH of Fig. 22-39, the circuit becomes
a Clapp oscillator. What is the frequency
of oscillation?

22-14. A Hartley oscillator like Fig. 22-16*b*, on
page 871, has $L_1 = 1\,\mu H$ and $L_2 = 0.2\,\mu H$.
What is the feedback fraction? The frequency of oscillation if $C = 1000$ pF?
The minimum voltage gain needed to
start oscillations?

22-15. An Armstrong oscillator has $M = 0.1\,\mu H$
and $L = 2\,\mu H$. What is the feedback
fraction? What is the minimum voltage
gain needed to start the oscillations?

Sec. 22-6 Quartz Crystals

22-16. A crystal has a fundamental frequency
of 5 MHz. What is the approximate value
of the first overtone frequency? The
second overtone? The third?

22-17. A crystal has a thickness of t. If you
reduce t by 1 percent, what happens to
the frequency?

22-18. A crystal has these values: $L = 1$ H,
$C_s = 0.01$ pF, $R = 1$ kΩ, and $C_m = 20$ pF. What is the series resonant frequency? The parallel resonant fre-
quency? The Q at each frequency?

TROUBLESHOOTING PROBLEMS

22-19. Does the output voltage of the Wien-
bridge oscillator (Fig. 22-38) increase,
decrease, or stay the same for each of
these troubles?
   a. Lamp open
   b. Lamp shorted
   c. Upper potentiometer shorted
   d. Supply voltages are 20 percent low
   e. 10 kΩ open

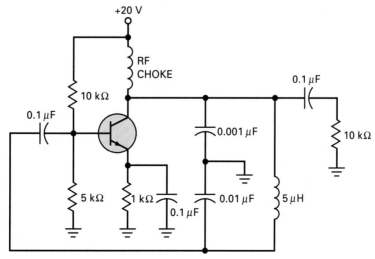

FIGURE 22-39

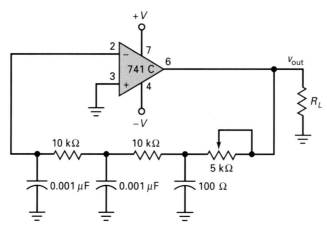

FIGURE 22-40

22-20. The Colpitts oscillator of Fig. 22-39 will not start. Name at least three possible troubles.

22-21. You have designed and built an amplifier. It does amplify an input signal, but the output looks fuzzy on an oscilloscope. When you touch the circuit, the fuzz disappears, leaving a perfect signal. What do you think the trouble is, and how would you try to eliminate it?

ADVANCED PROBLEMS

22-22. Design a Wien-bridge oscillator similar to Fig. 22-38a that meets these specifications: three decade frequency ranges covering 20 Hz to 20 kHz with output voltage of 5 V rms.

22-23. Select a value of L in Fig. 22-39 to get an oscillation frequency of 2.5 MHz.

22-24. Figure 22-40 shows an op-amp phase-shift oscillator. If $f_{2(CL)} = 1$ kHz, what is the loop phase shift at 15.9 kHz?

22-25. A 555 timer is connected for monostable operation. If $R = 10$ kΩ and $C = 0.022$ μF, what is the width of the output pulse?

22-26. An astable 555 timer has $R_A = 10$ kΩ, $R_B = 2$ kΩ, and $C = 0.0047$ μF. What are the output frequency and duty cycle?

22-27. Design a 555 timer circuit that free-runs at a frequency of 1 kHz and a duty cycle of 75 percent.

T-SHOOTER PROBLEMS

Use Fig. 22-41 for the remaining problems. In this T-shooter you are troubleshooting at the system level, meaning that you are to locate the most suspicious block for further testing. The system shown here is called a *superheterodyne receiver*. It is the typical block diagram of an ordinary radio. The antenna picks up a radio signal, which goes to the RF amp. This amplifier is a medium-frequency amplifier operating between 535 and 1605 kHz. The local oscillator is a Colpitts oscillator whose frequency is 455 kHz higher than the frequency of the RF amp. The mixer is a circuit that produces an output signal with a frequency of 455 kHz. This mixer output signal is amplified by the IF amps. The detector converts the signal to audio frequencies. The audio signal is amplified by the audio amps. Finally, the signal drives a loudspeaker to produce the sound.

In this exercise, all you have to do is locate the defective block. Because we have not discussed this system in detail, we will limit the measurements to Y or N, stands for "Yes, a signal is present at this point" or "No, there is no signal at this point." Even in complicated systems, a simple yes or no is often all you need to pin down the defective block. Then you can troubleshoot the block at the component level if desired. In the following problems,

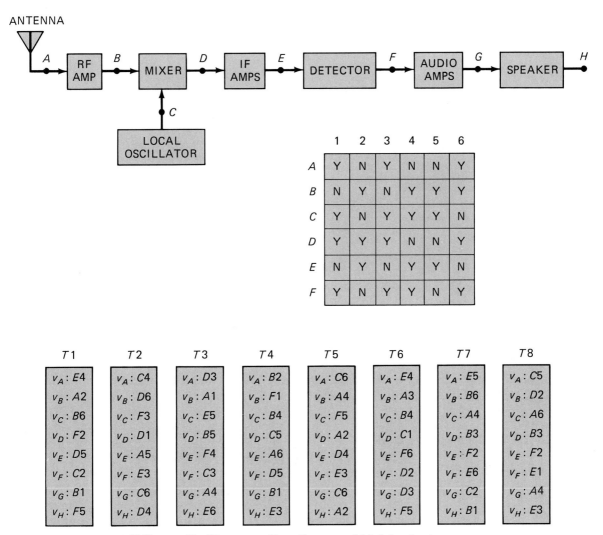

FIGURE 22-41 T-Shooter™. (*Patent pending: Courtesy of Malvino Inc.*)

the possible answers are antenna, RF amp, mixer, local oscillator, IF amps, detector, audio amps, and speaker.

22-28. Find Trouble 1.

22-29. Find Trouble 2.

22-30. Find Troubles 3 and 4.

22-31. Find Troubles 5 and 6.

22-32. Find Troubles 7 and 8.

REGULATED POWER SUPPLIES

Chapter 5 introduced the zener diode, a device with a constant voltage in the breakdown region. With a zener diode, we are able to build simple voltage regulators that can hold the load voltage constant. Now we discuss ways to improve voltage regulation. In this chapter, negative feedback is used to hold the output voltage almost constant despite relatively large changes in line voltage and load current. Although the discussion begins with discrete circuits, it ends with integrated circuits because they are commonly used to regulate supply voltages. The chapter concludes with switching regulators, the type of voltage regulator preferred for large load currents at low voltages, like those needed in personal computers.

23-1 VOLTAGE-FEEDBACK REGULATION

The temperature coefficient of a zener diode changes from negative to positive between 5 and 6 V. Because of this, zener diodes with breakdown voltages in this area will have temperature coefficients of approximately zero. In critical applications, therefore, zener voltages near 6 V are used because the zener voltage is constant when temperature changes over a large range. This highly stable zener voltage, sometimes called a *reference voltage*, can be amplified with a negative-feedback amplifier to get a higher output voltage that has the same temperature stability as the reference voltage.

Basic Idea

Figure 23-1a shows a discrete voltage regulator. Transistor Q_2 is an emitter follower. Therefore, its base voltage is one V_{BE} drop higher than the output voltage across the load. Transistor Q_2 is called a *pass transistor* because all the load current passes through it. Notice that a voltage divider samples the output voltage V_{out} and delivers a feedback voltage V_F to the base of Q_1. This transistor operates in the active region as a CE amplifier. The feedback voltage V_F controls the Q_1 collector current. Since the output voltage is being sampled, it is stabilized against changes in open-loop gain, load resistance, line voltage, and so on. The larger the feedback voltage, the greater the Q_1 collector current.

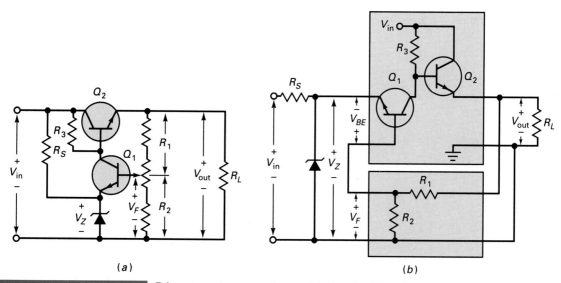

FIGURE 23-1   Discrete voltage regulator: (*a*) circuit; (*b*) redrawn to emphasize gain and feedback sections.

The dc input voltage V_{in} comes from an unregulated power supply, such as a bridge rectifier and a capacitor-input filter. Typically, V_{in} has a peak-to-peak ripple of about 10 percent of the dc voltage. But the output voltage V_{out} has almost no ripple and is almost perfectly constant, even though the input voltage and load current may change. Why? Because any change in output voltage is fed back through the voltage divider to the base of Q_1. This produces an error voltage that automatically compensates for the attempted change.

For instance, if V_{out} tries to increase, more V_F is fed back to the base of Q_1, producing a larger Q_1 collector current through R_3 and less base voltage at Q_2. The reduced base voltage to the Q_2 emitter follower results in less output voltage. Similarly, if the output voltage tries to decrease, there is less base voltage at Q_1, more base voltage at Q_2, and more output voltage. In either case, the attempted change in V_{out} produces an amplified output change in the opposite direction. The overall effect is to almost eliminate the attempted changes in output voltage.

Output Voltage

Figure 23-1*b* shows the circuit redrawn to allow easy recognition of the amplifier and the negative-feedback circuit. The reference voltage V_Z is the input to the emitter of the first stage, which drives an emitter follower. The output voltage V_{out} is applied to a voltage divider to produce a feedback voltage V_F for the base of Q_1. The feedback fraction is

$$B = \frac{R_2}{R_1 + R_2}$$

The closed-loop voltage gain is

$$A_{CL} = \frac{1}{B}$$

or

$$A_{CL} = \frac{R_1}{R_2} + 1 \qquad (23\text{-}1)$$

In Fig. 23-1a, the feedback voltage is applied to the base of Q_1. As you can see, this voltage may be written as

$$V_F = V_Z + V_{BE}$$

Since $V_F = BV_{\text{out}}$,

$$BV_{\text{out}} = V_Z + V_{BE}$$

or

$$V_{\text{out}} = \frac{V_Z + V_{BE}}{B}$$

or

$$V_{\text{out}} = A_{CL}(V_Z + V_{BE}) \qquad (23\text{-}2)$$

Therefore, the regulated output voltage equals the closed-loop voltage gain times the sum of the zener voltage and the base-emitter voltage.

Because of the closed-loop voltage gain, we can use a low zener voltage (5 to 6 V) where the temperature coefficient approaches zero. The amplified output voltage then has the same temperature coefficient. The potentiometer of Fig. 23-1a allows us to adjust the output voltage to the exact value required in a particular application. In this way, we can adjust for the tolerance in zener voltages, V_{BE} drops, and feedback resistors. Once the potentiometer is adjusted to the desired output voltage, V_{out} remains almost constant, despite changes in line voltage or load current.

Power Dissipation in the Pass Transistor

The pass transistor (the emitter follower) is in series with the load. This is why the circuit is called a *series regulator*. The main disadvantage of a series regulator is the power dissipation in the pass transistor:

$$P_D = V_{CE}I_C \qquad (23\text{-}3)$$

As long as the load current is not too large, the pass transistor does not get too hot. But when the load current is heavy, the pass transistor has to dissipate a lot of power. This implies heavier heat sinks and a bulkier power supply. In some cases, a fan may be needed to remove the excess heat. Depending on the application, a designer may prefer to use a switching regulator (discussed later) for heavy load currents.

EXAMPLE 23-1

You often see a Darlington pair used for the pass transistor in a series regulator. This allows the regulator to drive smaller load resistors. What are the minimum and maximum output voltages in Fig. 23-2?

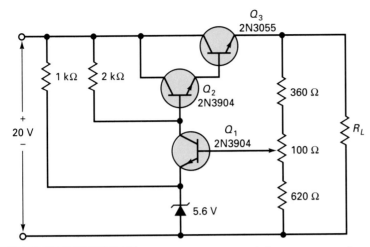

FIGURE 23-2 Darlington pair used for a pass transistor to increase maximum load current.

SOLUTION

To begin, the closed-loop voltage gain is still given by $A_{CL} = R_1/R_2 + 1$. If the wiper is turned all the way up, $R_1 = 360\ \Omega$ and $R_2 = 720\ \Omega$. In this case, the closed-loop voltage gain is

$$A_{CL} = \frac{360}{720} + 1 = 1.5$$

and the regulated output voltage is

$$V_{\text{out}} = 1.5(5.6\text{ V} + 0.7\text{ V}) = 9.45\text{ V}$$

However, when the wiper is moved all the way down, $R_1 = 460\ \Omega$ and $R_2 = 620\ \Omega$. Now,

$$A_{CL} = \frac{460}{620} + 1 = 1.74$$

and

$$V_{\text{out}} = 1.74(5.6\text{ V} + 0.7\text{ V}) = 11\text{ V}$$

So we can adjust the regulated output voltage between 9.45 and 11 V.

Incidentally, there is another way to find the output voltage by using only Ohm's law. In Fig. 23-2, visualize the wiper all the way up. Then the voltage across the R_2 of the voltage divider is

$$V = 5.6\text{ V} + 0.7\text{ V} = 6.3\text{ V}$$

The current through R_2 is

$$I = \frac{6.3\text{ V}}{720\ \Omega} = 8.75\text{ mA}$$

This current times the total divider resistance gives the output voltage:

$$V_{\text{out}} = (8.75 \text{ mA})(360 \ \Omega + 720 \ \Omega) = 9.45 \text{ V}$$

In a similar way, you can calculate an output voltage of 11 V when the wiper is all the way down.

This second method is very direct, accurate, and easy to remember since it is nothing more than Ohm's law applied in an intelligent way. The intelligent use of Ohm's law always beats the blind use of formulas. Many professionals feel that the simpler you can do something, the better. This principle is true in art, writing, and almost every other field. Why not electronics? Why not do things in the simplest way possible and fall back on formulas only as a last resort?

EXAMPLE 23-2

In Fig. 23-2, $V_{\text{out}} = 10$ V and $R_L = 5 \ \Omega$. If the 2N3055 has a β_{dc} of 50 and the 2N3904 has a β_{dc} of 100, what is the current through the zener diode?

SOLUTION

Since V_L is 10 V and R_L is 5 Ω, the load current is

$$I_L = \frac{10 \text{ V}}{5 \ \Omega} = 2\text{A}$$

The current through the voltage divider is

$$I = \frac{10 \text{ V}}{1.08 \text{ k}\Omega} = 9.26 \text{ mA}$$

This is very small compared with 2 A. Therefore, the total emitter current through the 2N3055 is approximately 2 A, and the base current is approximately

$$I_B = \frac{2 \text{ A}}{50} = 40 \text{ mA}$$

The 2N3904 has to supply the 40 mA of base current for the 2N3055. This means that the base current of the 2N3904 is

$$I_B = \frac{40 \text{ mA}}{100} = 0.4 \text{ mA}$$

Because of the two V_{BE} drops, the voltage at the bottom of the 2 kΩ is

$$V = 10 \text{ V} + 2(0.7 \text{ V}) = 11.4 \text{ V}$$

The current through the 2 kΩ is

$$I = \frac{20\,V - 11.4\,V}{2\,k\Omega} = 4.3\,mA$$

This current splits into the base current for the upper 2N3904 and the collector current for the lower 2N3904. Therefore, the lower 2N3904 has a collector current of

$$I_C = 4.3\,mA - 0.4\,mA = 3.9\,mA$$

The current through the 1 kΩ is

$$I = \frac{20\,V - 5.6\,V}{1\,k\Omega} = 14.4\,mA$$

The total current through the zener diode is the sum of the current through the 1 kΩ and the Q_1 emitter current:

$$I_Z = 14.4\,mA + 3.9\,mA = 18.3\,mA$$

Incidentally, the 2N3055 is a workhorse power transistor. This widely used industry standard can handle up to 15 A of continuous current, has breakdown voltages of 60 V or more, and can dissipate 115 W at 25°C with heat sinking.

EXAMPLE 23-3

Calculate the power dissipation of the 2N3055 in Fig. 23-2 for a load voltage of 10 V and a load current of 2 A.

SOLUTION

The collector voltage of the 2N3055 is 20 V, and the emitter voltage is 10 V. The voltage between the collector and the emitter is

$$V_{CE} = 20\,V - 10\,V = 10\,V$$

The power dissipation is the product of V_{CE} and I_C. Since the load current is 2 A,

$$P_D = (10\,V)(2\,A) = 20\,W$$

This is a fair amount of power if you think of the nearest standard light bulb (25 W). Because of the large amount of heat being dissipated by this transistor, the 2N3055 would need a heat sink to keep its junction temperature at a safe level.

23-2 CURRENT LIMITING

The series regulator of Fig. 23-2 has no *short-circuit protection*. If we accidentally short the load terminals, we get a large load current that will destroy the pass transistor. It may also destroy one or more diodes in the unregulated power supply that is driving the series regulator. To avoid this possibility, regulated supplies usually include some form of *current limiting.*

Another Look at the Diode Knee

Figure 23-3 shows the transconductance curve of a silicon transistor. If the linear part of this curve is extended downward, it intersects the V_{BE} axis at a value of approximately 0.7 V. This is typical of silicon transistors and is the basis for the diode approximations discussed earlier. In the ideal approximation, we ignore the knee voltage. In the second approximation, we include the knee voltage. In the third approximation, we add the voltage drop across the bulk resistance. To discuss current limiting, we have to go one step further. We need to use a fourth approximation.

In Fig. 23-3, you can see that a small current exists between approximately 0.6 and 0.7 V. In this region, the diode is undergoing a transition from nonconducting to conducting. In the fourth approximation of a diode, we consider the diode off below 0.6 V, on above 0.7 V, and in transition between 0.6 and 0.7 V. The current in this transition region is very small but is important in the discussion that follows.

Simple Limiting

Figure 23-4 shows one way to limit the load current to safe values even though the output terminals are shorted. Resistor R_4 is called a *current-sensing resistor,* which is a small resistor that has almost no effect on the load current but whose voltage is used as a measure of current. The current-sensing resistor produces a voltage that is applied to the base-emitter terminals of Q_3. If the load current is too large, the voltage across R_4 will turn on Q_3. In turn, this will produce current limiting.

Here are the details of how the current limiting works. When the load current is less than 600 mA, the voltage across R_4 is less than 0.6 V. In

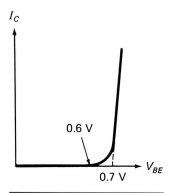

FIGURE 23-3

Diode characteristic.

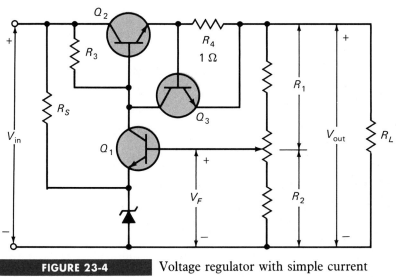

FIGURE 23-4   Voltage regulator with simple current limiting.

this case, Q_3 is cut off, and the regulator works as previously described. When the load current is greater than 600 mA, the voltage across R_4 is more than 0.6 V, which turns on Q_3. The collector current of Q_3 flows through R_3 and decreases the base voltage to Q_2. In turn, this decreases the load voltage. In effect, we have negative feedback because the original increase in load current is producing a decrease in load voltage. This negative feedback becomes very heavy for a V_{BE} between 0.6 and 0.7 V. The exact value depends on the transistor type and the collector current.

Here is the summary explanation: Assume R_L is being decreased from infinity to zero in Fig. 23-4. This means the load current is increasing from zero to higher values. Somewhere in the vicinity of 600 to 700 mA, Q_3 turns on, and this decreases the base voltage of Q_2. Since Q_2 is an emitter follower, the load voltage decreases. Further decreases in R_L increase the load current and decrease the load voltage. Because the load voltage is decreasing at approximately the same rate as the load resistance is increasing, the load current increases only slightly.

Figure 23-5 summarizes the current limiting. When R_L is infinite, the output voltage is regulated and has a value of V_{reg}. The load current is zero for this condition. When R_L decreases, the load current increases up to the point where R_L equals R_{min}. At this minimum load resistance, I_L equals 600 mA, and V_{BE} equals 0.6 V. Beyond this point, Q_3 turns on and current limiting sets in. Further decreases in R_L produce decreases in load voltage, and regulation is lost. When R_L is zero, the load current is limited to a value between 600 and 700 mA.

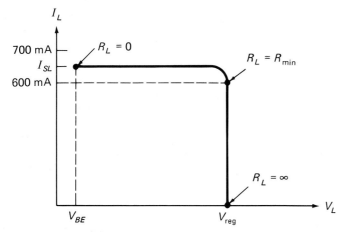

FIGURE 23-5 Load current versus load voltage with simple current limiting.

The load current with *shorted-load* terminals is symbolized as I_{SL}. When the load terminals are shorted in Fig. 23-4 (equivalent to $R_L = 0$), the voltage across R_4 is

$$V_{BE} = I_{SL}R_4$$

or

$$I_{SL} = \frac{V_{BE}}{R_4} \qquad (23\text{-}4)$$

where V_{BE} is typically between 0.6 and 0.7 V. For convenience, use $V_{BE} = 0.7$ V when you use Eq. (23-4). But remember, the exact value of I_{SL} may be slightly less or slightly more than what you calculate with Eq. (23-4) because the exact value depends on the transistor and the load current.

The minimum load resistance where regulation is lost can be estimated with this equation:

$$R_{\min} = \frac{V_{\text{reg}}}{I_{SL}} \qquad (23\text{-}5)$$

The exact value of $R_{\min}$ will be slightly smaller or greater than this, but Eq. (23-5) is excellent for troubleshooting and preliminary design.

Disadvantage of Simple Limiting

The simple current limiting just described is a big improvement because it will protect the pass transistor and rectifier diodes in case the load terminals are accidentally shorted. But it has the disadvantage of a large power dissipation in the pass transistor when the load terminals are shorted. With a short across the load, almost all the input voltage appears across the pass transistor. Therefore, the pass transistor has to dissipate approximately

$$P_D = (V_{\text{in}} - V_{BE})I_{SL} \qquad (23\text{-}6)$$

where V_{BE} is the base-emitter voltage of Q_3, the current-limiting transistor. Depending on the values of V_{in} and I_{SL}, this power may be small or large. In those applications where the value of P_D is acceptable, a designer may decide to use simple current limiting.

But if the value of P_D becomes a design problem, the next level of design is *foldback current limiting*. Figure 23-6 illustrates the advantage of this type of current limiting. As before, the load voltage is regulated for a load resistance between infinity and $R_{\min}$. When the load resistance decreases to less than $R_{\min}$, current limiting sets in. But with foldback limiting, small values of R_L decrease the load current as well as the load

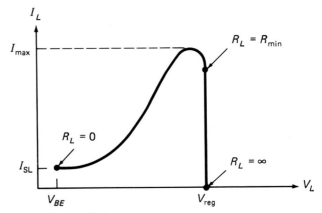

FIGURE 23-6 Load current versus load voltage with foldback current limiting.

voltage. When R_L is zero, the load current is much smaller than the maximum load current. Because of this, the value of P_D in Eq. (23-6) is much smaller. As a result, a voltage regulator with foldback current limiting can handle much larger load currents than one with simple current limiting. (See the "Optional Topics" if you are interested in a circuit with foldback current limiting.)

Dominant Bypass Circuit

The Q_1 stage of Fig. 23-4 has a phase shift of $-180°$. At high frequencies, this stage has a base bypass circuit and a collector bypass circuit, each producing a phase shift between $0°$ and $-90°$. Also the Q_2 stage has some phase shift at higher frequencies. Therefore, the loop gain may be greater than 1 when the phase shift around the loop becomes $-360°$, equivalent to $0°$. To avoid the possibility of oscillations, we can add a large capacitor to the base circuit of the Q_1 stage as shown in Fig. 23-7. This gives the voltage regulator a dominant bypass circuit that prevents oscillations, as discussed in Sec. 22-7.

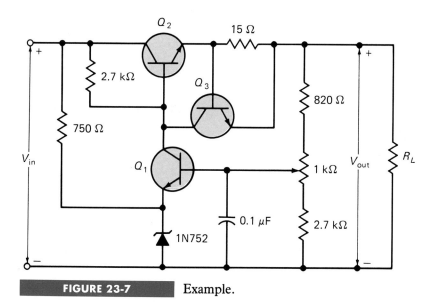

FIGURE 23-7 Example.

EXAMPLE 23-4

The wiper of Fig. 23-7 is at the middle of its travel. If the input voltage is 15 V, what is the output voltage of the regulator?

SOLUTION

The resistance below the wiper is

$$R_2 = 2.7 \text{ k}\Omega + 500 \text{ }\Omega = 3.2 \text{ k}\Omega$$

The feedback fraction is

$$B = \frac{3.2 \text{ k}\Omega}{820 \ \Omega + 1 \text{ k}\Omega + 2.7 \text{ k}\Omega} = 0.708$$

The closed-loop voltage gain is

$$A_{CL} = \frac{1}{0.708} = 1.41$$

Since a 1N752 (see the Appendix) has a zener voltage of

$$V_Z = 5.6 \text{ V}$$

With Eq. (23-2), the output voltage is

$$V_{\text{out}} = 1.41(5.6 \text{ V} + 0.7 \text{ V}) = 8.88 \text{ V}$$

Here is another way to solve the problem. After you find the feedback fraction of 0.708, the next step is to set up this equation:

$$\frac{V_F}{V_{\text{out}}} = 0.708$$

which can be written as

$$V_{\text{out}} = \frac{V_F}{0.708}$$

In Fig. 23-7, you can see that the zener voltage plus the base-emitter voltage of Q_1 is

$$V_F = 5.6 \text{ V} + 0.7 \text{ V} = 6.3 \text{ V}$$

Now, solve for V_{out}:

$$V_{\text{out}} = \frac{6.3 \text{ V}}{0.708} = 8.9 \text{ V}$$

Roundoff error aside, this is the same answer as before.

Here is a third way to solve the problem. The feedback voltage is given by

$$V_F = 5.6 \text{ V} + 0.7 \text{ V} = 6.3 \text{ V}$$

The feedback voltage is between the wiper and the common side of the circuit. Since this feedback voltage appears across

$$R_2 = 500 \ \Omega + 2.7 \text{ k}\Omega = 3.2 \text{ k}\Omega$$

the current through the voltage divider is

$$I = \frac{6.3 \text{ V}}{3.2 \text{ k}\Omega} = 1.97 \text{ mA}$$

Since the total resistance in the voltage divider is

$$R = 820 \ \Omega + 1 \text{ k}\Omega + 2.7 \text{ k}\Omega = 4.52 \text{ k}\Omega$$

the output voltage is

$$V_{out} = (1.97 \text{ mA})(4.52 \text{ k}\Omega) = 8.9 \text{ V}$$

As usual, you can use whichever method feels comfortable. The important thing is not how to solve this particular problem today, but whether you can solve similar problems by any method two months from now, a year from now, etc. This is why you have to trust your intuition about which method to use. You can use any method you want. That's what a professional does. The reason you have this freedom is because electronics is an art as well as a science. This allows you to do things any way you want, provided you don't violate the fundamental laws of electricity.

EXAMPLE 23-5

What is the shorted-load current in Fig. 23-7? With the wiper at the center of the potentiometer, what is the minimum load resistance where regulation is lost? If the input voltage is 15 V, what is the power dissipation of the pass transistor when the load resistance is shorted?

SOLUTION

Look at current-sensing resistor of Fig. 23-7. Its resistance is 15 Ω. When the voltage across this resistor is between 0.6 and 0.7 V, current limiting starts. As mentioned earlier, we will use $V_{BE} = 0.7$ V in our calculations for shorted-load current. Ohm's law applied to the current-sensing resistor gives

$$I_{SL} = \frac{0.7 \text{ V}}{15 \Omega} = 46.7 \text{ mA}$$

In Example 23-4, we found that $V_{out} = 8.9$ V. Ohm's law applied to the load resistor gives

$$R_{min} = \frac{8.9 \text{ V}}{46.7 \text{ mA}} = 191 \Omega$$

This is the smallest load resistance that can be used without losing voltage regulation.

When the load resistor is shorted, current limiting occurs and approximately 0.7 V is across the current-sensing resistor. The voltage across the pass transistor is

$$V_{CE} = 15 \text{ V} - 0.7 \text{ V} = 14.3 \text{ V}$$

The current through the pass transistor is equal to shorted-load current of 46.7 mA. Therefore, the power dissipation is

$$P_D = (14.3 \text{ V})(46.7 \text{ mA}) = 668 \text{ mW}$$

23-3 POWER SUPPLY CHARACTERISTICS

The quality of a power supply depends on its load voltage, load current, voltage regulation, and other factors. In this section, we will look at some characteristics of regulated power supplies.

Load Regulation

Figure 23-8 shows a block diagram of a bridge rectifier with a capacitor-input filter driving a voltage regulator. The quality of the load voltage can be specified in several ways. To begin, the *load regulation*, abbreviated *LR* (also called the *load effect*), is the change in regulated output voltage when the load current changes from minimum to maximum:

$$\text{LR} = V_{NL} - V_{FL} \tag{23-7}$$

where LR = load regulation
V_{NL} = load voltage with no load current
V_{FL} = load voltage with full load current

This is a defining formula. In this equation, V_{NL} occurs when the load resistance is infinite. And V_{FL} occurs when the load resistance is the minimum value where voltage regulation is lost. For instance, if the load voltage is 10 V at zero load current and 9.9 V at full load current, then

$$\text{LR} = 10 \text{ V} - 9.9 \text{ V} = 0.1 \text{ V}$$

As another example, the Hewlett-Packard 6214A is a regulated power supply with a maximum load voltage of 10 V and a maximum load current of 1 A. Its data sheet lists a load regulation of 4 mV. This means that the load voltage changes only 4 mV when the load current varies from 0 to 1 A.

Load regulation is often expressed as a percentage by dividing the load regulation by the full-load voltage and multiplying the result by 100 percent:

$$\%\text{LR} = \frac{V_{NL} - V_{FL}}{V_{FL}} \times 100\,\% \tag{23-8}$$

For instance, if the no-load voltage is 10 V and the full-load voltage is 9.9 V, then the percentage of load regulation is

$$\%\text{LR} = \frac{10 \text{ V} - 9.9 \text{ V}}{9.9 \text{ V}} \times 100\% = 1.01\%$$

As another example, if the change in load voltage is 4 mV and the no-load voltage is 10 V, then

$$\%\text{LR} = \frac{4 \text{ mV}}{9.996 \text{ V}} \times 100\% = 0.04\%$$

In the last two calculations, the denominator of the fraction is very close to the no-load voltage. Many engineers and technicians use this

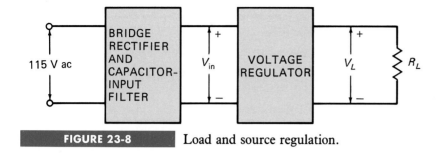

FIGURE 23-8   Load and source regulation.

approximation for Eq. (23-8):

$$\%\text{LR} \approx \frac{V_{NL} - V_{FL}}{V_{NL}} \times 100\% \qquad (23\text{-}9)$$

This is easier to use and still accurate when %LR is less than 5 percent, which is most of the time with modern regulated power supplies. With this approximation, the two preceding calculations look like this:

$$\%\text{LR} = \frac{10\,\text{V} - 9.9\,\text{V}}{10\,\text{V}} \times 100\% = 1\%$$

and

$$\%\text{LR} = \frac{4\,\text{mV}}{10\,\text{V}} \times 100\% = 0.04\%$$

Suggestion: Use Eq. (23-9) because it is easier. If you get a %LR that is greater than 5 percent, switch to Eq. (23-8) to calculate a more accurate value of %LR.

Source Regulation

In Fig. 23-8, the input line voltage has a nominal value of 115 V. Depending on the demand for electricity in any given area of the country, this line voltage may be different from 115 V. In fact, the line voltage at your power outlet can vary a lot between peak and quiet hours. Since this line voltage is the input to a bridge rectifier, the filtered output of the bridge rectifier is almost directly proportional to the line voltage. As you can see in Fig. 23-8, the filtered output of the bridge rectifier is the input to the voltage regulator.

Another way to specify the quality of a regulated power supply is by its *source regulation* (also called *source effect* or *line regulation*). Abbreviated SR, the source regulation is defined as the change in regulated load voltage for a specified range of line voltage, typically 115 V ± 10 percent. The defining equation is

$$\text{SR} = V_{HL} - V_{LL} \qquad (23\text{-}10)$$

where SR = source regulation
$\quad V_{HL}$ = load voltage with high line voltage
$\quad V_{LL}$ = load voltage with low line voltage

As an example, if the load voltage is 10 V $\pm$ 0.3 V for a line voltage of 115 V $\pm$ 10 percent, then

$$\text{SR} = 10.3 \text{ V} - 9.7 \text{ V} = 0.6 \text{ V}$$

As a comparison, a quality power supply like the Hewlett-Packard 6214A has SR = 4 mV.

The percentage of source regulation is

$$\%\text{SR} = \frac{\text{SR}}{V_{\text{nom}}} \times 100\% \qquad (23\text{-}11)$$

where V_{nom} is nominal load voltage, the output voltage under typical operating conditions. For instance, if the change in the load voltage is 0.6 V and the nominal load voltage is 10 V, the percentage of source regulation is

$$\%\text{SR} = \frac{0.6}{10 \text{ V}} \times 100\% = 6\%$$

Output Impedance

A regulated power supply is a very stiff dc voltage source. This means that the Thevenin or output resistance is very small. In the voltage regulator discussed earlier, an emitter follower supplies the load voltage. The emitter follower already has a low output impedance. The use of voltage feedback further reduces this output impedance because

$$z_{\text{out}(CL)} = \frac{z_{\text{out}}}{1 + AB} \qquad (23\text{-}12)$$

Regulated power supplies have output impedances in milliohms. Because of this, they are very stiff voltage sources. Even though you vary the load resistance, you will see almost no change in the load voltage. An ideal voltage source has an output impedance of zero. Today's modern regulated power supplies approach ideal voltage sources.

Ripple Rejection

Voltage regulators stabilize the output voltage against changes in input voltage. Ripple is equivalent to a periodic change in the input voltage. Therefore, a voltage regulator attenuates the ripple that comes in with the unregulated input voltage. Since a voltage regulator uses negative feedback, the improvement can be found with the desensitivity or sacrifice factor discussed in Chap. 19. As you recall,

$$D = 1 + AB$$

The output ripple of a voltage regulator is given by

$$V_{R(\text{out})} = \frac{V_{R(\text{in})}}{1 + AB} \qquad (23\text{-}13)$$

Data sheets sometimes list the ripple rejection (RR). It is defined as

$$\text{RR} = \frac{V_{R(\text{out})}}{V_{R(\text{in})}} \qquad (23\text{-}14)$$

Often, you will see the ripple rejection specified in decibels:

$$RR' = 20 \log \frac{V_{R(\text{out})}}{V_{R(\text{in})}} \qquad (23\text{-}15)$$

For example, an RR' of 80 dB means that the output ripple is 80 dB less than the input ripple. In ordinary numbers, this means that the output ripple is 10,000 times smaller than the input ripple.

EXAMPLE 23-6

The regulator of Fig. 23-9 has a no-load voltage of 9 V and a full-load voltage of 8.75 V. What is the load regulation? The percentage of load regulation?

SOLUTION

With Eq. (23-7), the load regulation is

$$LR = 9 \text{ V} - 8.75 \text{ V} = 0.25 \text{ V}$$

With Eq. (23-9), the percentage of load regulation is

$$\%LR = \frac{0.25 \text{ V}}{9 \text{ V}} \times 100\% = 2.78\%$$

EXAMPLE 23-7

In Fig. 23-9, the input voltage to the voltage regulator comes from a bridge rectifier with a capacitor-input filter. This input voltage

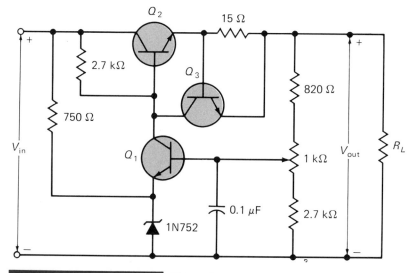

FIGURE 23-9   Example.

has a nominal value of 15 V. If the line voltage is 115 V ± 10 percent, what are the minimum and maximum values of the input voltage?

SOLUTION

When the knee voltage is small compared to the peak voltage, the output of a bridge rectifier is directly proportional to the line voltage. Because of this, a change of 10 percent in the line voltage produces a change of 10 percent in the input voltage of Fig. 23-9. This means the input voltage can be as low as

$$V_{\text{in}} = 15 \text{ V} - 0.1(15 \text{ V}) = 13.5 \text{ V}$$

or as high as

$$V_{\text{in}} = 15 \text{ V} + 0.1(15 \text{ V}) = 16.5 \text{ V}$$

EXAMPLE 23-8

In Fig. 23-9, suppose you measure a minimum load voltage of 8.9 V and a maximum load voltage of 9.1 V when the line voltage varies by ±10 percent. What is the source regulation? The percentage of source regulation?

SOLUTION

The source regulation is

$$\text{SR} = 9.1 \text{ V} - 8.9 \text{ V} = 0.2 \text{ V}$$

and the percentage of source regulation is

$$\%\text{LR} = \frac{0.2 \text{ V}}{9 \text{ V}} \times 100\% = 2.22\%$$

In this calculation, we used a nominal or average load voltage of 9 V.

EXAMPLE 23-9

Suppose the input ripple to Fig. 23-9 is 1 V peak to peak. If $1 + AB = 50$, what is the output ripple? What is the ripple rejection in decibels?

SOLUTION

The output ripple is

$$V_{R(\text{out})} = \frac{1 \text{ V}}{50} = 20 \text{ mV}$$

The ripple rejection is

$$RR = \frac{20\,\text{mV}}{1\,\text{V}} = 0.02$$

In decibels, the ripple rejection is

$$RR' = 20 \log 0.02 = -34\,\text{dB}$$

23-4 THREE-TERMINAL IC REGULATORS

In the late 1960s, IC manufacturers began producing a voltage regulator on a chip. First-generation devices like the μA723 and the LM300 included a zener diode, a high-gain amplifier, current limiting, and other useful features. The disadvantage of these early IC regulators was the need for many external components, plus the eight or more pins that had to be connected in various ways to get optimum performance.

The latest generation of IC voltage regulators has devices with only three pins: one for the unregulated input voltage, one for the regulated output voltage, and one for ground. The new devices can supply load current from 100 mA to more than 5 A. Available in plastic or metal packages, these three-terminal regulators have become extremely popular because they are so inexpensive and easy to use. Aside from a couple of bypass capacitors, the new three-terminal IC voltage regulators require no external components.

The LM340 Series

The LM340 series is typical of the new breed of three-terminal voltage regulators. Figure 23-10 shows the block diagram. The built-in reference voltage V_{ref} drives the noninverting input of an amplifier. The voltage regulation is the same as described earlier with a few minor differences. The amplifier consists of several stages of voltage gain. Because of the high gain of the amplifier, the error voltage between the two amplifier input terminals approaches zero. Therefore, the inverting input voltage is approximately equal to V_{ref}. This means the current through the voltage divider is

$$I = \frac{V_{\text{ref}}}{R_2'}$$

Incidentally, the prime attached to R_2' indicates that this resistor is inside the IC itself, rather being an external resistor. Since the same current flows through R_1', the output voltage is

$$V_{\text{out}} = \frac{V_{\text{ref}}}{R_2'}(R_1' + R_2') \tag{23-16}$$

This shows that the output voltage of the regulator can be precisely controlled by setting up the desired values of R_1' and R_2'. The LM340 series is available with preset output voltages of 5, 12, and 15 V. For

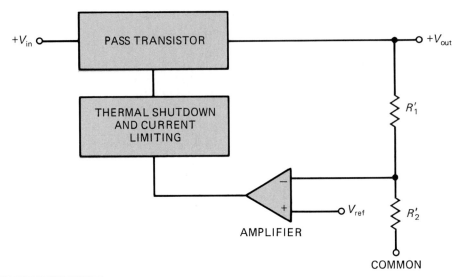

FIGURE 23-10   Functional block diagram of typical three-terminal IC voltage regulator.

instance, the LM340-5 produces a regulated output voltage of 5 V, the LM340-12 produces an output of 12 V, and the LM340-15 produces an output of 15 V. The tolerance of the regulated output voltage is ±4 percent. (The LM340A is also available with a tolerance of ±2 percent.)

The chip includes a pass transistor that can handle more than 1.5 A of load current provided that adequate heat sinking is used. Also included are *thermal shutdown* and current limiting. Thermal shutdown means that the chip will automatically turn itself off if the internal temperature becomes dangerously high, around 175°C. This is a precaution against excessive power dissipation, which depends on the ambient temperature, type of heat sinking, and other variables. Because of thermal shutdown and current limiting, devices in the LM340 series are almost indestructible.

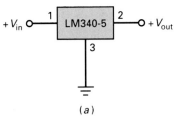

Fixed Regulator

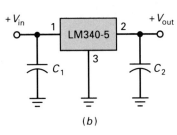

FIGURE 23-11

(a) LM340 connected as voltage regulator; (b) input bypass capacitor prevents oscillation, and output bypass capacitor improves transient response.

Figure 23-11a shows an LM340-5 connected as a *fixed voltage regulator*. Pin 1 is the input, pin 2 is the output, and pin 3 is ground. The LM340-5 has an output voltage of +5 V ± 4 percent, a maximum load current of 1.5 A, a load regulation of 10 mV, a source regulation of 3 mV, and a ripple rejection of 80 dB. With an output impedance of approximately 0.01 Ω, the LM340-5 is a very stiff voltage source to all loads within its maximum current rating.

When an IC is more than a few inches from the filter capacitor of the unregulated power supply, the lead inductance may produce oscillations within the IC. This is why you often see a bypass capacitor C_1 on pin 1 (Fig. 23-11b). To improve the transient response of the regulated output voltage, a bypass capacitor C_2 is sometimes used. Typical values for either bypass capacitor are from 0.1 to 1 μF. (The data sheet of the LM340 series suggests 0.22 μF for the input capacitor and 0.1 μF for the output capacitor.)

Any device in the LM340 series needs a minimum input voltage at least 2 to 3 V greater than the regulated output voltage. Otherwise, it stops regulating (sometimes it is called *brownout*). Furthermore, there is a maximum input voltage because of excessive power dissipation. For instance, the LM340-5 will regulate over an input range of approximately 8 to 20 V. The data sheet gives the minimum and maximum input voltages for the other preset output voltages.

Two More Applications

Figure 23-12a shows two external resistors R_1 and R_2 added to an LM340 to get an *adjustable output voltage*. The common terminal of the LM340 is not grounded. Rather, it is connected to the top of R_2. This means that the regulated output V_{reg} is across R_1. This produces a current of

$$I = \frac{V_{\text{reg}}}{R_1} \qquad (23\text{-}17)$$

This current flows through R_2. Therefore, the total output voltage is

$$V_{\text{out}} = \frac{V_{\text{reg}}}{R_1}(R_1 + R_2) \qquad (23\text{-}18)$$

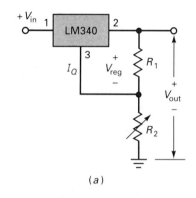

(a)

The foregoing equation is an approximation because it ignores the quiescent current I_Q out of pin 3. In Fig. 23-12a, the current I_Q that normally would flow to ground (see Fig. 23-11a) now has to flow through R_2 to reach ground. Because of this, there is an additional voltage drop across R_2. Many designers swamp out the effects of I_Q to prevent it from having any effect on the regulated output. They do this by making it much smaller than the current given by Eq. (23-17). In symbols,

$$I_Q \ll \frac{V_{\text{reg}}}{R_1} \qquad (23\text{-}19)$$

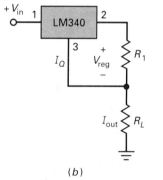

(b)

FIGURE 23-12

LM340 circuits: (a) adjustable output voltage; (b) load current is regulated.

For the LM340 series, the data sheet gives a worst-case I_Q of 8 mA. If V_{reg}/R_1 is greater than 160 mA, Eq. (23-18) is accurate to within 5 percent. If Eq. (23-19) is not satisfied, you will need to use the exact equation:

$$V_{\text{out}} = \frac{V_{\text{reg}}}{R_1}(R_1 + R_2) + I_Q R_2 \qquad (23\text{-}20)$$

The second term in this equation is the additional voltage drop produced by the quiescent current out of the IC regulator.

Figure 23-12b is another application, this time a current source (or current regulator). In this circuit, a load resistance R_L is used in place of R_2. As before, the current through R_1 is given by

$$I_{\text{out}} = \frac{V_{\text{reg}}}{R_1} \qquad (23\text{-}21)$$

This current flows through the load resistor. Since the value of load resistance does not appear in Eq. (23-21), the circuit appears like a current source to the load. As before, a designer can swamp out I_Q by using Eq.

(23-19). If the highest accuracy is required, you should add the quiescent current to get the exact value:

$$I_{\text{out}} = \frac{V_{\text{reg}}}{R_1} + I_Q \qquad (23\text{-}22)$$

The LM320 Series

The LM320 series is a group of negative voltage regulators with preset voltages of -5, -12, and -15 V. For instance, an LM320-5 produces a regulated output voltage of -5 V. At the other extreme, an LM320-15 produces an output of -15 V. With the LM320 series, the load-current capability is approximately 1.5 A with adequate heat sinking. The LM320 series is similar to the LM340 series and includes current limiting, thermal shutdown, and excellent ripple rejection.

By combining an LM320 and an LM340, we can regulate the output of a split supply (see Fig. 23-13). The LM340 regulates the positive output, and the LM320 handles the negative output. The input capacitors prevent oscillations, and the output capacitors improve transient response. The manufacturer's data sheet recommends the addition of two diodes to ensure that both regulators can turn on under all operating conditions.

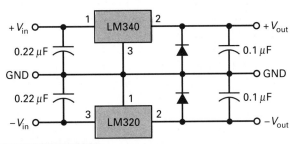

FIGURE 23-13 Using LM340 and LM320 to produce split-supply voltages.

Adjustable Regulators

A number of IC regulators, like the LM317, LM338, and LM350, are adjustable. These have maximum load currents from 1.5 to 5 A. For instance, the LM317 is a three-terminal positive voltage regulator that can supply 1.5 A of load current over an adjustable output range of 1.25 to 37 V. The load regulation is 0.1 percent. The line regulation is 0.01 percent per volt. This means that the output voltage changes only 0.01 percent for each volt of input change. The ripple rejection is 80 dB, equivalent to 10,000.

Figure 23-14 shows an unregulated supply driving an LM317 circuit. The data sheet of an LM317 gives this formula for output voltage:

$$V_{\text{out}} = 1.25 \left(\frac{R_2}{R_1} + 1 \right) \qquad (23\text{-}23)$$

which is valid from 1.25 to 37 V. Typically, the filter capacitor is selected to get a peak-to-peak ripple of about 10 percent. Since the regulator has

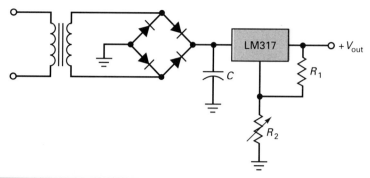

Bridge rectifier and capacitor-input filter produce unregulated input for LM317.

about 80 dB of ripple rejection, the final peak-to-peak ripple is smaller by four factors of 10, or

$$\frac{10\%}{10^4} = 0.001\%$$

In other words, a voltage regulator also filters the input ripple. This eliminates the need for RC and LC filters in most power supplies.

Dual-Tracking Regulators

When a split supply is needed, dual-tracking regulators like the RC4194 and RC4195 are convenient. These voltage regulators produce equal positive and negative output voltages. The RC4194 is adjustable from ± 0.05 to ± 32 V, while the RC 4195 produces two fixed output voltages of ± 15 V. For instance, Fig. 23-15 shows an RC4195. It needs two un-regulated input voltages. The positive input may be from $+18$ to $+30$ V, and the negative input from -18 to -30 V. As indicated, the two outputs are ± 15 V. The data sheet of an RC 4195 lists a maximum output current of 150 mA for each supply, a load regulation of 3 mV, a line regulation of 2 mV, and a ripple rejection of 75 dB.

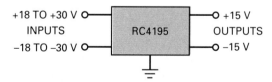

Dual-tracking voltage regulator produces split-supply voltages.

Regulator Table

Table 23-1 lists data for some popular IC regulators. For instance, the LM309 is a fixed positive regulator with an output of $+5$ V, a maximum load current of 1 A, a load regulation of 15 mV, a source regulation of

Table 23-1 IC Voltage Regulators

Number	V_{out}, V	I_{max}, A	LR, mV	SR, mV	RR, dB	Drop-out, V	Comment
LM309	+5	1	15	4	75	2	Fixed positive
LM317	—	1.5	0.1%	0.2%	65	2.5	Adjustable: 1.2 to 32 V
LM320-5	−5	1.5	50	10	65	2	Fixed negative
LM320-15	−15	1.5	30	5	80	2	Fixed negative
LM338	—	5	0.1%	0.1%	60	2.7	Adjustable: 1.2 to 32 V
LM340-5	+5	1.5	10	3	80	2.3	Fixed positive
LM340-15	+15	1.5	12	4	80	2.5	Fixed positive
LM350	—	3	0.1%	0.1%	65	2.5	Adjustable: 1.2 to 33 V
RC4194	—	0.15	0.2%	0.2%	75	3	Dual-tracking: 0 to 32 V
RC4195	±15	0.15	3	2	70	3	Dual-tracking

4 mV, and a ripple rejection of 75 dB. For the adjustable regulators, LR and SR are given in percentage rather than millivolts.

The table also includes the drop-out voltage, or the minimum allowable difference between the input voltage and the output voltage. For example, an LM340-5 has a drop-out voltage of 2.3 V. This means that the input voltage must be at least 2.3 V greater than the output voltage. Since the output is 5 V, the input must be at least 7.3 V.

EXAMPLE 23-10

What is the load curent in Fig. 23-16? The output ripple?

SOLUTION

The first item is easy. The LM340-12 produces a regulated output voltage of +12 V. Therefore, the load current is

$$I = \frac{12 \text{ V}}{100 \text{ }\Omega} = 120 \text{ mA}$$

The next item, the output ripple, will take a little longer to calculate. Notice that the unregulated +18 V is the input to the LM340-12. The filter capacitor charges and discharges as discussed in Chap. 4. The ripple across this filter capacitor is given by

$$V_R = \frac{I}{fC}$$

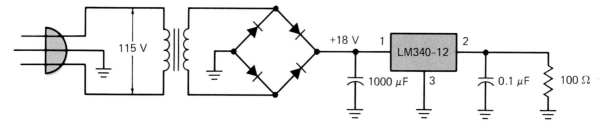

FIGURE 23-16 Example.

The value of current in this equation is approximately the same as the load current, 120 mA. The frequency is 120 Hz, and the capacitance is 1000 µF. The peak-to-peak ripple is

$$V_R = \frac{120 \text{ mA}}{(120 \text{ Hz})(1000 \text{ µF})} = 1 \text{ V}$$

This is the input ripple to the LM340-12. If you look at its data sheet in the Appendix, you find a worst-case ripple rejection of 55 dB. As an approximation, round this to 60 dB, which represents three factors of 10. Then calculate the output ripple as

$$V_R = \frac{1 \text{ V}}{1000} = 1 \text{ mV}$$

This is how a troubleshooter would estimate the output ripple.

Here is how to get a more accurate answer. Set up this decibel equation:

$$-55 \text{ dB} = 20 \log \frac{V_{R(\text{out})}}{V_{R(\text{in})}}$$

or

$$\frac{V_{R(\text{out})}}{V_{R(\text{in})}} = \text{antilog} \frac{-55 \text{ dB}}{20}$$

If you have a calculator, enter -55 and divide by 20 to get -2.75. Then take the inverse log of -2.75 to get 0.00178. This means

$$\frac{V_{R(\text{out})}}{V_{R(\text{in})}} = 0.00178$$

or

$$V_{R(\text{out})} = 0.00178(1 \text{ V}) = 1.78 \text{ mV}$$

EXAMPLE 23-11

What are the minimum and maximum output voltages in Fig. 23-17?

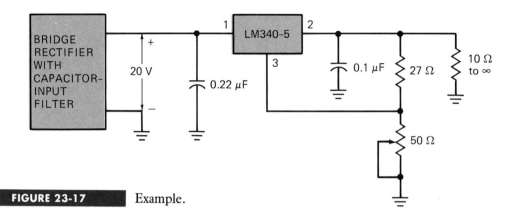

FIGURE 23-17   Example.

SOLUTION

You don't need formulas when you have Ohm's law and know what you are doing. In Fig. 23-17, the output of the LM340-5 appears across the 27 Ω, producing a current of

$$I = \frac{5\,V}{27\,\Omega} = 185\,mA$$

When the wiper is all the way up, the 27 Ω is grounded, which means that

$$V_{out} = 5\,V$$

When the wiper is all the way down, 185 mA flows through the 50 Ω as well as the 27 Ω. In this case, the output voltage is

$$V_{out} = (185\,mA)(27\,\Omega + 50\,\Omega) = 14.2\,V$$

So, we have a circuit that produces a regulated output voltage that is adjustable from 5 to 14.2 V. If you want to include the effect of quiescent current, here is how to proceed: If you look at the data sheet of the LM340 series in the Appendix, you will see that the worst-case quiescent current is 8 mA. With the wiper all the way up, the quiescent current has no effect. With the wiper all the way down, the quiescent current flows through the 50 Ω and produces a voltage of

$$V = (8\,mA)(50\,\Omega) = 0.4\,V$$

This has to be added to the maximum voltage calculated earlier. This means the output is adjustable from 5 to 14.6 V.

23-5 CURRENT BOOSTERS

Three-terminal regulators have a maximum load current that they can pass before thermal shutdown occurs. The value of this load current depends on whether a heat sink is used. As shown on the data sheet of the LM340 series (in the Appendix), an LM340 device has a maximum power dissipation of

$$P_{D(\text{max})} = 3 \text{ W}$$

when no heat sink is used and the ambient temperature is 25°C. With an infinite heat sink, $P_{D(\text{max})}$ increases to approximately 20 W. The power dissipation of the regulator is approximately equal to

$$P_D = (V_{\text{in}} - V_{\text{reg}})I_L \qquad (23\text{-}24)$$

If the load current is too large, P_D becomes greater than $P_{D(\text{max})}$ and increases the internal temperature of the device past the safe level. In this case, the device does not burn out, it shuts down. Recall that the LM340 series has a special feature called *thermal shutdown*, which means the pass transistor will be nonconducting until the device cools down. In the following discussion, let us assume that 1 A is our design limit for the maximum load current through the LM340-XX, where XX represents 5, 12, or 15 V.

The Outboard Transistor

What do we do when we want a load current greater than 1 A? One solution is to use an *outboard transistor* as shown in Fig. 23-18. In this circuit, 0.6 Ω is used as a current-sensing resistor. When the current is less than 1 A, the voltage across the 0.6 Ω is less than 0.6 V and the transistor is off. In this case, the voltage regulation works as before. The LM340-XX holds the output voltage constant. Furthermore, since the transistor is off, all the load current has to pass through the LM340.

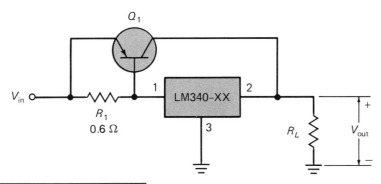

FIGURE 23-18 Outboard transistor increases load current capability.

When the load current is greater than 1 A, the voltage across the 0.6 Ω is greater than 0.6 V and the transistor turns on. This outboard transistor will supply the extra load current above 1 A. The current through the LM340 is approximately equal to the voltage across the current-sensing resistor R_1 divided by its resistance. For this example,

$$I = \frac{V_{BE}}{0.6 \ \Omega}$$

When the V_{BE} of the pass transistor is greater than 0.6 V, the pass transistor produces current on demand, that is, as needed by the load. This is a beautiful circuit because the transistor current automatically adjusts to the value of the excess load current. In other words, the current through the LM340 is only slightly more than 1 A. The outboard transistor handles the rest of the current. If the load current is 10 A, the outboard transistor supplies about 9 A of the load current.

Current Limiting

One improvement is to add current limiting as shown in Fig. 23-19. Here Q_1 and R_1 work as previously described, and Q_2 and R_2 produce current limiting when the load current becomes too large. Here is how it works: When the outboard current is 10 A, the voltage across R_2 is

$$V = (10 \text{ A})(0.06 \text{ } \Omega) = 0.6 \text{ V}$$

This means the current-limiting transistor Q_2 is on the verge of turning on. For any outboard current less than 10 A, the circuit works as previously described because the voltage across R_1 is the base-emitter drive for Q_1.

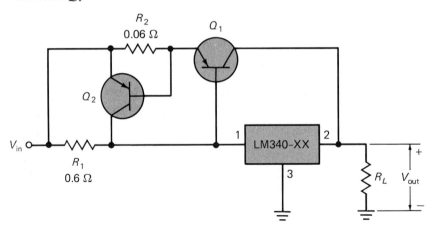

FIGURE 23-19 Outboard transistor with current limiting.

Current limiting starts when the outboard current is greater than 10 A because the voltage across R_2 is greater than 0.6 V, which turns on Q_2. Its collector current passes through the LM340 and produces thermal shutdown.

The V_{BE} drive for Q_1 is given by the *difference* in voltages across the current-sensing resistors:

$$V_{BE} = I_1 R_1 - I_2 R_2 \tag{23-25}$$

If there is no outboard current I_2, then V_{BE} is simply $I_1 R_1$. But when the outboard current is greater than zero, V_{BE} equals the difference of the two current-sensing voltages.

Equation (23-25) tells us something important. When a short is accidentally placed across the load resistance, I_2 tries to increase to infinity, which forces I_1 toward infinity. Since I_1 flows through R_1, the input voltage to pin 1 decreases toward zero. But the voltage regulator needs at least 2 to 3 V more than the output voltage to work properly. Therefore, a short across the load resistance forces the regulator to stop regulating, and the load voltage drops to zero.

EXAMPLE 23-12

What is the power dissipation of the LM340-15 if the load current is 25 mA in Fig. 23-20?

SOLUTION

When the load current is 25 mA, the voltage across R_1 is

$$V_1 = (25 \text{ mA})(20 \text{ } \Omega) = 0.5 \text{ V}$$

This means there is no outboard current. It also means that the current through the LM340-15 equals the load current.

The pass transistor inside the LM340 is where the bulk of the power dissipation appears. For this reason, we can approximate the total power dissipation of the device with Eq. (23-24), which says the power dissipation equals the difference between the input and output voltages of the device times the load current. Ignoring the 0.5 V across R_1, we get an approximate answer of

$$P_D = (23 \text{ V} - 15 \text{ V})(25 \text{ mA}) = 200 \text{ mW}$$

This is much less than the 3-W maximum without a heat sink at 25°C. Because of this, the LM340-15 shown in Fig. 23-20 can operate without a heat sink.

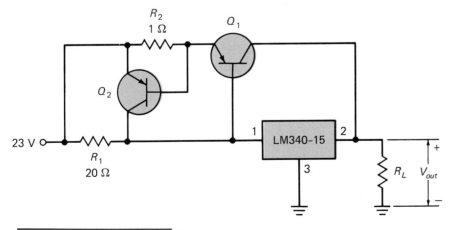

FIGURE 23-20 Example.

EXAMPLE 23-13

In Fig. 23-20, estimate the load current where the outboard current appears. Estimate the load current where current limiting begins.

SOLUTION

Now Q_1 starts to turn on at approximately 0.6 V. This means the current R_1 is

$$I_1 = \frac{0.6\,V}{20\,\Omega} = 30\,mA$$

This is the approximate load current where the outboard current appears.

Current limiting starts when the V_{BE} of Q_2 is approximately 0.6 V. This voltage appears when the current through R_2 is

$$I_2 = \frac{0.6\,V}{1\,\Omega} = 600\,mA$$

The load current is the sum of the currents through R_1 and R_2. Therefore, the load current where current limiting starts is approximately

$$I_{max} = 600\,mA + 30\,mA = 630\,mA$$

You may be wondering why an outboard transistor is being used in this circuit. Since an LM340-15 can easily handle more than 1 A with adequate heat sinking, why are we using an outboard transistor? Why not just use an LM340 with a heat sink? Here is the answer: The heat sink for a three-terminal regulator is usually larger and more expensive than the heat sink for an outboard transistor. With a circuit like Fig. 23-20, a designer can run the LM340 without a heat sink. The outboard transistor still needs a heat sink, but it is simpler and less expensive.

23-6 DC-TO-DC CONVERTER

Sometimes we want to convert a dc voltage to another dc voltage of a different value. For instance, if we have a system with a positive supply of +5 V, we can use a dc-to-dc converter to produce an output of +15 V. Then we would have two supply voltages for our system: +5 and +15 V. All kinds of designs are possible for dc-to-dc converters. In this section, we discuss a hypothetical design to get an idea of how a dc-to-dc converter works.

Basic Idea

In most dc-to-dc converters, the input dc voltage is applied to a square-wave oscillator whose output drives a transformer, as shown in Fig.

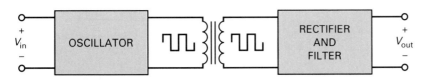

FIGURE 23-21 Output of relaxation oscillator is transformed to a different peak value before rectifying and filtering.

23-21. The frequency is usually between 1 and 100 kHz. The higher the frequency, the smaller the transformer and filter components. On the other hand, if the frequency is too high, it is difficult to produce a square wave with vertical sides. As a rule, 20 kHz works out as the best compromise, and you will see this frequency used a lot in practical circuits.

By selecting different turns ratios, we can get a smaller or larger secondary voltage. To improve the efficiency, a special kind of transformer is used. It has a toroidal core with a rectangular hysteresis loop. This produces a secondary voltage that is a square wave. The secondary voltage can then be rectified and filtered to get a dc output voltage. It is relatively easy to filter this voltage because it is a rectified square wave at a high frequency. If you visualize this correctly, you can see that it is almost a dc voltage.

One of the most common dc-to-dc conversions is $+5$ to $+15$ V. In digital systems, $+5$ V is a standard supply voltage for most ICs. But a few ICs, for example op amps, may require $+15$ V. In a case like this, you commonly find a low-power dc-to-dc converter producing $+15$ and -15 V. These voltages are used for the few ICs that require the higher voltages.

One Possible Design

There are many ways to design a dc-to-dc converter, depending on whether the voltage is stepped up or down, the maximum load current, and other factors. Figure 23-22 shows a dc-to-dc converter. This design uses only circuits that have already been discussed, so that you can follow the action.

Here is how the circuit works: A relaxation oscillator produces a square wave, whose frequency is set by R_3 and C_2. Typically, this frequency is in kilohertz. The square wave drives a *phase splitter* Q_1, whose outputs are equal and opposite square waves. These square waves are the input to the class B switching transistors Q_2 and Q_3. Transistor Q_2 conducts during one half-cycle and Q_3 during the other half-cycle. The primary current is therefore a square wave, which induces a square wave of voltage across the secondary winding. The square wave of voltage out of the secondary winding drives a bridge rectifier and a capacitor-input filter. Because the signal is a rectified square wave in kilohertz, it is easy to filter and get an unregulated dc voltage for the input to the three-terminal regulator. The final output is then a dc voltage at some level different from the input.

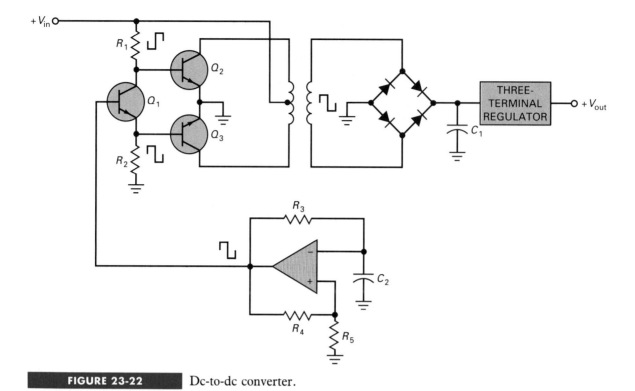

FIGURE 23-22 Dc-to-dc converter.

23-7 SWITCHING REGULATORS

Series regulators are very popular and fill many of our needs. Their big disadvantage is the power dissipation of the pass transistor. As the load current increases, the pass transistor has to dissipate more power, which implies larger heat sinks. Because of this, series regulators tend to get bulky at low voltages and high currents. In some cases, a fan may be needed to remove the heat generated by the pass transistor. Series regulators are sometimes called *linear regulators* because the pass transistor operates in the active region.

One way to reduce the power dissipation of the pass transistor is by using a *switching regulator*. This type of regulator does not allow the pass transistor to operate in the active or linear region. Instead, it alternately saturates and cuts off the pass transistor. This results in much less power dissipation in the pass transistor. Switching regulators can provide large load currents at low voltages, precisely what is needed in personal computers (PCs). This is why the switching regulator is commonly used in a PC.

Three Basic Configurations

Switching regulators come in three basic configurations or shapes. First, there is the *stepdown* version shown in Fig. 23-23a. The rectangular pulses on the base alternately saturate and cut off the pass transistor during each cycle. This produces a rectangular voltage at the input to LC filter. This filter blocks the ac component but passes the dc component to the output. Because of the on-off switching, the average value is always less than the input voltage. This is why the circuit is the stepdown variety.

Figure 23-23*b* shows the *step-up* version of the switching regulator. Again, the transistor is alternately saturated and cut off. When the transistor is saturated, current flows through the inductor. (Conventional flow is to the right, and electron flow to the left.) When the transistor suddenly cuts off, the magnetic field around the coil collapses and induces a large voltage across the coil of opposite polarity. This keeps the current flowing in the same direction. Furthermore, the inductive kickback voltage is larger than the input voltage, which is why the circuit is the step-up configuration.

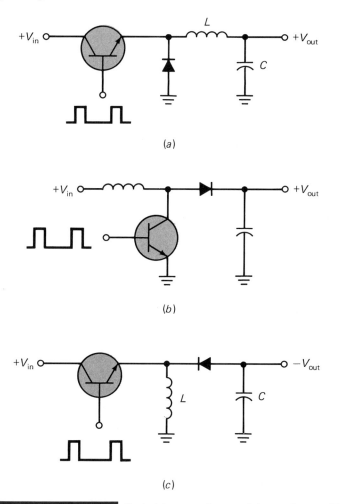

(a)

(b)

(c)

FIGURE 23-23 Switching regulators: (*a*) stepdown; (*b*) step-up; (*c*) inverting.

Figure 23-23*c* shows the *inverting regulator*. When the transistor is saturated, current flows through the inductor (conventional down, electron up). When the transistor cuts off, the magnetic field collapses, and the inductive kickback keeps current flowing in the same direction. Since the transistor is off, the only path is through the capacitor. If you check the direction of charging current through the capacitor, you can see that the output voltage is negative.

More About the Stepdown Version

The stepdown switching regulator is extremely important because it is widely used in computer power supplies. This is the one we want to examine in more detail. In Fig. 23-24, a string of pulses drives the base of the pass transistor. When the base voltage is high, the transistor is saturated. When the base voltage is low, the transistor is cut off. The main idea is that the transistor acts as a switch. Ideally, a switch dissipates no power when it is closed or open. In reality, a transistor switch is not perfect, and so it does dissipate some power, but this power is much less than that dissipated by a series or linear regulator.

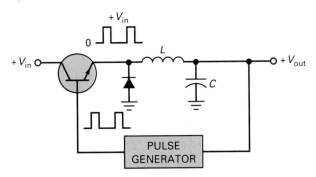

FIGURE 23-24 Transistor switch controls the duty cycle and final dc output voltage.

A diode is connected from the emitter to ground. This is necessary because of inductive kickback. An inductor will try to keep the current through it constant. When the transistor cuts off, the diode continues to provide a path for current through the inductor. Without the diode, the inductive kickback would produce enough reverse voltage to destroy the transistor.

The duty cycle D is the ratio of the on time W to the period T. By controlling the duty cycle out of the pulse generator, we control the duty cycle of the input voltage to the LC filter. Ideally, this input voltage swings from 0 to V_{in}, as shown. Although it is almost obsolete in ordinary power supplies, the LC filter is very popular in switching regulators because the switching frequency is typically around 20 kHz. This means that a smaller inductor and capacitor can be used. The output of the LC filter is a dc voltage with only a small ripple. This output voltage is directly proportional to the duty cycle and is given by

$$V_{out} = DV_{in} \tag{23-26}$$

Since D can vary from 0 to 1, V_{out} can vary from 0 to V_{in}.

The output voltage is fed back to the pulse generator. In most switching regulators, the duty cycle is *inversely proportional* to the output voltage.

If the output voltage tries to increase, the duty cycle will decrease. This means that narrower pulses will drive the LC filter, and its output will decrease. In effect, we have negative feedback. Since the output voltage is being sampled and fed back, the output voltage is the quantity being stabilized. Therefore, if the loop gain is high, we can produce a very stable output voltage.

One Possible Design

To give you a concrete idea of how a switching regulator works, Fig. 23-25 shows a low-power design, using circuits that you are already familiar with. The relaxation oscillator produces a square wave whose frequency is set by R_5 and C_3. The square wave is integrated to get a triangular wave, which drives the noninverting input of a triangle-to-pulse converter. The pulses out of this circuit then drive the pass transistor, as previously described. The duty cycle of these pulses will determine the output voltage. Notice that the output of the LC filter is sampled by a voltage divider, which returns a feedback voltage to the comparator. This feedback voltage is compared with a reference voltage V_{ref} from a zener diode or other source. The output of the comparator then drives the inverting input of the triangular-to-pulse generator.

Here is how the regulation works: If the regulated output voltage tries to increase, the comparator produces a higher output voltage, which increases the inverting input voltage to the triangular-to-pulse converter. This results in narrower pulses at the base of the pass transistor. Since the duty cycle is lower, the filtered output voltage is less, which tends to cancel almost all the original increase in output voltage. In other words,

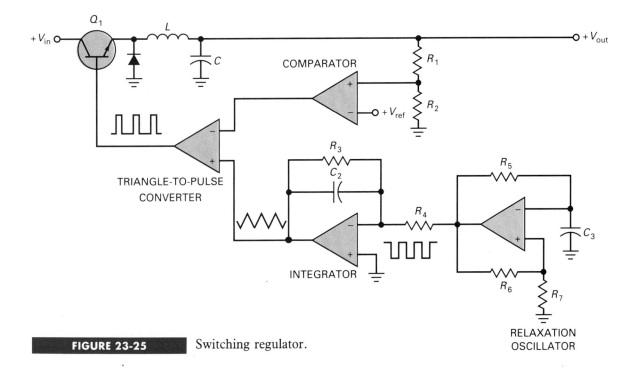

FIGURE 23-25   Switching regulator.

the attempted increase in output voltage produces a negative-feedback voltage that almost eliminates the original increase.

If the regulated output voltage tries to decrease, the output of the comparator decreases the inverting input voltage to the triangle-to-pulse converter. This results in a larger duty cycle, so that more voltage comes out of the LC filter. The final effect is a much smaller increase in output voltage than would occur without the negative feedback.

There is enough open-loop gain in the system to ensure a well-regulated output voltage. Since the error voltage to the comparator is near zero, the voltage across R_2 is approximately equal to V_{ref}. Therefore, the current through R_2 is

$$I = \frac{V_{ref}}{R_2}$$

This current flows through R_1, which means that the output voltage is given by

$$V_{out} = \frac{V_{ref}}{R_2}(R_1 + R_2) \qquad (23\text{-}27)$$

EXAMPLE 23-14

In Fig. 23-25, $V_{ref} = 1.25$ V, $R_1 = 3$ kΩ, and $R_2 = 1$ kΩ. What is the regulated output voltage?

SOLUTION

Because the noninverting and inverting input voltages to a comparator are approximately equal, the voltage across R_2 equals V_{ref}, or 1.25 V. This means the current through R_2 is

$$I = \frac{1.25\text{ V}}{1\text{ k}\Omega} = 1.25\text{ mA}$$

This current also flows through R_1. Therefore, the output voltage is

$$V_{out} = (1.25\text{ mA})(3\text{ k}\Omega + 1\text{ k}\Omega) = 5\text{ V}$$

EXAMPLE 23-15

If the unregulated input voltage is $+25$ V, what is the duty cycle of the pulses at the base of Q_1 in Example 23-14?

SOLUTION

Rearrange Eq. (23-26) to get

$$D = \frac{V_{out}}{V_{in}}$$

Now, substitute and solve as follows:

$$D = \frac{5 \text{ V}}{25 \text{ V}} = 0.2$$

If you multiply this by 100 percent, you get a duty cycle of 20 percent.

OPTIONAL TOPICS

The following material continues the earlier discussions at a more advanced and specialized level. All the topics are optional because they are not used in any of the basic discussions in earlier chapters. This section will be a useful reference when you are in industry because then you will probably want more advanced viewpoints.

23-8 FOLDBACK LIMITING

One way to reduce the high power dissipation of a series regulator under shorted-load conditions is with the circuit shown in Fig. 23-26a. The load current I_{out} flows through R_4, producing a voltage drop of approximately $I_{\text{out}}R_4$. This means that a voltage of $I_{\text{out}}R_4 + V_{\text{out}}$ is fed to a voltage divider (R_5 and R_6) whose output controls Q_3. The feedback fraction of the voltage divider is approximately

$$K = \frac{R_6}{R_5 + R_6} \tag{23-28}$$

The mathematical analysis of this circuit is complicated. The following formulas can be derived. When the load terminals are shorted, the output current is

$$I_{SL} = \frac{V_{BE}}{KR_4} \tag{23-29}$$

When the load terminals are not shorted, the maximum output current is

$$I_{\text{max}} = I_{SL} + \frac{(1 - K)V_{\text{out}}}{KR_4} \tag{23-30}$$

This equation says that the maximum load current is higher than the shorted-load current. Typically, K is selected to produce a maximum load current of 2 to 3 times the shorted-load current. The main advantage of foldback current limiting is the reduced power dissipation in the pass transistor when the load terminals are accidentally shorted.

Figure 23-26b shows how the output current varies with load resistance. When R_L is large, I_{out} is small. When R_L decreases, I_{out} increases until it reaches a maximum value of I_{max}. The circuit still has a regulated output voltage at this maximum load current. Beyond this point, foldback

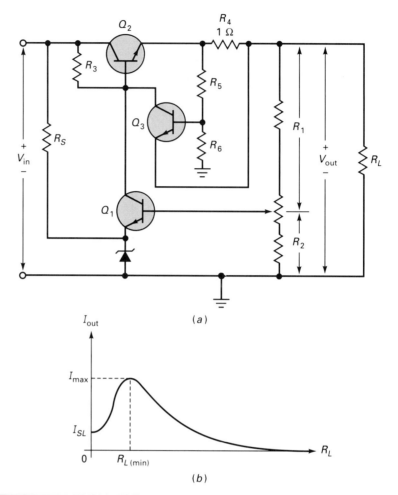

FIGURE 23-26 (a) Voltage regulator with foldback current limiting; (b) foldback means that shorted-load current is less than maximum load current.

current limiting takes over. Any further decrease in R_L forces I_{out} to decrease. When R_L is zero, I_{out} equals I_{SL}.

EXAMPLE 23-16

The output voltage of Fig. 23-27 is adjusted to 10 V. If V_{BE} is 0.7 V for the current-limiting transistor, what are the shorted-load current and the maximum load current?

SOLUTION

The feedback fraction of the current-limiting voltage divider is

$$K = \frac{180}{20 + 180} = 0.9$$

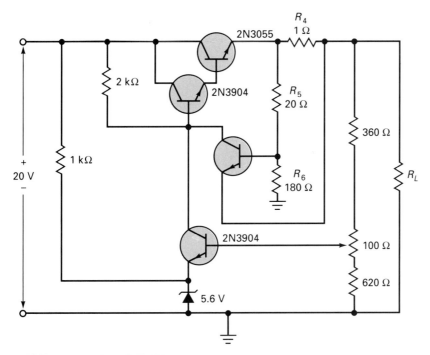

FIGURE 23-27 Voltage regulator uses Darlington
transistor and foldback current limiting.

From Eq. (23-29),

$$I_{SL} = \frac{0.7\,\text{V}}{0.9(1\,\Omega)} = 0.778\,\text{A}$$

From Eq. (23-30),

$$I_{max} = 0.778\,\text{A} + \frac{(1 - 0.9)(10\,\text{V})}{0.9\,(1\,\Omega)} = 1.89\,\text{A}$$

The regulator can supply a maximum load current of 1.89 A.
The minimum load resistance is 10 V/1.89 A, or 5.29 Ω. If the load
resistance is less than this, the load current is less than 1.89 A.
When the load terminals are shorted, the load current drops to
0.778 A.

EXAMPLE 23-17

For Example 23-16, calculate the power dissipation of the pass
transistor when the load terminals are shorted.

SOLUTION

With a shorted load, the load voltage is zero, and the emitter voltage
of the 2N3055 is

$$V_E = (0.778\,\text{A})(1\,\Omega) = 0.778\,\text{V}$$

Since the input voltage is 20 V, the voltage across the 2N3055 is

$$V_{CE} = 20 \text{ V} - 0.778 \text{ V} = 19.2 \text{ V}$$

and the power dissipation is

$$P_D = (19.2 \text{ V})(0.778 \text{ A}) = 14.9 \text{ W}$$

This power dissipation is much lower than it would be with simple current limiting.

23-9 REGULATED OUTPUT WITH FOLDBACK LIMITING

Figure 23-28 shows an improved version of Fig. 23-20. The earlier circuit used an LM340 with two transistors and two current-sensing resistors to produce simple current limiting. By adding another transistor, we can produce foldback current limiting. As before, R_1 is the current-sensing resistor that turns on Q_1, the outboard transistor. The voltage across R_2 is used to turn on Q_2. But the situation is more complicated now because of the voltage divider driving the base of Q_2. At some point, Q_2 turns on and drives the voltage divider for Q_3. The overall action of this circuit is too complicated to analyze here. The main advantage of this circuit is that the shorted-load current is less than the maximum load current. The decreases the power dissipation in the outboard transistor under shorted-load conditions.

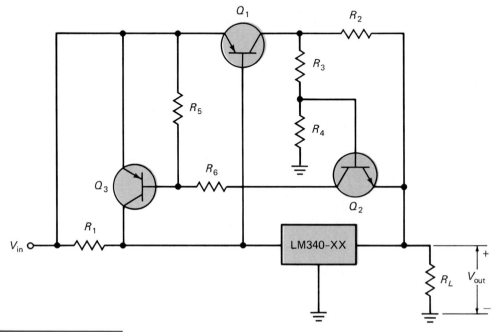

FIGURE 23-28 Regulator with foldback current limiting.

The following study aids will help to reinforce the ideas discussed in this chapter. For best results, use these study aids within 6 hours of reading the earlier material. Then review these study aids a week later and a month later to ensure that the concepts remain in your long-term memory.

SUMMARY

Sec. 23-1 Voltage-Feedback Regulation
The temperature coefficient of zener diodes is approximately zero between 5 and 6 V. This highly stable zener voltage is used as the reference voltage for regulator circuits with negative feedback. The output voltage is sampled and fed back to an amplifier. This produces a change in output voltage that opposes the original change. The overall effect is an output voltage that changes much less than it would without the negative feedback.

Sec. 23-2 Current Limiting
If there is any possibility of a short being accidentally placed across the load resistor, the designer will include current limiting in a voltage regulator. Load current flows through a current-sensing resistor. This produces a voltage that is directly proportional to the load current. This voltage can then be used to turn on a transistor and to decrease the load voltage. As the load resistance approaches zero, the load voltage approaches zero at the same rate. Therefore, the load current remains approximately constant and limited to the value set by the current-limiting resistance.

Sec. 23-3 Power Supply Characteristics
Load regulation is defined as the change in load voltage when the load current changes from its minimum to maximum specified value. Source regulation is the change in load voltage when the source voltage changes from its minimum to maximum specified value. Both load and source regulation may be expressed as percentages. Voltage regulators are equivalent to stiff voltage sources with extremely small output impedances. Input ripple is equivalent to a change in source voltage. The voltage regulator reduces this ripple, the same as a change in source voltage.

Sec. 23-4 Three-Terminal IC Regulators
The latest generation of IC voltage regulators has only three pins: one for the unregulated input voltage, one for the regulated output voltage, and one for ground. In some applications, the third pin is not grounded, but instead is tied to a voltage divider to change the value of the regulated output voltage. The LM340-XX series is typical of three-terminal voltage regulators. It is available with preset voltages of 5, 12, and 15 V and a tolerance of ± 4 percent. The LM340A series has a tolerance of ± 2 percent. Negative IC regulators and adjustable IC regulators are also available.

Sec. 23-5 Current Boosters
To extend the current range of a three-terminal regulator, we can add an outboard pass transistor in parallel with the voltage regulator. The maximum current through the voltage regulator is set up with a current-sensing resistor. The voltage across this current-sensing resistor is used for the V_{BE} of the outboard transistor. When the maximum regulator current is exceeded, the extra current passes through the outboard transistor to the load. By adding a second transistor and current-sensing resistor, we can get simple current limiting. More advanced designs include foldback current limiting.

Sec. 23-6 DC-to-DC Converter
Sometimes we need a power supply that can convert dc voltage of one value to dc voltage of another value. All kinds of designs are possible. The usual starting idea is to create a rectangular wave from the dc input voltage. This voltage is then stepped up or down with a special kind of transformer. The output of this transformer is also rectangular, but its peak-to-peak value is different from the input voltage. The secondary voltage can then be rectified and filtered to get the new value of dc voltage.

Sec. 23-7 Switching Regulators
There are three basic configurations for the output transistor and filter: stepdown, step-up, and inverting. The stepdown is the most widely used because it is the best choice when you need a low voltage and a high current. Personal computers use $+5$ V at high current, so the stepdown switching

regulator is widely used with PCs. One way to build a stepdown switching regulator is to use a pass transistor followed by an *LC* filter. The pass transistor is turned on and off by a series of rectangular pulses. The output voltage is sampled and fed back to circuits that control the duty cycle of these pulses. The negative feedback produces a regulated output voltage.

VOCABULARY

In your own words, explain what each of the following terms means. Keep your answers short and to the point. If necessary, verify your answer by rereading the appropriate discussion or by looking at the end-of-book Glossary.

current limiting	reference voltage
current-sensing resistor	ripple rejection
dc-to-dc converter	series regulator
foldback current limiting	short-circuit protection
linear regulator	source regulation
load regulation	switching regulator
outboard transistor	thermal shutdown
phase splitter	voltage regulator

IMPORTANT EQUATIONS

Here are some important equations. Say each of the following equations in symbols, then say each in words. Try to explain what the equation means and how it is used. Then read the description that follows.

Eq. 23-2 Output Voltage of Discrete Series Regulator

$$V_{\text{out}} = A_{CL}(V_Z + V_{BE})$$

This says the regulated output voltage equals the closed-loop voltage gain times the sum of the zener voltage and the diode drop. And A_{CL} is greater than 1, which means the regulated output voltage is as stable as the zener voltage and the diode drop.

Eq. 23-7 Load Regulation

$$LR = V_{NL} - V_{FL}$$

This is a defining formula. It says the load regulation equals the difference between the no-load and full-load voltages.

Eq. 23-8 Percentage of Load Regulation

$$\%LR = \frac{V_{NL} - V_{FL}}{V_{FL}} \times 100\%$$

This is the historical definition of load regulation, expressed as a percentage. You can simplify the calculation by using V_{NL} in the denominator. This introduces some error, but it is negligible when %LR is small.

Eq. 23-10 Source Regulation

$$SR = V_{HL} - V_{LL}$$

This is another defining formula. It says the source regulation equals the change in load voltage for the highest and lowest line voltages specified for the regulator.

Eq. 23-11 Percentage of Source Regulation

$$\%SR = \frac{SR}{V_{\text{nom}}} \times 100\%$$

This says the percentage of source regulation equals the source regulation divided by the nominal line voltage times 100 percent.

Eq. 23-12 Output Impedance

$$z_{\text{out(CL)}} = \frac{z_{\text{out}}}{1 + AB}$$

This shows what happens to the output impedance of a voltage regulator. Because of the negative feedback, the closed-loop output impedance approaches zero. This is why a voltage regulator acts as a stiff voltage source with an output or Thevenin impedance that is near zero.

Eq. 23-14 Ripple Rejection

$$RR = \frac{V_{R(\text{out})}}{V_{R(\text{in})}}$$

Besides regulating the load voltage, a voltage regulator cleans up the ripple. This eliminates the need for heavy filtering in the unregulated power supply. Usually, a bridge rectifier and a capacitor-input filter are all that is needed to produce the unregulated input voltage for a voltage regulator. It then produces a regulated output with a much smaller ripple.

Eq. 23-24 Power Dissipation

$$P_D = (V_{in} - V_{reg})I_L$$

This is not exact, but it's close enough. It says the power dissipation of a three-terminal regulator equals the difference of the input and regulated voltages times the load current. It is based on the idea that most of the power dissipation takes place in the internal pass transistor.

Eq. 23-26 Power Dissipation

$$V_{out} = DV_{in}$$

This is the key to how a stepdown switching regulator controls the output voltage. By changing the duty cycle D, we can control the value of output voltage.

IMPORTANT PROCESSES

Each of the following processes is important for analyzing op-amp circuits with negative feedback. If you can remember the basic ideas in each step, you will be able to solve problems more easily.

Output Voltage of a Regulator

1. Locate the voltage divider that samples the output voltage.
2. Find the current through the feedback resistor.
3. Multiply this current by the total resistance of the voltage divider.

Power Dissipation of the Pass Transistor

1. Visualize the unregulated input voltage at one end of the pass transistor.

2. Visualize the regulated output voltage at the other end of the pass transistor.
3. Multiply the difference in voltages by the load current.

Shorted-Load Current with Simple Current Limiting

1. Locate the current-sensing resistor.
2. Divide 0.7 V by the current-sensing resistance.

Estimate of Ripple Rejection

1. Round the ripple rejection to nearest multiple of 20 dB.
2. Count a factor of 10 for each 20 dB.
3. Divide the input ripple by factors of 10 to get the output ripple.
4. For a better answer, include a factor of 2 for each additional 6 dB.

Startup Level of Current Booster and Current Limiting

1. Locate the current-sensing resistor that turns on outboard transistor.
2. Divide 0.7 V by the current-sensing resistance.
3. Locate the current-sensing resistor that turns on current-limiting transistor.
4. Divide 0.7 V by the current-sensing resistance.

Output Voltage of Stepdown Switching Regulator

1. Locate the voltage divider that samples the output voltage.
2. Find the current through the feedback resistor.
3. Multiply this current by the total resistance of the voltage divider.

STUDENT ASSIGNMENTS

QUESTIONS

The following questions may have more than one right answer. Select the best answer. This is the one that is always true, or covers more situations, or fits the context, etc.

1. The temperature coefficient of a zener diode is approximately zero when the zener voltage is
   a. 0
   b. 6 V
   c. 15 V
   d. 45 V

2. The zener voltage in a voltage regulator is called the
   a. Current-sensing voltage
   b. Reference voltage
   c. Load voltage
   d. Foldback voltage

3. Voltage regulators use
   a. Negative feedback
   b. Positive feedback
   c. No feedback
   d. Current feedback

4. The closed-loop voltage gain of a voltage regulator is usually greater than
   a. 1
   b. The zener voltage
   c. The load voltage
   d. 10

5. The power dissipation of the pass transistor equals the collector-emitter voltage times the
   a. Base current
   b. Load current
   c. Zener current
   d. Foldback current

6. Without current limiting, a shorted load will probably
   a. Produce zero load current
   b. Destroy diodes and transistors
   c. Have a load voltage equal to the zener voltage
   d. Have too little load current

7. A current-sensing resistor is usually
   a. Zero
   b. Small
   c. Large
   d. Open

8. Simple current limiting produces too much heat in the
   a. Zener diode
   b. Load resistor
   c. Pass transistor
   d. Ambient air

9. With foldback current limiting, the load voltage approaches zero, and the load current approaches
   a. A small value
   b. Infinity
   c. The zener current
   d. 1 A

10. A capacitor may be needed in a discrete voltage regulator to prevent
    a. Negative feedback
    b. Excessive load current
    c. Oscillations
    d. Current sensing

11. If the output of a voltage regulator varies from 15 to 14.7 V between the minimum and maximum load current, the load regulation is
    a. 0.2 V   c. 0.7 V
    b. 0.3 V   d. 14.7 V

12. In Question 11, the percentage of voltage regulation is closest to
    a. 0   c. 2%
    b. 1%   d. 5%

13. If the output of a voltage regulator varies from 20 to 19.8 V when the line voltage varies over its specified range, the source regulation is
    a. 0.2 V   c. 0.8 V
    b. 0.3 V   d. 19.8 V

14. In Question 13, the percentage of source regulation is closest to
    a. 0   c. 2%
    b. 1%   d. 5%

15. The output impedance of a voltage regulator is
    a. Very small
    b. Very large
    c. Equal to the load voltage divided by the load current
    d. Equal to the input voltage divided by the output current

16. Compared to the ripple into a voltage regulator, the ripple out of a voltage regulator is
    a. Equal in value
    b. Much larger
    c. Much smaller
    d. Impossible to determine

17. A voltage regulator has a ripple rejection of -60 dB. If the input ripple is 1 V, the output ripple is
    a. -60 mV
    c. 10 mV
    b. 1 mV
    d. 1000 V

18. Thermal shutdown occurs in an IC regulator if
    a. Power dissipation is too high
    b. Internal temperature is too high
    c. Current through the device is too high
    d. All the above occur

19. If an IC regulator is more than a few inches from the filter capacitor, you may get oscillations inside the IC unless you use
    a. Current limiting
    b. A bypass capacitor on the input pin
    c. A coupling capacitor on the output pin
    d. A regulated input voltage

20. The LM340 series of voltage regulators produces an output voltage that is
    a. Positive
    b. Negative
    c. Either positive or negative
    d. Unregulated

21. The LM340-12 produces a regulated output voltage of
    a. 3 V
    c. 12 V
    b. 4 V
    d. 40 V

22. A current booster is a transistor in
    a. Series with the IC regulator
    b. Parallel with the IC regulator
    c. Either series or parallel
    d. Shunt with the load

23. To turn on a current booster, we can drive its base-emitter terminals with the voltage across
    a. A load resistor
    b. A zener impedance
    c. Another transistor
    d. A current-sensing resistor

24. The heat sink of a three-terminal regulator is usually larger and more expensive than the heat sink of
    a. An IC regulator
    b. A zener diode
    c. An outboard transistor
    d. Power supply

25. A phase splitter produces two output voltages that are
    a. Equal in phase
    b. Equal in amplitude
    c. Opposite in amplitude
    d. Very small

26. A series regulator is an example of a
    a. Linear regulator
    b. Switching regulator
    c. Shunt regulator
    d. DC-to-DC converter

27. The most widely used switching regulator is the
    a. Stepdown
    c. Inverting
    b. Step-up
    d. Linear

28. To get more output voltage from a step-down switching regulator, you have to
    a. Decrease the duty cycle
    b. Decrease the input voltage
    c. Increase the duty cycle
    d. Increase the linearity

29. The stepdown regulator is useful when you want to produce low voltages at
    a. Low currents
    b. High currents
    c. High impedances
    d. Low impedances

BASIC PROBLEMS

Sec. 23-1 Voltage-Feedback Regulation

23-1. What is the minimum regulated output voltage in Fig. 23-29? The maximum?

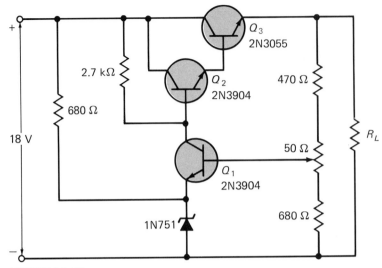

FIGURE 23-29

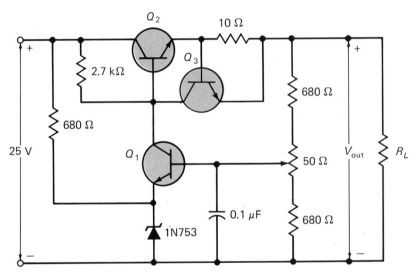

FIGURE 23-30

23-2. The wiper is at the middle of its range in Fig. 23-29. The load resistance is 7 Ω. The 2N3055 has a dc gain of 60, and the 2N3904 has a dc gain of 200. What is the current through the zener diode?

23-3. Calculate the power dissipation in the 2N3055 of Fig. 23-29 for the wiper all the way up and for a load resistance of 10 Ω.

Sec. 23-2 Current Limiting

23-4. In Fig. 23-30, the wiper is at the middle of its range. What is the output voltage of the regulator? What is the load current where current limiting starts?

23-5. What is the minimum load resistance in Fig. 23-30 where regulation is lost?

23-6. What is the power dissipation of the pass transistor when the load resistor is shorted in Fig. 23-30?

Sec. 23-3 Power Supply Characteristics

23-7. A voltage regulator has a full-load voltage of 27 V and a no-load voltage of 27.5 V. What is the voltage regulation? What is the percentage of voltage regulation?

23-8. A regulated power supply has these specifications: $V_{\text{out}} = 20.3$ V when $V_{\text{line}} =$

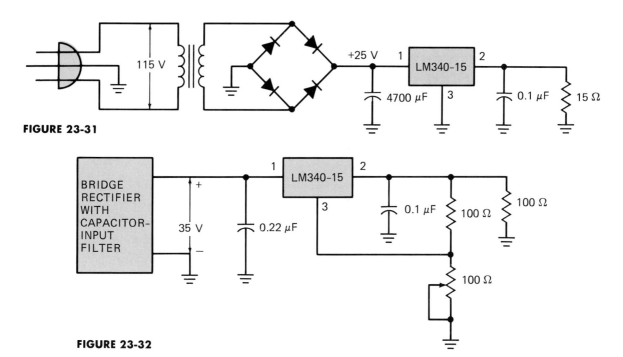

FIGURE 23-31

FIGURE 23-32

127 V and $V_{out} = 19.7$ when $V_{line} = 103$ V. What is the source regulation? The percentage of source regulation?

23-9. A voltage regulator has a sacrifice factor of 200. If the input ripple is 1 V, what is the output ripple? The ripple rejection, expressed in decibels?

23-10. The data sheet of a voltage regulator gives a ripple rejection of 74 dB. If the input ripple to this regulator is 2 V, what is the output ripple?

Sec. 23-4 Three-Terminal IC Regulators

23-11. What is the load current in Fig. 23-31? The output ripple?

23-12. If the line voltage is 115 V $\pm$ 10 percent in Fig. 23-31, what is the maximum load current? The maximum ripple?

23-13. If the load resistance is changed to 180 Ω in Fig. 23-31, what is the output ripple?

23-14. What are the minimum and maximum voltages in Fig. 23-32?

Sec. 23-5 Current Boosters

23-15. What is the power dissipation of the LM340-12 in Fig. 23-33?

23-16. If the input voltage of Fig. 23-33 has a tolerance of $\pm$ 10 percent, what is the worst-case power dissipation for the LM340-12?

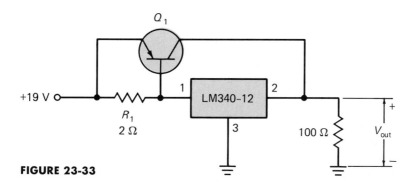

FIGURE 23-33

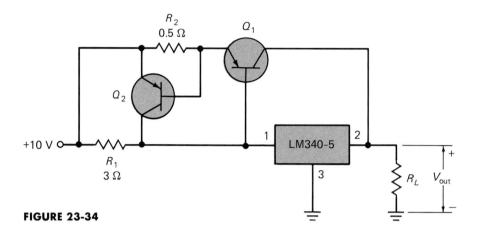

FIGURE 23-34

23-17. In Fig. 23-34, estimate the load current where the outboard current and current limiting begin.

Sec. 23-7 Switching Regulators

23-18. In Fig. 23-25 a stepdown switching regulator has a voltage divider sampling the output voltage. If $V_{ref} = 1.25$ V, $R_1 = 820$ Ω, and $R_2 = 330$ Ω, what is the output voltage?

23-19. The duty cycle is 50 percent. If the input voltage to a stepdown switching regulator is 10 V, what is the output voltage? If the input voltage were 15 V, what duty cycle would we need to maintain the output voltage as before?

TROUBLESHOOTING PROBLEMS

23-20. In Fig. 23-29, does the output voltage increase, decrease, or remain the same for each of these troubles?
   a. 2N3055 shorted
   b. 2N3055 open
   c. Zener diode shorted
   d. Zener diode open

23-21. In Fig. 23-30, the shorted-load current is approximately 7 mA. Which of the following is the trouble?
   a. 10 Ω shorted
   b. 680 Ω open
   c. Potentiometer open
   d. 100 Ω used instead of 10 Ω

23-22. Is the output voltage of Fig. 23-14 likely to increase, decrease, or remain about the same for each of these?
   a. One diode open
   b. Filter capacitor shorted
   c. LM317 defective
   d. R_1 open

23-23. The switching regulator of Fig. 23-25 has the following symptoms: the output voltage almost equals the input voltage, the triangular wave out of the integrator is all right, and V_{ref} is normal. Which of these is a possible trouble?
   a. Q_1 open
   b. R_1 open
   c. R_3 shorted
   d. R_7 open
   e. Triangle-to-pulse inputs reversed

ADVANCED PROBLEMS

23-24. In Fig. 23-34, what is the load current where current limiting begins?

23-25. Figure 23-35 shows an LM317 regulator with electronic shutdown. When the shutdown voltage is zero, the transistor is cut off and has no effect on the operation. But when the shutdown voltage is approximately 5 V, the transistor saturates. What is the adjustable range of output voltage when the shutdown voltage is zero? What does the output voltage equal when the shutdown voltage is 5 V?

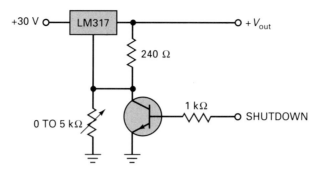

FIGURE 23-35

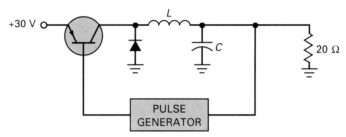

FIGURE 23-36

23-26. The transistor of Fig. 23-35 is cut off. To get an output voltage of 15 V, what value should the adjustable resistor have?

23-27. When a bridge rectifier and a capacitor-input filter drive a resistance, the discharge is exponential. But when they drive a voltage regulator, the discharge is almost a perfect ramp. Explain why this is true and what significance this has for Eq. (4-8).

23-28. Figure 23-36 shows a switching regulator in which the *LC* filter is collector-driven instead of emitter-driven. The transistor still acts as a switch, and the circuit regulates as previously described. If the output voltage is 15 V, what is the direct current through the inductor? If the duty cycle is 25 percent, what is the direct current through the diode?

T-SHOOTER PROBLEMS

Use Fig. 23-37 for the remaining problems. In this T-shooter you are troubleshooting a switching regulator. Before you start, look at the OK box and

decode the tokens to see the normal waveforms with their correct peak voltages. In this exercise, most of the troubles are IC failures rather than resistors. When an IC fails, anything can happen. Pins may be internally open, shorted, etc. No matter what the trouble is inside the IC, the most common symptom is a *stuck output*. This refers to the output voltage being stuck at either positive or negative saturation. If the input signals are okay, an IC with a stuck output has to be replaced because something is wrong on the inside. The following problems will give you a chance to work with outputs that are stuck at either $+13.5$ or -13.5 V.

23-29. Find Trouble 1.

23-30. Find Trouble 2.

23-31. Find Trouble 3.

23-32. Find Trouble 4

23-33. Find Trouble 5.

23-34. Find Trouble 6.

23-35. Find Trouble 7.

23-36. Find Trouble 8.

23-37. Find Trouble 9.

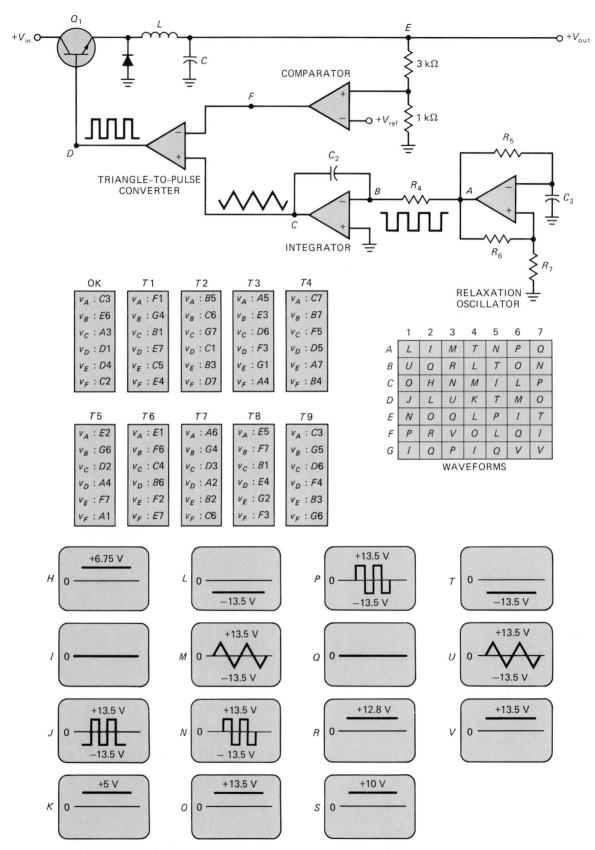

FIGURE 23-37 T-Shooter™. (*Patent pending: Courtesy of Malvino Inc.*)

APPENDIX
(DATA SHEETS)

MOTOROLA Semiconductors

BOX 20912 • PHOENIX, ARIZONA 85036

Designer's Data Sheet

1N4001 thru 1N4007

LEAD MOUNTED SILICON RECTIFIERS

50-1000 VOLTS
DIFFUSED JUNCTION

"SURMETIC"▲ RECTIFIERS

. . . subminiature size, axial lead mounted rectifiers for general-purpose low-power applications.

Designers Data for "Worst Case" Conditions

The Designers▲ Data Sheets permit the design of most circuits entirely from the information presented. Limit curves — representing boundaries on device characteristics — are given to facilitate "worst case" design.

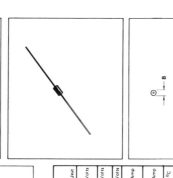

CATHODE BAND

| DIM | MILLIMETERS | | INCHES | |
	MIN	MAX	MIN	MAX
A	5.97	6.60	0.235	0.260
B	2.79	3.05	0.110	0.120
D	0.76	0.86	0.030	0.034
K	27.94	–	1.100	–

CASE 59-04
Does Not Conform to DO-41 Outline.

© MOTOROLA INC. 1975

DS 6015 R3

*MAXIMUM RATINGS

Rating	Symbol	1N4001	1N4002	1N4003	1N4004	1N4005	1N4006	1N4007	Unit
Peak Repetitive Reverse Voltage Working Peak Reverse Voltage DC Blocking Voltage	V$_{RRM}$ V$_{RWM}$ V$_R$	50	100	200	400	600	800	1000	Volts
Non-Repetitive Peak Reverse Voltage (halfwave, single phase, 60 Hz)	V$_{RSM}$	60	120	240	480	720	1000	1200	Volts
RMS Reverse Voltage	V$_{R(RMS)}$	35	70	140	280	420	560	700	Volts
Average Rectified Forward Current (single phase, resistive load, 60 Hz, see Figure 8, T$_A$ = 75°C)	I$_O$	1.0							Amp
Non-Repetitive Peak Surge Current (surge applied at rated load conditions, see Figure 2)	I$_{FSM}$	30 (for 1 cycle)							Amp
Operating and Storage Junction Temperature Range	T$_J$, T$_{stg}$	-65 to +175							°C

*ELECTRICAL CHARACTERISTICS

Characteristic and Conditions	Symbol	Typ	Max	Unit
Maximum Instantaneous Forward Voltage Drop (i$_F$ = 1.0 Amp, T$_J$ = 25°C) Figure 1	v$_F$	0.93	1.1	Volts
Maximum Full-Cycle Average Forward Voltage Drop (I$_O$ = 1.0 Amp, T$_L$ = 75°C, 1 inch leads)	V$_{F(AV)}$	–	0.8	Volts
Maximum Reverse Current (rated dc voltage) T$_J$ = 25°C T$_J$ = 100°C	i$_R$	0.05 1.0	10 50	µA
Maximum Full-Cycle Average Reverse Current (I$_O$ = 1.0 Amp, T$_L$ = 75°C, 1 inch leads)	I$_{R(AV)}$	–	30	µA

*Indicates JEDEC Registered Data

MECHANICAL CHARACTERISTICS

CASE: Void free, Transfer Molded

MAXIMUM LEAD TEMPERATURE FOR SOLDERING PURPOSES: 350°C, 3/8" from case for 10 seconds at 5 lbs. tension

FINISH: All external surfaces are corrosion-resistant, leads are readily solderable

POLARITY: Cathode indicated by color band

WEIGHT: 0.40 Grams (approximately)

▲Trademark of Motorola Inc.

951

ELECTRICAL CHARACTERISTICS ($T_A = 25°C$, $V_F = 1.5$ V max at 200 mA for all types)

Type Number (Note 1)	Nominal Zener Voltage V_Z @ I_{ZT} (Note 2) Volts	Test Current I_{ZT} mA	Maximum Zener Impedance Z_{ZT} @ I_{ZT} (Note 3) Ohms	*Maximum DC Zener Current I_{ZM} (Note 4) mA	Maximum Reverse Leakage Current $T_A=25°C$ I_R@V_R=1V µA	Maximum Reverse Leakage Current $T_A=150°C$ I_R@V_R=1V µA
1N4370	2.4	20	30	150	100	200
1N4371	2.7	20	30	135	75	150
1N4372	3.0	20	29	120	50	150
1N746	3.3	20	28	110	10	100
1N747	3.6	20	24	100	10	30
1N748	3.9	20	23	95	10	30
1N749	4.3	20	22	85	2	30
1N750	4.7	20	19	75	2	30
1N751	5.1	20	17	70	1	20
1N752	5.6	20	11	65	1	20
1N753	6.2	20	7	65	0.1	20
1N754	6.8	20	5	55	0.1	20
1N755	7.5	20	6	50	0.1	20
1N756	8.2	20	8	45	0.1	20
1N757	9.1	20	10	40	0.1	20
1N758	10	20	17	35	0.1	20
1N759	12	20	30	30	0.1	20

Type Number (Note 1)	Nominal Zener Voltage V_Z (Note 2) Volts	Test Current I_{ZT} mA	Maximum Zener Impedance Z_{ZT} @ I_{ZT} Ohms	Maximum Zener Impedance Z_{ZK} @ I_{ZK} Ohms	Maximum Zener Impedance I_{ZK} mA	*Maximum DC Zener Current I_{ZM} (Note 4) mA	Maximum Reverse Current I_R µA	Test Voltage Vdc 5% V_R	Test Voltage Vdc 10% V_R
1N957	6.8	18.5	4.5	700	1.0	47	150	5.2	4.9
1N958	7.5	16.5	5.5	700	0.5	42	75	5.7	5.4
1N959	8.2	15	6.5	700	0.5	38	50	6.2	5.9
1N960	9.1	14	7.5	700	0.5	35	25	6.9	6.6
1N961	10	12.5	8.5	700	0.25	32	10	7.6	7.2
1N962	11	11.5	9.5	700	0.25	28	5	8.4	8.0
1N963	12	10.5	11.5	700	0.25	26	5	9.1	8.6
1N964	13	9.5	13	700	0.25	24	5	9.9	9.4
1N965	15	8.5	16	700	0.25	21	5	11.4	10.8
1N966	16	7.8	17	700	0.25	19	5	12.2	11.5
1N967	18	7.0	21	750	0.25	17	5	13.7	13.0
1N968	20	6.2	25	750	0.25	15	5	15.2	14.4
1N969	22	5.6	29	750	0.25	14	5	16.7	15.8
1N970	24	5.2	33	750	0.25	13	5	18.2	17.3
1N971	27	4.6	41	750	0.25	11	5	20.6	19.4
1N972	30	4.2	49	1000	0.25	11	5	22.8	21.6
1N973	33	3.8	58	1000	0.25	9.2	5	25.1	23.8
1N974	36	3.4	70	1000	0.25	8.5	5	27.4	25.9
1N975	39	3.2	80	1000	0.25	7.8	5	29.7	28.1
1N976	43	3.0	93	1500	0.25	7.0	5	32.7	31.0
1N977	47	2.7	105	1500	0.25	6.4	5	35.8	33.8
1N978	51	2.5	125	1500	0.25	5.9	5	38.8	36.7
1N979	56	2.2	150	1500	0.25	5.4	5	42.6	40.3
1N980	62	2.0	185	2000	0.25	4.9	5	47.1	44.6
1N981	68	1.8	230	2000	0.25	4.5	5	51.7	49.0
1N982	75	1.7	270	2000	0.25	4.0	5	56.0	54.0
1N983	82	1.5	330	3000	0.25	3.7	5	62.2	59.0
1N984	91	1.4	400	3000	0.25	3.3	5	69.2	65.5

*Left column based upon JEDEC Registration, right column based upon Motorola rating.

NOTE 1. TOLERANCE AND VOLTAGE DESIGNATION

Tolerance Designation

The type numbers shown have tolerance designations as follows:

1N4370 series: ±10%, suffix A for ±5% units.
1N746 series: ±10%, suffix A for ±5% units.
1N957 series: ±20%, suffix A for ±10% units, suffix B for ±5% units.

NOTE 2. ZENER VOLTAGE (V_Z) MEASUREMENT

Nominal zener voltage is measured with the device junction in thermal equilibrium at the lead temperature of 30°C ±1°C and 3/8'' lead length.

NOTE 3. ZENER IMPEDANCE (Z_Z) DERIVATION

Z_{ZT} and Z_{ZK} are measured by dividing the ac voltage drop across the device by the ac current applied. The specified limits are for $|z(ac)| = 0.1 \, |z(dc)|$ with the ac frequency = 60 Hz.

NOTE 4. MAXIMUM ZENER CURRENT RATINGS (I_{ZM})

Maximum zener current ratings are based on the maximum voltage of a 10% 1N746 type unit or a 20% 1N957 type unit. For closer tolerance units (10% or 5%) or units where the actual zener voltage (V_Z) is known at the operating point, the maximum zener current may be increased and is limited by the derating curve.

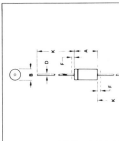

1N746 thru 1N759
1N957 thru 1N984
1N4370 thru 1N4372

GLASS ZENER DIODES

500 MILLIWATTS
2.4–91 VOLTS

MOTOROLA Semiconductors
BOX 20912 • PHOENIX, ARIZONA 85036

Designers Data Sheet

500-MILLIWATT HERMETICALLY SEALED
GLASS SILICON ZENER DIODES

- Complete Voltage Range — 2.4 to 91 Volts
- DO-35 Package — Smaller than Conventional DO-7 Package
- Double Slug Type Construction
- Metallurgically Bonded Construction
- Nitride Passivated Die

Designer's Data for "Worst Case" Conditions

The Designers Data sheets permit the design of most circuits entirely from the information presented. Limit curves — representing boundaries on device characteristics — are given to facilitate "worst case" design.

MAXIMUM RATINGS

Rating	Symbol	Value	Unit
DC Power Dissipation @ $T_L \leq 50°C$ Lead Length = 3/8''	P_D		
*JEDEC Registration		400	mW
Derate above $T_L = 50°C$		3.2	mW/°C
Motorola Device Ratings		500	mW
Derate above $T_L = 50°C$		3.33	mW/°C
Operating and Storage Junction Temperature Range	T_J, T_{stg}		°C
*JEDEC Registration		−65 to +175	
Motorola Device Ratings		−65 to +200	

*Indicates JEDEC Registered Data.

MECHANICAL CHARACTERISTICS

CASE: Double slug type, hermetically sealed glass

MAXIMUM LEAD TEMPERATURE FOR SOLDERING PURPOSES: 230°C, 1/16'' from case for 10 seconds

FINISH: All external surfaces are corrosion resistant with readily solderable leads.

POLARITY: Cathode indicated by color band. When operated in zener mode, cathode will be positive with respect to anode.

MOUNTING POSITION: Any

STEADY STATE POWER DERATING

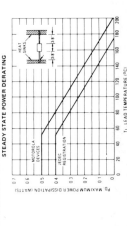

All JEDEC dimensions and notes apply

CASE 299-01
DO-35

NOTE:
1. POLARITY DENOTED BY CATHODE BAND
2. LEAD DIAMETER IS NOT CONTROLLED WITHIN DIMENSION "F"

©MOTOROLA INC. 1977

DS 7021 R

▲Trademark of Motorola Inc.

MOTOROLA Semiconductors

BOX 20912, PHOENIX, ARIZONA 85036

2N3903 2N3904

NPN SILICON SWITCHING & AMPLIFIER TRANSISTORS

NPN SILICON ANNULAR▲ TRANSISTORS

...designed for general purpose switching and amplifier applications and for complementary circuitry with types 2N3905 and 2N3906.

- High Voltage Ratings — BV_{CEO} = 40 Volts (Min)
- Current Gain Specified from 100 μA to 100 mA
- Complete Switching and Amplifier Specifications
- Low Capacitance — C_{ob} = 4.0 pF (Max)

MAXIMUM RATINGS

Rating	Symbol	Value	Unit
*Collector-Base Voltage	V_{CB}	60	Vdc
*Collector-Emitter Voltage	V_{CEO}	40	Vdc
*Emitter-Base Voltage	V_{EB}	6.0	Vdc
*Collector Current	I_C	200	mAdc
Total Power Dissipation @ T_A = 60°C	P_D	250	mW
**Total Power Dissipation @ T_A = 25°C	P_D	350	mW
Derate above 25°C		2.8	mW/°C
**Total Power Dissipation @ T_C = 25°C	P_D	1.0	Watts
Derate above 25°C		8.0	mW/°C
**Junction Operating Temperature	T_J	150	°C
*Storage Temperature Range	T_{stg}	-55 to +150	°C

THERMAL CHARACTERISTICS

Characteristic	Symbol	Max	Unit
Thermal Resistance, Junction to Ambient	$R_{\theta JA}$	357	°C/W
Thermal Resistance, Junction to Case	$R_{\theta JC}$	125	°C/W

*Indicates JEDEC Registered Data
**Motorola guarantees this data in addition to the JEDEC Registered Data
▲ Annular Semiconductors Patented by Motorola Inc.

CASE 29-02 / TO-92

DIM	MILLIMETERS MIN	MAX	INCHES MIN	MAX
A	4.450	5.200	0.175	0.205
B	3.180	4.190	0.125	0.165
C	4.320	5.330	0.170	0.210
D	0.407	0.533	0.016	0.021
F	0.407	0.482	0.016	0.019
K	12.700	—	0.500	—
L	1.150	1.390	0.045	0.055
N	1.270	—	0.050	—
P	—	6.350	—	0.250
Q	2.430	—	0.135	—
R	2.410	2.610	0.095	0.105
S	2.030	2.610	0.080	0.105

STYLE 1: PIN 1. EMITTER 2. BASE 3. COLLECTOR

*ELECTRICAL CHARACTERISTICS (T_A = 25°C unless otherwise noted)

Characteristic	Fig. No.		Symbol	Min	Max	Unit
OFF CHARACTERISTICS						
Collector-Base Breakdown Voltage (I_C = 10 μAdc, I_E = 0)			BV_{CBO}	60	—	Vdc
Collector-Emitter Breakdown Voltage (1) (I_C = 1.0 mAdc, I_B = 0)			BV_{CEO}	40	—	Vdc
Emitter-Base Breakdown Voltage (I_E = 10 μAdc, I_C = 0)			BV_{EBO}	6.0	—	Vdc
Collector Cutoff Current (V_{CE} = 30 Vdc, $V_{EB(off)}$ = 3.0 Vdc)			I_{CEX}	—	50	nAdc
Base Cutoff Current (V_{CE} = 30 Vdc, $V_{EB(off)}$ = 3.0 Vdc)			I_{BL}	—	50	nAdc
ON CHARACTERISTICS						
DC Current Gain (1)	15	2N3903 2N3904	h_{FE}			
(I_C = 0.1 mAdc, V_{CE} = 1.0 Vdc)		2N3903 2N3904		20 / 40	—	
(I_C = 1.0 mAdc, V_{CE} = 1.0 Vdc)		2N3903 2N3904		35 / 70	—	
(I_C = 10 mAdc, V_{CE} = 1.0 Vdc)		2N3903 2N3904		50 / 100	150 / 300	
(I_C = 50 mAdc, V_{CE} = 1.0 Vdc)		2N3903 2N3904		30 / 60	—	
(I_C = 100 mAdc, V_{CE} = 1.0 Vdc)		2N3903 2N3904		15 / 30	—	
Collector-Emitter Saturation Voltage (1)	16, 17	2N3903 2N3904	$V_{CE(sat)}$			Vdc
(I_C = 10 mAdc, I_B = 1.0 mAdc)				—	0.2	
(I_C = 50 mAdc, I_B = 5.0 mAdc)				—	0.3	
Base-Emitter Saturation Voltage (1)	17	2N3903 2N3904	$V_{BE(sat)}$			Vdc
(I_C = 10 mAdc, I_B = 1.0 mAdc)				0.65	0.85	
(I_C = 50 mAdc, I_B = 5.0 mAdc)				—	0.95	
SMALL-SIGNAL CHARACTERISTICS						
Current-Gain — Bandwidth Product (I_C = 10 mAdc, V_{CE} = 20 Vdc, f = 100 MHz)		2N3903 2N3904	f_T	250 / 300	—	MHz
Output Capacitance (V_{CB} = 5.0 Vdc, I_E = 0, f = 100 kHz)	3		C_{ob}	—	4.0	pF
Input Capacitance (V_{BE} = 0.5 Vdc, I_C = 0, f = 100 kHz)	3		C_{ib}	—	8.0	pF
Input Impedance (I_C = 1.0 mAdc, V_{CE} = 10 Vdc, f = 1.0 kHz)	13	2N3903 2N3904	h_{ie}	0.5 / 1.0	8.0 / 10	k ohms
Voltage Feedback Ratio (I_C = 1.0 mAdc, V_{CE} = 10 Vdc, f = 1.0 kHz)	14	2N3903 2N3904	h_{re}	0.1 / 0.5	5.0 / 8.0	X 10⁻⁴
Small-Signal Current Gain (I_C = 1.0 mAdc, V_{CE} = 10 Vdc, f = 1.0 kHz)	11	2N3903 2N3904	h_{fe}	50 / 100	200 / 400	-
Output Admittance (I_C = 1.0 mAdc, V_{CE} = 10 Vdc, f = 1.0 kHz)	12		h_{oe}	1.0	40	μmho
Noise Figure (I_C = 100 μAdc, V_{CE} = 5.0 Vdc, R_S = 1.0 k ohms, f = 10 Hz to 15.7 kHz)	9, 10	2N3903 2N3904	NF	—	6.0 / 5.0	dB
SWITCHING CHARACTERISTICS						
Delay Time (V_{CC} = 3.0 Vdc, $V_{BE(off)}$ = 0.5 Vdc,	1, 5, 6		t_d	—	35	ns
Rise Time I_C = 10 mAdc, I_{B1} = 1.0 mAdc)	1, 5, 6		t_r	—	35	ns
Storage Time (V_{CC} = 3.0 Vdc, I_C = 10 mAdc,	2, 7	2N3903 2N3904	t_s	—	175 / 200	ns
Fall Time I_{B1} = I_{B2} = 1.0 mAdc)	2, 8		t_f	—	50	ns

(1) Pulse Test: Pulse Width = 300 μs, Duty Cycle = 2.0%.
*Indicates JEDEC Registered Data

FIGURE 1 — DELAY AND RISE TIME EQUIVALENT TEST CIRCUIT

$+10.9$ V $+3.0$ V ϕ $C_s < 4.0$ pF*
DUTY CYCLE = 2%
300 ns → < 1.0 ns
-0.5 V
*Total shunt capacitance of test jig and connectors.

FIGURE 2 — STORAGE AND FALL TIME EQUIVALENT TEST CIRCUIT

$+3.0$ V ϕ $C_s < 4.0$ pF*
$10 < t_1 < 500$ μs, DUTY CYCLE = 2%
$+10.9$ V
-9.1 V < 1.0 ns
10 k   1N916

MOTOROLA Semiconductor Products Inc.

DS 5127 R2
© MOTOROLA INC. 1973

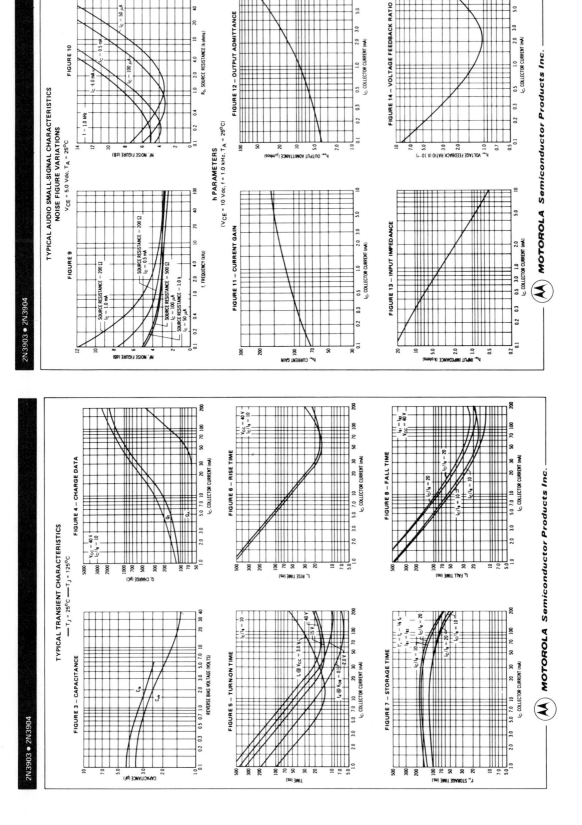

MOTOROLA *Semiconductor Products Inc.*

LM140A/LM140/LM340A/LM340 Series 3-Terminal Positive Regulators

National Semiconductor

LM140A/LM140/LM340A/LM340 Series 3-Terminal Positive Regulators

General Description

The LM140A/LM140/LM340A/LM340 series of positive 3-terminal voltage regulators are designed to provide superior performance as compared to the previously available 78XX series regulator. Computer programs were used to optimize the electrical and thermal performance of the packaged IC which results in outstanding ripple rejection, superior line and load regulation in high power applications (over 15W).

With these advances in design, the LM340 is now guaranteed to have line and load regulation that is a factor of 2 better than previously available devices. Also, all parameters are guaranteed at 1A vs 0.5A output current. The LM140A/LM340A provide tighter output voltage tolerance, ±2% along with 0.01%/V line regulation and 0.3%/A load regulation.

Current limiting is included to limit peak output current to a safe value. Safe area protection for the output transistor is provided to limit internal power dissipation. If internal power dissipation becomes too high for the heat sinking provided, the thermal shutdown circuit takes over limiting die temperature.

Considerable effort was expended to make the LM140-XX series of regulators easy to use and minimize the number of external components. It is not necessary to bypass the output, although this does improve transient response.

Input bypassing is needed only if the regulator is located far from the filter capacitor of the power supply.

Although designed primarily as fixed voltage regulators, these devices can be used with external components to obtain adjustable voltages and currents.

The entire LM140A/LM140/LM340A/LM340 series of regulators is available in the metal TO-3 power package and the LM340A/LM340 series is also available in the TO-220 plastic power package.

Features

- Complete specifications at 1A load
- Output voltage tolerances of ±2% at $T_j = 25°C$ and ±4% over the temperature range (LM140A/LM340A)
- Fixed output voltages available 5, 6, 8, 10, 12, 15, 18 and 24V
- Line regulation of 0.01% of V_{OUT}/V ΔV_{IN} at 1A load (LM140A/LM340A)
- Load regulation of 0.3% of V_{OUT}/A ΔI_{LOAD} (LM140A/LM340A)
- Internal thermal overload protection
- Internal short-circuit current limit
- Output transistor safe area protection

Typical Applications

Fixed Output Regulator

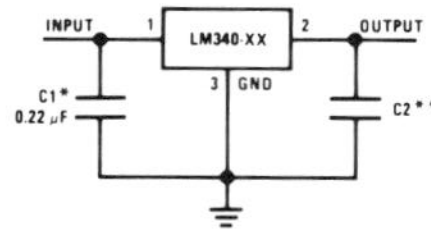

Adjustable Output Regulator

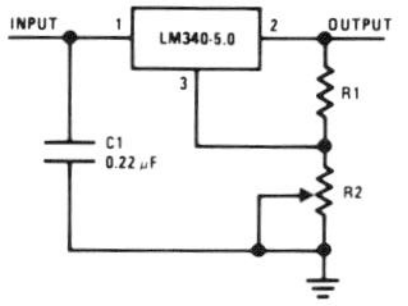

$V_{OUT} = 5V + (5V/R1 + I_Q) R2$
$5V/R1 > 3 I_Q$, load regulation $(L_r) = ((R1 + R2)/R1) (L_r \text{ of LM340-5})$

Current Regulator

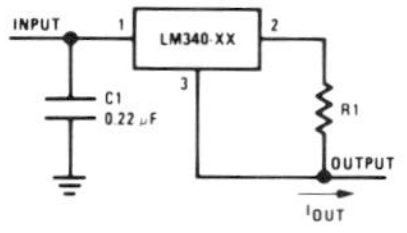

$I_{OUT} = \dfrac{V2-3}{R1} + I_Q$
$\Delta I_Q = 1.3 \text{ mA over line and load changes}$

* Required if the regulator is located far from the power supply filter

** Although no output capacitor is needed for stability, it does help transient response. (If needed, use 0.1 μF, ceramic disc)

TYPICAL STATIC CHARACTERISTICS

FIGURE 15 – DC CURRENT GAIN

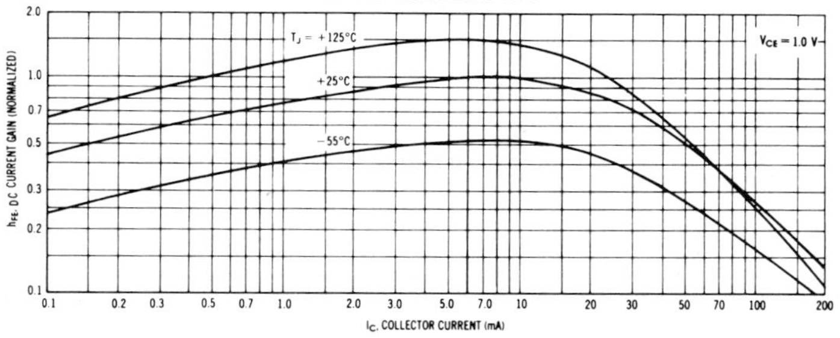

FIGURE 16 – COLLECTOR SATURATION REGION

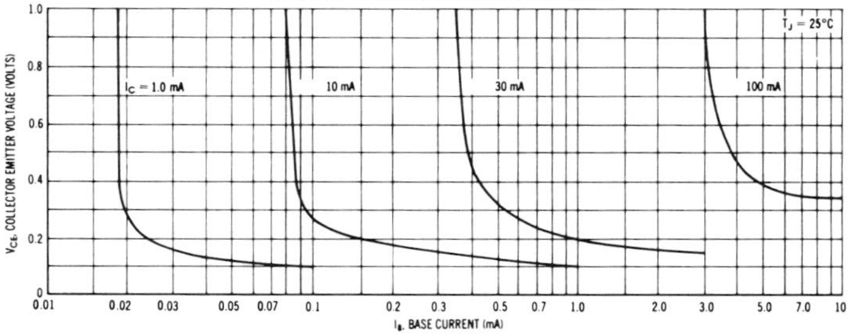

FIGURE 17 – "ON" VOLTAGES

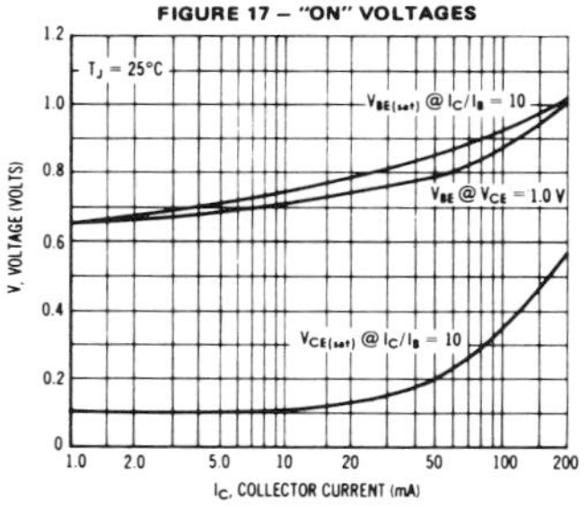

FIGURE 18 – TEMPERATURE COEFFICIENTS

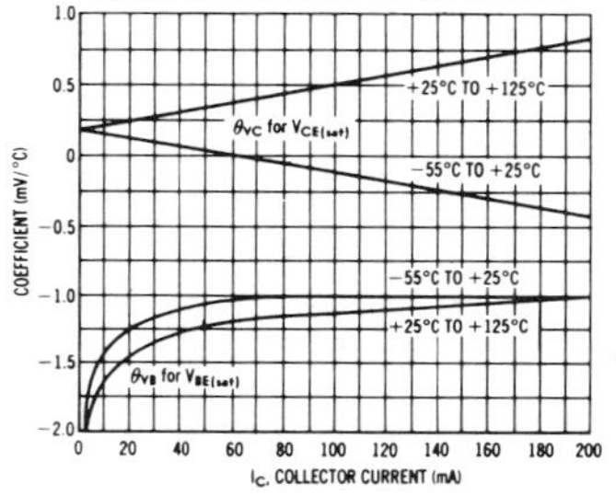

2N3903 ● 2N3904

MOTOROLA *Semiconductor Products Inc.*

Typical Performance Characteristics

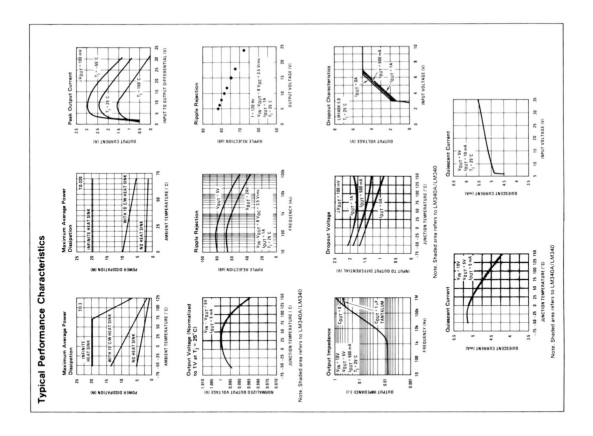

Electrical Characteristics LM340 (Note 2)

$0°C \leq T_j \leq +125°C$ unless otherwise noted.

OUTPUT VOLTAGE			5V			6V			8V			10V			12V			15V			18V			24V			
INPUT VOLTAGE (unless otherwise noted)			10V			11V			14V			17V			19V			23V			27V			33V			UNITS
PARAMETER	CONDITIONS		MIN	TYP	MAX	MIN	TYP	MAX	MIN	TYP	MAX	MIN	TYP	MAX	MIN	TYP	MAX	MIN	TYP	MAX	MIN	TYP	MAX	MIN	TYP	MAX	

Note 2: All characteristics are measured with a capacitor across the input of 0.22 μF and a capacitor across the output of 0.1 μF. All characteristics except noise voltage and ripple rejection ratio are measured using pulse techniques ($t_w \leq 10$ ms, duty cycle $\leq 5\%$). Output voltage changes due to changes in internal temperature must be taken into account separately.

2N4441 thru 2N4444

ELECTRICAL CHARACTERISTICS (T_C = 25°C unless otherwise noted)

Characteristic	Symbol	Min	Typ	Max	Unit
*Peak Forward Blocking Voltage (T_J = 100°C) Note 1 2N4441 2N4442 2N4443 2N4444	V_{DRM}	50 200 400 600	— — — —	— — — —	Volts
Peak Forward Blocking Current (Rated V_{DRM}, T_J = 100°C, gate open)	I_{DRM}	—	—	2.0	mA
Peak Reverse Blocking Current (Rated V_{DRM}, T_J = 100°C, gate open)	I_{RRM}	—	—	2.0	mA
Gate Trigger Current (Continuous dc) (Anode Voltage = 7.0 Vdc, R_L = 100 Ohms) T_C = 25°C * T_C = -40°C	I_{GT}	—	7.0	30 60	mA
Gate Trigger Voltage (Continuous dc) (Anode Voltage = 7.0 Vdc, R_L = 100 Ohms) T_C = 25°C * (Anode Voltage = 7.0 Vdc, R_L = 100 Ohms) T_C = -40°C * (Anode Voltage = Rated V_{DRM}, R_L = 100 Ohms) T_J = 100°C	V_{GT}	— — 0.2	0.75 — —	1.5 2.5 —	Volts
Peak On-State Voltage (Pulse Width = 1.0 to 2.0 ms, Duty Cycle ≤ 2.0%) * (I_{TM} = 5.0 A peak) * (I_{TM} = 15.7 A peak)	V_{TM}	— —	1.0 —	1.5 2.0	Volts
Holding Current (Anode Voltage = 7.0 Vdc, gate open) T_C = 25°C * T_C = -40°C	I_H	— —	6.0 —	40 70	mA
Gate Controlled Turn-On Time (I_{TM} = 5.0 A, I_{GT} = 20 mA)	t_{gt}	—	1.0	—	µs
Circuit Commutated Turn-Off Time (I_{TM} = 5.0 A, I_R = 5.0 A) (I_{TM} = 5.0 A, I_R = 5.0 A, T_J = 100°C)	t_q	— —	15 20	— —	µs
Critical Rate of Rise of Off-State Voltage (Rated V_{DRM}, Exponential Waveform, T_J = 100°C, Gate Open)	dv/dt	—	50	—	V/µs

*Indicates JEDEC Registered Data

Note 1.   Ratings apply for zero or negative gate voltage but positive gate voltage shall not be applied concurrently with a negative potential on the anode. When checking forward or reverse blocking capability, thyristor devices should not be tested with a constant current source in a manner that the voltage applied exceeds the rated blocking voltage.

Note 2   Torque rating applies with use of torque washer (Shakeproof WD19522 #6 or equivalent). Mounting torque in excess of 8 in. lbs. does not appreciably lower case-to-sink thermal resistance. Anode lead and heatsink contact pad are common.

For soldering purposes (either terminal connection or device mounting), soldering temperatures shall not exceed +225°C.

 MOTOROLA *Semiconductor Products Inc.*

 **MOTOROLA** *Semiconductors*

BOX 20912 • PHOENIX, ARIZONA 85036

2N4441 thru 2N4444

PLASTIC SILICON CONTROLLED RECTIFIERS

8.0 AMPERES RMS
50 thru 600 VOLTS

HEAT SINK CONTACT AREA (BOTTOM)

DIM	MILLIMETERS		INCHES	
	MIN	MAX	MIN	MAX
A	15.95	16.71	0.628	0.658
B	12.45	13.21	0.490	0.520
C	3.05	3.81	0.120	0.150
D	1.09	1.25	0.043	0.049
F	3.51	3.76	0.138	0.148
G	4.22 BSC		0.166 BSC	
H	3.18	3.18	0.125	0.125
J	0.76	0.86	0.030	0.034
K	14.99	16.51	0.590	0.650
Q	4.50	5.00	0.177	0.197
R	1.91	2.16	0.075	0.085

CASE 90-04
STYLE 1: PIN 1. CATHODE
2. ANODE
3. GATE

©MOTOROLA INC. 1973   DS 6533 R1

PLASTIC THYRISTORS

. . . designed for high-volume consumer phase-control applications such as motor speed, temperature, and light controls and for switching applications in ignition and starting systems, voltage regulators, vending machines, and lamp drivers requiring:

● Small, Rugged, Thermopad ▲ Construction — for Low Thermal Resistance, High Heat Dissipation, and Durability.

● Practical Level Triggering and Holding Characteristics @ 25°C
I_{GT} = 7.0 mA (Typ)
I_H = 6.0 mA (Typ)

● Low "On" Voltage — V_{TM} = 1.0 Volt (Typ) @ 5.0 Amp @ 25°C

● High Surge Current Rating — I_{TSM} = 80 Amp

MAXIMUM RATING (T_J = 100°C unless otherwise noted.)

Rating	Symbol	Value	Unit
*Repetitive Peak Reverse Blocking Voltage (Note 1) 2N4441 2N4442 2N4443 2N4444	V_{RRM}	50 200 400 600	Volts
*Non-Repetitive Peak Reverse Blocking Voltage (t = 5.0 ms (max) duration) 2N4441 2N4442 2N4443 2N4444	V_{RSM}	75 300 500 700	Volts
*RMS On-State Current (All Conduction Angles)	$I_{T(RMS)}$	8.0	Amp
Average On-State Current, T_C = 73°C	$I_{T(AV)}$	5.1	Amp
*Peak Non-Repetitive Surge Current (1/2 cycle, 60 Hz preceded and followed by rated current and voltage)	I_{TSM}	80	Amp
Circuit Fusing Considerations (T_J = -40 to +100°C; t = 1.0 to 8.3 ms)	I^2t	25	A^2s
*Peak Gate Power	P_{GM}	5.0	Watts
*Average Gate Power	$P_{G(AV)}$	0.5	Watt
*Peak Forward Gate Current	I_{GM}	2.0	Amp
*Peak Reverse Gate Voltage	V_{RGM}	10	Volts
*Operating Junction Temperature Range	T_J	-40 to +100	°C
*Storage Temperature Range	T_{stg}	-40 to +150	°C
Mounting Torque (6-32 screw) (Note 2)	—	8.0	in. lb.

THERMAL CHARACTERISTICS

Characteristic	Symbol	Typ	Max	Unit
*Thermal Resistance, Junction to Case	$R_{\theta JC}$	—	2.5	°C/W
Thermal Resistance, Junction to Ambient	$R_{\theta JA}$	40	—	°C/W

*Indicate JEDEC Registered Data.
*Trademark of Motorola Inc.

MPF102

ELECTRICAL CHARACTERISTICS ($T_A = 25°C$ unless otherwise noted)

Characteristic	Symbol	Min	Max	Unit
OFF CHARACTERISTICS				
Gate–Source Breakdown Voltage ($I_G = -10$ μAdc, $V_{DS} = 0$)	BV_{GSS}	-25	—	Vdc
Gate Reverse Current	I_{GSS}			
($V_{GS} = -15$ Vdc, $V_{DS} = 0$)		—	-2.0	nAdc
($V_{GS} = -15$ Vdc, $V_{DS} = 0$, $T_A = 100°C$)		—	-2.0	μAdc
Gate–Source Cutoff Voltage ($V_{DS} = 15$ Vdc, $I_D = 2.0$ nAdc)	$V_{GS(off)}$	—	-8	Vdc
Gate–Source Voltage ($V_{DS} = 15$ Vdc, $I_D = 0.2$ mAdc)	V_{GS}	-0.5	-7.5	Vdc
ON CHARACTERISTICS				
Zero-Gate-Voltage Drain Current* ($V_{DS} = 15$ Vdc, $V_{GS} = 0$ Vdc)	I_{DSS}*	2	20	mAdc
DYNAMIC CHARACTERISTICS				
Forward Transfer Admittance* ($V_{DS} = 15$ Vdc, $V_{GS} = 0$, f = 1 kHz)	$\|y_{fs}\|$*	2000	7500	μmhos
Input Capacitance ($V_{DS} = 15$ Vdc, $V_{GS} = 0$, f = 1 MHz)	C_{iss}	—	7	pF
Reverse Transfer Capacitance ($V_{DS} = 15$ Vdc, $V_{GS} = 0$, f = 1 MHz)	C_{rss}	—	3	pF
Forward Transfer Admittance ($V_{DS} = 15$ Vdc, $V_{GS} = 0$, f = 100 MHz)	$\|y_{fs}\|$	1600	—	μmhos
Input Conductance ($V_{DS} = 15$ Vdc, $V_{GS} = 0$, f = 100 MHz)	$Re(y_{is})$	—	800	μmhos
Output Conductance ($V_{DS} = 15$ Vdc, $V_{GS} = 0$, f = 100 MHz)	$Re(y_{os})$	—	200	μmhos

*Pulse Test: Pulse Width ≤ 630 ms, Duty Cycle ≤ 10%

MOTOROLA Semiconductor Products Inc.
BOX 955 • PHOENIX, ARIZONA 85001 • A SUBSIDIARY OF MOTOROLA INC

Si FIELD-EFFECT TRANSISTOR
MPF102
DS 5203

THE **RF** LINE
MPF102

JUNCTION FIELD-EFFECT TRANSISTOR

SYMMETRICAL SILICON N-CHANNEL

SEPTEMBER 1966 — DS 5203

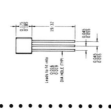

TO-92
BOTTOM VIEW
Drain and Source may be Interchanged.

MOTOROLA

SILICON N-CHANNEL
JUNCTION FIELD-EFFECT TRANSISTOR

... designed for VHF amplifier and mixer applications.

• Low Cross-Modulation and Intermodulation Distortion
• Guaranteed 100-MHz Parameters
• Drain and Source Interchangeable
• Low Transfer and Input Capacitance
• Low Leakage Current
• Unibloc* Plastic Encapsulated Package

MAXIMUM RATINGS

Characteristic	Symbol	Rating	Unit
Drain-Source Voltage	V_{DS}	25	Vdc
Drain-Gate Voltage	V_{DG}	25	Vdc
Gate-Source Voltage	V_{GS}	-25	Vdc
Gate Current	I_G	10	mAdc
Total Device Dissipation @ 25°C	P_D	200	mW
Derate above 25°C		2	mW/°C
Operating Junction Temperature	T_J	125	°C
Storage Temperature Range	T_{stg}	-65 to +150	°C

*Trademark of Motorola Inc.

MOTOROLA Semiconductor Products Inc.   A SUBSIDIARY OF MOTOROLA INC

LM741/LM741A/LM741C/LM741E

National Semiconductor

LM741/LM741A/LM741C/LM741E operational amplifier

general description

The LM741 series are general purpose operational amplifiers which feature improved performance over industry standards like the LM709. They are direct, plug-in replacements for the 709C, LM201, MC1439 and 748 in most applications.

The amplifiers offer many features which make their application nearly foolproof: overload pro-

tection on the input and output, no latch-up when the common mode range is exceeded, as well as freedom from oscillations.

The LM741C/LM741E are identical to the LM741/LM741A except that the LM741C/LM741E have their performance guaranteed over a 0°C to +70°C temperature range, instead of −55°C to +125°C.

schematic and connection diagrams (Top Views)

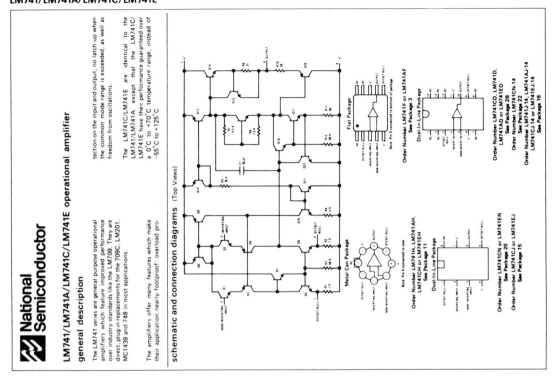

Flat Package

Note: Pin 5 connected to bottom of package

Order Number LM741F or LM741AF
See Package 3

Dual-In-Line Package

Order Number LM741CD, LM741D,
LM741AD or LM741ED
See Package 28
Order Number LM741CN-14
See Package 22
Order Number LM741-14, LM741AJ-14
LM741C-14 or LM741EJ-14
See Package 16

Metal Can Package

Note: Pin 4 connected to case

Order Number LM741H, LM741AH,
LM741CH or LM741EH
See Package 11

Dual-In-Line Package

Order Number LM741CN or LM741EN
See Package 20
Order Number LM741CJ or LM741EJ
See Package 15

GLOSSARY

absolute value The value of an expression without regard for its sign. Sometimes called the *magnitude*. Given $+5$ and -5, the absolute value is 5.

acceptor A trivalent atom, one that has three valence electrons. Each trivalent atom produces one hole in a silicon crystal.

ac current gain With a transistor, the ratio of ac collector current to ac base current.

ac cutoff The lower end of the ac load line. At this point, the transistor goes into cutoff and clips the ac signal.

ac equivalent circuit All that remains when you reduce the dc sources to zero and short all capacitors.

ac ground A node that is bypassed to ground through a capacitor. Such a node will show no ac voltage when it is probed by an oscilloscope, but it will indicate a dc voltage when it is measured with a voltmeter.

ac load line The locus of instantaneous operating points when an ac signal is driving the transistor. This load line is different from the dc load line whenever the ac load resistance is different from the dc load resistance.

ac resistance The resistance of a device to a small ac signal. The ratio of a voltage change to a current change. The key idea here is changes about an operating point.

ac saturation The upper end of the ac load line. At this point, the transistor goes into saturation and clips the ac signal.

active current gain The current gain in the active region of a transistor. That is what you usually find on a data sheet and what most people mean when they talk about current gain. (See *saturated current gain*.)

active filter In the good old days, filters were made out of passive components like inductors and capacitors. Some filters are still made this way. The problem is that at low frequencies, inductors become very large in passive filter designs. Op amps give another way to build filters and eliminate the problem of bulky inductors at low frequencies. Any filter using an op amp is called an active filter.

active loading This refers to using a bipolar or MOS transistor as a resistor. It's done to save space or to get resistances that are difficult with passive resistors.

active-load resistor A FET with its gate connected to the drain. The resulting two-terminal device is equivalent to a resistor.

active region Sometimes called the *linear region*. It refers to that part of the collector curves that is approximately horizontal. A transistor operates in the active region when it is used as an amplifier. In the active region, the emitter diode is forward-biased, the collector diode is reverse-biased, the collector current almost equals the emitter current, and the base current is much smaller than either the emitter or collector current.

amplifier A circuit that can increase the peak-to-peak voltage, current, or power of a signal.

amplitude The size of a signal, usually its peak value.

analogy A likeness in some ways between dissimilar things that are otherwise unlike. The analogy between bipolar transistors and JFETs is an example. Because the devices are similar, many of the equations for them are identical except for a change of subscripts.

approximation A way to retain your sanity with semiconductor devices. Exact answers are tedious and time-consuming and almost never justified in the real world of electronics. On the other hand, approximations give us quick answers, usually adequate for the job at hand.

audio amplifier Any amplifier designed for the audio range of frequencies, 20 Hz to 20 kHz.

avalanche effect A phenomenon that occurs for large reverse voltages across a *pn* junction. The free electrons are accelerated to such high speeds that they can dislodge valence electrons. When this happens, the valence electrons become free electrons that dislodge other valence electrons.

bandwidth The difference between the two dominant critical frequencies of an amplifier. If the amplifier has no lower critical frequency, the bandwidth equals the upper critical frequency.

barrier potential The voltage across the depletion layer. This voltage is built into the *pn* junction because it is the difference of potential between the ions on both sides of the junction. It equals approximately 0.7 V for a silicon diode.

base The middle part of a transistor. It is thin and lightly doped. This permits electrons from the emitter to pass through it to the collector.

base bias The worst way to bias a transistor for use in the active region. This type of bias sets up a fixed value of base current.

BIFET op amp An IC op amp that combines FETs and bipolar transistors, usually with FET source followers at the front end of the device, followed by bipolar stages of gain.

bipolar transistor A transistor where both free electrons and holes are necessary for normal operation.

breakdown region For a diode or transistor, it is the region where either avalanche or the zener effect occurs. With the exception of the zener diode, operation in the breakdown region is to be avoided under all circumstances because it usually destroys the device.

breakdown voltage The maximum reverse voltage a diode can withstand before avalanche or the zener effect occurs.

breakover When a transistor breaks down, the voltage across it remains high. But with a thyristor, breakdown turns into saturation. In other words, breakover refers to the way a thyristor breaks down and then immediately goes into saturation.

bridge rectifier The most common type of rectifier

circuit. It has four diodes, two of which are conducting at the same time. For a given transformer, it produces the largest dc output voltage with the smallest ripple.

buffer amplifier This is an amplifier that you use to isolate two other circuits when one is overloading the other. A buffer amplifier usually has a very high input impedance, a very low output impedance, and a voltage gain of 1. These qualities mean that the buffer amplifier will transmit the output of the first circuit to the second circuit without changing the signal.

Butterworth filter This is a filter designed to produce as flat a response as possible up to the cutoff frequency. In other words, the output voltage remains constant almost all the way to the cutoff frequency. Then it decreases at $20n$ dB per decade, where n is the number of poles in the filter.

bulk resistance The ohmic resistance of the semiconductor material.

bypass capacitor A capacitor used to ground a node.

capacitor input filter Nothing more than a capacitor across the load resistor. This type of passive filter is the most common.

carrier A free electron or a hole.

cascaded stages Connecting two or more stages so that the output of one stage is the input to the next.

case temperature This is the temperature of the transistor case or package. When you pick up a transistor, you are in contact with the case. If the case is warm, you are feeling the case temperature.

chip This has two meanings. First, an IC manufacturer produces hundreds of circuits on a large wafer of semiconductor material. Then the wafer is cut into individual chips, each containing one monolithic circuit. In this case, no leads have been connected to the chip. The chip is still an isolated piece of semiconductor material. Second, after the chip has been put inside a package and external leads have been connected to it, you have a finished IC. This finished IC is also referred to as a *chip*. For instance, we can call a 741C a chip.

class A operation This means the transistor is conducting throughout the ac cycle without going into saturation or cutoff.

class B operation Biasing of a transistor in such a way that it conducts for only half of the ac cycle.

closed-loop quantity The value of any quantity such as the voltage gain, input impedance, and output impedance that is changed by negative feedback.

CMOS inverter A circuit with complementary MOS transistors. The input voltage is either low or high, and the output voltage is either high or low.

collector The largest part of a transistor. It is called the collector because it collects or gathers the carriers sent into the base by the emitter.

collector cutoff current The small collector current that exists when the base current is zero in a *CE* connection. Ideally, there should be no collector current. But there is because of the minority carriers and the surface leakage current of the collector diode.

collector diode The diode formed by the base and collector of a transistor.

common-emitter circuit A transistor circuit where the emitter is common or grounded.

Colpitts oscillator One of the most widely used *LC* oscillators. It consists of a bipolar transistor or FET and an *LC* resonant circuit. You can recognize it because it has two capacitors in the tank circuit. They act as a capacitive voltage divider that produces the feedback voltage.

common-collector amplifier This is an amplifier whose collector is at ac ground. The signal goes into the base and comes out of the emitter.

common-mode signal A signal that is applied with equal strength to both inputs of a diff amp or an op amp.

comparator A circuit or device that detects when the input voltage is greater than a predetermined limit. The output is either a low or a high voltage. The predetermined limit is called the *trip point*.

compensating capacitor A capacitor inside an op amp that prevents oscillations. Also, any capacitor that stabilizes an amplifier with a negative-feedback path. Without this capacitor, the amplifier will oscillate. The compensating capacitor produces a low critical frequency and decreases the voltage gain at a rate of 20 dB per decade above the midband. At the unity-gain frequency, the phase shift is in the vicinity of 270°. When the phase shift reaches 360°, the voltage gain is less than 1 and oscillations are impossible.

compensating diodes These are the diodes used in a class B push-pull emitter follower. These diodes have current-voltage curves that match those of the emitter diodes. Because of this, the diodes compensate for changes in temperature.

coupling capacitor A capacitor used to transmit an ac signal from one node to another.

coupling circuit A circuit that couples a signal from a generator to a load. The capacitor is in series with the Thevenin resistance of the generator and the load resistance.

covalent bond The shared electrons between the silicon atoms in a crystal represent covalent bonds because the adjacent silicon atoms pull on the shared electrons, just as two tug-of-war teams pull on a rope.

critical frequency Also known as the *cutoff frequency, break frequency, corner frequency,* etc. This is the frequency where the total resistance of an *RC* circuit equals the total capacitive reactance.

crowbar The metaphor used to describe the action of an SCR when it is used to protect a load against supply overvoltage.

crystal The geometric structure that occurs when silicon atoms combine. Each silicon atom has four neighbors, and this results in a special shape called a crystal.

current booster A device, usually a transistor, that increases the maximum allowable load current of an op-amp circuit.

current feedback This is a type of feedback where the feedback signal is proportional to the output current.

current limiting Electronically reducing the supply voltage so that the current does not exceed a predetermined limit. This is necessary to protect the diodes and transistors, which usually blow out faster than the fuse under shorted-load conditions.

current mirror A circuit that acts as a current source whose value is a reflection of current through a biasing resistor and a diode.

current source Ideally, this is an energy source that produces a constant current through a load resistance of any value. To a second approximation, it includes a very high resistance in parallel with the energy source.

cutoff frequency Identical to the critical frequency. The name *cutoff* is preferred when you are discussing filters because that's what most people use.

cutoff point Approximately the same as the lower end of the load line. The exact cutoff point occurs where base current equals zero. At this point, there is a small collector leakage current, which means the cutoff point is slightly above the lower end of the dc load line.

cutoff region The region where the base current is zero in a *CE* connection. In this region, the emitter and collector diodes are nonconducting. The only collector current is the very small current produced by minority carriers and surface leakage current.

Darlington transistor Two transistors connected to get a very high value of β. The emitter of the first transistor drives the base of the second transistor.

dc equivalent circuit What remains after you open all capacitors.

dc return This refers to a path for direct current. Many transistor circuits won't work unless a dc path exists between all three terminals and ground. A diff amp and an op amp are examples of devices that must have dc return paths from their input pins to ground.

dc-to-dc converter A circuit that converts dc voltage of one value to dc voltage at another value. Usually, the dc input voltage is chopped up or changed to a rectangular voltage. This is then stepped up or down as needed, rectified, and filtered to get the output dc voltage.

dc value The same as the average value. For a time-varying signal, the dc value equals the average value of all the points on the waveform. A dc voltmeter reads the average value of a time-varying voltage.

decade A factor of 10. Often used with frequency ratios of 10, as in a decade of frequency referring to a 10:1 change in frequency.

decibel voltage gain This is a defined voltage gain given by 20 times the logarithm of the ordinary voltage gain.

defining formula A formula or an equation used to define or give the mathematical meaning of a new quantity. Before the defining formula is used for the first time, the quantity does not appear in any other formula.

depletion layer The region at the junction of p- and n-type semiconductors. Because of diffusion, free electrons and holes recombine at the junction. This creates pairs of oppositely charged ions on each side of the junction. This region is depleted of free electrons and holes.

depletion-mode MOSFET A FET with an insulated gate that relies on the action of a depletion layer to control the drain current.

derating factor A value that tells you how much to reduce the power rating for each degree above the reference temperature given on the data sheet.

derived formula A formula or an equation that is a mathematical rearrangement of one or more existing equations.

diff amp A two-transistor circuit whose ac output is an amplified version of the ac input signal between the two bases.

diode A pn crystal. A device that conducts easily when forward-biased and poorly when reverse-biased.

direct coupling Using a direct wire connection instead of a coupling capacitor between stages. For this to succeed, the designer has to make sure the dc voltages of the two points being connected are approximately equal before the direct connection is made.

discrete circuit A circuit whose components, such as resistors, transistors, etc., are soldered or otherwise connected mechanically.

donor A pentavalent atom, one that has five valence electrons. Each pentavalent atom produces one free electron in a silicon crystal.

doping Adding an impurity element to an intrinsic semiconductor to change its conductivity. Pentavalent or donor impurities increase the number of free electrons, and trivalent or acceptor impurities increase the number of holes.

duty cycle The width of a pulse divided by the period between pulses. Usually, you multiply by 100 percent to get the answer as a percentage.

efficiency The ac load power divided by the dc power supplied to the circuit multiplied by 100 percent.

emitter The part of a transistor that is the source of carriers. For *npn* transistors, the emitter sends free electrons into the base. For *pnp* transistors, the emitter sends holes into the base.

emitter bias The best way to bias a transistor for operation in the active region. The key idea is to set up a fixed value of the emitter current.

emitter diode The diode formed by the emitter and base of a transistor.

emitter follower Identical to a *CC* amplifier. The name *emitter follower* caught on because it describes the action better. The ac emitter voltage follows the ac base voltage.

enhancement-mode MOSFET A FET with an insulated gate that relies on an inversion layer to control its conductivity.

error voltage The voltage between the two input terminals of an op amp. It is identical to the differential input voltage of the op amp.

experimental formula A formula or an equation discovered through experiment or observation. It represents an existing law in nature.

extrinsic Refers to a doped semiconductor.

failure A necessary step to success. You have to make mistakes to grow. If you never fail, your goals are set too low. If you always fail, your goals are set too high. The trick is to set goals that are just outside your comfort zone, goals that produce both failure and success. As a wise person once said, "Show me someone who has never failed, and I'll show you a failure."

field-effect transistor A transistor that depends on the action of an electric field to control its conductivity.

flag A voltage that indicates an event has taken place. Typically, a low voltage means the event has not occurred, while a high voltage means that it has. The output of a comparator is an example of a flag.

floating load This is load that has nonzero node voltages on each. You can spot it on a schematic diagram by the fact that neither end of the load is grounded.

foldback current limiting Simple current limiting allows the load current to reach a maximum value while the load voltage is reduced to zero. Foldback current limiting takes this one step further. It allows the current to reach a maximum value. Then further decreases in the load resistance reduce both the load current and the load voltage. The main advantage of foldback limiting is less power dissipation in the pass transistor under shorted-load conditions.

forward bias Applying an external voltage to overcome the barrier potential.

free electron One that is loosely held by an atom. Also known as a *conduction-band electron* because it travels in a large orbit, equivalent to a high energy level.

frequency response The graph of voltage gain versus frequency for an amplifier.

full-wave rectifier A rectifier with a center-tapped secondary winding and two diodes that act as back-to-back half-wave rectifiers. One diode supplies one-half of the output, and the other diode supplies that other half. The output is a full-wave rectified voltage.

gate-source cutoff voltage The voltage between the gate and the source that reduces the drain current of a depletion-mode device to approximately zero.

germanium One of the first semiconductor materials to be used. Like silicon, it has four valence electrons.

go/no-go test A test or measurement where the readings are distinctly different, really high or really low.

ground loop If you use more than one ground point in a multistage amplifier, the resistance between the ground points will produce small unwanted feedback voltages. This is a ground loop. It can cause unwanted oscillations in some amplifiers.

h parameters An early mathematical method for representing transistor action. Still used on data sheets.

half-wave rectifier A rectifier with only one diode in series with the load resistor. The output is a half-wave rectified voltage.

hard saturation Operating a transistor at the upper end of the load line with a base current that is one-tenth of the collector current. The reason for the overkill is to make sure the transistor remains saturated under all operating conditions, temperature conditions, transistor replacement, etc.

heat sink A mass of metal attached to the case of a transistor to allow the heat to escape more easily.

high-frequency border The frequency above which a capacitor acts as an ac short. Also, the frequency where the reactance is one-tenth of the total series resistance.

holding current The minimum current through a thyristor that can keep it latched in the conducting stage.

hole A vacancy in the valence orbit. For instance, each atom of a silicon crystal normally has eight electrons in the valence orbit. Heat energy may dislodge one of the valence electrons, producing a hole.

hysteresis The difference between the two trip points of a Schmitt trigger. When used elsewhere, hysteresis refers to the difference between the two trip points on the transfer characteristic.

ideal approximation The simplest equivalent circuit of a device. It includes only a few basic features of the device and ignores many others of less importance.

ideal diode The first approximation of a diode. The viewpoint is to visualize the diode as an intelligent switch that closes when forward-biased and open when reverse-biased.

ideal transistor The first approximation of a transistor. It assumes a transistor has only two parts: an emitter diode and a collector diode. The emitter diode is treated as an ideal diode, while the collector diode is a controlled current source. The current through the emitter diode controls the collector current source.

initial slope of sine wave The earliest part of a sine wave is a straight line. The slope of this line is the initial slope of the sine wave. This slope depends on the frequency and peak value of the sine wave.

input bias current The average of the two input currents to a diff amp or an op amp.

input offset current The difference of the two input currents to a diff amp or an op amp.

input offset voltage If you ground both inputs of an op amp, you will still have an output offset voltage. The input offset voltage is defined as the input voltage needed to eliminate the output offset voltage. The cause of input offset voltage is the difference in the V_{BE} curves of the two input transistors.

instrumentation amplifier This is a differential amplifier with high input impedance and high CMRR. You find this type of amplifier as the input stage of measuring instruments like oscilloscopes.

integrated circuit A device that contains its own transistors, resistors, and diodes. A complete IC using these microscopic components can be produced in the space occupied by a discrete transistor.

integrator A circuit that performs the mathematical operation of integration. One popular application is generating ramps from rectangular pulses. This is how the time base is generated in oscilloscopes.

intrinsic Refers to a pure semiconductor. A crystal that has nothing but silicon atoms is pure or intrinsic.

intuition Thinking processes that cannot be explained verbally because they occur in the nonverbal right brain. In fact, Sperry says it is impossible to explain in words how the right brain works. If he is right, it probably means that computers will never be able to duplicate the processes in right brain, which border on the mystical.

inverting input The input to a diff amp or an op amp that produces an inverted output.

inverting voltage amplifier As the name implies, the

amplified output voltage is inverted with respect to the input voltage.

junction The border where p- and n-type semiconductors meet. Unusual things happen at a *pn* junction such as the depletion layer, the barrier potential, etc.

knee voltage The point or area on a graph of diode current versus voltage where the forward current suddenly increases. It is approximately equal to the barrier potential of the diode.

lag circuit Another name for a bypass circuit. The word *lag* refers to the angle of the output phasor voltage, which is negative with respect to angle of the input phase voltage. The phase angle may vary from 0 to $-90°$ (lagging).

latch Two transistors connected with positive feedback to simulate the action of a thyristor.

lead circuit Another name for a coupling circuit. The word *lead* refers to the angle of the output phasor voltage, which is positive with respect to angle of the input phase voltage. The phase angle may vary from 0 to $+90°$ (leading).

lead-lag circuit A circuit that combines a coupling and a bypass circuit. The angle of the output phasor voltage may be positive or negative with respect to the input phasor voltage. The phase angle may vary from -90 (lagging) to $+90°$ (leading).

leakage current Often used for the total reverse current of a diode. It includes thermally produced current as well as the surface leakage current.

LED driver A circuit that can produce enough current through a LED to get light.

left brain The left half of the human brain. It is like a computer that processes words and numbers. It operates sequentially, meaning it uses the results of each step to perform the next step. Ideally, it is characterized by the following adjectives: *verbal, mathematical, linear, logical, conscious, sequential, time-oriented*, etc.

lifetime The average amount of time between the creation and recombination of a free electron and a hole.

light-emitting diode A diode that radiates colored light such as red, green, yellow, etc. or invisible light such as infrared.

linear Usually refers to the graph of current versus voltage for a resistor.

linear op-amp circuit This is a circuit where the op amp never saturates under normal operating conditions. This implies that the amplified output has the same shape as the input.

linear regulator The series regulator is an example of a linear regulator. The thing that makes a linear regulator is the fact that the pass transistor operates in the active or linear region. Another example of a linear regulator is the shunt regulator. In this type of regulator, a transistor is shunted across the load. Again, the transistor operates in the active region, so the regulator is classified as a linear regulator.

line voltage The voltage from the power line. It has a nominal value of 115 V rms. In some places, it may be as low as 105 or as high as 125 V rms.

load power The ac power in the load resistor.

load regulation The change in the regulated load voltage when the load current changes from its minimum to its maximum specified value.

loop gain The product of the differential voltage gain A and the feedback fraction B. The value of this product is usually very large. If you pick any point in an amplifier with a feedback path, the voltage gain starting from this point and going around the loop is the loop gain. The loop gain is usually made up of two parts: the gain of the amplifier (greater than 1) and the gain of the feedback circuit (less than 1). The product of these two gains is the loop gain.

LSI Large-scale integration. Integrated circuits with more than 100 integrated components.

majority carrier Carriers are either free electrons or holes. If the free electrons outnumber the holes, the electrons are the majority carriers. If the holes outnumber the free electrons, the holes are the majority carriers.

maximum forward current The maximum amount of current that a forward-biased diode can withstand before burning out or being seriously degraded.

measured voltage gain The voltage gain that you calculate from the measured values of input and output voltage.

metaphor Use of words in a way that the literal left brain cannot understand, but the wholistic right brain can. *Example:* "The ugly duckling became a beautiful swan." The left brain takes this literally and thinks about a duck and a swan. The right brain sees through the words and knows what they really mean. Being able to think in metaphors is critical to invention. Aristotle said, "The greatest thing by far is to be a master of metaphor."

midband We have defined this as $10f_1$ to $0.1f_2$. In this range of frequencies, the voltage gain is with 0.5 percent of the maximum voltage gain.

Miller's theorem It says a feedback capacitor is equivalent to two new capacitances, one across the input and the other across the output. The most significant thing is that the input capacitance is equal to the feedback capacitance times the voltage gain of an amplifier. This assumes an inverting amplifier.

minority carrier The carriers that are in the minority. (See the definition of majority carrier.)

mixer An op-amp circuit that can have a different voltage gain for each of several input signals. The total output signal is a superposition of the input signals.

monolithic IC An integrated circuit that is entirely on a single chip.

motorboating A low-pitched putt-putt sound that comes out of a loudspeaker. It indicates that an amplifier is oscillating at a low frequency. The cause is usually the power supply having too large a Thevenin impedance.

MPP value Also called the *output voltage swing*. This is the maximum unclipped peak-to-peak output of an amplifier. With an op amp, the MPP value is ideally equal to the difference of the two supply voltages.

MSI Medium-scale integration. Circuits with 10 to 100 integrated components.

multivibrator A circuit with positive feedback and two

active devices, designed so that one device conducts while the other cuts off. There are three types: a free-running multivibrator, a flip-flop, and a one-shot. The free-running or astable multivibrator produces a rectangular output, similar to a relaxation oscillator.

n-type semiconductor A semiconductor where there are more free electrons than holes.

negative feedback Feeding a signal back to the input of an amplifier that is proportional to the output signal. The returning signal has a phase that opposes the input signal.

noninverting input The input to a diff amp or an op amp that produces an in-phase output.

nonlinear device A device that has a graph of current versus voltage that is not a straight line. A device that cannot be treated as an ordinary resistor.

normalized variable A variable that has been divided by another variable with the same units or dimensions.

notch filter A filter that blocks a signal with at most one frequency.

octave A factor of 2. Often used with frequency ratios of 2, as in an octave of frequency referring to a 2:1 change in frequency.

ohmic region The part of the drain curves that starts at the origin and ends at the proportional pinchoff voltage.

op amp A high-gain dc amplifier that provides usable voltage gain for frequencies from 0 to over 1 MHz.

open Refers to a component or connecting wire that has an open circuit, equivalent to a high resistance approaching infinity.

optimum Q point The point where the ac load line has equal maximum signal swings on both half-cycles.

optocoupler A combination of a LED and a photodiode. An input signal to the LED is converted to varying light which is detected by the photodiode. The advantage is very high isolation resistance between the input and output.

oscillations The death of an amplifier. When an amplifier has positive feedback, it may break into oscillations, which is unwanted high-frequency signal. This signal is unrelated to the amplified input signal. Because of this, oscillations interfere with the desired signal. Oscillations make an amplifier useless. This is why a compensating capacitor is used with an op amp; it prevents the oscillations from occurring.

outboard transistor A transistor placed in parallel with a regulating circuit to increase the amount of load current that the overall circuit can regulate. The outboard transistor kicks in at a predetermined current level and supplies the extra current needed by the load.

output impedance Another term used for the Thevenin impedance of an amplifier. It means the amplifier has been Thevenized, so that the load sees only a single resistance in series with a Thevenin generator. This single resistance is the Thevenin or output impedance.

output offset voltage Any deviation or difference of the output voltage from the ideal output voltage.

overloading Using a load resistance so small that it decreases the voltage gain of an amplifier by a noticeable amount. In terms of the Thevenin theorem, overloading occurs when the load resistance is small compared to the Thevenin resistance.

parasitic oscillations Oscillations of a very high frequency that cause all sorts of strange things to happen. Circuits act erratically, oscillators may produce more than one output frequency, op amps will have unaccountable offsets, supply voltage will have unexplainable ripples, video displays will contain snow, etc.

peak detector The same as a rectifier with a capacitor input filter. Ideally, the capacitor charges to the peak of the input voltage. This peak voltage is then used for the output voltage of the peak detector, which is why the circuit is called a peak detector.

peak inverse voltage The maximum reverse voltage across a diode in a rectifier circuit.

peak value The largest instantaneous value of a time-varying voltage.

periodic An adjective that describes a waveform that repeats the same basic shape for cycle after cycle.

phase shift The difference in phase angle between phasor voltages at points A and B. For an oscillator, the phase shift around the amplifier and feedback loop at the resonant frequency must equal 360°, equivalent to 0°, for the oscillator to work.

phase splitter A circuit that produces two voltages of the same amplitude but opposite phase. It is useful for driving class B push-pull amplifiers. If you will visualize a swamped CE amplifier with a voltage gain of 1, then you will have a phase splitter because the ac voltages across the collector and emitter resistances are equal in magnitude and opposite in phase.

photodiode A reverse-biased diode that is sensitive to incoming light. The stronger the light, the larger the reverse minority-carrier current.

pinchoff voltage The border between the ohmic region and the current-source region of a depletion-mode device when the gate voltage is zero.

plug and chugger Someone who always reaches for a calculator and a book of formulas. One who substitutes values into one formula after another, hoping to get the right answer. One who is seldom sure of the answer because he or she hasn't thought the problem through. Don't walk on any bridges built by this person, and try to be in another part of the laboratory when she or he powers up the circuits.

pnp transistor A semiconductor sandwich. It contains an n region between two p regions.

positive clamper A circuit that produces a positive dc shift of a signal by moving all the input signal upward until the negative peaks are at zero and the positive peaks are at $2V_p$.

positive feedback Feedback where the returning signal aids or increases the effect of the input voltage.

positive limiter A circuit that clips off the positive parts of the input signal.

power bandwidth The highest frequency that an op amp can handle without distorting the output signal. The power bandwidth is inversely proportional to the peak value.

power dissipation The product of voltage and current in a resistor or other nonreactive device. Rate at which heat is produced within a device.

power gain The ratio of output power to input power.

power transistor A transistor that can dissipate more than 0.5 W. Power transistors are physically larger than small-signal transistors.

predicted voltage gain The voltage gain you calculate from the circuit values on a schematic diagram. For a *CE* stage, it equals the ac collector resistance divided by the ac resistance of the emitter diode.

process A step-by-step routine followed to solve a problem. It is like a map that gets you from the given data to the final solution.

proportional pinchoff voltage The border between the ohmic region and the current-source region for any gate voltage.

prototype A basic circuit that a designer can modify to get more advanced circuits.

***p*-type semiconductor** A semiconductor where there are more holes than free electrons.

pullup resistor A resistor that the user has to add to an IC device to make it work properly. One end of the pullup resistor is connected to the device, and the other end is connected to the positive supply voltage.

push-pull connection Use of two transistors in a connection that makes one of them conduct for half a cycle while the other is turned off. In this way, one of the transistors amplifies the first half-cycle, and the other amplifies the second half-cycle.

r' parameters One way to characterize a transistor. This model uses quantities like β and r_e'.

recombination The merging of a free electron and a hole.

rectifier diode A diode optimized for its ability to convert ac to dc.

reductio ad absurdum A trick used when a device may be operating as a current source or as a resistor. You assume a current source and proceed with the calculations. If any contradictory answers turn up, you know your original assumption was wrong. Then you can change to the resistor model and finish off the calculations. Reductio ad absurdum usually works whenever you have a two-state system and don't know which state it is in.

reference voltage Usually, a very precise and stable voltage derived from a zener diode with a breakdown voltage between 5 to 6 V. In this range, the temperature coefficient of the zener diode is approximately zero, which means its zener voltage is stable over a large temperature range.

relaxation oscillator A circuit that creates or generates an ac output signal without an ac input signal. This type of oscillator depends on the charging and discharging of a capacitor through a resistor.

resonant frequency The frequency of a lead-lag circuit or the frequency of an *LC* tank circuit where the voltage gain and phase shift are suitable for oscillations.

reverse-bias Applying an external voltage across a diode to aid the barrier potential. The result is almost zero current. The only exception of when you can exceed the breakdown voltage. If the reverse voltage is large enough, it can produce breakdown through either avalanche or the zener effect.

reverse saturation current The same as the minority-carrier current in a diode. This current exists in the reverse direction.

right brain The right half of the human brain. It contains a visual and nonverbal processing unit unlike any computer currently in existence. It solves problems simultaneously rather than sequentially. It ignores all rules and creates all possibilities, both good and bad, useful and useless, etc. It works best when you suspend judgment and allow the mind to go where it will. It is characterized by these adjectives: *intuitive, artistic, emotional, timeless, subconscious, creative, playful, metaphorical,* etc.

ripple With a capacitor input filter, this is the fluctuation in load voltage caused by the charging and discharging of the capacitor.

ripple rejection Used with voltage regulators. It tells you how well the voltage regulator rejects or attenuates the input ripple. Data sheets usually list it decibels, where each 20 dB represents a factor-of-10 decrease in ripple.

rms value Used with time-varying signals. Also known as the *effective value* and the *heating value*. This is the equivalent value of a dc source that would produce the same amount of heat or power over one complete cycle of the time-varying signal.

safety factor The leeway between the actual operating current, voltage, etc. and the maximum rating specified on a data sheet.

saturated current gain The current gain of a transistor in the saturation region. This value is less than the active current gain. For soft saturation, the current gain is slightly less than the active current gain. For hard saturation, the current gain is approximately 10.

saturation point Approximately the same as the upper end of the load line. The exact location of the saturation point is slightly lower because the collector-emitter voltage is not quite zero.

saturation region The part of the collector curves that starts at the origin and slopes upward to the right until it reaches the beginning of the active or horizontal region. When a transistor operates in the saturation region, the collector-emitter voltage is typically only a few tenths of a volt.

Schmitt trigger A comparator with hysteresis. It has two trip points. This makes it immune to noise voltages, provided their peak-to-peak values are less than the hysteresis.

Schottky diode A special-purpose diode with no depletion layer, extremely short reverse recovery time, and the ability to rectify high-frequency signals.

second approximation An approximation that adds a few more features to the ideal approximation. For a diode or transistor, this approximation includes the barrier potential in the model of the device. For silicon diodes or transistors, this means 0.7 V is included in the analysis.

self-bias The bias you get with a JFET because of the voltage produced across the source resistor.

series regulator This is the most common type of linear regulator. It uses a transistor in series with the load. The regulation works because a control voltage to the base of the transistor changes its current and voltage as needed to keep the load voltage almost constant.

series switch A type of JFET analog switch where the JFET is in series with the load resistor.

short One of the common troubles that may occur. A short occurs when an extremely small resistance is approaching zero. Because of this, the voltage across a short approaches zero, but the current may be very large. A component may be internally shorted, or it may be externally shorted by a solder splash or miswire.

short-circuit output current The maximum output current that an op amp can produce for a load resistor of zero.

short-circuit protection A feature of most modern power supplies. It usually means the power suppy has some form of electronic current limiting that prevents excessive load currents under shorted-load conditions.

shunt switch A type of JFET analog switch where the JFET is in shunt with the load resistor.

silicon The most widely used semiconductor material. It has 14 protons and 14 electrons in orbit. An isolated silicon atom has four electrons in the valence orbit. A silicon atom that is part of a crystal has eight electrons in the valence orbit because the four neighbors share one of the electrons.

silicon controlled rectifier A thyristor with three external leads called the *anode*, *cathode*, and *gate*. The gate can turn the SCR on, but not off. Once the SCR is on, you have to reduce the current to less than the holding current to shut off the SCR.

sink If you visualize water disappearing down a kitchen sink, you will have the general idea of what engineers or technicians mean when they talk about a current sink. It is the point that allows current to flow into ground or out of ground.

slew rate The maximum rate that the output voltage of an op amp can change. It causes distortion for high-frequency large-signal operation.

small-signal operation This refers an input voltage that produces only small fluctuations in the current and voltage. Our rule for small-signal transistor operation is a peak-to-peak emitter current less than 10 percent of the dc emitter value.

small-signal transistor A transistor that can dissipate 0.5 W or less.

soft saturation Operation of the transistor at the upper end of the load line with just enough base current to produce saturation.

source follower The leading JFET amplifier. You see it used more than any other JFET amplifier.

source regulation The change in the regulated output voltage when the input or source voltage changes from its minimum to its maximum specified voltage.

SSI Small-scale integration. Refers to integrated circuits with 10 or fewer integrated components.

stepdown transformer A transformer with more primary turns than secondary turns. This results in less secondary voltage than primary voltage.

stiff current source A current source whose internal resistance is at least 100 times larger than the load resistance.

stiff voltage divider A voltage divider whose loaded output voltage is within 1 percent of its unloaded output voltage.

stiff voltage source A voltage source whose internal resistance is at least 100 times smaller than the load resistance.

stray wiring capacitance The unwanted capacitance between connecting wires and ground.

summer An op-amp circuit whose output voltage is the sum of the two or more input voltages.

superposition When you have several sources, you can determine the effect produced by each source acting alone and then add the individual effects, to get the total effect of all sources acting simultaneously.

surface-leakage current A reverse current that flows along the surface of a diode. It increases when you increase the reverse voltage.

surge current The large initial current that flows through the diodes of a rectifier. It is the direct result of charging the filter capacitor, which initially is uncharged.

swamped amplifier A *CE* stage with a feedback resistor in the emitter circuit. This feedback resistor is much larger than the ac resistance of the emitter diode.

switching regulator A linear regulator uses a transistor that operates in the linear region. A switching regulator uses a transistor that switches between saturation and cutoff. Because of this, the transistor operates in the active region only during the short time that it is switching states. This implies that power dissipation of the pass transistor is much smaller than in a linear regulator.

temperature coefficient The rate of change of a quantity with respect to the temperature.

thermal energy Heat energy.

thermal runaway As a transistor heats, its junction temperature increases. This increases the collector current, which forces the junction temperature to increase further, producing more collector current, etc., until the transistor is destroyed.

thermal shutdown A feature found in modern three-terminal IC regulators. When the regulator exceeds a safe operating temperature, the pass transistor is cut off and the output voltage goes to zero. When the device cools, the pass transistor is again turned on. If the original cause of the excessive temperature is still present, the device again shuts off. If the cause has been removed, the device works normally. This feaure makes the regulator almost indestructible.

thermistor A device whose resistance experiences large changes with temperature.

Thevenin's theorem A fundamental theorem that says any circuit driving a load can be converted to a single generator and series resistance.

third approximation An accurate approximation of a

diode or transistor. Used for designs that need to take into account as many details as possible.

threshold voltage The voltage that turns on an enhancement-mode MOSFET. At this voltage, an inversion layer connects the source to the drain.

thyristor A four-layer semiconductor device that acts as a latch.

transconductance The ratio of ac output current to ac input voltage. A measure of how effectively the input voltage controls the output current.

transfer function The inputs and outputs of an op-amp circuit may be voltages, currents, or a combination of the two. When you use complex numbers for the input and output quantities, the ratio of output to input becomes a function of the frequency. The name for the ratio is the transfer function.

triac A thyristor that can conduct in both directions. Because of this, it is useful for controlling alternating current. It is equivalent to two SCRs in parallel with opposite polarities.

trial and error Suppose you have a problem involving two simultaneous equations. Instead of solving this in the usual left-brain mathematical way, you can guess an answer and then calculate all the unknowns. One of the calculated unknowns is the very answer that you guessed. You compare the calculated and guessed answers to see how different they are. Then you guess another answer that will close the gap between the guessed and calculated answers. After several trials, the gap becomes small enough that you have an approximate answer.

trigger A sharp pulse of voltage and current that is used to turn on a thyristor or other switching device.

trigger current The minimum current needed to turn on a thyristor.

trigger voltage The minimum voltage needed to turn on a thyristor.

trip point The value of the input voltage that switches the output of a comparator or Schmitt trigger.

two-state output This is the output voltage from a digital or switching circuit. It is referred to as two-state because the output has only two stable states, low and high. The region between the low and high voltages is unstable because the circuit cannot have any value in this range except temporarily when switching between states.

unijunction transistor Abbreviated UJT, this low-power thyristor is useful in electronic timing, waveshaping, and control applications.

unity-gain frequency The frequency where the voltage gain of an op amp is 1. It indicates the highest usable frequency. It is important because it equals the gain-bandwidth product.

universal curve A solution in the form of a graph that solves a problem for a whole class of circuits. The universal curve for self-biased JFETs is an example. In this universal curve I_D/I_{DSS} is graphed for R_D/R_{DS}.

unwanted bypass circuit A circuit that appears in the base or collector sides of a transistor because of internal transistor capacitances and stray wiring capacitances.

upside-down *pnp* bias When you have a positive power supply and a *pnp* transistor, it is customary to draw the transistor upside-down. This is especially helpful when the circuit uses both *npn* and *pnp* transistors.

varactor A diode optimized for a reverse capacitance. The larger the reverse voltage, the smaller the capacitance.

varistor A device that acts like two back-to-back zener diodes. Used across the primary winding of a power transformer to prevent line spikes from entering the equipment.

virtual ground A type of ground that appears at the inverting input of an op amp that uses negative feedback. It's called virtual ground because it has some of, but not all, the effects of a mechanical ground. Specifically, it is ground for voltage but not for current. A node that is a virtual ground has 0 V with respect to ground, but the node has no path for current to ground.

voltage amplifier An amplifier that has its circuit values selected to produce a maximum voltage gain.

voltage-controlled device A device like a JFET or MOSFET whose output is controlled by an input voltage.

voltage feedback This is a type of feedback where the feedback signal is proportional to the output voltage.

voltage follower An op-amp circuit that uses noninverting voltage feedback. The circuit has a very high input impedance, a very low output impedance, and a voltage gain of 1. It is ideal for use as a buffer amplifier.

voltage gain This is defined as the output voltage divided by the input voltage. Its value indicates how much the signal is amplified.

voltage regulator A device or circuit that holds the load voltage almost constant, even though the load current and source voltage are changing. Ideally, a voltage regulator is a stiff voltage source with an output or Thevenin resistance that approaches zero.

voltage source Ideally, an energy source that produces a constant load voltage for any value of the load resistance. To a second approximation, it includes a small internal resistance in series with the source.

voltage-to-current converter A circuit that is equivalent to a controlled current source. The input voltage controls the current. The current is then constant and independent of the load resistance.

Wien-bridge oscillator An *RC* oscillator consisting of an amplifier and a Wien bridge. This is the most widely used low-frequency oscillator. It is ideal for generating frequencies from 5 Hz to 1 MHz.

zener effect Sometimes called *high-field emission*, this occurs when the intensity of the electric field becomes high enough to dislodge valence electrons in a reverse-biased diode.

zener resistance The bulk resistance of a zener diode. It is very small compared to the current-limiting resistance in series with the zener diode.

zener voltage The breakdown voltage of a zener diode. This is the approximate voltage out of a zener voltage regulator.

ANSWERS TO ODD-NUMBERED PROBLEMS

CHAPTER 1

1-1. 50 Ω or more 1-3. 100 Ω or more 1-5. 0.1 V 1-7. 200 kΩ or less 1-9. 1 kΩ
1-11. 4.8 mA; no 1-13. 6 mA, 4 mA, 3 mA, 2.4 mA, 2 mA, 1.71 mA, and 1.5 mA 1-15. Nothing;
4 kΩ 1-17. 100 V and a series resistance of 10 kΩ 1-19. R_1 is shorted 1-21. One possible trouble
is an open connecting lead between the voltage source and the top of R_1. Another possible trouble is a defective
voltage source, one that has no voltage. 1-23. 0.04 Ω 1-25. Disconnect the load resistor; then, measure
the Thevenin voltage. 1-27. Older methods work but they are not as efficient as Thevenin's theorem when
the only change is the load resistance. 1-29. A voltage source of 1000 V and a series resistance of 1 MΩ will
simulate a stiff current source of 1 mA. 1-31. The two resistances in the voltage divider have 30 kΩ and
15 kΩ. This produces a Thevenin resistance of 10 kΩ. 1-33. Measure the Thevenin voltage. Connect a
resistor. Measure the load voltage. Calculate the load current. Subtract the load voltage from the Thevenin voltage
to get the voltage across the Thevenin resistance. Divide this internal voltage by the load current to get the
Thevenin resistance. 1-35. Trouble 1: R_1 shorted 1-37. Trouble 3: R_2 shorted; Trouble 4: R_3 open;
Trouble 5: R_3 shorted; Trouble 6: R_2 open

CHAPTER 2

2-1. -3 2-3. *a*. Semiconductor *b*. Conductor *c*. Semiconductor *d*. Conductor 2-5. *a*. *p*-type *b*. *n*-type
c. *p*-type *d*. *n*-type *e*. *p*-type 2-7. 1.74 nA and 320 nA

CHAPTER 3

3-1. 18.2 mA 3-3. 500 mA 3-5. 10 mA 3-7. 16 mA 3-9. 19.3 mA, 19.3 V, 373 mW,
13.5 mW, 386 mW 3-11. 30.4 mA, 14.3 V, 435 mW, 21.3 mW, 456 mW 3-13. 0, 15 V
3-15. 9.65 mA 3-17. 15.2 mA 3-19. Open 3-21. The resistor is open, the diode is shorted, etc.
3-23. No supply voltage, R_1 open, R_2 shorted, open connection between the supply voltage and the measured
junction 3-25. Cathode band; diode arrow points toward this band 3-27. 1N914: $R_F = 100 \ \Omega$,
$R_R = 800$ MΩ, 1N4001: $R_F = 1.1 \ \Omega$, $R_R = 5$ MΩ, 1N1185: $R_F = 0.095 \ \Omega$, $R_R = 21.7$ kΩ
3-29. 22.9 Ω. The nearest standard value is 22 Ω. 3-31. 4.47 mA 3-33. When the source of 15 V is
normal, the upper diode conducts and produces a load voltage of 14.3 V. The lower diode is open during this
normal operation. If the 15 V fails, the upper diode opens up. Then the battery takes over and forces the lower
diode to turn on. 3-35. An increase in R_1 has no effect on the value of V_A, which is an independent quantity,
but it does produce a decrease in all the dependent quantities. 3-37. V_A, V_B, V_C, I_1, I_2, P_1, and P_2

CHAPTER 4

4-1. 35.4 V 4-3. 651 V 4-5. 208 mA 4-7. 54.2 V and 53.5 V 4-9. 14.9 V and 17.7 V
4-11. 6.54 mA (ideal) and 6.27 mA (second) 4-13. 0.245 V 4-15. 18 V, 18 mA, 9 mA, 2.7 mA, 18 V,
and 9.45 4-17. Ideal: 28.3 V and 0.236 V; second: 26.9 V and 0.224 V 4-19. Possible troubles include
an open diode or an open connection in one of the diode branches. 4-21. You should check the load
resistance to see if it is being shorted out. 4-23. Ideal and ignore ripple: $V_L = 33.9$ V, $C = 252$ μF, $I_O = 51$ mA,
and PIV = 33.9 V; second and ignore ripple: $V_L = 32.5$ V, $C = 252$ μF, $I_O = 49.2$ mA, and PIV = 33.9 V;

969

second and include ripple: V_L = 30.9 V, C = 252 μF, I_O = 46.8 mA, and PIV = 33.9 V   4-25. We can look up the sine of the angle every 5 degrees between 0° and 90°. There are 19 samples including the sine of 0°. By adding up the sine values and dividing by 19, we get 0.629. This is close to the exact value of 0.636. If a more accurate answer is needed, we could use a smaller interval, say every degree.   4-27. 44.2 A   4-29. Trouble 2: Diode open; Trouble 3: Load resistor shorted   4-31. Trouble 6: Load resistor open; Trouble 7: Secondary winding open

CHAPTER 5

5-1. 24.2 mA   5-3. 26.9 mA   5-5. 14.6 V   5-7. 19.6 mA   5-9. P_S is 194 mW, P_L is 96 mW, and P_Z is 195 mW   5-11. 33.7 mV   5-13. 5.91 mA   5-15. 13 mA   5-17. 200 mW   5-19. 11.4 V, 12.6 V   5-21. a. 0 b. 16.4 V c. 0 d. 0   5-23. Check for a short across the 330 Ω.   5-25. 12.2 V   5-27. Many designs are possible here. One design is a 1N754, a series resistance of 270 Ω, and a load resistance of 220 Ω. This design results in a series current of 48.9 mA, a load current of 30.9 mA, and a zener current of 18 mA.   5-29. 26 mA   5-31. 7.98 V   5-33. Trouble 2: Wire BD open   5-35. Trouble 5: No supply voltage; Trouble 6: R_L open

CHAPTER 6

6-1. 0.05 mA   6-3. 3.75 mA   6-5. With the second approximation 28.2 μA   6-7. The minimum base resistance is 314 kΩ, so the base current is 29.6 μA.   6-9. 0.35 W   6-11. Ideal: 13.1 V and 20.9 mW; second: 13.4 V and 18.4 mW   6-13. −55 to 150°C   6-15. It is destroyed or seriously degraded.   6-17. a. Increase b. Increase c. Increase d. Decrease e. Increase f. Decrease   6-19. 166   6-21. 0.01 V, 0.1 V, and 0.5 V   6-23. With the second approximation, 5.64 mA   6-25. 0.35 V   6-27. V_C increases and P_D increases. All other variables show no change.   6-29. The variables that show no change are V_A, V_B, V_D, I_B, I_C, P_B.

CHAPTER 7

7-1. Approximately 18   7-3. The load line passes through 6.06 mA and 20 V.   7-5. The load line passes through 20 mA and 20 V.   7-7. The load line passes through 10.6 mA and 5 V.   7-9. The load line passes through 5 mA and 5 V.   7-11. Second: 10.8 V and 19.2 V   7-13. 4.7 V (second)   7-15. 3.95 V and 5.38 V   7-17. a. No b. No c. Yes d. No   7-19. 5 V and 0.2 V   7-21. 13.2 V   7-23. 3.43 V   7-25. 8.34 V   7-27. 11 mA and 3 V   7-29. D, D, U   7-31. N, N, U   7-33. D, D, U   7-35. N, N, U   7-37. Ideally, 10 V at both the base and the collector   7-39. Transistor open, emitter resistor open, no ground on emitter, no base supply voltage, etc.   7-41. A solder bridge between the collector and emitter, collector resistance much higher than 3.6 kΩ producing saturation, emitter resistance much smaller than 1 kΩ producing saturation, open 3.6 kΩ, no 10 V, etc.   7-43. It sounds impractical for mass production. You can hand-select but the yield is going to be low. Another more serious problem is the effect of temperature changes causing β_{dc} to change.   7-45. a. If the troubleshooter uses the typical voltmeter whose common lead is grounded, the emitter will be shorted to ground. What happens next depends on how stiff the + 1.8 V is at the base. If the base voltage holds at + 1.8 V, the emitter diode will be destroyed because the emitter is grounded. The voltmeter then will read + 10 V. b. If a VOM with 20 kΩ per volt is used, the input impedance is 200 kΩ on the 10-V range. This input impedance is much higher than the 3.6 kΩ and 1 kΩ, so the reading will be valid and slightly less than the theoretical 4.94 V because of the small loading error.   7-47. 7.2 μA   7-49. 11.9 mA   7-51. 1.13 V   7-53. 0 V   7-55. V_{CC} affects only two variables, V_C and P_D. It causes both to increase.   7-57. V_B, V_E, I_E, I_C, I_B, and P_E

CHAPTER 8

8-1. 3.8 V, 11.3 V   8-3. 1.63 V, 5.22 V   8-5. 4.12 V, 6.14 V   8-7. The dc load line passes through 5.89 mA and 21.2 V. The Q point is at 3.8 mA and 7.5 V.   8-9. The dc load line passes through 55.8 μA

and 8.37 V. The Q point is at 31.9 μA and 3.59 V.   8-11. 52 μA, 60 μA   8-13. 1.43 mA, 8.28 V
8-15. 7.57 V, 8.92 V   8-17. −4.94 V or 4.94 V with the emitter more positive than the collector
8-19. −1.1 V, −6.04 V   8-21. *a*. Decrease *b*. Increase *c*. Decrease *d*. Increase *e*. Increase *f*. Unchanged
8-23. *a*. Zero *b*. 7.57 V *c*. Zero *d*. 14.6 V *e*. Zero   8-25. The 2N3904 will be destroyed.   8-27. One
way to destroy the transistor is by increasing the supply voltage until it exceeds the breakdown voltage. Another
possibility is to short R_C. This increases the transistor power to around 2.6 W, which may exceed the power rating
of the transistor.   8-29. Q_1: 9 V. Q_2: 8.97 V. Q_3: 8.45 V   8-31. 8.8 V   8-33. 27.5 mA
8-35. Trouble 1: R_1 short   8-37. Trouble 3: R_C short; Trouble 4: *CES*, all transistor terminals shorted
8-39. Trouble 7: R_E open; Trouble 8: R_2 shorted   8-41. Trouble 11: No V_{CC}; Trouble 12: Base-emitter diode
open

CHAPTER 9

9-1. 0.155 mA, 1.27 V   9-3. Unchanged   9-5. Doubles   9-7. 0.435 Hz; 4.35 Hz   9-9. Drops
in half; unchanged; drops in half; drops in half   9-11. 9.57 V; 2.65 kΩ   9-13. Draw ac equivalent
circuits similar to Fig. 9-7*a* and *b*. In the base circuit, 381 Ω. In the collector circuit, 543 Ω   9-15. 0.5 mA
9-17. 0.978 mA   9-19. 8.62 Ω   9-21. 1 kΩ   9-23. 320 Ω, 174 Ω   9-25. 5.8 Ω. Slightly
larger   9-27. The leads of the capacitor have a small amount of inductance. The reactance is given by $2\pi fL$.
At a high enough frequency, the capacitor is an ac short but its leads have significant inductive reactance. This is
why a voltage appears across the capacitor at high enough frequencies.   9-29. 14.6 Ω   9-31. 19.9 MHz;
nothing   9-33. Draw the ac equivalent of the first stage driving the second. The input impedance of the
second stage then is part of the ac collector resistance of the first stage. The input impedance of the base is 3.85
kΩ. This is in parallel with 1 kΩ and 4 kΩ to give an input impedance of the stage of 662 Ω.   9-35. 30.6 μF;
nearest standard size is 33 μF   9-37. An increase in R_1 increases r'_e, has no effect on r_c, increases $\beta r'_e$, and
increases z_{in}.   9-39. None of them decrease.   9-41. r_c. The ac collector resistance increases because it
equals R_C in parallel with R_L.   9-43. An increase in supply voltage will produce more dc emitter current. This
reduces re' and related quantities like $\beta r'_e$ and z_{in}.

CHAPTER 10

10-1. 1.28 kΩ   10-3. 1.28 kΩ   10-5. 2.58 kΩ   10-7. 119   10-9. 500   10-11. 160 mV
10-13. 89.9 mV   10-15. 117   10-17. 0.386 V   10-19. 13.1   10-21. 3.8 V   10-23. 2.48 V;
2.55 V   10-25. *a*. Increase *b*. Increase *c*. Increase *d*. Increase   10-27. An open coupling capacitor at
the second collector, an open connection anywhere between the second collector and the load resistor, a shorted
load resistor, and a shorted collector resistor   10-29. 297 mV; higher   10-31. 3.6 kΩ; nothing
happens; 3.24 kΩ   10-33. −268, −104   10-35. Trouble 2: R_E open   10-37. Trouble 5: C_2
open; Trouble 6: R_2 open   10-39. Trouble 9: No V_{CC}; Trouble 10: *BE* diode open

CHAPTER 11

11-1. 3.66 V   11-3. 13.2 V; 86.8 mV   11-5. 207 Ω; the nearest standard value is 200 Ω   11-7. 15.9 mA
11-9. 239 mW   11-11. 2.18%   11-13. 60.5 mW   11-15. 2 W   11-17. *a*. Increase
b. Increase *c*. Increase *d*. Decrease   11-19. The input signal is too large, so that clipping occurs on both
output peaks.   11-21. You probably would not want to buy the book because the maximum possible
efficiency is 100 percent. You cannot get more power out of a system than you put into it.   11-23. The dc
load line has a cutoff voltage of 12.9 V and a saturation current of 18.9 mA. The Q point is 9.77 mA and 6.2 V.
The ac load line passes through the Q point and also passes through a cutoff voltage of 11.5 V.   11-25. 5.2 V
11-27. The first stage: 2.12 V; the second stage: 11 V   11-29. It increases everything except the
efficiency. The change in efficiency is less than 1 percent at this signal level.   11-31. P_D and MPP. An
increase in R_2 increases the dc emitter voltage, increases the dc emitter current, decreases the collector-emitter
voltage, and decreases the transistor power. Also, since the stage is optimized for maximum MPP value with the
original values, the increase in R_2 decreases the MPP value.   11-33. P_L and efficiency. An increase in R_C
increases the voltage gain and the load voltage, which means more load power. In turn, this increases the

efficiency. 11-35. An increase in R_G decreases the load power and efficiency. It has no effect on transistor dissipation, dc supply power, MPP value, and efficiency. 11-37. P_L and efficiency. The increase in β increases the input impedance of the stage, producing more ac base voltage, which increases load power and efficiency.

CHAPTER 12

12-1. 77.1 kΩ   12-3. 0.956 V   12-5. 0.995, 0.951 V   12-7. 0.954 V   12-9. 10.2 V
12-11. 21.5 V   12-13. 330   12-15. 8.58 V   12-17. 220 mV   12-19. 48 kΩ, 7.2 MΩ
12-21. 9.36 mV   12-23. 16.6 V   12-25. 3.83 Ω   12-27. 1 kΩ   12-29. At the positive peak, we lose approximately 1.5 V across the last 1 kΩ because the base current for Q_3 has to flow through this 1 kΩ. At the negative peak, the voltage across the 100 Ω prevents the base of Q_2 from being pulled all the way down to ground. The approximate effect is to lose about 3 V in total swing. Therefore, the MPP value drops from the ideal 30 V to approximately 24 V.   12-31. The zener voltage should be 6.2 V. Assume β is 100. Then the base current is 5 mA. Let the zener current also equal 5 mA. This gives a series current of 10 mA and a series resistance of 880 Ω. 12-33. The two ac output signals each have a peak-to-peak value of 5 mV. The circuit produces two sine waves which are equal in magnitude and opposite in phase.   12-35. Trouble 1: C_4 open   12-37. Trouble 3: C_1 open   12-39. Trouble 5: Wire BC open   12-41. Trouble 7: Base-emitter diode of Q_2 open

CHAPTER 13

13-1. $4(10^{12})$ Ω   13-3. 0, 16 mA, −3 V   13-5. 50 Ω   13-7. The K factors are 1, 0.64, 0.36, 0.16, 0.04, and 0. The drain currents are 1 mA, 0.64 mA, 0.36 mA, 0.16 mA, 0.04 mA, and 0.   13-9. 13.8 V
13-11. 9 V   13-13. 16.7 V   13-15. 2.5 V   13-17. 7.5 V   13-19. 2.35 V   13-21. 10.6 V
13-23. 13.5 V   13-25. 17.5 V   13-27. 30 V   13-29. 3.89 V   13-31. 10 V   13-33. 7.5 MΩ
13-35. 20 mA, 5 V, 5 V to 30 V   13-37. 7.68 mA   13-39. 13 V   13-41. 10 V
13-43. It has no effect on any of the variables   13-45. I_D and R_{DS}. If the gate-source cutoff voltage increases, the K factor will increase in value, producing more drain current. Also, an increase in gate-source cutoff voltage for the same I_{DSS} produces an increase in R_{DS}.   13-47. The only variable that decreases is R_{GS}. An increase in I_{GSS} implies a decrease in R_{GS}.

CHAPTER 14

14-1. 12.5 V, 0.675 V, and −0.675 V   14-3. −1 V and 10.3 V   14-5. −0.77 V and 7.7 mA
14-7. Ideally 2.5 kΩ; or 2.4 kΩ for the nearest standard value   14-9. 5.2 mA and −1.4 V   14-11. 1.17 V
14-13. 1050 μmhos   14-15. 5.4 mV   14-17. Approximately 1.8 mV   14-19. 0.453 mV and 50 mV
14-21. 15 V and 0.577 V   14-23. 15 V and 0.938 V   14-25. Open coupling capacitor, no ac input voltage, or an open connection in the ac signal path   14-27. a. 283 Ω; the nearest standard value is 270 Ω b. 667 Ω; the nearest standard value is 680 Ω   14-29. 8.47 V; 16.2 mV   14-31. 181 Ω
14-33. 16 mA and 30 V; 2.94 mA and 0.588 V   14-35. Trouble 1: Wire EF open   14-37. Trouble 3: R_L open   14-39. Trouble 5: C_2 open; Trouble 6: R_G open   14-41. Trouble 9: Wire BC open; Trouble 10: R_S open

CHAPTER 15

15-1. 4.7 V   15-3. 160 mA   15-5. 7.3 V   15-7. 14.7 V   15-9. 18 V   15-11. 58 mA
15-13. Approximately 0 V and 15 V   15-15. 5 A   15-17. 0.98 A   15-19. Ideally, 12.6 V; To a second approximation, 13.3 V   15-21. Ideally, 14 V; To a second approximation, 14.7 V   15-23. 160 mA
15-25. 0.01 μF   15-27. 1 kHz   15-29. 16 mA and 1 A   15-31. 90.9 Hz and 1 kHz
15-33. Trouble 2: Power outlet   15-35. Trouble 4: Load shorted   15-37. Trouble 6: Bridge rectifier and filter   15-39. Trouble 8: Fuse

CHAPTER 16

16-1. 2.5 kHz to 500 kHz   16-3. 405 Hz   16-5. 15.8 Hz   16-7. 260 Hz   16-9. 505 Hz,
531 Hz   16-11. Ignoring R_E: 607 Hz, 1.71 kHz   16-13. 528 Hz   16-15. 79.6 MHz
16-17. 1000 pF and 5.03 pF   16-19. 52.9 kHz   16-21. 255 pF   16-23. 2.31 MHz
16-25. 44.7, 24.2, 9.95, 0.999   16-27. 22.4 dB   16-29. 100 dB   16-31. 98 dB, 79432
16-33. Draw a horizontal line between 50 Hz and 100 kHz. Then draw sloping lines below 50 Hz and above 100
kHz. The rate of drop is 20 dB per decade.   16-35. The cutoff frequency associated with this coupling
capacitor will be 100 times higher than it should be.   16-37. It degrades it because the long leads produce
more stray capacitance.   16-39. Since the stray capacitance is not specified, we will ignore it. Then, the cutoff
frequencies are 194 kHz and 10.6 MHz.   16-41. 251 W   16-43. 40 dB, 0 dB   16-45. 0.44 μs
16-47. The amplifier with a cutoff frequency of 1 MHz   16-49. It decreases f_1, but has no effect on the other
variables.   16-51. The only variable that decreases is f_1. The emitter bypass capacitor will affect the lower
cutoff frequency because it produces one of the critical frequencies.   16-53. It decreases f_1 and f_2. The dc
collector resistance affects the output coupling circuit and it also affects the collector bypass circuit.
16-55. It only increases f_2. The higher the f_T, the higher the upper cutoff frequency.

CHAPTER 17

17-1. 7.5 V   17-3. −0.05 V and −0.0409 V   17-5. 18.5 nA and 21.5 nA   17-7. 18.5 pA and
21.5 pA   17-9. 667 kΩ   17-11. 10 V, 10.4 V, and 9.6 V   17-13. ±2.25 V   17-15. ±1.35 V
17-17. Desired signal is 300 mV and a common-mode signal is 2.5 mV.   17-19. 10   17-21. 14.3 V
17-23. Ideally: 2 mA and 1 mA; second: 1.95 mA and 0.975 mA   17-25. 6.75 V   17-27. 1.86 mA, 25.4 V,
25.4 V   17-29. It increases V_{B2}, V_{in}, and V_{out}. It has no effect on V_{B1} and I_T.   17-31. None

CHAPTER 18

18-1. 67.7 Hz and 10.2 kHz   18-3. 10 MHz   18-5. 10 Hz   18-7. 75 Hz   18-9. 20 pF
18-11. No   18-13. 0.188 V/μs and 0.377 V/μs   18-15. No   18-17. 95.5 kHz   18-19. 31.8 kHz
and 255 kHz   18-21. 1.11 MHz   18-23. 239 kHz, 23.9 kHz, and 2.39 kHz   18-25. 6 V
18-27. 1 μV, 10 μV, and 100 μV   18-29. ±9.4 mV and ±15 V (ideal) because of saturation
18-31. 4.14 MHz and 414 kHz   18-33. 5 MHz, 1 MHz, 1 MHz, and 4 MHz   18-35. LM11C
18-37. Approximately 22.9 mV   18-39. 80 nA, 20 nA, 70 μV, and 7 V   18-41. 20 μV and 2 V
18-43. It increases V_1, V_{in}, and V_{out}. It has no effect on V_2, MPP, and f_{max}.   18-45. V_1, V_2, and MPP. The
increase in supply voltages will increase both base currents. Also, more supply voltage means the signal can swing
further before clipping occurs, which is why MPP increases.   18-47. The only variable that decreases is f_{max}.
It decreases because it is inversely proportional to peak voltage.

CHAPTER 19

19-1. 50   19-3. 0.5 μV and 2 μV   19-5. 150 mV and 3 μV   19-7. 101, 5, and 2   19-9. 0.15 V
19-11. 1 mV and 5 μA   19-13. 20,000 MΩ and 0.0125 Ω   19-15. 1982 MΩ and 0.0757 Ω; 40,000 MΩ
and 0.00375 Ω; 100,000 MΩ and 0.0015 Ω   19-17. 0.454 V   19-19. 2 V   19-21. 0.1 V
19-23. 2 kΩ   19-25. 5 kHz   19-27. 15 kHz   19-29. 2.65 V   19-31. c   19-33. a
19-35. No supply voltages, load resistor is shorted, no input voltage, open connection between op-amp output and
load resistor   19-37. 41 mV (negative polarity)   19-39. A sine wave that swings between 0.9 V and
1.1 V   19-41. 1 mV, 10 mV, 100 mV, 1 V, and 10 V   19-43. Trouble 2: R_2 shorted
19-45. Trouble 4: R_2 open   19-47. Trouble 6: R_3 open   19-49. Trouble 8: R_3 shorted

CHAPTER 20

20-1. 92.4 mV, 21.5 kHz, 1.59 Hz, 1.59 Hz, and 10.3 Hz   20-3. 81.7 mV to 101 mV   20-5. 611 mV,
2.7 kΩ, 8.18 kHz   20-7. Either change R_S to 3.3 kΩ or change R_F to 270 kΩ.   20-9. 200 mV;

0 to 15.2 kHz   20-11. 200 mV; 0 to 17.1 kHz   20-13. 55 mV   20-15. 110 mV   20-17. 2.45 V, 24.5 mA   20-19. 0.5 mA   20-21. 165 kΩ   20-23. Change 3.3 kΩ to 2 kΩ.   20-25. 1.56 times v_{in}; 7.96 kHz   20-27. The 741C will saturate with an output of either $+13.5$ V or -13.5 V. The final output will be 0.7 V less, either $+12.8$ V or -12.8 V.   20-29. The minimum voltage gain is zero. The maximum voltage gain is 10 with an output of either polarity.   20-31. The first stage has a voltage gain of 194, and the second stage has a voltage gain of 48. The total voltage gain is 9312, which is equivalent to 79.4 dB. 20-33. 0.667 mA   20-35. 143 Ω   20-37. Trouble 2: R_2 open   20-39. Trouble 5: Wire CD open; Trouble 6: Wire JA open   20-41. Trouble 9: R_3 shorted; Trouble 10: R_2 shorted

CHAPTER 21

21-1. 106 mV dc   21-3. -106 mV dc   21-5. 213 Hz   21-7. 1.13 Hz; approximately 10 mV rms 21-9. $+0.291$ V, -0.291 V, and 0.582 V   21-11. 0.528 V   21-13. 5 mA   21-15. 10 V, 1 V, 0.1 V, and 0.01 V   21-17. A triangular wave of 3.91 mV peak-to-peak; 39.1 mV peak-to-peak 21-19. 50%; zero   21-21. 9.23 Hz   21-23. First solution: Change capacitance to 4615 pF. Second solution: Change capacitance to 1000 pF and change resistance from 2 kΩ to 9.1 kΩ.   21-25. First solution: Use an oscilloscope to test the output. Second solution: Use a Schmitt trigger with a hysteresis of 10 V. Third solution: Use a peak-to-peak detector and a comparator with a trip level of 5 V.   21-27. Many solutions are possible. One solution is to peak detect the line voltage and use this as the input to a comparator with a trip level of 148 V. When the trip level is passed on the low side, the comparator turns on the alarm. 21-29. Trouble 1: Positive clamper   21-31. Trouble 3: Relaxation oscillator; Trouble 4: Peak detector   21-33. Trouble 7: Integrator; Trouble 8: Peak detector

CHAPTER 22

22-1. 9 V rms, or 25.4 V peak-to-peak   22-3. A: 35.8 Hz and 362 Hz. B: 358 Hz and 3.62 kHz. C: 3.58 kHz and 36.2 kHz. D: 35.8 kHz and 362 kHz   22-5. At least 3.62 MHz   22-7. 39.8 Hz   22-9. 2.36 MHz, 0.1, 10   22-11. 1.67 MHz   22-13. 10.4 MHz   22-15. 0.05 and 20   22-17. It increases 1%. 22-19. a. Decrease b. Increase c. Increase d. Same e. Same   22-21. Definitely sounds like parasitic oscillations. Insert a small resistor in series with the base or use a ferrite bead on the base lead.   22-23. 4.46 μH 22-25. 242 μs   22-27. Nearest standard values are $R_A = 22$ kΩ, $R_B = 11$ kΩ, and $C = 0.033$ μF 22-29. Trouble 2: IF amp   22-31. Trouble 5: Antenna; Trouble 6: Loudspeaker

CHAPTER 23

23-1. 9.53 V and 10.2 V   23-3. 8.07 W   23-5. 230 Ω, assuming the wiper is at the middle 23-7. 0.5 V and 1.8%   23-9. 5 mV, 46 dB   23-11. 1 A, 0.56 mV   23-13. 46.7 μV 23-15. 0.811 W   23-17. 0.2 A and 1.2 A   23-19. 5 V, 33.3%   23-21. d 23-23. b 23-25. 1.25 V to 27.3 V; 1.25 V   23-27. The load current is approximately constant. This implies that the input current to the voltage regulator is approximately constant. Since the filter capacitor has to supply an approximately constant current, its voltage resembles a ramp rather than an exponential. The significance of this for equation 4-8 is that the equation can be used for large ripple, as well as small.   23-29. Trouble 1: Triangle-to-pulse converter   23-31. Trouble 3: Q_1 23-33. Trouble 5: Relaxation oscillator 23-35. Trouble 7: Triangle-to-pulse converter   23-37. Trouble 9: Triangle-to-pulse converter

INDEX